MW00560696

ALBATROSSES

ALBATROSSES

W.L.N. Tickell

YALE UNIVERSITY PRESS
NEW HAVEN AND LONDON

Published in the United Kingdom by Pica Press (an imprint of Helm Information Ltd) and in the United States by Yale University Press.

ISBN 0-300-08741-1

Library of Congress Catalog Card Number: 00-102414
Printed in Hong Kong.

A catalogue record for this book is available from the British Library.

The paper in this book meets the guidelines for permanence and durability of the Committee on Production Guidelines for Book Longevity of the Council on Library Resources.

10 9 8 7 6 5 4 3 2 1

CONTENTS

Preface		5
Acknowledgements		7
INTRODUCTION		11
Chapter 1	Names and Naturalists	13
Chapter 2	Tube-nosed Seabirds	22
SOUTHERN ALBATROSSES		39
Chapter 3	The Southern Ocean	41
Chapter 4	Islands of the Southern Ocean	46
Chapter 5	Mollymawks	71
Chapter 6	Sooties	112
Chapter 7	Great Albatrosses	130
TROPICAL ALBATROSSES		169
Chapter 8	The Equatorial Pacific Ocean	171
Chapter 9	The Galápagos Albatross	175
NORTHERN ALBATROSSES		187
Chapter 10	The North Pacific Ocean	189
Chapter 11	Gooneys	194
Chapter 12	Steller's Albatross	233
COMPARATIVE BIOLOGY		247
Chapter 13	Moult	249
Chapter 14	Flight	260
Chapter 15	Behaviour	276
Chapter 16	Ecology	323
VICTIMS AND VERSE		355
Chapter 17	The Sea of Man	357
Chapter 18	The Mariner Syndrome	371
Appendices		383
Photographic section		416
Bibliography		417
Index		439

In Memoriam

L. E. Richdale	1900–1983
J. H. Sorensen	1905–1982
H. I. Fisher	1916–1994
J. D. Gibson	1926–1984
P. A. Prince	1948–1998

PREFACE

However remote the chance of seeing a live albatross, most people are familiar with the icon of retribution. As a paradigm of the greatest seabirds, it has always been represented as larger than life and its undoubted powers of flight exaggerated, but when the real bird soars past, no-one lucky enough to be watching is disappointed.

Of course, there is no such bird as *the* albatross, there are a number of species and this book is about that diversity. It has been gleaned from the work of many people who have studied specimens in museums, encountered albatrosses at sea and lived with them on remote islands. Scientists are usually preoccupied with the latest research, but time lends perspective to discovery and the achievements of those who have gone before are part of the story.

The book is divided into six self-contained parts and readers may begin with whichever ocean or subject interests them. In the Introduction, Chapter 1 has to begin by addressing the current problem of conflicting scientific names. Users of nomenclatures often have different needs and expectations from those of the taxonomists who construct evolutionary trees. As the science of systematics advances, ornithologists and birders encounter difficulties when their measurements, or the plumage and behaviour they observe, cease to be the characters used by taxonomists to determine species. Not long ago, most biologists could take part in taxonomic debate with some confidence. Now, short of re-training in biochemistry and cladistics, many are obliged to adopt an act of faith in the latest technology.

Chapter 2 outlines the relationship of albatrosses with their nearest relatives and deals with common topics that might otherwise have to be repeated in later chapters. It explains what I mean by the terms juvenile, subadult and adult (p.36).

Throughout the book, groups of species are reviewed from a geographic perspective. Albatross distribution is, ultimately, a response to the hydrography, hydrology and climatology of oceans, which I have introduced very briefly in Chapters 3, 8 and 10.

Many problems on albatross breeding grounds today, originated in events that followed the discovery of islands, so I have outlined their history and geography. To avoid repetition, the many islands of the southern hemisphere are all placed in Chapter 4, while those of the Pacific Ocean have been mentioned with the species.

The albatrosses of the Southern Ocean (Chapters 5, 6, 7) are followed by the one species of the tropical Pacific. In a family of beautiful birds, the Galápagos Albatross (Chapter 9) is the dowdy exception, but it is a unique albatross in more ways than one. The North Pacific albatrosses (Chapters 11, 12) experienced periods appalling destruction, but all three have survived and revealed remarkable potential for recovery.

Four topics of albatross biology are discussed. The complex patterns of albatross moult (Chapter 13) began to be understood in the early 1980s, but were not analysed in any detail until ten years later. The so-called *dynamic* soaring of albatrosses is still quoted, long after measurements of flight at sea put that theory in perspective (Chapter 14). In spite of television exposure, albatross behaviour (Chapter 15) is still of minor academic interest, while long-term research into the ecology of a few populations (Chapter 16) has attracted much more attention.

Albatrosses have always suffered at the hand of man, but saving them from drowning is now the prevailing objective of many people at sea and in offices around the world (Chapter 17). However, it was the *Rime of the Ancient Mariner* that established the public conception of albatross, so I hope readers will enjoy the few other poems and doggerel that bring the book to a close (Chapter 18).

ACKNOWLEDGEMENTS

In the early 1980s, desk-top computers were installed at the University of Malawi. I had finished writing-up my own South Georgia research, so I began compiling a bibliography of albatross literature. One publication tempted me to start this monograph, *The flight of Petrels and Albatrosses (Procellariiformes) observed at South Georgia and its vicinity* (1982). I thank Colin Pennycuick, who has been my guide to bird flight since the early 1960s.

The first chapters were drafted at the University of Bradford in 1985, and I thank Sue Norman who typed them, but most of the book has been written in Bristol. I am grateful to Brian Follett, Tom Thompson and the University of Bristol for an honorary research fellowship in the School of Biological Sciences, and to Colin Mapes, Richard Tinsley and Paul Hayes for their continued support. I thank the librarians of the School of Biological Sciences and Special Collections, Linda Birch at the Alexander Library (Edward Grey Institute) in the University of Oxford, and Michael Wilson for translating occasional Russian extracts. Also I thank Robert Prys-Jones and the the staff of the Natural History Museum at Tring. In 1990-92 the work benefited from a Leverhulme Research Fellowship, for which I thank the Trustees.

I am indebted to the publishers of many scientific journals who are cited throughout the book, and to all who contributed to my earlier published maps of albatross distribution at sea, including Simon Godden and Jonathan Tooby for their graphics. Up-dated editions of those maps appear in Chapters 5, 6, 7 and 9.

Many albatrosses winter off South Africa and I thank the owners of I. & J. South Africa, for allowing me to go to sea in July 1984 on their trawler *Pioneer* IV. Skipper Langley and his crew I remember for their hospitality throughout nine days on the fishing grounds off the Cape, among thousands of seabirds.

Over the years many people involved in albatross research have talked openly to me about their work and responded to my letters and telephone calls, none more so than John Croxall and Peter Prince of British Antarctic Survey. They constantly kept me up to date with progress at South Georgia, providing many unpublished manuscripts and reprints. I am deeply indebted to them and to Simon Berrow, Dirk Briggs, Norman Cobley, Jon Cooper, Julian Hector, Nick Huin, Simon Pickering, Keith Reid, Paul Rodhouse and Gerry Thomas.

In 1957, on a voyage home from the Antarctic, my ship, RRS *John Biscoe* called at Tristan da Cunha and I talked about seabirds with Gerry Stableford. A few years later, during another brief call, J.H. Flint told me about the status of albatrosses. For later information about Tristan da Cunha and Gough Islands I thank John Cooper, Robert Furness, Martin Holdgate, Michael Swales and Barry Watkins. Aldo Berruti and John Cooper provided unpublished material on the Prince Edward Islands.

I first heard about albatrosses on the Iles Crozet from Roger Tufft in 1960, then from Jean-François Voisin and Jean-Louis Mougin. In 1968, at the Conference on Antarctic Ecology in Cambridge, Benoît Tollu related his long journeys on foot about Kerguelen. At another Cambridge conference in 1982, Pierre Jouventin told me of unusual Wandering Albatrosses on the Ile Amsterdam, a discovery that heralded the present taxonomic shake-up. Soon after the first satellite tracking of Wandering Albatrosses from the Ile de la Possession in 1989, I visited the Centre d'Etude Biologiques de Chizé. I am grateful to Pierre Jouventin, Henri Weimerskirch and Benoît Lequette for keeping me informed of their work.

I thank Knowles Kerry who told me about Macquarie Island and Light-mantled Sooty Albatrosses while on safari in Uganda, Gavin Johnstone who wrote of Heard and McDonald Islands and Eric Woehler who talked of Antarctic seabirds over breakfast in a California diner. In 1989 Graham Robertson drove a long way to meet me near Sydney, he later made sure I received a volume of *Albatross Biology and Conservation* and writes evocative notes from islands I should have visited myself! I pay attention to what Geoff Copson has to say about Macquarie Island, but I always think of Bob Tomkins, alone with the Wanderers at Caroline Cove. Nigel Brothers first talked to me about fishermen and Shy Albatrosses in a Hobart bar, and has since responded to my many queries about Tasmanian islands. I remember them all, as well as the new generation of albatross workers I met at the 1995 Hobart Conference: Trudy Disney, Rosemary Gales, April Hedd and Jenny Scott.

Long after ringed Wandering Albatrosses brought us into contact in 1959, Doug Gibson continued to send me reports from the coast of New South Wales. By the time I visited Australia he had died, but Harry Battam introduced me to Malabar and Bellambi. With Lindsay Smith, he brought together decades of data that heralded a new era of seabird reporting off Australia. David Nicholls, a pioneer in albatross telemetry, still comes up with innovative ideas. I thank him for his forbearance and Karin Reinke for exciting displays of satellite tracking.

Forty years ago, L.E. Richdale sent me copies of his three albatross monographs from New Zealand and later we met, one afternoon in London Zoo! In 1989, I spent ten days at his Taiaroa Head colony and I am grateful to the rangers for their warm welcome and discussions. I learnt much about the early colony from Stan Sharpe, Kaj Westerskov and Alan Wright. Over the years, many New Zealanders have answered my lengthy queries. I thank Sandy Bartle, Paul Dingwall, Graeme Elliott, Fred Kinsky, Rick McGovern Wilson, David Medway, Peter Moore, Aalbert Rebergen, Chris Robertson, Rowley Taylor, Alan Tennyson, Kath Walker, Susan Waugh and others.

I am grateful to Michael Salisbury, who commissioned my BBC television documentary, *Marathon Birds* and to Rob Chappell, who master-minded the expedition to film Northern Royal Albatrosses in the Chatham Islands. Alastair Fothergill gave me the opportunity to work again on South Georgia and joined those who are excited by the island.

I am grateful to Roddy Napier of West Point Island who always has something interesting to say about Falkland Islands albatrosses and to Richard Hill for allowing me access to Steeple Jason Island. I thank Peter Prince, Kate Thompson and Robin Woods for details of their unpublished counts. Also Jérôme and Sally Poncet, who have always been generous with their knowledge of South America, the Falklands and South Georgia. I treasure memories of the occasions that I sailed their schooner, *Damien II.*

The New York Zoological Society funded my visit to the Galápagos Islands in 1988. My thanks are due to William Conway and to the staff of the Charles Darwin Research Station, especially to Johnny Navarrete Torres, who camped with me above the beach on Isla Española. Since then I have received information from Dave Anderson, Catie David (neé Rechten), Hector Douglas, Michael Harris, Guy Houvenaghel, Bryan Nelson and Chela Vásquez. I thank them all.

The two 1962 papers on Laysan and Black-footed Albatrosses by Dale Rice and Karl Kenyon, reached me at South Georgia when I was in the final year of my own field-work. I thank them and other members of the Fish & Wildlife Service for that work and advice in later years. In 1973, I was a guest of the US Navy at Midway Atoll and remember the generous hospitality. A decade of outstanding albatross research was coming to an end and I went on to visit Harvey Fisher who had directed it. His students, Peggy Dooley (née Van Ryzin), Eugene LeFebvre, Earl Meseth and Don Sparling were still at the University of Southern Illinois. I remember their welcome and thank them for help in later years. As Fish & Wildlife Service activities have extended through the Hawaiian Islands I have received data from Heidi Auman, Stewart Fefer, Beth Flint, Craig Harrison and Nanette Seto. I thank them as well as Hubert Frings, Patrick Gould, Louis Hitchcock James Ludwig, Rocky Strong and G. Causey Whittow. As I was finishing this book, the Black-footed Albatross Population Biology Workshop was convened. I am grateful to Kathy Cousins and acknowledge their contribution, especially the data of Chan Robbins.

The Royal Navy played a pivotal role in Steller's Albatross conservation. In 1973, funds negotiated by ICBP from the New York Zoological Society made it possible for me to visit the Far East where HMSs *Brighton* and *Antrim* supported my expedition to Torishima. Afterwards, as I was returning home through Japan, I was entertained one evening by students of the University of Kobe. In that group was Hiroshi Hasegawa who became Japan's guardian of Steller's Albatross. I am deeply indebted to him for decades of co-operation and friendship.

I thank Gary Nunn for reading Chapter 1 and explaining his research philosophy, Storrs Olson for details of albatross fossils (Chapter 2) and Robert Headland for some history of southern hemisphere islands (Chapter 4). At the James Rennell Division of Ocean Circulation, University of Southampton, I thank Beverley de Cherfas and NERC for FRAM images of the Southern Ocean, and David Cotton for satellite data of Pacific Ocean winds and waves. The Scott Polar Research Institute enabled me to use the drawings and notebook of Edward Wilson while Surrey Beatty & Sons allowed me to reprint maps from *Albatross Biology and Conservation.*

I first talked about albatross moult with Richard Brooke in Cape Town. Many years later, Peter Prince read Chapter 13 in Cambridge. I thank them and Sievert Rohwer of the Burke Museum, Seattle for the diagram of moult options.

I am grateful to Colin Pennycuick and Thomas Alerstam for reading Chapter 14. Also to John Wilson and Colin Woods who patiently talked me through the aerodynamics of soaring and to Jeremy Rayner who directed me towards many rewarding papers in the literature of animal flight.

My thanks are due to Pierre Jouventin, Benoit Lequette, Earl Meseth and Simon Pickering for their models of albatross behaviour. Also to my wife, Willow for help in translating from the French – one Christmas vacation in Africa, on a veranda overlooking the Indian Ocean. Hector Douglas, Peter Corkhill, Gareth Jones, Leslie McPherson, Richard Ranft, Don Sparling and John Warham all provided sonagrams or facilities. I thank them and Academic Press for allowing me to reproduce sonograms that first appeared in John Warham's book *The Petrels.*

I am grateful to John Croxall, Julian Hector and Peter Prince for comments on Chapter 16, and to John Coulson and Peter Rothery for revisions. Mike Imber, Malcolm Clarke, Keith Reid and Paul Rodhouse all made corrections and up-dates to the lists of prey (Appendices 6-9).

Thanks also to David Medway and Storrs Olsen for confirming historical episodes in Chapter 17, to Nigel Brothers for his views on fishing hazards, Bob Furness, Heidi Auman and James Ludwig for comments on pollution and John Croxall for checking points of conservation legislation.

After reading Peter Alexander's biography of Roy Campbell, I sent him my anthology of albatross verse (Chapter 18). I am grateful for his encouragement and to Ad Donker Publishers for allowing me to reprint *Albatross* from *Roy Campbell's Collected Works,* by F.C. Custódio. *Punch* gave permission for me to include R.P. Lister's amusing poem, while the Society of Authors, as the literary representative of the estate of John Masefield, allowed me to end the book with *Sea Change.*

In 1958, Bill Bourne came to share a room with me at the Edward Grey Institute in Oxford, when I was about to embark on my first albatross expedition. I thank him for many years of co-operation and for reading several chapters and extracts. I met John Warham in Paris, at the 1962 Symposium of Antarctic Biology and we have exchanged seabird gossip ever since. His book, *The Petrels* has been a frequent reference in recent years, and I thank him also for checking several of my chapters. Robin and Anne Woods first read the whole book and I am grateful for their diligent labour on behalf of an old friend. Jeffery Boswall has always been ready when called upon for support and Nigel Marvin has cheered from the sidelines.

My own tracings of albatross behaviour from 16 mm film were improved by Robin Prytherch, who then did all the remaining species and wings. I thank him for his warm interest and meticulous drawing. Many friends and acquaintances responded generously to my requests for their best colour pictures. I thank them and an unknown Chinese photographer for an aerial photograph of the the Senkaku Islands.

I am indebted to Christopher Helm and Nigel Redman for accepting the book, and to John Coulson for his constructive editing. Lastly I thank Julie Dando and Marc Dando for their dedication and skill in putting the book together. Working with them in Looe has been an unexpected and wonderful ending.

INTRODUCTION

A mixed flock of petrels off South Georgia (B. Osbourne).

1
NAMES AND NATURALISTS

'I now belong to a higher cult of mortals, for I have seen the albatross!'

Robert Cushman Murphy, 28 October 1912
on board the whaling brig *Daisy* in the South Atlantic[1]

For a long time there was a widely accepted consensus that these were the albatross species (Peters 1931, Jouanin & Mougin 1979). I have arranged them in the groups they occupy throughout this book:

Black-browed Albatross
Grey-headed Albatross
Yellow-nosed Albatross
Buller's Albatross
Shy Albatross

Sooty Albatross
Light-mantled Sooty Albatross

Wandering Albatross
Royal Albatross

Galápagos Albatross

Laysan Albatross
Black-footed Albatross

Steller's Albatross

They included a number of subspecies, but these may soon be swept away by a revolution that aims to create 11 new species (Appendix 1).

In the 1990s, albatross systematics was overtaken by a cladistic methodology applied to molecular (genetic) characters, which provided new insights into phylogeny and evolution (Fig 1.1) (Sibley & Ahlquist 1990, Viot *et al.* 1993, Nunn *et al.* 1996). However, the definition of species and the usefulness of subspecies has been a long-running debate (Ridley 1986, Mallet 1995, Zink 1997, Snow 1997) and Jeremy Greenwood (1997) spoke for many when he said:

> 'Taxonomy and related fields are battle grounds onto which the non-combatant ventures at his peril, liable to be shot at from all sides. Even the definition of the subject is one on which its practitioners clearly disagree.'

Species remain the fundamental units of taxonomy, but generations of naturalists and scientists have used trinomials to label variations within species. In some groups, including albatrosses, named subspecies are so numerous that for practical purposes they have taken over the lowest rank. Most ornithologists have no difficulty with this, but taxonomists employing modern analytical methods need a consistent unit (terminal taxon) at the end of branching diagrams. There are, for example, two royal albatrosses, the Northern Royal Albatross and the Southern Royal Albatross; their separate morphologies and breeding distributions are well defined. Some ornithologists believe

[1] Murphy 1947.

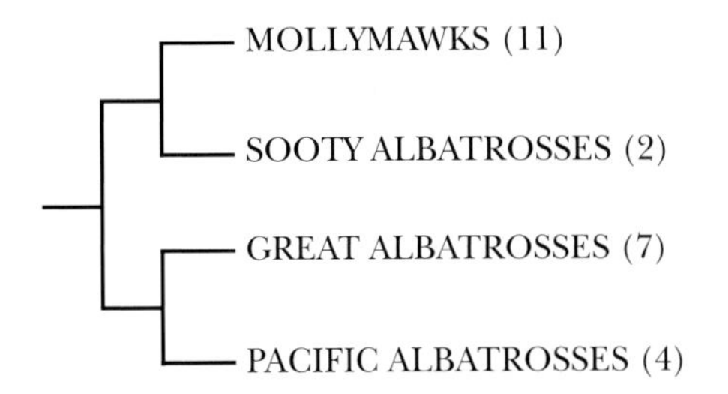

Fig. 1.1 Cladistic analysis of mitochondrial cytochrome-*b* genes from 22 albatross taxa. The numbers in parentheses are the terminal taxa (species or subspecies) in each group (after Nunn & Stanley 1998).

they are both subspecies of *Diomedea epomophora*; others, with equal conviction, say there are two species, *Diomedea sanfordi* and *Diomedea epomophora* (Robertson & Nunn 1997). A similar dilemma attends other subspecies (Bourne & Warham 1999).

Scientific nomenclatures are listed in Appendix 1 and readers may choose for themselves whether the albatrosses[2] they see or read about are best considered as species or subspecies. Common usage is the final arbiter, but it takes time for a new consensus to become apparent and it is still too early to anticipate how the majority of scientists and birders will react. In the meantime, I hope that English names will be adequate for the purposes of this book. To some, I have added new geographic prefixes and others I identify by their island breeding grounds.

ALBATROSS

The word albatross has a curious etymology. It was traced to its classical origin by Alfred Newton (1896) in his *Dictionary of Ornithology*. An Arabic name for the pelican, *al - câdous* seems to have come from a Greek word κάδος for the leather scoop used on ancient water wheels. After the Moors had been driven out of Spain, the word evidently remained as *Alcatraz, Alcatràz, Alcatraza, Alcatraze, Alcatràç* and *Algatraz*, names used by Spanish and Portuguese mariners around the world and widened to include boobies, gannets[3] and frigatebirds.

Sixteenth century English mariners used the familiar names of home for strange seabirds. Some of those encountered at sea by Sir Richard Hawkins in 1593 were clearly Wandering or Royal Albatrosses:

> 'During this storme, certain great fowles as big as swannes, soared about us... the hooke in his upper beake... is like to a faulcons bill, but that the poynt is more crooked...and from the poynt of one wing to the poynt of the other, both stretched out, was about two fathomes'
>
> (Markham 1878).

Almost a hundred years later the long-winged 'eagles' and 'hawks' that 'suffered themselves to be taken up' by the crew of the *Welfare* at the Falkland Islands could only have been Black-browed Albatrosses (Richard Simson in Cawkell *et al.* 1960). Other mariners had long since adopted the Spanish names; Peter Mundy on a voyage from Mauritius to Madagascar in 1638 identified the birds well enough:

> 'Allcatrazes is againe the biggest of any Seaffowle I have yett seene, spreading Near 6 or 7 Foote with his wings, which hee seemeth not to Move att all as hee Flyeth leisurely and close to the Rymme of the water...'
>
> (Temple 1919).

English dialects (possibly those spoken by seamen of southwest England) appear to have softened the hard Arabic *c* to *g* or *b*. John Fryer (1909), sailing to India in 1673 with a fleet of East Indiamen, included '*Albetrosses*' among the 'feathered Harbingers of the Cape' and mentioned that their wings were twice as long as their bodies. William Dampier (1697) also noted that '*Algatrosses*' were very large long-winged fowl. In 1700, Edmond Halley crossed the South Atlantic between latitudes 27°S and 44°S on board the *Paramore* and recorded '*Alcatrosses*', but William Dalrymple who edited the journal for publication, corrected the spelling to *Alcatrasses*, almost back to the Spanish (Thrower 1981). During his circumnavigation of 1719–22, George Shelvocke (1726) knew '*Albitrosses*' for what they were: 'the largest sort of sea-fowls, some of them extending their wings 12 or 13 foot'. As late as 1769, Joseph Banks on the *Endeavour* recorded '*Albatross* or *Alcatrace*'; but ever after used only the former (Beaglehole 1963).

[2] The singular – albatross – is sometimes used also as a plural. It is an eccentric habit that has not been widely adopted.
[3] In Spain and some parts of Portugal, Alcatraz is still the name for the North Atlantic Gannet.

French mariners first knew the large white birds as *Mouton* or *Mouton du Cap*[4], but by the mid-1700s, Parisian savants had adopted *L'Albatros* which became their binomial, *Albatros mouton*. The name Albatros(s) has since been absorbed into many other languages and is understood worldwide.

A skull acquired by the Royal Society in London was described by Nehemiah Grew (1681) as 'The HEAD of the MAN of WAR; called also Albitrosse'. The specimen was clearly from an albatross, but the Man of War[5] synonym was evidence of an ambiguity that confused collectors of the time. The misconception was still around 57 years later (Albin 1738).

During the 18th century, serious interest in natural history was encouraged by a wider dilettante curiosity among fashionable society. Displays of collections became an entertaining diversion at well-to-do salons and led to the publication of large, illustrated books. It was in such expensive volumes, illustrated by the best artists of the day and destined for aristocratic libraries, that pictures of albatrosses were first seen. George Edwards's *Natural History of Birds* (1747) featured a convincing reconstruction with useful bill detail (Fig. 1.2).

Fig. 1.2 'The Albatross' illustrated in George Edwards's *Natural History of Birds* Part II (1747).

DIOMEDEA

For over 200 years the name *Diomedea* has prevailed as the genus of almost all albatrosses. It has been split many times, but new species were usually accommodated within the single genus. Nunn *et al.* (1996) have again split the genus and *Diomedea* is about to shed most of its former species, but it will probably continue as the collective name of the group.

[4] Cape sheep after the Cape of Good Hope.
[5] Now more commonly frigatebirds *Fregata* spp.

Diomedes[6], King of Thrace, was one of Homer's heroes. With Odysseus and Pelomedes, he commanded the Greek army under Agamemnon that took part in the siege and sack of Troy. During another campaign he offended the goddess Athene, who sent a mighty storm on his fleet. Some of his followers, instead of being chastened by the hardships, taunted the goddess, who became even more enraged and transformed them all into large white birds – 'gentle and virtuous'. In another version, Diomedes and a few followers survived the storm and the transformations to live out exile in the land of King Daunus where Diomedes married Euippe, the king's daughter and became a founder of great cities. In his old age he was murdered by the king and buried on one of the Diomedan Islands[7], later becoming a god worshipped in Verletia and throughout southern Italy (Graves 1955).

The story appears in Ovid's *Metamorphoses*, and Pliny the Elder wrote of rather fanciful seabirds he called *Diomedeae* that were said to occur only near the tomb and temple of Diomedes, on an island off the coast of Apulia. Friderici Lachmund's 1674 dissertation on Pliny's birds is quite a convincing, illustrated description of a petrel[8]. John Ray (1713) included it in his *Synopsis methodica Avium*. He had two major divisions of birds: Land Fowl and Water Fowl and in the latter, *Diomedea* was the first of his Web-footed Birds. He identified the breeding ground as the Tremiti Islands off the Adriatic coast of Italy, where Isabel Winthrope found them in the mid-1960s.

Jacob Theodore Klein (1750) separated the three-toed, web-footed seabirds into a family which included auks, penguins, skuas, storm petrels and albatrosses, most of them he named *Plautus*[9]. In the tenth edition of *Systema Naturae*, Linnaeus[10] (1758) created a much larger group of waterbirds; within it, his genus *Diomedea* comprised an albatross and a penguin! Shortly afterwards, Mathurin Jaques Brisson (1760) created an apt substitute, *Albatrus*, from the sailors' name and placed it in a separate order. However, Linnaeus's *Diomedea* became so widely known that Johann Reinhold Forster (1785) considered it would be 'pure affectation' to substitute another name.

Since that time there have been many such affectations, but *Diomedea* will probably endure as the exclusive genus of the largest albatrosses (Nunn *et al.* 1996). Some writers have had misgivings about calling them GREAT ALBATROSSES and have felt constrained to use the name in parentheses, but they are significantly bigger and, like Murphy (1936), I have always found the name apt and unambiguous. It follows that all other species are SMALL ALBATROSSES.

MOLLYMAWKS

The old seafarers' name for the small albatrosses of the southern hemisphere is well established in New Zealand and among the islanders of the Falklands and Tristan da Cunha. It has always been variously written and spoken as mollymauk, mollimauk, mollymoke, mollyhawk, mollymawk and very often contracted to molly (Murphy 1936, Tickell 1976). In this book it is a collective term for the group of albatrosses in the proposed genus *Thalassarche* (Nunn *et al.* 1996), and never used as a substitute for 'Albatross' in an English name.

> 'Mollymawk... is a sailors' corruption of an English word which was corrupted from a German word corrupted in turn from the Dutch! It was originally a name of the Arctic Fulmar, the vast flights of which reminded the Dutch whalemen of "malle-mugge", the tiny midges that whirl around a lamp; but today it is applied by mariners of nearly all nations to the smaller albatrosses of the southern oceans.'

Murphy's (1936) brisk derivation came from Friderich Martens who was at sea with whalers off Greenland and Svalbard in 1671 (White 1855); but 'Mallemucke' was also said to have been derived from the Dutch *mal* (foolish) and *mok* (gull) (Armstrong 1958). John Ray adopted the name in 1713 for the Northern Fulmar and it is still current in Denmark as *Mallemuk*. The name was later applied by sealers and whalers in the southern oceans to the smaller albatrosses that fed on discarded blubber and carcasses in the same frenzied manner as the Fulmars of the north.

[6] Greek: (*Dio*) – god, (*medes*) – counselled.
[7] Not to be confused with the Diomede Islands in the Bering Strait; named by Russian navigators or cartographers after Saint Diomede (304 AD) on whose day, 16 August 1728, they were discovered (Berg 1926).
[8] Cory's or Mediterranean Shearwater which Scopoli (1769) described as *Procellaria diomedea*, taking his specific name not from Lachmund but from Linnaeus. It is now *Calonectris diomedea*.
[9] Latin: *plautus* – flat or broad; among the Umbrians flat-footed.
[10] Carl von Linné (1707–1778) usually known by his Latin name.

The English dialect of the Tristan da Cunha islanders contains many early 19th century seafaring words that are now no longer heard elsewhere. Among these people, mollymawks have always been considered quite different birds from albatrosses.

At the Falkland Islands, Lieutenant S.W. Clayton (1774) of the Royal Navy and the whaling master Edmund Fanning (1833) knew albatrosses when they saw them and referred to them as such, but in 1855, when W.P. Snow (1857) visited Keppel Island, the birds were spoken of locally as 'mollimauks'. After two years commanding the Falklands garrison, Captain C.C. Abbott (1861) included 'Molly-mawk' in his list of birds and by the end of the century this name was firmly established (Vallentin 1904). Later writers with ornithological pretensions adopted the style 'Black-browed Albatross (Mollymauk)' (Bennett 1926, Cobb 1933).

European emigrants to New Zealand evidently picked up sailors' vernacular on the long sea voyages. F.W. Hutton, who arrived in 1866, later catalogued three New Zealand *Diomedea* species as 'Molly-mawks' (Hutton 1871); but Walter Buller (1873), the pioneer ornithologist who had been born in New Zealand, ignored the name altogether. By the 1920s it was being used often enough to be adopted as an alternative to Albatross (Oliver 1930) and from about that time Mollymawk was resolutely used in New Zealand as a proper name (Marples 1946; Falla *et al.* 1966), while across the Tasman Sea it was just as strongly eschewed by Australian ornithologists.

SOOTIES

A collective term for the two sooty albatrosses of the genus *Phoebetria.* For a long time, they were thought to be systematically remote from all other albatrosses, but recent genetic evidence indicates a common ancestry with the mollymawks (Nunn *et al.* 1996). Nevertheless, the proposed name 'Sooty Mollymawks' (Robertson & Nunn 1997) is misleading. The other characters that clearly distinguish sooties from mollymawks remain significant for field-workers.

GOONEYS

Newly discovered islands were soon visited by fishermen, sealers and whalers. Many of these hardy men were shipwrecked or marooned for years in makeshift huts with meagre supplies. When they were about to take eggs or kill the occupants of albatross nests, the birds' disregard of human intruders was seen as dumb or stupid. The bewildered behaviour of captured albatrosses on the decks of sailing ships (p.377) also contributed to this impression and the name *gony* or *gooney* followed. The latter eventually became attached to the Laysan and Black-footed Albatrosses of the North Pacific Ocean. Some present-day admirers of the birds find the name embarrassing, but ever since the early 19th century, when seamen sometimes described albatrosses as 'gulls' (p.211), it has served posterity. It was preserved at Midway Atoll in the everyday speech of US Navy personnel and their families who were, none the less, admirers of the birds and who daily went to endless effort to avoid harming them. Thousands of Americans have lived among the gooneys of Midway, where the name had long since lost its pejorative connotation. It became an affectionate familiarity that diminished no-one, least of all the birds, and remains as a convenient collective term that distinguishes Laysan and Black-footed Albatrosses from all other species.

NATURALISTS[11]

The folklore of the sea contains many fragments of albatross natural history, but it is often difficult to separate them from the myths. Most passengers sailing between Europe and Australia in the 19th century heard sailors' tales of the seabird islands passed *en route.* Only the unlucky ones achieved such perilous intimacy with albatrosses as Florence Wordsworth (1876), castaway on the notorious Ilots des Apôtres.

> 'I was stunned with cold, and almost fainting, so that it seemed only a few minutes to me till Charlie came with the reeking-hot skins of two albatrosses and wrapped my feet in them.'

Heinrich Gottlieb Ludwig Reichenbach (1850) was the *éminence grise* who split the albatross genus *Diomedea,*

[11] Including many who were no more than collectors and others who might prefer to be called scientists.

creating three new genera without a word of justification. Clear logical argument was the strength of Elliott Coues (1866), who argued from anatomy that all albatrosses should remain in the one genus. Comparative anatomy was at the forefront of 19th century biology, but well-preserved seabirds were hard to come by. Petrels and albatrosses collected especially for this purpose during the voyage of the *Challenger* (1873–76) were dissected by the brilliant young anatomist, William Alexander Forbes (1882). Tragically he died soon afterwards, at the age of 27.

Throughout the late 19th century, Osbert Salvin was the acknowledged expert on petrels and albatrosses at the British Museum. He was the friend of Lord Rothschild, who amassed a huge collection of seabirds (p.199). Gregory Macalister Mathews in Australia became the gadfly of seabird taxonomy. For 40 years, throughout the first half of the 20th century, he constantly got ahead of more conservative ornithologists, exasperating them with his new taxa (Serventy 1950). His crowning achievement was an edifice of nine albatross genera (Mathews & Hallstrom 1943), but in the end he was prevailed upon to recant (Mathews 1948):

> 'There is a growing desire to place the Albatrosses and Mollymawks in one genus and we must be consistent...'

Nunn *et al.* (1996) have recently championed Mathews and implied that Alexander *et al.* (1965) ignored valid taxa. In reality it was Mathews's contemporaries three to four decades earlier in the RAOU[12], BOU[13] and AOU[14] who had opposed his splitting. In the first pocket field-guide, *Birds of the Ocean*, W.B. Alexander (1928) reflected a pragmatic consensus of two albatross genera. It had the support of Murphy (1936) and served ornithology well for over half a century.

The best seabird artist of the period was Edward Wilson, the heroic doctor naturalist of Captain Scott's two Antarctic expeditions (Roberts 1967). He was an acute seabird observer (Fig. 1.3) who sketched on the decks of ships in all weathers and put together a personal seabird guide. His remarkable life drawings (Fig. 1.4) give a vivid stylised impression of flight at a time when most illustrations of seabirds were lifeless re-creations from dead specimens.

In 1912–13 the brig *Daisy*, the last of the old New England whalers, visited South Georgia. The owners had contracted with the American Museum of Natural History and the Brooklyn Museum to carry and assist the young Robert Cushman Murphy in the collection of seabirds. His account of the voyage, *Logbook for Grace*, written for his newly married wife, is one of the classics of sea adventure (Murphy 1947). He returned from the one year voyage with about 500 specimens of 55 species, more than 100 sets of eggs, many photographs and other data. Murphy saw many albatrosses and visited Wandering Albatross nests in the Bay of Isles for ten weeks. The voyage was a time-warp that left him with a unique ability to find natural history entries in old whaling log books. It was also the foundation of an eminent career in marine ornithology. Well funded with endowments, the American Museum of Natural History commissioned the best collectors of the day, men like Roland Howard ('Rollo') Beck, who spent years at sea. When Lord Rothschild's collection unexpectedly came onto the market[15], there was no hesitation, the large sum was found and Murphy was sent to Tring to oversee the packing. He was tall, confident and well connected (Amadon 1974, Brooks 1984), commanding the scholarship across disciplines needed to draw together the great synthesis, *Oceanic Birds of South America* (Murphy 1936) that launched seabird science and became the source book for future generations of seabird researchers.

Modern seabird guides began to appear in the 1970s (Watson 1975, Harper & Kinsky 1978, Tuck & Heinzel 1978). Nevertheless, the identity of great albatrosses at sea remained obscure. Peter Harrison (1978) gave us the diagnostic features. He worked for two years as cook, winchman and pot-baiter on fishing boats off New Zealand to gain the experience needed for his admirable identification guide, *Seabirds* (Harrison 1983).

The first field-work on albatrosses was that of Henri Filhol (1885) who watched Royal Albatross chicks at Campbell Island in 1874 (p.58). Two years later, William Froude (1888) measured the flight of albatrosses from

[12] Royal Australasian Ornithologists' Union; formerly Australian Ornithologists' Union.
[13] British Ornithologists' Union.
[14] American Ornithologists' Union.
[15] His lordship was being blackmailed by a former mistress (Rothschild, M. 1983).

the deck of a warship off South Africa (p.261) and C.C. Dixon (1933) accumulated some 3,500 days of observations during 27 years at sea (1892–1919). Lancelot Eric Richdale became interested in New Zealand albatrosses in the late 1930s. He responded to challenging opportunities with determination and throughout his career as an educational instructor in agriculture, his enlightened employers granted generous leave of absence for academic study at home and abroad. A shy, solitary man, he attracted the warm regard and respect of many ordinary people who co-operated with him, but harboured resentments and remained aloof from the ornithological and conservation establishments of New Zealand (Fleming 1984, Warham 1984, K. Westerskov pers. comm., S. Sharpe pers. comm.).

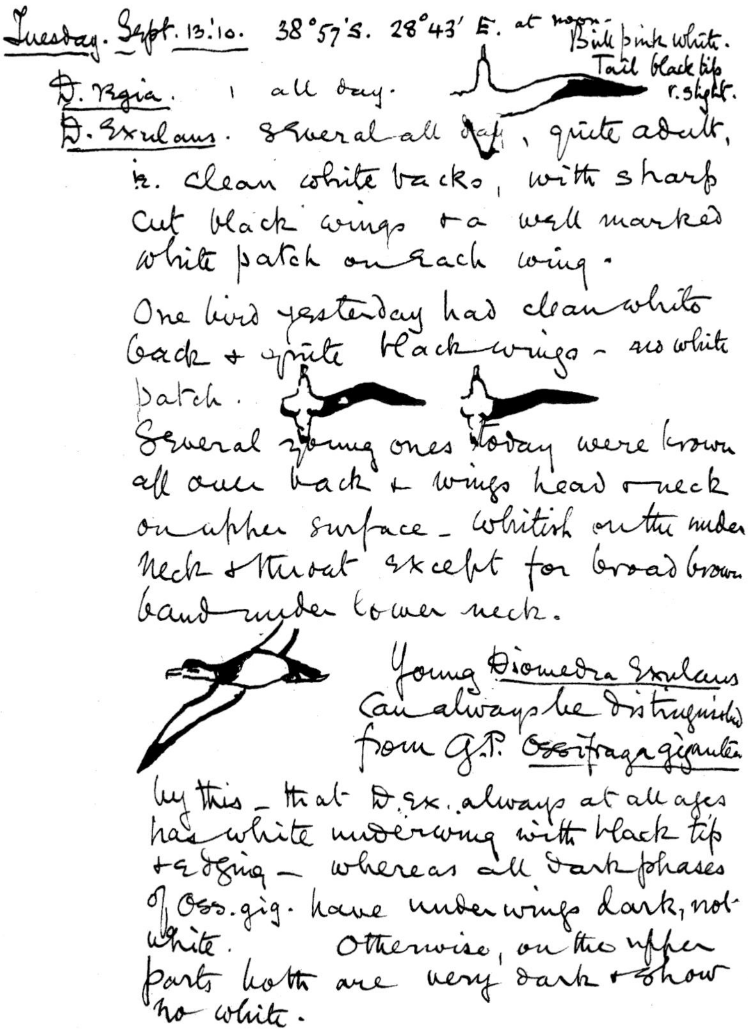

Tuesday. Sept. 13.'10. 38°57'S. 28°43' E. at noon.
Bill pink white. Tail black tip r. slight.
D. regia. 1 all day.
D. exulans. Several all day, quite adult, i.e. clean white backs, with sharp cut black wings & a well marked white patch on each wing.
One bird yesterday had clean white back & quite black wings – no white patch.
Several young ones today were brown all over back & wings head & neck on upper surface – whitish on the under neck & throat except for broad brown band under lower neck.
Young Diomedea exulans can always be distinguished from G.P. Ossifraga gigantea by this – that D. ex. always at all ages has white underwing with black tip & edging – whereas all dark phases of Oss. gig. have underwings dark, not white. Otherwise, on the upper parts both are very dark & show no white.

Fig. 1.3 A page from Edward Wilson's sketchbook with fieldnotes on Royal Albatrosses, Wandering Albatrosses and Giant Petrels written in 1910 (Roberts 1967).

Richdale's seven weeks long pioneering enterprise on the Snares Islands (Richdale 1949), and the long study of Northern Royal Albatrosses at Taiaroa Head, secured his position as the founder of albatross research. He personally guarded the first pair of Northern Royal Albatrosses to breed successfully in New Zealand and monitored the growing colony for 16 years. When editors of journals declined his long papers, he had them printed by his local newspaper as private monographs. *The Pre-egg Stage in the Albatross Family* (1950) and *Post-egg Period in Albatrosses* (1952) remain classics of albatross biology.

Fig. 1.4 Sketches of Black-browed Albatrosses at sea by Edward Wilson (Roberts 1967).

Military imperatives decreed that the United States Navy should commission albatross research on Midway Atoll in the mid-1950s and scientists of the Fish & Wildlife Service led by Dale W. Rice and Karl W. Kenyon (1962a&b) made the first comparative study and world census of two closely related species. They were followed by Harvey I. Fisher (1976), who kept field-work on these birds going for 13 years (1960–73), taking albatross research to a new level.

There were few logistic problems at Midway, but most albatross islands are uninhabited and far beyond scheduled shipping routes. The main impediment to independent research has been the high cost of chartering vessels. Over the years, warships, cruise liners, tankers, whale-catchers, sealers, trawlers and crab boats have all put field parties on islands. Since the 1980s, Jérôme Poncet in the schooner *Damien II* and Gerry Clark in *Totorore* have not only landed small expeditions, but assumed a role of their own, running seal and seabird surveys of remote island coasts.

The presence of a whaling and sealing industry at South Georgia made it possible for me to start albatross research at Bird Island in 1958 with very little money and the initiative was later supported by the United States Antarctic Research Program (USARP) and British Antarctic Survey (BAS) (Tickell 1968). With substantial and easily accessible colonies of four albatross species the island had outstanding potential. Field-work could not be prolonged beyond 1964, but the 50,000 albatrosses ringed in those years contributed to decades of later research. BAS took over the base and study colonies in 1972 and teams of field-workers led by John Croxall and Peter Prince pursued far-sighted objectives centred on prolonged comparative demography. Innovative field-technology has replaced many labour-intensive tasks (Prince & Walton 1984). The tolerance of incubating albatrosses to serial blood sampling led to remarkable achievements in endocrinology (Hector 1984), and genetic fingerprinting of individual birds may soon expose kinship-related variation within and between colonies.

In the late 1960s, French scientists began similar albatross research, first from a new TAAF[16] base on the Ile de la Possession in the Iles Crozet (Voisin 1969, Mougin 1970), then at the Iles Kerguelen and the Ile Amsterdam. In the late 1970s, Pierre Jouventin and Henri Weimerskirch initiated comparative ecological studies of several species at sea (Jouventin *et al.* 1981) and on the breeding ground (Weimerskirch *et al.* 1986). The application of satellite telemetry to albatross ecology was a notable advance (Jouventin & Weimerskirch 1990), and when later coupled with physiological instrumentation (Wilson *et al.* 1992), yielded undreamt-of insights into the activities of albatrosses at sea. Satellite tracking is now widely available to any field-workers who have the money to purchase microtransmitters and pay for operating services. Smaller and less costly position loggers known as 'archival tags' are coming into use. They store the times of sunset and sunrise, from which geographic positions can later be computed (Tuck *et al.* 1999). High resolution remote sensing may also be expected to play an increasing role in the surveillance of breeding grounds (Croxall 1997).

In the early 1990s, it was finally confirmed that large fisheries were responsible for the deaths of many albatrosses. Government agencies began monitoring fishing gear, and private expeditions began yearly inspections of colonies on hitherto little-known islands (Elliott 1997). Amid increasing publicity, the Australian Antarctic Division and Tasmanian Parks and Wildlife Service hosted the First International Conference and Workshop on the Biology and Conservation of Albatrosses, which was held at Hobart in 1995 (Alexander *et al.* 1997, Robertson & Gales 1998).

[16] Territoire des Terres Australes et Antarctiques Françaises.

2

TUBE-NOSED SEABIRDS

'...these melancholy water birds – one does not know what to do about them,
and their number is overwhelming!'[1]

Buffon[2] (1779)

This chapter is concerned with the general similarities and differences between albatrosses and their closest relatives, the petrels (fulmars, prions, shearwaters, gadfly-petrels)[3], storm petrels and diving petrels. All belong to one large group of pelagic seabirds characterised by the possession of tubular nostrils. There are various speculations about the function of these tubes, but it cannot be said that we understand what they give to their possessors, that other seabirds do well enough without.

Most of the petrels are medium-sized birds; there are many diverse species in all oceans and most seas of the world. In comparison, there are fewer albatrosses and large areas of the oceans where they are never seen. Many of the apparent differences between albatrosses and petrels have to be qualified by exceptions, and some at least disappear on close scrutiny.

EARLY TUBE-NOSES

A few fossil seabird bones from the Lower Eocene epoch, around 50 million years ago are believed to have come from extinct petrels or albatrosses. The earliest identified petrel and albatross fossils come from the Oligocene, about 32 million years ago.

In the early Tertiary, new marine faunas were replacing those lost in the Cretaceous mass extinction. The squids and bony fish were radiating together with new hierarchies of predators. Most living albatrosses are squid-eaters and it is plausible to imagine that early albatrosses may have co-evolved with proliferating squid. Middle Miocene fossils indicate that albatrosses similar to living genera were well established about 16 million years ago (Olson 1985).

A fossil bone from the Middle Eocene of Nigeria was once claimed to have been part of an extinct albatross much bigger than any living species (Brodkorb 1963), but that identification has been dismissed (Olsen 1985). No acknowledged albatross fossils have been as large as the bones of present day great albatrosses, although some have been as big or bigger than those of Steller's Albatross. Some fossils may have been from albatrosses smaller than any living species.

Fossils from the west coast of the North America and Japan indicate that there were albatrosses throughout the North Pacific in the mid-Miocene to late Pliocene, while others from the east coast of North America and lower Pliocene deposits of Europe suggest that similar albatrosses also inhabited the North Atlantic about five million years ago. At that time, the Atlantic was open to the Pacific Ocean through the Caribbean Sea (Olson 1985). Some time after the Panama isthmus formed, two to three million years ago, albatrosses disappeared from the North Atlantic other than as vagrants; nevertheless, the two immigrants that became resident (p.77) prove that albatrosses can make a living in the North Atlantic.

Asymmetry of land masses between the northern and southern hemispheres ensures that there are more fossil deposits in the northern hemisphere. A few fossils from late Miocene to early Pliocene deposits of Australia, South Africa and Argentina confirm that there were albatrosses in the Southern Ocean from five to nine million years ago (Wilkinson 1969, Olson 1983, 1984).

1 Translated from the French '...ces tristes oiseaux d'eau, dont on ne sais quois dire, et dont le nombre et accablant!' (Stresemann 1975).
2 George-Louis Leclerc comte de Buffon.
3 A wider definition also includes the albatrosses, storm petrels and diving petrels (Warham 1990,1996).

In the distant past, some albatrosses must have crossed the equator, but the tropical ocean has apparently been a barrier to albatrosses for a long time and they have evolved independently in the two hemispheres (Nunn *et al.* 1996). The one tropical species has a very restricted distribution just to the south of the equator (p.178).

Trans-equatorial migration occurs among living petrels (Warham 1996). Albatrosses also cross the equator in both directions, but they are few and some may have followed ships into latitudes beyond those they would normally enter alone.

CHANGING OCEANS

The drifting of continents and the reshaping of oceans in the Tertiary Period was approaching the present configuration by the time of the first albatrosses, but the climate, hydrology and productivity of the oceans some 30 million years ago were quite different. Warheit (1992) has reviewed the fossil seabirds in the North Pacific in relation to plate tectonics, palaeoceanography and faunal change. Seabirds moved with the changing ocean systems and the biological properties of both northern and southern oceans followed physical patterns driven by events in the Antarctic.

The Drake Passage was already open when Australia and Antarctica began moving apart in the Eocene-Oligocene, 25–30 million years ago. The Antarctic Circumpolar Current began flowing and established the polar fronts that insulated Antarctica from the warmer seas of lower latitudes. From the Late Miocene to the Early Pliocene, 5–15 million years ago, Antarctic and Arctic ice-sheets expanded causing progressive cooling of the adjacent oceans.

Even in warmer seas towards the equator, extension of the Antarctic ice sheet five million years ago is believed to have been the source of cold water that upwelled off SW Africa to start the Benguela Current (Olson 1983). These waters are highly productive today and have probably always attracted foraging seabirds.

The marine ecology of the Quarternary oceans has been determined by ice. At the last glacial maximum, 18,000 years ago, winter pack-ice covered twice as much of the Southern Ocean as it does today, but not so much of it melted and ten times as much remained throughout the summer as today. The Arctic had almost twice as much ice as Antarctica, but while the Arctic lost about 90% and was left with just the Greenland Ice Cap, the Antarctic Continent lost only 11% of its ice sheet.

During the many alternating glacials and interglacials, shifts of polar fronts in both the northern and southern oceans repeatedly changed the physical and biological characteristics of the oceans. The difference in mean annual air temperature between glacial and interglacial periods is believed to have been about 14°C. Changes of this order had greatest impact on the mid-latitudes favoured by albatrosses today. The tropical oceans were still comparatively hot and the subtropical fronts of the Southern Ocean remained in about the same position, but the polar fronts moved closer, so the area of subantarctic surface water was reduced.

In the North Pacific, the dynamics of the ocean have undoubtedly changed many times during the millions of years that albatrosses have been part of its ecosystem. About 20,000 years ago, sea-ice filled the whole of the Bering Sea and extended south of the Aleutian chain, from Kamchatka to the Gulf of Alaska. The retreat of the northern continental ice-sheets in the last 10,000 years opened up these waters to seabirds, notably auks, but albatrosses may have been taking larger prey in the North Pacific for millions of years.

Albatrosses today are separated geographically. The variety and abundance in each region is roughly proportional to the areas of ocean which they occupy:

Southern Ocean:	113 million km^2
North Pacific:	10 million km^2
Galápagos-Peru:	1 million km^2

ISLANDS

Tectonic uplifting of the ocean floor was often accompanied or followed by volcanic activity. New volcanoes were eroded by sea and weather to become irregular mountainous islands which acquired vegetation and become the breeding grounds of seabirds.

Many islands bordering the Arctic and Antarctic acquired glaciers and ice-caps that repeatedly changed their suitability as seabird breeding grounds. South Georgia is today a mountainous, glaciated island with enough exposed vegetation to enable four species of albatross to build nests. These populations are comparatively recent. During the most severe glaciation, an ice-sheet covered all of South Georgia except the highest mountains and extended over the surrounding sea (Sugden & Clapperton 1977); there would have been nowhere for albatrosses to breed. The last glaciation came to an end about 15,000–10,000 years ago and vegetation began to accumulate peat in mires and bogs from about 9,500 years ago. Albatrosses do not usually nest in bogs at South Georgia, but in valleys and on slopes dominated by tussock grass. The oldest tussock peat was deposited 4,300 years ago (Smith 1981), so colonisation of this significant albatross island has been comparatively recent.

Many of today's ice-free albatross islands experienced only partial glaciation; the Falklands had only small glaciers which would not have prevented seabirds breeding. The Andean ice-cap overlaid most of South America's offshore islands, but the Islas Diego Ramirez appear to have been beyond its southerly point and could have retained albatross colonies. Far to the north of the Antarctic Polar Front there is no evidence of glaciation at Tristan da Cunha and several other albatross islands (Denton & Hughes 1981).

Changes in sea-level have exposed or submerged albatross islands (Olson 1983). Although the sea-level rose 120 m as Arctic ice melted 10,000–13,000 years ago, the time scale of these events would have been long enough for albatross colonies to have moved gradually to higher ground, if any was available.

In the sub-tropical Pacific Ocean, tectonic movements gradually submerged many volcanic islands, but coral reefs maintained some as atolls which may thus have remained albatross breeding grounds for millions of years.

LIVING TUBE-NOSES

The name Procellaria[4] (Linnaeus 1758) has been used at several taxonomic levels for groups of petrels. Johann Carl Illiger (1811) of the Berlin Museum first recognised the significance of tubular nostrils and created a family he called the Tubinares; it included petrels, prions, diving petrels and albatrosses. Elliott Coues (1866) had misgivings about using tubular nostrils as a principal distinguishing character and Max Fürbringer, in his massive anatomical work[5], created a suborder Procellariiformes, in which Tubinares was a synonym of his 'gentes' Procellariae. Hans Friedrich Gadow (1892) raised the Procellariiformes to an order, retaining Tubinares as the lower category, but Tubinares prevailed at the British Museum (Salvin 1896, Godman 1907–10). The two names thus became synonyms that have vied for prominence ever since.

Order Procellariiformes/Tubinares
Family Diomedeidae (albatrosses)
Procellariidae (petrels)
Hydrobatidae (storm petrels)
Pelecanoididae (diving petrels)

This arrangement was generally accepted (Godman 1907–10, Peters 1931, Jouanin & Mougin 1979, Warham 1990) and, apart from a brief lumping of petrels and storm petrels (Murphy 1936, Mayr & Amadon 1951), remained unchanged until Charles Sibley and Jon Ahlquist (1990) launched their controversial revision, *Phylogeny and Classification of Birds* based on DNA-DNA hybridisation. The tube-nosed seabirds lost their exclusive order:

Order Ciconiiformes
Family Procellariidae
Sub-family Diomedeinae (albatrosses)
Procellariinae (petrels and diving petrels)
Hydrobatinae (storm petrels)

Seabird researchers have tended to ignore this revision.

[4] Latin: *procella* – storm, hurricane, tempest.
[5] more accessible in Fürbringer (1902).

Peter Harper (1978) identified ancient traits in the plasma proteins of storm petrels, later confirmed by Viot *et al.* (1993), and Sibley & Ahlquist (1990) also put the storm petrels ancestral (basal) in branching diagrams (Fig. 2.1). Nunn *et al.* (1994) at first believed sequences of mitochondrial cytochrome-*b* indicated that albatrosses were the earliest family, but later suggested two ancestral (paraphyletic) lines of storm petrels (Nunn & Stanley 1998).

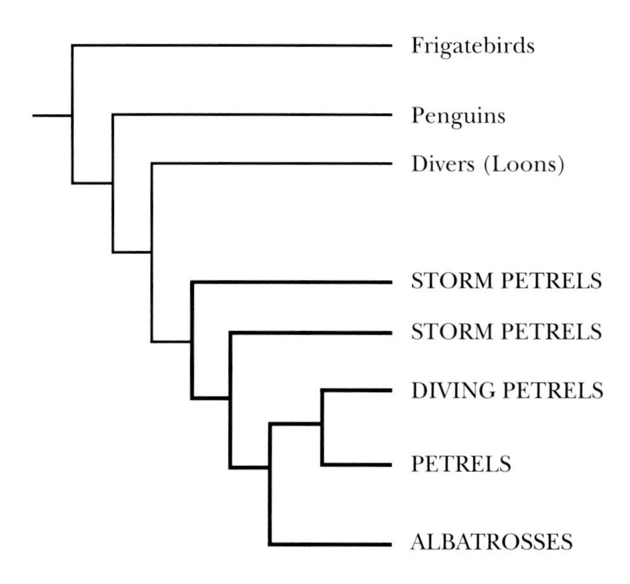

Fig. 2.1 Relationships between tube-nosed seabirds and some other seabirds based upon data from DNA-DNA hybridisation (Sibley and Ahlquist 1990) and mitochrondrial DNA sequencing (Nunn & Stanley 1998).

SIZE

Living albatrosses are clearly distinguished by their large size. The great albatrosses, with wing spans of about three metres are by far the biggest birds of the ocean. Other albatrosses have wing spans of about two metres, but there is great variation in mass and other dimensions (Table 2.1).

	Mass (kg)	Wing span (m)	Tail (cm)	Bill (mm)
Mollymawks	1.7–5.3	1.90–2.03	19–25	102–141
Sooty albatrosses	1.8–3.7	2.18	23–32	97–121
Great albatrosses	4.6–11.9	2.71–3.23	18–25	130–188
Galápagos Albatross	2.8–4.1	2.00–2.50	13–16	134–160
Gooneys	2.4–4.3	2.00–2.20	13–16	94–113
Steller's Albatross	5.1–7.5	2.17	16–18	129–141

Table. 2.1 The size of albatrosses: extreme ranges of dimensions of the main group of albatrosses. For more detailed measurements see Appendix 3.

Albatrosses are not just scaled-up petrels, two species of giant petrels fulfil that paradigm. They equal or exceed small albatrosses in mass and white phase individuals are often mistaken for albatrosses. However, in spite of their large size, giant petrels have remained conspicuously petrels in other characteristics. They are unique specialists and their size may have co-evolved with the habit of scavenging seal carcasses on land. Other petrels only rarely feed on marine organisms cast up on beaches.

There is a marked absence of petrels or albatrosses weighing around 1.5 kg with wing spans of about 1.5 m (Warham 1996). Relatively few petrels are heavier than 1 kg and more than 1 m in wing span, but there are many smaller species. At the lower end of the range the prions are 100–200 g and have wing spans of about 60 cm. Diving petrels are just over 100 g with wings spanning about 40 cm, while the smallest storm petrels average only 20 g with wing spans of 31 cm (Warham 1990).

Sexual dimorphism in size varies greatly in different groups of tube-nosed seabirds. In albatrosses and some petrels, males are larger than females, while in storm petrels and some diving petrels the reverse occurs. In some species the difference may be conspicuous in the field when the male and female are together at a nest, but often it is apparent only after statistical tests on measurements.

PLUMAGE AND MOULT

Most albatrosses have black and white plumage resembling black-backed gulls. Some are almost completely white with black wing tips rather like gannets and quite unlike anything found among petrels. Petrels have a great range of plumage patterns ranging from pure white to all black, and sooty albatrosses resemble some of the darkest petrels. Galápagos Albatrosses have plumage remarkably similar to adult giant petrels. Other colour is absent from the plumage of albatrosses, except on the heads of Steller's and Galápagos Albatrosses, which are tinged with yellow, like those of gannets.

In most albatrosses and petrels, plumage is similar in both sexes; there are sometimes slight differences, recognisable when pairs are seen together. The juvenile plumage of most petrels is similar to that of adults, but in albatrosses, it is usually distinguishable, sometimes conspicuously darker and gradually changing to the adult condition over many years.

The pattern of body feathers (pterylosis) in most albatrosses differs slightly from those of petrels. The ventral tract in albatrosses is like that of petrels, but except for sooty albatrosses, the dorsal tract of the divided anterior portion is composed of larger feathers separated by a space from the posterior, lumbar portion (Nitzsch 1867, Forbes 1882).

The gross structure of feathers is similar. Contour feathers are double, but the second vanes (aftershafts) are very small, even vestigial in albatrosses. Albatross flight feathers (remiges) were considered by Chandler (1916) to be more specialised than those of petrels; the barbules having more 'frills' than in any other known feathers. Moult is an energy demanding activity. Most birds do not breed and moult at the same time. Petrels complete their moult between breeding seasons. Feathers may begin to be dropped during the latter part of the summer, while birds are still visiting young, but most are replaced during the winter at sea, when flight in some species may be impeded (Voous 1970). The two giant petrels are a notable exceptions; they begin moulting during the breeding season and complete it afterwards (Hunter 1984).

Feathers grow at roughly similar absolute rates, so small birds replace their plumage more rapidly than big birds (Langston & Rohwer 1996). The smaller albatrosses, which are similar in size to giant petrels, cannot replace all their flight feathers in the time available between two successful breedings. They have to preserve flight capabilities, and although sometimes seen with a projecting feather or a narrow slot in one wing, they do not expose the conspicuous gaps and hatchet-shaped wing tips often seen in giant petrels. Scavenging seal carcasses on beaches perhaps minimises flight costs for giant petrels.

After the first year, moult in albatrosses becomes more prolonged and complex than in most petrels. Several albatrosses gradually change their plumage from a dark to a lighter condition over many years and a similar sequence also occurs in giant petrels.

ANATOMY

The internal organs of petrels and albatrosses: heart, liver, kidneys, testes, ovaries and intestines are surrounded by air-sacs, which appear to form shock absorbers around each organ (Hamlet & Fisher 1967).

The skeletons of albatrosses are generally more pneumatised than those of petrels, perhaps reflecting size differences. Air enters the dorsal vertebrae of petrels from small holes in depressions on the sides of each vertebra; in albatrosses there is a single hole in the middle of each centrum. The large wing bones (humeri) of albatrosses contain extensions of air sacs which are lacking in those of petrels.

The posterior margins of albatross skulls are only slightly convex compared with the strongly arched crania of the smaller petrels (Condon 1939 in Warham 1990). There are differences in the shape of the lacrymal bone. It is connected to the palatine by a ligament which in some petrel skulls has a minute sesamoid nodule, but in albatrosses there is a long ossicle (Forbes 1882). All petrels and albatrosses have longitudinally cleft (schizognathous) palates, allowing air to pass between the mouth and nasal chamber. In albatrosses this cleft is very narrow and in front it is filled by the end of a downward curved vomer (Forbes 1882, Warham 1990).

The dorsal vertebrae of petrels have ventral, spinous processes (hypapophyses) which are absent from albatrosses (Kuroda 1954). The breastbones (sterna) of albatrosses are relatively shorter and broader than in most petrels, but those of some gliding petrels approach the same shape (Forbes 1882, Warham 1996).

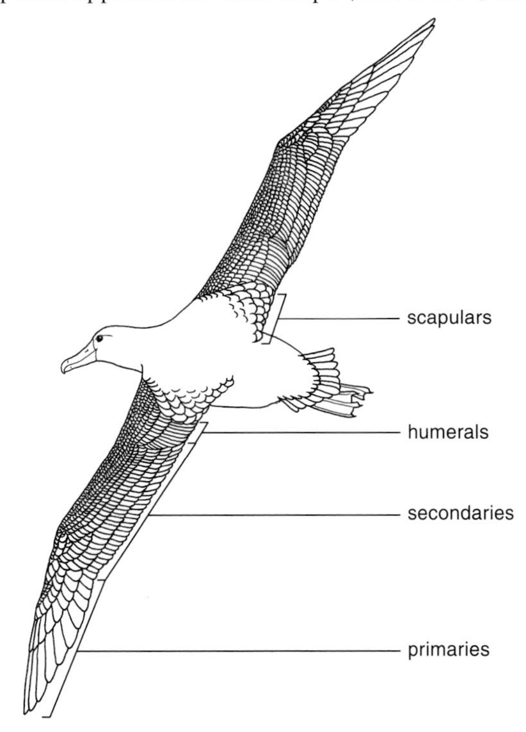

Fig. 2.2 Steller's Albatross showing the dorsal wing feather tracts (drawn by Robin Prytherch from a photograph by Hiroshi Hasegawa).

WINGS

Petrels and albatrosses all have long, narrow wings (Fig. 2.2, Table 2.2). Pennycuick (1989) has described how they should be measured for flight studies and stressed the importance of including the body width.

The shape and flight characteristics of many albatross and petrel wings have been discussed by John Warham (1996). Body mass and wing area were shown to approximate to a logarithmic function in which great albatrosses and giant petrels have the highest wing loadings.

Elongation of the wings has involved equal lengthening of the humerus and radius/ulna, with the result that the outer wings (manus plus primaries) become proportionally shorter in the larger petrels and albatrosses (Table 2.3). When folded, the skeletal outer wing (manus) of birds is tucked under and inside the radius/ulna, but the secondaries and secondary coverts of most birds only partially cover the folded primaries which remain visible along the side of the body. Most petrels, whose outer wings are long relative to the folded inner wings, are like other birds in this respect. In the folded wings of albatrosses, the relatively long bones project backwards and

Group	Mass (kg)	Wing span (m)	Wing area (m^2)	Aspect ratio
Great albatrosses	8.46	3.01	0.592	15.3
Small albatrosses	3.53	2.18	0.344	13.8
Giant petrels	3.25	1.98	0.326	12.0
Large petrels	1.08	1.41	0.167	11.9
Fulmars	0.82	1.13	0.119	10.7
Shearwaters	0.53	0.87	0.077	10.2
Prions	0.16	0.63	0.047	9.8
Storm petrels	0.03	0.35	0.016	7.9

Table 2.2 Wing characteristics of albatrosses and petrels (Pennycuick 1999).

	Mass (kg)	Ratio
Wandering Albatross	8.68	1 : 1.4
Black-browed Albatross	3.51	1 : 2.0
Southern Giant Petrel	4.55	1 : 2.0
Grey Petrel	1.02	1 : 2.5
Northern Fulmar	0.76	1 : 2.8
Sooty Shearwater	0.85	1 : 3.0
Dove Prion	0.16	1 : 3.4
Storm Petrel	0.03	1 : 4.5

Table 2.3 Ratios of the lengths (mm) of wing bones in albatrosses and petrels. Ratio = radius/ulna : manus+primaries (after Warham 1996).

upwards at the elbow, so that the secondaries and their coverts cover the primaries except for the ends crossed over the tail. The same arrangement is also found in giant petrels. It is most conspicuous in great albatrosses where the folded primaries are almost hidden by the inner wings (Warham 1990).

The wings of petrels and albatrosses each have ten functional primary remiges and a small alula. Albatrosses have 25–34 secondary remiges, giant petrels have 26–29 (Hunter 1984), most medium sized petrels about 20, the smallest prions as few as 13 and storm petrels only ten. Humeral feathers (overlaid by scapulars) form the innermost wing surfaces of albatrosses and giant petrels, but are probably few or absent in most petrels.

The long bones of petrel and albatross wings differ. In many petrels the radius and ulna are bowed to accommodate muscles between them; in albatrosses these muscles are poorly developed and the bones are long, thin and close together (Kuroda 1954).

Structures on the long bones of petrels and albatrosses appear to assist the prolonged extension of the wings for gliding flight. At the elbow joint of the extended wing the two heads of the radius lock into corresponding articular depressions on the ulna and introduce a rigidity that is believed to relieve muscles of that function (Joudine 1955, Yudin 1957).

The triangular shaped skin (propatagium) in front of the elbow is held taut by a fan of tendons, tensioned by the extended bones of the wing. In many petrels the tendons alone are adequate to the task, but in all albatrosses and some petrels, extra support is present. Very small bones (sesamoid ossicles) articulate with a forward-pointing (ectepicondylar) process on the outer head of the humerus. They appear to give a strut-like support to the long tendon (tensor patagii longus) that forms the leading edge of the wing (Mathews 1936, Brooks 1937, Warham 1996).

Flight muscles in most birds average about 16% of the body mass. Small petrels that do a lot of flapping have almost as much (diving petrels 13%, prions 14%), but the flight muscles of most albatrosses and giant petrels amount to less than nine percent of their body mass and in great albatrosses it is less than six percent (Pennycuick 1999).

Petrels and albatrosses have the divided pectoral muscle characteristic of gliding birds. The larger superficial muscle provides power for flapping, while the smaller, deeper muscle is a slow, long-acting (tonic) muscle responsible for extending the wing (Kuroda 1961, Pennycuick 1982b). The muscle that elevates the wing (supra-coracoideus) is well developed but varies in length. In petrels it is homogeneous, but in albatrosses it is divided into four separate parts which unite before passing through the shoulder-pulley (foramen triosseum).

The biceps muscle is very much reduced. It is best developed in albatrosses where it arises by two heads, on the coracoid and humerus, but they soon unite into one muscle. From the coracoid head a very long narrow tendinous biceps-slip reaches the edge of the propatagium, where it joins the tensor patagii longus near the elbow. In petrels, the coracoid head alone becomes the main muscle while the humeral head becomes detached and the very thin muscle inserts as a short cylindrical tendon in the tensor patagii longus near the shoulder. Thus in albatrosses, the biceps-slip is derived solely from the coracoid head while in petrels, the slip represents the shorter or humeral head of the biceps (Forbes 1882).

Albatrosses and giant petrels have tendon mechanisms that lock their wings at the shoulders and prevent them lifting above the horizontal when gliding. These shoulder locks are believed to reduce strain on the tonic muscles (Pennycuick 1982b).

When landing, most birds lower the rear of the body, so that the longitudinal axis ceases to be horizontal and the wings come to beat forward, applying braking forces (Rüppell 1980). Some birds remain horizontal for longer, rotating their forewings (radius and ulna) along their axes (supination), thus imparting a downward and forward fanning motion to the feathers. The twisting occurs at the elbow leaving the humeral feathers, next to the body, in the horizontal plane (Fig. 14.12). This action is probably common in petrels (Pennycuick & Webbe 1959) and it is particularly conspicuous in albatrosses (Scholey 1982).

TAILS

Most albatrosses and petrels have relatively short tails. Twelve feathers (rectrices) are normal, but at least two fulmarine petrels have 14 and giant petrels have 16 (Forbes 1882).

FLYING

Petrels and albatrosses all soar on sea-surface winds, but petrels, especially the smaller shearwaters and prions, employ powered flight (by flapping) much more than albatrosses. The slow flapping of broad-winged storm petrels and the fast whirring wings of diving petrels are quite unlike anything in albatross flight. Albatrosses, petrels and storm petrels all spend more time flying across wind than in any other direction (Spear & Ainley 1997).

The distances petrels and albatrosses are capable of flying was appreciated long ago; birds with distinguishing marks were seen to follow vessels for thousands of miles. Many were captured alive and released carrying labels inscribed with messages. On 30 December 1847, Captain Hiram Luther of the whaler *Cachalot*, shot a great albatross off the coast of Chile at 43° 24'S, 79° 05'W. Tied around its neck was a vial containing the message:

> 'Dec. 8th, 1847. Ship "Euphrates", Edwards, 16 months out, 2300 barrels of oil, 150 of it sperm. I have not seen a whale for 4 months. Lat. 43°S., long. 148° 40'W. Thick fog, with rain.'

The bird had flown[6] at least 5,466 km in 22 days or 248 km per day and until the advent of satellite telemetry in the 1990s, it remained the best record of albatross flight between two locations at sea (Murphy 1936, Snyder 1958).

Some petrels, including the small storm petrels, have long migrations equal to those of many albatrosses. In contrast, the Galápagos Albatross remains within a small area of ocean and travels much less than many petrels.

[6] Great circle distance recomputed at the US Naval Academy, Annapolis.

Swimming

Most petrels and albatrosses feed on the surface of the sea, but some feed underwater. In addition to the specialist diving petrels, the shearwaters include accomplished divers capable of swimming deep, using flexed wings and feet (Warham 1996).

Albatrosses are buoyant; they float high on the water and have long been characterised as picking or scooping food from the surface (Ashmole 1971). Nevertheless, mollymawks are able to dive and swim, their eyes functioning equally well under water as in air (Martin 1998), and sooty albatrosses appear to approach the underwater proficiency of shearwaters (Prince *et al.* 1994).

Standing and Walking

The feet of storm petrels, petrels and albatrosses are webbed between three toes. The hallux is a vestige, showing at least a claw in petrels, but is hardly visible in albatrosses. The thigh muscles of albatrosses have one minor muscle which is not found in petrels (Forbes 1882).

On land, most petrels remain underground in burrows, hidden in dense vegetation or high-up on cliff ledges. Much of the time they are sitting down, with the tarsometatarsi flat on the surface in line with the feet. When moving, they usually raise the tarsometatarsi just off the ground, open the tarsometatarsi-tibiotarsi angle slightly and shuffle forwards; a mode quite adequate for most of their needs. They can usually raise themselves higher for short bursts of activity. Giant petrels are far more adept on land than other petrels, but they have typical petrel build and move with ungainly prancing steps, unless helped by extended wings.

Albatrosses are much better equipped to stand poised and to walk. They may appear rather determined, compared with the confident sauntering of gulls, but their gait is well balanced and there is no doubt that they are capable walkers.

Salt-excreting Glands

Many species of birds have paired glands in curved grooves above the orbits of the skull (Shoemaker 1972). When these birds take in more salt than can be handled by the kidneys the glands secrete hypertonic saline. The size of the glands varies with the amount of salt processed and are important in seabirds, but even within the albatrosses some species have larger salt glands than others.

Ducts discharge the fluid into the first nasal chamber (anterior concha). It flows forward and out through the tubes of petrels, but in albatrosses there is a way out under the folded naricorn and into the lateral grooves just below and outside the tubes i.e. a functional separation from the airway (Bang & Wenzel 1985). In sitting albatrosses discharged saline runs down grooves between the culminicorn and latericorns to drip off the tip of the hooked bill. In flight, airflow over the head may be expected to impede such passive flow. The saline spray photographed leaving a storm petrel nasal tube (Schmidt-Nielsen 1960) was evidently produced by 'snorting' air out through the nasal tubes; a venturi effect would draw saline from the ducts as an aerosol, to be carried away in the slipstream of the flying bird; other petrels and albatrosses may do the same in flight.

Albatrosses may secrete less salt than storm petrels, but all secrete proportionally more than other seabirds such as penguins and gulls (Schmidt-Nielsen 1960). They may obtain enough water from prey with body fluids hypotonic to seawater (Costa & Prince 1987), but captive Laysan Albatrosses fed fish isotonic with their own blood, needed to drink seawater and died if it was not available (Frings & Frings 1959).

Bills

All petrels and albatrosses have sharply hooked bills. The horny sheath (rhamphotheca) of petrels is composed of eleven separate plates. In albatrosses there is an extra piece, a thin wedge (inter-ramicorn) between the lower mandibles near the symphysis. The rhamphotheca is soft in chicks and hardens with age, but may retain fleshy strips between the plates. Ridging at both ends of the culmen in some albatrosses may indicate periodic accumulation of keratin; but they are quite different from the encrusting laminations at the base of giant petrel bills which look like the remains of keratin that has flaked off over years.

In most young petrels and albatrosses, the rhamphotheca is black or dark brown and remains more or less so in many adults, but in some species the dark pigmentation is a juvenile condition that is replaced as the birds age. Albatrosses have more conspicuously coloured bills than most petrels. Colour is present on the bill plates and strips of skin; unique coloured margins at the base of the mandibles are continuous with the edges of the gape. In some species the base of the brightly coloured culminicorn is surrounded by contrasting black skin. In a few species, the rhamphotheca is unpigmented in the chick and remains so throughout life.

TUBES

Tubular nostrils are the conspicuous characteristics of all albatrosses and petrels[7]. In diving petrels they are much reduced. In petrels the tubes are fused together on top of the bill behind the culminicorn; in albatrosses they are on either side of the bill separated by the culminicorn. From both positions, grooves slope down on either side of the culmen towards the hooked tip (unguis). The lateral position of the two separate tubes in albatrosses unequivocally distinguishes them from all petrels.

The fused nasal tubes of most petrels are superficially soft with the naricorns embedded in skin whereas albatross tubes are hard exposed keratin held in place at their periphery by thin strips of skin connected to the edges of the latericorns and culminicorn.

At the time of hatching, albatross naricorns are bulbous, but not completely formed. In adult Wandering Albatrosses, the airway of each tube is separated by a vertical fold (Forbes 1882). These folds appear during the first two months after hatching and are pushed slowly upwards during growth. They do not achieve their final shape until the fledglings are about to fly (Voisin 1969).

The positions of nasal tubes in petrels and in albatrosses are so different that questions arise about their position in the common ancestor. There are no bony formats, so nasal tubes are not present in fossils. Skulls have large external nares; those of petrels are close together near the culmen ridge while those of albatrosses are in a lateral position. Forbes (1882) assumed that in the ancestral condition, separate tubes evolved above each nostril, as in present-day albatrosses, and Cracraft (1981) considered that the morphology of albatrosses was primitive compared with storm petrels. Cladistic analysis of molecular characters, however, puts storm petrels basal to petrels and albatrosses in branching diagrams (Fig. 2.1). During the early development of Kerguelen Petrel chicks, lateral flaps of skin grow over the nostrils and fuse above (Studer in Warham 1996). In ancestral albatrosses, during the evolution of gigantism, superior fused tubes could have parted to accommodate a rearward extension of the culminicorn. The separate tubes would then be free to migrate forward, between the culminicorn and latericorns to their present lateral positions. Indeed, some albatrosses still have areas of skin at the base of the bill that may represent the remains of an ancestral state after migration of the tubes and before the extension and butting of the latericorns with the expanded culminicorn. It is perhaps significant that when gigantism evolved within the fulmarine petrels, the fused tubes remained on top of the bills of giant petrels and became proportionately longer than in normal sized fulmars. The resulting naricorns are comparatively massive compared with those of albatrosses.

SENSE OF SMELL

Seamen have long believed that seabirds are attracted from a distance by the smell of heated galley fat (Murphy 1936). Field and laboratory experiments reviewed by Warham (1996) confirm olfaction in petrels, but are less positive for albatrosses. Black-footed Albatrosses distinguish bacon fat from turpentine controls (Miller 1942), but several other species show little or no response to olfactory cues (Lequette *et al.* 1989, Verheyden & Jouventin 1994).

Petrels and albatrosses have large olfactory lobes in the brain and histology of these suggest that the birds have an acute but not very discriminating sense of smell (Wenzel & Meisami 1990). In a rank order of birds with

[7] Similar structures are found in oilbirds.

olfactory equipment, petrels and storm petrels are at the top with albatrosses ranked below them (Bang & Wenzel 1985). Experiments at sea confirm that albatrosses do not respond to dimethyl sulphide which attracts some petrels and storm petrels (Nevitt *et al.* 1995).

Anatomy of the nasal passages suggests that nasal tubes funnel air into anterior chambers (conchae), where each air stream is divided towards septate scrolls of olfactory and respiratory surfaces. When the bill is submerged, a valve prevents water entering the olfactory spaces (Bang 1966).

RESPIRATORY AIRWAYS

In some petrels the trachea is divided by a medial vertical septum into left and right airways that extend forward from the bronchial divide and end a short distance before the larynx (McLelland 1989). It is not known whether albatross tracheae are similarly divided, but in some, at least, the bronchi of males are elongated with convolutions and/or bulges while in females they are short and simple (Swinhoe 1863, Forbes 1882).

The internal organs are surrounded by air sacs. Only minor differences are apparent between those of albatrosses and petrels. The large air sacs seen in the throat of a Wandering Albatross (Warham 1996) were discovered to arise from the bronchi of each lung (Hamlet & Fisher 1967).

MOUTHS

The lining of the mouth in petrels and albatrosses may be comparatively smooth or furnished with longitudinal rows of pointed papillae-like spines. These spines evidently serve as barbs to maintain a grip on slippery prey and vary between species in length and distribution. The tongue likewise differs in shape and the quantity of papillae it carries. Nestling albatrosses may have more papillae than adults (Forbes 1882, Warham 1996).

STOMACHS AND OIL

The oesophagus of petrels and albatrosses opens without constriction into a ridged, glandular proventriculus. When full, it expands to becomes a great bag of water, oil and solids occupying most of the abdominal cavity. The gizzard is small.

At one time, the well known pink-orange oil was believed to have been secreted by the wall of the proventriculus, but several investigations concluded that it is of dietary origin (Clarke & Prince 1976, Warham 1977). It is present in quantities commensurate with the diet and the length of time the food is retained before delivery to a chick. Water and oil separate by gravity in the proventriculus; the oil remains there longer than the water and acts as an energy store. The composition of stomach oils are similar in petrels and albatrosses (Warham 1996). Thirty-one samples from 11 species had a mean calorific value slightly below the value for commercial diesel oil (Warham 1977). The oil accumulating in the proventriculi of albatrosses and petrels contains wax esters, which are not found in diving petrels. Wax esters occur in some other birds, but are not digested.

REGURGITATION

It is often claimed that petrels eject oil through the nasal tubes. Oil may indeed seep up from the mouth into the nasal passages, but it does not take much experience to establish that a fulmar shoots through its open bill. The oil may contain fragments of food that could not be ejected through the narrow tubes. Petrels and albatrosses vary in the readiness with which they let go their stomach contents.

Early naturalists noted fragments of fish, squid and crustaceans in these regurgitations. Albatross adults rarely disgorge, but when alarmed, their chicks vomit up considerable quantities. It is a simple matter to approach a nestling, grab its throat before it spews and up-end it over a bucket. During the first collections at Bird Island in 1958–59 and 1960–61, 120 kg of regurgitated food was obtained from 220 sympatric Black-browed and Grey-headed Albatross chicks (Tickell 1964), but although taxonomists could identify crustacean exoskeletons, they could not do the same for the horny beaks of squids or jumbles of fish bones, so analysis was limited. By the time Peter Prince (1980) had devised quantitative methods, Malcolm Clarke was constructing keys for the identification of squid beaks and hundreds of kilograms of regurgitations were analysed. Later it became possible to identify digested fish from their otoliths (Reid *et al.* 1996).

BEHAVIOUR

Most petrels nest in burrows or rock crevices which they visit only at night. They communicate vocally and perhaps by olfaction (Bretagnolle 1990). Visual communication is apparent in the few species of petrels that are active in the open during daylight; the signals are simple and movement is restricted on cliff ledges. Giant petrels nesting on gentle slopes and flat ground have much more mobility, but their visual signals and vocalisations remain typically fulmarine (Warham 1996).

Albatross communication is clearly more complex, involving repertoires of quick, synchronised postures, exposing conspicuous head patterns and bill colours, accompanied by a range of sounds, both dramatic and subtle.

BREEDING

Albatrosses and petrels usually return to the areas in which they were reared (philopatry), although there have been notable exceptions. Fidelity to one nest or nest area supports monogamy. Experienced pairs often remain together for many years and until one member dies. Separation and re-pairing of both partners (divorce) is infrequent, especially in albatrosses. Juvenile petrels and albatrosses spend some years at sea before returning to their birth place for the first time. They then visit the breeding grounds for several years before acquiring a mate and breeding.

Early sealers on many islands knew that great albatross chicks remained in their nests throughout the winter (Earle 1832, Goodridge 1832). Most of them probably guessed that the adults did not breed every year, but one at least imagined a different scenario.

Richard Harris was a seaman on the *Betsey and Sophia* when it was wrecked at Kerguelen in March 1831 (Savours 1961). He visited Wandering Albatross nests at intervals throughout the winter until he was rescued the following December. Many years later, when he was an engineer officer of the *Adventure*, he related these events to F.W. Hutton (1865), a passenger on the vessel. Harris claimed that Wandering Albatross chicks were abandoned by their parents sometime between February and June – to survive on stored fat. When these parents returned in October to breed again, they greeted their chicks, but did not feed them.

More convincing observations discredited this notion. In September 1874, Henri Filhol (1885) watched a Southern Royal Albatross[8] feeding a fledgling at Campbell Island (p.58) and in the late 1920s Harrison Matthews (1929,1951) saw Wandering Albatrosses feeding chicks at South Georgia. Nevertheless, Robert Cushman Murphy (1936) clung to Harris's story in support of annual breeding. The so-called 'starvation theory' was not finally demolished until Richdale (1939) weighed a Northern Royal Albatross chick throughout the winter of 1939 and also confirmed that its parents did not breed the following season.

Birds that rear no more than one young every two years are known as *biennial* breeders, but the name obscures the fact that some pairs defer breeding for two, three or even four years. In the late 1930s V.C. Wynne-Edwards (1939) noticed that some Northern Fulmars did not breed every year, and the habit has since been found in other petrels (Zotier 1990, Chastel *et al.* 1993, Mougin *et al.* 1997).

BREEDING SEASONS

Most petrels and albatrosses lay their eggs in the spring-summer of the hemisphere in which they breed. On islands in high latitudes with cold winters, eggs are usually laid in the spring, but in lower latitudes, and especially on sub-tropical islands some populations lay much later, even in winter. North Pacific albatrosses lay their eggs in the autumn-winter.

Southern hemisphere breeding seasons usually begin in one calendar year and end in the next, for example 1958–59[9], but several populations are exceptions to this rule and complete their breeding within one calendar year.

[8] At the time Filhol believed they were Wandering Albatrosses.
[9] When identified by a single calendar year, it is usually taken as the year in which the egg was laid i.e. 1958 in the above example.

PRE-EGG PERIOD

Within the albatrosses and petrels, there is great variation in nest attendance early in the breeding season (Warham 1990). Nevertheless, the male generally spends more time ashore than the female before the egg is laid.

Female absence at this time is usually explained as a nutritional necessity associated with egg production. Perhaps coincidentally, males of some albatrosses appear to benefit by having their mates at sea, safe from the attentions of other males.

NEST BUILDING

Warham (1990, 1996) has reviewed the wide range of nests and nest-building behaviour in petrels. All albatrosses are surface nesters, active in daylight and at night. Apart from those that favour cliff ledges, most nest on steep ground or exposed open slopes, sometimes almost hidden in tall tussock grass or shrubs. Some albatrosses nest under trees.

Neither petrels nor albatrosses carry nest material. Albatrosses pick up material with their bills, but they do not walk with it, nor is there any proof that petrels do so. It is gathered within reach of a bird on a nest or standing/ sitting nearby. The vegetation or mud is pulled up from in front and tossed or placed to one side or the other in a stereotyped over-the-shoulder action. It may be shuttled backwards in this manner for several body lengths to the nest. These simple actions can move considerable quantities of material and even result in the entrances to petrel burrows being blocked. It should not be imagined that this is primitive behaviour. There is evidently strong selection in favour of not wasting time and energy carrying material that can be had close by. Penguins, frigatebirds, grebes and divers (loons) are all committed transporters of nest material, but some phylogenetically more distant groups, such as wildfowl, do not carry nest material and move it in the same way as petrels and albatrosses (Harrison 1967, Wilmore 1974, Ogilvie 1978).

Providing there is material available, all albatrosses build nests; many of them are large and persist for years. Some colonies are limited by lack of material, but fragments of nests can usually be found and nest building actions detected. Where there is no vegetation, stones and bones are used by both petrels and albatrosses, but fulmarine petrels are much more proficient at using stones than albatrosses.

COPULATION

Copulation occurs on land and has been seen more frequently in open-nesting albatrosses, giant petrels and fulmars than in burrow-nesting species. Shearwaters copulating on the ground outside burrows are believed to have been birds that had failed to acquire burrows in crowded colonies (Warham 1990,1996). In experienced mated pairs, copulation usually occurs without conspicuous preliminary behaviour by either participant, but soliciting female postures have been recognised in both albatrosses and fulmars (Fisher 1971, Hatch 1987).

Extra-pair copulations have been seen many times in albatrosses (p.304) and fulmars. Co-operation is essential if insemination is to follow; female birds remain in control and are capable of thwarting males even when overpowered (Birkhead & Møller 1993). In the Northern Fulmar, 2.4% of all observed copulations were with birds other than a mate; but although extra-pair copulations had the same probability of insemination as those of pairs, mated females copulated far more often with their own males. DNA fingerprinting confirmed that the chicks from females that had copulated with other males were the genetic offspring of the male of the pair (Hatch 1987, Hunter *et al.* 1992).

EGGS

Albatross and petrel females lay only one egg in a season, but a few Manx Shearwaters (Perrins *et al.* 1973) and storm petrels (Davis 1957) are believed to have laid replacement eggs. The presence of two eggs in an albatross nest indicates the activity of a second female (Tickell & Pinder 1966, Fisher 1968), however, one giant petrel female is believed to have laid a clutch of two (Warham 1996). Two follicles may develop in a female, but in one Western Yellow-nosed Albatross, the oviduct solved the physiological problem by packaging the two yolks in a single abnormally long shell instead of two separate shells. Very small eggs are occasionally laid by albatrosses and probably by petrels.

Females appear to lay eggs only after establishing pair-bonds and copulating with mates, but one Southern Royal Albatross is known to have laid eggs over a number of years, during which time no mate was seen (Robertson 1993b). A solitary Steller's Albatross at Midway Atoll is also believed to have laid eggs in three seasons (Elizabeth Flint pers. comm.).

The eggs are white, but in albatrosses, some storm petrels and less often in other petrels there is usually a ring of peppered red-brown spots around the broad end (Warham 1990); these are believed to be dried blood. On some eggs the spots are so fine as to be barely visible without a lens, on others they form a dense cap over the whole of the broad end and may also include larger blotches (Verrill 1895, Frings 1961, Fisher 1969).

Egg mass increases with body mass (Rahn *et al.* 1984). The eggs of albatrosses, petrels, storm petrels and diving petrels are relatively larger than those of other seabirds except auks. Although albatross eggs are the largest, they are relatively small (<7% of body mass) compared with those of smaller petrels, particularly the storm petrels (>16% of body mass).

In shape, albatross eggs tend to be proportionally longer and thinner than petrel eggs; long sub-elliptical to oval in the terminology of Palmer (1956). The shells are thicker than those of petrels, but relatively thin for their size and mass; those of great albatrosses are quite fragile. Calcium in the diet of albatrosses may be low or conserved for other purposes (Houston 1978).

Smaller eggs tend to have proportionally larger yolks. The yolk of a Northern Royal Albatross was 26% of the 408 g egg, while the average of two White-faced Storm Petrel yolks was 39% of the 12 g eggs (Warham 1985).

Petrel and shearwater eggs concealed in burrows and crevices often survive cooling during interrupted incubation. Albatross eggs are so conspicuous that unattended eggs are usually broken by predators, such as skuas, gulls, sheathbills, caracaras and mocking birds. Neither albatrosses nor petrels retrieve eggs that have rolled out of their nests. After hatching, shell fragments may be pushed out by the movements of parents or chicks, but usually they are trampled into the nest.

The period over which eggs are laid may be curtailed or prolonged. Individual females in successive years probably all lay over much shorter periods than that characteristic of the population.

FORAGING

Foraging seabirds are constrained by the need to return to their island nests from time to time. Albatrosses habitually traverse huge areas of ocean, but White-chinned Petrels fly even greater distances, as far as 3,495 km from their nests (Catard & Weimerskirch 1998).

Ranges are determined by how long chicks can withstand fasting and the imperative for adults to maintain their own physical condition (Weimerskirch 1998). Trade-offs between these conflicting demands, alternating long and short flights, were first demonstrated in albatrosses and have subsequently been discovered in some petrels (Weimerskirch *et al.* 1994).

Most albatrosses and petrels associate with other seabirds in multi-species flocks at marine resources; they often accompany marine mammals or large fish and frequently congregate at fisheries. Many forage at frontal zones and over bathymetric features such as continental shelves. Some species of albatrosses, petrels and storm-petrels are significantly more numerous above seamounts[10] that reveal no obvious surface clues to their location hundreds of metres below the surface, whereas other species are no more common than over the surrounding deep ocean (Blaber 1986, Haney *et al.* 1995).

FEEDING CHICKS

Albatrosses and petrels use the same characteristic technique for feeding their young. The adult regurgitates food by muscular contraction of the abdominal wall accompanied by retching. As the stomach contents are

[10] Undersea mountains rising above the ocean floor.

pushed up into the bird's throat, the chick places its open bill crosswise as a funnel inside the parent's open mandible. The jet of semi-liquid food goes straight down the chick's throat. Only if there are large undigested prey is there interruption or spillage.

CHICK GROWTH

Chicks of albatrosses, petrels and storm petrels usually become heavier than their parents (Fig. 2.3), due to an accumulation of fat. It is oil rendered from chick carcasses that has largely contributed to the value of petrels and albatrosses in the human economies.

Excesses of 111% to 155% adult mean mass have been measured in petrels, 115% to 166% in albatrosses and 138% to 170% in storm petrels. These peaks are achieved at 60% to 85% of the fledging periods in petrels, 57% to 81% in albatrosses and 60% to 95% in storm petrels. Several hypotheses have been proposed to explain this phenomenon. An anatomical study of the Northern Fulmar has recently concluded that fat helps fledglings survive their early days at sea, before they become proficient at foraging for themselves (Phillips & Hamer 1999). In albatrosses, Reid *et al.* (in press) have reached a similar conclusion, but with an earlier crucial role in feather development (p.352).

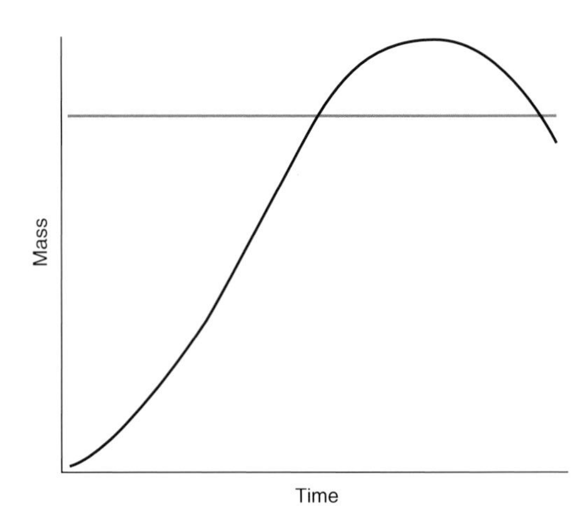

Fig. 2.3 Growth curve of a petrel/albatross chick from hatching to fledging. The straight line represents the average adult body mass.

Throughout their development, fulmar chicks lose body water, and during their last weeks ashore this accounts for a net loss of body mass (Phillips & Hamer 1999). Most petrel and albatross fledglings are still being fed, but less frequently, and with smaller meals. Only at a few locations are they definitely known to be deserted by both parents. By the time they fly, fledglings are about as heavy as adults, but storm petrels remain consistently heavier (Fig. 2.3).

BREEDING SUCCESS

The usual measure of breeding success is the proportion of eggs laid by pairs that yield fledglings which fly. It may vary from 100% in some small colonies, to total failure (0%) in large colonies (Prince *et al.* 1994; Saether *et al.* 1997). Sample size, colony location, age, experience, physical fitness, progress of moult, past and present nutrition all contribute to breeding success and interact with variation in the availability of food. In both petrels and albatrosses, individual determinants have greater influence than environmental factors (Croxall & Saether 1998). Although experience contributes to breeding success in petrels, which fail more often in their early breedings; at least some albatrosses breeding for the first time raise as many young as more experienced birds (Berrow *et al.* (in press)).

Scientific study may itself influence breeding success. In one year Northern Fulmars that were handled for ringing during the pre-egg period were 8% less successful in the same season than birds that were handled after hatching (Ollason & Dunnet 1978). Experience with some albatrosses suggests that they are less sensitive (Tickell 1968).

JUVENILES, SUBADULTS AND ADULTS

When fledgling petrels and albatrosses leave their nests and make their first flights seawards, they do not turn back. Most of them disperse widely over great distances. As juveniles they remain at sea continuously for some years. Sooner or later, though, their neuro-endocrine systems respond to photoperiodic cues and initiate a seasonal return to land. There is no way of knowing when and where this physiological change occurs, but the first return to a colony is evidence that it has already happened. It is an observable event and convenient to regard young petrels and albatrosses which have returned to the breeding grounds as no longer juveniles. They are subadults whose sex may not be obvious, but can be discerned.

The state of becoming adult is also difficult to determine. Physiological competence to breed precedes the acquisition of behavioural proficiency, which may take several years to acquire. It is impossible to judge exactly when the physiology matures, or when the behaviour is perfect, so field-workers adopt the next observable event. In petrels and albatrosses, subadults become adults when they share in the production of an egg; it may not produce a chick, or even hatch, but it is the beginning of breeding.

POPULATIONS

In this book a *breeding population* is defined as the total number of pairs in a defined area that have laid an egg at least once, including those that are not currently breeding. A *seasonal population*[11] is the number of pairs in a defined area that lay eggs in one season. Breeding populations are always larger than seasonal populations, conspicuously so in biennial breeding species and to a lesser degree in annual species.

There are difficulties in counting small to medium sized nocturnal, burrowing petrels. Large diurnal species like albatrosses and giant petrels nesting in the open are easier. World populations of tens, or hundreds of thousands are usual among petrels and albatrosses, but some, such as the Bermuda Petrel and Steller's Albatross, occur in very much smaller numbers. The most numerous species of petrels numbers more than twenty million birds; only one albatross species has over a million birds (Warham 1996).

LONGEVITY

Albatrosses and petrels are long-lived seabirds. Some storm petrels were still alive at 29 years or age and shearwaters at over 40 (Warham 1996). The oldest known fulmar was said to have been over 50 years old and an albatross over 60 when they ceased returning to their nests (Robertson 1993c, Anon 1997).

PARASITES AND DISEASE

Chewing lice (Phthiraptera: Ischnocera & Amblycera) and feather mites (Acarina: Astigmata: Analgoidea) are common avian ectoparasites (Appendix 5). In addition to their significance as parasites, some chewing lice have long been believed to reflect the systematics of birds (Harrison 1916, Hopkins & Clay 1952, Paterson *et al.* 1993). Two genera, *Austromenopon* and *Saemundssonia* have been found on albatrosses, giant petrels, fulmars, gadfly-petrels, shearwaters, prions and storm petrels. *Perineus* occurs on albatrosses, giant petrels and fulmars; *Docophoroides* and *Paraclisis* on albatrosses and giant petrels, while *Harrisoniella* and *Episbates* are unique to albatrosses. Within these genera, no species found on albatrosses have been reported on any petrels, shearwaters or storm petrels and *vice versa.* There are other genera of chewing lice on petrels that have not been found on albatrosses (Pilgrim & Palma 1982, Palma & Pilgrim 1984, 1987).

Among the feather mites, *Brephosceles* occurs on albatrosses, giant petrels and prions as well as on other water birds and waders. *Alloptellus* has been found on albatrosses, petrels, frigatebirds and pelicans while *Diomedacarus* occurs only on albatrosses. No species within these genera of feather mites is confined to just one host species (Atyeo & Peterson 1970, Cerny 1973, Peterson & Atyeo 1972).

Ticks and fleas are cosmopolitan. Ticks of the genus *Ixodes* occur widely on albatrosses, several petrels, prions and diving petrels as well as on penguins (Wilson, 1970). Fleas of the genus *Parapsyllus* have been found in the nests of albatrosses, various petrels, prions and diving petrels as well as of penguins, shags, skuas and gulls (Smit 1970).

Petrels and albatrosses suffer from similar virus infections, whose symptoms include blisters on the webbed feet and around the base of the bill (Warham 1990).

[11] Before the details of deferred breeding was fully understood, I coined the term *demi-population* for the Wandering Albatross (Tickell 1968). It now needs to be more precisely defined. A *seasonal population* can be counted on the ground, but the *demipopulation* is the average proportion of *breeding population* that breeds each season. It is determined from the records of ringed birds over several seasons.

PREDATORS

The natural predators of seabirds are mostly other birds. At sea, predation may be incidental to piracy (kleptoparasitism), but on the breeding grounds large numbers of petrels are killed by skuas, gulls and raptors. In the northern hemisphere foxes and otters may be locally significant while on some tropical islands snakes, skinks and sand crabs take chicks and eggs. Snakes take underground chicks on some Australian islands, as does the indigenous tuatara *Sphenodon* in New Zealand.

Albatrosses suffer less, largely because of their size but also because of their aggressive posture; giant petrels have the size advantage, but on some islands they are intimidated by skuas and leave their eggs.

Alien mammals, notably cats and rats introduced by man onto islands that have never known them, devastated many populations of petrels, but had little impact on albatrosses.

Marine predators – leopard seals, furseals, sealions and sharks – are opportunistic, taking a variety of seabirds. Only two albatrosses are known to be particularly vulnerable (p.232).

Man himself has been the most dangerous predator. Millions of petrels have been killed for food, bait and feathers. Albatrosses have also suffered, but their numbers could never have sustained equivalent levels of exploitation. Huge numbers of seabirds die as the direct or indirect result of marine industry, be it fishing, oil extraction or shipping. Most of these birds are auks, cormorants and petrels, but it is the hooked albatrosses that attract most publicity.

SOUTHERN ALBATROSSES

Wandering Albatross in flight (B. Osbourne).

3
THE SOUTHERN OCEAN

'This is the belt of the great west winds – the "Roaring Forties, Furious Fifties, and Shrieking Sixties." Throughout this zone, however, the most important climatic truth is the inconsistency of the weather. We speak as a matter of course of the "westerlies" of these latitudes, but in reality the winds are shifting all the time.'

Robert Cushman Murphy (1936)

The South Atlantic Ocean, Indian Ocean, and South Pacific Ocean each have their own separate characteristics, but they flow together around the Antarctic Continent. Captain Cook called this continuous expanse of water the Southern Ocean, a name that is still used, although some authorities insist upon maintaining the integrity of the separate oceans by writing Southern Oceans or southern oceans. Alternative names such as Antarctic circumpolar ocean or Antarctic Ocean are sometimes preferred. The northern boundary is not commonly agreed and scientists adopt whichever best suits their interests. For some it has been the Antarctic Front and for others the Subantarctic Front[1] or even the Subtropical Convergence.

The Southern Ocean may be considered one massive ecosystem. The distribution of water masses and main circulation were established millions of years ago. Many marine organisms are circumpolar in distribution, but productivity is greater in some regions, for instance at the Southern Boundary of the Antarctic Circumpolar Current where regional influences are conspicuous (Tynan 1998).

The bathymetry of the ocean (Fig. 3.1) influences events at the surface, both in the long and short term. The huge numbers of albatrosses breeding at the Falkland Islands owe their existence to the broad continental shelf from which the islands rise and areas of ocean above some seamounts are known to attract seabirds. It is a fair guess that the exceptional speciation among albatrosses around New Zealand is the consequence of varied hydrology to which bathymetry has contributed in no small part.

Atmospheric pressure remains high over the Antarctic and katabatic winds off the ice-cap join easterly winds that circle the continent. Farther north a succession of low pressure systems (cyclones), with associated fronts circle the Southern Ocean at a rate of about 7–10 per month in summer and slightly more in winter. They are generated in mid-latitudes and spiral inwards towards the continent. Over colder water, they fill and disappear. In summer, they are mainly between the continent and latitude 45°S, while in winter this area expands to about 30°S. Anticyclones in lower latitudes also move east, but more slowly. In summer, most of them are north of the low pressure belt, but in winter they become more frequent to the south, sometimes blocking the easterly progress of cyclones and causing notable changes in wind direction (Jury 1991). Such zones of opposing pressure systems with changeable winds are favoured by Wandering Albatrosses (Nicholls *et al.* 1997).

Tropical cyclones from SE Asia cross the equator before filling over the the southern Indian Ocean and the Tasman Sea. Although they are more frequent in summer than winter, there is little monthly variation and the highest winds of the Southern Ocean remain in the eastern hemisphere between longitudes 75°E and 150°E.

The effect of winds upon the ocean is cumulative over a time scale of decades. Even when interrupted, westerlies maintain the west to east flow of surface water, formerly known as the West Wind Drift and now usually referred to as the Antarctic Circumpolar Current (ACC). It is a deep ocean current with many conspicuous mesoscale[2] eddies and meanders (Fig. 3.2).

[1] The historic Antarctic Convergence of surface waters (Deacon 1984) has been discovered to consist of several convergencies or fronts. Separate temperature and salinity fronts occur at the Antarctic (Polar) Front and the Subantarctic Front which mark the southern and northern boundaries of the Antarctic Polar Frontal Zone (APFZ) (Open University 1989).

[2] 50–200 km long and lasting from one to several months (Open University 1989).

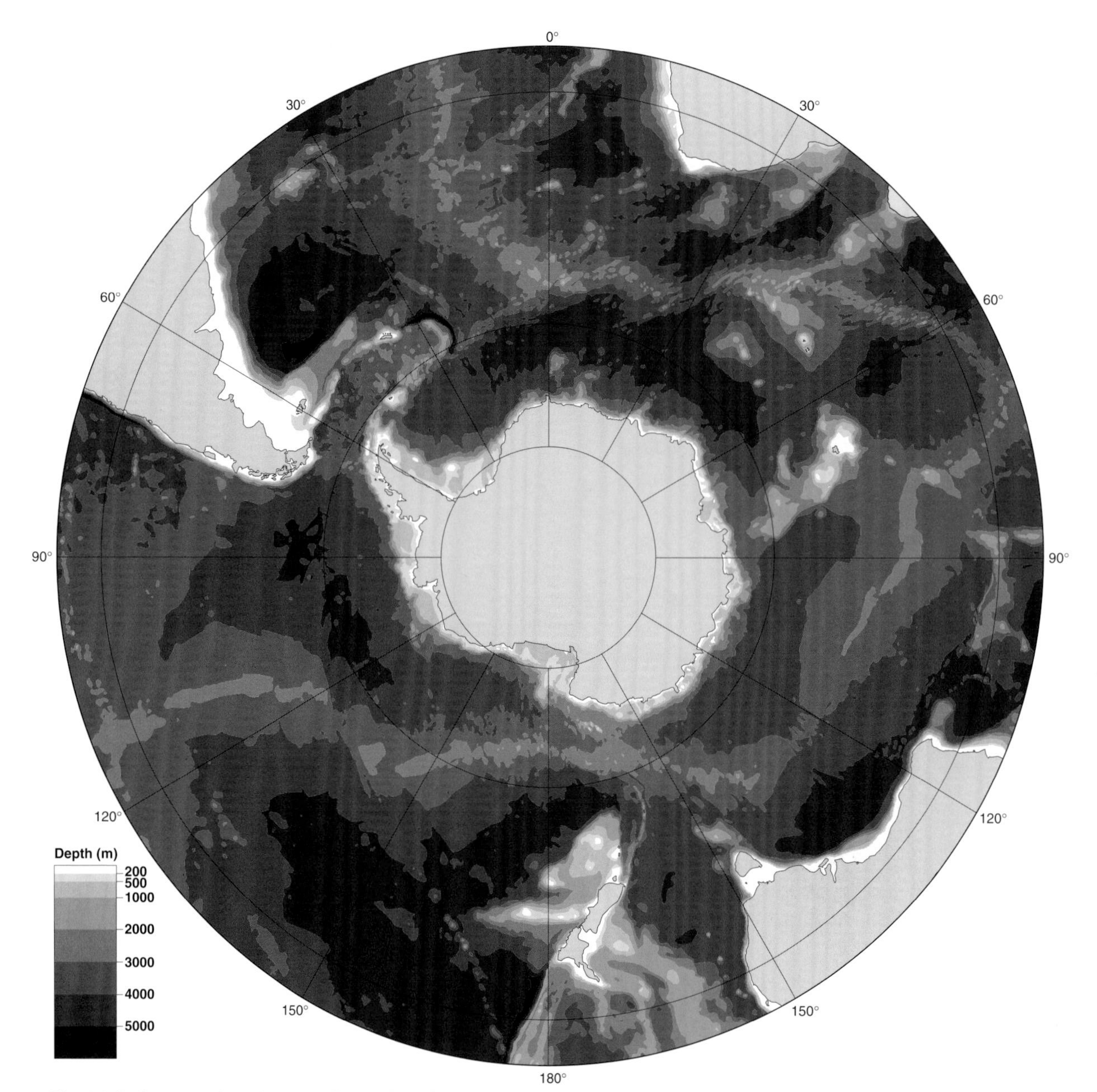

Fig. 3.1 Bathymetry of the Southern Ocean: Fine Resolution Antarctic Model (FRAM) (Webb *et al.* 1991).

The speed of the ACC has been measured by thousands of drift bottles and buoys, but a natural event has provided similar results. In 1962–63, a huge field of pumice from an eruption in the South Sandwich Islands was identified as it was swept around the Southern Ocean. Fine particles in the current reached Tasmania at a mean speed of 11–17 km per day (Deacon 1984). Much faster jet currents of 43–86 km per day occur at the Antarctic and Subantarctic Fronts.

In winter, pack-ice covers 17–20 million km^2 of the ocean in high latitudes (Fig. 3.3). Biological activity below and within the ice is minimal, but in November–December 75% to 80% of it breaks up, disperses and melts[3], leaving 3–5 million km^2 close to the Antarctic Continent. Large areas of the marginal ice zone (a region of transition from completely ice-covered water to water uninfluenced by the effects of pack-ice) become biologically productive. Phytoplankton and then grazing zooplankton proliferate.

[3] Melting takes less time than freezing.

Cold polar water of low salinity flowing from the region of melting ice, meets the slightly warmer and nutrient-rich upper layer of the Circumpolar Deep Water at the Southern Boundary of the Antarctic Front. The surface temperature of the Antarctic water averages 1° to 2°C in winter and 3° to 5°C in summer. The front varies by about 100 km either side of a mean position. From high latitudes south of New Zealand and west of the Drake Passage, it pushes north to about 49°S in the South Atlantic (Fig. 3.2). It is usually described in spare scientific language, but in one lecture to the Royal Geographical Society, Dilwyn John (1934) created a poetic and almost pastoral image:

> 'But we, whether sailors or scientists, know and will remember the convergence best....as the line to the north of which we felt, at the right season, after months in the Antarctic, genial air again and soft rain like English rain in the spring. I can remember a number of those days vividly. It was like passing at one step from winter to spring.'

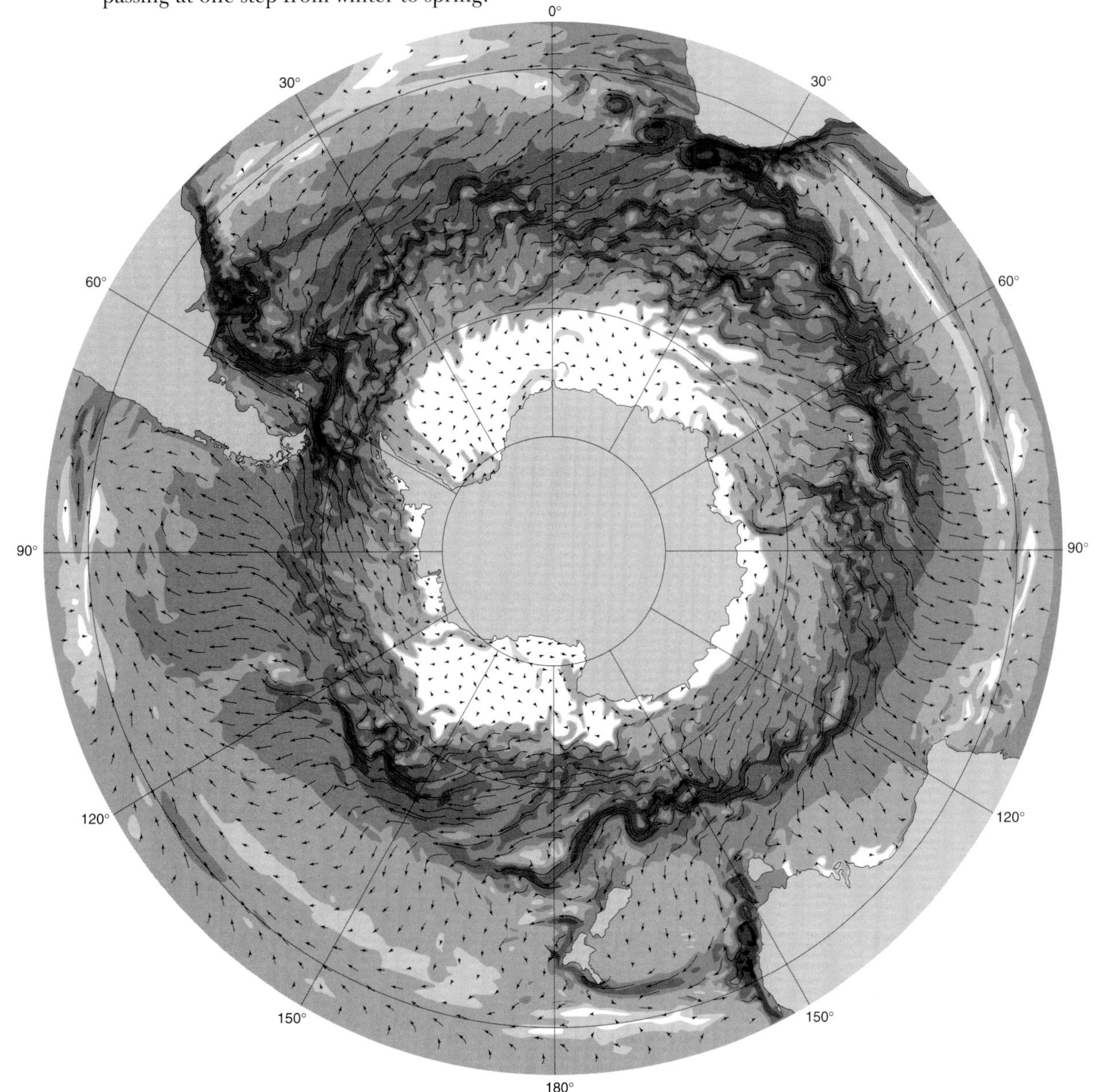

Fig. 3.2 Surface water currents of the Southern Ocean: Fine Resolution Antarctic Model (FRAM). Arrows represent the simulated movements of particles over a period of 50 days (Webb *et al.* 1991).

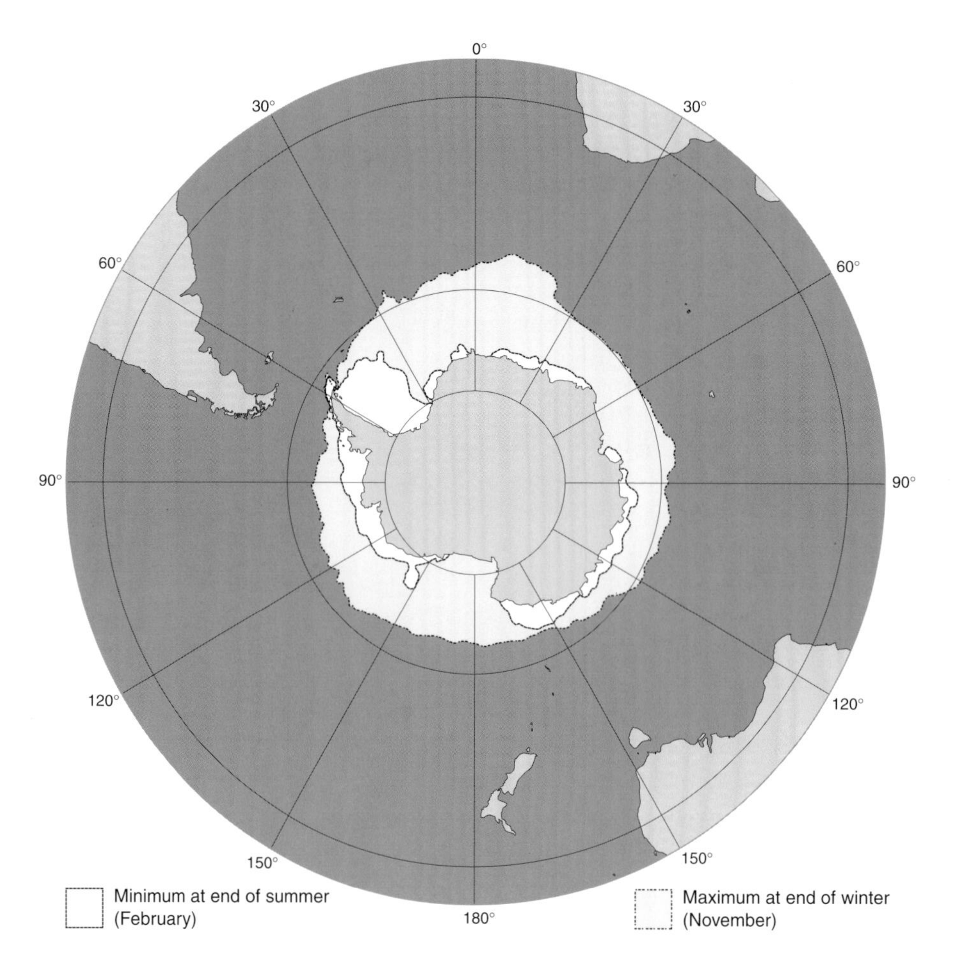

Fig. 3.3 Extent of Antarctic pack-ice in winter (September) and summer (March).

The continental shelf along the west of the Antarctic Peninsula is an important spawning ground for Antarctic krill *Euphausia superba*, whose swarming larval stages are swept northeast by water streaming up from the Bellingshausen Sea, through the Bransfield Strait and into the Scotia Sea, where it joins water emerging from the Weddell Sea and encounters the submarine ridge of the Scotia Arc. This polar water is highly productive; within it, plankton and nekton mature to become the prey of squid, fish, seals, cetaceans and seabirds. However, complex circulation in the Scotia Sea varies from year to year (Rodhouse *et al.* 1996) and the extent of winter sea-ice may determine the quantity of, or direction that growing krill follows. In at least four seasons over the last 20 years krill has been much reduced off South Georgia where huge populations of predators are largely dependent upon it (Murphy *et al.* 1998).

The main flow out of the Drake Passage is deflected north and carries with it the Antarctic Polar Frontal Zone (APFZ). Water emerging from the north side follows the continental slope to become the Falkland Current, flowing north around the Patagonian Shelf (Fig. 3.2). The shelf itself is a prolific spawning ground for squid that supports the largest numbers of albatrosses in the southern hemisphere. The warm, southward flowing Brazil Current is comparatively weak and the Subtropical Convergence is deflected north by the Falkland Current, approaching the Argentine coast at about latitude 33°S. Over the continental slope, complex mixing evidently determines productivity at all trophic levels. The physical properties of these waters, for instance, appears to be linked to the quantity of squid that will later be available off the Falkland Islands (Waluda *et al.* 1999). Albatrosses from the South Atlantic as well as others from as far away as New Zealand are attracted by these resources.

The Subtropical Convergence, APFZ and ACC all pass south of South Africa, but the warm Agulhas Current flowing southwest and west along the continental slope from the Mozambique Channel, mixes with eddies of the ACC (Fig. 3.2). Considerable turbulence is apparent, and on one occasion the German research ship *Meteor* traversed visibly disturbed surface water SE of the Cape, where the surface temperature rose by 5.6°C in 2 km and 9.1°C in 10 km (Deacon 1984).

Cool water that sinks at the Subtropical Convergence south of the Cape continues to flow north below this disturbance, and SE trade winds off the land force water away from the coast of South Africa, allowing the cooler deep water to upwell and flow northwards. Productivity within the Benguela Current is higher than in the surrounding warmer water and is the basis of a considerable fishery which attracts many seabirds, including albatrosses (Summerhayes *et al.* 1974).

The APFZ and Subtropical Convergence are so close together in the south Indian Ocean that they may be imagined as one frontal zone spanning three degrees, 43° to 46°S near the Iles Kerguelen and 41° to 43°S near the Iles Crozet (Gamberoni *et al.* 1982). It was believed to be due to a northerly projection of the APFZ west of the Iles Crozet, perhaps caused by the shape of the ocean floor. The disposition of albatrosses in the south Indian Ocean may be linked to this phenomenon.

The APFZ and ACC pass well south of Australia, but cold water is diverted north into the southern Tasman Sea. Many albatrosses fly great distances to Australian waters. The continental shelf is broad across the Great Australian Bight to Bass Strait, but comparatively narrow along the coasts of New South Wales and southern Queensland (Fig. 3.1). The warm East Australian Current flowing south meets cooler water flowing east around Tasmania and generates areas of turbulence and mixing in the south Tasman Sea that are productive and attractive to many seabirds.

The Macquarie Ridge is an obstruction to the ACC. Two streams flow through passes north and south of Macquarie Island, but the greater volume is deflected south in a sharp bend before flowing on eastwards along the southern slope of the New Zealand continental shelf. In the course of these convolutions, fronts within the APFZ separate. The divided shelf south and west of New Zealand (Fig. 3.1,2) contributes to complex hydrology that has had a profound influence on the albatrosses of the region.

In the South Pacific, the ACC and APFZ remain far south. Water from the immense South Pacific basin backs-up as it funnels towards the Drake Passage, creating the highest waves in the Southern Ocean. Cold water is forced northwards off the coast of Chile (Fig 3.2). The Subtropical Convergence also turns north, probably flanking the Peru (Ocean) Current and reaching the coast at about 38°S. Many albatrosses cross the South Pacific from New Zealand waters and are found along the continental slope off Chile and in bays such as the Golfo de Penas.

4

ISLANDS OF THE SOUTHERN OCEAN

'I have had one of the most fabulous days of the expedition. The only thing wrong with it was that there was nobody to share it with me ... this group of Apostles is perhaps the most exciting and the most incredible of all the islands that I have visited ... I have spent the whole day wandering around through little narrow gaps between the rocks surveying the incredible coast.'

Gerry Clark in the yacht *Totorore*
at the Ilots des Apôtres, 1 March 1986[1]

Albatrosses breed on scattered islands between latitudes 37°S (Tristan da Cunha) and 57°S (Islas Diego Ramírez), but there are no islands in these latitudes of the South Pacific between longitudes 75°W (Chile) and 176°W (Chatham Islands). Many albatross islands have more than one species present, often nesting close to each other. The numbers that survive today have been greatly influenced by historical events.

Southern hemisphere albatrosses range over Antarctic waters, although none breed on typically polar islands. Others may sometimes be found in very low latitudes, but do not breed on tropical islands. The cool, damp climates of Southern Ocean islands promote peat formation. Large tussock-forming grasses are often perched on top of tall, peat pedestals. They are an important source of building material for albatross nests. The common name 'tussock grass' includes a number of species which are listed in Appendix 2.

In Fig. 4.1, numbered spots correspond to the following islands or island groups. Maps of the islands themselves appear in later chapters.

1. PRINCE EDWARD ISLANDS (46° 50'S, 37° 45'E)

Marion Island (area 290 km^2, maximum altitude 1,230 m a.s.l.)
Prince Edward Island (44 km^2, 672 m)

These two volcanic islands, separated by 22 km of sea, were first seen in 1663 from a Dutch East Indiaman driven south of the Cape. In 1772, they were rediscovered by two French warships commanded by Marc Macé Marion du Fresne. Captain Cook took *Resolution* and *Discovery* between them in 1776 and named them for the young Prince Edward[2] (Dunmore 1965, Beaglehole 1967).

Sealers were there by 1799; ships were wrecked and large numbers of men spent months or even years on Marion Island (Goodridge 1832, Savours 1961). Sealing reached a peak about 1840, but after 1870 very few furseals remained. A revival in 1909 lasted barely two years.

Scientists from the *Challenger*, went ashore briefly at Marion Island in 1873, where they took the first photographs of Wandering Albatrosses (Fig. 4.2).

In 1947–48, South Africa annexed the group and established a meteorological station on Marion Island, which has remained occupied ever since. Six field huts, built around the coast in 1975, have been replaced from time to time.

Early exploration (Bennets 1949, Rand 1952, 1957) revealed several species of albatross breeding on both islands. In the mid-1960s there was a significant initiative in biology and geology, including field work on the

[1] Gerry Clark (1988) was an intrepid yachtsman with a passion for seabird islands.This day at the Apôtres was one of many similar episodes in his epic circumnavigation of the Southern Ocean. He died in June 1999 when *Totorore* was wrecked at the Antipodes Islands.

[2] later Duke of Kent and father of Queen Victoria.

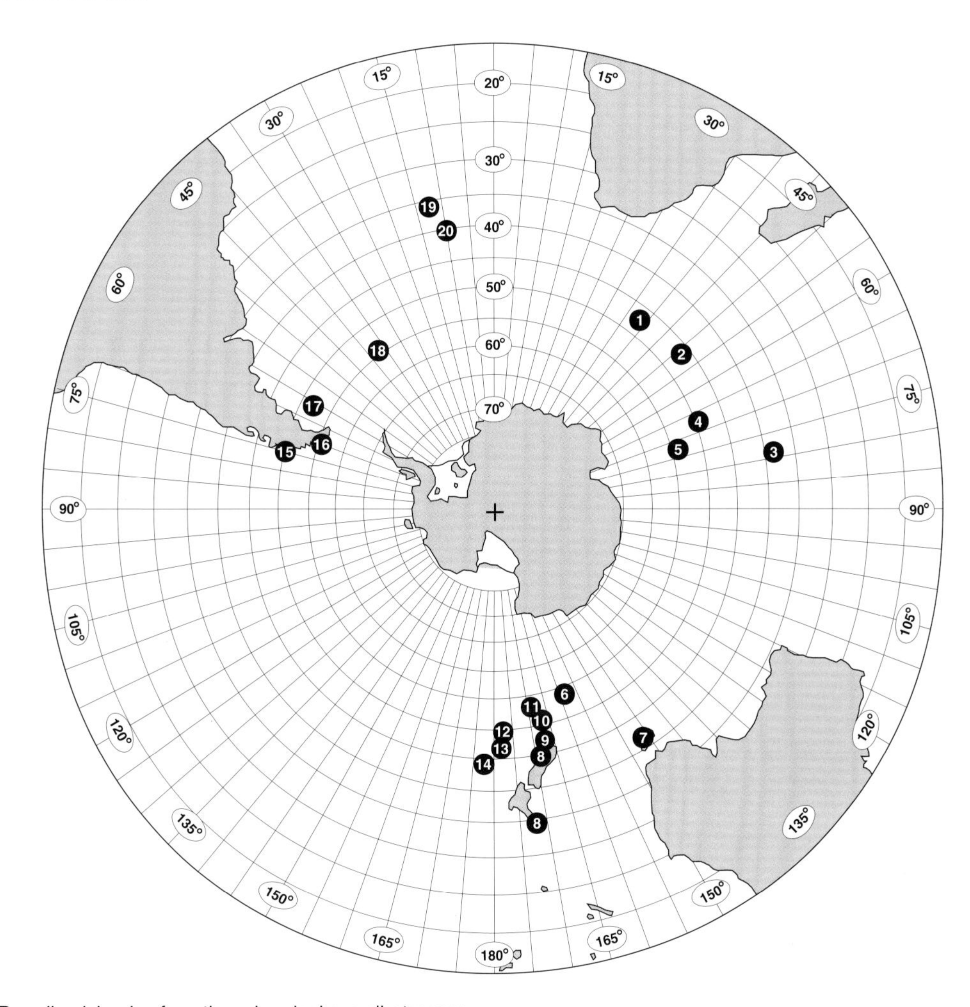

Fig. 4.1 Breeding islands of southern hemisphere albatrosses.

1 Prince Edward Islands
2 Iles Crozet
3 Iles Amsterdam & St Paul
4 Iles Kerguelen
5 Heard & McDonald Islands
6 Macquarie Island
7 Tasmanian offshore islands
8 New Zealand & offshore islands
9 Snares Islands
10 Auckland Islands
11 Campbell Island
12 Antipodes Islands
13 Bounty Islands
14 Chatham Islands
15 Isla Diego de Almagro
16 Islas Diego Ramirez & Ildefonso
17 Falkland Islands
18 South Georgia
19 Tristan da Cunha
20 Gough Island

albatrosses of Marion Island (van Zinderen Bakker 1971). The Seabirds and Seals Protection Act 46 came into force in 1973 and periodic surveillance increased, with some colony mapping and counting on both islands (Berruti 1979, Grindley 1981, Williams 1984).

Marion Island has a low, rolling profile of old, eroded grey lava, but most of it is covered by more recent black lava, much of which is exposed inland with volcanic cinders and numerous extinct scoria cones. The last eruption was in 1980. There is a small ice-cap on the summit, with snow surrounding it for eight months of the year. Several fault scarps and ridges with cliffs radiate outward, they rise to several hundred metres. Prince Edward Island is much smaller and lower, but with a steeper profile and no ice-cap.

Coastal cliffs and steep slopes subject to sea-spray support associations of tussock grass and *Cotula plumosa*. The relatively flat and poorly drained coastal plains above are dominated by mires of the grass *Agrostis magellanica* with

Fig. 4.2 Wandering Albatrosses on nests at Marion Island in the Prince Edward Islands, December 1873 (Thomson 1885).

herbfields of burnet *Acaena.* Inland, drained slopes with the fern *Blechnum penna-marina* and vegetated lava hummocks gradually give way at about 300 m to a feldmark of mosses, lichens and cushions of *Azorella selago* (Huntley 1971).

House mice probably came to Marion Island in sealers' provisions early in the 19th century and spread throughout the island. They invaded the new base in 1949 and domestic cats were introduced in an attempt to control them. The cats escaped and proliferated to become a feral predator throughout the island. They killed immense numbers of seabirds, but few albatross chicks were taken (Williams 1984). By 1977, there were about 2,100 cats and control measures began with the introduction of feline panleucopaenia virus, later accompanied by shooting, trapping and poisoning. The last cats on Marion Island were killed in 1991. Prince Edward Island has no introduced mammals.

In 1987, a South African Government proposal to construct a hard runway for aircraft on Marion Island met with strong opposition from conservation groups and after an unfavourable environmental impact assessment, the proposal was abandoned. The Prince Edward Islands are now a nature reserve under Parliamentary Act with a published management plan.

2. ILES CROZET (45° 57'–46° 30'S, 50° 20'–52° 20'E)

Ile de la Possession (145 km², 934 m)
Ile de l'Est (132 km², 1,090 m)
Ile aux Cochons (70 km², 770 m)
Ile des Pingouins (3 km², 360 m)
Ilots des Apôtres (2 km², 289 m)

This group of islands in the south Indian Ocean was also discovered in 1772 by Marion du Fresne, soon after he had left the Prince Edward Islands. His officers, Julien Marie Crozet and Jean Roux, landed on the Ile de la Possession (Dunmore 1965). Since 1955, the group has been part of the Territoire des Terres Australes et Antarctiques Françaises (TAAF).

Henry Fanning, a sealing master, arrived in 1805 in the *Catherine* and other sealers soon followed. They introduced pigs, cats, rabbits, rats and house mice (Jouventin *et al.* 1984). Some of their ships were wrecked and men lived on the islands for months or years. Early information on the fauna and flora first came from the stories of shipwrecked mariners (Goodridge 1832) and specimens brought back from sealing voyages (Layard 1867). Coastal vegetation is dominated by tussock grass. Inland, smaller tussocks give way to a heath of short grass *Deschampsia*, with burnet *Acaena* and ferns *Blechnum*. There are no ice-caps; feldmark with cushions of *Azorella* predominates on the higher interiors.

The Ile de la Possession is the largest of the group, with green hills and valleys rising from huge sea-cliffs. *Discovery*[3] called briefly in 1929, when Robert Falla (1937) went ashore for a few hours and described colonies of several albatross species. French scientists visited briefly in 1939 and again in 1957 when a photo-reconnaissance was undertaken. In December 1959, the cutter *Mischief* arrived and H.W. Tilman (1961) spent ten days exploring the island with Roger Tufft, who counted and ringed Wandering Albatrosses. Two years later, Port Alfred, above the Crique du Navire, was selected as the site for a permanent French base and the following season a cable-lift was constructed. Building was completed in 1963–64 and since 1966, the island has been the scene of continuous seabird research.

The Ile de l'Est is more rocky, with jagged ridges and peaks rising above wide valleys with rivers. In 1825, *l'Aventure* was wrecked on the north coast; its crew spent two years ashore during which time the master made a chart of the island. French scientists have visited the island from time to time (Prévost 1970, Despin *et al.* 1972, Jouventin *et al.* 1984).

The Ile aux Cochons, an ancient volcanic cone, no longer has any pigs. Several field parties of French biologists have been ashore (Dreux & Milon 1967, Derenne *et al.* 1976, Voisin 1984).

The Ilots des Apôtres were notorious hazards to sailing ships. On the night of 30 June 1875, the *Strathmore* (Capt.C. McDonald) sailing from London to New Zealand was wrecked there. Forty people were lost, but 49 including a woman (p.17), got ashore on Grande Ile (3 km x 1 km), where they survived for almost seven months before being rescued by an American whaler. After the rescue, C.F. Wordsworth (1876) wrote:

> '... the food we lived upon was young and old albatross; the young gave more eating than the old, being large heavy birds, with beautiful white down upon them about three inches long...We got very hard up for anything to eat at one time...Black Jack's tent had had nothing to eat all day. We were very weak and low spirited...I went down by the side of the island, where I remembered to have seen a large quantity of nests, built of mud, smooth and round, about a foot from the ground, looking at a distance like the turrets of a small castle. Down the rocks I went, and saw, to my great delight a quantity of beautiful white birds. We named them "Freemasons", but we afterwards discovered that their real name was mollyhawk. I killed about fourteen of these, as they let me come quite close to them...Many of the others went out, and killed about a hundred in all. Such a feast of tails we had then! That appendage was cut off close to the back, the long feathers pulled out, and being burnt for a time in the fire, was considered a great delicacy, and one of the perquisites of the hunter...The young albatrosses were on the island when we landed in July, and just before we left the old birds returned and built their nests and laid their eggs, so we presumed we had seen the round of the sea-birds. We never took any albatross eggs as we looked forward to depending on the young for food later on...We used to see parts of fish in the big gut of the albatross when they had young to feed. I remember once killing an albatross and, as is often the case just before dying, it vomited up the contents of its bag, and amongst the mess was an eel quite perfect, having the appearance of being cooked. I took it and ate it, and it tasted quite like stewed eel.'

[3] British, Australian and New Zealand Antarctic Research Expedition 1929–1931

It was well over a hundred years before French scientists landed on the Ilots des Apôtres and the Ile des Pingouins (Jouventin *et al.* 1984). They returned to spend somewhat longer ashore in 1986–87 to count seabirds (Jouventin 1990).

3. ILE AMSTERDAM[4] (37° 50'S, 77° 31'E)

This island (55 km^2, 881 m) was sighted in 1618 by Adriaen de Wale and in 1633 Anthoonnij van Diemen named it after his ship *Nieuw Amsterdam.* Together with the Ile St Paul[5], it was claimed for France in 1792 and charted from warships commanded by Antoine-Raymond-Joseph de Bruni d'Entrecasteaux.

Sealers arrived in 1789 and were often resident for years at a time. Pigs and domestic fowl were put ashore soon afterwards and mice date from 1825. Cats and rats were introduced in 1931, but while Segonzac (1972) identified black rats, Micol & Jouventin (1995) have since referred to brown rats.

There were wrecks and attempted settlements. One family and four employees from Réunion began farming the island in 1870, but gave up and left within two years. The animals they had introduced were left behind. The sheep have gone, but the feral cattle that graze the island are now of considerable zoological interest (Micol & Jouventin 1995).

In 1892, castaway depots were put ashore and commercial concessions issued. A meteorological station was established near Pointe Hosken in 1949 and has remained operational ever since. In 1955, both islands were incorporated within TAAF (p.21).

The climate is cool with little seasonal difference in temperature, but strong winds blow throughout the year. In 1769 John Wane, a midshipman of the East India Company, described the island as mostly forested. Forest fires, started by sealers, date from 1791; they often burned for months at a time (Dunmore 1965). In 1974, the last fire burnt across the whole island, except for the western cliffs, in a year. Repeated burning with constant grazing, browsing and trampling by cattle destroyed the lowland forest, which now survives only as small copses separated by dense stands of introduced grasses and ferns (Micol & Jouventin 1995). Around the coast, tussock grass and *Spartina arundinacea* predominate, with a few *Phylica nitida* trees in sheltered gullies. Above 100 m, the sedge *Scirpus nodosus* forms a dense cover that extends to the peat bog of the central plateau (Jouventin *et al.* 1984).

Edgar Aubert de la Rüe listed some seabirds in 1928 and 1931; others were collected in the early 1950s when Patrice Paulian (1953) spent several days ashore. Segonzac (1972) twice visited the island, spending more than three months producing the first report on the fauna and reporting large numbers of albatrosses. It was not until the early 1980s that regular field-work revealed a small number of Amsterdam Wandering Albatrosses (Roux *et al.* 1983). Their breeding ground has been fenced to exclude feral cattle, which are now confined to the northern sector of the island (Micol & Jouventin 1995).

3. ILE ST. PAUL (38° 43'S, 77° 32'E)

This small island (7 km^2, 272 m) was discovered at the same time as the Ile Amsterdam. Its crater is a natural harbour for small vessels. Sealers arrived in 1789 and charted the anchorage. Five men led by P.F. Péron remained for more than three years, after which French, Polish and Mauritian settlers fished there between 1819 and 1857. A British troopship bound for Australia was beached in 1871 and 500 soldiers camped ashore for three months until taken off by a liner and two warships (Headland 1989).

Fishing interests revived in 1928 and 1938, when short-lived companies took sea crayfish ('rock lobster'). After a more reliable market had been established in the early 1950s, fishing off St Paul and Amsterdam Islands thrived until 1957–58 when it was suspended to allow stocks to recover. The industry was again busy from 1965 to 1974 (Headland 1989).

There are no trees on the island and all slopes are covered by a dense and sometimes impenetrable growth of

[4] Long ago common usage abbreviated the translated name to Ile Amsterdam and in 1967 this was accepted by the French Institut Géographique National.

[5] For many years the names of the two islands were transposed on British charts (Richards 1984).

the sedge *Scirpus nodosus* (Segonzac 1972). The island has been devastated by fire and introduced animals. Pigs, goats and house mice were reported as early as 1823 and were later joined by rats and cats (Goodridge 1832, Jouventin *et al.* 1984).

Several expedition ships called at the island in the late 19th century. The first scientific investigation and charting of the island was in 1857 by Austrian scientists from the *Novara.* A French scientific expedition to St Paul observed the transit of Venus in 1874, when natural history observations were made by Charles Vélin and M. De L'Isle. Aubert de la Rüe surveyed the island in 1930 and, in 1971, Segonzac (1972) spent about three weeks ashore assessing the status of the fauna. A few albatrosses have bred there.

4. ILES KERGUELEN (48° 35'–49° 44'S, 68° 43' –70° 35'E)

The Iles Kerguelen are quite unlike the other islands of the Southern Ocean. The mainland is much larger and, with about 300 islands, occupies about 7,215 km^2. Numerous long, finger-like fjords increase the length of coastline and enclose sheltered inland waters far from the sea. There are mountains rising to 1,850 m and two small icecaps.

In 1772, two French ships under the command of Yves-Joseph de Kerguelen-Trémarec sighted land. Gales, heavy seas and worn rigging forced Kerguelen to leave, but *Gros-Ventre* entered a bay on the south coast and François Alesno de Saint-Allouarn made a brief landing. The two vessels lost sight of each other and Kerguelen returned to France making rash claims that he had found a great land. He was sent back with three ships to found a colony. This time they approached from the northwest and on 6 January 1774 one brief landing was made from the *Oiseau,* after which the expedition departed without further exploration (Dunmore 1965).

Almost three years later, Cook's ships, *Resolution* and *Discovery* also approached from the northwest. They sighted the Iles Nuageuses and sailed around the adjacent peninsula where they entered a large bay. On 25 December 1776, they worked their way into an inner cove, Christmas Harbour, where the men enjoyed time ashore and found the French claim. Before leaving, Cook charted the east coast and confirmed his suspicion that there was open sea to the south. His name, *Desolation Island,* was used for some years, but J. Douglas, posthumously editing the journals for publication in 1784 (Beaglehole 1967), arrived at a form of words that Cook might have written had he lived:

> 'an island of no great extent; which from its sterility, I should, with great propriety, call the Island of Desolation, but that I would not rob Monsieur de Kerguelen of the honour of its bearing his name'

Sealers arrived in 1791 and for more than a century had a profound influence on the islands. Vessels sometimes came in fleets and numbers peaked in the 1860s; wrecks were frequent. It was seasonal work, but gangs of men with boats were left in makeshift camps where they sometimes remained for years. Several whaling stations operated intermittently between 1908 and 1956, after which sealing was revived for a few years (1956–60).

Kerguelen became an early focus for scientists. The *Erebus* and *Terror* put into Christmas Harbour in 1840 and *Challenger* in 1874. The islands were selected as one of the locations from which to make measurements of the 1875 transit of Venus. Later, many other expedition ships called at the islands. Specimens, including albatrosses, were collected for various museums from Betsy Cove, Molloy Point and Observatory Bay (Eaton 1875). Other naturalists (Hall 1900) reached the islands with sealers and whalers.

In 1893, a fifty-year commercial concession was granted to René and Henri Emile Bossière, who created the *Compagnie Générale des Iles Kerguelen.* The concession was later sub-contracted to *Irvin and Johnson (S Africa) Ltd,* which operated a sealing/whaling station in Port-Couvreux (1912–1931). René Bossière was appointed the French Resident. In 1938, a decree included the Iles Kerguelen in a French national park and since 1955, they have been part of TAAF.

Charting was an immense task and some sealing masters contributed to the early exploratory work of Cook and Ross. The main coastline was completed by Rallier du Baty in 1914, using two small vessels, *J.B.Charcot* and *La Curieuse.* A mining syndicate was set up by the Bossières who contracted the geologist, Aubert de la Rüe to survey the islands. During field-work from the trawler *Austral* in 1928–29 and 1930–31, he examined most of the coast and although no workable deposits were found, he produced a comprehensive study of the geology and geography of the islands. He returned to Kerguelen in 1949 as an advisor to TAAF and wintered in 1952. Altogether, he journeyed 2,500 km on foot within the islands and his several contributions to natural history included the first analytic description of the Kerguelen vegetation (Aubert de la Rüe 1964).

A French scientific base with a small farm was built in 1951, next to the whaling station at Port-aux-Français that closed in 1956. Today, 120 people may be on Kerguelen in the summer and 70 overwintering. Patrice Paulian (1953), the first zoologist to winter there, explored much of the Peninsule Courbet and Presqu'île du Prince de Galles.

Albatrosses had long been known in this area and in 1958 an extensive coastal survey was carried out by launch. From 1964 to 1966 a search for seabird colonies was made along more distant coasts. Even with helicopter support, the small field parties were obliged to make arduous journeys over difficult country (Tollu 1967). It was many more years before the Ile de Croy was visited and discovered to have five species of albatross (Jouventin & Stonehouse 1985). Other potential albatross islands remain to be explored (Weimerskirch *et al.*1989a).

The two ice-caps occupy only about a tenth of the mainland. Mountains, plateaux and plains are subject to wind erosion, creating desert-like conditions adjacent to valleys with braided rivers and wetlands with thousands of pools.There are 28 species of native vascular plants, of which the Kerguelen cabbage *Pringlea antiscorbutica* is best known. Coastal and cliff vegetation near the open sea is a tussock grass and *Cotula* association; around the fjords burnet *Acaena* and *Azorella* are widespread. *Colobanthus* with stunted *Azorella*, moss and lichen characterise the higher feldmark. Agricultural experiments have included the planting of trees which did not survive.

Twelve species of mammals have been introduced to the Iles Kerguelen and seven remain (Mawson 1934, Weimerskirch *et al.* 1989a). House mice arrived with the early sealers and were common along the coast and elsewhere by 1875, but rats were not seen until 1956. Rabbits from a convict settlement near Cape Town were shipped to the islands in 1875 by orders of the British Admiralty and liberated about the Golfe de Morbihan. They proliferated and devastated the vegetation. Two introductions of *Myxoma* virus in 1955 and 1956 were not accompanied by vectors and the disease has remained locally endemic in the population. Since 1952, repeated introductions of domestic sheep and Mouflon (wild sheep) have been confined to several islands in the Golfe de Morbihan. Herds of reindeer, released in 1955, roam the mainland. Feral dogs (1903–29) have not survived (Headland 1989), but the progeny of two cats introduced in 1956 have proliferated and remain serious predators of burrow-nesting birds.

5. HEARD ISLAND (53° 05'S, 73° 30'E)

This island (375 km^2, 2,745 m) was sighted in 1833, but named for Captain John Heard, of the merchant barque *Oriental* who sailed past in 1853. Britain claimed the island in 1908 and in 1947 transferred sovereignty to Australia. Sealers landed in 1855 and began wintering in 1857 (Keage 1982). Scientists landed briefly from passing ships. When the *Challenger* called in 1874, 40 sealers were scattered along the coast. The Kerguelen Sealing and Whaling Company built a small hut in 1929, which scientists from *Discovery* occupied for a week. ANARE[6] established a scientific base in 1947 and occupied it until 1955. Since then, there have been intermittent visits and field shelters have been installed about the island. There were overwinterings in 1969 and 1992.

Heard Island is an active volcano which has erupted many times, the latest in 1992–93. About three-quarters of the island is covered with ice; the Laurens Peninsula and a long shingle spit project northwest and east of the main ice dome. The vegetation is generally poor, with 11 species of vascular plant, but tussock grass, Kerguelen Cabbage and *Azorella* are locally prolific, while cushions of *Colobanthus* are abundant. There are small numbers of Black-browed Albatrosses (Downes *et al.* 1959, Johnstone 1982, Woehler 1989).

The flora and fauna of the islands have been protected by Australian ordinances dating back to 1918 and new conservation ordinances came into force in 1980. In 1983, Heard and McDonald Islands were made Australian Heritage Sites. The Australian *Territory of Heard Island and McDonald Islands Environment Protection and Management Ordinance 1987* provides for the protection and use of both islands and the surrounding sea within 200 nautical miles. Since 1992, there has been a management plan.

[6] Australian National Antarctic Research Expeditions.

5. McDONALD ISLANDS (53° 02'S, 73° 36'E)

Although this small volcanic[7] island (2.6 km^2, 212m) and its associated islets (total 4.8 km^2) were seen by sealers in 1854 and named for one of them, it is thought that the coast was too dangerous for sealing. The islands were included in the 1908 British territorial claim to Heard Island, 43 km to the east. The first reconnaissance did not take place until 1971, when field-workers were ashore for about an hour. In 1980, four days were spent surveying the islands (Johnstone 1982). There is an impoverished vegetation of five vascular species and a few Black-browed Albatrosses are among the breeding seabirds.

6. MACQUARIE ISLAND (54° 30'S, 158° 55'E)

This is a long, thin island (34 km x 5 km; 128 km^2 ,433 m) has two groups of rocky islets; the Judge and Clerk Islets 11 km to the north and the Bishop and Clerk Islets 37 km to the south. They are the exposed crest of an undersea ridge at the boundary of two major tectonic plates. Deep water separates the Macquarie Ridge from the Campbell plateau to the east. Earthquakes are common and in 1980 there were nine events of magnitude greater than five on the Richter scale (Selkirk *et al.* 1990). There is no permanent ice.

The island was discovered in 1810 by Captain Frederick Hasselburg of the sealing brig *Perseverance*, and named after the Governor of New South Wales. Ten years later, von Bellingshausen's Antarctic Expedition in the *Vostok* and *Mirny* called at the island. In 1825, Macquarie came under the jurisdiction of the new colony of Van Diemen's Land and continued as a dependency after 1890, when the colony adopted the name Tasmania.

Sealing gangs were left ashore and many men wintered. Furseals were exterminated early in the 19th century, but elephant seals survived on remote beaches. Accessible beaches were worked out by 1829 and afterwards few ships visited the island, apart from a brief renewal of elephant sealing in 1874–78, when European rabbits were liberated as food for men. From 1889, men were again resident, killing enormous numbers of penguins and occasional elephant seals for oil.

From December 1911, a party from Douglas Mawson's Australasian Antarctic Expedition stayed on Macquarie for two years operating a radio link between Antarctica and Australia. The island was surveyed and H. Hamilton made biological collections and observations. The station was manned for another two years by the Commonwealth Meteorological Service, but closed in 1915. Mawson and former members of his expedition successfully campaigned for an end to the oiling industry. In 1919, the killing of penguins and seals ceased and all men left the island.

Dogs introduced by sealers, died out, but cats proliferated to become a serious predator of ground-nesting birds. In the early 1970s, it was estimated that 250 to 500 cats, whose main prey had become the introduced rabbit, were killing about 60,000 petrels per year. Shooting started in 1974 and, since 1985, has been accompanied by trapping, gassing and poisoning. By 1995 the cat population had accommodated to this increased mortality and stabilised at a lower level, with no immediate prospect of extinction (Copson 1995). Wekas from New Zealand were released on several occasions in the 1840s to 1870s; they also preyed on burrowing petrels but were completely eradicated between 1985 and 1988. Black rats and house mice arrived in the 1890s; they feed mainly on vegetation and invertebrates, but the rats also prey on small petrels.

The vulnerability of burrowing petrels to introduced cats and rats depends upon predator/prey interactions with rabbits (Copson 1995). By 1978 about 150,000 wintering rabbits were having a marked effect on vegetation. Numbers have been brought down to about 10,000 since the introduction of rabbit fleas (1968) and *Myxoma* virus (1978).

Horses, donkeys, cattle, goats, sheep, pigs, domestic geese, ducks and fowl have all been free on Macquarie, but none remain.

European Redpolls, Starlings and Mallards, originally introduced to Australia and New Zealand have found their own way to Macquarie Island (Copson 1995).

[7] Volcanic activity was seen in 1997 (E.J. Woehler pers. comm.).

The first inventory of flora and fauna was made in 1880, two years after the introduction of rabbits. Today there are 46 species of native vascular plants including a megaherb, the Macquarie Cabbage *Stilbocarpa polaris*. About half of the island is an exposed plateau with feldmark dominated by cushion forming *Azorella macquariensis* and mosses (Selkirk *et al.* 1990). At lower altitudes the vegetation includes a mosaic of tall tussock grassland, meadow-like short grassland of *Agrostis*, *Uncinia* and *Deschampsia* or *Festuca*. Rabbits greatly reduced the area of tall tussocks, which were replaced by short grassland, but since reduction of the rabbit population tall tussocks have spread back into the short grassland. Mires with rushes are common in valleys and there is a variety of herbfield on higher ground.

Macquarie Island was declared a wildlife sanctuary in 1933 (Tasmanian Animals and Birds Protection Act 1928). In 1947–48 ANARE established a permanent scientific station at the north end of the island and six field huts have since been put up about the island. Albatrosses have been under surveillance since the late 1940s and ringing started in the early 1950s. When the Tasmanian National Parks and Wildlife Service was set up in 1971, the island was designated a conservation area and in the following year upgraded to a State Reserve (National Parks and Wildlife Act 1970). The Macquarie Island Nature Reserve includes the two groups of distant islets. According to a management plan gazetted in 1991, it is a restricted area managed by the Tasmanian Parks and Wildlife Service. Caroline Cove, with the few remaining Wandering Albatrosses, is closed from 1 November to 30 April, but permits may be issued for approved research.

7. TASMANIAN OFFSHORE ISLANDS

Albatross Island (40° 22'S, 144° 40'E. 33 ha, 35 m).
Mewstone (43° 44'S, 146° 23'E. 7 ha, 133 m).
Pedra Branca (43° 51'S, 146° 59'E. 2.5 ha, 60 m).

Bass Strait was named after George Bass, a Royal Navy surgeon who, in 1797, explored the northern coast in an open whaleboat. The following year, he set out with Matthew Flinders in the 25 ton sloop *Norfolk* and completed a circumnavigation of Van Dieman's Land (Tasmania). On 9 December 1798, they came upon an island where Bass landed and brought off the first boatload of seals and Shy Albatrosses.

> '...there are vast numbers of albatrosses on the isle to which their name is given, which were tending young in the beginning of December; and being unacquainted with the power or disposition of man, did not fear him: we taught them their first lesson of experience.'
>
> Matthew Flinders (1801)

Albatross Island is within comparatively easy reach of mainland Tasmania and in the decades that followed, sealers destroyed the furseals. They then killed the albatrosses for their feathers, which could be sold in Launceston for up to a shilling a pound. Thousands of carcasses were left to rot on the island. By 1909, numbers had been reduced to about 300 pairs and have not yet fully recovered (Plomley in Green 1974, N. Brothers pers. comm.). The island is dry and its sparse vegetation is mainly a tussock grass association with about 20 other vascular plants. It was unallocated Crown Land until 1981, when it was made a Nature Reserve of the Tasmania Department of Parks and Wildlife (N. Brothers pers. comm.). In 1960, Macdonald & Green (1963) ringed albatrosses, but regular field-work on albatrosses did not begin until 1980.

Two rock stacks, Mewstone and Pedra Branca, south of Tasmania, were named by Abel Janszoon Tasman in 1642, during his famous voyage of discovery. There are no records of sealers or fowlers and landings are hazardous. The first albatross counts were made in 1977 and 1978 (Brothers 1979a,b). Since 1951, the stacks have been part of the SW Tasmania National Park which, in the 1980s, was renamed the Tasmanian Wilderness and adopted as a World Heritage site.

8. NEW ZEALAND AND OFFSHORE ISLANDS

Three Kings Islands (34° 09'S, 172° 03'E)
South Island – Otago Peninsula (45° 46'S, 170° 44'E)
Solander Islands (46° 55'S, 166° 55'E)

New Zealand had been inhabited by Maoris for several hundred years before the first Europeans arrived in the 17th century. These seagoing people had been familiar with albatrosses – *toroa* in their language. Bones of albatrosses have been found in prehistoric middens around New Zealand; many of them had been worked into artifacts – awls, needles, hooks, toggles, necklaces and flutes. The albatrosses from which they originated may have been caught at sea when attending fishermen or picked up from beaches after being driven ashore in storms. In historic times, *toroa* featured prominently in Maori culture; canoes decorated with white albatross feathers were symbolic of swift flight – skimming over the ocean. They were also used in burial rites (R. McGovern-Wilson pers. comm.). Men and women wore strips of feathered skin through pierced ears (Forster 1968) and tufts of albatross down, called *pohoi* were worn in the hair like tropical flowers. *Te Hiwi o Toroa* (Shoal of the Albatross) in Lake Rotorua is named not for the bird itself, but for the *pohoi* from the beautiful *Kura* that floated on the surface when she dived (Pomare & Cowan 1930). These tufts of white down later acquired political significance as the emblem of the *Te Whiti* (Taranaki Pacifist) movement (Robertson 1991).

Three Kings Islands

A group of islands lying 65 km off Cape Reinga, the NW extremity of New Zealand. They were named by Tasman who approached them on 4 January 1643, the feast of Epiphany, at the end of his voyage around North Island in *Heemskerck* and *Zeehaen*. Attempts to land for freshwater were prevented by heavy surf. Maoris were seen ashore. There is one main island, three smaller islets and a chain of rocky stacks called the Princes Islands, which have a few Northern Buller's Albatrosses.

Away from steep coastal cliffs, the larger islands are covered with a dense scrub of *Phormium* and other plants which in wetter sheltered areas, give way to luxuriant coastal forest of puka *Meryta sinclairii* and other trees. On Great Island, the forest has been cleared and cultivated from time to time. There have been large numbers of goats which were shot out in 1945 (Buddle 1948; Wilson 1959).

South Island

At the end of the Otago Peninsula, overlooking the entrance to Otago Harbour, the natural features of Taiaroa Head make it a good defensive position and it is said to have been a *Pukekura* (significant site) for Maoris from about the mid-17th century. In the early decades of the 19th century, the sheltered harbour was an anchorage for whalers and since 1865 there has been a lighthouse and signal station on Taiaroa Head. In the 1880s, it was fortified and by the turn of the century the garrison and families numbered about a hundred people. Most of them departed in the early 1900s, leaving only the lighthouse families, but during two world wars, units of soldiers returned.

The slopes of Taiaroa Head are carpeted with European grasses and herbs but the flora is gradually reverting to native species (C.J.R. Robertson pers. comm.). When the Northern Royal Albatross began breeding there, the headland was owned by the Otago Harbour Board. In 1951, an area of 5.3 ha (now 9 ha) was fenced and a ranger appointed. People still took eggs and were prosecuted, but the small fines were not considered a deterrent. Dogs, cats, stoats and ferrets also continued to gain access and were shot or trapped.

Taiaroa Head was made a Wildlife Refuge under the New Zealand Wildlife Act of 1953 and the Otago Harbour Board began negotiations to transfer responsibility to the Government. In the meantime, no-one was allowed into the colony except the ranger, members of the British Royal Family and the Governor General; some other distinguished visitors were turned away. From 1963, the Department of Internal Affairs (Wildlife Service) became responsible for the site, which was gazetted a Flora and Fauna Reserve (1964). In 1967, the newly formed Otago Peninsula Trust (OPT) began lobbying for the albatross colony to be opened for public viewing, but in the Wildlife Service there were concerns that the colony was not yet ready for such potential disturbance. A small hut had just been built for the ranger on a rocky promontory overlooking the west end of the colony and this served to test the reactions of the birds. By 1972, there was enough confidence for a public observatory to be opened on the same site. Five years later, the colony became a Nature Reserve (Reserves Act 1977). In 1989, a substantial visitors' centre was built and the number of visitors to the colony rapidly increased (Robertson 1997).

Taiaroa Head is remarkable for being only 25 km from the City of Dunedin. Protection of the birds is paramount; rangers destroy feral predators and monitor breeding. Since a grass fire in 1967, fire breaks have been maintained by mowing and herbicide spraying. Broken eggs are replaced by artificial eggs and deserted eggs fostered onto the still incubating birds. Nests with sick chicks may be screened from the wind. On windless days, albatrosses at Taiaroa Head have suffered from heat exhaustion. Three adults died and blowflies were attracted to newly hatched chicks. A vet with antibiotics is on call when birds are seriously ill and six deserted fledglings have been hand-reared (Robertson & Wright 1973).

Outside the pre-egg and laying periods (September to November), visitors are introduced to the reserve by OPT guides and conducted to the observatory. On windy days the great birds sweep past within a few metres of the windows. It is an exciting experience well worth the fee, but at other times visitors may see very little activity. Long grass can obscure albatrosses on the ground and new breeders have tended to nest out of sight of the observatory (Robertson 1993a, 1997).

The sensitivities of albatrosses were not apparent when the observatory was built, but they will determine future viewing strategy. The birds could clearly see and hear many people through the large windows of the observatory; these have been tinted. Eventually, concealed observation chambers and passages will probably be required. There remains the eventuality of albatrosses occupying sites outside the reserve fence, which has always been too close to the colony. The buildings, car park and fenced approaches are out of sight of the birds on the breeding ground, but albatrosses in flight are aware of them and the activity of people.

Solander Islands

The group lies in the western entrance of the Foveaux Strait, about 40 km south of South Island. It consists of Solander Island (1km^2, 340 m), Little Solander Island (8 ha, 180 m) and some stacks (Cooper *et al.* 1986). They were discovered on 11 March 1770, when the *Endeavour* was circumnavigating New Zealand, and named by Cook after Daniel Solander, Joseph Banks' greatly respected naturalist and a former student of Linnaeus. Sealers were at the islands in the first decade of the 19th century and one gang was marooned from 1808 to 1813 (McNab 1909).

The islands were visited by the New Zealand Government steamer in 1908, when Captain Bollons and his wife landed to collect plants. Edgar Fraser Stead got ashore for an hour during a mid-winter gale in 1933 and in the last months of 1947, Robert Falla and A.J. Black in the *Alert* spent 12 days mapping and investigating the birds. Graham Wilson (1973) camped ashore for four days early in 1973 and towards the end of that year, a party of biologists spent 16 days at the islands. Both islands were visited again briefly in February 1984 (Cooper *et al.* 1986).

Away from the steep cliffs, where Southern Buller's Albatrosses nest, both islands are covered with blanket bog and tussock grasses. The plateau of Solander Island is dominated by *Olearia* forest, with an understory of ferns and megaherbs. Little Solander has a less diverse vegetation.

In 1959, the Solander Islands were incorporated into the Fjordland National Park and since 1973, they have been a Specially Protected Area.

9. SNARES ISLANDS (48° 02'S, 166° 36'E)

A group of islands about 100 km southwest of Stewart Island, New Zealand. North East Island (2.8 km^2, 130 m) and Broughton Island (90 ha, 86 m) have a number of offlying stacks and rocks. About five kilometres to the SW lie the five islets of the Western Chain: Tahi, Rua, Toru, Wha and Rima.

Captain George Vancouver in the *Discovery* and Lieutenant William Robert Broughton in the *Chatham*, independently found the islands on 23 November 1791. Sealers arrived at about the same time, but the Snares never attracted many vessels; four men who were put ashore against their will in 1810 did not see another ship for seven years (Richdale 1949b).

The coasts include vegetated cliffs of 50 m and above the steep ground, tussock grasses become meadows, with clearings of the megaherb *Stilbocarpa robusta.* There is a unique forest of tree daisies *Olearia lyalli* and *Brachyglottis stewartiae* with a canopy at six to eight metres.

The islands were incorporated as one of the Dependencies of New Zealand in 1842. Between 1867 and 1927 New Zealand Government steamers called regularly to look for castaways and replenish a depot. During these and other voyages various collectors and naturalists landed briefly. Robert Falla and R.C. Murphy stayed on the Snares for two weeks in 1947 and in the following year, L.E. Richdale spent seven weeks on the island with one companion during his pioneer study of Southern Buller's Albatrosses (Richdale 1949b). In 1961, biologists from the University of Canterbury spent four weeks converting an old castaways hut into a field station, but they did not return until 1967 (Warham 1967). Albatross research revived again in the late 1990s and received funding for several years (Sagar 1999).

Falla landed on Rua for an hour in 1947 and there were several more brief visits to the Western Chain before exploration of the group was completed by the University of Canterbury expeditions of 1976–77, 1983–84, 1984–85 and 1985–86 (Miskelly 1984, 1997).

In 1961, the Snares Islands were gazetted as a nature reserve and in 1986 upgraded to National Nature Reserve. In 1998, they were designated a World Heritage Site.

10. AUCKLAND ISLANDS (50° 44'S, 166° 06'E)

Auckland Island (510 km^2, 644 m)
Adams Island (101 km^2, 705 m)
Enderby Island (7 km^2, 43 m)
Disappointment Island (5 km^2, 316 m)

Abraham Bristow, master of the whaler *Ocean*, sighted islands in 1806, which he roughly charted and named 'Lord Auckland's Groupe'. The *Ocean* was owned by S. Enderby and Sons of London, enterprising merchants in the seal trade who encouraged geographic explorations. The following year, Bristow was sent back in the *Sarah*; he claimed the islands for Britain and released pigs. In the years that followed, many vessels took seals and whales or laid-up in the sheltered bays. House mice and cats were probably introduced around these times (Taylor 1975).

In 1840, American, British and French expedition vessels called at the Aucklands on their way to the Antarctic. Captain James Clark Ross and the officers of the HMSs *Erebus* and *Terror* charted the islands and released pigs and rabbits. Natural history collections were made by the surgeons, among them Robert McCormick, who found albatrosses on nests near Port Ross.

Forty Maori with 26 Moriori slaves arrived from the Chatham Islands in 1842 and eked out a meagre existence for 13 years. In that period, the British Southern Whale Fishery Co. (Enderby and Sons) built a shore station at Port Ross, only to discover that there were no more whales. Within three years, the enterprise was abandoned.

Introduced rabbits survived on Enderby and Rose Islands, but have now been eradicated. Cattle released on Enderby Island in 1895 thrived for almost a hundred years, until eventually removed in 1993. These animals had a profound effect on the vegetation, which may have influenced the nesting of Southern Royal Albatrosses.

Two thousand sheep were released on Auckland Island in 1904, but the climate was evidently too wet for them and they soon died out. Introduced goats never moved far from Port Ross and have all been shot. Feral pigs do well and continue to affect the vegetation (Challies 1975). Feral cats are still present and house mice are widespread (Taylor 1975, P. Dingwall pers. comm.). Adams and Disappointment Islands have no introduced mammals.

From 1864, at least ten sailing ships were wrecked on the Aucklands and over a hundred lives lost. After the earlier disasters, the New Zealand Government provisioned castaway huts and from 1882 visited them regularly until 1929. Under Captains Fairchild and Bollons, these voyages became popular tours for Governors and their guests.

The barque *Dundonald* was wrecked on Disappointment Island in 1907. Fifteen of the crew got ashore and were able to build shelters of peat and tussock grass. Against all odds, they brought ashore a dry match and throughout a winter of great hardship were able to cook albatross meat and megaherb roots. With the departure of the albatross fledglings, they became anxious and built a rickety coracle of twisted rata branches covered with canvas. Several of them made a desperate crossing to Auckland Island and reached the castaway depot. There they found a boat and were able to rescue their shipmates.

Naturalists and collectors occasionally joined steamers charted by the New Zealand Government on voyages to rescue castaways and provision depots (1882 to 1927). Captain Fairchild evidently approved of the Western Arm of Carnley Harbour and often landed passengers on the adjacent coast of Adams Island. In January 1890, F.R. Chapman and others named the place Fairchild's Garden. The most important scientific endeavour to benefit from these voyages was the 1907 expedition of the Philosophical Institute of Canterbury that sailed in the *Hinemoa.* Twelve scientists and their assistants worked in the Carnley Harbour area and also landed on Disappointment Island (Chilton 1909).

At the beginning of the second world war, the activities of German surface raiders around New Zealand prompted the NZ Government to set up observation posts on its remote islands; they were known collectively as the 'The Cape Expedition'. On the Auckland Islands, there were outposts at Port Ross in the north and Carnley Harbour in the south, both within reach of albatross colonies. Towards the end of the war, survey parties traversed the three main islands before all observers were withdrawn.

New Zealand scientists visited Enderby Island and the vicinity of Port Ross several times between 1954 and 1966, when the nests of the Southern Royal Albatross on Enderby Island were mapped. A field party with a small boat also spent three weeks on Adams Island in 1966. The Auckland Islands Expedition of 1972–73 had two vessels, *Acheron* and *St Michael*, which landed many parties, totalling 29 scientists and their assistants. Albatross breeding grounds throughout the islands were located (Robertson 1975b). During the early 1990s, Kath Walker led several yearly expeditions to Adams Island, to count and ring albatrosses. Landings on Disappointment Island have been few, but photography from the air and sea has been used to make a census of albatrosses (A. Rebergen pers. comm.).

The land is well vegetated with 196 indigenous and 37 introduced vascular plants. Tussock grasses occur along exposed coasts. Away from sea-spray, there are large-leaved megaherbs and narrow strips of gnarled trees. Within sheltered bays forests of southern rata, *Metrosideros umbellata* are at sea level. As one ascends, the stature of the forests soon diminishes to a sub-alpine shrubland, with dense, almost impenetrable tangles of *Dracophyllum*, *Coprosoma* and *Myrsine divaricata* often overlying deep bogs. On Adams Island above 200 m, wet tussock grass is dominated by *Chionochloa antarctica.* On the higher slopes, sparse fellfield with flat rosettes of dwarf *Pleurophyllum hookeri*, give way to bare rock on the most exposed ridges; there is no permanent ice.

A New Zealand Government declaration protecting the flora and fauna of Adams Island was made in 1910, and in 1934 extended to the other islands. In 1986, the group was declared a National Nature Reserve and a management plan was published. In 1998, the Auckland Islands were designated a World Heritage Site.

11. CAMPBELL ISLAND (52° 33'S, 169° 09'E)

Frederick Hasselburg, master of the sealing brig *Perseverance*, discovered this island (114 km^2, 569 m) in 1810 and named it for Robert Campbell of Sydney, the owner of the vessel. From that year, sealing gangs lived ashore and by 1842, when the island became a Dependency of New Zealand, seal stocks were depleted. Towards the end of the century, seals became slightly more numerous and although killing was prohibited, cheap sealing licences were issued up to 1923–24 and the occasional closed seasons were not enforced. In 1909 and 1911, licences were also issued for small whaling stations at Northwest Bay and Northeast Harbour. Sixty right whales had been taken by 1916, when all the men enlisted for the war.

Many naturalists and collectors visited the island (Westerskov 1960). Most of them were from ships that anchored in sheltered harbours for a day or so on their way to or from the Antarctic. The exception was the French transit of Venus expedition of 1874–75. The *Vire* was at Perseverance Harbour from 9 September to 28 December and Henri Filhol, a versatile natural scientist of devoted industry, had a camp ashore, not far from Southern Royal Albatross nests (Filhol 1885).

Between 1883 and 1927, New Zealand Government steamers visited the island each year and during one visit William Dougall (1888) took the first photographs of the Southern Royal Albatross. At the beginning of 1958, a party from the Denver Museum of Natural History spent six weeks on Campbell Island (Bailey & Sorensen 1962). Kaj Westerskov, who accompanied them as scientific liaison officer of the NZ Wildlife Service, later produced his own history of naturalists and checklist of the birds of the island (Westerskov 1960).

On the lower slopes there is a shrubland of *Dracophyllum*, *Coprosoma* and *Myrsine*. Above about 200 m the natural vegetation is tussock grass interspersed with deep peat-bogs. In 1894, the New Zealand Government offered Campbell Island as a sheep run. Over 3,000 sheep were released and thrived, increasing to 8,500 by 1910. Shetland shepherds were engaged and labour was also available from among whalers during their off season. World depression eventually made the island wool unprofitable and, in 1931, the shepherds departed leaving some 4,000 sheep and 30 cattle to fend for themselves. Feral sheep and the earlier burning of pasture had considerably modified the vegetation; tussock grass and many of the megaherbs had been replaced by extensive stands of unpalatable *Bulbinella rossii*. (Westerskov 1959).

During the second world war, the Cape Expedition had a base at Tucker Cove in Perseverance Harbour. Jack Sorensen (1950a,b), who served in different capacities between 1942 and 1947, completed pioneer studies of Southern Royal and Light-mantled Sooty Albatrosses. At the end of the war the huts became a meteorological station until 1958, when new buildings were erected at Beeman Cove. From 1941 to 1995, the island was inhabited by technicians and scientists; it is now visited only intermittently.

Between 1943 and 1970, some 20,000 Southern Royal Albatrosses were ringed, but sustained field-work on the more distant mollymawks was not initiated until the late 1980s (Moore & Moffat 1990a). Westerskov (1963), who worked on the ecology of the Southern Royal Albatross during the summer of 1957–58, drew attention to changes in vegetation and recommended that the sheep be removed. This led to an investigation into the ecology of the feral sheep (Wilson & Orwin 1964, Taylor *et al.* 1970). Experimental management began in 1970, when a fence was erected across an isthmus separating the island into two, and 1,100 sheep were removed from the northern part. After the effects on vegetation had been noted, a second fence was erected (1984), confining sheep to the SW corner of the island and by 1992 the last sheep had been removed.

Pigs, goats and domestic fowl were released in the 1860s, but did not survive. Brown rats probably arrived during the period of sheep farming and at least one cat was photographed with the shepherds at that time. Today rats are widespread, but there are said to be fewer than 50 cats (Fraser 1986).

After a brief call in 1907, Edgar Waite claimed that albatross eggs were being sold as curiosities. He suggested that measures should be taken to protect these birds while breeding, but it was not until 1953 that Campbell Island was declared a flora and fauna reserve. Campbell Island is now administered by the New Zealand Department of Conservation (DoC). Since 1983, a management plan has been in operation, and a boardwalk has been laid to one of the albatross nests habitually visited by people from cruise ships. In 1998, Campbell Island was designated a World Heritage Site.

12. ANTIPODES ISLANDS (49° 41'S, 178° 48'E)

Antipodes Island (21 km^2, 366 m)
Bollons Island (50 ha, 200 m)

These islands were first sighted by Captain Henry Waterhouse of the *Reliance* on 26 March 1800 and named the 'Penantipodes' for their opposite global position relative to England. Sealing gangs were ashore in 1804, and by 1842, when the islands became a Dependency of New Zealand, there were few seals left. From 1882, New Zealand Government steamers visited the islands regularly in search of castaways and a hut, built at Anchorage Bay in 1886, is still standing.

Around the coast, tussock grass is intersperced with ferns and the megaherbs *Stilbocarpa*, *Pleurophyllum* and *Anisotome*. Tussocks are tall on the coast and shorter inland. The only woody plants are four species of *Coprosoma*; their interlaced trunks and branches can be impenetrable. There are 68 indigenous vascular plants. Introduced house mice are said to have had little impact on the vegetation.

Birds were collected for the British Museum in 1901, and in 1926 Rollo Beck collected for the American Museum of Natural History. A biological survey was carried out by the New Zealand Government in November 1950 and in 1969 a University of Canterbury Expedition was on the island from January to March and made the first census of the Antipodes Wandering Albatross (Warham & Johns 1975, Warham & Bell 1979). A New Zealand Wildlife Service Expedition was on the Antipodes for six weeks in 1978–79, when a hut was built and a landing achieved on Bollons Island (C.J.R. Robertson pers. comm.).

Interest in the islands revived in 1994, when the small yacht *Totorore* landed expeditions to count the Antipodes Wandering Albatross. Landings were also made on Bollons Island (Clark & Robertson 1996, Tennyson *et al.* 1998).

The Antipodes Islands were protected by the New Zealand Government in 1961 and declared a National Nature Reserve in 1986, under the Reserves Act 1977. In 1998 they were designated a World Heritage Site.

13. BOUNTY ISLANDS (47° 45'S, 179° 03'E)

This group of about 20 low islands and rocks of total area about 135 ha is nowhere more than 73 m above sea level (Fraser 1986). They were discovered on 9 September 1788 by Captain William Bligh of the *Bounty*, who named them for his ship, but did not approach closer than about ten miles. Sealing gangs were ashore from 1808.

Although they were designated a Dependency of New Zealand in 1842, it was not until 1870 that Captain George Palmer of the *Rosario* took possession for Britain (Headland 1989). From 1882 to 1927 Government steamers visited the islands regularly to service a depot and search for castaways. William Dougall's (1888) photographs clearly show the characteristic rocky nature of the islets, confirming earlier reports of lack of vegetation; but this did not prevent the New Zealand Government from offering the islands as a sheep run in 1894! A biological survey was carried out in 1950 and in 1978 three scientists spent two weeks on the islands and made the first study of Salvin's Albatross (Robertson & van Tets 1982). In 1998, the Bounty Islands were designated a World Heritage Site.

14. CHATHAM ISLANDS (44° 00'S, 176° 30'W)

Moriori people had migrated to the Chatham Islands by the 15th century. There were said to have been some 2,000 of them on 29 November 1791 when Lieutenant Broughton landed from the *Chatham* and claimed the islands for Britain. In 1835, Maoris from mainland New Zealand invaded and enslaved the Morioris.

Sealers began working the outer islands in the 1790s and European settlement dates from about 1827 when provisioning stations for whalers were set up. After the Treaty of Waitangi (1840)[8], the New Zealand Company tried to acquire the Chatham Islands privately from Maori chiefs, but the British Government intervened in 1842, making the islands a Dependency of New Zealand.

Sheep-farming began in 1842 and most accessible land has long since been changed by animal husbandry and agriculture. Many alien plants and animals have been introduced. Today there are serious conservation concerns (Ritchie 1970). In recent times, fishing for sea crayfish and paua *Abalone* has been the most profitable industry. After 1840, collectors,naturalists and scientists visited the Chathams from time to time, but few went beyond the main islands. The remote islets were first approached in 1872.

Morioris were nomadic hunter-gatherers with no agriculture. The men made periodic voyages to collect albatrosses from the outer islets. The wash-through boats they are said to have used (Warham 1990) were capable of carrying about 40 albatrosses in addition to the eight-man crew. Albatross expeditions were continued by Maoris in boats that could carry hundreds of birds (Buller 1888). From 1841, preserved albatross meat was exported to mainland New Zealand for Maori celebrations, but the tradition declined in the 1880s. This ethnic delicacy has recently acquired wider political significance with some islanders claiming cultural rights to kill albatrosses. However, many islanders are content for the birds to remain protected (C.J.R. Robertson pers. comm.)

[8] signed 22 January 1840 by more than 400 Maori chiefs and the representative of Queen Victoria, making New Zealand a British dominion (Headland 1989).

Maori ownership of the offshore albatross islets was allocated by the Maori Land Court of 1870 (R. Chappell pers. comm.). Albatrosses have been protected throughout New Zealand since the Animal Protection and Game Acts of 1921–22 and 1931. The Wildlife Act of 1953 confirmed this protection. Since then, albatross meat has come from birds that have been picked up 'dead' off the sea (flotsam), killed illegally or washed up on main Chatham Island beaches. During storms, fledglings making their first flights from the islets are vulnerable. At such times, rather fine legal interpretations of flotsam are possible. Islanders have been prosecuted in the Chatham Islands Court for possessing albatross carcasses (R. Chappell pers. comm.), but few convictions have been obtained (Robertson 1991).

Petitions to the New Zealand Parliament by Chatham Islands Maoris claiming a traditional right to take albatrosses have been made from time to time since 1933, but have never been successful (Robertson 1991).

The Forty Fours (43° 58'S, 175° 50'W) lie 68 km east of the main Chatham Island and comprise one island, *Motuhara* (14 ha, 100 m) and three large stacks. Steep cliffs make landings hazardous in heavy seas. The fairly flat top is rocky, with no water more permanent than small shallow rain pools and seeps. A thin soil occurs, mainly on the northwest part of the island, and supports a sparse vegetation dominated by *Cotula renwicki*; few other plant species have been found. Soil and vegetation are used for nest building. A photograph taken by Wotherspoon in the 1930s (Falla 1938) shows fledgling Northern Royal Albatrosses among quite well grown *Cotula*. It was much the same in 1954 (Dawson 1955) and 1973 (C.J.R. Robertson pers. comm.), but by 1989, *Cotula* growth was regarded as sparse and vegetation limited (R. Chappell pers. comm.). By 1994, a substantial reduction in vegetation had been accompanied by soil erosion (Robertson & Sawyer 1994). This is believed to have been caused by a series of at least seven exceptional storms and hurricanes in the decade 1985–96 (Robertson 1997).

Low altitude air-photography began in 1989, but it was not until 1991 that anyone remained on the islet long enough to attempt a thorough survey. On that occasion Graham Robertson and Rex Page spent 13 exciting days there. Since then, Department of Conservation field-workers have camped there several times (C.J.R. Robertson pers. comm.).

The Sisters (43° 34'S, 176° 49'W) are 20 km to the north of the main island. They were seen and named by Broughton as the *Chatham* approached in 1791, but they already had the Moriori name *Rangitutahi*. There are two islets, stacks and an exposed reef. Little (Middle) Sister (6.5 ha) is steep-sided with cliffs rising to 40–80 m. The centre of the island is a shallow trough where some soil has accumulated. There is no water other than shallow rain puddles. However the salt dampened soil once supported a significant vegetation of *Cotula featherstonii*, *Senecio radiolatus* and *Lepidium oleraceum*; at least ten other plants have been recorded (R. Chappell pers. comm., C.J.R. Robertson pers. comm.). This vegetation, favoured by Northern Royal Albatrosses, is now impoverished (Robertson 1997).

In reasonable weather, a landing can be made on rock platforms at sea level from where a moderate rock-climb leads to the centre of the islet. Big Sister (8 ha), several hundred metres to the west, is long and flat-topped at about 90 m, with cliffs rising steeply from the sea.

The Sisters were visited sometime in the 1930s (Falla 1938) and by N Z Wildlife Service biologists in 1953 (Bell 1955), 1954 and 1964 (Dawson 1955, 1973). During the early 1970s, DoC field-workers visited Little Sister seven times, spending five to 32 days ashore (C.J.R. Robertson pers. comm.). In 1990, a party was on Little Sister for six days (R. Chappell pers. comm.) and since 1991, field-workers have been there on 12 occasions. Landing on Big Sister are more hazardous, but the steep ascent was made in 1973, 1975 and 1996; no-one has stayed on the islet for more than a few hours (C.J.R. Robertson pers. comm.).

The Pyramid or *Tarakoikoia* (44° 26'S, 176° 15'W) is a solitary rock nine kilometres south of Pitt Island. It rises to 177 m and occupies about two hectares (R. Chappell pers. comm.). In 1924, during the Whitney South Seas Expedition, Rollo Beck shot seabirds from a small boat just off the rock, but he could not get ashore. Charles Fleming (1939) landed on 16 December 1937 and spent two hours among the albatrosses:

> "Its northern aspect is steep, giving a perfect 'pyramid' shape, but on the southern side its precipitous cone flattens out in a concave steeply sloping ledge of several acres which we were able to reach after a 50-foot climb from sea level."

In 1974, Christopher Robertson landed twice with Alan Wright and camped with Rodney Russ for six days, a considerable feat of endurance. Graham Robertson and Rex Page were there in 1991 and Graham's brief note, written shortly afterwards, tells it all:

> 'Managed 10 days on Pyramid and 13 days on the 44s...Weather on both stacks atrocious. We spent much (most) time and energy just trying to stay alive. We did! Wind and rain in volumes. In spite of this both stacks absolutely brilliant and the birds breathtaking in courtship and behaviour...I'm rapt.'

15. ISLA DIEGO DE ALMAGRO (51° 26'S, 75° 15'W)

This is one of the outer islands of the Chilean fjordland. North of the Pacific entrance to the Magellan Strait is the Estrecho Nelson; the southern end of Isla Diego de Almagro at Cabo Jorge guards the northern entrance to this strait. It is a militarily sensitive area and nothing was known about its seabirds until January 1984 when the yacht *Totorore* made a thrilling inshore run under its impressive western cliffs and discovered thousands of Black-browed Albatrosses (Clark 1984, 1988).

16. ISLAS ILDEFONSO (55° 44'S, 69° 26'W)

Richard Hough (1971) has suggested that these were the islands that Sir Francis Drake (1589) encountered as the *Golden Hind* headed northwest from the region of Cape Horn, where it had been driven by storms in September 1578. The Captain-General wrote:

> 'We returning hence Northward againe, found the 3 of October three islands, in one of which was such plentie of birds as is scant credible to report.'

His nephew, John Drake, age 15, who was later captured and interrogated by a Tribunal of the Inquisition in Lima, Peru 1587 (Nuttall 1914) remembered an island 'covered with ducks' that 'made a provision of their meat', while in 1628 the chaplain, Francis Fletcher (1854) remembered:

> 'In this course we chanced...with two Ilands, being, as it were, storehouses of most liberall pro[v]ision of victualls for [u]s, of birds; yielding not onely sufficient and plentifull store for [u]s who were present, but enough to ha[v]e ser[v]ed all the rest also which were absent.'[9]

The Islas Ildefonso are steep, rocky islands covered with tussock grass, situated 96 km W of Isla Hermite, Tierra del Fuego. There are two main islands (130 m) with smaller islets, stacks and rocks extending about six kilometres NW-SE. In December 1914, when Rollo Beck was collecting near Cape Horn, he landed from a small sealing cutter and spent two hours ashore among the penguins and Black-browed Albatrosses. Seventyone years later, Gerry Clark (1988) sailed his small yacht *Totorore* to the Ildefonsos and has vividly described putting Peter Harrison ashore.

Shortly before, *Totorore* had been at the Islas Evout (55° 34'S, 66° 47'W), where P.W. Reynolds (1935) had been told there were Black-browed Albatrosses. Military restrictions prevented landing, but a careful offshore search from the sea convinced Clark that the species did not occur there (G. Clark *pers. comm.*).

In November 1901 a meteorological observatory and lighthouse was established on Isla Observatorio, in the Año Nuevo group just north of Isla de los Estados (54° 47'S, 64° 20'W). An egg from the collection of S. Venturi at the British Museum, Tring is labelled 'Black-browed Albatross, Año Nuevo (?)' and dated 12 December 1902. There is nothing to indicate that Venturi himself went to the islands (Hartert & Venturi 1909). The lighthouse ceased operating in 1917, and Castellanos (1935) saw no albatrosses on Isla Observatorio. Meade Cadot (pers. comm.), who spent 27 days on board the *Hero*, circumnavigating Estados at the height of the 1971–72 breeding season, saw no colonies; neither did Juan Carlos Chebez and Claudio Bertonatti (1994), who were there in the summer of 1981–1982, nor Jérôme and Sally Poncet (pers. comm.), who later explored the northern coast in their schooner *Damien II*.

[9] refers to the two ships that had disappeared earlier when leaving the Magellan Strait.

16. ISLAS DIEGO RAMÍREZ (56° 31'S, 68° 44'W)

These two small groups of islets and rocks, about 112 km SW of Cape Horn, lie on the southern edge of the continental shelf and extend over 10 km in a NNW-SSE direction. They were discovered in 1619 by the Portuguese brothers Bartolomé and Gonzálo de Nodal who named them for their cosmographer. Their caravels, *Nuestra Señora de Achoa* and *Nuestra Señora de Buen Succeso,* made the first circumnavigation of Tierra del Fuego (Hough 1971). The islands are in the direct line of vessels keeping well south of Cape Horn and, in the days of sailing ships, acquired a notorious reputation.

The larger southerly group consists of Isla Bartolomé (93 ha, 190 m) separated by a narrow channel from Isla Gonzalo (38 ha, 139 m). These two main islands are hilly, with cliffs on the west and fairly steep rocky slopes on the east. There are only eight species of vascular plants. Coastal tussock grass, often 1–2 m high becomes lower inland and is replaced on drier ground by a mosaic of cushion plants *Plantago barbata/ Colobanthus quitensis.* The smaller group, about four kilometres to the north, consists of Isla Norte and several smaller islets. They have less vegetation than the larger southerly islands.

Sealers were there by the 1820s and the group was surveyed in 1826–30 by the *Adventure* and *Beagle.* The islands were confirmed Chilean territory in the 1881 treaty with Argentina. A small meteorological station was established on Isla Gonzálo in 1951 and has been occupied ever since. The French geographer, Edgar Aubert de la Rüe (1959) was ashore in 1958 for one day of scientific reconnaissance and Roberto Schlatter spent three months there in 1980–81 (Schlatter & Riveros 1997).

17. FALKLAND ISLANDS (51° 00'–52° 54'S, 57° 42'–61° 27'W)

The Falklands lie about 500 km northeast of Tierra del Fuego. In addition to the two main islands, there are at least 746 smaller offshore islands and islets, altogether totalling about 12,200 km^2 (Woods 1988). The coastline has many deep, sheltered bays while the interiors of the main islands are treeless heathland and bogs interrupted by areas of shattered rock known locally as 'stone-runs'. Craggy hills rise to 705 m and on the higher slopes feldmark predominates. The flora of 292 species has strong natural affinities with Tierra del Fuego, and includes 125 alien plants introduced by man. The vegetation has been drastically changed by sheep farming. In particular there has been a widespread loss of coastal tussock grass[10] due to over-grazing and burning (Strange *et al.* 1988). The grass survives as dense stands, two metres high on some albatross islands, but fire has reached one albatross colony in recent years (T. Chater pers. comm.).

The islands were sighted by English mariners in 1590s and in 1690 John Strong landed from the *Welfare* and named Falkland Sound, separating the two main islands which became known as East Falkland and West Falkland. In 1701, Jacques Gouin de Beauchêne with two French vessels, discovered the most detached island to the south, now named after him. A French settlement was established at Port Louis, East Falkland in 1764 by Louis Antoine de Bougainville. Three years later, the French territory was formally transferred to Spain. In the meantime, at the beginning of 1765, John Byron of the *Dolphin* had found a safe harbour in West Falkland, which he called Port Egmont. The islands were claimed for Britain and in 1766 settlers were brought out in the *Jason* and *Carcass.* A Spanish force in five ships expelled them in 1770, but the following year they were reinstated. The political consequences of these events are still with us.

Whaling and sealing vessels visited the Falkland Islands before 1771, and from 1775 New England fleets began to lay-up in the sheltered bays, where they put cattle and pigs ashore. New Island and West Point Island were well known for their albatross eggs. Sealers probably landed at most albatross islands, but customs duty, levied on seal skins from 1904, eventually brought an end to old-style sealing in the islands.

French naturalists visited Port Louis in the early 1820s. The *Beagle,* with Charles Darwin, was there twice (1833,1834) and in 1842, Joseph Hooker and Robert McCormick in the *Erebus* and *Terror* stayed for six months.

By the 1830s, large numbers of feral cattle were roaming East Falkland. Goats, pigs, hares, rabbits, brown rats,

[10] known in the Falklands as Tussac.

black rats, house mice and cats had all been introduced and some were widespread. A unique endemic fox, known locally as the 'Warrah', had been exterminated by the 1870s, but Patagonian foxes and South American guanacos were introduced to a few islands in the early 1900s. Sheep farming was not generally successful until the mid-19th century, but from then on sheep gradually replaced cattle to become the main grazing animals of the islands.

British colonisation of the Falkland Islands did not get under way until the 1840s when a new settlement, Stanley, was built on the shore of Port William and farm leases were offered on East Falkland. Apart from an evangelical mission on Keppel Island opened in 1856, leases in West Falkland were not offered until 1869.

Black-browed Albatross eggs were collected near Port Egmont in the late 1760s and preserved in casks of sand (Clayton 1774, Gower 1803). American sealers, who anchored at New Island in 1797 kept eggs by coating them in whale oil and throughout the first half of the 19th century intermittent egging probably occurred at several colonies.

From 1855, the Keppel Island Mission was within reach of one colony and by 1869, when West Falkland was opened to sheep farming, homesteads at Saunders Island, West Point Island and New Island all had ready access to albatrosses. Uninhabited albatross islands also passed into private ownership or were leased and stocked with sheep.

Albatross eggs for home use were probably collected each year. They were considered superior to penguin eggs and from time to time boat crews harvested the distant colonies and sold eggs to the people of Stanley. In 1900, one sealer shipped 10,000 albatross eggs from Beauchêne Island and obtained 12 shillings a hundred for them in Stanley (Vallentin 1904). Since 1914, egging has been subject to government control, but licences have not recorded the species involved. Most were penguin eggs, with comparatively few from albatrosses.

Between 1880 and 1930, over 2,000 albatross eggs were taken each year at West Point Island and at least one colony had declined. The Devil's Nose, which yielded 1,500 eggs in 1907, was down to 200 pairs by 1949. Collecting from this colony ceased and by 1959 numbers had increased to about 550 pairs (Strange 1969). From 1965 to 1988, 11,000 albatross eggs were collected at Saunders and West Point Islands. They amounted to less than 300 per year on each island, just enough for the island farms and well within sustainable levels.

There is no evidence that such levels of egging had any detrimental effect upon the enormous numbers of Black-browed Albatrosses in the Falklands; even an occasional large haul was probably well within the expected loss from natural causes. The albatrosses themselves do not appear to have been eaten, perhaps because there were so many geese available. Life styles and tastes have since changed. After the 1982 Argentine/Falkland War, imported hens eggs have largely replaced those of wild birds, which are now eaten only by older people.

Wild animal and bird protection ordinances were enacted by the Falkland Islands Government in 1908 and 1913, but had little impact other than to licence sealing and egging. The Wild Animals and Birds Protection Ordinance of 1964 protected all birds and restricted the taking of albatross eggs without a licence. Legislation towards setting up nature reserves on government land was also enacted. Beauchêne Island (1.7 km^2) was among the first albatross islands to be declared a sanctuary (1964) and Bird Island (1.2 km^2) was similarly protected in 1969. High priority was given to saving tussock grass and among the important tussock islands made reserves, South Jason (3.7 km^2) has an albatross colony. Sheep were removed from Steeple Jason (7.9 km^2) and Grand Jason (13.8 km^2) in 1968; two years later these important albatross islands were sold by Dean Brothers to Leonard Hill of Gloucestershire, England, for £5,500. Both islands have since remained private nature reserves. In 1993, they were put on the market at an asking price that was said to be $300,000, and were bought by Michael H. Steinhardt of New York; whether he paid anything like that amount was not revealed.

Seabird islands were well known to seafaring islanders long before anyone attempted to estimate the numbers of birds. From 1919 to 1949, J.E. Hamilton served as naturalist to the F.I. Government, but after he retired, the post was abolished and natural history in the islands has been largely the province of amateurs with occasional visits by professionals. Since the formation of the Falkland Islands Foundation[11] in 1979, several seabird projects have been funded.

The surrounding sea, over the Patagonian continental shelf, is highly productive and oil exploration is under

[11] In 1991 merged with the Falkland Islands Trust as Falklands Conservation.

way. The only indigenous fishery has been a decade of whaling (1908–17) when two whale-catchers operated out of New Island. From the 1960s to the present, squid and finfish fisheries have attracted fleets from all over the world, but since 1982, fishing in the Falkland Islands Conservation Zone (FICZ), within 150 to 200 miles of the islands, has been confined to foreign vessels licensed by the F.I. Government. A ship was wrecked on Beauchêne Island in 1992, and for a while there was concern that rats might have got ashore (Thompson 1992). Optimistic forecasts continue to be made about oil extraction. All these industries have significant implications for seabirds of the Falklands.

18. SOUTH GEORGIA (53° 30'–55° 00'S, 35° 30'–38° 40'W)

South Georgia is a large mountainous island (3,755 km^2, 2,934 m) with several prominent offshore islands and many islets. Glaciers descending to sea level occupy 57% of the land and in winter there is total snow cover. The *Resolution* and *Adventure* sailed through the Stewart Strait on 17 January 1775; later that day, Captain Cook landed at Possession Bay and claimed the land for Britain.

Vegetation extends from sea level to about 250 m, with 26 indigenous vascular plants and 54 alien plants, the latter mostly in the vicinity of the whaling stations. Close to the coast, on cliffs exposed to sea-spray, tussock grass is often very dense, sometimes growing to two metres. Inland it is less prolific and interspersed with meadows of short Antarctic hair-grass *Deschampsia antarctica*, herbs and mosses. Grassy heath formations with burnets are extensive on the better drained ground. Wet meadows and bogs with pools and rushes occur wherever the water table approaches the surface.

Sealers arrived in 1786 and by 1791 there were over 100 vessels with perhaps 3,000 men at the island. From the earliest times albatrosses and their eggs were taken for food.

> '...with regard to meat, we were supplied with young albatrosses, that is to say, about a year old: the flesh of these is sweet, but not sufficiently firm to be compared with that of any domestic fowl.'
>
> James Weddell (1825)

In 1800–01, a single vessel, *Aspasia*, obtained 57,000 skins but after 1825 few furseals remained. The 170 taken in 1908–09 by Benjamin Cleveland, master of the brig *Daisy*, were thought to have been the last commercial kill.

The first of several whaling stations was built at Grytviken in 1904–05 and by 1907 there were 720 whalers working at South Georgia, almost all of them Scandinavian. Apart from pauses during times of war, they operated continuously until 1964, when shore-based whaling at South Georgia came to an end. In sixty years, 175,250 whales had been killed. Elephant sealing for oil began in 1909–10 and was part of the whaling industry; it also ceased in 1964, by which time 260,950 seals had been killed (Headland 1984).

The whaling stations have probably been infested with introduced house mice and brown rats since they were built, and horses, pigs, dogs and cats have all lived in or close to the factories. Away from whaling stations, house mice have been found in only one isolated location where they evidently came ashore in early sealers' provisions. Brown rats are widespread, but absent from most offshore islands and there is no evidence that they prey on any albatrosses (Pye & Bonner 1980). In 1905, an attempt to farm sheep was abandoned after three months and introduced Upland Geese from the Falkland Islands did not survive. Two herds of reindeer, the first introduced in 1911, are still grazing separate areas.

A meteorological station was set up on King Edward Point in 1907 and the first magistrate and customs officer arrived from the Falkland Islands in 1909. A radio station was installed in 1925 and a small government community with families became a permanent feature until the end of whaling. From 1969 to 1982, British Antarctic Survey was established at King Edward Point, but during the Argentine/Falkland war this became a military garrison. From 1985, South Georgia and the South Sandwich Islands ceased to be a Dependency of the Falklands and became a designated territory with a Commissioner resident in the Falkland Islands.

During the International Polar Year of 1882–83, German scientists lived at South Georgia. They took the first photographs of the island and made biological collections, but remained in the vicinity of Royal Bay. The Swedish

Antarctic Expedition on board the *Antarctic* was at South Georgia in 1902, when Carl Skottsberg extended botanical knowledge of the island. In 1912–13, Murphy (1947) camped for a while next to a few Wandering Albatross nests at the Bay of Isles, while the *Daisy* was sealing.

The next person to find Wandering Albatrosses at South Georgia was Sir Ernest Shackleton. On 10 May 1916, the small boat *James Caird* with its exhausted crew, approached the forbidding and uncharted south coast after an epic voyage from Elephant Island. They negotiated the coastal hazards and discovered a sheltered cove with a dry cave.

> 'Crean and I climbed the tussock slope behind the beach and reached the top of a headland overlooking the sound. There we found the nests of albatrosses, and, much to our delight, the nests contained young birds. The fledglings were fat and lusty, and we had no hesitation about deciding that they were destined to die at an early age...what a stew it was...The young albatrosses weighed about fourteen pounds each fresh killed, and we estimated that they weighed at least six pounds each when cleaned and dressed for the pot. Four birds went into the pot for six men, with a Bovril ration for thickening. The flesh was white and succulent, and the bones, not fully formed, almost melted in our mouths. That was a memorable meal. When we had eaten our fill we dried our tobacco in the embers of the fire and smoked contentedly.'
>
> (Shackleton 1919).

Concern for the state of whale stocks led to the *Discovery* Investigations of 1925–31. Scientists of the marine biological laboratory built near the whaling station at Grytviken, worked primarily on the biology and ecology of whales, krill and elephant seals. Leonard Harrison Matthews (1929, 1951) accompanied sealers about much of the coast, becoming familiar with the island and its fauna. His personal narrative is a vivid reminder of a hardy way of life, before modern sensitivities, when biologists shared the lives of whalers and sealers. He confirmed Wilkins' (1923) report that large numbers of albatrosses bred at the northwest, where whalers habitually took thousands of eggs and fledglings.

The ships and facilities of the whaling industry made it possible for a few naturalists to visit South Georgia and the captains of the sealers *Petrel*, *Albatros* and *Diaz*, who steamed around the island each October in the decades after the second world war, had an unrivalled knowledge of the coast. Carl Gibson-Hill (1950), curator of the Raffles Museum in Singapore, spent the early months of 1946 at Leith Harbour. He was taken to Elsehul, Bird Island and the Bay of Isles. Niall Rankin (1951), a well-known bird photographer shipped his boat *Albatross* south on a whale factory-ship and spent the summer of 1946–47 exploring the coast and islands with a large plate camera. His photographs remain classics of South Georgia birds. Since the early 1980s, Jérôme and Sally Poncet have sailed the coast of South Georgia many times in their schooner *Damien II*, and have acquired an intimate knowledge of the island's natural history.

The almost extinct furseals of South Georgia, have recovered in a spectacular manner. A few had been seen at Bird Island in 1927, and in 1936, 36 were counted (Deacon 1984). By 1956, 4,500 breeding females were present, and they have continued to increase exponentially, spreading inland from the beaches. Many areas of formerly lush tussock grass have been eroded.

A small field hut was put on Bird Island in 1958, and, in 1962, a larger one was built for overwintering. Since 1982 several more buildings have been added. Today, furseal and seabird research continues from a modern, permanent base.

19. TRISTAN DA CUNHA ISLANDS (37° 03'–37° 25'S, 12° 13'–12° 42'W)

Tristan da Cunha (86 km², 2,062 m)
Inaccessible Island (14 km², 511 m)
Nightingale Island (3 km², 366 m)

In 1506, a Portuguese fleet commanded by Tristão da Cunha and bound for India, discovered this group of islands. Ships began calling at Tristan da Cunha[12] early in the 17th century and by 1656 it was known as a watering place. Sealers arrived in the late 18th century, releasing goats and pigs. By the early 1800s, few furseals were left, but sealers and whalers continued to visit the islands.

Three Americans were making a living on Tristan in 1810 by cultivating vegetables. One of them, Tomaso Corri, was still there with a companion on 14 August 1816, when Captain Robert Festing of the *Falmouth* took possession of the islands for Britain. Later a garrison was landed to prevent the islands being used to rescue Napoleon, then exiled on St Helena. Captain Dugald Carmichael (1818) accompanied the expedition and on 4 January 1817, climbed to the crater summit with a small party. On the way down they encountered three species of albatrosses on nests.

The following year, after a wreck in which 55 men were lost, the garrison was evacuated, leaving behind three men, a woman, two children, cattle and horses. Later, more men joined the community and in 1827, at their request, a visiting ship brought five women and four children from St Helena. Farming produced enough surplus for this community to trade meat and vegetables with passing vessels and briefly (1880–82) to export cattle to St Helena.

The coastal plains of Tristan are now pastures of introduced grasses grazed by domestic stock, but on the 750 m cliffs and 'the Base' above, stunted trees of indigenous *Phylica arborea* and dwarf tree ferns, *Blechnum palmiforme* persist, interspersed with and giving way to a wet heath of grasses, sedges and *Empetrum* to about 900 m. Above this, the cinder cone rises to the summit.

The islanders have traditionally kept a range of introduced domestic animals. The inventory of livestock on Tristan in 1974 was: 48 donkeys, 447 cattle, 55 sheep, 10 pigs and 610 hens and ducks. Inaccessible and Nightingale Islands have had pigs and sheep in the past, but are now free of all alien animals (Richardson 1984).

House mice probably arrived with the early sealers and are believed to be harmless. Black rats were said to have come ashore at Tristan in 1882 from the wreck of the cattle ship and to have soon infested the settlement. In 1962, it was estimated that there were 400,000 on Tristan which killed many small birds, but they had not reached the other islands (Richardson 1984).

Feral cats on Tristan were said to be capable of killing well grown mollymawk chicks and since the 1950s, the islanders have been encouraged to shoot them on sight (Elliott 1953). There have been none on the other islands. Domestic cats in the settlement were implicated in the spread of toxoplasmosis and all 54 were killed. Feral dogs were all shot in 1961–62, but the 70 or so domestic dogs at the settlement in 1974 were considered a potential menace to birds (Richardson 1984). In 1993, there were more than 100, but they were rarely taken beyond the coastal plateaux (M.K. Swales pers. comm.).

The islanders cared for the survivors of many wrecks and, in their frail boats, rescued castaways from Inaccessible and Nightingale Islands. In 1857, the British Government resettled 45 of them at the Cape of Good Hope, leaving 28 on the island. By 1885, after 15 men had been drowned in a boating accident, only three adult males remained, but from time to time castaways or absconding seamen joined the community. After the accident, ships called more regularly with stores and mail, but it was not until the 1930s that colonial responsibilities were fully implemented; a magistrate was appointed and the first Island Council was officially established. Tristan da Cunha became a Dependency of St Helena in 1938 and Norwegian scientists made a study of the islands in 1937–38 (Hagen 1952).

During the second world war, a Royal Navy meteorological and radio station was established on Tristan. In 1946, it was transferred to the South African Weather Bureau. These services continue to the present day. Since 1948, colonial administrators have been appointed. The Tristan da Cunha Development Company began fishing for sea crayfish *Jasus tristani* in 1949 and Margaret Rowan, one of the marine biologists, visited Nightingale Island several times. A cannery and later a freezer were built near the village on Tristan.

Members of the Gough Island Scientific Survey (1955–56) spent some time on Tristan and have since remained influential in scientific and environmental issues.

On 9 October 1961, a volcanic eruption occurred close to the village and the whole population was evacuated.

[12] the name Tristan da Cunha here refers to the whole group, while 'Tristan' is the largest island.

Two years later 198 islanders returned. The cannery was rebuilt and a harbour constructed. Trade links were strengthened and the community is now largely self-sufficient. There have never been permanent settlements on the smaller islands of the group.

Nightingale Island, 38 km to the SSW of Tristan, is dominated by tussock grass with a few scattered trees and bushes. Between high ground, there are several boggy valleys with standing water known as the Ponds. In spite of a century of egging, fowling, guano digging and fires, the island is said to have been little damaged by man or stock.

Inaccessible Island, 40 km SW of Tristan, is steep-sided. The lower slopes are dominated by tussock grass, but above the 200 m cliffs fern bush (*Histiopteris incisa* and *Phylica arborea*) gives way to wet heath on the plateau. Goats and pigs were put ashore in 1821, the year that the *Belenden Hall* was wrecked, leaving the crew and passengers on the island for six months (Thomson 1885). Gustav and Friedrich Stoltenhoff (1873), who lived on the island for two years (1871–73), saw numbers of albatrosses and shot out all the goats. Sheep were introduced in 1923 and an attempt to farm the island in 1937–38 failed. The domestic animals were removed, but feral pigs were still numerous in the 1950s (Booy 1957). They have not been seen since and their impact on the vegetation has been slight (Wace & Holdgate 1976, Fraser *et al.* 1988).

Tristan da Cunha had the longest history of sustained albatross exploitation. Augustus Earle (1832), an artist who was marooned on Tristan for eight months in 1824, saw Gough Wandering Albatrosses killed on the mountain. It took almost a hundred years for the islanders to eliminate them. In 1852, many nests were scattered around the 'Base' (of the summit cone) at 600–900 m (Bourne & David 1981) and some were still to be found in 1873 (Moseley 1879). The last pair to have bred on Tristan was said to have been in 1907.

For over a hundred years, thousands of Western Yellow-nosed Albatrosses, Sooty Albatrosses and their eggs were taken for food. The people were aware that replacement eggs were not laid in the same season and Katherine Barrow (1910) wrote 'In time I fear that these beautiful birds will be driven from the island.'

For five years, Bob Glass kept a record of the number of mollymawks killed:

1923 – 4,807
1924 – 3,764
1925 – 4,947
1926 – 4,645
1927 – 3,742

The Norwegian scientific expedition of 1937–38, which had studied albatrosses at Nightingale Island (Hagen 1952), was alarmed by these numbers and tried to persuade the islanders to adopt conservation measures (Munch 1946), but in the subsistence economy that prevailed, the needs of the growing human population increased. Margaret Rowan (1951) worked out that an average family needed six albatrosses for a meal. If every family had only one albatross dinner each month and one egg each week, the community would have consumed about 4,000 fledglings and 3,000 eggs a year. Without a more assured alternative to missionary charity, the islanders could not afford to neglect albatrosses.

The devastation of seabirds on Tristan during the last century, obliged later generations of islanders to go to the other islands. The dangers surrounding voyages in such frail craft, imposed constraints that worked indirectly to the benefit of the albatrosses which survived over-exploitation longer than they might otherwise have done.

During the Second World War a Naval Chaplain, C.P. Lawrence, discovered how few albatrosses remained on Tristan and realised that the islanders were taking proportionally more birds from Nightingale Island. In 1942, the military authorities imposed absolute control on all fowling. This was possible largely because the wartime garrison had introduced new elements into the economy. Islanders were employed, they could buy imported supplies and sell their produce. A total ban on fowling was enforced on Tristan and yearly quotas were placed upon the numbers of eggs and birds taken from Nightingale Island. Each family was allowed 80 albatross eggs and 30 to 40 chicks; there were 50 to 60 families so at least 5,500 eggs/chicks were taken each year.

The wartime ban was lifted in 1949–50. Numbers of albatrosses on Tristan had increased; killing of adults was discouraged, but quotas of chicks and eggs were allowed and a reserve area, Joey's Garden, agreed informally (Elliott 1953). By 1961, albatross numbers had declined again; the two year evacuation allowed a slight and temporary recovery, but in 1965 the quota was about 400 and only half that number in 1968. The Wild Life (Tristan da Cunha) Protection Ordinance of 1950 did not include albatrosses on its schedule of protected species, but nests in the SE quarter of the island were said to be protected 'in theory' (Flint 1967). Civil Administrators could not, or chose not, to use the legal powers available to them.

After the war, there were substantial numbers of Western Yellow-nosed Albatrosses on Nightingale Island and licences were issued for eggs and chicks. The quotas were rarely observed by the islanders; 2,500 chicks were taken at Nightingale in 1951 and 1,500 eggs in 1965. By 1968, the Ponds colony was visibly smaller and only 150 chicks were taken (Wace & Holdgate 1976).

Sooty Albatrosses used to nest in colonies high on Tristan (Carmichael 1818), where they were particularly vulnerable. They survived in nests scattered along steep slopes and cliffs, where egg and chick collecting required much more effort. In the early 1950s, about half the estimated 1,500 fledglings on Tristan were killed each year (Table 4.1) and unknown (small) numbers were still being taken in 1968 (Wace & Holdgate 1976).

		1950	1951	1973	1974
Western Yellow-nosed Albatross	eggs	2 000	2 400	100	156
	chicks	1 200	1 700	1 550	1 745
	adults	200	200	100	128
Sooty Albatross	chicks	150	100	325	210
	adults	0	0	7	7

Table 4.1 Numbers of eggs, chicks and adult albatrosses taken in four years from Tristan da Cunha Islands by Tristan Islanders (Richardson 1984).

Although the wartime garrison had been followed by commercial development, the people were not yet ready to dispense with seabird food and evidently considered the albatross quotas inadequate. Dr M.E. Richardson (1984), a medical officer on the island, calculated the UK Sterling value of the 1973 seabird harvest to the islanders as £9,671. Most of it was from penguins and Great Shearwaters, but 100 eggs and 1,550 chick carcasses of Western Yellow-nosed Albatrosses were valued at £10 and £775 respectively, while 325 Sooty Albatross chick carcasses were worth £244. At market value, albatross eggs and carcasses were slightly cheaper than local domestic produce.

In the 1960s, more powerful boats became available and by 1964, following two visits to Nightingale Island, J.H. Flint (1967) suggested that it might become necessary to control the number of albatrosses taken. Ten years later, they were still being taken (Table 4.1). In spite of being less dependent upon seabirds for food, voyages to the other islands were enjoyed by the islanders and had become a tradition in which fowling or egging was still an essential ingredient.

The same year that IUCN[13] published *Man and Nature in the Tristan da Cunha Islands* (Wace & Holdgate 1976), the Tristan da Cunha Conservation Ordinance 1976 was passed by the Government of St Helena and Dependencies. It protected the Wandering Albatross throughout the islands, but the Western Yellow-nosed and Sooty Albatrosses were not protected outside the SE reserve on Tristan, and the Sooty Albatrosses could be taken on the other islands under licence. In 1984, the Conservation (Protected Birds) Amendment Order made both the Western Yellow-nosed and Sooty Albatrosses fully protected species, but only on Tristan. Two years later, another amendment order was passed, specifically to remove the Sooty Albatross from the list of those birds that could be killed under licence at the other islands. Only then did all three albatross species finally achieve full protection throughout the Tristan da Cunha Islands (Cooper & Ryan 1993). By 1996, 44% of the land area of the group had become protected including all of Inaccessible Island.

[13] International Union for the Conservation of Nature and Natural Resources.

20. GOUGH ISLAND (40° 21'S, 9° 53'W)

This small island (65 km^2, 910 m), 350 km SSE of the Tristan da Cunha group, was discovered by the Portuguese navigator, Gonçalo Alvarez in 1505 and named by Charles Gough, master of the *Richmond* in 1732. It was formally claimed in 1938 and, like Tristan da Cunha became a Dependency of the British Dependent Territory of St Helena.

Sealing gangs lived ashore from 1806. They took whatever they could obtain and by 1891 there were no furseals left so penguins were boiled for oil. Ships went to the island for guano only to be disappointed. A collier was wrecked in 1878 and some survivors stayed on the island for six months before being rescued.

George Comer, one of an American sealing gang, made meteorological observations in 1888 and collected bird specimens (Verrill 1895). Collections were later made by scientists from the *Scotia* (1904), *Quest* (1922) and *Discovery* (1927). In 1919, South African diamond prospectors built huts in the Glen and worked the island. Commercial fishing for crayfish about Gough Island began in 1950 as part of the Tristan da Cunha fishery.

The Gough Island Scientific Survey spent six months on the island in 1955–56 and completed the first thorough investigation (Holdgate 1958). The base hut, built at the foot of the Glen, was handed over to the South African Weather Bureau at the end of the expedition. It was rebuilt in 1972–73 and relocated in 1984–85. This station is still occupied by meteorologists, and is used every year by visiting biologists.

Gough Island has no permanent ice and winter snow on high ground does not accumulate. The coastal vegetation of tussock grass rises to 300 m on the cliffs of the west, but only to about 100 m on the east where fern bush (*Histiopteris incisa* and *Phylica arborea*) is prominent, especially in sheltered gullies. On higher ground above about 500 m, wet heath with peat bogs predominates and gives way to feldmark on exposed ridges. There are 63 vascular plants, including 25 ferns and six probable aliens.

House mice are the only alien mammals. They were probably introduced by 19th century sealers and are now widespread over the whole island (Cooper & Ryan 1993).

Although albatrosses and their eggs were once taken by sealers and the crews of occasional ships, the numbers of Wandering Albatrosses breeding at Gough Island do not appear to have been seriously depleated. South African and Tristanian fishermen attending crayfish traps around the coast also 'fished' for albatrosses (Elliott 1953).

The Wild Life (Tristan da Cunha) Protection Ordinance of 1950 which did nothing for albatrosses, was replaced by the Tristan da Cunha Conservation Ordinance, 1976. Gough Island was declared a wildlife reserve in which all native fauna and flora was protected, and in 1984, it was designated a World Heritage Site (Cooper & Ryan 1993).

5
MOLLYMAWKS

'Their nests are little hollows on small mounds of earth, about a foot above the surface; and this earth the birds collect and lay in regular lines, the nests being equal distance from each other. Few streets are more regularly formed, for which reason we gave these assemblages the name of towns.'

Erazmus Gower
Port Egmont, Falkland Islands 1770[1]

Mollymawks are albatrosses with wing spans of just over two metres and average body masses of two to four kilograms. All have black and white plumage, similar to those of black-backed gulls and although they may be strikingly different from each other in head plumage, bill colour and aspects of their biology, there are common features and habits that clearly distinguish them as a group from all other albatrosses.

They are the most numerous and frequently seen albatrosses of the Southern Ocean and may be encountered in large numbers. Their island colonies are sometimes very large, with nests at maximum possible density.

The 11 names below, identify the terminal taxa. If you favour the biological species concept, they represent five species with ten subspecies. If you have adopted the phylogenetic approach of Robertson & Nunn (1997), there are 11 species (Appendix 1). Here and subsequently, I avoid the question of rank.

Black-browed Albatross (BBA)
Campbell[2] Black-browed Albatross (CBBA)

Grey-headed Albatross (GHA)[3]

Western[4] Yellow-nosed Albatross (WYNA)
Eastern[4] Yellow-nosed Albatross (EYNA)

Northern[5] Buller's Albatross (NBA)
Southern[5] Buller's Albatross (SBA)

Tasmanian Shy Albatross (TSA)[6]
Auckland Shy Albatross (ASA)[6]
Salvin's Albatross (SLA)[6]
Chatham Albatross (CHA)[6]

DESCRIPTION

Adult mollymawks have mainly white bodies, but at all ages the black or dark grey upper surfaces of the wings are linked by a dark band of similar width and colour across the back. Pale primary shafts are often visible at the tips of outstretched wings.

[1] Gower 1803 – a 'journal' entry written more than 30 years after the event.
[2] Robertson & Nunn (1997).
[3] The Grey-headed Albatross measured at South Georgia is slightly larger than that at the Iles Kerguelen (Waugh *et al.* 1999d), but although average differences are statistically significant, no separate taxon has been named.
[4] Bourne & Casement (1993).
[5] Turbott *et al.* (1990).
[6] Formerly known collectively as the *cauta* mollymawks.

The white under-surfaces of the wings have black tips formed by the primaries, and black margins of varying width formed by black-tipped coverts on the leading edges and black-tipped secondaries on the trailing edges. Wings are subject to abrasion and new, blackish feathers fade to shades of brown at different rates in different positions. Tails are dark grey and likewise fade to brown when worn.

The head is white or shades of grey, sometimes with contrasting white forehead and crown. The juvenile may have a darker or lighter head than in the adult. All mollymawks have dark areas immediately in front of the eyes, mostly below horizontal supercilia and shading to white through the lores and cheeks, while those birds with grey heads also have flashes of minute white feathers close behind and below the eyes in the quadrant 90° to 180°[7]. The shape and colour of the bill plates are important diagnostic characters (Fig. 5.1). A conspicuous yellow-orange-pink gape-stripe (rictus oris) extends back horizontally from each side of the open mouth. This is normally hidden under minute contour feathers and is revealed only during display. Tarsi and feet vary from whitish-blue to grey, often suffused with pink.

The two black-browed albatrosses[8] have white heads with grey-black lores and black supra-orbital 'brows' extending back over the ear coverts in a groove or 'parting' of the feathers. Juveniles may have a pale grey collar extending from the nape to upper mantle and around the sides of the neck, but this disappears by the time they are five years old. The dark underwing margins are broader than in other mollymawks, particularly on the leading edges and even more so in juveniles, The bill is yellow and ends with a rosy-orange tip (unguis), but there is considerable variation in shade. Surrounding the base of the bills, at the junction with the plumage, there is a very fine and contrasting strip of black fleshy skin. Fledglings have horny-brown bills with black tips which gradually pale and assume the adult colour over five or six years, but once again there is considerable variation (Prince & Rodwell 1994). The Black-browed Albatross has dark brown irides (Plate 1). The Campbell Black-browed Albatross can be distinguished by pale irises (Plate 4), white in the juvenile and yellow in the adult (Barton 1979), dark patches in the white inner underwings between the axilliaries and greater coverts, and more prominent darker 'brows' in the region of the lores (Harrison 1983). Black-browed and Campbell Black-browed Albatrosses interbreed and rear chicks; subadult hybrids have not yet been confirmed, but birds with anomalous iris colours have been seen (Moore *et al.* 1997a).

Grey-headed, yellow-nosed and buller's albatrosses all have grey on the head, neck, throat and chin as well as extending down the nape and darkening over the mantle to merge with the grey-black back. The underwings of the juvenile Grey-headed Albatross are almost completely grey-black, darker than those of any other mollymawks. They have dark grey heads which become white before acquiring the blue-grey of the adult at about the age of five, but there is considerable variation (Prince & Rodwell 1994). The grey heads of adult buller's albatrosses are distinguishable by their contrasting white foreheads.

The bills of the Grey-headed, two buller's and two yellow-nosed albatrosses are all black with yellow culminicorns becoming reddish over the tip (maxillary unguis). In the yellow-nosed albatrosses, the culminicorns are narrow; the posterior ends have characteristic shapes and are surrounded by extensive black skin (Fig. 5.1). The Grey-headed Albatross (Plate 8) has a rather broader culminicorn with less black skin, while in the two buller's albatrosses they are very broad, exposing no skin. There are thin strips of conspicuous yellow-orange skin at the base of the mandibles between the posterior edge of the ramicorns and the fringing feathers. The colour continues horizontally along the bottom of the mandibles. In both buller's albatrosses it is prominent along the whole length, in the Grey-headed Albatross along about two-thirds and in the two yellow nosed albatrosses it is absent altogether.

Both the yellow-nosed albatrosses appear slight compared with other mollymawks. The Western Yellow-nosed Albatross has a pale grey head with a white forehead in contrast to the Eastern Yellow-nosed Albatross (Plate 7), whose head is so pale that for all practical purposes it is seen as white-headed, like the juvenile (Brooke *et al.* 1980, Harrison 1983).

Southern and Northern Buller's Albatrosses (Plate 9 & 11) breed in different latitudes at different seasons. Morphological differences have been described (Marchant & Higgins 1990), but they have not been distinguished at sea (Harrison 1983).

[7] A vertical line through the eye is 0°.
[8] names without capitals are collective terms incorporating two or more taxa.

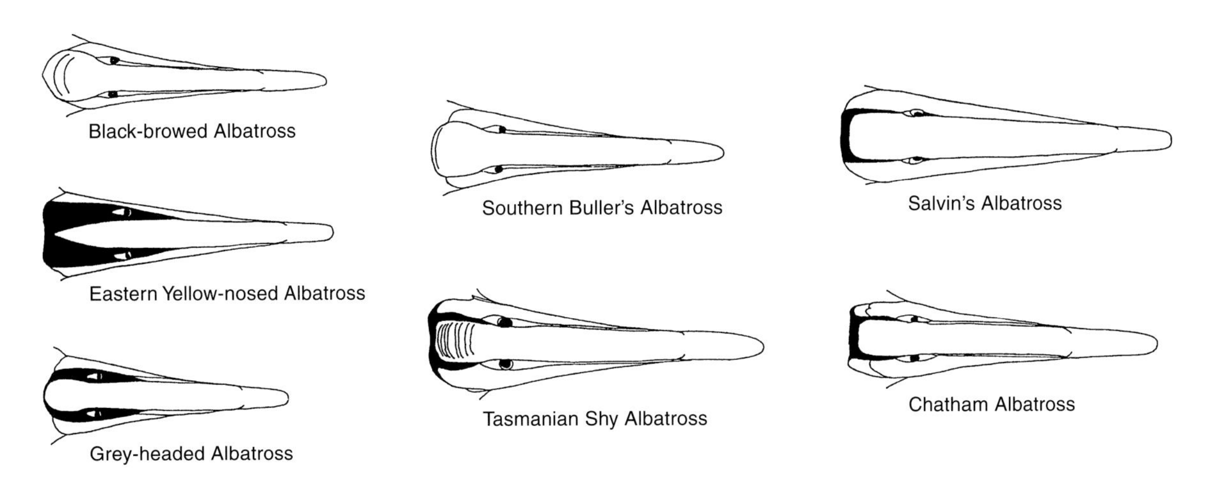

Fig. 5.1 The dorsal view of bills of mollymawks, illustrating the shape of the culminicorns and the extent of exposed skin (black) surrounding their bases (drawn by R. Holdaway in *The Petrels* (Warham 1990), published by permission of Academic Press).

The four *cauta* mollymawks (p.71) are the largest mollymawks. Their white underwings have very narrow black margins and distinct black 'thumb marks' on the axilliaries of the leading edges. The bills are deeper than in other mollymawks. The two shy albatrosses (Plates 12&13) are pale grey on the sides of the head and throat, with the forehead, crown and hind neck white. Starting from various shades of brown in the juvenile, the bill become light metallic-grey sometimes with suffused yellow at the base of the culminicorn and on the tip (unguis). The fleshy black skin between the culminicorn and the latericorns extends around the base of the bill as a black line. The gape-stripes are bright yellow and continue along the bases of the ramicorns. Juveniles have dark spots on the ends of the mandibles (mandibular unguis) that become yellow in adults. Auckland and Tasmanian Shy Albatrosses breed on different islands, and their breeding cycles are out of phase (Robertson 1985). Slight morphological differences are apparent in museum skins, but the two have not yet been distinguished at sea (Harrison 1983, Marchant & Higgins 1990, Brothers *et al.* 1998a&b). Salvin's Albatross (Plate 14) is closely related to the shy albatrosses and morphologically similar. The head and neck is said to be rather brownish-grey; the bill is darker grey with a more contrasting pale culminicorn, becoming yellow towards and around the tip; the dark spots on the ends of the juvenile mandibles are retained in adults (Harrison 1983). Nevertheless it is frequently mis-identified as a shy albatross. The Chatham Albatross (Plate 15) is also closely related to the shy albatrosses, but this is a distinctly grey mollymawk with a conspicuous orange-yellow bill, slightly greyish on the latericorns, and, like Salvin's Albatross, retaining dark mandibular spots in the adult.

HISTORY

Joseph Banks had been seeing albatrosses for several weeks in January 1769 as the *Endeavour* approached Cape Horn during Captain Cook's first voyage (1768–1771) and on 3 February 1769 he shot '...*diomedea profuga*[9] Lesser Albatross with a party coloured bill...'. His comments, together with Daniel Solander's Latin description and Sydney Parkinson's drawing and notes, tell us that it was a Grey-headed Albatross (Salvin 1876). By the time the ship had passed through the Drake Passage into the South Pacific, he was often recording 'Lesser Albatrosses' in his journal.

Johann Reinhold Forster visited Banks and Solander after he had been appointed naturalist to Captain Cook's second expedition (1772–1775), but because of ill feelings[10] he had been shown few if any of their descriptions or drawings before he and his son George sailed in the *Resolution* (Hoare 1982). They first encountered albatrosses in the South Atlantic as the ship was approaching South Africa in October 1772 and Forster shot three 'Yellow-

[9] Latin: *profugio* – to flee, escape.

[10] Banks had taken umbrage and withdrawn from the expedition when asked by the Admiralty to reduce the number of his party; his extra accommodation had made the ship top-heavy.

billed Albatrosses' (our Western Yellow-nosed Albatross), which he described as *Diomedea chrysostoma*[11]. Over a year later, in February 1774, another albatross was obtained in the SE Pacific which Forster believed was similar to the earlier birds, but which was in fact a different species (our Grey-headed Albatross). His Latin descriptions (published posthumously by Lichtenstein in 1844) and the drawings of his son George reveal that they had collected two different mollymawks, but remained unaware of the second (Lysaght 1959). None of his specimens were preserved and Forster pooled the descriptive notes. In his *Memoire sur les Albatros* (Forster 1785), '*L'Albatros a bec doré Diomedea chrysostoma*' was accompanied by a description of the SE Pacific bird, not the South Atlantic one to which the Latin name had originally been applied. Forster published his important albatross paper in the *Mémoire de Mathématique et de Physique* of the *Academie des Sciences* (Paris) where it remained largely unnoticed by ornithologists for over a century (Medway 1997a).

The manuscripts of Banks, Solander and Forster did not became generally available until long after their deaths, but John Latham was acquainted with Banks and, for a time at least, was helped by Forster. He evidently saw George Forster's drawings and used some of them to prepare coloured plates for his 1785 *General Synopsis of Birds*. He had no experience of the live birds, but Medway (1997a) believes that he had access to a specimen. He did not detect that the Forsters had confused two species and his illustration of a 'Yellow-nosed Albatross', although crude, was close enough to the birds we know by that name today.

Gmelin incorporated Latham's bird into his 1789 edition of Linnaeus's *Systema Naturae* and, apparently unaware of either Banks's *profuga* or Forster's *chrysostoma*, created a new name, *Diomedea chlororhynchos*[12], which remained fundamentally ambiguous, but was widely adopted (Medway 1997a).

It was not until John Gould went to Australia in 1838 that the two species were separated. Gould spent weeks at sea off Tasmania and in Bass Strait where he saw many mollymawks of several species. His 'Culminated Albatros', which he confidently distinguished from Latham's yellow-nosed albatross, later became known as the Grey-headed Albatross. It concealed yet another unsuspected species that in turn remained undetected for another 50 years. Albatrosses with grey heads were well known off New Zealand where they were believed to be Grey-headed Albatrosses, but a specimen seen by Osbert Salvin at the British Museum was clearly different, and in 1893 Rothschild named it *Diomedea bulleri* after the New Zealand ornithologist, Walter Buller.

When Forster's *Memoire sur les Albatros...* was eventually discovered, Mathews (1913) reinstated *chrysostoma* in place of Gould's *culminata*, so that the Grey-headed Albatross is known today by the Latin name *Diomedea chrysostoma* that Forster first applied to a Western Yellow-nosed Albatross!

On 11 April 1770, the *Endeavour* was in the Tasman Sea approaching Australia. It was a calm day and Banks went out in a small boat to shoot seabirds. Among several albatrosses he obtained was one that he called '*diomedea impavida*'[13] (Beaglehole 1962). Solander's Latin notes were not consulted for 143 years. Meanwhile, in 1820, Coenraad Jacob Temminck was approached by the Baron Laugier de Chartrouse to co-operate in an additional set of coloured plates for the latest edition of Buffon's celebrated *Histoire Naturelle des Oiseaux* (1786). It included an *Albatros sourcils noirs Diomedea melanophris*[14]; the plate looks nothing like a black-browed albatross, but the description, attributed to a Dr Boie, identifies the species (Temminck & Laugier 1838).

Mathews (1913) eventually found Solander's notes in the British Museum. They recorded that Banks' albatross had yellow eyes which unequivocally identifies it as a Campbell Black-browed Albatross. Mathews applied *impavida* as a subspecific trinomial to that population.

Among the mollymawks John Gould obtained in Bass Strait, was a specimen which he called the 'Cautious Albatross *Diomedea cauta*'; in time it became known as the Shy Albatross. Later writers substituted 'White-Capped' claiming that the species was not at all shy, but use of this synonym has been erratic and is avoided here. Scientists on board the steamer *Hinamoa*, which rescued survivors of the *Dundonald* disaster at the Auckland Islands, visited Disappointment Island shortly afterwards. Edgar Waite (1909) wrote of black-browed albatrosses on the island,

[11] Greek: (*chrysos*) – gold, (*stoma*) – mouth.
[12] Greek: (*chloro*) – green, (*rhynch*) – snout.
[13] Latin: *im-pavidus* – fearless, undaunted.
[14] Greek: (*melan*) – black, (*phrys*) – eyebrow. Temminck wrote *melanophris*, but others including Marchant (Tickell 1976), insist that –*phrys* is the correct Latin representation of the Greek.

but an accompanying photograph clearly identifies an Auckland Shy Albatross. Robert Falla (1933) at the Dominion Museum, New Zealand, described specimens from this island as a distinct subspecies, but as late as 1990, the *Handbook of Australian, New Zealand and Antarctic Birds* (Marchant & Higgins 1990) claimed that there was no difference between individuals from the Auckland Islands and Tasmania.

Throughout the late 19th century similar mollymawks about New Zealand had been identified as shy albatrosses, but a specimen sent to the British Museum was seen to be different and Rothschild (1893) named a new species, *Diomedea salvini*, after the museum's seabird specialist. Murphy (1930) considered it a subspecies (*Diomedea cauta salvini*) of the Shy Albatross, a view that long prevailed (Marchant & Higgins 1990). In New Zealand itwas promoted as a separate species in a popular field guide (Falla *et al.* 1966), but remained a subspecies in the OSNZ[15] checklist (Turbott *et al.* 1990). Its breeding ground remained unknown for many years, but the matter was eventually resolved by a Mr. Bethune, who had served for many years as Chief Engineer on government ships visiting castaway depots on the New Zealand islands. He had made a collection of albatross heads which, with Dougall's (1888) photographs, convinced Buller (1905) that the Bounty Islands were the home of Salvin's Albatross.

In 1924, Rollo Beck shot a quite different mollymawk off the Chatham Islands. At the American Museum of Natural History, Murphy (1930) distinguished it as another subspecies (*Diomedea cauta eremita*[16]) of the Shy Albatross. New Zealand ornithologists later raised it to a species (Falla *et al.* 1966), then reverted to a subspecies (Turbott *et al.* 1990).

In the middle of the 19th century, Ludwig Reichenbach of the Dresden Museum published a set of coloured plates in many volumes. The albatrosses were all captioned within the genus *Diomedea*, but in his systematic list he created a new genus, *Thalassarche*[17] for the Black-browed Albatross (Reichenbach 1850). Elliott Coues (1866) contested this splitting by putting all mollymawks in the one genus *Diomedea*; no-one else at that time was so inhibited. After the first southern voyage of the Italian Navy (1865–68), Enrico Giglioli added the Grey-headed Albatross to *Thalassarche* and Robert Ridgeway (Baird *et al.* 1884) at Harvard created another genus *Thalassogeron*[18]. Godman (1910) in his *Monograph of the Petrels*, did not accept *Thalassarche*, but was content to keep *Thalassogeron*, while Mathews (1912–13) accepted both and created others: *Nealbatrus*, *Diomedella* and *Thalasseus*. When the British Ornithologists' Union published *Systema Avium Australasianarium* (Mathews 1927), *Diomedella* was still the genus of the shy albatrosses, while the other mollymawks were all included in *Thalassarche* – together with the Laysan and Black-footed Albatrosses!

In 1930, Murphy wrote 'In my opinion it is impracticable to attempt generic subdivision among the smaller albatrosses...' He put them all in *Thalassarche*, but six years later, in *Oceanic Birds of South America*, he had returned them to *Diomedea*. Conservatism prevailed; revision of Peter's Check List (Jouanin & Mougin 1979) preserved *Thalassarche* as a subgenus containing just the Black-browed Albatross and put all other mollymawks in a subgenus *Thalassogeron*.

In 1996, Nunn *et al.* presented molecular evidence that the mollymawks were monophyletic and deserved a single genus of their own – *Thalassarche* was reinstated.

OCEANIC DISTRIBUTION

The data upon which our knowledge of oceanic distribution is based come from various sources. Collecting at sea with shotgun and baited line may have ended, but the substantial samples of birds killed on fishing gear again provide museum specimens. Ringing recoveries have yielded good data in some species. Satellite transmitters provide large numbers of locations at sea, but they come from relatively few individuals. Sightings from ships remain the main source for mapping, but the reliability of identifications has always been uncertain and less amenable to fine discrimination. The distribution maps (Figs. 5.2–5.8) represent pooled data from all sources. The 5° x 5° shaded squares are not measures of abundance; they may represent just one observation of a single

[15] Ornithological Society of New Zealand.
[16] Latin: *eremite* – hermit.
[17] Greek: (*thalassa*) – sea, (*-archy*) – leader.
[18] Greek: (*thalassa*) – sea, (*-geron*) – aged.

bird at the edge of its range or many reports totalling thousands of albatrosses throughout the year.

Mollymawks are all seabirds of middle to high latitudes, but some move into tropical seas and a few find their way across the equator into the northern hemisphere.

Continental and island shelves or slopes with their influence on productive water masses, are constant determinants of mollymawk oceanic distribution (Thompson 1992, Jehl 1974, Summerhayes *et al.* 1974, Harrison *et al.* 1991, Stahl *et al.* 1985, Weimerskirch *et al.* 1988, Barton 1979).

Black-browed albatrosses

The Black-browed Albatross occurs throughout the Southern Ocean, but if numbers are taken into account it is mainly a bird of the South Atlantic (Fig. 5.2). In mid-ocean the northern limit is at about 35°S; but near the continents they are found farther north, particularly over colder, northerly moving waters such as the Benguela and Humboldt Currents.

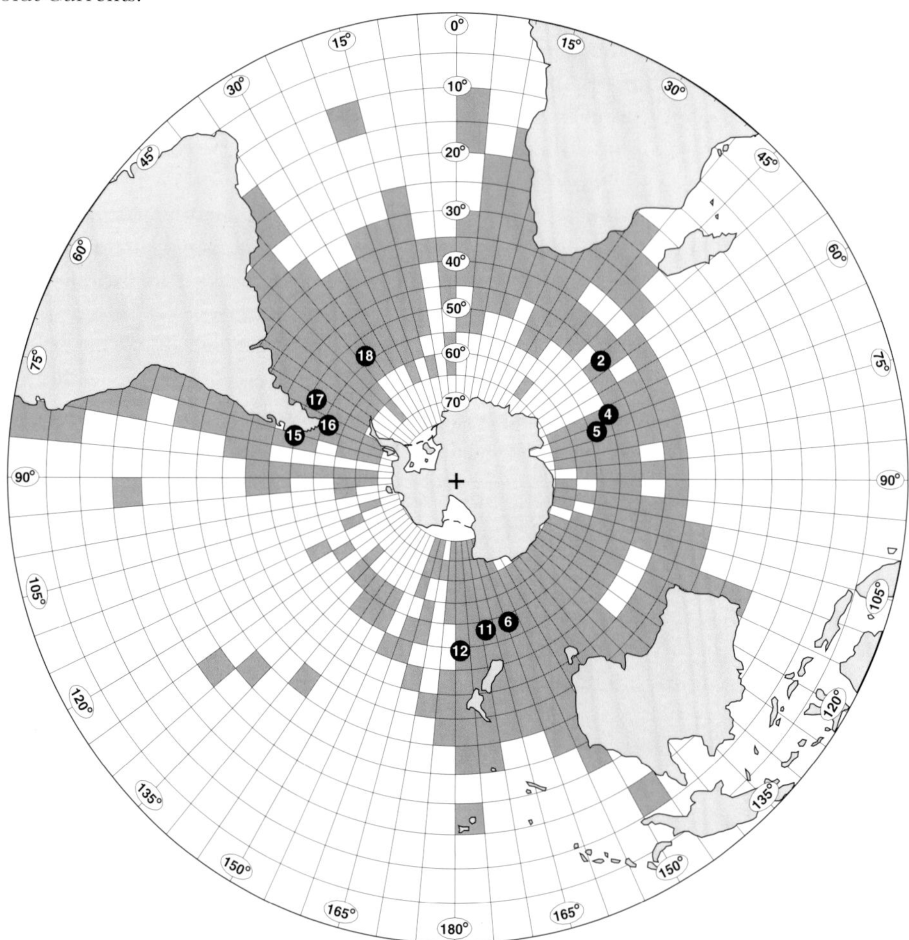

Fig. 5.2 Oceanic distribution and breeding islands of the Black-browed and Campbell Black-browed Albatrosses. The numbered black spots indicate the breeding islands as coded in Fig. 4.1.

The cold water of the Falkland (Malvinas) Current flowing north over the broad Argentine continental shelf and meeting the warmer, southward-flowing Brazil Current off the Rio de la Plata offers the classic scenario for high productivity sustaining large numbers of predators.

Northerly movements, with many juveniles, occur in the South Atlantic at the end of the breeding season. In winter, the Black-browed Albatross is abundant off the Rio de la Plate (Stagi *et al.* 1997) and become the commonest albatross off SE Brazil as far north as 23°S (Olmos *et al.* 1995). It is a regular winter visitor off Tristan da Cunha, being most numerous towards the end of August (Elliott 1957) and is sometimes seen near St. Helena (16°S) (Rowlands *et al.* 1998).

The Black-browed Albatross occasionally crosses the equator into the North Atlantic, where it is the commonest vagrant albatross, occasionally driven inland by storms (Fisher & Lockley 1954, Bourne 1967). There were 41 records between 1958 and 1994 off the British Isles, Denmark, France, Germany, Iceland, Mauretania, Morocco, Norway, Spain, Sweden and as far north as 80° N off Svalbard (Lubbock 1937, Harrop 1994). At least two individuals became resident. One visited the colony of Northern Gannets on Myggenaes Holm, Faroes, for 25 years (1860–94) and was known to the local people as *Sulkonge*, the gannet king, until it was shot (Andersen 1895). Another, that visited the Bass Rock in 1967 (Waterston 1968), may have been the bird that settled among the gannets of Hermaness, Shetland in the early 1970s and was seen on an empty nest most years until 1996 (Bourne 1977a, Sutherland & Brooks 1979, B. Tulloch pers. comm., J.L. Johnston pers. comm.). Vagrant Black-browed Albatrosses also occur in the eastern North Atlantic. Two were collected off the Windward Islands and two were seen together in the Caribbean Sea, northeast of the Islos Los Roques (Bruijine 1970).

The Black-browed Albatross can been seen throughout the year on the fishing grounds off South Africa, where it assembles in large rafts and scavenges around ships hauling in trawls. Numbers increase during autumn, reaching a peak in winter (May–June); most of them 50–100 km offshore and varying in abundance from year to year (Liversidge & Le Gras 1981). Some of these birds may be from the Iles Crozet and Iles Kerguelen (Weimerskirch *et al.* 1985), but most are juveniles that left South Georgia as recently as four weeks earlier (Prince *et al.* 1997). They appear to spend their first three years mainly off the Cape of Good Hope, but some move north off the coast of Namibia and few go east as far as Durban (Tickell 1967, Sladen *et al.* 1968). Very few Black-browed Albatrosses are found dead on South African shores; between 1977 and 1986 the average was less than one juvenile per 100 km, but in May 1984, during severe storms, 58 juveniles per 100 km were wrecked along the coast of the western Cape (Ryan & Avery 1987, Jury 1991).

A few South Georgia juveniles go west to join Falkland birds off Argentine and Brazil and some fly east as far as Australia. Most Black-browed Albatrosses from the Falkland Islands remain in the SW Atlantic, moving up the adjacent coast of Argentina and Brazil; a few cross the ocean northeast as far as South African waters (Jehl 1974, Lashmar 1990, Tickell 1967).

The Black-browed Albatross is rarely encountered in the central South Pacific (Szijj 1967, Clark 1986). Those individuals seen off the coast of Chile probably all come from the South American breeding islands (Islas Diego de Almagro, Ildefonso and Diego Ramírez). In winter, large numbers occur north of these islands to 33°S, but a few follow the coast north into tropical latitudes off Peru and Ecuador (Jehl 1973, Clark 1986). They regularly enter the fjords and channels of southern Chile and Tierra del Fuego, penetrating as far as Lago Argentino (R.P. Schlatter pers. comm.) and have also been seen on Lago Fagnano (Cami), a long freshwater lake east of Lago Argentino (Stiles 1974).

Black-browed Albatrosses from the Crozet, Kerguelen, Heard and Macquarie Islands winter in south Australian coastal waters, some going as far as New Zealand (Lashmar 1990, Howard 1954, Hindwood 1955). Ocean currents south of Australia bring cold productive water up against the warm sub-tropical water of the continent. In the open sea away from coastal influences, the Black-browed Albatross is common throughout the year. In the southern autumn and winter, rough seas carry cold water over the continental shelf, which is narrowest at both ends of the Great Australian Bight. Many pelagic birds, including the Black-browed Albatross, then approach the coast, especially off Kangaroo Island, South Australia (Cox 1973, 1976, Swanson 1973). It is also common off the coasts of Victoria and New South Wales. In winter, it may reach as far north as 24°S (Shuntov 1974). Ringed Black-browed Albatrosses from as far away as South Georgia have been recovered throughout the year in Australian and New Zealand waters, with many juveniles arriving in October–January (Tickell 1967, Barton 1979, Blaber 1986, Wood 1992).

Occasionally the Black-browed Albatross moves into high latitudes off the west coast of the Antarctic Peninsula, as far as 66°S (P. Kinnear pers. comm.), but it does not enter the Bellingshausen Sea and is absent from the adjacent Antarctic waters of the Pacific (Zink 1981, Wanless & Harris 1988). In the Ross Sea, the Black-browed Albatross reaches 66°S in December and 68° to 70°S in January and February (Lowe & Kinnear 1930, Siple &

Lindsey 1937, Darby 1970, Watson *et al.* 1971, Hicks 1973). Between longitudes 60°W and 65°E, few have been seen south of 60°S. Thurston (1982) commented upon their absence in the Weddell Sea even when the ice-edge was 200 km distant.

The Campbell Black-browed Albatross is not often identified at sea or even when washed up dead on New Zealand beaches (Powlesland 1985, Taylor 1997). Few approach the coasts of New South Wales and Victoria, but many have been seen in April–May off Kangaroo Island, South Australia (Swanson 1997). Ringing recoveries indicate that juveniles, in their first winter, disperse into the Tasman Sea. Older juveniles, subadults and adults have been recovered between latitudes 30°S and 50°S from the coastal waters of Western Australia to the South Pacific, as far east as longitude 140°W, and north to the sub-tropical waters of French Polynesia, Tonga, Fiji and New Caledonia (Bartle 1974, Clark 1986, Battam & Smith 1993, Waugh *et al.* 1999b).

Campbell Black-browed Albatrosses, colour-dyed when incubating at their colonies in 1995–96 and 1996–97, were later seen along the east coast of New Zealand as far north as 43°S, south to almost 64°S and west beyond Macquarie Island. None approached the west coast of New Zealand. In winter one individual was seen beyond East Cape at about 36°S and another crossed the Tasman Sea to the east coast of Tasmania (Waugh 1998).

The Grey-headed Albatross

These are birds of the open ocean, mostly south of 40°S (Fig 5.3). Away from the immediate surroundings of breeding islands, they are usually encountered in ones and twos (Harrison *et al.* 1991, Jehl 1973, Weimerskirch *et*

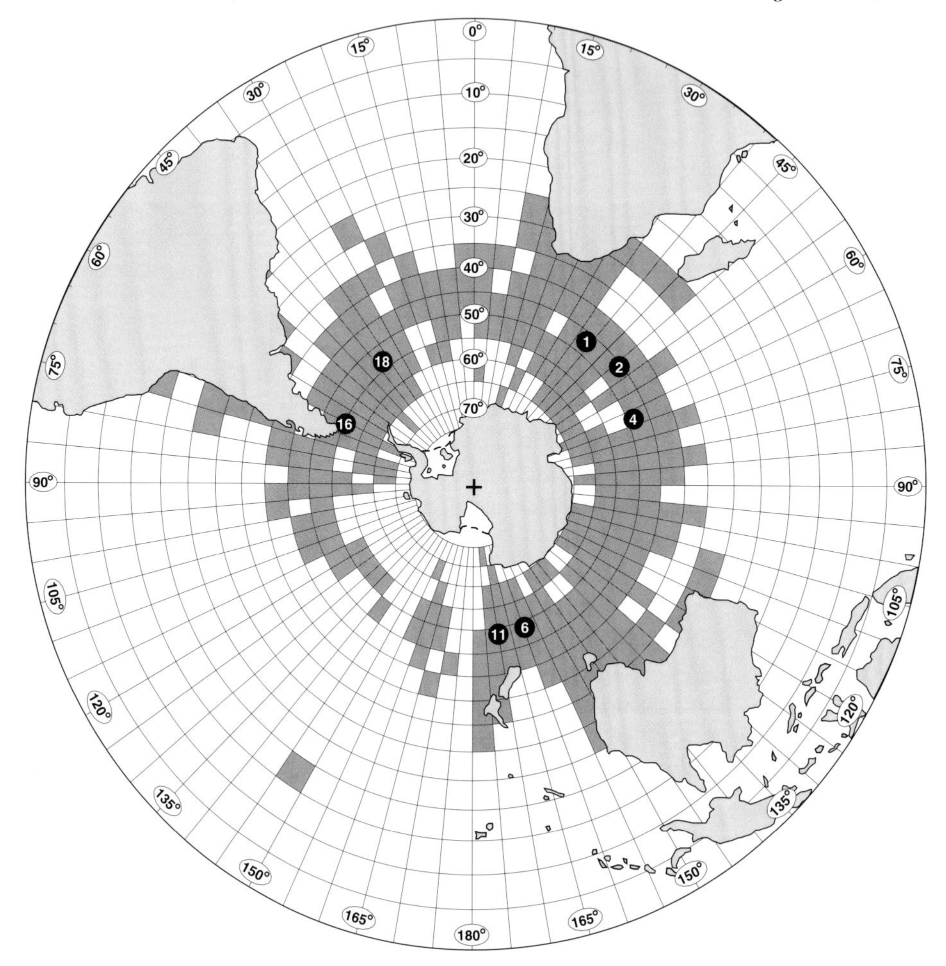

Fig. 5.3 Oceanic distribution and breeding islands of the Grey-headed Albatross. The numbered black spots indicate the breeding islands as coded in Fig. 4.1.

al. 1988). An adult that left South Georgia in March 1996 and was tracked by satellite, flew southwest to the continental shelf off the South Shetlands and followed it down the Antarctic Peninsula as far as 67°S before flying north into the South Pacific off Chile (Prince *et al.* 1997).

Ringed juveniles have much lower recovery rates than juvenile Black-browed Albatrosses; the few recoveries from fledglings ringed at South Georgia have been from South Africa, Australia and New Zealand (Tickell 1967). Many Grey-headed Albatrosses die in storms each winter off New Zealand, increasing from a low in April to a peak in September. Powlesland (1985) has suggested that these are mainly juveniles from Campbell Island, but some are from other islands as distant as South Georgia (Tickell 1967). South of New Zealand in December and February, the Grey-headed Albatross remains north of the Antarctic Polar Front, but in January, when pack-ice breaks out, it enters the Ross Sea, sometimes reaching 68°S (Darby 1970, Hicks 1973).

Grey-headed Albatrosses rarely stay with ships on passage, but can be enticed to bait (Wood 1992). Bourne & Curtis (1985) have reported large numbers following fishing boats south and east of the Falklands in winter and Gales *et al.* (1998) include this species among the albatrosses killed on longlines south of Australia.

The Grey-headed Albatross is seldom among the species that congregate at continental shelves off South Africa and Australia (Cox 1976), but it was seen more often off South Africa during exceptional storms in the winter of 1984 (Ryan *et al.* 1989) and many were washed ashore on the coast of Victoria, Australia in 1975–76 (Carter 1977).

In the SW Indian Ocean and central South Pacific it appears to be more common than the Black-browed Albatross (Szijj 1967, Clark 1986). Incubating Grey-headed Albatrosses colour-dyed at Campbell Island were not seen at sea away from the island (Waugh 1998) and few rings have been recovered (Waugh *et al.* 1999b). Off the coast of southern Chile at 50°S, it may outnumber the Black-browed Albatross 3:2 in winter, while farther north, at 41°S, the situation may be reversed, with Black-browed Albatrosses dominating by 90:1 (Jehl 1973).

Yellow-nosed albatrosses

The two yellow-nosed albatrosses have rarely been distinguished at sea or even among dead birds on beaches. In mid-ocean they occur mostly between 30°S and 40°S (Fig. 5.4).

In the South Atlantic, ringed adult and juvenile Western Yellow-nosed Albatrosses breeding at Tristan da Cunha and Gough Island have been recovered off the Atlantic coast of southern Africa, so it is usually inferred that all yellow-nosed albatrosses seen throughout the year in warm oceanic water, and avoiding the upwelling inshore waters of the Benguela Current, are likewise Western Yellow-nosed Albatrosses. During November, small numbers are widespread in the South Atlantic, north of 26°S. From March to April, they are numerous off the African continental shelf or slope north of 30°S, and locally abundant north of 22°S. South and west of Cape Point small numbers occur over the open ocean. By May, there are few at the shelf-edge north of 26°S and none to the south. There appears to be a winter migration, that begins with a southerly movement out of Angolan and Namibian coastal waters (Summerhayes *et al.* 1974), possibly the beginning of a return to the breeding islands. Some move into the southern Indian Ocean, making long forays as far as Australia (Reid & Carter 1988) and the western Tasman Sea (Smith 1997), perhaps as far as New Zealand (Powlesland & Powlesland 1994).

Western Yellow-nosed Albatrosses are common off Brazil throughout the year between 23°S and 25°S (Olmos *et al.* 1995), and sometimes hundreds occur on the fishing grounds off the Rio de la Plata (Curtis 1994). They are infrequent south of 45°S.

Western Yellow-nosed Albatrosses are sometimes seen around St. Helena (16°S) (Rowlands *et al.* 1998) and occasionally find their way into the North Atlantic (Bourne 1967). Most have been in the western ocean; one was off Tobago (Brackenridge 1971) and specimens have been collected along the coast of North America as far north as the Gulf of St. Lawrence (Squires 1952). There have been sightings at sea in the eastern North Atlantic off Spain, Britain and Norway (Curtis 1993, Harrop 1994).

Eastern Yellow-nosed Albatrosses from the Prince Edward Islands and Iles Crozet occur off the south to southeast coasts of South Africa. They may be in company with Western Yellow-Nosed Albatrosses between Cape Point and Cape Agulhas (Harrison *et al.* 1997). Over the Agulhas Bank, they are the most numerous mollymawks in winter.

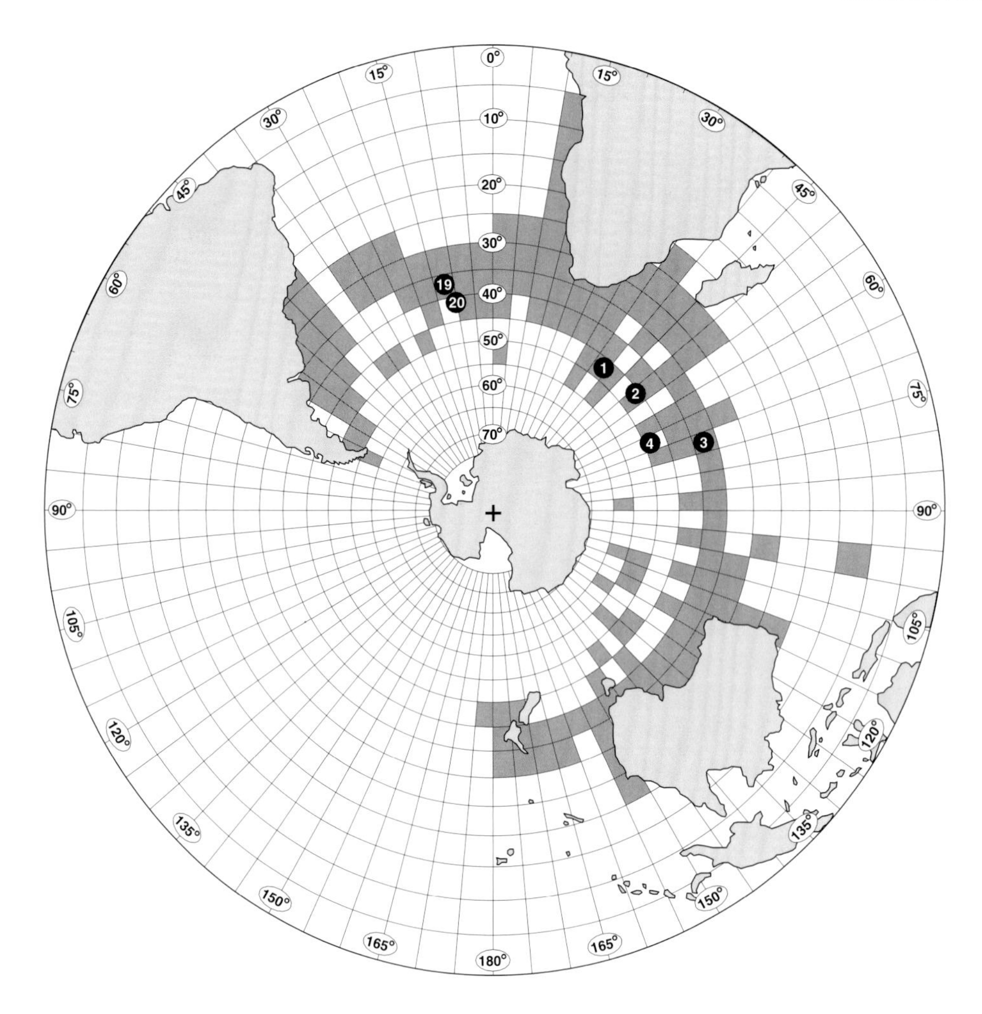

Fig. 5.4 Oceanic distribution and breeding islands of the Western and Eastern Yellow-nosed Albatrosses. The numbered black spots indicate the breeding islands as coded in Fig. 4.1.

The Eastern Yellow-nosed Albatross from the the Ile Amsterdam, migrates southeast during November (Jouventin *et al.* 1983, Weimerskirch *et al.* 1985). Large numbers are present throughout the year off Australia, where they are the most numerous inshore mollymawk, exceeded by black-browed albatrosses only in rough weather. Substantial numbers arrive off Western Australia in autumn, as far north as 26°S, and increase through the winter, particularly in the Great Australian Bight (Shuntov 1974) and off Kangaroo Island. Many enter Bass Strait and the western Tasman Sea where they have been seen to 26°S off Queensland. They are absent for only about one month (January–February). There are fewer juveniles in May–June, but they become more numerous in September–October (Barton 1979, Wood 1992, Battam & Smith 1993).

Since 1921, small numbers of unspecified yellow-nosed albatrosses have been seen around New Zealand as far as the Chatham Islands (Robertson 1975a). Nineteen carcasses were collected by beach patrols between 1981 and 1992 (Powlesland & Powlesland 1994).

Buller's albatrosses

Published records of the Grey-headed Albatross probably include misidentified buller's albatrosses. Experienced observers are now more confident of their identifications, but Northern and Southern Buller's Albatrosses have not been distinguished at sea (Fig. 5.5).

Most recoveries of ringed Southern Buller's from the Snares have been on the south and east coasts of South Island, New Zealand (Stahl *et al.* 1998), and in March 1996, colour-dyed Southern Buller's Albatrosses from the Snares foraged off the south coast of New Zealand (Waugh *et al.* 1996).

The home range of both buller's albatrosses appears to be over the New Zealand continental shelf or slope, where large numbers attend fishing vessels. Many are storm-wrecked on New Zealand beaches in winter, reaching a peak in June (Powlesland 1985, Taylor 1996). A few cross the Tasman Sea to the coasts of New South Wales, Tasmania and South Australia. Around Tasmania, they are seen mainly between January and June, with most during March–April (Marchant & Higgins 1990, Stahl *et al.* 1998).

Long ago, Rollo Beck and others collected unspecified buller's albatrosses off the coasts of Chile and Peru between latitudes 12°S and 35°S (Murphy 1936). During 15 cruises off South America between 1980 and 1995, Spear *et al.* (1995) logged 86, most of them over the continental slope between 30°S and 40°S. An adult Southern Buller's ringed at the Snares Islands was recovered in the subtropical South Pacific at 12°S 105°W (Warham 1982) and satellite tracking has confirmed these migrations (D.G. Nicholls pers. comm.).

Some buller's albatrosses pass through the Drake Passage into the SW Atlantic (Curtis 1988b) and several have been reported from high latitudes off Antarctica (Woehler *et al.* 1990, E.J. Woehler pers. comm.).

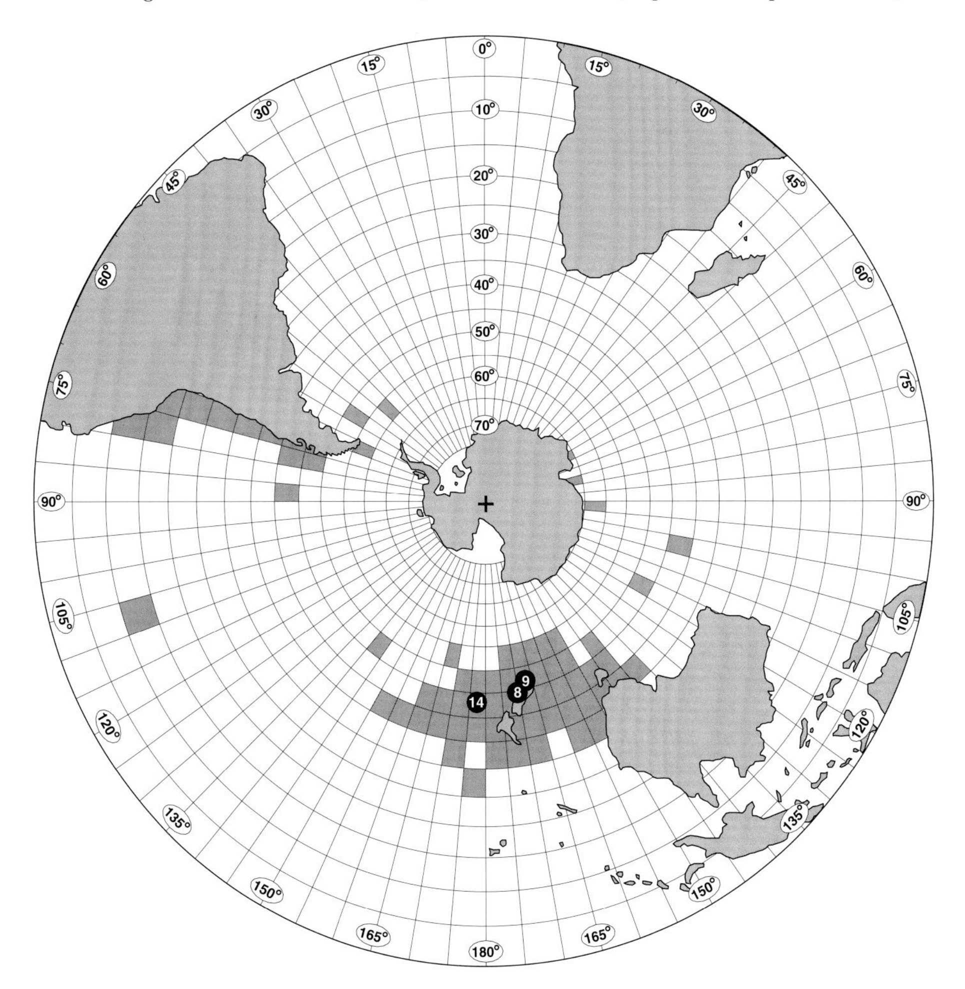

Fig. 5.5 Oceanic distribution and breeding islands of the Northern and Southern Buller's Albatrosses. The numbered black spots indicate the breeding islands as coded in Fig. 4.1.

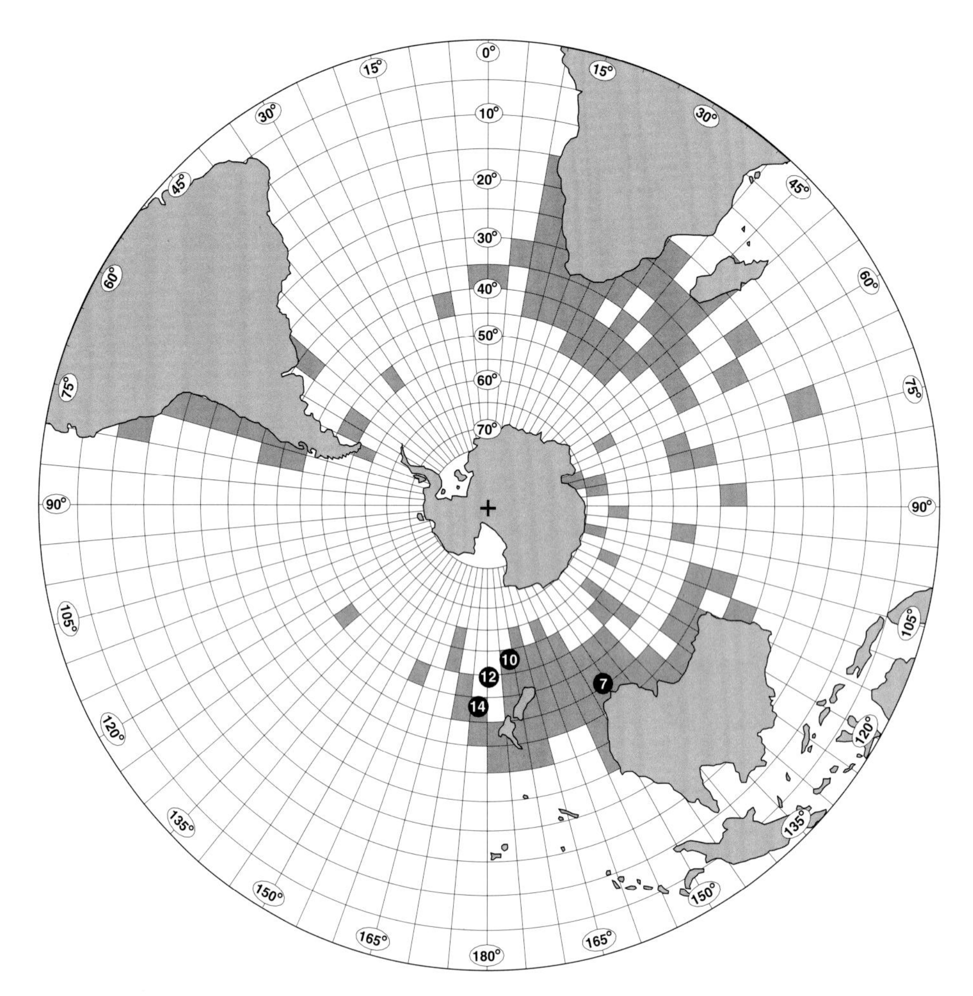

Fig. 5.6 Oceanic distribution and breeding islands of the Auckland and Tasmanian Shy Albatrosses. The numbered black spots indicate the breeding islands as coded in Fig. 4.1.

Shy albatrosses

Auckland and Tasmanian Shy Albatrosses have rarely been distinguished at sea, even by experienced observers. In the Tasman Sea, they occur together, sometimes in comparable numbers (Hansbro 1996). Fledglings leave the Auckland Islands in August (Robertson 1985) and may be the juveniles seen in the western Tasman Sea during September–October (Barton 1979). Those that die during storms and are washed ashore on New Zealand beaches are assumed to have come from the Auckland Islands (Taylor 1996). The shy albatrosses that cross the Pacific Ocean eastwards and seen off the coast of South America, may also have come from the Auckland Islands (Meeth & Meeth 1983, P. Harrison pers. comm., L.B. Spear & D.G. Ainley pers. comm.).

In Australian waters, shy albatrosses are seen throughout the year. They are assumed to be the Tasmanian Shy Albatross, but some may be Auckland Shy Albatrosses. Tasmanian Shy Albatrosses, foraging from their islands, are largely segregated by the shallow waters of Bass Strait, with some mixing off the west coast of Tasmania. In summer, the birds from Albatross Island remain mostly over the continental shelf of the western Bass Strait while Pedra Branca birds favour the continental slope southeast of Tasmania. Ringed or painted adults and subadults have mostly been reported off Tasmania and the adjacent mainland of Australia, up to 587 km from their breeding islands (Brothers *et al.* 1997).

Tasmanian Shy Albatrosses are widespread over the continental shelf south of Kangaroo Island, where they are sometimes as numerous as the two black-browed albatrosses (Swanson 1973, Barton 1979). Inshore it may be locally abundant, outnumbering all black-browed and yellow-nosed albatrosses combined (Swanson 1973), but elsewhere off South Australia, they approach the coast only in spring and autumn (Cox 1976). The latter including Albatross Island breeders whose young have recently flown (Fig. 5.7) (Brothers *et al.* 1997, 1998c).

From March to April 1995, after breeding on Pedra Branca, adult Tasmanian Shy Albatrosses follow the shelf northwards off the east coasts of Tasmania, Flinders Island and Victoria (Fig. 5.7), and are rare in the open sea southeast of the sub-tropical front (Blaber 1986, Brothers *et al.* 1998c).

During the winter months of June to August, ringed Tasmanian Shy Albatrosses from all three islands are in the western Tasman Sea. With many other seabirds, they follow the subtropical front north (Barton 1979, Wood 1992).

Ringed juveniles from Albatross Island follow the continental shelf west, some as far as West Australia. Nevertheless, others have been recovered off the east coast of Tasmania and New South Wales, several going as far north as Queensland and one crossing the Tasman Sea to New Zealand. Although most ringed juveniles have been recovered in Australian waters, many Mewstone juveniles fly west, clearing Australian continental waters and some have reached South Africa within 60 days (Brothers 1988, Marchant & Higgins 1990, Battam & Smith 1993, Brothers *et al.* 1997).

Many shy albatrosses remain around South Africa all year, about 50 to 100 km offshore; Harrison *et al.* (1997) state that almost all of them are Tasmanian Shy Albatrosses. They are more numerous in winter. Peak numbers off the south coast occur in June to July, but off Namibia they may peak later (August to October). They congregate over the continental slope, but unlike other mollymawks they often approach close inshore. In five years of observations east of the Cape, they outnumbered the Black-browed Albatross every year, sometimes by as much as six times. North of 25°S, off Namibia, they may also be more numerous than Black-browed Albatrosses, but over the deep ocean west of the continental slope they are rare north of 35°S (White

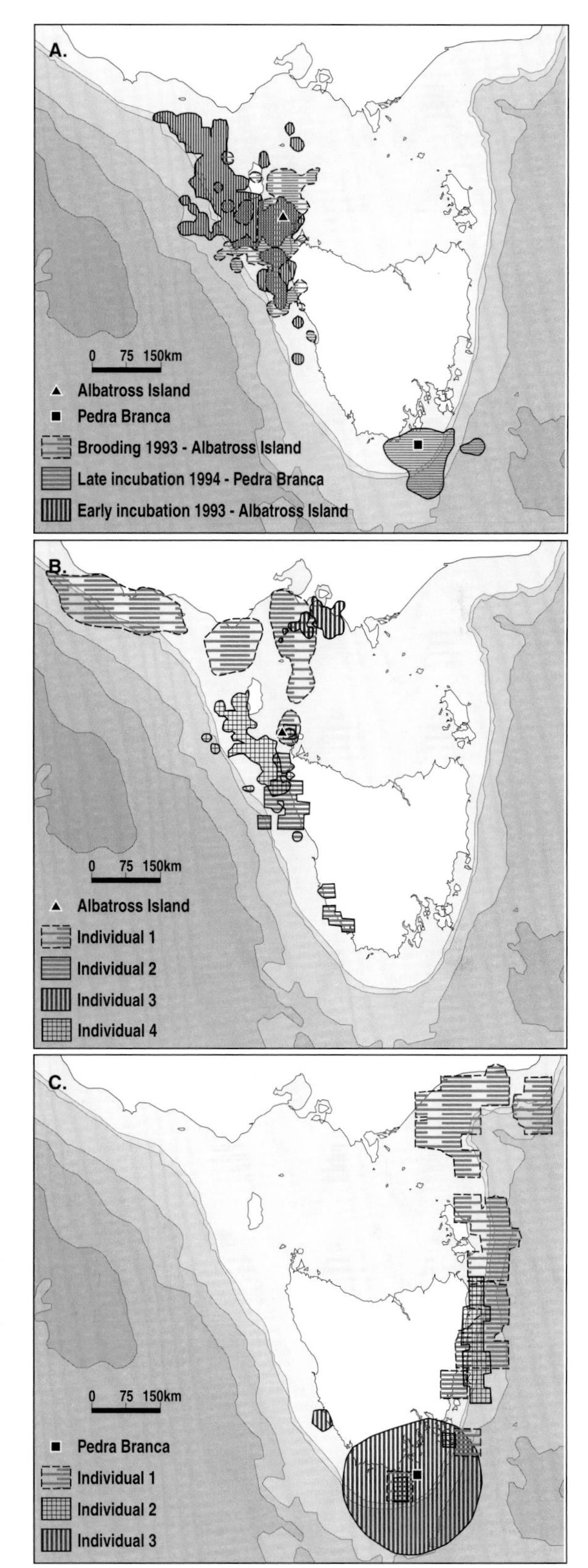

Fig. 5.7 The Tasmanian Shy Albatross: distributions at sea off Tasmania and SE Australia, based on satellite tracking in 1993–94 and 1994–95. Home ranges of both breeding populations during incubation and brooding (A). Areas of four individuals in April–May 1994, immediately after breeding at Albatross Island (B) and areas of three individuals in March–April 1995 after breeding at Pedra Branca (C) (Brothers *et al.* 1998c) (Fig. 5.16).

1973, Summerhayes *et al.* 1974, Bourne 1977b, Liversidge & Le Gras 1981, Marchant & Higgins 1990, Gales 1993).

Unspecified shy albatrosses have been seen and collected in the tropical Indian Ocean, off Africa. There were 14 sightings off southern Kenya in 1989–90 (Zimmerman *et al.* 1996), and north of the equator, one was seen off Somalia. Another, alive in the Gulf of Aqabah, was later found dead (Meeth & Meeth 1988).

One Tasmanian Shy Albatross has been collected in the North Pacific, off the coast of North America (Slipp 1952).

Salvin's Albatross

This albatross has usually been misidentified as a shy albatross and is under-represented in records; the map (Fig. 5.8) probably does not adequately represent it's distribution. It is common over the New Zealand shelf. The number of dead Salvin's Albatrosses washed-up on New Zealand beaches increases briefly in March, at the time when fledglings are going to sea for the first time. Throughout winter, the number of dead Salvin's on beaches remains low and declining just when storms are killing many Grey-headed and Auckland Shy Albatrosses. Towards spring, storm-wrecked Salvin's again increase in numbers as they return to their breeding islands (Powlesland 1985).

Some Salvin's Albatrosses cross the Tasman Sea and join Tasmanian Shy Albatrosses (Barton 1979, Wood 1992); many fly east across the South Pacific to the coasts of Chile and Peru (Jehl 1973, Meeth & Meeth 1983, P.

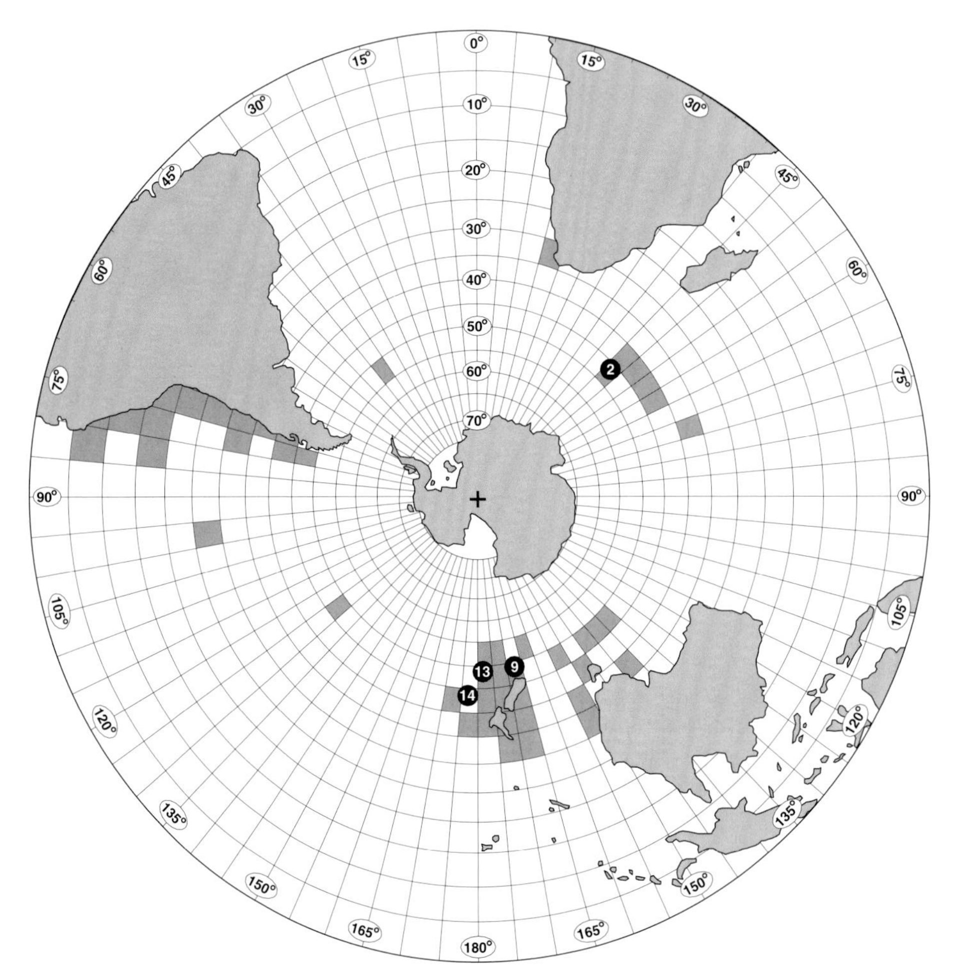

Fig. 5.8 Oceanic distribution and breeding islands of Salvin's Albatross. The numbered black spots indicate the breeding islands as coded in Fig. 4.1.

Harrison pers. comm.). In 15 cruises off South America between 1980 and 1995, Spear *et al.* (1995) logged 475 sightings, mostly over the continental slope. They were found farther north in winter than in summer; specimens have been collected at 6°S (Murphy 1936). They are rarely seen in the Drake Passage, but some do take that route into the South Atlantic.

Salvin's Albatross has seldom been confirmed amongst the numerous shy albatrosses off South Africa (Harrison *et al.* 1997).

The Chatham Albatross

This albatross migrates across the south Pacific Ocean (Fig. 5.9), Ben Haase (1994) saw one 305 km off the coast of Peru (8°S 82°W). In 15 cruises off South America between 1980 and 1995, Spear *et al.* (1995, unpublished) logged 27, mostly in pelagic waters between 14°S and 41°S. They were scarce in summer and more southerly in winter. In 1997, one was tracked by satellite from its nest on the Pyramid, off the Chatham Islands, across the ocean to Chile (C.J.R. Robertson pers. comm.).

Chatham Albatrosses are very rarely reported around New Zealand. A few have been seen at the Snares (Miskelly 1997) and some fly west to the south of Tasmania (Reid & James 1997). Others pass to the north of New Zealand and cross the Tasman Sea westwards (Battam & Smith 1993), where they may enter Bass Strait.
There has been an unconfirmed report of one off South Africa (Harrison *et al.* 1997).

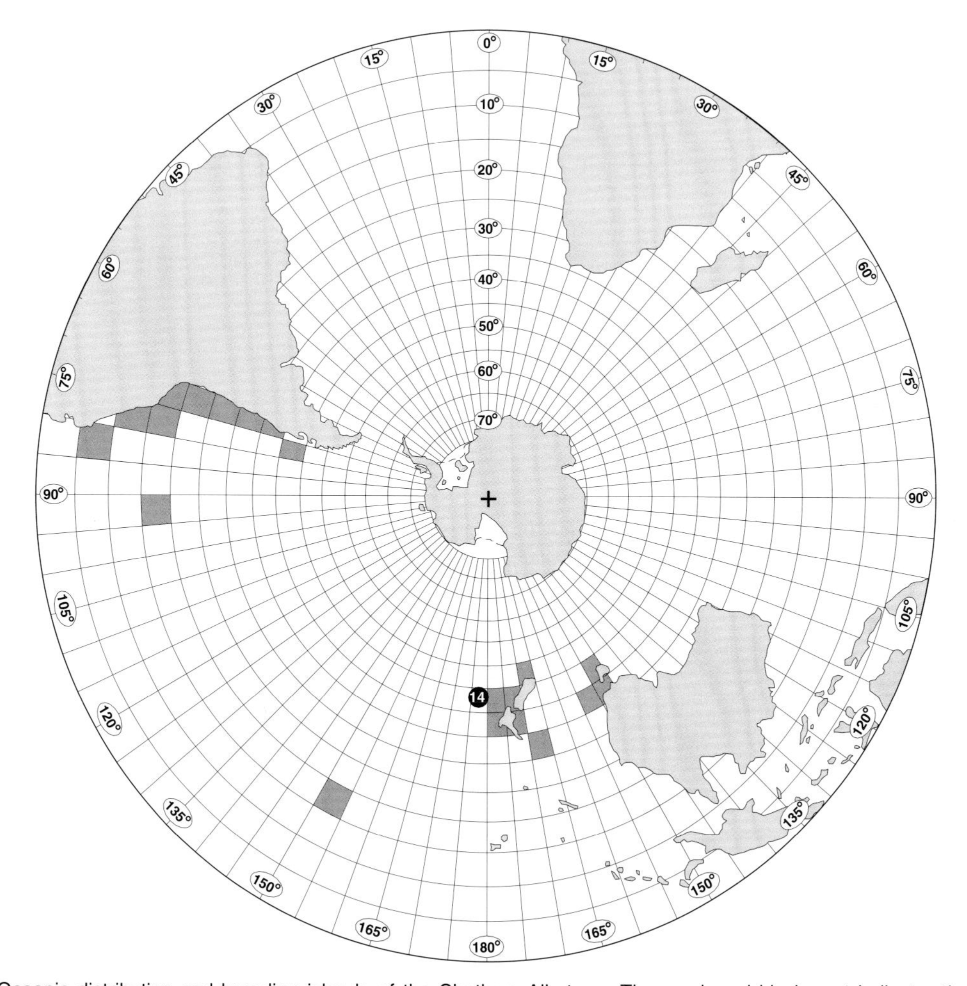

Fig. 5.9 Oceanic distribution and breeding islands of the Chatham Albatross. The numbered black spot indicates the Chatham Islands.

BREEDING ISLANDS

All mollymawks breed on isolated islands and where those islands are in groups or archipelagos, they often favour offshore or detached islets. On large islands they are usually at the ends of peninsulas or on capes and headlands.

Prince Edward Islands

In 1951, Robert Rand (1954) discovered colonies of the Grey-headed Albatross on high inland cliffs overlooking Goodhope and Rook's Bays, on the south coast of Marion Island inland (Fig. 5.10). It also breeds along the northern cliffs of Prince Edward Island, adjacent to colonies of Eastern Yellow-nosed Albatrosses (van Zinderen Bakker Jr. 1971, Grindley 1981).

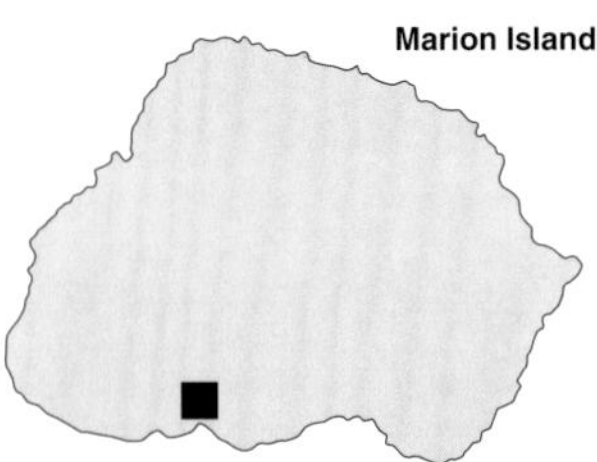

Fig. 5.10 The Prince Edward Islands showing the breeding sites of the Grey-headed Albatross (■) and Eastern Yellow-nosed Albatross (△) (Rand 1954, Grindley 1981).

Iles Crozet

Most mollymawk colonies are on the smallest islands to the west of the group (Fig. 5.11). Eastern Yellow-nosed, Black-browed and Grey-headed Albatrosses all breed at the Ilots des Apôtres and Ile de Pingouins (Jouventin *et al.* 1984). Four pairs of Salvin's Albatrosses were discovered on the Ile de Pingouins in 1986–87 (Jouventin 1990). At the eastern end of the group, Grey-headed Albatrosses are well established on the Ile de l'Est together with a smaller number of Black-browed Albatrosses (Prévost 1970, Despin *et al.* 1972). There are no mollymawks on the Ile aux Cochons and only few Grey-headed Albatrosses nest on the Ile de la Possession (Jouventin *et al.* 1984).

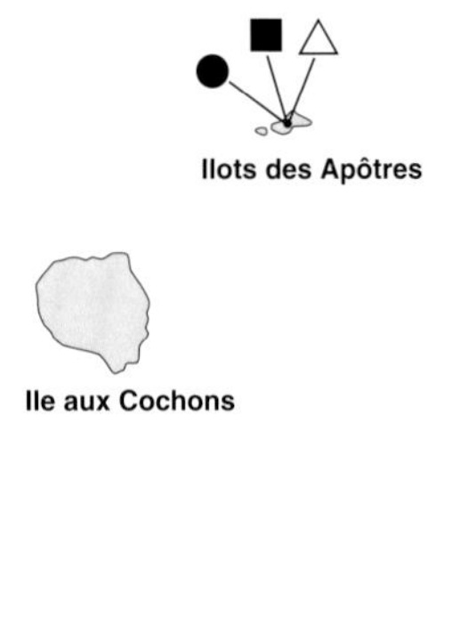

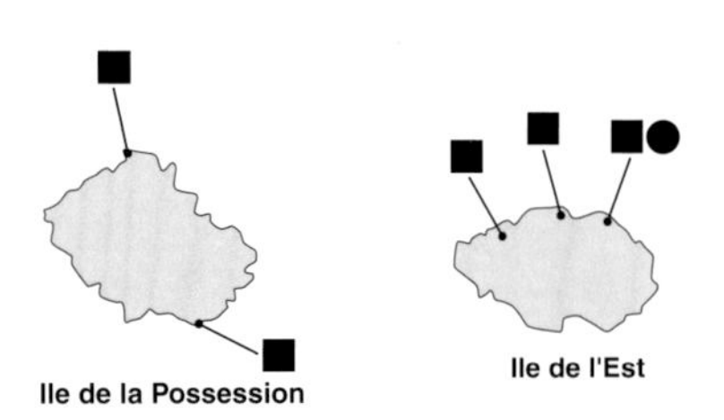

Fig. 5.11 The Iles Crozet showing the breeding sites of the Black-browed Albatross (●), Grey-headed Albatross (■), Eastern Yellow-nosed Albatross (△) and Salvin's Albatross (◇) (Despin *et al.* 1972).

Ile Amsterdam

Patrice Paulien (1953) saw many Eastern Yellow-nosed Albatrosses at sea and over the cliffs of the west coast in 1951–52, but it was not until 1970 that the huge colonies were visited. They are on tussock grass slopes 50–400 m above the west coast below the Grand Balcon, from the Pointe de la Rookerie to the Pointe del Cano (Segonzac 1972). At least one Western Yellow-nosed Albatross has been seen among breeding Eastern Yellow-nosed Albatrosses. Mapping of these colonies was accomplished in 1981–82 (Jouventin *et al.* 1983) (Fig. 5.12).

Ile Saint Paul

Vélain saw several mollymawks on the island in 1874 (Segonzac 1972), and Mougin found a downy mollymawk chick on a nest in 1969. The following year other French biologists, who spent 12 days on the island, located seven nests of the Eastern Yellow-nosed Albatross on the slopes of the Grand-Morne (Fig. 5.12). Since then a few more nests have been found (Jouventin *et al.* 1984).

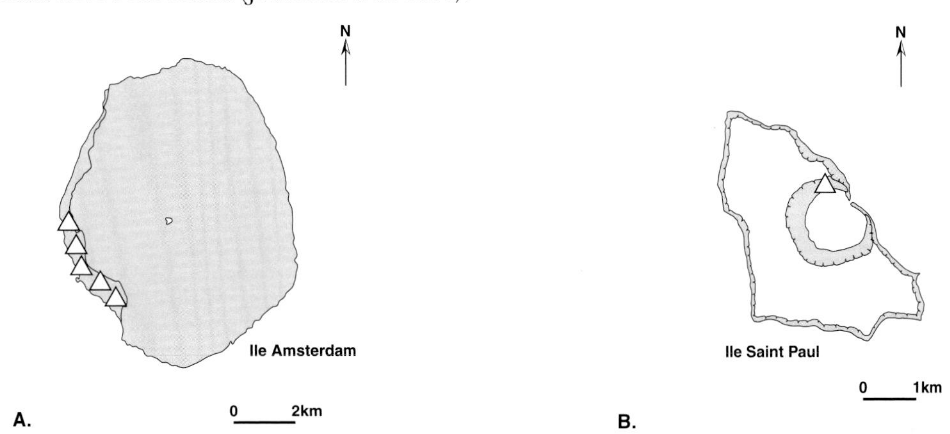

Fig. 5.12 The Ile Amsterdam (A) and Ile St Paul (B) showing breeding sites of the Eastern Yellow-nosed Albatross (△) (Jouventin *et al.* 1983, Segonzac 1972).

Iles Kerguelen

On 15 May 1840, when the *Erebus* and *Terror* arrived at Christmas Harbour, Joseph Hooker[19] landed and climbed a ridge to the north coast.

> 'In the Eastwards the cliffs were of immense height, and some huge sea birds were soaring like eagles around the top of it.'

A few days later he followed the ridge to the summit above Cape Français. By that time the birds had gone, but he found nests that he thought were those of albatrosses:

> '...it consists of a raised cylindrical pillow of earth contracted above and below, swelling out in the centre, with a depression at the top. They were situated, huddled together as many as fifty or sixty of them, and were built on the grassy ledges above the precipice 7 or 800 feet above the sea. A good deal of straw and stubble were mixed with them, or rather plastered up with the clay to give it consistency. The height was about one and a half feet and their breadth much the same; from a distance they looked like so many Cheshire cheeses.'

Hooker assumed they were Wandering Albatrosses (Sharpe 1879), but they have never been seen in this location. The Black-browed Albatross was reported at this site in 1914, by a French sailor (Loranchet 1915). Benôit Tollu (1967) and Brian Roberts found colonies in 1964–65, along the north cliffs of the Peninsule Loranchet, adjacent to the Iles Nuageuses. Colonies also extended along the south-facing cliff-tops of the Table d'Oiseau for about two kilometres between Cap Français and the entrance to Port Christmas; among them were a few Grey-headed Albatrosses. Large colonies of the Black-browed, Grey-headed and Eastern Yellow-nosed Albatrosses were later found on Ile du Croy in the Iles Nuageuses (Jouventin & Stonehouse 1985) (Fig. 5.13).

Black-browed Albatrosses nesting on steep cliffs of the Presque ile Jeanne d'Arc near the Baie Greenland, were seen by R. Hall (1900) from a ship close inshore. It was not until 1966 that Benôit Tollu (1967) and his colleagues approached from the land and confirmed two colonies on steep, loose slopes facing east-southeast high on the

[19] Antarctic Journal (unpublished MS at Kew).

cliffs of the Cañyon des Sourcils Noirs, northeast of Cap George. Several pairs have bred elsewhere: on the Ile Longue above Port Bizet and near Molloy in the north of the Golfe du Morbihan (Pascal 1979).

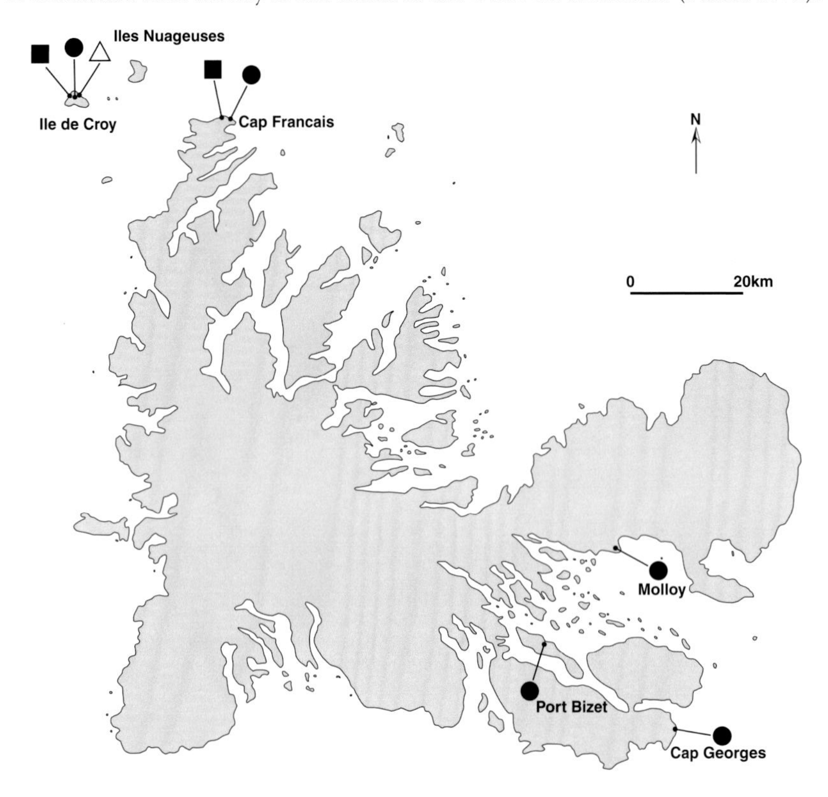

Fig. 5.13 The Iles Kerguelen showing the breeding sites of the Black-browed Albatross (●), Grey-headed Albatross (■) and Eastern Yellow-nosed Albatross (△) (Tollu 1967, Jouventin & Stonehouse 1985).

Heard and McDonald Islands

Between 1949 and 1954, ANARE[20] personnel visited several small colonies of Black-browed Albatrosses above the northeast cliffs of the Laurens Peninsula and on Henderson Bluffs (Downes *et al.* 1959). Among the birds breeding there in 1988 (Kirkwood & Mitchell 1992), was a ringed immigrant from the Iles Kerguelen (Woehler 1989).

In 1980, Black-browed Albatrosses were found breeding above the southern cliffs of McDonald Island and on Meyer Rock (Johnstone 1982) (Fig. 5.14).

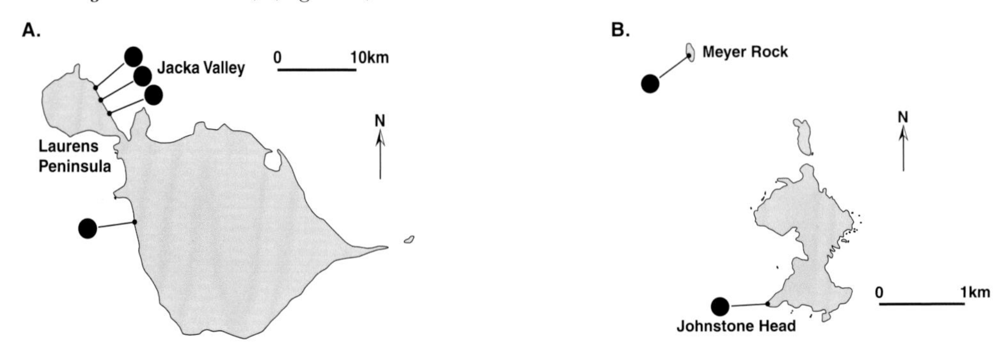

Fig. 5.14 Heard Island (A) and McDonald Islands (B) showing breeding sites of the Black-browed Albatross (●) (Kirkwood & Mitchell 1992).

[20] Australian National Antarctic Research Expedition.

Macquarie Island

A few Grey-headed Albatrosses were breeding on the southwest side of Petrel Peak at the southern end of the island in 1911–12 and in the mid-1950s there were about 50 nests. By the mid-1980s they had increased to about 90 pairs with some 25 pairs of the Black-browed Albatross next to them (Copson 1988) (Fig. 5.15). Similar numbers of Black-browed Albatrosses have been seen on the Bishop and Clerk Islets, 37 km south of Petrel Peak (Mackenzie 1968, Lugg *et al.* 1978).

A small number the Black-browed Albatross settled on the west side of North Head sometime after 1914. They were breeding in 1947–48, but from 1968 to 1972 a maximum of eight pairs were counted and only one fledgling was reared (G.W. Johnstone pers. comm., Copson 1988). Since 1988, they have disappeared altogether.

Tasmanian Islands

The main colony of the Tasmanian Shy Albatross is on Mewstone, with a smaller one nearby on Pedra Branca (Brothers 1979a&b). A separate colony has long been known on Albatross Island in Bass Strait (Johnstone *et al.* 1975) (Fig. 5.16). Since the early 1980s, two Chatham Albatrosses have been regular visitors to Albatross Island (Brothers & Davis 1985).

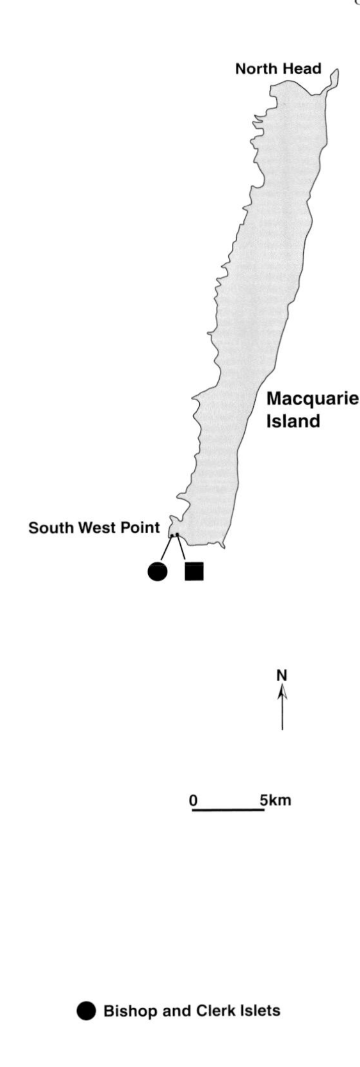

Fig. 5.15 (above) Macquarie Island and its offlying islets showing breeding sites of the Black-browed Albatross (●) and Grey-headed Albatross (■) (Mackenzie 1968, Copson 1988).

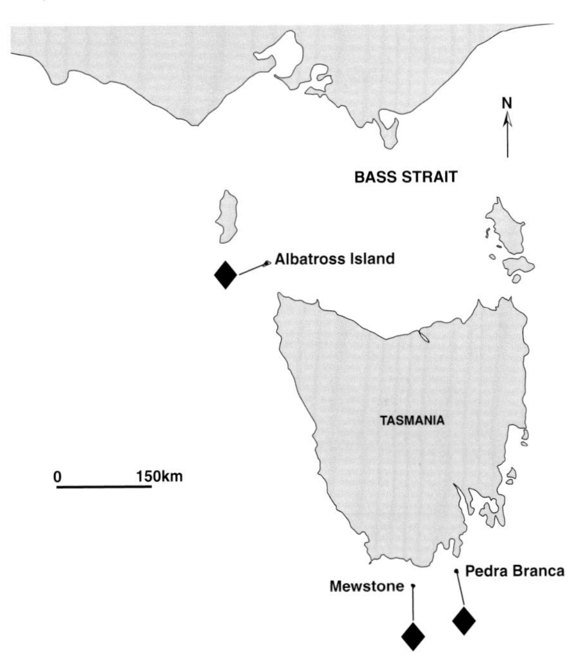

Fig. 5.16 (left) Tasmania showing the locations of Mewstone, Pedra Branca and Albatross Island, all breeding sites of the Tasmanian Shy Albatross (◆) (Johnstone *et al.* 1975, Brothers 1979a,b).

New Zealand Islands

Several pairs of Northern Buller's Albatrosses breed on Rosemary Rock in the Princes Islands of the Three Kings group (Wright 1984) (Fig. 5.17).

Large numbers of Southern Buller's Albatrosses nest from 30 m to 212 m above sea level, all around the coast of Solander Island and its off-lying stacks to the southwest; there are fewer of them on Little Solander Island (Cooper *et al.* 1986, I.G. McLean pers. comm.) (Fig. 5.17).

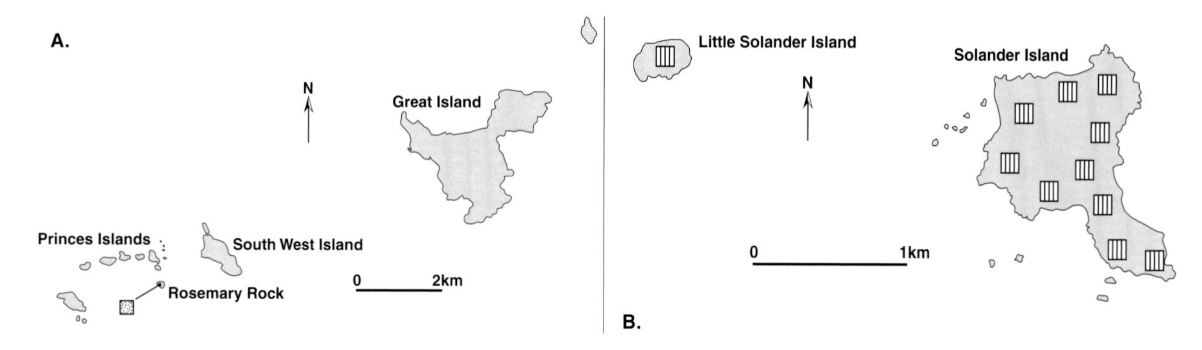

Fig. 5.17 New Zealand breeding locations of the Northern Buller's Albatross (▩) at the Three Kings Islands (A) and Southern Buller's Albatross (▥) at the Solander Islands (B) (Wright 1984, I.G. McLean pers. comm.)

Snares Islands

Southern Buller's Albatrosses nest around most of the main island, but most occur towards the south and west (Fig. 5.18). Some cliff-top colonies extend into the *Olearia* forest (Plate 11). A few pairs nest on Toru islet in the Western Chain.

Salvin's Albatross is well established on Toru and Rima islets in the Western Chain. An Auckland Shy Albatross and several Chatham Albatrosses have visited the colonies on Rima and Toru (Miskelly 1984, 1997). Six Black-browed Albatrosses were also seen on Toru in 1983–84 and a pair was incubating an egg in the following year. In January 1986, the same nest contained a chick (Miskelly 1997).

Fig. 5.18 The Snares Islands showing the breeding sites of the Southern Buller's Albatross (▥) Salvin's Albatross (◇) and Black-browed Albatross (●) on Main Island and the Western Chain (Miskelly 1984, 1997, Sagar *et al.* 1994).

Auckland Islands

Many colonies of Auckland Shy Albatrosses cover the lower tussock slopes of Disappointment Island (Fig. 5.19) (Plate 13). Smaller colonies are situated on the South West Cape of Auckland Island and on the southwest coast of Adams Island (Robertson 1975b).

Campbell Island

Large colonies of Campbell Black-browed and Grey-headed Albatrosses extend along the south side of the Courrejolles Peninsula, on the north coast near Hooker's Finger and Hooker's Peninsula, and south of Bull Rock North (Fig. 5.20) (Plate 5). The Isle de Jeanette Marie has just Campbell Black-browed Albatrosses nesting. All are distant from the main anchorage. Although there has been a history of human occupation of the island, these mollymawks were probably visited only occasionally by shepherds before 1931. Sorensen's notes for the 1940s contain the first descriptions of the colonies and Alfred Bailey photographed them in 1958 (Bailey & Sorensen 1962). Christopher Robertson (1980) began mapping them in 1975–76, a task that was completed in 1987–88 (Moore & Moffat 1990b).

In 1975, two Black-browed Albatrosses were discovered among the endemic Campbell Black-brow Albatrosses at Bull Rock South. They have increased and are now more numerous in the Courrejolles colonies (Tennyson *et al.* 1998). At least two-thirds of them interbreed with Campbell Black-browed Albatrosses (Moore *et al.* 1997a).

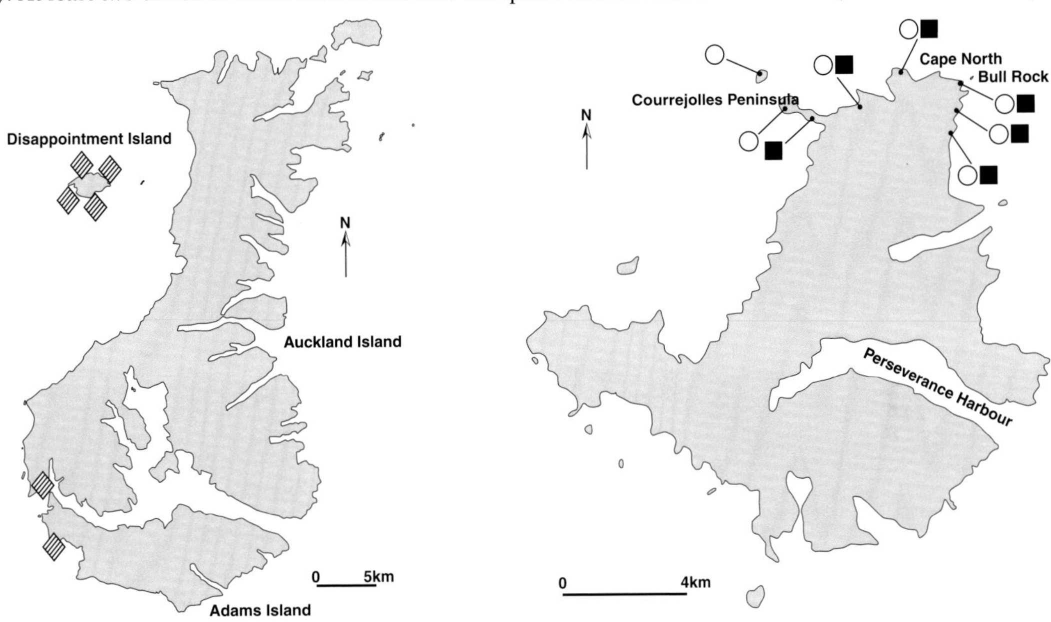

Fig. 5.19 The Auckland Islands showing the breeding sites of the Auckland Shy Albatross (◈) on Auckland, Adams and Disappointment Islands (Robertson 1975b, A. Rebergen pers. comm.).

Fig. 5.20 Campbell Island showing the breeding sites of the Campbell Black-browed Albatross (○) and Grey-headed Albatrosses (■) (Moore & Moffat 1990b).

Antipodes Islands

Small numbers of mollymawks breed in tussock grass high on the steep south to east cliffs of Bollons Island (Fig. 5.21). They have recently been confirmed as mainly Black-browed Albatrosses, with a few Auckland Shy Albatrosses (Clark & Robertson 1996, Tennyson *et al.* 1998).

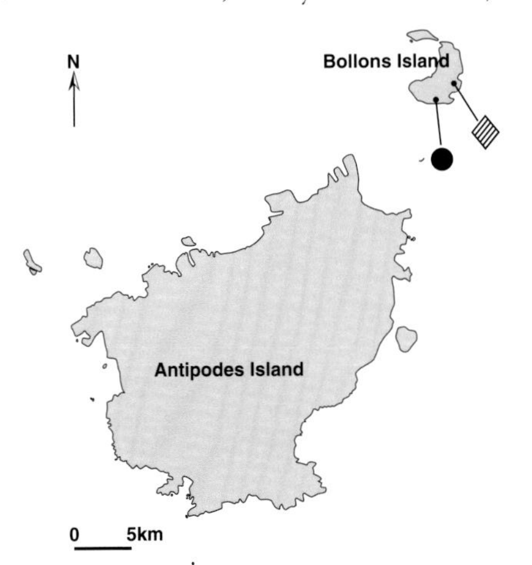

Fig. 5.21 The Antipodes Islands showing the breeding sites of the Black-browed Albatross (●) and Auckland Shy Albatross (◈) on Bollons Island (Clark & Robertson 1996).

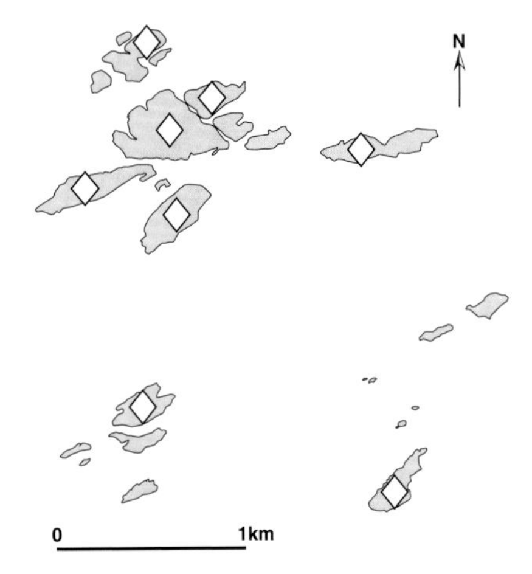

Fig. 5.22 The Bounty Islands showing the mixed colonies of Salvin's Albatrosses (◇) and Erect-crested Penguins (Robertson & van Tets 1982).

Bounty Islands

Large numbers of Salvin's Albatrosses nest among penguins and prions on the eight rocky islets (Robertson & van Tets 1982) (Fig. 5.22) (Plate 14).

Chatham Islands

Northern Buller's Albatrosses (Fig. 5.23) nests in clefts on steep cliffs, on rock slabs, in caves and on the flat tops of the Sisters and Forty-Fours islets (Plate 10). A single pair of the Auckland Shy Albatross breeds on the Forty-Fours (Robertson *et al.* 1997).

Chatham Albatrosses (Fig. 5.23) nest only on the steep rocky flanks and cave of the Pyramid (Plate 15) (Fleming 1939). A single Salvin's Albatross occupied a site among these birds in December 1991 (G. Robertson pers. comm.).

Fig. 5.23 The Chatham Islands showing the Forty Fours and Sisters Islets with their colonies of the Northern Buller's Albatross (▩) and the Pyramid with the Chatham Albatross (◈) (C.J.R. Robertson pers. comm., Fleming 1939).

Fig. 5.24a Isla Diego de Almagro, Chile, with breeding sites of the Black-browed Albatross (●) (G. Clark 1984, 1988, pers. comm.).

Isla Diego de Almagro

In October 1955, when the frigate HMS *Mounts Bay* was steaming north off the Chilean coast, R. Gibbs saw a large colony of Black-browed Albatrosses on an island (Tuck 1956). No location was given, but it may have been one of the colonies later discovered by Gerry Clark (1984) on the Isla Diego de Almagro (Fig. 5.24a). They are at at several sites along the west coast: Cabo Taplas and an offshore islet; Cabo Gruta and an offshore islet; Punta Negra – two sites to the southwest and at Cabo Jorge – two sites and two islets to the northwest. The nests are scattered among patches of vegetation or in small colonies on very steep rocky cliffs 15–170 m above sea-level (G. Clark 1984, 1998, pers. comm.).

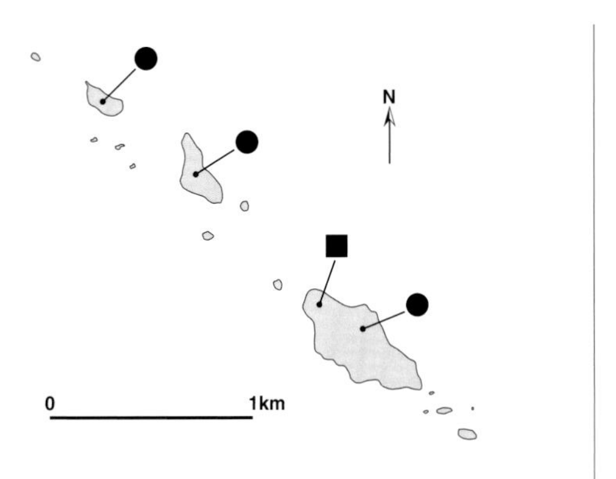

Fig. 5.24b Islas Ildefonso, Chile; the Black-browed Albatross (●) breeds on all three main islets and offshore stacks. A few Grey-headed Albatrosses (■) nest among the Black-browed Albatrosses at the NW point of the most southeasterly islet (G. Clark 1988, pers. comm.).

Islas Ildefonso

In December 1914, Beck took a photograph of Black-browed Albatrosses that appeared in *Oceanic Birds of South America* (Murphy 1936). The species breeds on all three main islets among eroding vegetation between the exposed rock of cliff-tops or seaward slabs and inland tussock grass. Beck saw a Grey-headed Albatross amongst them (Murphy 1936) and in 1985 Peter Harrison found six nesting at the northwest end of the southern islet (G. Clark pers. comm.) (Fig. 5.24b).

Islas Diego Ramírez

Chilean seamen knew that there were mollymawks on these islands long before 1958 when Edgar Aubert de la Rüe (1959) found Grey-headed Albatrosses nesting on Isla Gonzalo. It was another 20 years before they were reported nesting with Black-browed Albatrosses on other islands of the southern group. The largest numbers were on Isla Bartolomé (Schlatter & Riveros 1997). Black-browed Albatrosses breed in conspicuous colonies, among eroding vegetation, along the crests of the islands. They outnumber Grey-headed Albatrosses (P. Harrison pers. comm.) which are more widely scattered among the tussock grass. No mollymawks were seen when *Damien II* sailed up the west side of the northern islets (S. Poncet pers. comm.) (Fig. 5.24c).

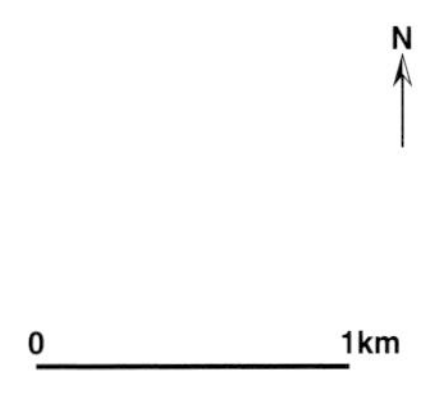

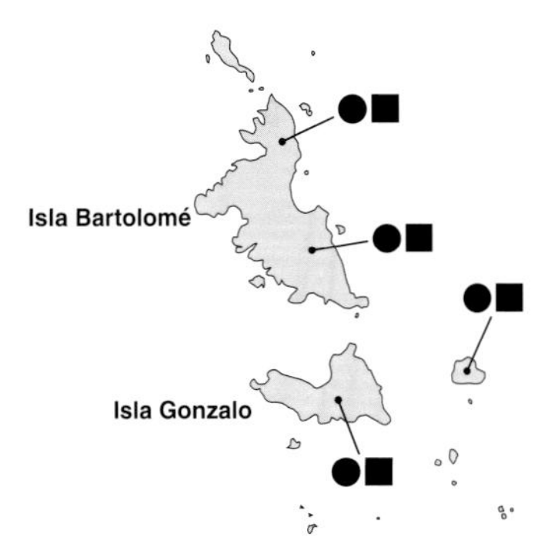

Fig. 5.24c Islas Diego Ramírez, Chile; the Black-browed Albatross (●) and Grey-headed Albatross (■) have not been seen on the Grupo Norte (S. Poncet pers. comm.).

Falkland Islands

After 1765, several warships lay at Port Egmont and sailors ashore on Saunders and Keppel Islands came across mixed colonies of albatrosses and penguins. One of the seamen, Erasmus Gower (1803) wrote many years later that they were on rising ground near the sea and numbered four to six thousand nests. Sealers and whalers discovered many other 'rookeries'.

From 1869, Black-browed Albatross colonies were well known on Keppel Island, Saunders Island, New Island

and West Point Island, all of which had permanent sheep farms. Bird Island and the Jason Islands remained uninhabited, but were stocked with cattle and/or sheep so the presence of colonies would have been well known. Beauchêne Island, however was never more than a haunt of sealers (Fanning 1833).

The Black-browed Albatross breeds at 12 sites in the Falkland Islands (Fig. 5.25). George Reid landed on Beauchêne Island in January 1959 and found the huge colony of albatrosses and penguins later described and photographed by Ian Strange (1965). Soon afterwards Len Hill drew attention to the even greater numbers on Grand and Steeple Jason Islands (Hill & Wood 1976) (Plate 2).

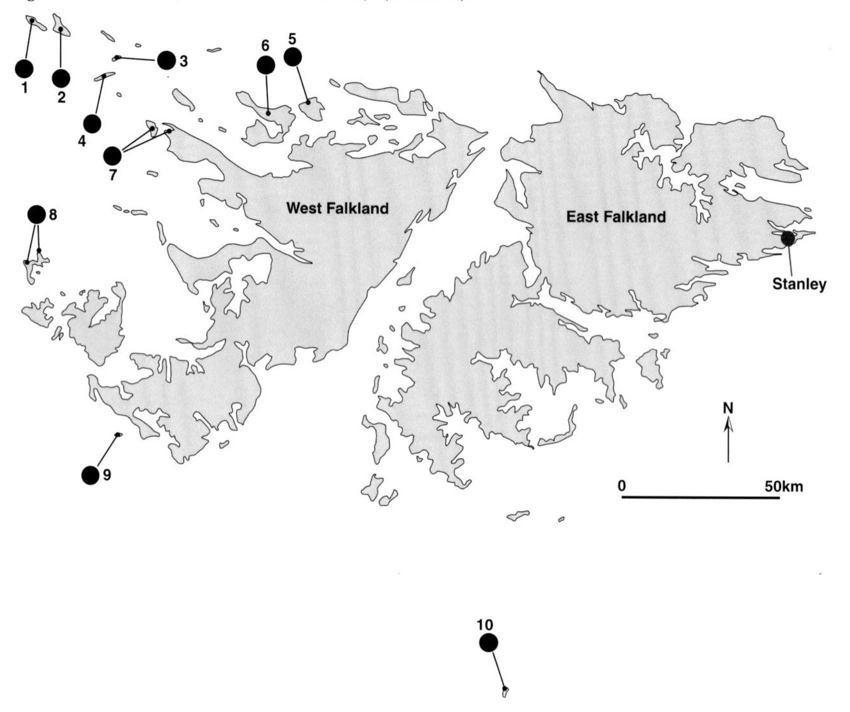

Fig. 5.25 The Falkland Islands showing breeding sites of the Black-browed Albatross (●) (Woods & Woods 1997, R.W. Woods pers. comm.). 1) Steeple Jason I. 2) Grand Jason I. 3) Elephant Jason I. 4) South Jason I. 5) Keppel I. 6) Saunders I. 7) West Point I. & Grave Cove. 8) New I. & North I. 9) Bird I. 10) Beauchêne I.

South Georgia

The mollymawk colonies of South Georgia were known to many generations of sealers and whalers. They understood that while 'white mollymawks' (Black-browed Albatrosses) bred at both the northwest and southeast ends of the island, 'blue mollymawks' (Grey-headed Albatrosses) were found only at the northwest (Harrison Matthews 1929, 1951). Colonies were first mapped at Bird Island in 1958–59 (Tickell & Pinder 1975), then gradually located around the the rest of South Georgia (Tickell 1976, Prince & Payne 1979, Prince *et al.* 1994, S. Poncet pers. comm.) (Fig. 5.26).

Tristan da Cunha

Colonies of Western Yellow-nosed Albatrosses were described at Tristan da Cunha in 1817 (Carmichael 1818). It breeds on all five islands in the group. Nests are built on tussock grass, on rocks and under trees/bushes. (Wace & Holdgate 1976) (Fig. 5.27). On Tristan itself, nesting grounds are mainly on the southeast and south-southeast slopes of the Base at 200–800 m, with fewer to the north above the settlement (Richardson 1984). At Inaccessible Island they breed on the plateau near Long Ridge, at Cairn Peak, Molly Bog and Dick's Bog (Fraser *et al.* 1988). On Nightingale Island, nests are widely scattered from the slopes of High Ridge and the Humps to the Ponds, which contain nests in colonies (Rowan 1951, Wace & Holdgate 1976, M.K. Swales pers. comm.).

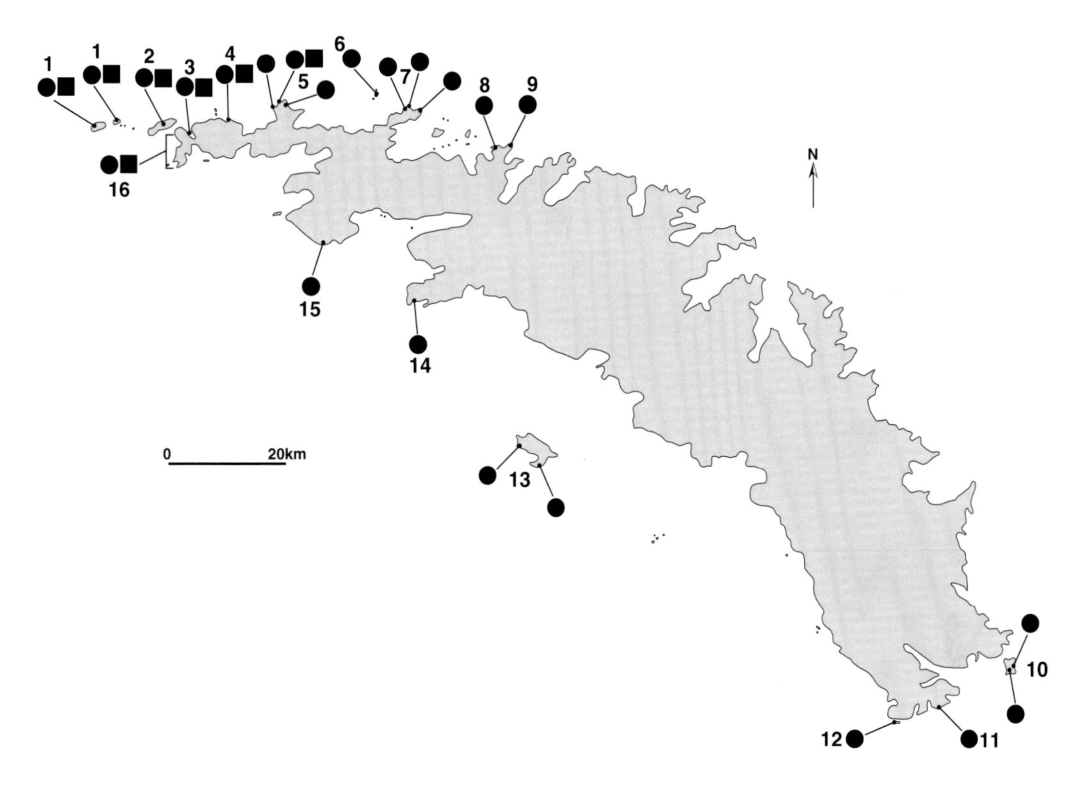

Fig. 5.26 South Georgia showing the breeding sites of the Black-browed (●) and Grey-headed Albatrosses (■) (Prince *et al.* 1994, S. Poncet pers. comm.). 1) Willis Is. 2.) Bird I. 3) Elsehul. 4) Sørn & Bernt & coast. 5) Cape North. 6) Welcome Is. 7) Cape Buller. 8) Cape Wilson. 9) Cape Crewe. 10) Cooper I. 11) Rumbolds Point. 12) Green I. 13) Annenkov I. 14) Cape Nuñez. 15) Klutschak Point. 16) Cape Alexandra-Cape Paryadin & the Jomfruene.

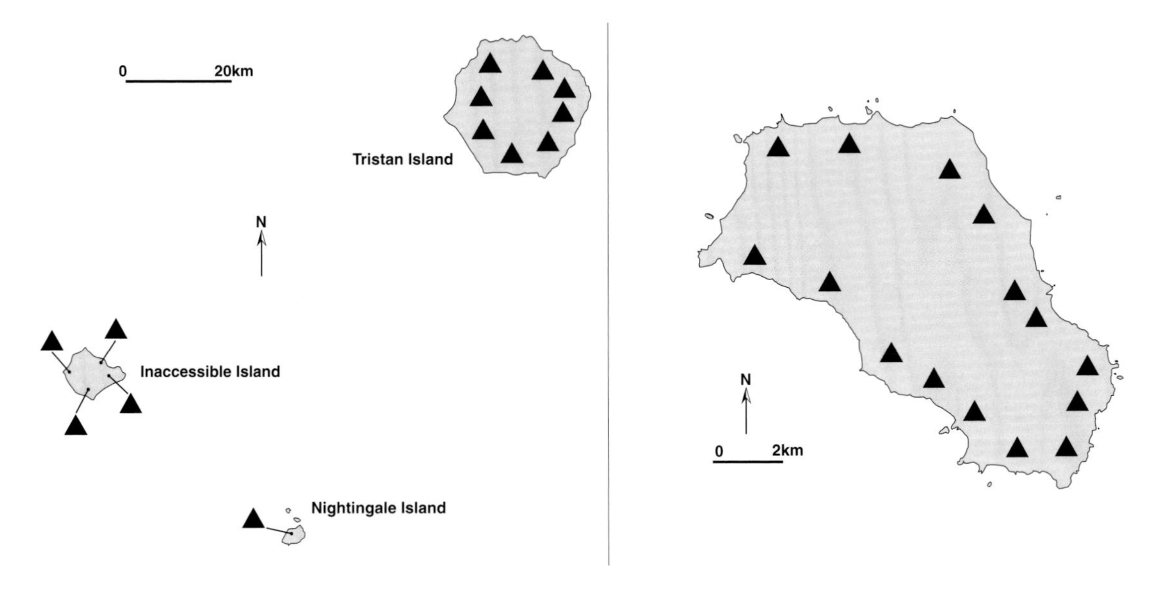

Fig. 5.27 The Tristan da Cunha Islands showing the breeding sites of the Western Yellow-nosed Albatross (▲).

Fig. 5.28 Gough Island showing the breeding sites of the Western Yellow-nosed Albatross (▲).

Gough Island

Western Yellow-nosed Albatrosses breed on steep slopes between 60 and 300 m, but the distribution is poorly understood (Fig. 5.28). Nests are scattered within dense fern bush on the sides of valleys such as The Glen and Sophora Glen, but are thought to be more numerous on high ground south of Little Glen and on the cliff-tops above Lot's Wife's Cove (Swales 1965, Voisin 1979, J. Cooper pers. comm.).

NUMBERS

Estimates of mollymawk numbers vary greatly in accuracy. Counting has often been *ad hoc*; the results one-off and difficult to interpret. After we had been told that there were more than a million pairs of the Black-browed Albatrosses on Beauchêne Island in the Falklands (Strange 1965), a more realistic figure of 137,000 was calculated (Woods 1966). Eleven years later, the hypnotic million was still being promoted (Strange 1976). However, colonies are entities that anchor even the most questionable figures; eventually it took only a few sample counts to bring the number down by an order of magnitude (Prince 1982). My own transfer of breeding success from one season to another resulted in an exaggeration of Black-browed and Grey-headed numbers at Bird Island, South Georgia in 1962 (Tickell 1976). Two decades of later, counts at Bird Island indicated how much annual variation exists in mollymawk colonies (Prince *et al.* 1994).

Most counts are of breeding adults. The cohorts of juveniles at sea are beyond counting and it is laborious to accumulate figures for subadults. Counts and estimates of the numbers of breeding pairs have been tabulated in Appendix 15. Tables 5.1–5.6 contain rounded summaries appropriate only for broad comparison.

	Number (pairs)
Black-browed albatrosses	701 000
cauta mollymawks	168 000
Grey-headed Albatross	157 000
Yellow-nosed albatrosses	71 000
Buller's albatrosses	31 000

Table 5.1 World numbers of breeding mollymawks based on most recent data (Appendix 15) and rounded to the nearest order of magnitude. (*cauta* mollymawks – Auckland Shy, Tasmanian Shy, Salvin's and Chatham Albatrosses).

	Number (pairs)
Falkland Islands	549 000
South Georgia	101 000
South America	20 000
Iles Kerguelen	3 000
Iles Crozet	1 000
Heard & McDonald Islands	700
Macquarie Island	200
Antipodes Islands	100
Campbell Island	30
Snares Islands	1
Total Black-browed Albatross	675 000
Campbell Black-browed Albatross	26 000

Table 5.2 The numbers of breeding pairs of Black-browed Albatrosses and Campbell Black-browed Albatrosses based upon the most recent data (Appendix 15) and rounded to the nearest order of magnitude.

There are probably over a million pairs of breeding mollymawks in the southern hemisphere. Well over half of them are Black-browed Albatrosses; no other mollymawk species approaches such numbers (Table 5.1). Almost all Black-browed Albatrosses breed in the SW Atlantic (Table 5.2). Two-thirds to three-quarters of of them are at the Falkland Islands, where most are on just two islands; Steeple Jason Island, with more than 200,000 nests (Thompson & Rothery 1991) and Beauchêne Island, with well over 100,000 pairs in an impressive single colony (Fig. 5.25).

Grey-headed Albatrosses are most numerous in the SW Atlantic and Drake Passage, with about half breeding at South Georgia, and much smaller numbers on other circumpolar islands (Table 5.3). The two northerly breeding yellow-nosed albatrosses achieve similar, relatively small numbers in both the South Atlantic and Indian Oceans (Table 5.5). There are only about half as many of the two Buller's albatrosses, perhaps rather more of the Northern than the Southern populations (Table 5.6).

Around New Zealand and SE Australia, mollymawk numbers are dominated by Salvin's Albatrosses and Auckland Shy Albatrosses (Table 5.4). Tasmanian Shy Albatrosses have increased substantially over the last 30 years; nevertheless, their numbers will probably remain small compared with those of the Auckland Shy Albatross (Croxall & Gales 1997). The Chatham Albatross is the least numerous of all mollymawks.

	Number (pairs)
South Georgia	81 000
South America	14 000
Iles Kerguelen	13 000
Prince Edward Islands	13 000
Campbell Island	11 000
Iles Crozet	10 000
Macquarie Island	100

Table 5.3 The numbers of pairs of biennial breeding Grey-headed Albatrosses rounded to the nearest order of magnitude have been computed by dividing recent counts of pairs breeding in one season (Appendix 15) by 0.592 to allow for those failing and trying again the following season (Prince *et al.* 1994).

		Number (pairs)
Auckland Shy Albatross		75 000
Tasmanian Shy Albatross		12 000
Salvin's Albatross		
	Bounty Islands	76 000
	Snares Islands	600
	Iles Crozet	10
Chatham Albatross		4 000

Table 5.4 The numbers of pairs of breeding *cauta* mollymawks (Auckland Shy, Tasmanian Shy, Salvin's and Chatham Albatrosses) based upon the most recent data (Appendix 15) and rounded to the nearest order of magnitude.

		Number (pairs)
Western Yellow-nosed Albatross		
	Tristan da Cunha	29 000
	Gough Island	7 000
Eastern Yellow-nosed Albatross		
	Amsterdam & St Paul Islands	25 000
	Prince Edward Islands	6 000
	Iles Crozet	4 000
	Iles Kerguelen	50

Table 5.5 The numbers of pairs of breeding Western Yellow-nosed Albatrosses and Eastern Yellow-nosed Albatrosses, based upon the most recent data (Appendix 15) and rounded to the nearest order of magnitude.

		Number (pairs)
Northern Buller's Albatross		
	Chatham Islands	18 000
	Three Kings Islands	20
Southern Buller's Albatross		
	Snares Islands	8 000
	Solander Islands	5 000

Table 5.6 The numbers of pairs of breeding Northern Buller's and Southern Buller's Albatrosses based upon the most recent data (Appendix 15) and rounded to the nearest order of magnitude.

BREEDING

Colonies and nests

At any particular island or group of islands, mollymawk colonies are scattered in an apparently unpredictable manner. It has rarely been possible to do more that guess at the significance of the locations. Nests are usually close enough for there to be no doubt that they form a colony. More or less isolated nests can usually be found, but solitary breeding is rare. Nesting habit is not species-specific. Typically, mollymawk colonies are on well-vegetated slopes above coastal cliffs (Plates 5&13), or on hills a short distance from the sea.

At Ile Amsterdam, some of the Eastern Yellow-nosed Albatrosses nest on slopes of 25° to 60°, but 63% of the nests are on the gentler slopes (Segonzac 1972). Mollymawks sometimes occupy narrow ledges on steeper broken crags or, in marked contrast, expanses of more or less level ground (Plate 2).

Colonies of two species may be adjacent to each other or even mixed where they meet. Nests of one species may be isolated within the colony of another and a pattern may be apparent; for instance, at South Georgia small numbers of Black-browed Albatrosses nest among Grey-headed Albatrosses, but not vice versa (Tickell & Pinder 1975). At the Iles Crozet, small groups of Grey-headed Albatrosses are scattered through two much larger colonies of Eastern Yellow-nosed Albatrosses and many Black-browed Albatross nests are also close to one or other of them (Weimerskirch *et al.* 1989a).

Penguins, cormorants and gannets are sometimes associated with mollymawk colonies. At Amsterdam Island, the Rockhopper Penguin nests on the lower slopes, but shares common overlapping areas with the higher nesting Eastern Yellow-nosed Albatrosses (Segonzac 1972). At the Falkland Islands, Rockhopper Penguins nests among the Black-browed Albatrosses even at the highest breeding densities. Erect-crested Penguin nest among Salvin's Albatrosses on the Bounty Islands (Plate 14) and the density of albatross nests remains the same whether or not there are penguins present (Robertson & van Tets 1982).

Cormorants nests alongside Black-browed Albatrosses at some locations in South Georgia, the Falklands and Kerguelen. At one Black-browed Albatross colony in Kerguelen, which was displaced by the slipping of an unstable slope, the Kerguelen Cormorant took over the abandoned mollymawk nests (Derenne *et al.* 1972; Pascal 1979). At Beauchêne Island, the King Shags congregate to excavate nest material and thus open up dense tussock grass that may benefit Black-browed Albatrosses (Smith & Prince 1985). Nests of Tasmanian Shy Albatrosses on Pedra Branca are scattered among those of Australian Gannets and benefit from the nest material that the latter carry onto the rock.

Mollymawk nests are typically columns made of excavated peat or mud (Fig. 5.29, Table 5.7). Peat is an excellent building material, soft and workable when wet, it dries into resilient structures that last for years, attracting surface growths of algae and moss. Gradually, they become eroded or the bases are excavated by neighbouring birds, so that they eventually topple over and may even roll down steep slopes.

Dimensions (Fig. 5.29)	BBA mean (range) (n=26)	SLA mean (range) (n=35)	NBA mean (range) (n=11)	WYNA (n=1)
Diameter of bowl (d_1)	29 (28-31)		22 (20-24)	31
Diameter of column (d_2)	40 (37-44)	37	31 (29-34)	41
Depth of bowl (d)	7 (5-8)	3		
Height on low side (h_1)	17 (10-28)	9 (1-25)	9 (6-18)	25/60
Height on high side (h_2)	36 (12-54)	15 (1-25/40)	9 (6-18)	25/60

Table 5.7 The dimensions (cm) of mollymawk nests: the Black-browed Albatross (BBA) at the Falkland Islands, Salvin's Albatross (SLA) at the Bounty Islands, Northern Buller's Albatross (NBA) at the Three Kings Islands and Western Yellow-nosed Albatross (WYNA) at Tristan da Cunha (Hagen 1952, Robertson & van Tets 1982, McCallum 1985, W.L.N. Tickell unpublished).

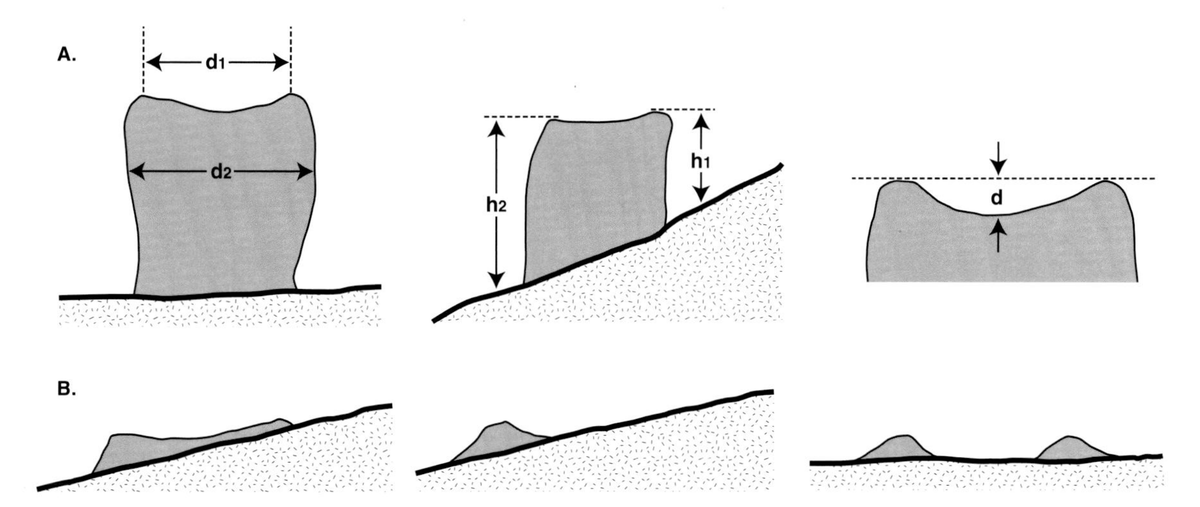

Fig. 5.29 Diagrams of Black-browed Albatross nests at West Point Island, Falkland Islands (A), where there is ample peat and tussock grass. Salvin's Albatrosses at the Bounty Islands (B) nest on bare rock where very little soil is available (Robertson & van Tets 1982). See Table 5.7 for nest measurements.

Although many Falkland Islands colonies are on tussock slopes, the huge colonies of the Black-browed Albatross at Beauchêne and Steeple Jason Islands are on flat or gently sloping rocky ground. These colonies were originally carved out from tussock grass, but little vestige of grass has survived the constant digging and trampling. The fibre content of the original peat has rotted away leaving a mineral soil. The birds cannot work this when it is dry, but after rain it becomes a soft mud and there is vigorous building everywhere in the colony. The newly built columns, up to 50 cm high, dry to a durable, brick-like texture and colour.

New nests at the edge of colonies, are often on the top of live tussocks. The crown growth eventually dies, but living grass may survive around the base, and the fibrous pedestals of dead tussock remain for many years. Nests are sometimes made from ragged piles of freshly pulled grass (Plate 6); those of the Western Yellow-nosed Albatross on Nightingale Island sometimes reach impressive proportions, but they may be so loosely constructed that by the end of the season they are trampled flat (Rowan 1951).

Most mollymawk islands are so wet and lush that nest building is seldom a problem. At the Snares, old nests average 23 cm in height when the Southern Buller's Albatrosses return and by mid-incubation they have been built up to 25 cm (Warham 1996). The Grey-headed Albatrosses returning to South Georgia at the end of a late winter may find their colonies drifted-up with snow, the ground frozen and nest bowls filled with ice. Mollymawks show preferences for damp or wet places. Carmichael (1818) noticed that Western Yellow-nosed Albatrosses nested near streams on Tristan da Cunha, while on the high slopes of West Point Island in the Falklands, the Black-browed Albatrosses prefer damp gullies rather than the surrounding well drained ground.

In drier climates where there is little vegetation, mineral soils are often so friable that nests crumble and disintegrate by the end of the season. Nests of Tasmanian Shy Albatrosses on Albatross Island are trampled flat by the time the chicks leave. In all mollymawk colonies, the bones of dead chicks become incorporated into nests, and where there is little nest material they are more conspicuous. At the Chatham Islands, Northern Buller's Albatrosses occupy the flat top of the Forty-Fours. In the few damp places, normal nests are constructed of soil, elsewhere stones and bones are incorporated into low nests averaging five to ten centimetres in height (Robertson & Sawyer 1994). On Little Sister, nests tend to be in sheltered places, often at the bottom of broad clefts in solid rock or under cave-like overhangs where soil accumulates. On the steep flanks of the Pyramid, quite substantial nests of Chatham Albatrosses are surrounded by clean, bare rock from which the birds have removed every trace of soil (Fleming 1939), while at the Bounty Islands the minimal rims of Salvin's Albatross nests (Fig. 5.29) point to a severe shortage of material.

Colonies vary in size from less than ten nests to more than 100,000 (Tickell & Pinder 1975, Prince 1981,

Weimerskirch *et al.* 1986, A. Rebergen pers. comm.). Within colonies, densities of Black-browed Albatrosses nesting on rocky slopes at South Georgia average 43 nests per 100 square metres (2.3 m^2/nest) while Grey-headed Albatrosses, which select less steep ground on the same island, create more even terrain on which densities of 60 nests per 100 square metres (1.7 m^2/nest) are possible (Tickell & Pinder 1975). At the Bounty Islands, in spite of being on bare rock with hardly any nesting material, Salvin's Albatrosses achieve densities of 53 nests per 100 square metres (Robertson & van Tets 1982). The highest nest densities are among Black-browed Albatrosses at Beauchêne and Steeple Jason Islands in the Falklands, which average 69 nests per 100 square metres (1.5 m^2/nest). One sample area contained 92 nests per 100 square metres (1.1 m^2/nest) (Prince 1981).

Most nests of Western Yellow-nosed Albatrosses at Tristan da Cunha and Gough Island are widely scattered and out of sight of each other in dense fern-bush or tussock grass (Plate 6). The only places where they aggregate is on the boggy areas of the Ponds on Nightingale Island, and, to a lesser extent, on the plateau of Inaccessible Island (Swales 1965, Voisin 1979). Eastern Yellow-nosed Albatrosses, at the Iles Crozet and Ile Amsterdam, form characteristic colonies with the largest exceeding 23,000 pairs (Jouventin *et al.* 1983).

At the Snares, nests of Southern Buller's Albatrosses are also scattered, but in small, loosely clumped aggregations on tussock ledges and cliff tops (Warham & Bennington 1983); 40% nest within the forest up to 400 m inland. The dense canopy inhibits growth of an understory and there is ample humus from which the birds construct tall nest columns, less ragged than those outside (Richdale 1949a) (Plate 11).

Breeding seasons

In most mollymawks, breeding is a summer activity, with adults arriving in the spring and leaving in the autumn. There are two exceptions; the Auckland Shy Albatross and Southern Buller's Albatross – both breeding on islands south of New Zealand.

Black-browed Albatrosses begin returning to South Georgia in late September and most fledglings leave around 2 May; all have gone by 14 May (Table 5.8). The small colony at Heard Island is similarly timed, but at the Iles Crozet, Iles Kerguelen, Macquarie Island and the Falkland Islands, the season begins about three weeks earlier.

Grey-headed Albatrosses at South Georgia begin arriving in mid-September and the build-up is about 14 days earlier than the sympatric Black-browed Albatrosses. At the end of the season, though, most Grey-headed fledglings leave in mid-May, 16 days later than those of the Black-browed Albatross. Almost all have gone by the end of the month, but some individuals may remain late into June (Table 5.8).

	Black-browed Albatross		Grey-headed Albatross	
	Mean±SD	n	Mean±SD	n
Arrival (male)	10 Oct±5.2	26	22 Sep±4.0	7
(female)	14 Oct±6.0	26	30 Sep±4.2	7
Copulation	14 Oct±2.6	22	28 Sep±3.3	64
Laying	27 Oct±4.0	302	19 Oct±3.4	278
Hatching	3 Jan±3.4	111	30 Dec±3.1	105
Fledging	2 May±3.7	110	18 May±5.3	102

Table 5.8 Breeding cycles of Black-browed and Grey-headed Albatrosses at Bird Island, South Georgia (Tickell & Pinder 1975, Croxall *et al.*1988).

Eastern Yellow-nosed Albatrosses have been seen back at the Ile Amsterdam as early as 21 August, but reoccupation of the colonies takes place at the beginning of September. Fledglings start flying on 20 March and most go to sea at the end of March (Jouventin *et al.* 1983). The timing of the South Atlantic colonies appears to be similar. At Tristan da Cunha, eggs of the Western Yellow-nosed Albatross were collected on Nightingale Island during the second week of September 1950 (Rowan 1951) and at Gough Island in 1981, the first eggs were recorded on 21 September (Williams & Imber 1982).

Tasmanian Shy Albatrosses begin laying on Pedra Branca early in September and one to two weeks later at Albatross Island where laying continues to mid-October (Hedd *et al.* 1997). Fledglings leave early in April (Johnstone *et al.* 1975). Sixteen hundred kilometres away, the Auckland Shy Albatross season is much later. Laying begins in late November and chicks remain in their nests throughout most of the southern winter, leaving in mid-August. (Table 5.9).

	Location	Arrival	Laying	Hatching	Departure
Tasmanian Shy Albatrosses	Albatross Island	August?	mid-Sept.	early Dec.	March?
	Pedra Branca		Sept.–Oct.		
	Mewstone		late Sept.?	late Dec.	early Apr.
Auckland Shy Albatross	Auckland Islands		Nov.–Dec.	Jan.–Feb.	mid-Aug.
Salvin's Albatross	Bounty Islands	Sept.		early Nov.	April
	Snares Islands	Sept.		early Nov.	April
Chatham Albatross	Pyramid	early Aug.	late Aug.	early Nov.	

Table 5.9 Breeding dates of Tasmanian Shy Albatrosses, Auckland Shy Albatrosses, Salvin's Albatrosses and Chatham Albatrosses (Fleming 1939, Buckingham *et al.* 1991, Clark & Robertson 1996, Dawson 1973, Robertson 1975, Johnstone *et al.* 1975, Robertson & van Tets 1982, Tennyson *et al.* 1998, N. Brothers pers. comm.).

Northern Buller's Albatrosses begin returning to the Chatham Islands towards the middle of September. Laying is from late October to late November (Robertson 1974). At the Snares Islands, 1,400 km to the southwest, Southern Buller's Albatrosses do not begin returning until early December and laying starts at the beginning of January. Fledglings leave from late August to late October (Horning & Horning 1974, Sagar & Warham 1997).

Philopatry

Most subadult mollymawks return to the islands and colonies where they were reared. Among 60 five-year-old Black-browed Albatrosses that had returned to South Georgia, 58% were seen in the colony where they had been reared and 80% were within 500 m. Of 124 Grey-headed Albatrosses aged 8–14 years, 85% were in the colony of birth and 93% within 250 m. At the Snares, 66% of 86 incubating Southern Buller's Albatrosses nested within 100 m of where they had been ringed as chicks and males showed much stronger philopatry than females (Sagar *et al.* 1998).

Pre-egg periods

After they return to their colonies, Black-browed and Eastern Yellow-nosed Albatrosses have16–18 days before the eggs are laid. In Grey-headed Albatrosses this pre-egg period lasts 26 days and in Southern Buller's Albatrosses it is 29 days (Table 5.10).

	mean±D (days)	male %	male longest	female %	female longest	empty %	n
Black-browed Albatross	16±4	83	18	10	2	21	30
Grey-headed Albatross	26±6	60	22	9	4	43	41
Southern Buller's Albatross	29	53	32	8	23	54	51

Table 5.10 Length of the pre-egg period (from the arrival of the male to laying by the female) and attendance of mated males and females of the Black-browed Albatrosses, Grey-headed Albatrosses and Southern Buller's Albatrosses. Percentages do not add up to 100% because pairs were often together at the nest (Richdale 1949a, Tickell & Pinder 1975, Sagar & Warham 1997).

Most mollymawks return repeatedly to the same site or nest. Ninety-three percent of Black-browed and Eastern Yellow-nosed Albatrosses used the same nest in consecutive seasons. At the Snares only 67% of Southern Buller's Albatross pairs reoccupy the same nests. Branches of trees sometimes prevent returning birds getting back to their nests, or the nests have collapsed after being undermined by burrowing petrels (Sagar & Warham 1997). One Southern Buller's Albatross at the Snares in 1993 was on the same nest that it had occupied in 1948 and 1971 (Richdale & Warham 1973, Sagar & Warham 1993). Nests of Black-browed and Grey-headed Albatrosses are occupied through most of the pre-egg period; usually the male is present, females returning from time to time, but generally spending only a day or less ashore. Nests of these species are rarely left unattended, but Southern Buller's Albatrosses at the Snares frequently vacate their nests, although individual males and females have been known to remain on their nests for much longer periods than either Black-browed or Grey-headed Albatrosses. Most female mollymawks return during the day before laying.

Eggs

Maturation of Black-browed and Grey-headed Albatross gonads begins several weeks before they return to land. Newly arrived males have enlarged testes with free sperm and can copulate as soon as they arrive. Copulation between established pairs occurs on the nest, but it has been seen away from the nest and between adults and subadults. Among pairs of Black-browed and Grey-headed Albatrosses at South Georgia, copulation was seen as late as two days before the egg was laid.

Female Black-browed and Grey-headed Albatrosses begin laying down yolk in several ova before returning, by which time one ovum is much bigger, though not yet full size. After copulation, sperm remain viable in the oviduct while the female goes off to feed and lay down the remainder of the yolk. Ovulation and fertilisation probably occur at sea and it takes another nine to twelve days before she is ready to lay. The total time for egg formation is 29 and 33 days respectively in Black-browed and Grey-headed Albatrosses (Astheimer *et al.* 1985). Availability of food during this period contributes to the size of Grey-headed Albatross eggs (Cobley *et al.* 1998), but it is not known what proportion of total variation is attributable to environmental factors.

Newly laid mollymawk eggs are 7% to 11% of the parents' body mass and the extremes between all species is: mass 186–333 g, length 93–110 mm, width 58–70 mm (Appendix 4). The smallest eggs are those of the two yellow-nosed albatrosses. The eggs of females laying for the first time are narrower than in subsequent years. Over a breeding lifetime of up to 50 years, eggs of individual Southern Buller's Albatrosses became shorter and broader (Richdale & Warham 1973, Sagar & Warham 1993).

Laying

Egg-laying in most mollymawk colonies lasts about three weeks. Eastern Yellow-nosed Albatrosses at the Ile Amsterdam continues laying for 17 days (Jouventin *et al.* 1983). Black-browed and Grey-headed Albatrosses at South Georgia lay for 23 and 19 days respectively (Tickell & Pinder 1975). At the Snares, laying by Southern Buller's Albatrosses is spread over 43–62 days; about 80% of the eggs appear within 35 days, but this remains a very prolonged laying period, and it is about twice as long as in the Northern Buller's Albatross, at the Chatham Islands (Richdale 1949a, Warham 1967).

Incubation

The average length of incubation among mollymawks is 68–73 days (Table 5.11). The temperature of the brood patch probably remains within very narrow limits between different mollymawk species and much of the variation in the duration of incubation probably reflects differences in the size of eggs. Two Black-browed Albatrosses at South Georgia incubated for 78 and 80 days respectively, well outside the normal range and indicating abnormal incubation behaviour (Huin 1997). In most mollymawks, males and females share incubation in five to ten shifts. After laying, the female is sometimes displaced within hours by her mate, especially if he was present at the time of laying. The first shift is always by the female; it is significantly shorter than subsequent shifts by both sexes. Nevertheless, females do sometimes remain on the egg for long periods after laying; one female Southern Buller's Albatross sat for 24 days before being relieved (Richdale 1949b). The next shifts by both male and female are

	Mass of egg mean (g)	Incubation (days) Mean±SD	n	Brooding (days) Mean±SD	n
Grey-headed Albatross	276	72±1.6	103	23±2.8	27
Black-browed Albatross	257	68±1.2	72	22±2.8	49
Eastern Yellow-nosed Albatross	200	71±1.5	40	21±2.4	42
Southern Buller's Albatross	250	69±1.3	61		

Table 5.11 Lengths of incubation and brooding in mollymawks: the Grey-headed Albatross, Black-browed Albatross, Eastern Yellow-nosed Albatross and Southern Buller's Albatross (Tickell & Pinder 1975, Jouventin & Weimerskirch 1984, Sagar & Warham 1997).

usually longer, before becoming progressively shorter towards the end of incubation (Fig. 5.30). Tasmanian Shy Albatrosses at Albatross Island are an exception to this sequence; all shifts after the first are of about equal length (N. Brothers pers. comm.).

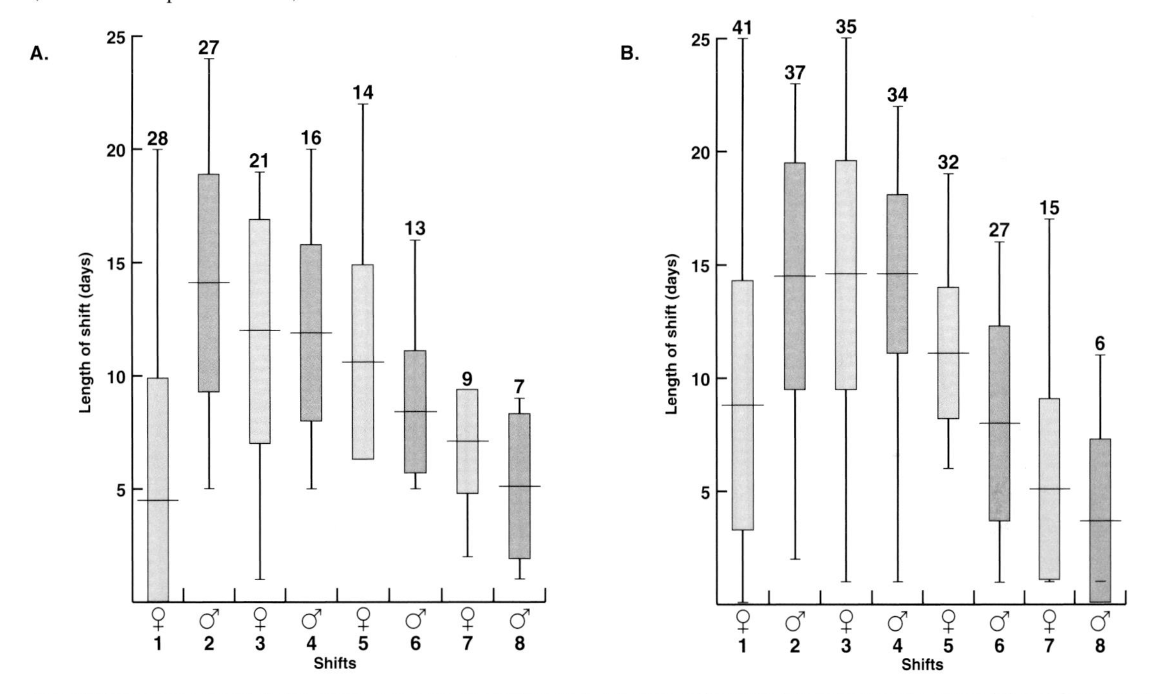

Fig. 5.30 The mean length (± SD, range & n) of male and female incubation shifts in the Black-browed Albatross (A) and Grey-headed Albatross (B) at South Georgia (Tickell & Pinder 1975).

Hatching and brooding

From the first signs of chipping, Black-browed and Grey-headed Albatross chicks take about four days to break out of their shells. Most are brooded by one or other parent for 22–23 days in short shifts of two to three days. Chicks of the Eastern Yellow-nosed Albatross are brooded for at least 18 days and, on average, are left alone by day 20, while those of the Southern Buller's Albatross are brooded for 23–25 days (Sagar & Warham 1997).

Chicks and fledglings

After they are left alone, all mollymawk chicks remain on their nests and are fed by both parents until departure. The fledging period has not been measured very often, but the variation of 115 to 167 days (Table 5.12) is five times greater than for incubation. In 1993, 1994 and 1995, selected pairs of Grey-headed Albatrosses at South Georgia, with histories of breeding success, fledged 13–15 young in 143, 133 and 121 days respectively, differences that were believed to reflect annual environmental conditions. Some pairs are consistently better providers than

	Islands	Incubation (days)	Fledging (days)	Total (days)
Eastern Yellow-nosed Albatross	Ile Amsterdam	71	115	186
Black-browed Albatross	South Georgia	68	116	184
	Iles Crozet		125	
Campbell Black-browed Albatross	Campbell Island		130	
Grey-headed Albatross	South Georgia	72	141	213
	Crozet Island	73	145	218
	Campbell Island		152	
Southern Buller's Albatross	Snares Islands	69	167	236

Table 5.12 Incubation and chick-rearing periods in mollyhawks (Tickell & Pinder 1975, Jouventin *et al.* 1983, Weimerskirch *et al.* 1986, Moore & Moffat 1990b, Sagar & Warham 1997).

others and this source of variation has to be superimposed upon environmental constraints when making generalisations (Cobley *et al.* 1998).

Mollymawk chicks are covered in smoky white to pale grey down that becomes matted under the belly. They fledge in a plumage that substantially resembles their parents' except on the head, neck and underwings, where intermediate stages prevail for some years. Chicks have black beaks that begin to turn brown by the time they fledge and then take several years to acquire adult colours.

Parents forage independently and there may be differences in the amounts of food brought to chicks by males and females. In years of adequate resources, most mollymawk chicks receive feeds at intervals of one to three days. The lengths of intervening fasts reflect foraging distance and food availability. Although fasts get longer, up to 12 days towards the end of the chick rearing period, most chicks receive food within one to three days of departure.

At the Iles Kerguelen, Black-browed Albatross parents feed their chicks mostly during the hours of daylight. The largest numbers fly in during the late afternoon. When they reach the island at night, they remain in rafts on the water offshore until dawn. If a chick has been fasting for some days, it can quickly consume all of the food that one adult can deliver, but if it has recently been fed, it may accept only a portion. At such times, an adult will take-off from the colony and rest on the sea offshore for some hours before coming in again to deliver the remainder of its load. To a degree, adults may adjust foraging strategy to their chick's needs. Parents that spend little time at the nest bring in larger meals more often. Parents delayed near the colony, take longer over foraging and deliver smaller meals (Weimerskirch *et al.* 1997c).

Single mollymawks can successfully rear chicks. In four known cases, the meals delivered by single, unsexed Black-browed Albatrosses were the same size as those of paired birds, but the foraging flights were on average 12% shorter and the total quantity of food delivered was 10% greater. On average, each single parent fed its chick 36% more than half the quantity delivered by a pair (Weimerskirch 1997).

In their first few days, the young chicks are fed liquid; this is mostly water with 6% to 9% suspended solids and oil derived from crustaceans (p.32). The energy content of the oil, 12 kJ per g, provides six times as much energy as the solids, so it could be particularly important in the early days. The delivered feeds of mollymawks all include substantial liquid components; among the Black-browed and Grey-headed Albatrosses at South Georgia they amounted to about half of the total mass of feeds (Prince 1980). The chicks need water; even the most experienced parents may sometimes be away for long periods and dehydration is probably a greater hazard to chicks than starvation. In February 1951 at Nightingale Island, large numbers of Western Yellow-nosed Albatross chicks died after three weeks of drought (Elliott 1957).

In February 1962, one Grey-headed Albatross nest at Bird Island, South Georgia contained two healthy chicks of the same size. Two females may have laid the eggs (p.34), but the rearing of two chicks – by two or three parents – is remarkable.

Growth of chicks

Since the early 1960s, there have been several studies of chick growth in the Black-browed and Grey-headed Albatrosses at Bird Island (Tickell & Pinder 1975, Clarke & Prince 1980, Prince & Ricketts 1981, Ricketts & Prince 1981, Huin *et al.* (in press)). Chicks of both species have sigmoid growth curves that differer in scale and shape. Inherited differences in growth patterns are influenced by variable provisioning. The mean mass of newly hatched Grey-headed chicks at South Georgia in 1963–65 was 197 g (range 150–260 g) (Tickell & Pinder 1975). Early growth of skeletal structures (bones and muscles) is followed by organs (liver, lung, heart and kidneys) and then fat and integument (skin, scales and plumage) (Reid *et al.* 2000).

Early Black-browed Albatross chicks are smaller, but grow faster than Grey-headed chicks, and attain higher peak mass. Krill contains more energy and calcium, so the larger quantities of it fed by the Black-browed Albatross may enhance differences in the rates of chick growth. In cross-fostering experiments, Grey-headed chicks gained weight faster when fed by Black-browed adults, while Black-browed chicks grew less well when fed by Grey-headed adults (Prince & Ricketts 1981).

There is considerable annual variation in growth, especially during the first 60 days. At the Iles Kerguelen, Black-browed Albatross chicks achieved mean growth rates of 65 and 39 g per day respectively in February 1994 and 1995 (Weimerskirch 1997). Over four years in the early 1990s, Black-browed Albatross chicks at South Georgia gained an average 42 g per day with the maximum rate of increase at 27 days from hatching, while Grey-headed chicks grew at 37 g per day and reached a maximum rate at 30 days. Black-browed and Grey-headed chicks reached mean peak mass of 4.61 kg and 4.47 kg at 88 days and 103 days respectively. The Black-browed chicks lost weight more rapidly than the Grey-headed chicks, but in years of adequate resources both achieved flight weights of 3.0–3.5 kg. at about 125 and 155 days respectively (Huin *et al.* (in press)). At Campbell Island, the opposite occurred; Grey-headed chicks grew faster and achieve greater peak and departure masses than those of the Campbell Black-browed Albatross (Moore & Moffat 1990b). Grey-headed Albatross pairs with histories of breeding success rear chicks that are heavier at hatching, grow faster and reach greater peak mass than those reared by parents with records of failure, even when food resources are poor (Cobley *et al.* 1998).

Departure

Most mollymawks vacate their colonies at the end of the southern summer (Table 5.9). The dates of departure vary between species and islands. At South Georgia, some adult Black-browed and Grey-headed Albatrosses keep returning with food up to the time that their young go to sea, but most successful breeding adults probably leave just before their fledglings. Subadults and failed breeders leave earlier. Black-browed Albatrosses that have lost eggs or young chicks continue to attend the colony for up to eight weeks, but Grey-headed Albatrosses leave soon after such losses (Tickell & Pinder 1975).

Two species of mollymawks are exceptional in rearing chicks through the southern winter. Auckland Shy Albatross fledglings depart slightly earlier in the following spring than those of Southern Buller's Albatross (Sagar & Warham 1997).

First breeding

Juvenile mollymawks remain at sea for several years before returning as subadults. At South Georgia, occasional Black-browed and Grey-headed Albatross subadults return to the island at two to three years of age. A few four-year-olds are usually to be found in the colonies, but most begin to come back when they are five years old (Tickell & Pinder 1975).

The modal age of first breeding in Black-browed Albatrosses at South Georgia is ten years (8–13 years). At Kerguelen and Macquarie breeding begins at six and seven years respectively. Campbell Black-browed Albatrosses also begin breeding at ten years (6–13 years) (Waugh *et al.* 1999a).

At South Georgia Grey-headed Albatrosses breed for the first time at a modal age of 12 years (10–14 years) and at Campbell Island at 13 years (10–17 years) (Prince *et al.* 1994, Waugh *et al.* 1999a).

FOOD

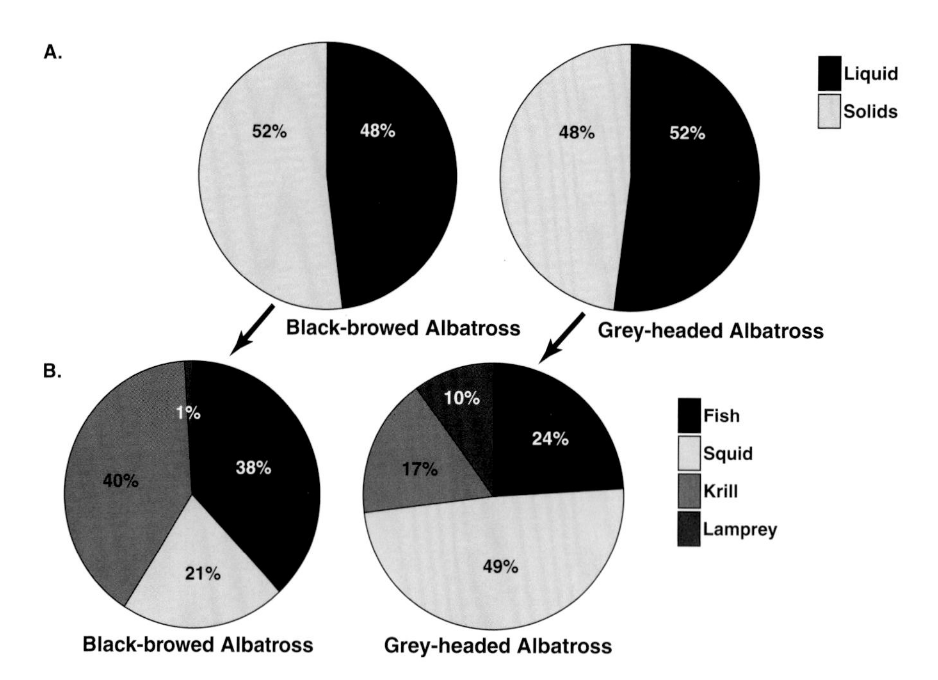

Fig. 5.31 The proportion of fluid (oil/water) and types of food delivered to 138 Black-browed and 132 Grey-headed Albatross chicks at South Georgia in February–March 1975 and 1976 (A). Percentage composition of solids by weight (B) (Prince 1980).

Mollymawks feed mainly on squid, fish and Crustacea (Fig. 5.31,32) (Appendix 6). Carrion from dead whales, seals and penguins may also be locally important. The tests of colonial tunicates such as *Salpa* and *Pyrosoma* have been identified in regurgitations (Tickell 1964, West & Imber 1986). Jellyfish and comb jellies *Ctenophora* have only been identified when birds have been seen feeding at sea, but they may, nevertheless, be more important than anyone has previously supposed, since they often occur at the surface of the ocean in great numbers.

The precise composition of mollymawk diets varies between different areas of ocean (Cherel & Klages 1997). On any one island, a diet may have some constants, but ocean resources are dynamic and the proportions of different prey brought to chicks change seasonally.

At South Georgia, Black-browed and Grey-headed Albatrosses take krill of similar size if it is available (Croxall *et al.* 1997a) and Black-browed bring in more krill than Grey-headed Albatrosses. Among a great variety of squid species eaten by both albatrosses, *Martialia hyadesi* is the single most important prey and is the predominant food of Grey-headed Albatrosses at

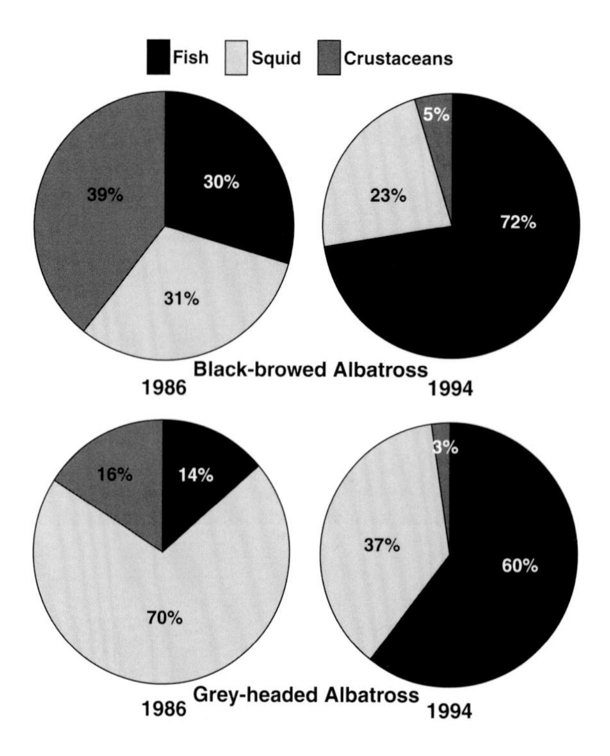

Fig. 5.32 The composition of food delivered to Black-browed and Grey-headed Albatross chicks at South Georgia in years when krill was abundant (1986) and scarce (1994) (Reid *et al.* 1996).

South Georgia and Campbell Island (Clarke & Prince 1981, Rodhouse & Prince 1993, Rodhouse *et al.* 1990, Waugh *et al.* 1999c). Both species at South Georgia eat a variety of fish, but the southern lamprey is taken almost exclusively by Grey-headed Albatrosses (Tickell 1964, Reid *et al.* 1996). As long as Grey-headed chicks receive enough squid and fish they grow well, but Black-browed chicks need some krill (Prince and Ricketts 1981).

In seasons when krill is short around South Georgia, Black-browed and Grey-headed Albatrosses eat more squid and fish, but the species of fish eaten may change. Both albatrosses continue to deliver meals of about the same size, but foraging takes longer, so the feeding rates decline (Croxall *et al.* 1999).

Food items eaten by other mollymawks have been identified (Appendix 6), but few have been sampled in a manner that permits convincing quantitative comparison between regions and years (Brooke & Klages 1986, Weimerskirch *et al.* 1986, Ridoux 1994). For example, the diet of Black-browed Albatross chicks at the Falkland Islands includes the squids *Loligo gahi* and *Illex argentinus*, hake and southern blue whiting. These albatrosses have also been seen feeding on lobster krill *Munida gregaria* (M.F. Holloway, in Bourne 1975). Food samples from chicks at New Island, Steeple Jason Island and Beauchêne Island had quite different constituents, but the samples were taken in different years (Fig. 5.33). In February 1990, food delivered to chicks at Beauchêne Island could have been obtained by scavenging from squid trawlers (Thompson 1992); so the apparent differences in diets at the three islands could reflect seasonal changes in fishing effort as well as geographical differences and hydrological fluctuations over the years and months of sampling.

At the Iles Crozet, a few Grey-headed Albatross food samples from the easternmost Ile de l'Est contained 89% squid, 11% crustaceans and a few fish, while at the Prince Edward Islands to the west the proportions were 34% squid, 58% fish and 3% crustaceans (Ridoux 1994, Hunter & Klages 1989).

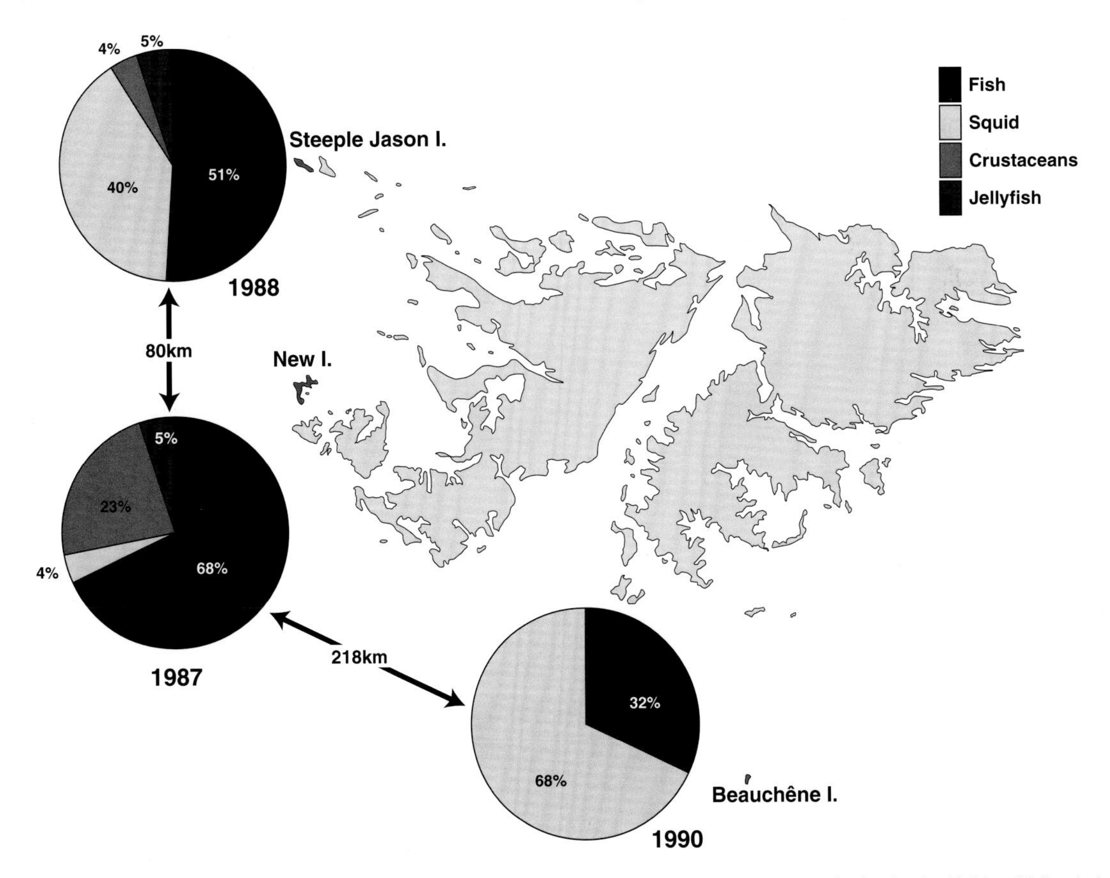

Fig. 5.33 The composition of food delivered to Black-browed Albatross chicks at three separate colonies in the Falkland Islands in different years. Segments are equivalent to percentage proportion by weight (Thompson 1992).

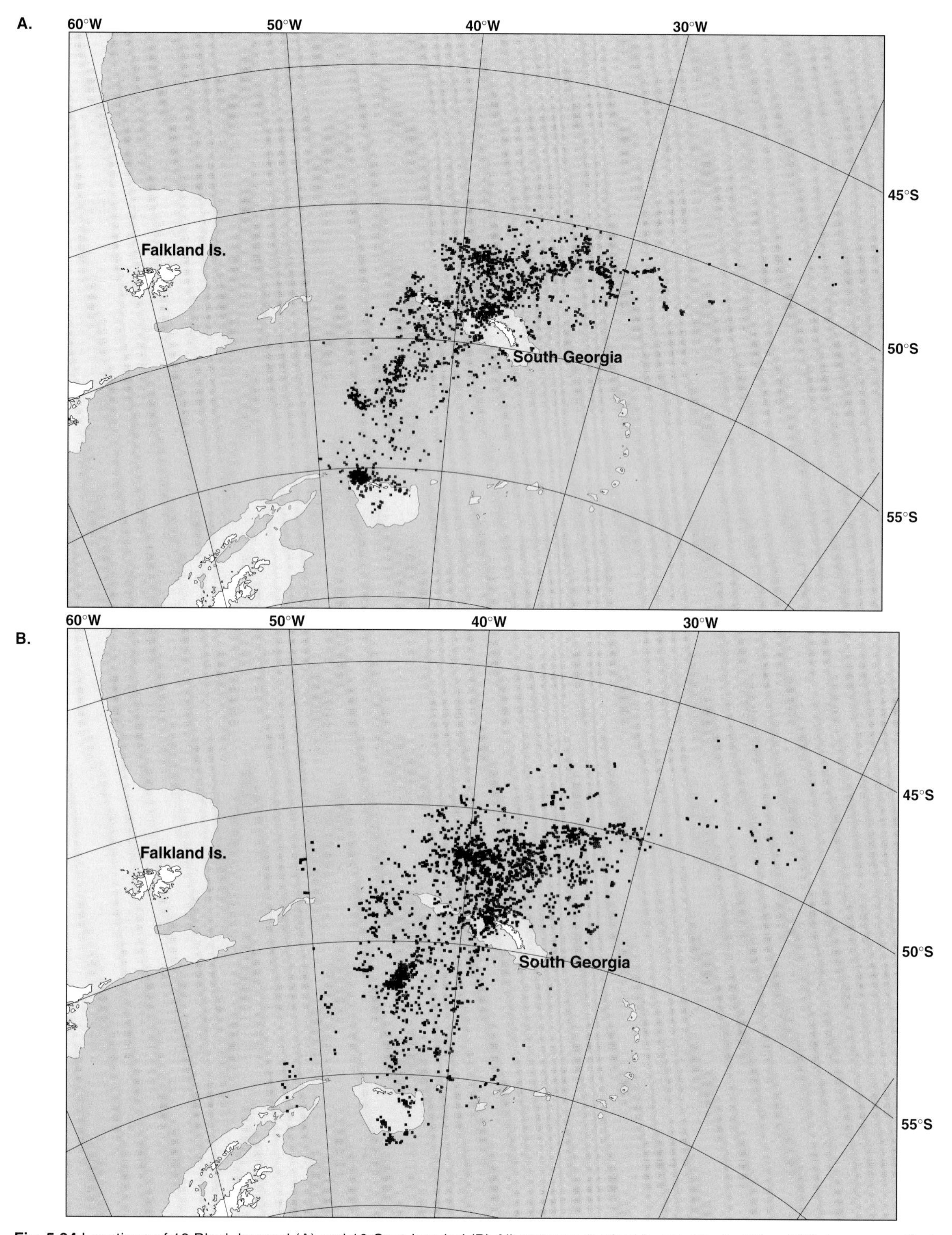

Fig. 5.34 Locations of 19 Black-browed (A) and 10 Grey-headed (B) Albatrosses tracked by satellite for 77 and 90 days respectively when foraging from Bird Island, South Georgia, January–March 1993. The position of the 1,000 m isobath is shown by change in shading and indicates the position of the continental shelf. Note that none of these birds approached the rich feeding grounds surrounding the Falklands Islands (Prince *et al.* 1997).

Eastern Yellow-nosed Albatross samples from the Ile aux Pingouins at the west end of the Iles Crozet contained 38% squid, 58% fish and 4% crustaceans (Ridoux 1994, Hunter & Klages 1989).

Black-browed and Grey-headed Albatrosses feeding chicks at South Georgia appear to forage over a broadly similar range extending to about 500 km from the island (Figs 5.34). Feeding areas have been detected, mainly over the continental shelf or slope surrounding South Georgia, Shag Rocks and the South Orkney Islands or associated with the Antarctic Polar Frontal Zone (APFZ). The Black-browed Albatross uses all areas, but the Grey-headed Albatross rarely lingers at the shelf and favours the APFZ (Prince *et al.* 1997).

A similar pattern is apparent south of New Zealand. Over the Campbell Plateau dense shoals of small juvenile southern blue whiting are an important resource for the Campbell Black-browed Albatross (Waugh *et al.* 1999c). It spends about half its time over the Campbell Plateau and the other half in forays of up to 2,000 km towards the APFZ. The Grey-headed Albatross breeding at Campbell Island spends most of its time over deep water and at the APFZ (Waugh *et al.* (in press)). In the mid-1990s, the Black-browed Albatross breeding on the south coast of the Iles Kerguelen, foraged at an average range of only 250 km, largely over a crescent of the shelf-edge, from north to south around the east of the islands. In 1994, some flew 500 km southeast to a distant seamount (Fig. 5.35).

Black-browed and Grey-headed Albatrosses at South Georgia in the early 1990s, returned to feed their chicks on average every 1.2 days and 1.3 days respectively. The probability of a Black-browed chick receiving meals from one or both of its parents in a single day was 58% and 28% compared with 51% and 24% for Grey-headed chicks. Provisioning rates varied from year-to-year and in any one year between successful and failed breeders. Black-browed Albatrosses at South Georgia delivered smaller meals, but more often than Grey-headed Albatrosses, so overall there was no significant difference between the two mollymawks.

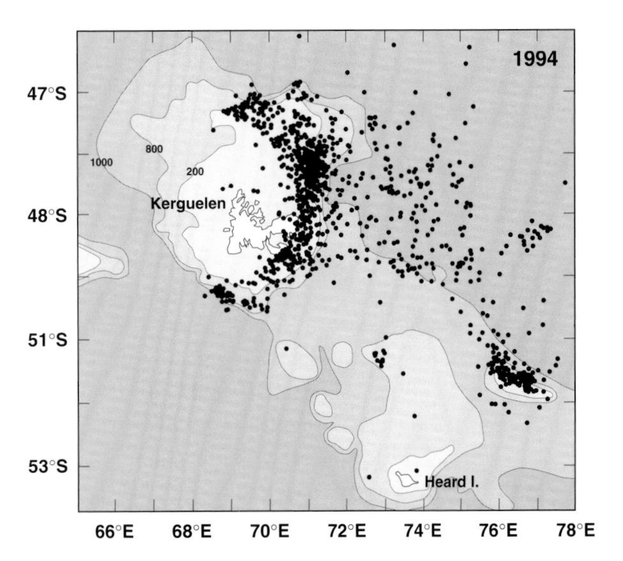

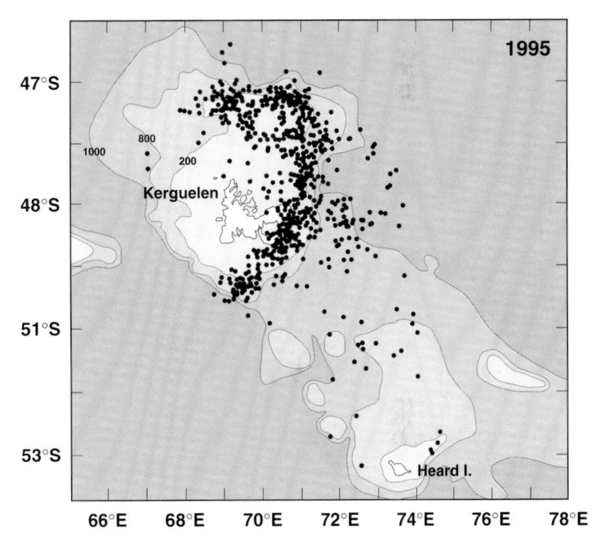

Fig. 5.35 Locations of adult Black-browed Albatrosses foraging from one small colony on the south coast of the Iles Kerguelen. In February 1994, 1,290 locations were obtained from 12 birds and in 1995, 857 locations from ten birds. Isobaths are in metres (Weimerskirch *et al.* 1997c).

At Campbell Island, provisioning rates also vary from year to year, and in 1996 and 1997, Campbell Black-browed Albatrosses had higher average provisioning rates (142 and 183 g per day) than the Grey-headed Albatrosses (108 and 157 g per day) (Waugh *et al.* (in press)).

At the Iles Kerguelen, foraging Black-browed Albatrosses carry food loads averaging 12% to 15% of their body mass, with the heaviest up to 30%. In February 1994, they collected significantly more food than in February 1995. The average feeding intervals were virtually the same, once every two days, but the meals were significantly larger (543 g) on average in 1994 than in 1995 (449g).

In both the Black-browed and Grey-headed Albatrosses at South Georgia, provisioning rates increased with the age of chicks as their gut capacities expanded; Grey-headed chicks were receiving full size meals by 45 days. In both species, rates peaked when the chicks were 60–69 days old and received, on average, 550–600 g per day. Provisioning rates declined with decrease in meal size and increase in the length of foraging flights, after 85 and 95 days in the Black-browed Albatross, and after 75 and 95 days in the Grey-headed Albatross. These declines were more rapid in the Black-browed Albatross (Huin *et al.* (in press)).

PARASITES AND DISEASE

Seven species of chewing lice (Phthiraptera: Ischnocera & Amblycera) have been found on mollymawks (Appendix 5); two on five host species, three on four host species and two on three host species. Fewer species have been found in any one host location. Hosts are sympatric on a number of islands, sometimes with colonies and nests close enough for external parasites to move from one species to another.

Five species of feather mites (Acarina: Astigmata: Analgoidea) have been collected. The Black-browed Albatross has *Brephosceles diomedei, Diomedacarus gigas* and *Echinacarus rubidus,* while the Grey-headed Albatross has *Brephosceles diomedei, B. gressitti* and *Promegninia pedimana.*

The incidence of ticks and fleas is more variable. Small numbers of the tick *Ixodes uriae* were sometimes seen on newly hatched Black-browed and Grey-headed chicks at South Georgia (Wilson 1970), but never fleas.

Ticks are fairly common on the feet of Black-browed Albatross chicks at the Falkland Islands, but rarely on their faces (R.B. Napier pers. comm.). These albatrosses have the flea *Parapsyllus longicornis,* which has also been recorded on the Campbell Black-browed Albatross (G.M. Dunnet 1964, pers. comm.).

On Tristan da Cunha, conspicuous infestations of the ticks *Ixodes rothschildi, I. percavatus* and *I. auritulus* have been seen under the bills and on the throats of nestling Western Yellow-nosed Albatrosses. Adults were free of ticks and Rowan (1951) considered them harmless. A single specimen of the flea *Parapsyllus longicornis* is believed to have come from a Western Yellow-nosed Albatross at Tristan da Cunha (G.M. Dunnet 1964).

The Tasmanian Shy Albatross on the cliffs of Mewstone apparently have no serious infestation, but climatic conditions on Albatross Island are evidently more favourable for ectoparasites; nests have been described as 'heaving with fleas and ticks'. The flea *Parapsyllus australiacus* is believed to be the vector of an avian pox virus that causes liver and kidney failure (Gales 1993). The ticks *Ixodes eudyptidis* (Green 1974) are seasonally variable (N. Brothers pers. comm.) and high chick mortality has been directly attributed to infestation (Johnstone *et al.* 1975). Ticks have also been found on Northern Buller's Albatrosses at the Chatham Islands (Robertson & Sawyer 1994) and on the faces of well grown chicks of Southern Buller's Albatrosses at the Solander Islands.

Terrestrial leeches *Ornithobdella* were discovered in two nests of the Southern Buller's Albatross on the Solander Islands (Cooper *et al.* 1986). Parasitic nematodes *Anisakis* have been regurgitated by Tasmanian Shy Albatrosses at Albatross Island and Southern Buller's Albatrosses at the Snares Islands (Green 1974, West & Imber 1986). The Tasmanian Shy Albatross also has the spiruoid worm *Tetramere.* They are are ingested with amphipod intermediate hosts and infections are apparently symptomless, perhaps because the worms become encased in nodules in the proventriculus (N. Brothers pers. comm.).

No serious disease was noticed among Black-browed or Grey-headed Albatrosses at South Georgia; but in the Falklands, blisters and warts on the feet and faces of Black-browed Albatrosses indicated pox-like virus infections. These lesions developed secondary bacterial infections and caused the death of some chicks (R.W. Woods pers. comm.). A few Black-browed and Grey-headed Albatrosses at South Georgia have parasitic Haematozoa *Heptozoon albatrossi* in their blood, associated with mysterious inclusion bodies, perhaps of viral or rickettsial origin (Peirce & Prince 1980).

A disease of unknown aetiology is endemic among chicks in the large colonies of the Eastern Yellow-nosed Albatross at the Ile Amsterdam. In some years it becomes epidemic with mortalities as high as 95% (Croxall & Gales 1997).

Syndromes associated with deficiency of melanin pigment are rare in most mollymawks. Over three decades of observations at South Georgia, one white Black-browed Albatross chick has been reported. Among Tasmanian Shy Albatrosses at Albatross Island about five white chicks are reared each season; so far, none have returned to the island. As many as 14 adults with deficient pigmentation of the back and upper wings have been recorded and varying eye and bill colours also occur (N. Brothers pers. comm.). An isabelline shy albatross has been seen in the Tasman Sea (Barton 1979).

PREDATORS

Sheathbills and skuas habitually scavenge in mollymawk colonies. Deserted eggs are usually not abundant and healthy chicks on tall nests can defend themselves, but squid, fish and krill are sometimes spilt during the chick-feeding period and skuas are known to harass feeding mollymawks and cause spillage. Feeding mêlées by skuas disturb nearby chicks, but normally skuas and sheathbills at a distance are tolerated. Giant petrels seldom enter mollymawk colonies at South Georgia and when they do, seem more intent upon getting out than molesting chicks. Black-browed and Grey-headed chicks throughout colonies at South Georgia always pay close attention to them, adopting synchronised defensive responses. At the Snares, Northern Giant Petrels have not been seen killing chicks of the Southern Buller's Albatross, but they were present in colonies at about the time that chicks were killed (Sagar & Warham 1997).

Raptors rarely occur on albatross islands, but Wedge-tailed or White-breasted Sea Eagles evidently reach Albatross Island from the adjacent mainland of Tasmania and occasional kill Tasmanian Shy Albatrosses. At the Falkland Islands considerable, numbers of Striated Caracaras nest in tussock grass alongside large mixed colonies of the Black-browed Albatross and Rockhopper Penguin, but most of their activity, and that of skuas in the same colonies, is probably directed towards the eggs and chicks of penguins.

At the Ile Amsterdam, unexplained losses of Eastern Yellow-nosed Albatross chicks may have been caused by feral cats or rats, whose burrows have been found near colonies (Segonzac 1972).

On Auckland Island in 1972–73, C.N. Challies (1975) visited colonies of the Auckland Shy Albatross. There were tracks of feral pigs in a part of one colony, but no signs of predation. Ten years later, only colonies inaccessible to pigs remained (Croxall & Gales 1997).

Gough Island Moorhens were introduced to Tristan da Cunha Island in 1956 and have proliferated, evidently re-occupying a niche formerly occupied by a similar extinct indigenous moorhen. There have been rumours that these moorhens have been preying upon eggs and chicks of the Western Yellow-nosed Albatross (M.W. Holdgate pers. comm.).

Fledgling mollymawks on the sea after their first flight are vulnerable. They may be taken by sea lions, furseals and leopard seals, giant petrels and skuas (Horning & Horning 1974). Even after they have flown thousands of miles and become competent in the air, juveniles have been attacked; one Black-browed Albatross, sleeping on the sea off South Australia was caught unawares, drowned and eaten by a giant petrel (Cox 1978). Awake and aware, yellow-nosed albatrosses in the western Tasman Sea can escape from mako sharks (Hunter 1997). One caught on a shark line off Wollongong bit a shark on the nose, causing the fish to swim off, before freeing itself.

6

SOOTIES[1]

'At 6 o'Clock, having but little Wind, we brought to among some loose Ice, hoisted out the boats and took up as much as filled all our empty Casks and compleated our Water to 40 Tons, the *Adventure* at the same time filled all her Empty Casks; while this was doing Mr Forster shott an Albatross whose plumage was of a Dark grey Colour, its head, uper sides of the Wings rather inclining to black with white Eye brows, we first saw these Birds about the time of our first falling in with these Ice Islands and they have accompanied us ever sence. Some of the Seamen call them Quaker Birds, from their grave Colour...'

Captain James Cook on board HMS *Resolution*,
Lat. 64°12'S. Long. 38°14E., 12 January 1773 (Beaglehole 1961)

The Sooty Albatross and Light-mantled Sooty Albatross are small brown albatrosses with conspicuous white crescents behind their eyes. Their flight has long been admired and their haunting calls, echoing around the cliffs of sea-washed coves, are the distinctive signature of subantarctic islands (p.371). The sealers' onomatopoeic vernacular, *Pee-oo* or *Pee-ah* has survived only among the islanders of Tristan da Cunha. An Australian initiative to shorten the name Light-mantled Sooty Albatross to Light-mantled Albatross is not common, and a translation of the French *L'Albatros fuligineux à dos sombre* has not replaced the name Sooty Albatross (Bourne & Casement 1993, 1996).

There are other albatrosses with dark plumage, but a world of difference separates the two sooties from all others. Dark plumage remains an enigma in all albatrosses, but in the two sooties there is an impression of

	Latitude (°S)	Sooty Albatross (pairs)	Light-mantled Sooty Albatross (pairs)
Tristan da Cunha Islands	37	3 000	
Ile Amsterdam	38	100	
Ile St Paul	39	20	
Gough Island	40	5 000	
Ile Crozet	46	3 000	2 000
Prince Edward Islands	47	5 000	200
Iles Kerguelen	49	10	4 000
Antipodes Islands	50		1 000
Auckland Islands	51		5 000
Campbell Island	53		2 000
Heard Island	53		300
South Georgia	54		6 000
Macquarie Island	55		1 000

Table 6.1 The approximate numbers of Sooty and Light-mantled Sooty Albatrosses breeding in a single season and rounded to orders of magnitude (Appendix 15).

[1] In the text Sooty Albatross written with capitals refers only to *Phoebetria fusca*, but in quotations from earlier literature it may refer to this and/or *P. palpebrata*. Without capitals 'sooty albatrosses' or 'sooties' are collective terms.

separation of habit and habitat, both at sea and on the breeding grounds. The wedge-shaped tails of these exquisite fliers may contribute more to other attributes, such as diving.

The two sooties have overlapping geographic distributions with the Light-mantled Sooty Albatross having a circumpolar distribution over cold Antarctic waters while the Sooty Albatross is characteristic of lower, less severe latitudes. These distributions are by no means mutually exclusive and sympatry is well established at some islands (Table 6.1). Both species are biennial breeders with very low productivities.

Less is known about sooties than other albatrosses. The dangers of working on steep cliffs[2], coupled with the small samples that scattered nests allow, has contributed to them being given low priority in the field.

DESCRIPTION

Sooty and Light-mantled Sooty Albatrosses weigh slightly less than most mollymawks and have thinner wings with spans of just over two metres. The central rectrices of the unique wedge-shaped tails are not only long, but very stiff. The heads appear stubby with smaller bills, but the pointed wings and tails contribute to a slender appearance, especially during flight. Males are slightly larger than females.

At both the Prince Edward Islands and Iles Crozet, the Sooty Albatross weighs less and has shorter wings but longer bills than the sympatric Light-mantled Sooty Albatross (Appendix 3).

The Sooty Albatross (Plate 16) is dark brown about the head. The back, mantle and belly are paler, but the contrast is slight and at sea the difference is often apparent only in good light. The primaries and tail feathers are dark brown, against which the whitish shafts (rachis) stand out. The concealed contour feathers are paler. Worn dark feathers are apparent as a lighter brown plumage and, with abrasion of exposed webs, the underlying paler webs are revealed, sometimes contributing to a mottled appearance.

To Harrison Matthews (1951), the Light-mantled Sooty Albatross was 'the Siamese cat of the albatross world'. As its name suggests, much of the nape, mantle, back, scapulars and rump is pale ashy-grey (Plate 19). In some lighting and/or in some stages of wear, the bird appears silvery-grey, an effect that may be due to the open (pennaceous) fringes of these feathers or to abrasion. Towards the head, tail and along the wings, the plumage becomes darker grey-brown merging to dark brown with contrasting whitish shafts to the primaries. Concealed parts of contour feathers are paler. The underparts are light greyish-brown.

In both species the conspicuous post-orbital crescents are formed by very short, white feathers immediately behind the eyes. They have characteristic shapes in each species (Berruti 1979b) (Fig. 6.1) and appear to be the

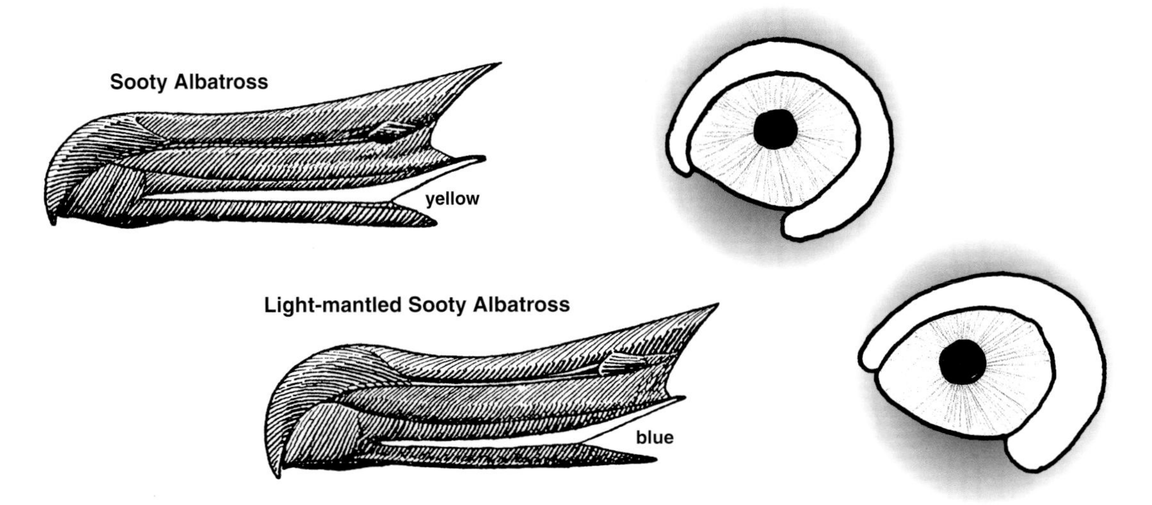

Fig. 6.1 Bill profiles and post-orbital crescents of the two sooty albatross species (Murphy 1936, Berruti 1979b).

[2] Roger Barker died after falling from a cliff on Macquarie Island (Kerry & Garland 1984).

remnants of the chick's white face-mask (Plate 17) (Sorensen 1950b). In the two species, the bills are black, but according to Murphy (1936), have slightly different profiles (Fig. 6.1), but Willis & Oniki (1993) have pointed out that the viewing angle is critical. Each ramicorn of the lower mandible is split on the outside by a longitudinal groove, called the sulcus, which contains a strip of coloured skin, continuous with the lower gape-stripe. In the adult Sooty Albatross it is cream or pale yellow and in the adult Light-mantled Sooty Albatross this skin is pale blue or violet[3]. The gape-stripe is of the same colour. In juveniles of both species, the skin of the sulcus and gape stripe is dark grey and gradually attains the characteristic colour. The legs and feet of both species vary from greyish-brown flesh, with a mauvish tinge in juveniles and a pinkish blue or lilac tinge in adults (Elliott 1957).

When they first go to sea, the plumage of fledglings is in pristine condition that is never apparent in the adults, where abrasion dilutes colours. During their juvenile years, Sooty Albatrosses develop a contrasting buff to grey or whitish collar, which extends down the nape onto the mantle. There seems to be some dispute as to whether the shafts of dark feathers at this time are dark or pale (Berruti 1979b, Jouventin & Weimerskirch 1984a, Harrison 1983). The post-orbital crescents of juveniles are at first thin and sometimes greyish white. By the time they return to the breeding islands, subadults of both species are indistinguishable from adults.

In worn or bleached plumage, some Sooty Albatrosses may resemble Light-mantled Sooties in the shade of grey colour on the nape and mantle, although the rump and underparts are always darker. In such condition the two species are indistinguishable at sea (Harrison 1983) and there have been difficulties even with specimens in the hand (Willis & Oniki 1993).

HISTORY

On 1 February 1769 when *Endeavour* was leaving the Drake Passage and sailing east into the South Pacific, Joseph Banks shot: '*Diomedea antarctica* ... the Black billd Albatross' (Beaglehole 1962). Two days later he obtained a 'Lesser black-billed Albatross'. Daniel Solander's description (Diment & Wheeler 1984) and Sydney Parkinson's drawing and note (Salvin 1876, Wheeler 1986), indicate Light-mantled Sooty Albatrosses, but they were never published at the time and Banks's name was overtaken by others.

The Forsters saw little if any of the Banks and Solander material before they sailed on Captain Cook's second expedition (1772–75) (p.73). At sea, and from a distance, they did not at first recognise these brown birds as albatrosses and used the sailors' name 'quakerbird' (p.112). They were first encountered on 5 December 1772 when the *Resolution* was south of the Cape and appeared in J.R. Forster's journal six times before he examined one in the hand and recognised it as an albatross. He named the bird that he shot amongst the ice, *Diomedea palpebrata*[4] for the white crescents about the eyes; it was a Light-mantled Sooty Albatross. Forster continued to enter 'Quackerbirds' in his journal until 28 January 1773, when he started writing 'sooty Albatross' (Hoare 1982). In 1775, at the end of the voyage, Forster had an acrimonious disagreement with the Admiralty[5], after which there was no further communication or co-operation between their Lordships and the learned but cantankerous doctor. It was ten years before the name of the new albatross appeared in his Mémoire sur les Albatros (1785), but as it was published in the mathematics and physics journal of the French Academy of Sciences, few naturalists saw it for over a hundred years.

John Latham (1785) read the two published accounts of Cook's second voyage and was thus able to include a 'Sooty Albatross' in his *General Synopsis of Birds*. J.F. Gmelin, unaware of Forster's 1785 *Mémoire...*, translated Latham's name literally as *Diomedea fuliginosa* for the 1789 edition of Linnaeus's *Systema Naturae*. This name became common usage long before anyone noticed Forster's name in the posthumous *Descriptiones Animalium* (Lichtenstein 1844).

Reichenbach (1850) also missed Forster's name when he created the new genus *Phoebetria*[6]. Elliott Coues (1866) disliked splitting *Diomedea*, but he nevertheless retained *Phoebetria*. There was still just one known species, *Phoebetria fuliginosa*.

[3] In dried museum specimens the yellow pigment of the Sooty Albatross remains distinct, but the blue of the Light-mantled Sooty Albatross disappears, leaving the sulcus brown and indistinguishable from the rest of the bill.
[4] Latin: *palpebra* – eyelid.
[5] The dispute concerned authorship of the official account of the voyage and led to the Forsters publishing their own account privately. It appeared in 1777, just six weeks before Cook's own account.
[6] Greek: (*phoibetria*) – a prophetess or soothsayer; (*phoebetron*) – an object of terror (Jobling 1991) (p.236).

About this time a few naturalists began to notice variations among the sooty albatrosses they came across at sea. F.W. Hutton (1867), with experiences of seven voyages between Europe and New Zealand, wrote:

> 'There is a very distinct variety of this bird, which from its resemblance to a Hooded Crow I have called var. *cornicoides*...'

By the turn of the century, other Antarctic expeditions included naturalists who adopted Hutton's varieties. Eagle Clarke (1913) of the *Scotia* expedition (1902–04) was convinced that there were two species: *Phoebetria cornicoides* in the Weddell and Scotia Seas, and *P. fuliginosa* in lower latitudes of the South Atlantic.

In 1902[7], Forster's name *palpebrata* at last came into circulation and Gregory Mathews (1912–13) undertook a revision of the *Phoebetria* albatrosses. Unaware of Banks' name, he claimed priority for Forster's *Diomedea palpebrata*; so the Light-mantled Sooty Albatross became *Phoebetria palpebrata* and *cornicoides* was discarded. Since Gmelin's redundant *D. fuliginosa* had been a synonym for the Light-mantled Sooty Albatross, it could not be used for the species of lower latitudes as proposed by the *Scotia* naturalists.

The Sooty Albatross was described and named from a specimen said to have been taken in the Mozambique Channel, but Hilsenberg's 1822 name *Diomedea fusca* went unnoticed and in 1839, John James Audubon independently gave the same name to another specimen. J.K. Townsend claimed to have collected this specimen in latitude 46°N, off the bar of the Columbia River, Oregon, USA.[8] Mathews (1912–13) had discovered Hilsenberg's name and took the opportunity to reinstate it as *Phoebetria fusca* with several subspecies.

In 1913, Robert Cushman Murphy brought back to the American Museum of Natural History the first good series of Light-mantled Sooty Albatross specimens from a single breeding ground, South Georgia. They became the foundation for another revision of the genus (Nichols & Murphy 1914), but in *Oceanic Birds of South America* (1936), Murphy discarded all subspecies.

Six years records of the Light-mantled Sooty Albatross at Campbell Island in the 1940s, indicated that individuals were not breeding every year (Sorensen 1950b), and in 1963–64 similar behaviour was evident at South Georgia. Biennial breeding was later confirmed in the Light-mantled Sooty Albatross at Macquarie Island (Kerry & Garland 1984) and in the Sooty Albatross at the Ile Amsterdam (Jouventin & Weimerskirch 1984a).

OCEANIC DISTRIBUTION

The Light-mantled Sooty Albatross has a widespread distribution in high southern latitudes, as far as 77° 50'S in the Ross Sea (Siple & Lindsey 1937). Individuals are frequently seen near the edge of the pack-ice and it is common from 40°S to 60°S (Johnstone & Kerry 1976, Thurston 1982, Weimerskirch *et al.* 1986). It has been reported as far north as 33°S (Fig. 6.2), but is rarely found off the coast of South Africa (Cooper 1974).

The Sooty Albatross occurs in warmer latitudes (Fig. 6.3), and is found north of the Subtropical Front to 25°S in the South Atlantic (Tickell & Woods 1972); one was seen near St. Helena (16°S) (Rowlands *et al.* 1998). In the Indian Ocean, it has been recorded as far north as 20°S (Weimerskirch *et al.* 1986). It is rare off the coast of South Africa, but more frequent than the Light-mantled Sooty (Harrison *et al.* 1997).

The Sooty Albatross makes forays eastwards, as far as the Tasman Sea and New Zealand. The species has been seen off South Georgia (Prince & Croxall 1996), near the Falkland Islands (Woods 1988), at almost 62°S 90°W in the South Pacific (Holgersen 1957) and at 65°S 38°E near the Antarctic continent (Weimerskirch *et al.* 1986). Some identifications might be questioned, but Martin Routh (1949) argued strongly that both species visit the ice-edge and Holger Holgersen had no doubt about his identification:

> 'Two birds came close to the ship, circled around us, and settled several times on the sea...they were adult Sooty Albatrosses. I got very good views of them...I could hardly believe my own eyes, finding the species so far east and south, but every possible doubt was swept aside when, only three

[7] C.D. Sherborn found *Mémoire sur les Albatross* when compiling his *Index Animalium.*

[8] J.T. Nichols and R.C. Murphy (1914) identified this specimen as a Light-mantled Sooty Albatross. W. Stone (1930) has argued that it was collected off South America and inadequately labelled when it was sent to Audubon (1839) with specimens of Black-footed Albatrosses.

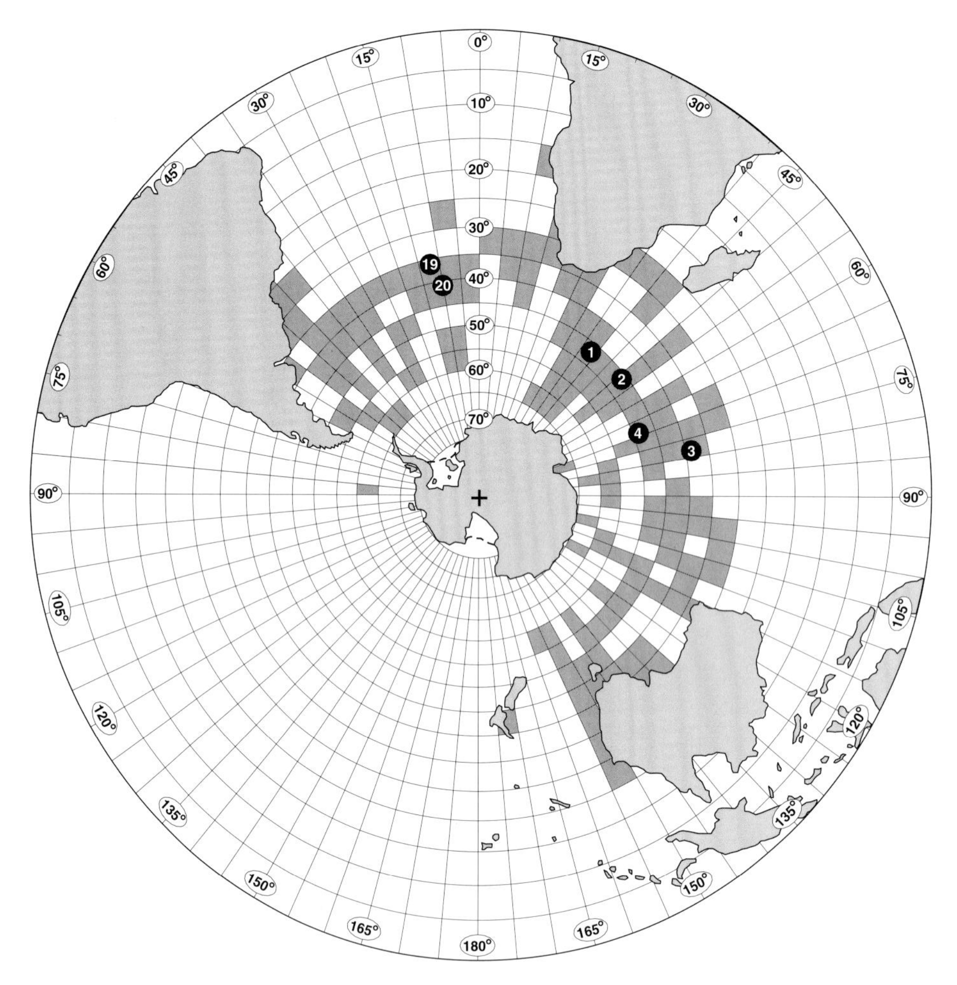

Fig. 6.2 The Sooty Albatross oceanic distribution (shaded) and breeding sites (numbered spots): Prince Edward Islands (1), Iles Crozet (2), Ile Amsterdam and Ile St Paul (3), Iles Kerguelen (4), Tristan da Cunha Islands (19) and Gough Island (20) (Tickell 1993).

> hours later and whilst we were still at work at the same station, one immature Light-mantled Sooty Albatross, and a short time afterwards also an adult bird of the same species, turned up for comparison. The difference between the two birds watched at so short a range was striking...'

Light-mantled Sooty Albatrosses are regularly washed up dead on New Zealand beaches. Numbers increase from April to July and most of them occur in the extreme northwest, as if they had approached North Island from the Tasman Sea. Powlesland (1985) has suggested that they may have been exhausted juveniles at the end of their first migration from distant breeding grounds. One fledgling, ringed in its nest at the Iles Crozet on 15 April 1970, was recovered dead in New Zealand on 7 July 1970 (Barrat *et al.* 1973).

Both the Sooty and Light-mantled Sooty Albatrosses breeding at Marion Island forage to the south of the island, but evidently are separated on either side of the Antarctic Polar Front (50°S), with the Sooty Albatross to the north and the Light-mantled Sooty to the south (Berruti 1979a).

South of the Indian Ocean, from longitude 50°E to 90°E, both species are common between latitudes 46°S and 56°S with a few Sooty and many Light-mantled Sooty Albatrosses south of 56°S. Both species are farther north in September than in February. Sooty Albatrosses are found north to 33°S in September and 41°S in February, while Light-mantled Sooties reach only 42°S and 46°S respectively (Jouventin *et al.* 1981). Although both species evidently share the middle latitudes of their joint distribution at sea, juveniles reveal a distinct polarity,

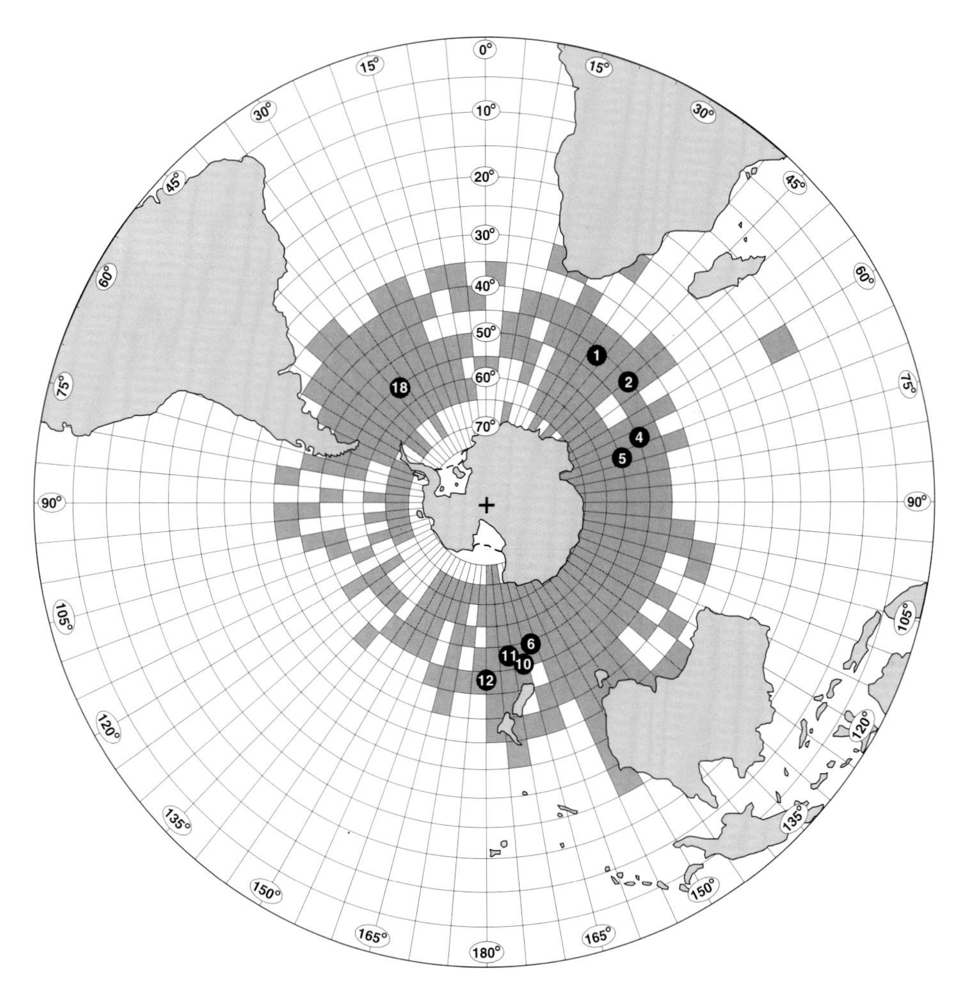

Fig. 6.3 The Light-mantled Sooty Albatross oceanic distribution (shaded) and breeding islands (numbered spots): Prince Edward Islands (1), Iles Crozet (2), Iles Kerguelen (4), Heard and McDonald Islands (5), Macquarie Island (6), Auckland Islands (10), Campbell Island (11), Antipodes Islands (12) and South Georgia (18) (Tickell 1993).

with young Sooty Albatrosses going north of their breeding islands into subtropical seas and young Light-mantled Sooty Albatrosses moving far south into cold polar waters, not far from the Antarctic continent (Stahl *et al.* 1985).

Sooty Albatrosses from the Iles Crozet forage over an area of about three million km^2, and have been seen in numbers up to 1,200 km from the islands, extending from about latitude 42°S to 57°S. Jouventin & Weimerskirch (1984a) have proposed the following disposition (Fig. 6.4). Birds of all ages are at sea during the winter, but by the time eggs are being laid in October, almost half are coming to land. Of those at sea, 22% bred last year and 33% are juveniles. As the season progresses, adults that lose eggs or young chicks return to sea; young subadults are still arriving, but older subadults, that have formed pair-bonds, are leaving; overall there are fewer ashore. By February, only about 17% of birds are associated with the breeding ground.

The Sooty Albatross occurs off Australia, but only two have been seen around New Zealand (Scofield 1994, C. Jowett pers. comm.), one to the west of North Island and the other over the continental shelf near the Bounty Islands. There are no records from the south Pacific Ocean.

Five Light-mantled Sooty Albatrosses tracked by satellite when off duty from incubation at Macquarie Island in November–December 1992, foraged south of the Polar Front, in Antarctic pelagic waters and at an average distance of 1,721 km from their nests. Two birds made circuits of 6,463 km and 6,975 km in ten and fifteen days respectively;

one flew directly to the foraging area and returned along the same route (Fig. 6.5) (Weimerskirch & Robertson 1994).

On 17 July 1994, a Light-mantled Sooty Albatross was seen and photographed on the Cordell Banks, off Marin County, California (Morlan 1994). The undoubted presence of one of the most southerly albatrosses off the coast of North America defies interpretation.

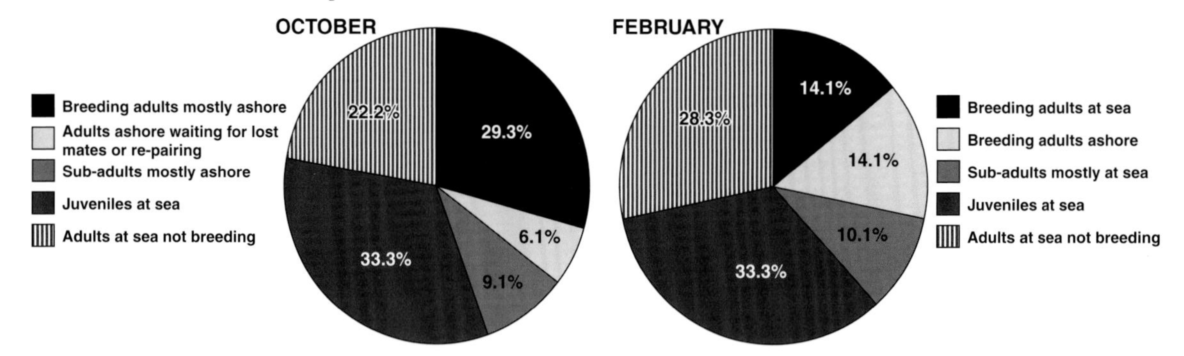

Fig. 6.4 The proportions (percentage) of Sooty Albatrosses in different activities, at sea and on the Ile de la Possession, Iles Crozet, in spring (October) and late summer (February) (Jouventin & Weimerskirch 1984a).

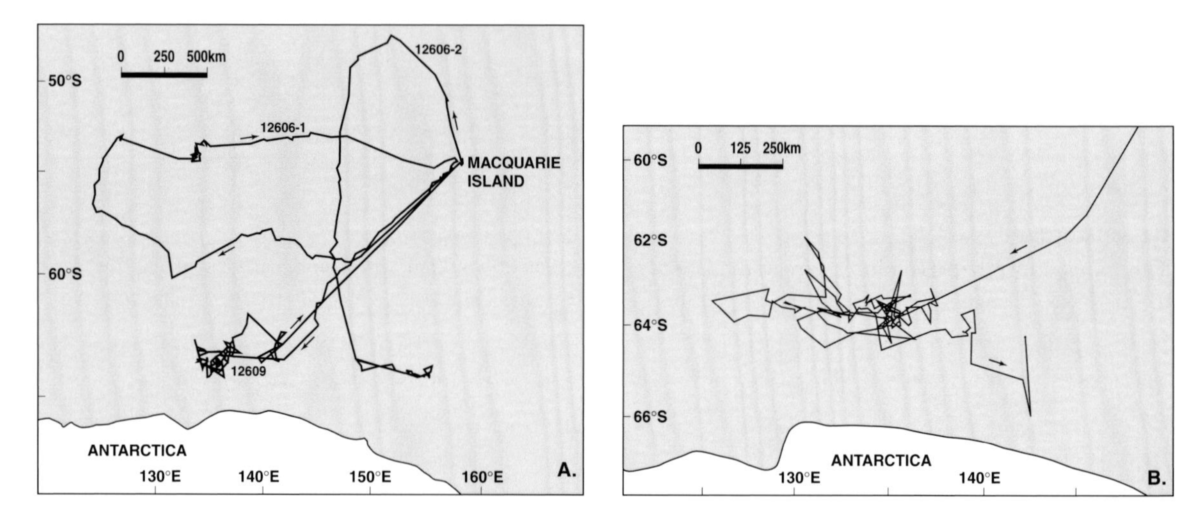

Fig. 6.5 Tracks of three Light-mantled Sooty Albatrosses off-duty from incubating at Macquarie Island November– December 1992. 'Commuting' and 'foraging' flights are apparent in track 12609 (A) and detail of foraging in B (Weimerskirch & Robertson 1994).

BREEDING ISLANDS

Prince Edward Islands

Sooty and Light-mantled Sooty Albatrosses were discovered breeding in 1951 (Rand 1954). Colonies of the Sooty Albatross are sited around the whole of Marion Island, usually on the smoother, grey larvas of the sheltered eastern and northern coast. The Light-mantled Sooty Albatross also nests on the coast, but some are inland (Berruti 1977). On Prince Edward Island, the Sooty Albatross is restricted to three areas of the northern cliffs, but the Light-mantled Sooty Albatross may be more widely distributed (Grindley 1981) (Fig. 6.6).

Iles Crozet

Sooty and Light-mantled Sooty Albatrosses breed on all five of the Iles Crozet (Jouventin *et al.* 1984). Both species nest in the same areas, but colonies are rarely mixed. The relative distribution of the two species is best known on Ile de la Possession, where they have been mapped in detail. Most are on the coast, but some Light-mantled Sooty Albatrosses breed on inland cliffs. On the Ile de l'Est, the Sooty Albatross appears to prefer the leeward northeast

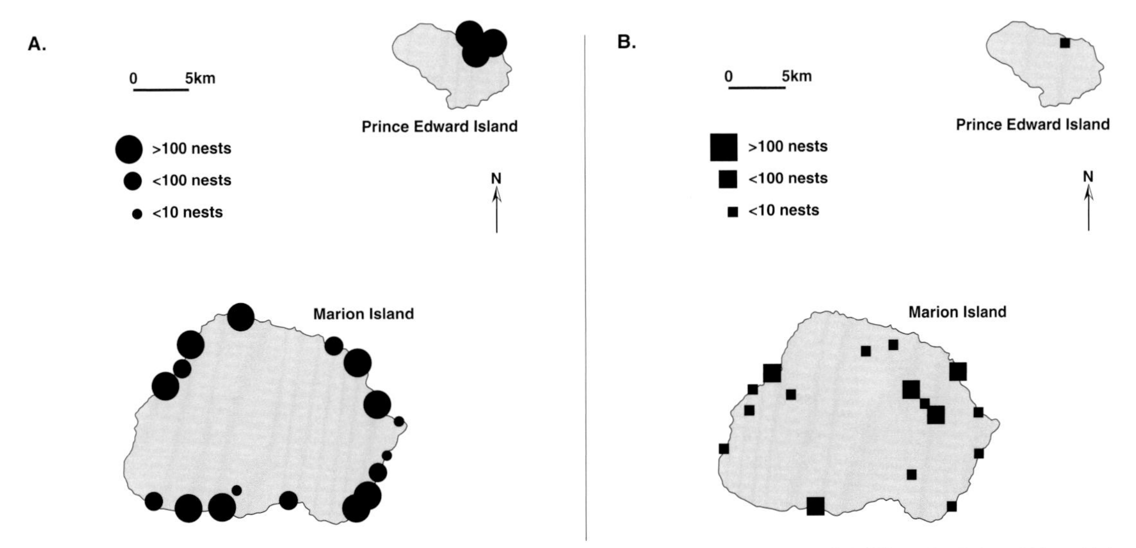

Fig. 6.6 Nesting areas of the Sooty Albatross (●) (A) and the Light-mantled Sooty Albatross (■) (B) on the Prince Edward Islands (Berruti 1977, Grindley 1981).

side of the island from the Baie du Nord to the Anse des Grandes Orgues. Light-mantled Sooties are also present, but tend to be farther from the sea, in valleys such as the Vallée Désolée (Despin *et al.* 1972). At the Ile aux Cochons, both species also prefer the northeast, in the region of the Cinq Géants and on cliffs to the east of Pointe Sud; very few nest on the west coast (Derenne *et al.* 1976). Likewise on the Ile des Pingouins and Iles des Apôtres, the Sooty Albatross appears to prefer the east coasts.

Ile Amsterdam

The main Sooty Albatross colonies are in the Grandes Ravines at 300–400 m, with some scattered along the upper d'Entrecasteaux cliffs between 400 m and 600 m and a few higher on Mont du Fernand (Segonzac 1972).

Ile Saint Paul

The Sooty Albatross breeds on the cliffs of the south coast. Formerly, they also bred on the upper cliffs of the crater above the thermal springs (Segonzac 1972).

Iles Kerguelen

The Light-mantled Sooty Albatross apparently breeds on the coastal cliffs all around the islands; some nest inland at altitudes up to 400 m (Loranchet 1915) and as far as ten kilometres from the shore and 20 km from the open sea (Weimerskirch *et al.* 1989a). A few Sooty Albatrosses nest at 150 m on the southern cliff of the Cañon des Sourcils Noires. At almost 50°S and 70°E, it is the most southerly and easterly colony of this species (Pascal 1978, Roux 1987).

Heard Island

The Light-mantled Sooty Albatross breeds all around the Laurens Peninsula. Elsewhere, they nest between glaciers on the NW Cornice, on Mount Andrée, and around the coast above Cape Gazert, Long Beach, Cape Lambert , Cape Lockyer, Gilcrist Beach and Rogers Head (Downes *et al.* 1959).

MacDonald Islands

The Light-mantled Sooty Albatross has been reported breeding on these islands, but no location was given (Johnstone 1980).

Macquarie Island

Nests of the Light-mantled Sooty Albatross are scattered along steep hills and sea cliffs around the whole 85 km of coastline. Field studies, started in 1951, have been located on the northeast coast in Gadget Gully, First Gully and on Wireless Hill between the ANARE base and North Head (Kerry & Colback 1972, Kerry & Garland 1984).

	Return	Laying	Hatching	Depart
Tristan da Cunha Islands	early Sept.	2–4 Oct.	mid-Dec.	16–20 May
Gough Island	late Aug.	late Sept.	mid-Dec.	mid-May
Prince Edward Islands	22 Aug.	8–11 Oct.	mid-Dec.	late May
Iles Crozet	17 Aug.	7 Oct.	18 Dec.	31 May
Ile Amsterdam			mid-Dec.	

Table 6.2 The breeding regime of the Sooty Albatross at different islands (Elliott 1957, Segonzac 1972, Berruti 1977, Williams & Imber 1982, Jouventin & Weimerskirch 1984, Richardson 1984).

Auckland Islands

The Light-mantled Sooty Albatross breeds along the great cliffs of the west coast (Auckland Island) and south coast (Adams Island). Elsewhere, occasional nests have been found on the north side of Carnley Harbour (Oliver 1955), in Waterfall Inlet (Reischek 1888b) and at the head of Smith Harbour (Bell 1975).

Campbell Island

The Light-mantled Sooty Albatross nests on steep sea cliffs all around Campbell Island and offshore on Jaquemart, Monowai, Wasp, Hook Keys, Dent, Jeanette Marie, and two unnamed islands. They nest along both sides of the outer Perseverance Harbour and a few breed inland on Mt. Lyall, Mt. Honey, Beeman Hill and above North East Harbour. Some nests are only 6 m above the shore, while those inland may be at an altitude of several hundred metres (Westerskov 1960, Robertson 1980).

Antipodes Islands

The Light-mantled Sooty Albatross breed on coastal cliffs of both Antipodes and Bollons Islands. They are particularly numerous around Ringdove and Stack Bays and near the creek opposite Ord Lees Islet (Warham & Bell 1979).

South Georgia

The Light-mantled Sooty Albatross nests on sea cliffs around much of mainland South Georgia and offshore on Cooper, Pickersgill, Annenkov, Jomfruene, Bird and Willis Islands. Glaciers, moraine and sheer rock walls without vegetation are avoided. Some pairs nest in sheltered fjords up to 13 km from the open sea (Thomas *et al.* 1983).

	Return	Laying	Hatching	Depart
Prince Edward Islands	4–9 Oct.	late Oct.	early Jan.	early June
Iles Crozet	9–21 Sept.	20 Oct.–13 Nov.	25 Dec.–7 Jan.	18 May–26 June
Iles Kerguelen		late Oct.	late Dec.	early June
Heard Island	1–4 Oct.	24 Oct.–early Nov.	early Jan.	late May
Macquarie Island	early Oct.	20 Oct.–5 Nov.	early Jan.	13–18 May
Campbell Island	4–9 Oct.	31 Oct.–6 Nov.	2–11 Jan.	21–22 May
South Georgia	early Oct.	30 Oct.–8 Nov.	31 Dec.–10 Jan.	15 May–1 June

Table 6.3 The breeding regimes of Light-mantled Sooty Albatrosses on different islands (Sorensen 1950b, Paulian 1953, Downes *et al.* 1959, Mougin 1970, Segonzac 1972, Tickell 1975, Berruti 1977, Thomas *et al.* 1983, Jouventin & Weimerskirch 1984, Kerry & Garland 1984, Weimerskirch *et al.* 1986).

Tristan da Cunha

The Sooty Albatross breeds on Tristan Island, using exposed ledges from 15 to 1,200 m. At Nightingale Island, they nest on the High Ridge, at Ned's Cave, Seahen Rock and at other scattered locations. They are also on Stoltenhoff Island and Inaccessible Island, where the largest colony is near Gony Ridge (Richardson 1984, Fraser *et al.* 1988).

Gough Island

The Sooty Albatross breeds on sea cliffs surrounding the island; Snug Harbour to Luff Point, Reef Point to Waterfall Point and Quest Bay to South Point have been named. Inland, they are on steep slopes and crags at 400–500 m. Areas mentioned include Gony Dale, South Peak, The Glen, Sophora Glen, Wild Glen, Deep Glen and Tavistock Crag. However, there is little confidence about the presence of this species over much of the island (Swales 1965, Shaughnessy & Fairall 1976, Voisin 1979, J. Cooper pers. comm., B.P. Watkins pers.comm.).

NUMBERS

Counts of Sooty and Light-mantled Sooty Albatrosses are seldom attempted even on small islands. The perceived difficulties of detecting nests from a distance and scanning long stretches of coastal cliffs, either from land or sea, deter serious attempts even by committed albatross researchers. Nevertheless, figures have been published for all breeding grounds. Counts are rarely explicit and without clear details of computations, 'estimates' are often so open-ended as to be little more than wild guesses. Questionable data have been cited repeatedly for decades and have acquired a veracity they do not deserve. Few of the figures available are convincing (Table 6.1, Appendix 15). There appear to be greater numbers of the Light-mantled Sooty Albatross than the Sooty Albatross, but the fact that there is longer total coastline available on the islands used by breeding Light-mantled Sooty Albatrosses may be more profound evidence than the counts.

At the Prince Edward Islands, it is not evident whether concurrent counts of both species on the two islands have ever been undertaken. The few published figures mostly lack details of areas and dates. In 1973, Grindley (1981) counted at least 892 pairs of the Sooty Albatross on Prince Edward Island, but saw only three Light-mantled Sooties. There was no corresponding count of Marion Island, but Berruti (1977) indicated the approximate sizes of colonies on Marion Island in 1974–75 (Fig. 6.6). The Sooty Albatross appears to be about ten times more numerous than the Light-mantled Sooty Albatross in this group (Cooper & Brown 1990).

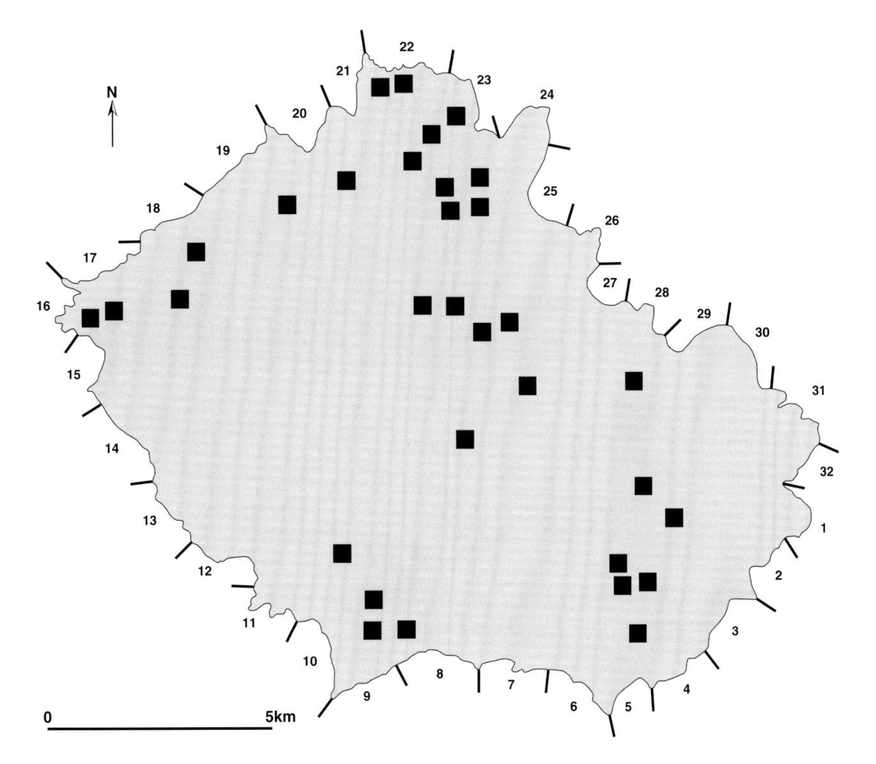

Fig. 6.7 Coastal counting zones of Sooty and Light-mantled Sooty Albatrosses at Ile de la Possession and inland nesting sites of the Light-mantled Sooty Albatross (■). See Table 6.4 for 1980–81 counts (H. Weimerskirch pers. comm.).

Both species breed at the Iles Crozet and numbers vary in relative abundance from island to island, most of which have been counted infrequently (Weimerskirch & Jouventin 1997). A count of the Ile de la Possession in 1980–81 indicated 639 pairs of the Sooty Albatross and 1,160 pairs of the Light-mantled Sooty Albatross. The Sooty Albatross nests were all on the coast and mostly concentrated in two localities, with 106–107 nests per km. By 1987, the Sooties had declined to 220 pairs (Gales 1993) and in 1995 there were still only 273 pairs (Gales 1997). In 1980–81, most of the Light-mantled Sooty Albatrosses were also on the coast, with 54 pairs inland, but they were far more scattered. Along 63 km of cliffs, 1,106 pairs varied from less than one to 68 nests per km, an overall average of 20 nests per km of occupied coast (Fig. 6.7, Table 6.4) (H. Weimerskirch pers. comm.).

Sooty Albatrosses are well established at the Ile Amsterdam and a few breed at the Ile St Paul (Segonzac 1972).

Estimates of 3,000–5,000 pairs of the Light-mantled Sooty Albatross at the Iles Kerguelen (Weimerskirch *et al.* 1989a) cannot be taken very seriously. The prospect of surveying at least 1,500 km of coast (Derenne *et al.* 1974) is still too intimidating for most researchers. On these islands the Sooty Albatross is at the southerly extreme of its breeding range. A few nests have been found at one location, but there remain hundreds of kilometres of unexplored cliffs on which others could be discovered.

The published figure of 200-500 pairs of the Light-mantled Sooty Albatross at Heard Island needs to be taken with caution. It dates from the 1950s and even then, 200 pairs was believed to be nearer the island total (Downes *et al.* 1959).

The Macquarie Island coastline is only about 85 km long and the Light-mantled Sooty Albatross has been studied there for longer than anywhere else, but counts of the whole coast were not undertaken. Sometime in the early 1970s, there were estimates of 500–700 pairs (Kerry & Colback 1972), but no details of areas or dates were given. The same figure evidently remained valid (Kerry & Garland 1984) until 1992–93, when a count of the whole island yielded 1,000–1,150 pairs (Gales 1993, 1997).

Fig 6.7		Sooty Albatross		Light-mantled Sooty Albatross	
coastal zone	length km	no.	nests/km	no.	nests/km
1	2.0	0		130	65
2	2.0	0		112	56
3	1.8	0		123	68
4	1.7	0		52	31
5	1.6	170	106	25	16
6	2.3	0		*5	
7	1.7	0		*2	
8	2.0	0		7	3
9	1.7	8	5	24	14
10	2.5	2	<1	10	4
11	2.1	0		4	2
12	2.1	8	4	34	16
13	1.8	12	7	7	4
14	2.2	0		*0	
15	2.4	0		20	8
16	1.8	31	17	37	21
17	2.3	0		34	15
18	1.9	29	15	12	6
19	1.8	0		1	<1
20	2.4	256	107	1	<1
21	2.0	73	37	52	26
22	2.3	48	21	71	31
23	2.2	0		23	10
24	1.7	0		17	10
25	2.6	2	<1	11	4
26	1.4	0		5	4
27	1.6	0		0	
28	1.3	0		0	
29	1.7	0		22	13
30	2.2	0		99	45
31	2.4	0		91	38
32	1.2	0		75	62
Coast	62.7	639		1106	
Inland		0		54	
Total		639		1160	

Table 6.4 Counts of the two sooty albatross species breeding at the Ile de la Possession, Iles Crozet in 1980-81. See Fig. 7.7 for coastal zones. *uncertain counts (H. Weimerskirch pers. comm.)

The Light-mantled Sooty Albatross has been seen in some numbers along the immense cliffs of the west and south coasts of the Auckland Islands, but only occasional nests have been mentioned along the eastern and northern coasts. No details whatsoever were given in 1972–73 to support the one-and-only estimate of 5,000 pairs along roughly 369 km of coast (Bell 1975).

In the early 1940s, Sorensen (1950b) estimated that there were no less than a thousand pairs of the Light-mantled Sooty Albatross breeding on Campbell Island. The coastline is about 116 km in length and of that, about 89 km has steep ground suitable for nesting. In 1995–96, a sample 19 km of coastline had 292 nests, about 15 nests per km. Together with nests on inland hills and offshore islands, it was estimated that the number breeding each season was at least 1,600 pairs (Moore 1996).

No-one has counted the Light-mantled Sooty Albatross at the Antipodes Islands. The figure that has crept into the literature was inferred from a statement that in 1969 they were less numerous than the Antipodes Wandering Albatross (Warham & Bell 1979). A census in 1994 revealed more than four times as many Wanderers as the 1969 estimate, so where does that leave the much cited figure of more than 1,000 pairs of the Light-mantled Sooty Albatross?

At South Georgia, the Light-mantled Sooty Albatross is scattered along 800 km of indented coastline, which is interrupted by glaciers. Sample counts over 10, 24 and 25 km of coast in different locations have given densities of four, six and seven nests per kilometre. A mean density of six nests per km indicates a total about 4,800 pairs breeding each season (Thomas *et al.* 1983).

In the Tristan da Cunha group, the Sooty Albatross was increasing on Tristan Island in the early 1950s (Elliott 1957) and was apparently still doing so in the early 1970s (Richardson 1984), so the 1972–74 estimate of 2,000–3,000 pairs is probably no longer realistic. The 2,000 pairs said to be present at Inaccessible Island in 1951 were also thought to be increasing (Elliott 1957), but an expedition in 1982–83 found only 60 pairs. By 1987, though, it was suggested that there was 'a considerably larger breeding population' (Fraser *et al.* 1988).

At Gough Island in 1974, Richardson (1984) counted 350 pairs of Sooty Albatrosses between Standoff Rock and the Admiral, which represented 5% of the coastline. If the birds were similarly scattered around the island that would indicate approximately 7,000 pairs for the coast. Richardson then added an unstated number of pairs which were 'abundant on the inland cliffs' and arrived at a total of 5,000–10,000 pairs. Since then, with even less explanation, Cooper & Ryan (1993) have published 5,000 pairs in an important management document!

BREEDING

Breeding Seasons

Sooty and Light-mantled Sooty Albatrosses take about seven months to complete a breeding cycle. After a fledgling has successfully flown, its parents have three or four winter months until the beginning of the next summer, but this is not long enough for either species. They remain at sea throughout the following summer and winter; with 14–15 months between the end of one successful breeding season and the beginning of the next.

Cliffs

Sooty and Light-mantled Sooty Albatrosses characteristically breed on much steeper ground than other albatrosses (Plate 18). Sheltered cliff sites may have microclimates that enhance the survival of chicks (Mougin 1970).

On high vertical cliffs both species need substantial ledges. At the Iles Crozet, cliffs are frequently crossed by wide terraces which provide ample room for nests and are ideal for take-offs. More often than not, nests are on broken cliffs where there are ledges, overhangs, and cracks providing all sorts of nest sites. Nests may not be on the cliff itself, but immediately below, at the

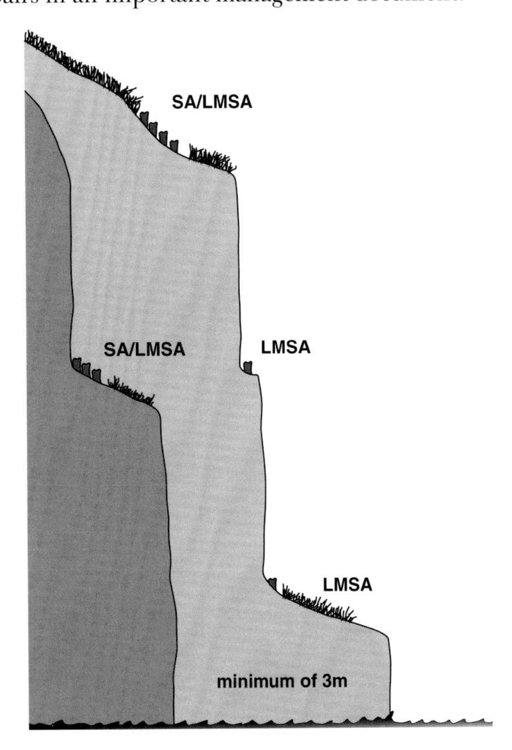

Fig. 6.8 Two cliff profiles with the positions of Sooty Albatross (SA) and Light-mantled Sooty Albatross (LMSA) nests.

foot of a rock wall or even on a slope immediately above the cliff face (Fig. 6.8). Aldo Berruti (1979a) observed that while Light-mantled Sooty Albatrosses at Marion Island always had a wall of rock or earth backing their nest, 80% of Sooty Albatrosses lacked such protection.

At Tristan da Cunha, some Sooty Albatrosses nest at an altitude of 1,200 m, but elsewhere on the same island other nests are no more than 15 m above the sea; the birds need at least 3 m vertical drop, to take flight and glide clear of the water (Berruti 1977).

Colonies and clusters

Although Sooty Albatrosses at Tristan da Cunha today nest in groups of two to five nests, there can be no doubt that formerly, before the depredations of man, some at least bred in much larger colonies. Captain Dugald Carmichael (1818) described one such colony he came across on 4 January 1817 when climbing to the summit crater of Tristan da Cunha.

> 'The black albatrosses (*D. \fuliginosa*) are at this season gregarious, building their nests close to each other. In the area of half an acre I reckoned upwards of a hundred. They are constructed of mud, raised five or six inches, and slightly depressed at the top. At the time we passed, the young birds were more than half grown and covered with a whitish[9] down. There was something extremely grotesque in the appearance of these birds standing on their respective hillocks motionless like so many statues, until we approached close to them, when they set up the strangest clattering with their beaks, and, if we touched them, squirted on us a deluge of foetid oily fluid from their stomach.'

At Marion Island, colonies of 50–60 Sooty Albatross nests still occur (Berruti 1977). Compared with typical mollymawk colonies, these are small, but we have no difficulty in thinking of them as colonies. In both species, however, most pairs nest in aggregations of several nests, which M.E. Richardson (1984) called 'clusters'.

At the Ile de la Possession, 49% of Light-mantled Sooty Albatross nests are solitary compared with only 12% among Sooty Albatrosses (Fig. 6.9). Clusters occur in both species, but are more frequent among Sooty Albatrosses. Only 10% of Light-mantled Sooty Albatrosses breed in clusters of 5–13 nests. In contrast, at South Georgia most Light-mantled Sooty Albatrosses breed in clusters of 5–15 nests (G. Thomas pers. comm.).

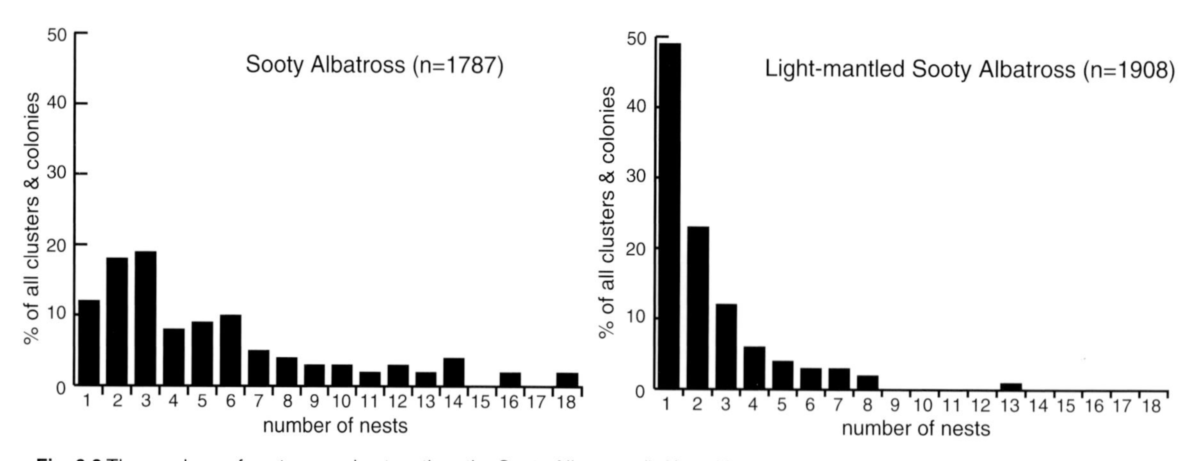

Fig. 6.9 The numbers of nests occurring together: the Sooty Albatross (left) and Light-mantled Sooty Albatross (right) (Weimerskirch *et al.* 1986).

Where the two species are sympatric, mixed clusters may occur. At the Iles Crozet, only about 3% of Sooty Albatrosses nest within Light-mantled Sooty Albatross clusters, while 7% of Light-mantled Sooty Albatrosses nest in Sooty Albatross clusters. The difference may reflect the proportionally greater number of Sooty Albatross clusters which attract Grey-headed Albatrosses (Berruti 1977, Weimerskirch *et al.* 1986).

[9] Sooty Albatross chick down is more appropriately described as grey (Marchant & Higgins 1990), but it should not be supposed that this indicates that the chicks were mollymawks. Carmichael accurately described *D. chlororhynchos* immediately after his *D.fuliginosa.*

Ninety-four percent of established Sooty Albatross pairs return each year to the same ledge or cluster at the Iles Crozet, but 74% of them use a different nest.

Nests

On most breeding grounds there is ample vegetation and nests of both species are truncated cones of plant material or pedestals of peaty soil about 20 cm high and 30–40 cm diameter. The bowl is six to eight cm deep and lined with grass pulled from surrounding plants. By the end of the season, it becomes trampled by the fledgling, but abundant guano promotes regeneration of the surrounding vegetation. At the Ile St Paul, where dry ledges have little vegetation, Sooty Albatross nests are very low, about five cm high, while at high altitude on Tristan da Cunha they may be no more than scrapes in volcanic soil or gravel (Segonzac 1972, Richardson 1984).

Pre-egg periods

At Tristan da Cunha and the Iles Crozet, Sooty Albatrosses return to the surrounding seas in August, and come ashore in September. In both species the pre-egg period is three to four weeks. Males return before females and visit nest sites, but the degree of attendance varies greatly. A few males seem to be at the nest almost all the time, but it is more usual for them to stay for only short spells until just before laying, the length of these visits is very variable. Females spend very little of the pre-egg period at the nest site (Mougin 1970, Richardson 1984, Weimerskirch *et al.* 1986). Copulation between Sooty Albatrosses at the Iles Crozet occurs on average 14 days before laying (n = 63) (Jouventin & Weimerskirch 1984a).

Eggs

There is an indication that Light-mantled Sooty Albatross eggs may be broader and heavier than those of Sooty Albatrosses (Berruti 1977), but some geographic intra-specific differences are of a similar order and there is no clear trend. Newly laid Sooty Albatross eggs from the Iles Crozet average 102 x 65 mm and 227 g (n = 74) (Appendix 4).

Laying

A high degree of synchrony is characteristic of both species. The peak of laying (2–4 October) among Sooty Albatrosses at Tristan da Cunha is about a week earlier than at the Iles Crozet (7–11 October) where sympatric Light-mantled Sooty Albatrosses are even later (23–27 October). At South Georgia, Light-mantled Sooty Albatrosses lay mainly from 31 October to 4 November. In both species laying is completed in ten to twelve days (Thomas *et al.* 1983, Weimerskirch *et al.* 1986).

Incubation

At the Iles Crozet, incubation in the Sooty Albatross (71 days) is longer than in Light-mantled Sooty Albatrosses (67 days), but at South Georgia, the Light-mantled Sooty Albatross incubates for 70 days (Table 6.5).

Male and female share incubation in seven to nine shifts. Sooty Albatrosses at the Iles Crozet share equally, even though males take an extra shift, while at the Prince Edward Islands both take the same number of shifts, but males average more time on the egg. In the Light-mantled Sooty Albatross, females sit for longer at the Iles Crozet and males for longer at South Georgia even though the latter completes the task in fewer shifts.

Incubation shifts at the Iles Crozet are 1–21 days in the Sooty Albatross and 1–29 days in the Light-mantled Sooty Albatross (Weimerskirch *et al.* 1986). In the Sooty Albatross, shift-lengths increase to an average maximum of 13 days by the third, female shift and decline progressively thereafter to about six days just before hatching; it

	Islands	mean (days)	n
Sooty Albatross	Prince Edward Islands	70±1.8	15
	Iles Crozet	71±1.5	40
Light-mantled Sooty Albatross	Iles Crozet	67±1.4	24
	South Georgia	70±1.3	22

Table 6.5 Length of incubation in the two sooty albatross species (Berruti 1977, Thomas *et al.* 1983, Weimerskirch *et al.* 1986).

is much the same at the Prince Edward Islands (Berruti 1979a). In the Light-mantled Sooty Albatross at the Iles Crozet, the fifth, female shift is about 14 days, and males are sitting for about eight days as hatching approaches. Elsewhere, they sit for much longer; an average of 24 days at the Prince Edward Islands and 20 days at South Georgia. These are the longest mean incubation shifts for any albatross.

Hatching

Variation in the dates of hatching correspond to geographic differences in dates of laying in the two species. Chicks of the Light-mantled Sooty Albatross take three to five days to break out of their egg shells (Thomas *et al.* 1983, Kerry & Garland 1984).

Chicks and fledglings

Throughout the first 19–21 days after hatching, chicks of the two species are brooded equally by both parents; shifts average two to three days, but may be up to nine days (Table 6.6). Two Light-mantled Sooty chicks at the Iles Crozet were 18 days and 21 days old when they became thermally stable (Mougin 1970). Chicks may be guarded intermittently for a few days more before being left alone (Thomas *et al.* 1983, Weimerskirch *et al.* 1986).

	Islands	mean (days)	n
Sooty Albatross	Prince Edward Islands	21±2.1	23
	Iles Crozet	21±2.4	42
Light-mantled Sooty Albatross	Iles Crozet	19±3.3	13
	South Georgia	20±2.6	–

Table 6.6 Length of brooding in the two sooty albatross species (Berruti 1977, Thomas *et al.* 1983, Weimerskirch *et al.* 1986).

There is considerable variation in the length of the fledging period. At the Iles Crozet a mean fledging time for Sooty Albatross chicks was 164±9 days, 135-178, n=28). A median time for Light-mantled Sooty Albatross chicks at the same island was 157 days, while at South Georgia and Macquarie Island chicks Light-mantled chicks flew at 139 and 149 days respectively (Thomas *et al.* 1983, Kerry & Garland 1984). Parents of both species continue to feed their young until they fledge, so adults and fledglings tend to leave their islands at about the same time. At the Iles Crozet, both species leave over a period of five weeks; the Sooty Albatross on average, at the end of May and most Light-mantled Sooties about a week later.

Sooty Albatrosses at the Iles Crozet feed their chicks once every 2.6 days (Jouventin & Weimerskirch 1984a) and those at the Prince Edward Islands every 2.4 days (Berruti 1979a). On the same islands, Light-mantled Sooty Albatrosses bring in food every 2.9 and 2.6 days respectively, and at South Georgia, at intervals of 2.1–2.6 days (Thomas 1982, Thomas *et al.* 1983).

From a mean hatching mass of 176 g, Light-mantled Sooty Albatross chicks at South Georgia have grown to about a kilogram by the end of brooding and attain a mean peak mass of about 3.4 kg, 120% of the mean adult mass, in about 80 days after hatching. Chick mass fluctuates around this level for 30 more days, reaching 140% of the mean adult mass before beginning to decrease. By the time they make their first flight, fledglings average 2.6 kg, or 91% of mean adult mass. At the Iles Crozet, Sooty and Light-mantled Sooty Albatrosses appear to have slightly different growth characteristics (Weimerskirch *et al.* 1986).

Primary feathers begin to grow shortly after the end of brooding and from 50 days to 120 days, the wings of both species grow at about 5 mm per day, 1% of the adult wing length (Berruti 1977, Thomas *et al.* 1983).

Juveniles

Many Sooty Albatross juveniles, possibly 64% of those which leave their islands, die in their first year at sea. Those that survive remain at sea from 5–14 years before returning to the Iles Crozet for the first time. Juvenile Light-mantled Sooty Albatrosses also spend many years away at sea.

Subadults

Sooty Albatrosses returning to the Iles Crozet between the ages of five and ten years, go back to the place where they were reared (n=352). The average age of first return is eight years (n= 67). The youngest birds arrive late in the season, but as they get older, subadults arrive progressively earlier each season (Jouventin & Weimerskirch 1984a).

Light-mantled Sooty Albatrosses return to Macquarie Island from the age of six years.

First breeding

Two Light-mantled Sooty Albatrosses at Macquarie Island are known to have bred at six and eight years of age (Kerry & Garland 1984), but there has been a decline in the average age of recruitment since the early 1970s. The mean age of first breeding among Sooty Albatrosses at the Iles Crozet is 12 years (9–15, n=22) (Fig. 6.10).

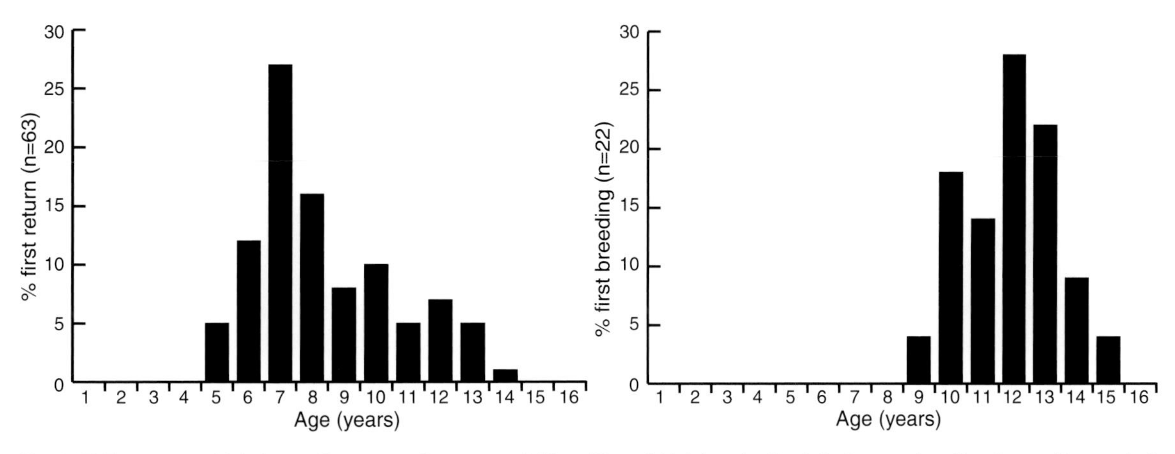

Fig. 6.10 The age at which Sooty Albatrosses first return (left) and breed (right) at the Ile de la Possession, Iles Crozet (Jouventin & Weimerskirch 1984a).

Adults

Both the Sooty and Light-mantled Sooty Albatrosses are biennial breeders. Following a successful season, Sooty Albatross pairs are away from the Iles Crozet for more than 15 months. However, pairs that fail in one season may return the following year and persistently unsuccessful pairs can produce eggs each season. One pair of Sooty Albatrosses at Nightingale Island was said to have had its chick taken each season for at least 11 years (Elliott 1957).

Sooty Albatross females at the Iles Crozet lay, on average, one egg every 1.6 years (0.6 eggs per year), and successfully rear one young every 2.8 years (0.36 fledglings per year). The Light-mantled Sooty Albatross at Macquarie Island rears only one young every three or four years, which is claimed to be the lowest for any bird species (Kerry & Garland 1984).

With few exceptions, Sooty Albatross pairs remain together until one of the partners dies. At Macquarie Island, one Light-mantled Sooty Albatross pair is known to have been together for 21 years (Kerry & Garland 1984). The annual mortality rate of Sooty Albatross adults averages 3.7% (Jouventin & Weimerskirch 1984a).

Breeding success

The mean breeding success of pairs of the Sooty and Light-mantled Sooty Albatrosses at the Iles Crozet is 43% and 52% respectively, but it varies greatly from year to year. Among Sooty Albatrosses 44% to 84% of eggs hatch and 32% to 70% of hatched chicks fledge. The overall performance evidently improves with age. The main cause of egg and chick loss is desertion by parents. It is rarely known why they leave. Sooty Albatrosses sometimes desert collectively; a whole colony of incubating or brooding birds suddenly leaving at the same time.

On narrow ledges of steep cliffs, fledglings of both species are prone to a particular danger rarely experienced by other albatrosses. Gusty winds sometimes unbalance fledglings as they are exercising their wings and cause them to fall from their ledges. Battered against cliff faces or the rocks below, they are soon prey to giant petrels, skuas and sheathbills (Jouventin & Weimerskirch 1984a).

At Macquarie Island, breeding success of the Light-mantled Sooty Albatross is locally variable, perhaps due to predation by cats, and there has been an overall decline from 60% to 70% in the early 1970s to 20% to 30% in the mid-1990s. There are also cats at Campbell Island, but they were not mentioned by Moore (1996) in the catalogue of disturbance likely to contribute to the low breeding success. In 1995–96, a sample of 60 nests at Campbell Island achieved a breeding success of less than 50%.

Longevity

Light-mantled Sooty Albatrosses at Macquarie Island include ringed birds still breeding at 32 years of age.

FOOD

The long pointed tongues of sooty albatrosses with their fleshy, backward-pointing gular spines (Forbes 1882) hint at different feeding habits from other albatrosses (p.30). Sooty and Light-mantled Sooty Albatrosses deliver meals of up to one and a half kilograms. As in mollymawks, about half is liquid (Cooper & Klages 1995).

Carrion is important; both species include more seabirds in their diet than other albatrosses. In addition to penguin skin and feathers, a variety of small petrels are eaten, some swallowed whole (Ridoux 1994, Cherel & Klages 1997).

The proportions of squid, fish and crustaceans in diets vary between the two species, and also in each species at different islands and in different years (Berruti & Harcus 1978, Thomas 1982, Cooper & Klages 1995) (Fig. 6.11, Appendix 7). Squid clearly predominates in diets at South Georgia and at the Iles Crozet, but at the Prince Edward Islands in 1989–90, when Sooty Albatrosses were feeding their chicks 42% squid and 33% fish, Light-mantled Sooties were bringing in 34% squid and 46% fish (Cooper & Klages 1995).

Because of inconsistencies in the collection and analyses of food samples, it is very difficult to make convincing generalisations about the squid and fish eaten by the two sooty albatrosses (Cherel & Klages 1997). The squid delivered to chicks are mostly about 100–300 g in mass, but the Sooty Albatross may take larger proportions of

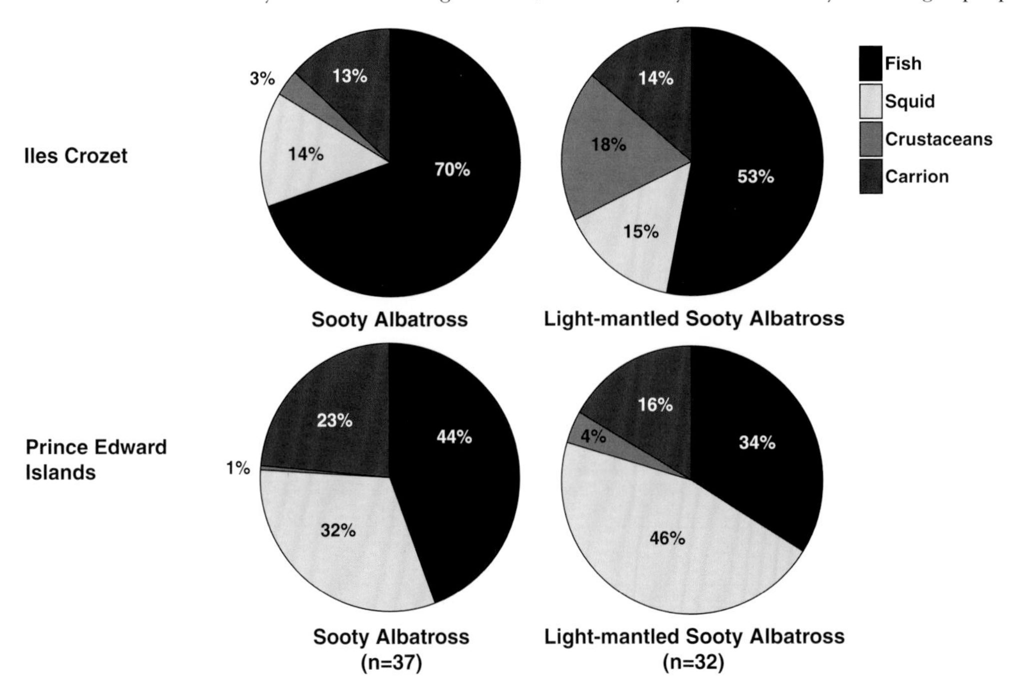

Fig. 6.11 Food delivered to chicks of the two sooty albatross species at the Iles Crozet (top) and the Prince Edward Islands (below), Feb–May 1990. Segments of the pie-diagrams indicate the percentage mass of the four major prey types in the samples (Weimerskirch *et al.* 1986, Cooper & Brown 1990).

smaller individuals than the Light-mantled Sooty. Both albatrosses also obtain fragments of much larger squid, up to five kilograms in mass, almost certainly fragments of dead or moribund spawners, perhaps scavenged from seals or whales (Berruti & Harcus 1978, Cooper & Klages 1995).

At the Prince Edward Islands and the adjacent Iles Crozet, where the two sooties are sympatric, there is a broad overlap in the species of squid and fish captured, and differences which exist are said to reflect different foraging areas; Sooty Albatrosses near the Subtropical Convergence and Light-mantled Sooties towards the Antarctic Polar Frontal Zone (APFZ).

The Sooty Albatross sometimes obtain a greater variety of squid than Light-mantled Sooty, but in 1974–75 at the Prince Edward Islands, only ten species each contributed more than one percent of those caught by either albatross species. Sooty diets at the Iles Crozet include large quantities of medium sized cranchid squids such as *Galiteuthis glacialis* and some big onychoteuthids, notably *Kondakovia longimana* and *Moroteuthis knipovitchi* (Ridoux 1994). To the west, at the Prince Edward Islands, *Galiteuthis glacialis* and *Kondakovia longimana* are also important together with *Chiroteuthis* sp. and *Histioteuthis eltaninae.*

Annual differences occur; in 1974–75 *Kondakovia longimana* contributed 48% and 64% of the squid prey of Sooty and Light-mantled Sooty respectively (Berruti & Harcus 1978), while in 1989–90 the proportions were 20% and 34% (Cooper & Klages 1995).

The Light-mantled Sooty Albatross at South Georgia in the late 1970s, fed chicks on a spectrum of squid in which 58% was the cycloteuthid *Discoteuthis* sp. and 17% the cranchid *Mesonychoteuthis hamiltoni,* but both were absent or negligible at the Prince Edward Islands and Iles Crozet (Thomas 1982, Cherel & Klages 1997).

Both species eat more pelagic crustaceans than other albatrosses, but they differ from each other in the relative quantities taken. At South Georgia in the late 1970s and at the Iles Crozet in 1982, krill amounted to 35% and 15% of Light-mantled Sooty Albatross meals (Thomas 1982, Ridoux 1994), while at the Prince Edward Islands in 1990, it was only 4% (Cooper & Klages 1995). Sooty Albatrosses at the Iles Crozet and Prince Edward Islands took negligible quantities of krill (Ridoux 1994).

Johannes van de Merwe, at Gough Island, claimed to have twice seen a Sooty Albatross break the egg of a Yellow-nosed Albatross and eat the contents, while the parent mollymawk stood by and did not react (Swales 1965). A clue to the interpretation of this bizarre story may be found in van de Merwe's character: '...depths of humour hidden beneath a placid exterior.' (Holdgate 1958).

PARASITES

Five species of chewing lice have been identified from the Light-mantled Sooty Albatross (Appendix 5), one of them, *Perineus circumfasciatus* has also been found on a Sooty Albatross (British Museum (Nat Hist) catalogue, Pilgrim & Palma 1982, Palma & Pilgrim 1987). A Light-mantled Sooty Albatross has been recorded as host of the flea *Parapsyllus magellanicus* at the Iles Kerguelen (Dunnet 1964).

PREDATORS

Giant petrels enter Sooty Albatross colonies at the Prince Edward Islands and although they have never been seen actually killing chicks, Berruti (1979a) believed they were important predators. These large birds are less agile than the sooties and do not land on many of the small ledges that both Sooty and Light-mantled Sooty Albatrosses use. On the other hand, giant petrels do patrol coastal cliffs and soon find the corpses of chicks and fledglings that have fallen from their nests. Skuas are certainly able to land on narrow ledges, but healthy chicks have quantities of oil in reserve. They can usually intimidate skuas, but anything short of fully aggressive postures will encourage skuas in their harassment; weak or starving chicks are eventually dragged or toppled from their nests.

Feral cats are potential predators at some islands, but remains of only one Sooty Albatross were found in the stomachs of 120 cats shot at Marion Island (Berruti 1977). Discharge of oil from a nestling Light-mantled Sooty Albatross at Macquarie Island was seen to hit one cat; it leapt out of the way and rolled over frantically on the grass as if trying to remove the contamination (K. Kerry pers. comm.).

7
GREAT ALBATROSSES

> 'In November, 1939 the German pocket battleship *Graf Spee* had sunk a small merchant ship off Madagascar, and then disappeared. She might have gone east into the Indian Ocean or doubled back into the South Atlantic. The ship[1] in which I was serving had been ordered to take up a strategic position south of the Cape to cover either eventuality, but the seas around remained empty. It was weary, monotonous work, and when the great white bird came sailing across the wake one evening it was a heart-lifting event.'
>
> William Jameson (1958)

The great albatrosses are the among the most spectacular sights of the Southern Ocean. The genus *Diomedea* formerly contained most albatross species, but Nunn *et al.* (1996) proposed that it be reserved exclusively for the great albatrosses. Notable differences distinguish the wandering albatrosses from the royal albatrosses, but on the high seas they are very often misidentified. Distinguishing the four different wandering albatrosses at sea is more difficult than separating the two royal albatrosses.

The seven names below, identify the terminal taxa. If you favour the biological species concept, they represent two species with seven subspecies. If you have adopted the phylogenetic approach of Robertson & Nunn (1997), they will represent seven species (Appendix 1). Here, again I avoid the question of rank.

Wandering Albatross (WA)
Gough[2] Wandering Albatross (GWA)
Amsterdam Wandering Albatross (ADWA)
Auckland Wandering Albatross (AKWA)
Antipodes Wandering Albatross (ATWA)

Northern Royal Albatross (NRA)
Southern Royal Albatross (SRA)

The longest wing span of 'an incredible number of albatrosses' measured, we are told accurately, by one sea captain over forty years, was 3.45 m (Murphy 1936). Murphy himself believed that a dead bird, with its wings pulled as tightly as possible, could not extend more than 3.5 m. Wing spans are important dimensions that are rarely obtained. They cannot be measured on folded museum specimens and are not standard taxonomic characters. A sample of 119 live, unsexed wandering albatrosses captured off Australia in the western Tasman Sea averaged 8.16 kg (5.89–11.34 kg, n=108) and had an average wing span of 3.0 m (2.72–3.23 m). The smallest weighed 5.44 kg and had a wing span of 2.56 m (Gibson & Sefton 1960). It was thought to be a 'runt', but could, perhaps, have been an Antipodes Wandering Albatross. Six female wing spans measured on Antipodes Island averaged 2.71 m (Warham & Bell 1979).

Claims have been made in favour of wanderers and royals being the biggest albatrosses of all, but shortage of wing span measurements, daily changes in the weight of individuals and a lack of comparable statistical treatment have not yet produced an unequivocal record.

Wandering albatrosses breed at twelve island groups around the Southern Ocean and are variable in size, plumage and breeding season. Four of these populations have been named sub-species/species and in this book are identified by island prefixes. The Wandering Albatross has sometimes been called the Snowy Albatross; it is a

[1] The aircraft carrier HMS *Ark Royal.*
[2] Bourne & Casement (1993) named this albatross after its main breeding ground.

misleading name as the plumage of most individuals is not completely white. Moreover, it comprises several allopatric populations, each of which could acquire specific status under a new taxonomy. In discussing them, I shall name the respective islands.

Most Southern Royal Albatrosses breed at Campbell Island with a small number on the Auckland Islands. The Northern Royal Albatross breeds at the Chatham Islands, with a very small colony on the east coast of New Zealand.

All great albatrosses take about a year to rear their young and if successful, do not breed in the following season. They are the classic biennial breeders; many pairs do indeed breed successfully every two years, but others sometimes defer breeding for longer (p.33).

The detailed features that distinguish wandering from royal albatrosses seen on their breeding grounds or as specimens in the hand, have been known since the late 1890s, but identification at sea remained a problem for over 70 years. Murphy (1936) doubted the ability of anyone to distinguish a royal albatross at sea from an old white male wandering albatross. Westerskov (1960) attempted to distinguish them by their flight, but it was not until the mid-1970s that Peter Harrison (1978) eventually gave us field-characters for separating them at sea.

DESCRIPTION

'Except for juveniles and downy young, no two Wandering Albatrosses look alike.'
Marchant & Higgins (1990)

The chicks of wandering albatrosses fledge from grey down (Plate 23) into an almost black plumage with contrasting white face masks[3] (Plates 24&25). At sea, the worn juvenile feathers become dark brown. The underwings are white with some dark-brown axilliaries just where the leading edge joins the body, thin dark-brown trailing edges (ends of secondaries) and almost black tips (ends of primaries). Patches of dark inner secondaries on the centre underwings have been seen twice; one was at sea (Cheshire & Carter 1992) and the other in the Bay of Isles, South Georgia. Underwing plumage remains substantially the same throughout life, but the dark feathers of the body, upper wings and tail gradually lose at least some pigmentation. Juvenile feathers are dark only on their exposed vanes; the concealed inner portions are white. Replacement feathers have progressively less pigment, so that white vanes may become exposed. In older birds, pigmentation persists as wavy pencilling (vermiculations) on the terminal fringes and may or may not eventually disappear altogether.

In 1929, Harrison Matthews sketched four plumage stages (Fig. 7.1) and Gibson (1967) later illustrated eight (Fig. 7.2). Gibson recognised that an infinite number of intermediate patterns were possible and with the help of friends he invented a versatile numerical system for recording any stages of wandering albatross plumage. Durno Murray (1989) described how it came about:

> '...at this time a sick Wandering Albatross appeared in Sydney Harbour. A meeting was arranged at short notice and held in the evening at Arthur Gwynn's home, attended by Doug Gibson, Allan Sefton, Clive Campion, Bill Lane, myself and the albatross. Doug explained his plan and at intervals we retired individually to the verandah to make our plumage scores, after which we compared notes and discussed our disagreements. From these Doug developed his scheme which became known as the "Gibson Code" [4]. The albatross recovered within two days from its unique experience and was last seen flying through Sydney Heads to the Tasman Sea.'

This innovation (Fig. 7.3) has been of great heuristic significance. Rapid and precise records have been amenable to statistical analysis and made possible the characterisation of sexes, age-groups and populations. It was refined for use with Amsterdam Wandering Albatrosses (Jouventin *et al.* 1989) (Fig. 7.4) and employed extensively by Battam & Smith (1993) who renamed it the Gibson Plumage Index (GPI).

The five wandering albatrosses have their own characteristic GPIs. Significant differences in climax plumage

[3] In the late 1980s, one Wandering Albatross chick at Bird Island, South Georgia fledged into white plumage just like an old adult (M. Jones pers. comm.).
[4] Tickell (1968).

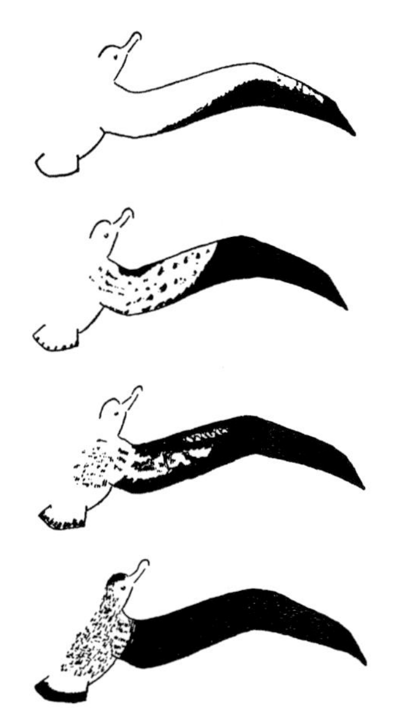

Fig. 7.1 Plumage development in the Wandering Albatross. From field-sketches made in 1925–27 by L. Harrison Matthews on the breeding grounds at South Georgia (Harrison Matthews 1929).

Fig. 7.2 Plumage development drawn by J.D. Gibson from observations of many different wandering albatrosses at sea off New South Wales, Australia (Gibson 1967).

and maturation time are also apparent between Wandering Albatrosses from the South Georgia, Prince Edward Islands and Iles Crozet populations.

When subadults begin returning to their birth places at the age of three, they have already lost considerable proportions of their feather pigmentation and males are conspicuously paler than females, but there is some overlap (Fig. 7.5). Up to about 20 years of age, South Georgia males are whiter than Crozet males (Fig. 7.6). Females never become quite as white as males (Table 7.1). Up to 10 years of age those from South Georgia and the Iles Crozet are indistinguishable; but from then on South Georgia females become whiter (Plate 20) than those from the Crozets, which retain more dark feathers on the head, nape and particularly on the back and inner wings (Prince *et al.* 1997). This stage in 25-year-old females corresponds to that of males at about nine to ten years of age (Weimerskirch *et al.* 1989b). Males breeding at seven and eight years of age had plumage significantly advanced over males of the the same age that were still subadults; but by nine years of age these difference had disappeared. Among breeding and subadult females, no difference was apparent at seven, eight and ten years of age (Tickell 1968, Weimerskirch *et al.* 1989b).

In the Antipodes Wandering Albatross (Plate 29), females breed in much darker plumage than males (mean GPI: males 8.7±1.6 SD, n=43, females 4.4±0.5, n=45) which is comparable to the condition of subadults or even juveniles of Wandering Albatrosses at South Georgia and the Iles Crozet (Robertson & Warham 1992). In the Amsterdam Wandering Albatross (Plate 30), breeding males are also dark (mean GPI 4.3, n=7) and females like juveniles (mean GPI 4.0, n=7) (Jouventin *et al.* 1989).

The bills of wandering albatrosses are narrower than those of royals and the nasal tubes are compressed laterally, pointing slightly upward (Fig. 7.7). The thin bill plates are translucent, like human finger nails, revealing the underlying dermal tissue and blood capillaries. Narrowing of the capillaries (vasoconstriction) leaves the bill whitish, while vasodilation produces various shades of pink. In most populations the cutting edges (tomia) of the latericorns are unpigmented, but in the Amsterdam Wandering Albatross they are pigmented (Plate 30) like those of royal albatrosses. The sharp hooked tip (maxillary unguis) is composed of dense keratin which scatters light and usually appears ivory-white, but in some populations pigmentation produces variable areas of dark shading. Sharp edges are maintained by the upper bill stropping against the mandibles (Warham 1996).

The unpigmented eyelids are bluish-white, turning pink with vasodilation (Tomkins 1983). Tarsi and feet are whitish-grey, with variable tinges of blue, also turning pink with vasodilation.

Variable crescents of pink-stained plumage just behind the concealed ear openings are conspicuous and common in wandering albatrosses (Plate 21). At South Georgia, they are present on approximately three-quarters of breeding adults at all times of the year, more prominent when they are at sea than when on the breeding grounds. Stains are more frequent and more prominent in males than females and constant in individuals over many months (Tickell 1980, Tomkins 1983). In museum specimens, they may still be apparent eight years after preparation (Warham & Bell 1979).

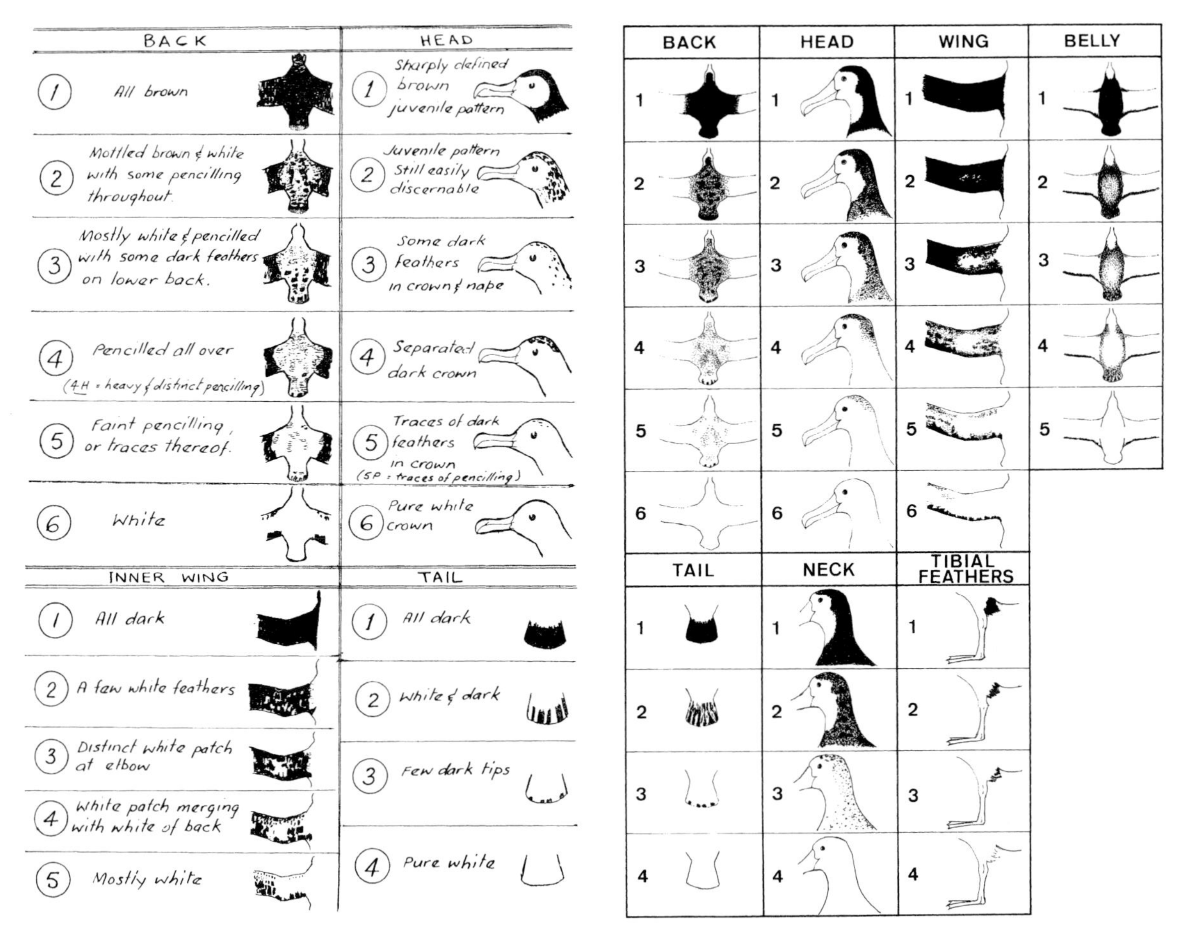

Fig. 7.3 The Gibson Plumage Index (GPI) devised by J.D. Gibson in collaboration with members of the New South Wales Albatross Study Group in the mid-1960s for describing plumage of wandering albatrosses. Four figure indices and totals vary from (4) for juveniles (1,1,1,1,) immediately after they have fledged to (21) among the oldest males (6,6,5,4) (Gibson 1967).

Fig. 7.4 Illustrated key of the expanded Gibson Plumage Index (GPI) devised by Jouventin *et al.* (1989) for the Amsterdam Wandering Albatross.

These stains are unique to wandering albatrosses; they taste salty (H. Battam pers. comm.) and ice crystals form on them in cold weather. In the absence of any obvious function, they can only be imagined as contaminants coloured by carotenoids of dietary origin. All birds go to great lengths to keep their feathers in perfect condition and it is particularly significant that no such stains discolour the white heads of royal albatrosses. Wanderers cannot preen the stains themselves and they have not been noted as a target for allopreening partners. It is a curious phenomenon that has yet to be explained.

George Comer, who was ashore at Gough Island for five months (1888–89), considered the Gough Wandering Albatross (Plate 28) to be smaller than the Wandering Albatross of South Georgia (Verrill 1895). This was later confirmed by measurements, and variation in size is apparent between other island populations (Appendix 3). Male are larger than females; the sex of adults and their advanced fledglings have been determined by discriminant function analyses of bill length on depth (Berrow *et al.* 1999).

At all ages, royal albatrosses are large, predominantly white birds (Plate 35). Chicks fledge directly into white plumages essentially like those of adults. The upper wings of Northern Royals remain black throughout life, while those of the Southern Royals become progressively whiter from the inner leading edge outwards (Plate 36). Peter Harrison (1983) illustrated this progression in two stages for Northern Royals and five for Southern Royals,

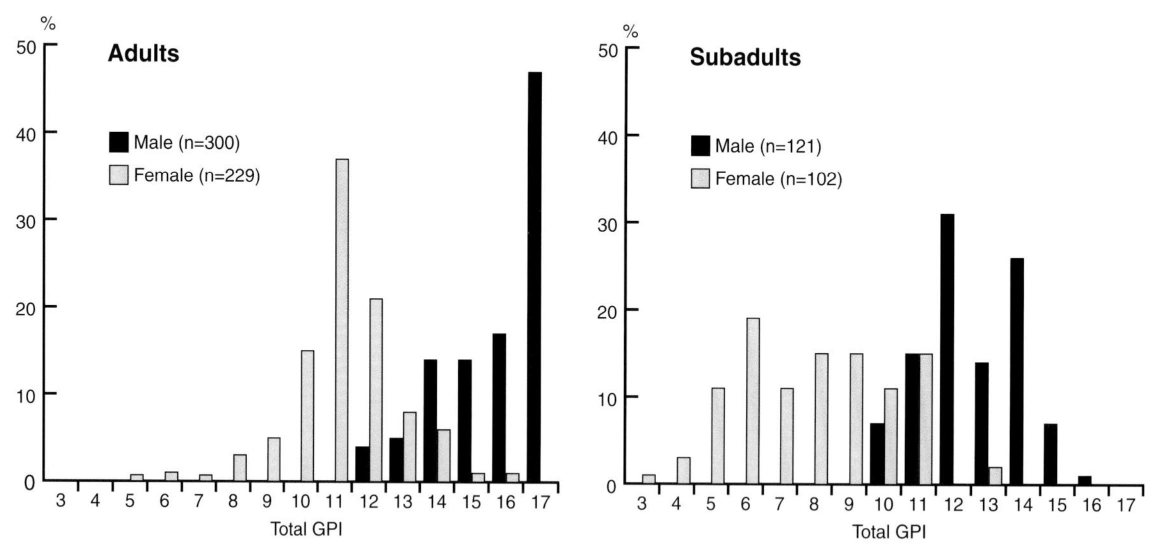

Fig. 7.5 Distribution of total Gibson Plumage Index (GPI) for the Wandering Albatross at the Iles Crozet (Weimerskirch *et al.* 1989).

but no GPI has yet been devised for these birds. Plumage progression proceeds more rapidly in males than females. In both royals, young birds have some dark feathers on the centre of the crown which also disappear more rapidly from males than females. The backs are white; in young birds some feathers have their outer vanes variably black or vermiculated, but these disappear with age. Tail feathers are black-tipped in young birds, becoming progressively whiter with age. Among 105 incubating Northern Royals that I examined at the Chatham Islands in January 1990, all had white crowns and backs, but only 43% had completely white tails, the remaining 57% had varying numbers of black-tipped tail feathers.

In the small colony at Taiaroa Head, New Zealand, the plumage of Northern Royal Albatrosses has diverged from that characteristic of the Chatham Islands (C.J.R. Robertson pers. comm.). This may reflect the limited gene-pool of the few founders and later Northern immigrants together with the contributions of an immigrant Southern Royal that has bred with a Northern Royal and successfully reared five hybrid offspring. By 1995, they were into the third generation and the colony contained various hybrid combinations including one pair in

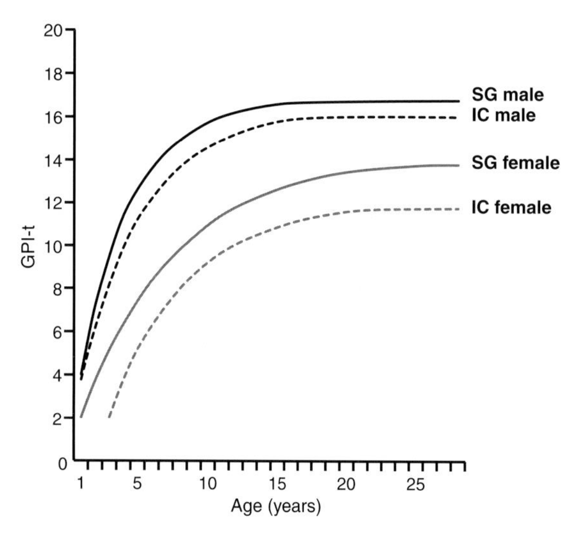

Fig. 7.6 Mean plumage scores excluding the tail (GPI - t) of known-age Wandering Albatrosses from South Georgia (SG) and the Iles Crozet (IC). Curves fitted by eye (Prince *et al.* 1997).

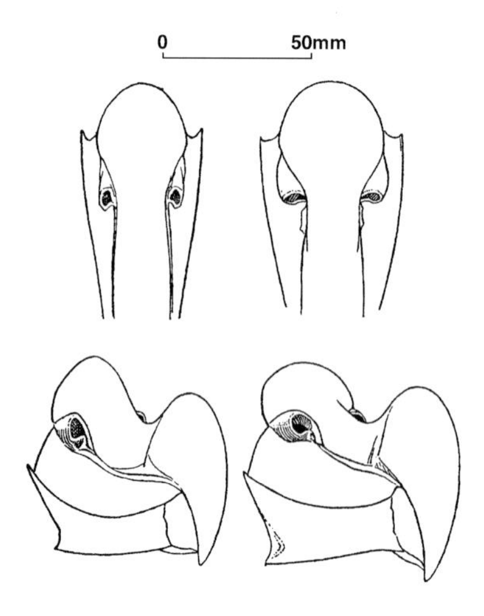

Fig. 7.7 The distinguishing characterisitics of great albatross bills and nasal tubes: wandering albatrosses (left) and royal albatrosses (right) (Murphy 1936).

which both partners were hybrids (C.J.R. Robertson 1993b, pers. comm.). Two immigrant Northern Royals have also paired with Southern Royals at Enderby Island in the Aucklands, but no offspring have been reported (Croxall & Gales 1997).

Both royal albatrosses have broad bills with bulbous nasal tubes directed forward (Fig. 7.7). As in the wandering albatrosses, vasoconstriction leaves the bill whitish, while vasodilation produces various shades of pink. The sharp hooked tip remains white.

	Total GPI-t at the age of:			
	>25 years		**>32 years**	
	mean (range)	**n**	**mean (range)**	**n**
Males	16 (13-17)	74	17 (16-17)	11
Females	12 (11-14)	32	12 (11-14)	6

Table 7.1 Sexual dimorphism in mature Wandering Albatrosses at the Iles Crozet; the maximum GPI-t score is 17 (Weimerskirch *et al.* 1989).

There are conspicuous black lines along the cutting edges of the lateral plates of the upper bill (Plate 31). Pigmentation extends through the keratin and is visible also from the inside. Internally, a second paletal ridge on the posterior roof of the mouth is also pigmented. These pigmented edges are distinctly sharper than the corresponding unpigmented edges of a Wandering Albatross bill from South Georgia.

HISTORY

After 1758, Linnaeus's name *Diomedea exulans* became available for albatrosses and William Turton's 1806 translation of *Systema Naturae* gave us 'wandering' as an English interpretation.[5] Linnaeus's description was quite inadequate; it has been doubted that he ever saw a specimen, but relied on Edwards's illustration which had been drawn from two specimens (Fig. 1.2). Much time has been spent trying to decide which population of wandering albatross Linnaeus's species represents (Robertson 1986, Bourne 1989, Medway 1993).

During Captain Cook's circumnavigation in the *Endeavour* (1768–71), Joseph Banks and Daniel Solander used Linnaeus's name from their first encounter with the great birds. On 23 December 1768, when they were in the SW Atlantic off the Rio de la Plata, Banks wrote in his journal (Beaglehole 1962):

> '...killed an albatross *Diomedaea exulans* who measured 9 ft 1 inch between the tipps of his wings'

This was the first wandering albatross to be described by naturalists present at the time the specimen was collected. Solander wrote the description and the bird was illustrated by Sydney Parkinson soon after it had been killed, but it was not preserved. It had dark plumage, but Banks and Solander were not deterred from including it in the same species as the much whiter birds they had already seen and later collected. In 1785 Latham, who had never seen the birds at sea, saw Parkinson's illustration and took it to be a different species. He called it the Chocolate Albatross to which Gmelin (1789) gave the Latin name *Diomedea spadicea*[6] (Medway 1993, Bourne 1993). Another dark brown specimen was later named *Diomedea adusta*[7].

On Cook's second voyage (1772–75), *Diomedea exulans* first appeared in Forster's journal on 24 October 1772. Subsequently he repeatedly referred to 'great white Albatrosses' or 'large common Albatrosses' and confirmed them as *Diomedea exulans* on 28 January and 11 February 1773. A year later, though, on 13 February 1774, the 'common Albatross' suddenly became *Diomedea Albatrus* and Forster never again used *exulans*, either in his journals or in his later *Mémoire sur les Albatros* (Forster 1785).

On 24 March 1773, as the *Resolution* was approaching New Zealand from the South, Forster appears to have remembered his visit to the St. Petersburg Museum in 1765–66, where he had examined the collection made by Steller in Kamchatka. The specimens included the albatrosses which Pallas (1769) had later described and named *Diomedea albatrus* (p.235).

Whatever Forster's reasons for discarding *exulans*, the substitution of Pallas's *albatrus* led him into curious contradictions. He stated that there was just one large species, which was found only in the southern hemisphere south of 26°–27°S and, on the authority of Captain Cook, affirmed that they did not cross the tropics. He knew *Diomedea*

[5] Latin: *ex* – out of, *solum* – one's country; *exsulare* – to live in banishment.
[6] Latin: *spadix* – date brown, nut-brown.
[7] Latin: *adustus* – burnt, scorched.

albatrus had already been given to specimens from the Bering Sea and that they were smaller than the 'common Albatross' of his experience. He even suggested that they may not have been albatrosses at all but giant petrels.

Forster's (1785) *Mémoire sur Les Albatros* did not attract the attention of anyone interested in seabirds for 128 years and his *Descriptiones Animalium...* was not published until after his death (Lichtenstein 1844). It was even longer before his journals were published; they are full of useful records and he guessed the significance of the brown plumage in wandering albatrosses. On 25 November 1772 he recorded (Hoare 1982):

> 'A great number of Albatrosses attended the ship. Some were caught...with a hook and line and a piece of Sheepskin for a bait so that the whole number was nine. I now plainly saw that the brown ones were young and small and that they grew whiter so as they increased in size and age.'

Coenraad Temminck indirectly came to the same conclusion. Two French specimens indicated that Steller's Albatross developed from a brown juvenile to a white adult and he became convinced that the same progression had also occurred in French specimens of *Albatros mouton* and *Mouton braun* (Temminck & Laugier 1838).

The novice naturalist Andrew Bloxam, sailing in the *Blonde* (1824–25), drew the same conclusion and was the first to define a plumage sequence, although he may have been seeing both royal and wandering albatrosses:

> 'They seem to have as great a variety of color as the common gull. Their heads however are almost invariably of a dull white, which is the case with the belly and under parts of their wing.The following were the colors of several varieties that I observed – age probably caused it.
>
> No.1. The upper surface of the back, wings, & tail of dark brown, the under parts white.
> No.2. The upper surface of the wings & tail brown, back white.
> No.3. The whole dull white color.
> No.4. The upper surface of wings alone brown, all the rest a dirty white.
>
> So great is the variety of color, that scarcely two in a whole flock resemble each other. The most beautiful were those which were purely white. These however were rare and seldom met with.'
>
> (Olson 1995).

In 1823 the French naturalist R.P. Lesson was crossing the South Pacific in the corvette *La Coquille* when he saw a great albatross that he believed was different from the species then recognised. One of the names used by French sailors for great albatrosses was *l'Amiraux*, because of the white marks on their wings, so he named the bird *L'Albatros à épaulettes Diomedea epomophora.* No specimen was collected and nothing was made of this observation in the scientific report of the expedition. Lesson's remarks (1825) were inadequate to distinguish the bird from wandering albatrosses and J.J. von Tschudi (1856), who translated it as *Der schulterfleckige Albatross*, seems to have been the only contemporary ornithologist to have noticed the observation.

In 1875, Filhol (1885) was convinced that there was a second species of albatross at sea more varied than the one nesting on Campbell Island. Later, Captain Fairchild who was constantly about the islands, had no doubt that there were two. Towards the end of the century Walter Buller also concluded that there was a second species, but he had never visited the breeding ground and had difficulty making a good case. Eventually, at a meeting of the Wellington Philosophical Society in 1891, he showed specimens and named a new species, *Diomedea regia.* Photographs of Southern Royal Albatrosses had been taken at Campbell Island by William Dougall (1888) and the species soon became widely accepted.

Twenty-one years later, Mathews came across Lesson's 1825 name and with unquestioning deference to priority, replaced *regia* in his *Birds of Australia* (Mathews 1912). It was not challenged at the time, so royal albatrosses became *Diomedea epomophora.* Murphy (1917) knew of *D. epomophora*, but a specimen collected off South America and distinguished by its lack of vermiculations was named *Diomedea sanfordi* within a new subgenus *Rhothonia.* Later Mathews & Hallstrom (1943) elevated this subgenus to a new genus, but it was never accepted.

In researching his history of royal albatrosses, Kaj Westerskov (1961) came to the conclusion that according to the rules of nomenclature *D. epomophora* could not be sustained against the claims of *D. regia.* So, undeterred by time and endorsed by Robert Falla, who had similar inclinations, he announced that together they were about to present a proposal to the International Commission of Zoological Nomenclature for Buller's name to be reinstated – but in fact they never did so. The *coda* to this bravado came two years later in a letter from Dean Amadon (1963) of the American Museum of Natural History:

> 'I am sure most of us would have preferred to have seen Buller's name used for this species. Still, are we not better off to continue to use the name that has been in circulation for half a century?...Since there are so few birds left to describe, perhaps ornithologists can be philosophical in promoting the opinion that naming species is mere drudgery and that it does not much matter who performs this chore.'

OCEANIC DISTRIBUTION

Because of their biennial reproductive cycles, great albatrosses in one seasonal population forage from the breeding ground, while those from the other are more distant. They change places each summer, but providing the population is reasonably stable (the two seasonal populations remain similar in size), little difference in at-sea abundance should be apparent, apart from that induced by climate, sea conditions and extent of sea-ice. Potential

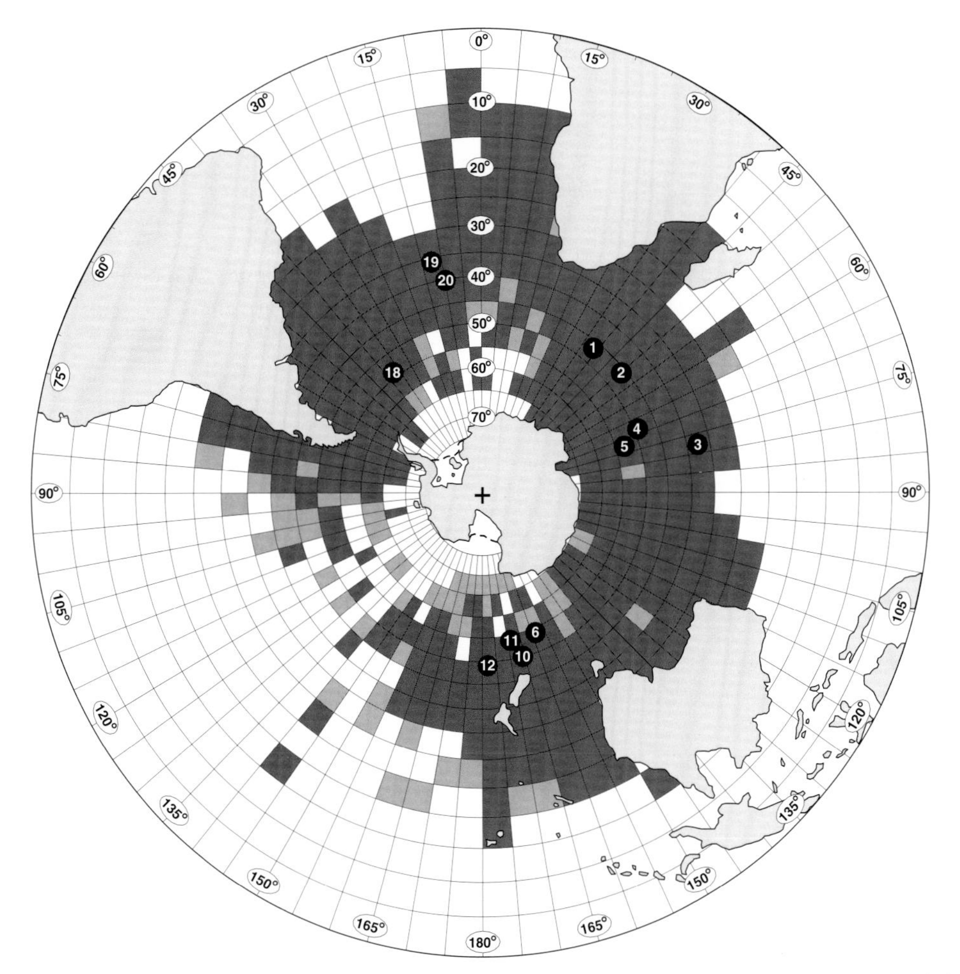

Fig. 7.8 The distribution of wandering albatrosses and their breeding islands throughout the southern oceans. The lighter shaded squares are based upon older data of unknown reliability. The numbered spots represent the Prince Edward Is. (1), Is. Crozet (2), I. Amsterdam (3), Is. Kerguelen (4), Heard I. (5), Macquarie I. (6), Auckland Is. (10), Campbell I. (11), Antipodes Is. (12), South Georgia (18), Tristan da Cunha (19) and Gough Island (20) (Tickell 1993).

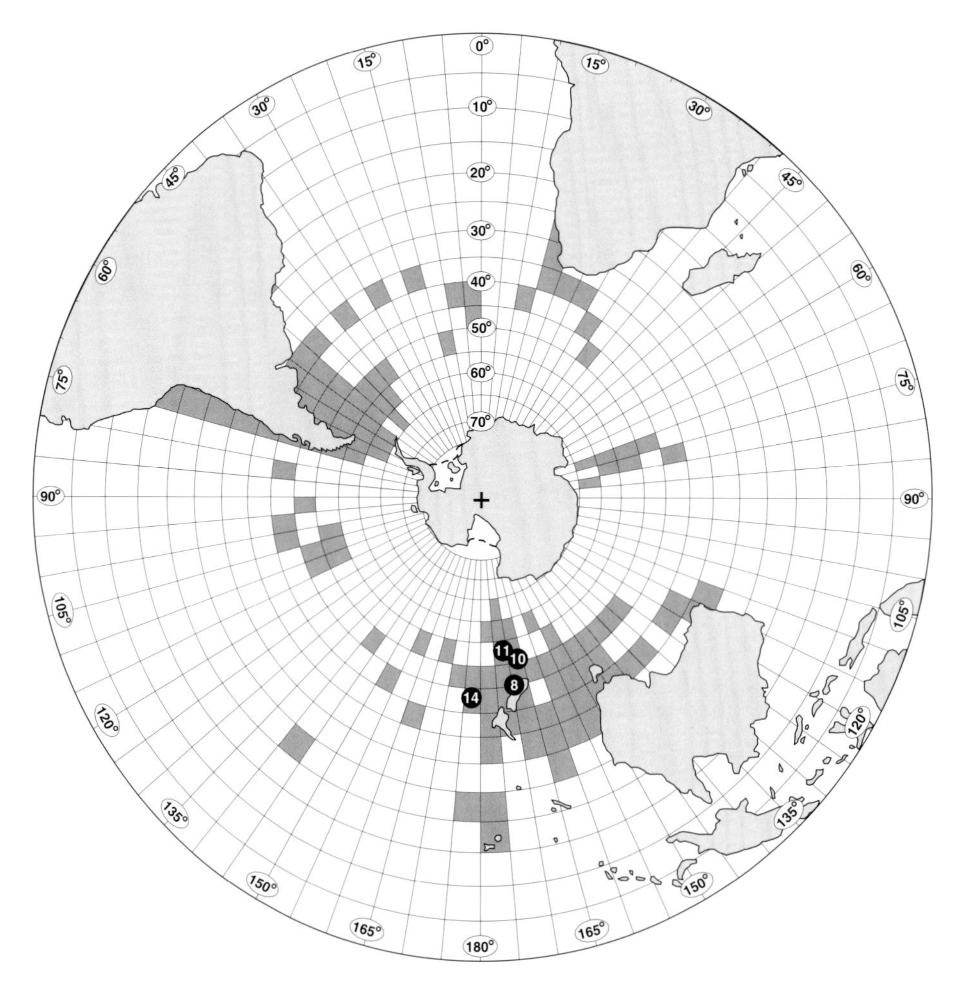

Fig. 7.9 The distribution of royal albatrosses and their breeding grounds throughout the southern oceans. The numbered spots represent New Zealand (8), the Auckland Is. (10), Campbell I. (11) and the Chatham Is. (14) (Tickell 1993).

changes in the number of birds at sea may be attributable to 1) the dispersion of juveniles, which for a short time in summer are close to the islands they have just left, and 2) several cohorts of subadults, that vacate distant places in time to arrive at their breeding islands later in the summer and join the breeding birds foraging in the surrounding sea.

At sea, wandering albatrosses (Fig. 7.8) have been recorded far more often than royal albatrosses (Fig. 7.9). They are more numerous and their breeding islands more widespread than those of royals, so there is a higher probability of seeing them. For a long time the published observations of wandering albatrosses by many respected ornithologists have included misidentified royal albatrosses, and great albatrosses seen at sea are still often assumed to be wanderering albatrosses.

Both wandering and royal albatrosses are in the seas about New Zealand all year (Robertson & Jenkins 1981), but less than half as many royals as wanderers are cast up dead on beaches and these are mostly on outer islands and south of Cook Strait (Powlesland 1985).

Around South America, both are present. Jehl (1973, 1974) saw wanderers in the Pacific off Chile, as far north as 27°S in June 1970, almost all of them more than 370 km from the coast. At the beginning of March in the Golfo de Penas, Chile, as many as 20 Northern and Southern Royal Albatrosses with up to 18 wandering albatrosses, have been seen at one time (P. Harrison pers. comm.). On one occasion, off Corral in October 1913, Rollo Beck killed a wanderer with one barrel of his shotgun and a royal with the other (Murphy 1936).

Both royal and wandering albatrosses have long been known to visit the Southwest Atlantic off Argentina, where royals have been reported to outnumber wanderers by as much as 23:1 (Dabbene 1926, Robertson & Kinsky 1972, Croxall & Prince 1990). Satellite telemetry indicates that after approaching from the west, through the Drake Passage, Northern Royal Albatrosses in non-breeding years turn north to forage over the Patagonian Shelf. Wandering Albatrosses breeding at South Georgia approach from the deep ocean to the east, but most remain near the shelf edge; it is there that they may be found in company with Northern Royals.

Wandering albatrosses

The separate island populations each have their own characteristic home ranges and distant migrations. Together they occupy a great expanse of the Southern Ocean (Fig. 7.8), but pack-ice excludes them from a considerable area during the southern winter.

During four summer voyages into the eastern Weddell Sea in the mid-1960s, Wandering Albatrosses were twice recorded close to the ice-edge at 59°S and 65°S (Thurston 1982). South of Heard Island, they have been seen for decades between 60° and 65°S; there are few in December, but numbers increase in January and February. Some reach 67°S off the coast of Antarctica between longitudes 65°E and 80°E (Routh 1949, Johnstone & Kerry 1976, Woehler *et al.* 1990). In the South Pacific at the the time of maximum ice (August to September 1964), Szijj (1967) was sailing south down the 150°W meridian on board the *Eltanin.* He saw wandering albatrosses about the ship to 59°S. The water temperature was 1°C and pack-ice was encountered three days later at 62°S. Returning north near the 125°W meridian, he met the first wanderers at 57°S. In summer (January to February), wandering albatrosses are absent from the polar waters of the South Pacific (65°S–75°S) (Zink 1981, Wanless & Harris 1988). Entering and leaving the western Ross Sea from the north, Darby (1970) saw them as far south as 62°S in December, while Hicks (1973) found them at 66°S in January. During the late 1960s, Russian fleets commonly saw them between 65°S and 67°S, sometimes as far as 69°S (Shuntov 1974).

Wandering Albatrosses breeding at South Georgia have a home range in the South Atlantic. They forage about the South Georgia shelf and west of a line running northeast from the South Shetland Islands through South Georgia to about 35°S 20°W (Fig. 7.10). There is a distinct attraction towards the edge of the South American continental shelf, but they rarely forage over the shelf itself (Croxall *et al.* 1999). They avoid the eastern basin of the Scotia Sea and Weddell Sea. Some males make forays westwards into the Drake Passage and towards the continental shelf west of the South Shetlands and Antarctic Peninsula (Wanless & Harris 1988, Croxall & Prince 1990, Prince *et al.* 1997). After breeding, their range increases. Between the 1950s and 1970s, many South Georgia Wanderers made forays east as far as Australia and into the Tasman Sea (Tickell & Gibson 1968), but few do so now (Croxall *et al.* 1999). One ringed bird was recovered at Mauritius (P.A. Prince pers. comm.) and two individuals tracked by satellite flew north from South Georgia before making their eastings between about 30°S and 40°S; they spent some time near South Africa before flying on eastwards (Fig. 7.11) (Prince *et al.* 1997).

Off the Atlantic coast of Brazil, wandering albatrosses are encountered as far north as 23°S in October where birds from South Georgia are joined by others from Gough Island. Sailing north from Tristan da Cunha in mid-ocean during May and June, wanderers accompany ships to 24°S (Tickell & Woods 1972) and are occasionally seen around St. Helena (16°S) (Rowlands *et al.* 1998).

Around southern Africa, there are seasonal variations in the numbers of wandering albatrosses seen at different localities. Ringed birds from Gough Island, South Georgia, the Prince Edward Islands and Iles Crozet have all been recovered in the same seas. West of the Cape, they are relatively common at the shelf-edge north of 30°S between March and October, with occasional birds as far north as 25°S (Summerhayes *et al.* 1974). Others have been seen at 18°S and the most northerly followed ships to about 10°S (Williamson 1975). East of the Cape (23°E to 28°E), adults and juveniles have been seen within 40 km of the coast. They occur throughout the year, being most numerous in December, January and May with fewer in July, but in some years very few arrive. Off the coast of Mozambique, the most northerly wandering albatross was at 26°S in June, when the water was 25°C (Rand 1962).

In the Indian Ocean, north of the Iles Crozet between 1971 and 1981, wandering albatrosses were seen at 31°S in September and 36°S in February (Jouventin *et al.* 1981). Sailing south in February 1982, Voisin (1983) did not see them until 39°S, but in January 1984, Stahl (1987) encountered them at 32°S. Returning north from Kerguelen

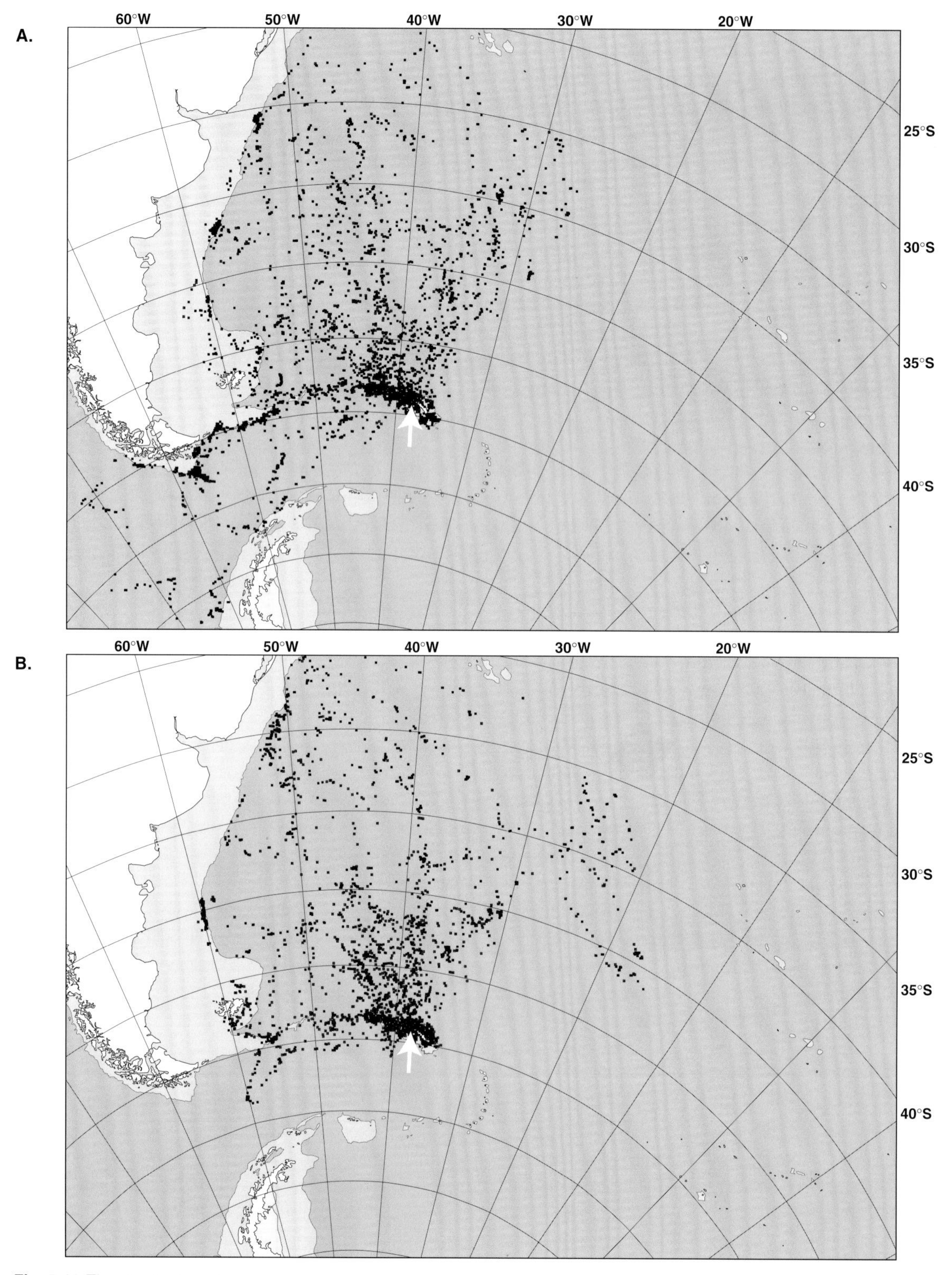

Fig. 7.10 The distribution of 10,581 satellite fixes from 34 male (A) and 22 female (B) Wandering Albatrosses, during 875 days foraging from Bird Island, South Georgia (arrow), August 1990–September 1994. The position of the 1,000 m isobath is shown by change in shading and indicates the edge of the continental shelf (Prince *et al.* 1997)

in February, they were seen until 34°S, but in March only as far as 42°S.

Wandering Albatrosses breeding at the Iles Crozet are birds of the southern Indian Ocean. In summer, they forage over an area bounded by latitudes 30°S–55°S and longitudes 30°E–70°E. Males tend to go south, some approaching Antarctica, while females tend to go north (Weimerskirch & Jouventin 1987, Weimerskirch *et al.* 1993). After breeding, they make more distant migrations, some flying east between latitudes 30°S and 45°S as far as the Tasman Sea and South Pacific (Nicholls *et al.* 1995, Tuck *et al.* 1999).

Wandering Albatrosses from the Iles Kerguelen occur around Australia and among the first ever ringed, by Loranchet in December 1913, was a bird that crossed the South Pacific to be recaptured three years later in the Drake Passage by the crew of a French four-masted sailing ship (Menegaux 1917).

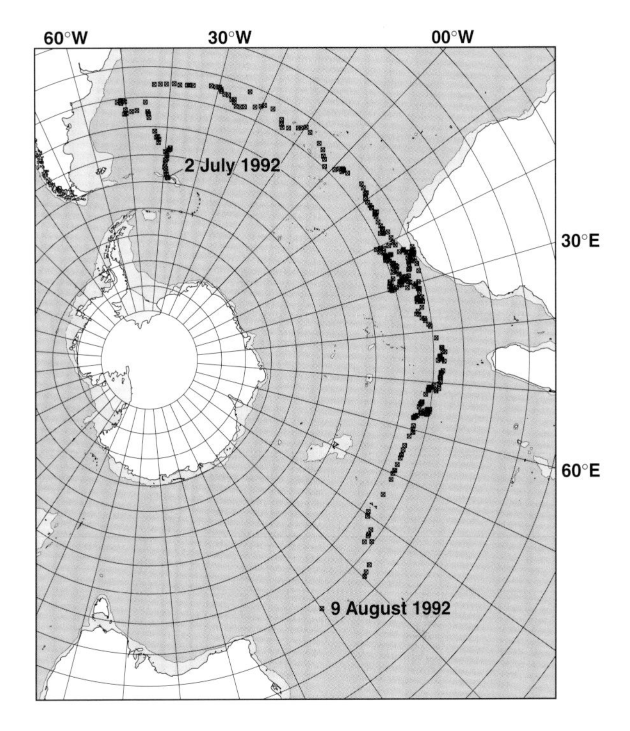

Fig. 7.11 The track of a Wandering Albatross on migration from South Georgia. After losing its chick, this female left Bird Island on 2 July 1992 and by the 14 August, when the battery of its transmitter expired, it had flown over 25,000 km. The 1,000 m isobath is indicated by change in shading and shows the edge of the continental shelf (Prince *et al.* 1997).

In the Indian Ocean, off West Australia, the northern limit of wandering albatrosses changes only slightly, from 29°S to 30°S between winter and summer (Shuntov 1974). Winter migrants off the east coast of New South Wales, Australia, have been known since 1912 (Smith 1998) and have been studied since the late 1950s (Gibson & Sefton 1959, Gibson 1963, Tickell & Gibson 1968). The area is a post-breeding destination for adults from several populations of the Wandering Albatross, as well as Auckland, Antipodes, Gough and possibly Amsterdam Wandering Albatrosses (Weimerskirch *et al.* 1985, Battam & Smith 1993, Smith 1997b, Palliser 1999). Some follow ships to 20°S off the coast of Queensland (Amiet 1958) and north of New Zealand, eventually reaching New Caledonia (Shuntov 1974) and Fiji (Jenkins 1986).

Auckland Wandering Albatrosses have home ranges in the Tasman Sea and SW Pacific (Walker *et al.* 1995b). Antipodes Wandering Albatrosses forage at the Chatham Rise and in the western South Pacific as far as 150°W. Males may forage south as far as 65°S and both sexes go north to 35°S. Antipodes Wanderers have post-breeding migrations traversing the South Pacific. One male, tracked by satellite, flew about 8,000 km eastwards in 17 days (Nicholls *et al.* 1996). Another completed a return circuit of some 50,000 km in about 200 days (Darby 1996).

Royal albatrosses

Southern and Northern Royal Albatrosses are seen in New Zealand waters (Robertson & Jenkins 1981). In the western Tasman Sea they are both rare (Barton 1979). Off Tasmania, Southern Royals have been identified more often (Blaber 1986, Enticott 1986).

Juvenile Southern Royal Albatrosses leave Campbell and the Auckland Islands in October and move into lower latitudes around New Zealand. Many cross the South Pacific in an easterly direction, apparently joining Northern Royals from the Chatham Islands (Clark 1986). Ringed Southern Royals have been recovered in the coastal waters of South America within weeks of leaving Campbell Island. Some were as far north as 21°S off Peru, but most were between about 30° and 40°S off Chile. Many have been seen in the Drake Passage (P. Harrison pers. comm.) and eastwards into the SW Atlantic.

Northern and Southern Royals occur over the Patagonian Shelf, off the coast of Argentina, and as far north as the Rio de la Plata (Robertson & Kinsky 1972, Curtis 1988a, 1994). Around the Falkland Islands, both have been seen in roughly equal numbers and one Northern Royal arrived there eight days after leaving the Chatham Islands (C.J.R. Robertson pers. comm.).

There have been scattered sightings elsewhere in the South Atlantic and southern Indian Ocean as far as Western Australia (Fig 7.9). Southern Royals have been seen north and south of the Antarctic Polar Frontal Zone (Woehler *et al.* 1991), but are considered vagrants off South Georgia (Prince & Croxall 1996).

Circumpolar migration

The notion that great albatrosses migrate continuously eastwards around the globe, crossing all the ocean basins, passing through the Drake Passage in the same direction and so returning to their breeding grounds was a plausible and persuasive speculation.

> 'There seems to be a progressive shifting of the centre of abundance from west to east each season as if the birds flew around the earth from west to east with the prevailing winds.' (Dixon 1933).

It lost some of its credibility when two Wandering Albatrosses were tracked westwards by satellite from Australia, one of them to its nest on the Iles Crozet (Fig, 7.12) (Nicholls *et al.* 1995). Since then, an Antipodes Wandering Albatross that had been tracked eastwards across the South Pacific to South American waters, turned around and flew back to the Tasman Sea (Darby 1996).

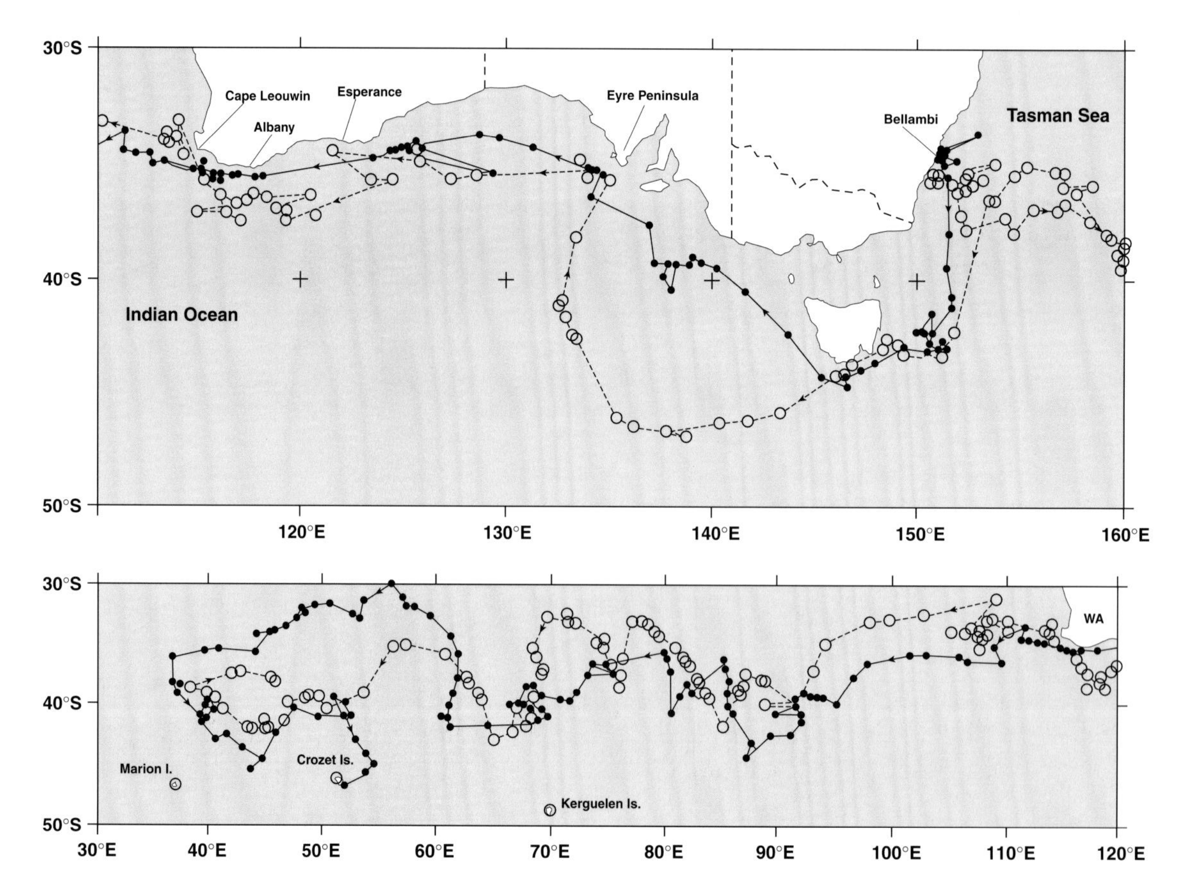

Fig. 7.12 The migration routes of two Wandering Albatrosses marked at Bellambi and tracked by satellite during the southern winter of 1992 as they returned to their breeding colony. In mid-August both birds were in the Tasman Sea; they set off south around Tasmania then westwards across the Great Australian Bight, past West Australia and across the southern Indian Ocean. One bird (○) was north of the Iles Crozet on 3 November when contact was lost, but the other (•) returned to a nest at the Iles Crozet on 3 December (Nicholls *et al.* 1995).

The northern hemisphere

Old records of albatrosses in the northern hemisphere include birds caught at sea in the southern hemisphere and transported by ships, to be later released in the north (Bourne 1967). The same suspicion still lingers over at least some more recent records.

A juvenile wandering albatross captured alive in the Bay of Panama was released later (Murphy 1938) while an adult or subadult, seen in a field in southern California, flew away over the Pacific Ocean (Paxton 1968). A male and a female said to have been caught by the crew of a fishing boat in the East China Sea in 1970 are now specimens in Japanese museums (Ornithological Society of Japan 1975). In the North Atlantic, a juvenile was encountered in 1963, 80 km off Portugal (Bourne 1966). Most bizarre of all, a wandering albatross somehow found its way through the Strait of Gibraltar into the Mediterranean. It crashed onto the coast road near Palermo, Sicily and was killed by a passing motorist (Orlando 1958).

BREEDING ISLANDS

Almost all great albatrosses breed on comparatively small islands, the exceptions being the Wandering Albatrosses on the mainlands of South Georgia and the Iles Kerguelen, and the Northern and Southern Royal Albatrosses on the Otago Peninsula of New Zealand.

Murphy (1936) unwittingly fostered a short-lived myth that royal albatrosses bred somewhere in the labyrinth of islands and channels of Tierra del Fuego. He had been told by Roberto Dabbene that P.W. Reynolds, who lived by the Beagle Channel, had discovered large white albatrosses nesting on mountain slopes above Lake Cami. Dabbene could not believe that the many royal albatrosses captured off South America had flown all the way from New Zealand breeding grounds, so the hearsay Reynolds had picked up was given more credence than it was worth.

Prince Edward Islands

Wandering Albatrosses nest on both islands (Fig. 7.13). Most breeding grounds were discovered in 1950–52 (Rand 1954, La Grange 1962) and later field-work filled in detail (van Zinderen Bakker 1971, Grindley 1981, Watkins 1987).

On Marion Island, they are almost confined to a flat, mossy coastal area up to about 100 m above sea level, although one nest was found at 300 m. Gony Plain on the north coast, west of Long Ridge, is a major colony. At Prince Edward Island, Wanderers are also on the coastal plain in several separate areas, but are most numerous to the east in Albatross Valley, where nests occur up to about 200 m.

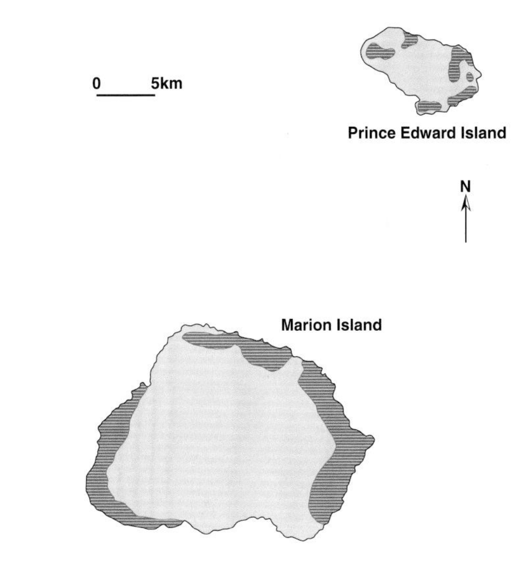

Fig. 7.13 Breeding grounds of the Wandering Albatross at the Prince Edward Islands. (Rand 1954, Grindley 1981, Percy Fitz-Patrick Institute unpublished).

Iles Crozet

After the sealing cutter *Princess of Wales* was wrecked in 1821, Charles Medyett Goodridge (1832) spent a year on the Ile de la Possession and wrote of the Wandering Albatross:

> 'They lay about Christmas, only one egg each, but their eggs are very large, the shell holding about a pint. Their period of incubation is about three months. The young when hatched are covered with down, and they grow wing feathers about May; they were then excellent for the table, and provided us with a very good dish for a long period, as they did not fly off until December.'

All three main islands have Wandering Albatrosses (Fig. 7.14). On the Ile aux Cochons, they nest at low altitude around most of the coast, with the major colony on the Plaine aux Albatros to the north and another important area on the Plaine Sud to the southwest (Voisin 1984).

The four colonies on the Ile de la Possession were counted in 1959–60 by Roger Tufft (pers. comm.). The smallest colony is near to the French base above the Baie du Marin and two are on the northeast coast, below the Morne Rouge (116 m) and above Pointe Max Douguet. The largest is at about 100 m above Pointe Basse on the north coast (Weimerskirch & Jouventin 1987).

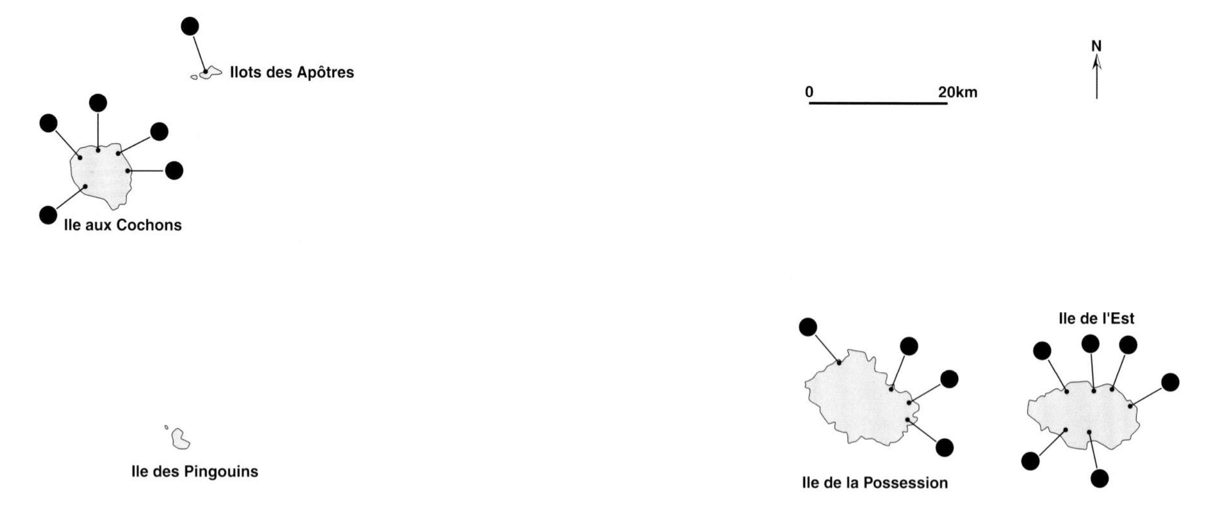

Fig. 7.14 Breeding locations of Wandering Albatrosses throughout the Iles Crozet (Voisin 1984, Despin *et al.* 1972, Weimerskirch & Jouventin 1987).

On the Ile de l'Est, there are separate colonies in the six main valleys, all below 100 m. The largest are in the Vallée des Phoquiers, Vallée de l'Abondance and l'Amphithéâtre (Despin *et al.* 1972).

The tiny Grande Ile of the Ilots des Apôtres also has a colony of Wanderers on its summit ridge at 139–289 m (Jouventin *et al.* 1984) (Fig. 7.14).

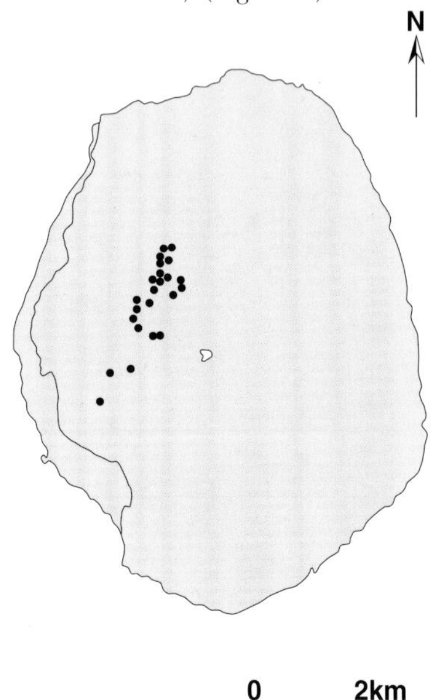

Fig. 7.15 Nest sites of the Amsterdam Wandering Albatross at the Ile Amsterdam, 1981–84 (Jouventin *et al.* 1989).

Ile Amsterdam

In 1951, members of the French meteorological station on the Ile Amsterdam found great albatrosses on the island. A published photograph appeared to be of a fledgling or juvenile wandering albatross (Paulian 1953). In the early 1980s, it was discovered that these birds all had very dark plumage and black lines on the cutting edges of the bills (like royal albatrosses). They were breeding six to seven weeks later than the Wandering Albatross at the Iles Crozet and Kerguelen. The French researchers believed there was reproductive isolation and a new species was proposed (Roux *et al.* 1983). However, Bourne (1989), Marchant & Higgins (1990) and Medway (1993) all referred to the population as a subspecies of the Wandering Albatross.

The Amsterdam Wandering Albatross nests only on the upper northwest moorland slopes at 400–600 m, in an area between les Events and the Plateau des Tourbières (Jouventin *et al.* 1989) (Fig. 7.15).

Iles Kerguelen

After the sealer *Betsy and Sophia* was wrecked on the Iles Kerguelen in March 1831, Richard Harris visited Wandering Albatross nests on the north coast of the Peninsule Courbet during the following winter (Hutton 1865) (p.33). Scientists from *Challenger* evidently saw Wandering Albatrosses near Mount Campbell in 1874 (Sharpe 1879) and J.H. Kidder (1875), of the American Transit of Venus Expedition, described their behaviour on the 'Prince of Wales Foreland'. Robert Hall (1900) found 30 nests on the Ile Howe and a few at other locations in the summer of 1897–98, but the coast of the Peninsule Courbet and Presqu'ile Prince de Galles (Paulian 1953, Angot 1954) remained the best known breeding ground until the 1980s.

Wandering Albatrosess are now known to be widely scattered around the coast of the Péninsule Courbet from below la Table de Diomede (408 m) to the Pointe de l'Ornithologue (26 m) and on the south coast of the Presqu'ile Prince du Galles, but the largest colony, on the southwest coast of the Péninsule Rallier du Baty, adjacent to the Ile Niñas, was not discovered until 1987 (Fig. 7.16). A few pairs nest on the Presqu'ile Joffre and nearby on Ile Sibbald, Ile Howe and Ile Briand. The Ile de Castries in the Iles Leygues has the most substantial colony in the northeast and a few nests have been found on the Iles de Ternay in the Iles Nuageuses to the far north of the group (Weimerskirch *et al.* 1989a).

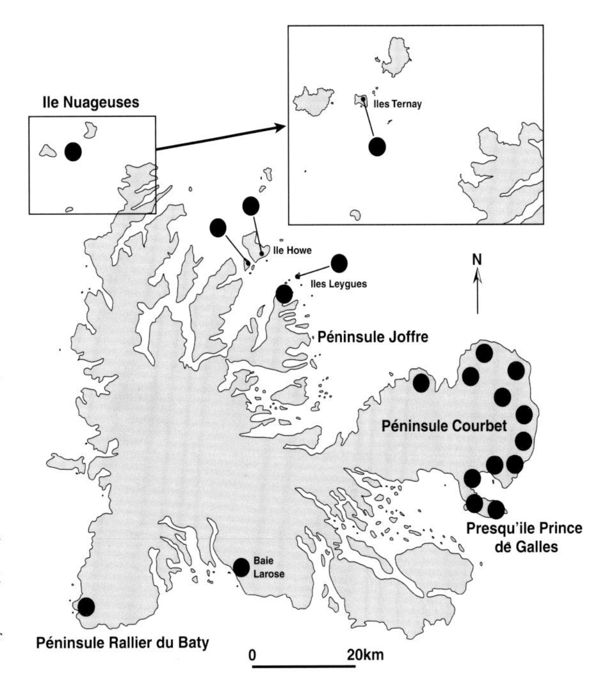

Fig. 7.16 The locations of Wandering Albatross breeding grounds at the Iles Kerguelen (Weimerskirch *et al.* 1989).

Heard Island

During the seven years (1947–55) that the Australian National Antarctic Research Expedition (ANARE) base on Heard Island was occupied, no Wandering Albatrosses were seen ashore. Scientists visiting the island in March 1980 were therefore surprised to find a male Wandering Albatross on the flats above Cape Gazert (90 m) (Fig. 7.17). It had been ringed as a subadult on Macquarie Island in 1967 and was incubating an egg which hatched shortly afterwards. There were two former nest mounds nearby (Johnstone 1982). Seven years later, it was seen again together with a female which was captured and ringed, but there was no egg or chick (E.J. Woehler pers. comm.).

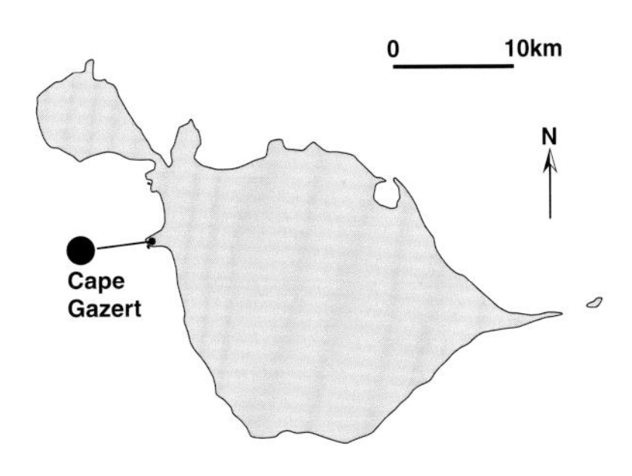

Fig. 7.17 The location of a single pair of the Wandering Albatross, above Cape Gazert at Heard Island (Johnstone 1982).

Macquarie Island

Albatrosses were known to the early sealers at Macquarie Island. A cave midden at Aurora Point, containing the bones of at least 65 wandering albatrosses, confirm that they were killed for food (de la Mare & Kerry 1994). They were also included among the birds seen or collected during Admiral Bellingshausen's brief visit in 1820.

During the Australasian Antarctic Expedition of 1911–14 (AAE) only two Wandering Albatrosses were found. One was killed above Caroline Cove with a tin of meat and another was seen on high ground (Falla 1937).

Nests were discovered when the ANARE base was established and since 1954, they have been counted yearly (de la Mare & Kerry 1994). They were at several locations along the west coast (Fig. 7.18). Most have been on tussock slopes up to 250 m above Caroline Cove in the southwest of the island and a few nested on the plateau edge at about 200 m above Cape Star, near Tiobunga Lake and above Rockhopper Point. At the northwest end of the island there were several nests on the low coastal flats at Handspike Corner, Eagle Bay, Langdon Point, Douglas Point to Boiler Rocks and south of Mawson Point. In 1996, an Antipodes Wandering Albatross was seen ashore at the south end of the island (Smith 1997a).

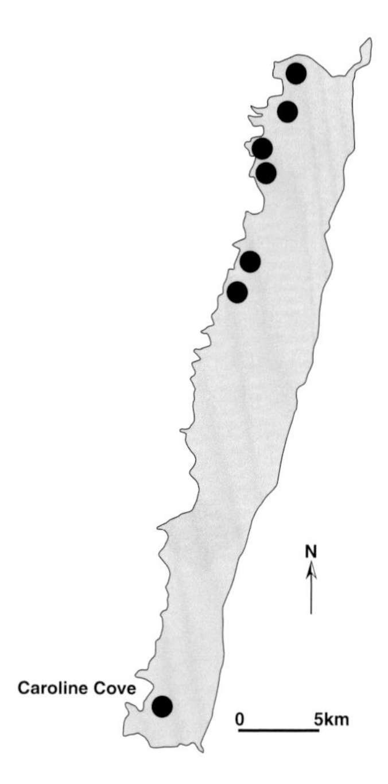

Fig. 7.18 Breeding locations of Wandering Albatrosses at Macquarie Island; the main colony has been at Caroline Cove with occasional nests scattered up the west coast (N. Brothers pers. comm.).

New Zealand

Northern Royal Albatrosses were prospecting the Otago coast in the late 19th century. During the early 1900s, albatrosses occasionally visited the grassy slopes of Taiaroa Head (Fig. 7.19), between the barracks and the gun positions. In 1918, six birds were seen together on the parade ground. Early in 1919, the garrison was withdrawn leaving only three men and their families at the lighthouse. Later that year, an egg was laid – it was fried and eaten. A similar fate befell others laid in the following years. The 1935 egg hatched, but the chick was killed and the next egg was again taken; Richdale knew of these losses and in 1937–38 guarded both egg and chick. The first Northern Royal Albatross to be reared on mainland New Zealand in recent times flew on 22 September 1938. A small colony became established on 3.7 ha of the northern slopes of Taiaroa Head, about 25 m above the mouth of Otago Harbour. The founders of this colony were Northern Royal

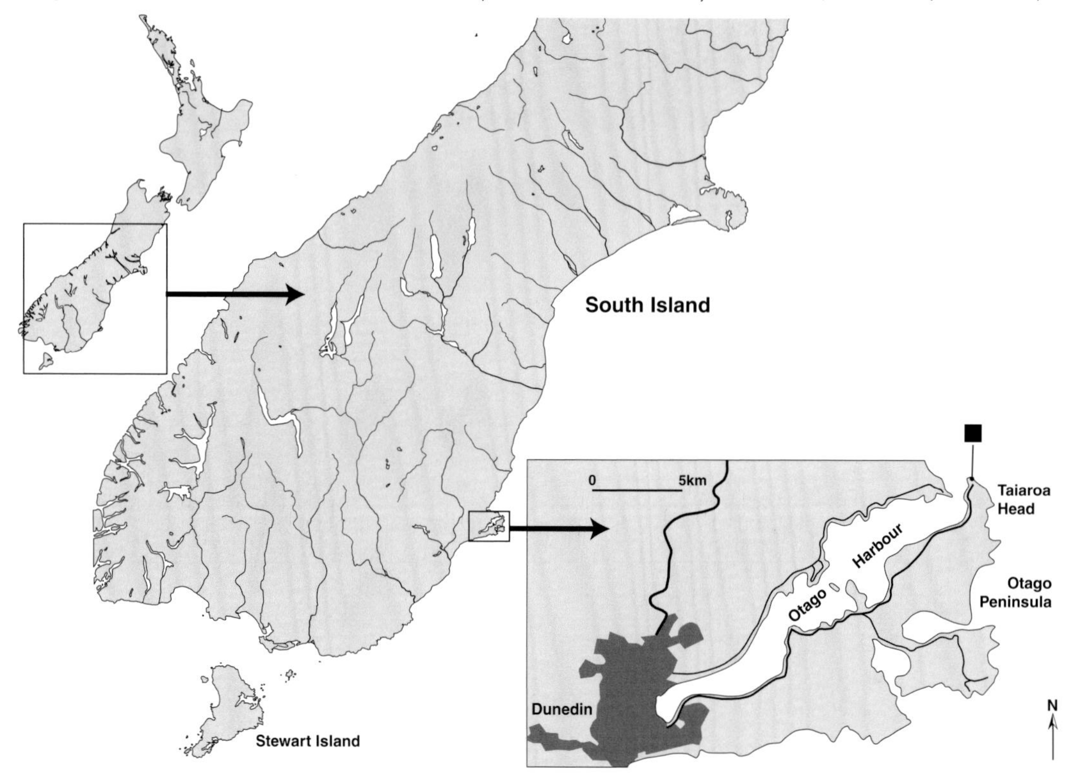

Fig. 7.19 The Northern Royal Albatross colony at Taiaroa Head, near Dunedin on the east coast of South Island, New Zealand.

Albatrosses. It is conceivable that one or two may have been aged survivors from an earlier breeding, but they were probably immigrants from the Chatham Islands. In 1950–51, a female Southern Royal arrived. It laid at the periphery of the colony for several years, but was never seen with a male and none of the eggs hatched. Some years later, another Southern Royal arrived and occupied a site outside the reserve fence. It was apparently a male for it attracted birds from inside the reserve, diverting them from potentially more successful breedings. The prospect of the colony expanding beyond the reserve fence and the unknown prognosis for hybrids was seen as a management dilemma; it was solved by shooting the bird (S. Sharpe pers. comm.). In 1962–63, another female Southern Royal took up residence in the reserve. This one was allowed to mate with a Northern Royal and over the following six years reared four hybrid chicks, some of which survived to become established breeders. The colony now contains a variety of hybrids and can no longer be imagined merely as an outpost of the Northern Royal Albatross. It has acquired a unique status (Robertson 1993b) (p.55).

Auckland Islands

In 1840, Robert McCormick (1884), the surgeon/naturalist on *Erebus* collected 'two old Albatrosses' near Port Ross at the north end of Auckland Island. One or two isolated nests were still there in the early 1970s. Nearby on Enderby Island, a small colony of the Southern Royal Albatross nests at about 40 m above sea-level. In 1962, sub-fossil skeletons of albatrosses and sealions were discovered in sand dunes about 10 m above Sandy Bay. They were buried between 1,000 and 3,000 years BP, but no cause was apparent (Otago Daily Times 1981, McFadgen & Yaldwyn 1984).

Auckland Wandering Albatrosses began to be mentioned in the 1880s. On 25 January 1888, Andreas Reischek (1888a) described hundreds of them flying about the *Stella* as it rounded the South Cape of Adams Island. Early in January 1890, Chapman (1891) climbed to nests above Fairchild's Garden on Adams Island and the following day crossed the island almost along the meridian 166°E. Just below the watershed on the southern slopes, large numbers of Auckland Wanderers were nesting. From March 1942 to January 1943, Charles Fleming was at the Cape Expedition base at the south end of Auckland Island. He found a solitary nest on high ground above the Musgrave Peninsula and visited the chick intermittently throughout the year. On several occasions he crossed Carnley Harbour to Adams Island, once climbing to the high saddle at the west end, where he found about 50 nests (Bailey & Sorenson 1962). In 1966, New Zealand Wildlife Service field-parties traversed the high ground of Adams Island on 26–29 January, a census of albatrosses was made, but no figures were published (Godley 1975).

Auckland Wandering Albatrosses (Plates 26&27) nest over much of the high tussock ground of Adams Island, at about 200–500 m (Fig. 7.20), but the largest concentrations are on the southern slopes, especially between Fly Harbour and Lake Turbott and above the coast between Astrolabe Point and Amherst Rock. They are also along the main ridges of Disappointment Island (316 m) and a few nests are widely scattered over the southern slopes of Auckland Island, on Mt. Raynal, Wilkes Peak and Mt. D'Urville (Robertson 1975b).

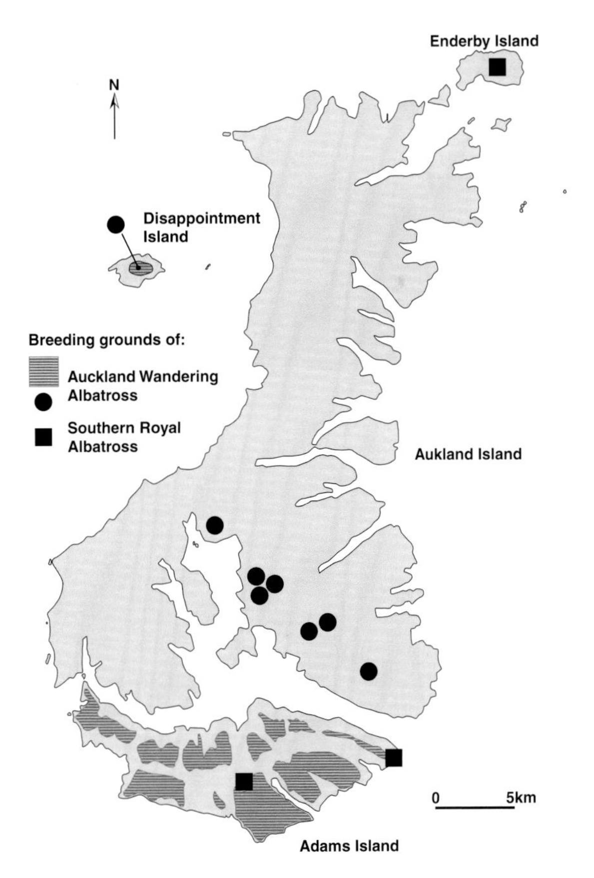

Fig. 7.20 The breeding grounds of the Auckland Wandering Albatross and Southern Royal Albatross at the Auckland Islands. (Robertson 1975b, Walker *et al.* 1991, K. Walker pers. comm.).

Breeding areas were mapped in 1972–73 and a few Southern Royals were discovered nesting near Auckland Wandering Albatrosses on the high slopes of Adams Island, notably above Gilroy Head and between Fly Harbour and Lake Turbott (Robertson 1975b). Wildlife biologists[8] were sent to look for the rare Auckland Islands Rail in 1989. They camped on Adams Island for five weeks and while there, counted other species including the two great albatrosses. After this initiative, the Auckland Wandering Albatross and Southern Royal Abatross were counted most years between 1991 and 1996 (Walker *et al.* 1991, 1995a, Walker & Elliott 1999).

Campbell Island

This is the main breeding ground of the Southern Royal Albatross, whose nests are scattered among tussock grass and stunted scrub on high ridges at 200–300 m over much of the island (Fig. 7.21). They avoid the dense stands of *Bulbinella rossii* (Westerskov 1963) (p.59).

During the few days that *Erebus* and *Terror* were at Campbell Island in 1840, McCormick (1884) saw albatrosses nesting on the hills above Perseverance Harbour. These were the birds that Filhol (1885) found in 1875 (p.58). Sorensen (1950) had seven pairs under observation between 1942 and 1946 and their chicks were weighed weekly throughout the winter of 1943. Richdale (1950b) considered the field-work inadequate, but it was never repeated and remained a useful source. Westerskov (1963) completed the first census of these birds during a brief visit in the summer of 1957–58.

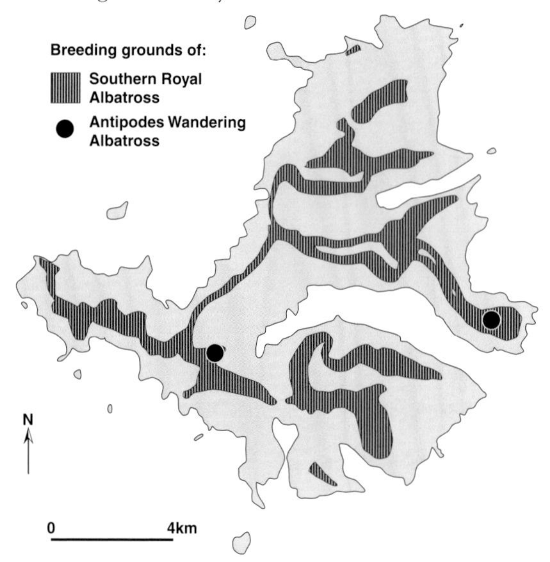

Fig. 7.21 The breeding grounds of the Southern Royal Albatross in 1957–58 and the locations of several Antipodes Wandering Albatross nests at Campbell Island (Westerskov 1963, Moore & Moffat 1990a).

From 1943 to 1970, staff of the meteorological station ringed over 20,000 Southern Royals from which 89 distant recoveries were obtained (Robertson & Kinsky 1972). Other data collected by meteorological staff during the 1960s has recently been published (Waugh *et al.* 1997).

Several pairs of Antipodes Wandering Albatrosses also nest at Campbell Island in the area of Moubray Hill and on the northeast slopes of Mt. Dumas (Bailey & Sorensen 1962, Moore & Moffat 1990a) (Fig. 7.21).

Antipodes Islands

The Antipodes Wandering Albatross nests throughout the interior of the main island, in tussock grass moorland from about 100 m above coastal cliffs to the summit of Mt. Galloway (404 m) (Warham & Bell 1979) (Fig. 7.22). When Captain Fairchild climbed to the high ground in 1886, he found many on nests. He and Captain Bollons passed the word that these birds laid their eggs much later than the Auckland Wanderers (Oliver 1955). Although ships called regularly in search of castaways, heavy seas often prevented landing and few of those who did get ashore reported on the albatrosses. Guthrie-Smith's (1936) photographs illustrated the dark plumage in which these birds breed (Robertson & Warham 1992). There was some ringing in 1950 and mapping in 1969. Recent field-work dates from 1993–94 when Jacinda Amey and Gus McAllister sailed to the island in Gerry Clark's small yacht *Totorore*.

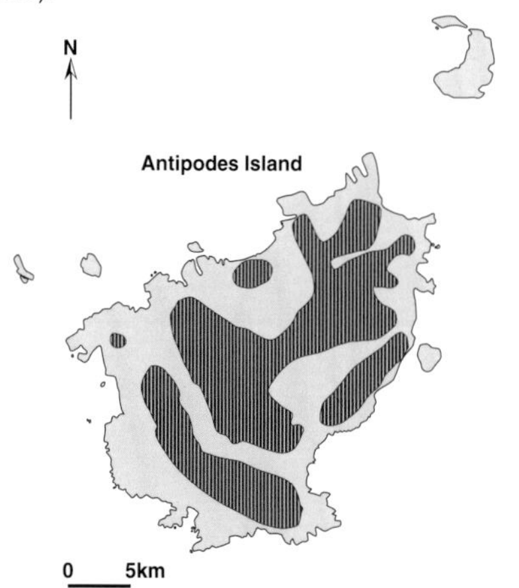

Fig. 7.22 The breeding grounds of the Antipodes Wandering Albatross at the Antipodes Islands (Warham & Bell 1979).

[8] New Zealand Department of Conservation (DoC).

Chatham Islands

The main breeding grounds of the Northern Royal Albatross are on three remote, rocky outliers. The largest colony is among the sparse vegetation or bare rock on top of Motuhara in the Forty-Fours group. In the Sisters group, Big Sister is a similar islet, but Little (Middle) Sister is lower and the nests are among *Cotula*, mostly on soil in a damp hollow surrounded by higher rock slabs (Plate 32) (Fig 7.23).

Field-work and air photography at the Chatham Islands in 1973–74 revealed nest densities far in excess of anything found elsewhere for either royal or wandering albatrosses (Robertson 1974). There were no further counts until 1989–90, when annual surveillance from the air was initiated (Robertson 1991).

South Georgia

The Wandering Albatross is scattered around South Georgia among open tussock grass, on headlands and small islands. Most are towards the northwest, but there are several to the southeast (Fig. 7.24). Some nests are little more than 15 m above sea-level and the highest at 140 m, but most colonies are at 30–100 m. The main breeding ground is at Bird Island and there are appreciable numbers on Annenkov Island and islands in the Bay of Isles (Croxall 1979).

In 1823, when James Weddell (1825) sailed north, out of the sea that now bears his name, the *Jane* and *Beaufoy* sheltered for a while at South Georgia. His crews

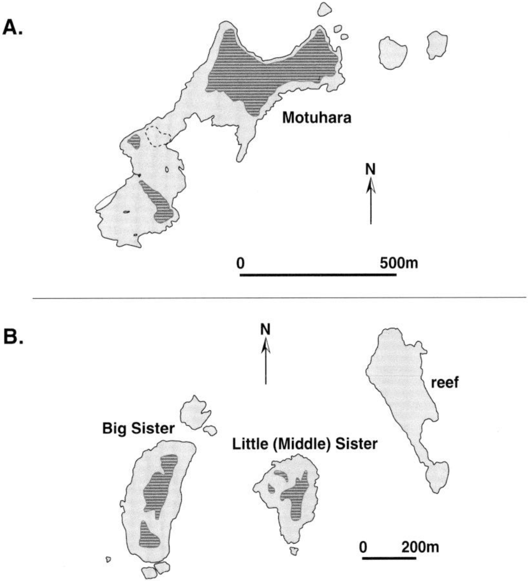

Fig. 7.23 Colonies of Northern Royal Albatrosses on the Forty Fours (A) and the Sisters (B) islets of the Chatham Islands (Fig. 5.23) (Robertson 1991 & 1994, C.J.R. Robertson pers. comm.)

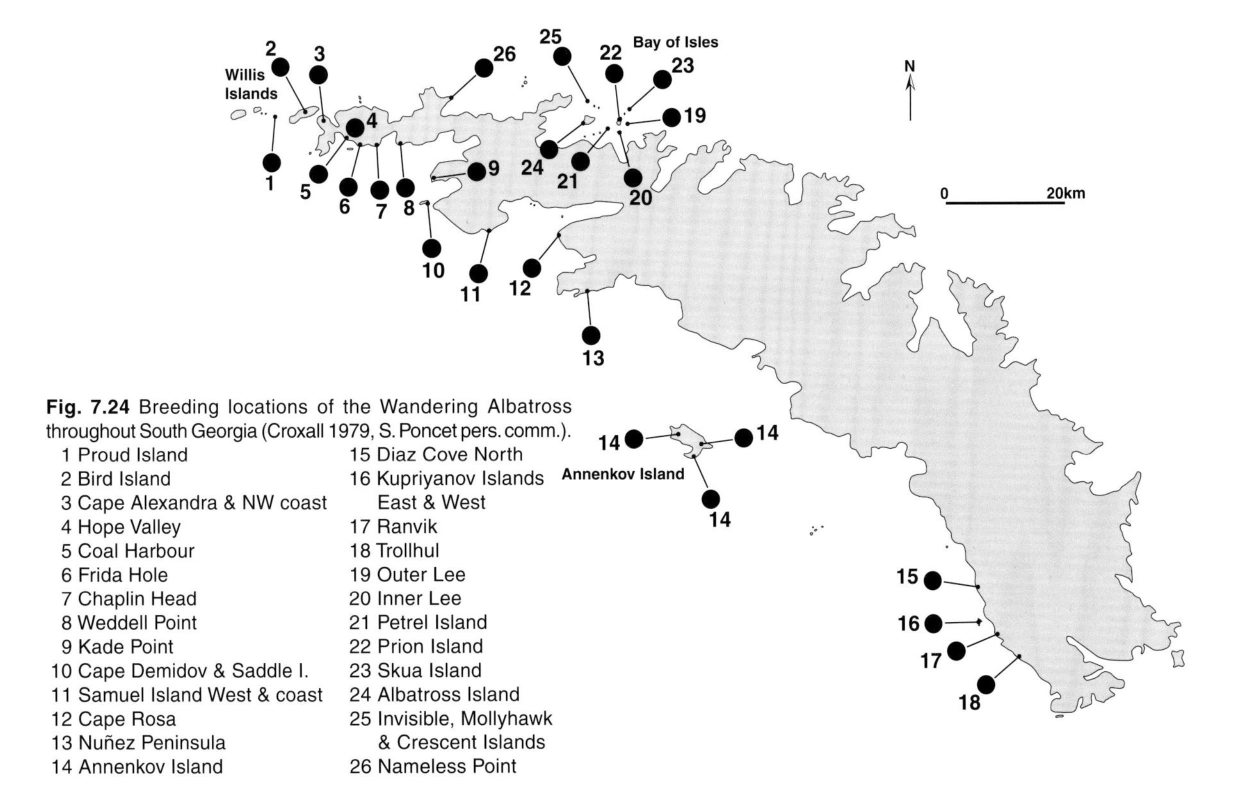

Fig. 7.24 Breeding locations of the Wandering Albatross throughout South Georgia (Croxall 1979, S. Poncet pers. comm.).

1 Proud Island
2 Bird Island
3 Cape Alexandra & NW coast
4 Hope Valley
5 Coal Harbour
6 Frida Hole
7 Chaplin Head
8 Weddell Point
9 Kade Point
10 Cape Demidov & Saddle I.
11 Samuel Island West & coast
12 Cape Rosa
13 Nuñez Peninsula
14 Annenkov Island
15 Diaz Cove North
16 Kupriyanov Islands East & West
17 Ranvik
18 Trollhul
19 Outer Lee
20 Inner Lee
21 Petrel Island
22 Prion Island
23 Skua Island
24 Albatross Island
25 Invisible, Mollyhawk & Crescent Islands
26 Nameless Point

found Wandering Albatrosses above Undine Harbour. A century later, Hubert Wilkins (1923) camped among these Wanderers and reported large numbers nesting on Bird Island. When the crew of the brig *Daisy* was working elephant seal beaches around the Bay of Isles (1912–13), Murphy had a camp ashore close to several Wandering Albatross nests and landed four times on Albatross Island. Niall Rankin (1951) counted all Wandering Albatrosses at the Bay of Isles in 1946–47 and there has been a gradual accumulation of information from these and other breeding locations (Croxall 1979)

Field-work on the Wandering Albatross at Bird Island began in 1958–59 (Tickell & Cordall 1960). Up to 1986, over 7,300 adults and 13,000 fledglings had been ringed and the subadults that returned formed many cohorts of known-age birds (Croxall & Prince 1990).

The Antarctic furseals at South Georgia have increased exponentially since the 1950s, and by the 1980s about half of the tussock grass at Bird Island had been more or less eroded. Beginning at lower altitudes, Wandering Albatrosses on their nests have been in contact with increasing numbers of fur seals. The birds have not been molested, but adult seals may sleep alongside nests and curious pups approach sitting birds closely. There is no evidence of Wanderers with established sites moving their nests to areas with fewer furseals. Subadult albatrosses are more active in areas with few furseals and new breeders therefore tend to occupy sites and establish nests away from furseals; natural losses of breeding birds in furseal areas are therefore not replaced. In time, Wandering Albatrosses will probably come to occupy only the higher areas of vegetation (Croxall *et al.* 1990a).

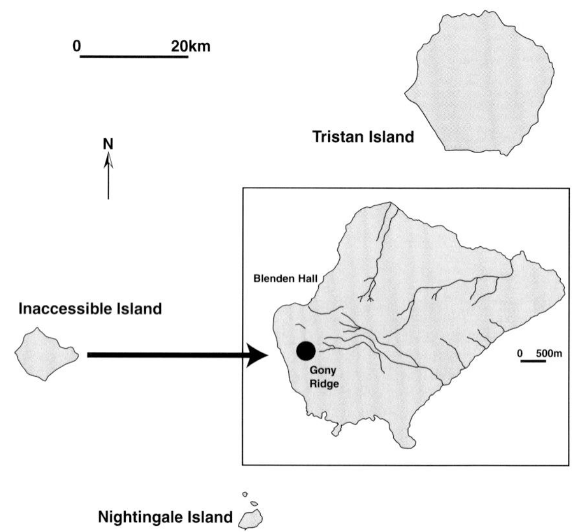

Fig. 7.25 The location of Gough Wandering Albatross nests on Inaccessible Island in the Tristan da Cunha group (Fig. 5.27) (Fraser *et al.* 1988).

Inaccessible Island (Tristan da Cunha)

Substantial numbers of Gough Wandering Albatrosses were seen on Inaccessible Island by the Stoltenhoff brothers during their 1871–73 sojourn (1873). They nested at about 400–500 m, but by 1938 visits of the Tristan islanders had reduced them to several pairs (Hagen 1952). They were no more numerous when the Denstone Expedition of 1982–83 spent three months ashore. The only birds remaining today nest at Gony Ridge (Fraser *et al.* 1988, Ryan *et al.* 1990) (Fig. 7.25).

Gough Island

The Gough Wandering Albatross nests over a large area of wet moorland between 400 and 750 m mainly to the southwest of the Triple Peak – Edinburgh Peak watershed, northwest towards Tavistock Crag and eastwards to Tafelkop (Fig. 7.26). Gony Dale and Albatross Plain have been visited many times (Holdgate 1958, Swales 1965). These albatrosses have been counted and ringed, but the rest of the island has been inspected only infrequently (Watkins 1987, Ryan *et al.* 1990).

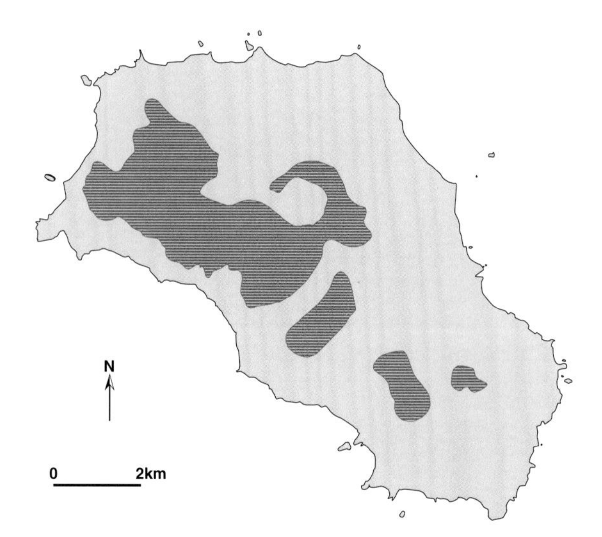

Fig. 7.26 Breeding grounds of the Gough Wandering Albatross at Gough Island (B.P. Watkins pers. comm.).

NUMBERS

	Islands	Pairs
WA	Prince Edward Islands	3 000
	South Georgia	2 000
	Iles Crozet	2 000
	Iles Kerguelen	1 000
	Macquarie Island	10
	Heard Island	1
GWA	Gough Island	1 000
	Tristan da Cunha	10
ADWA	Ile Amsterdam	10
AKWA	Auckland Islands	6 000
	Campbell Island	10
ATWA	Antipodes Islands	5 000

Table 7.2 The numbers of wandering albatrosses breeding in a single season. Figures rounded to the nearest order of magnitude (Appendix 15). Wandering Albatross (WA), Gough Wandering Albatross (GWA), Amsterdam Wandering Albatross (ADWA), Auckland Wandering Albatross (AKWA), Antipodes Wandering Albatross (ATWA).

	Islands	Pairs
NRA	Chatham Islands	5 000
	New Zealand	10
SRA	Campbell Island	7 000
	Auckland Islands	100

Table 7.3 The numbers of Northern (NRA) and Southern (SRA) Royal Albatrosses breeding in a single season. Figures rounded to the nearest order of magnitude (Appendix 15).

The figures considered here are mostly the numbers of pairs, eggs or chicks that can be counted on the ground in any one season (seasonal population). To find the total, it is necessary to make counts in at least two consecutive seasons and to know the breeding success. Seasonal variation may obscure increasing or decreasing trends at islands that are not counted each year. Incorporating the many cohorts of non-breeding juveniles and subadults is more difficult.

Available counts are listed in Appendix 15. In Tables 7.2 and 7.3 numbers have been rounded to orders of magnitude for comparison. The demography of better known colonies will be discussed in Chapter 16.

In this century, the numbers of great albatrosses breeding on islands throughout the southern hemisphere have probably varied from 50,000 to 60,000 pairs, about two-thirds of them wandering albatrosses and one-third royal albatrosses.

Wandering albatrosses

Before the 1980s, there had been only one or two counts of the Wandering Albatross at the Prince Edward Islands but three annual counts of fledglings at both islands gave totals of 2,282, 2,272 and 2,461 (Watkins 1987). Between 1982 and 1984, when other populations were in decline, the Marion Island population remained stable.

At the Iles Crozet, there have been annual counts of the Wandering Albatross on the Ile de la Possession since the late 1960s, but it has only about 13% of the total for Crozet. Other islands of the group have been visited less often. Pooled counts total 1,637 pairs (Weimerskirch *et al.* 1997a), but this population was once much larger (Jouventin *et al.* 1984).

The Amsterdam Wandering Albatross is recovering from near extinction. It is believed that about 30 pairs bred each season in the 1960s. At the time of the first census in 1984, about five pairs bred each year; by 1996 they had increased to 17 pairs. (Jouventin *et al.* 1989, Weimerskirch *et al.* 1997a, Weimerskirch & Jouventin 1997).

Wandering Albatross nests are widely scattered among the Iles Kerguelen, but counts have been relatively few, intermittent and incomplete. It was only in 1987 that the largest colony, at the Péninsule Rallier du Baty, was discovered. There is at least one breeding ground for which figures have never been published; so unless the Iles Leygues have an appreciable number of Wanderers, the Kerguelen population will probably remain at about 1,000 pairs (Weimerskirch *et al.* 1989a, 1997a).

It took the scattered Wandering Albatrosses of Macquarie Island about 50 years to reach 31 pairs (Fig 7.27). After the recent catastrophic decline (de la Mare & Kerry 1994), it is likely to take the few survivors equally as long to increase again to such numbers.

The largest numbers of wandering albatrosses are on islands of the New Zealand continental shelf. In the 1968–69 season, it was estimated that there were 750–900 pairs of Antipodes Wandering Albatrosses breeding on Antipodes Island (Warham & Bell 1979). Twenty-five years later, counts of 4,522, 5,757 and 5,148 pairs in three consecutive seasons (1994–1996) represent an increase far in excess of any known from other great albatross

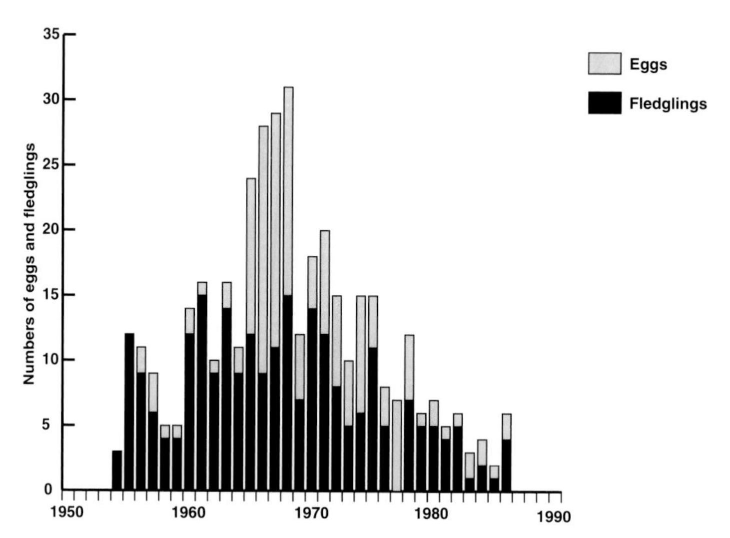

Fig. 7.27 The numbers of eggs laid (shaded) and fledglings reared (black) by the Wandering Albatross at Macquarie Island between 1954 and 1987 (after de la Mare & Kerry 1994).

populations (J. Amey pers comm.). The recent counts were more thorough, but the 1968–69 estimate was not a casual guess; care was taken to eliminate errors. It is conceivable that the Antipodes population in the late 1960s was out of balance and that 1968–69 was a low year, but It is also possible that the number of unseen birds in control plots may have been underestimated (J. Warham pers. comm.). The few pairs of immigrant Antipodes Wandering Albatrosses at Campbell Island have not increased noticeably since the 1940s.

Estimated numbers of the Auckland Wandering Albatross breeding each year, between 1990 and 1996, throughout the Auckland Islands, have varied from 4,826 to 7,417 pairs (Walker *et al.* 1991, 1995a, Walker & Elliott 1999).

At South Georgia in the early 1960s, there were probably almost 3,000 pairs of Wandering Albatross breeding each year (Tickell 1968). They have been counted on Bird Island each year since 1976, but smaller colonies, amounting to about 39% of the South Georgia population, have been visited infrequently. Pooled counts and estimates from many sources now indicate about 2,000 pairs (Croxall *et al.* 1990b, Gales 1997).

The breeding ground of the Gough Wandering Albatross at Gough Island has been divided into census areas, but they have rarely all been seen at one time. In November 1979, 792 chicks were counted in an almost complete survey (Williams & Imber 1982) and in October–November 1982 there were 798 (Cooper 1983). Gonydale, which is the most accessible area, has been counted more often and in the three years 1979–81, there were 63, 53 and 66 chicks respectively. From the little that is known of breeding success, it seems likely that there are no more than about 1,000 pairs breeding each season over the whole island (Watkins 1987, Cooper & Ryan 1993). The two or three pairs of Gough Wandering Albatrosses surviving at Tristan da Cunha, on Inaccessible Island, have not increased in 60 years.

Royal albatrosses

The 6,308 and 7,787 pairs of the Southern Royal Albatross on Campbell Island in 1994 and 1995 appear to represent an increase over several decades (Moore *et al.* 1997b), but Croxall & Gales (1997) have commented upon differences in counting methods. At the Auckland Islands, a small colony of 55 pairs on Enderby Island has apparently been slowly increasing (Gales 1997). There are only two or three widely separated pairs on the main island and 10–20 pairs at Adams Island (Robertson 1975b, K. Walker pers. comm.). At the Chatham Islands, Northern Royal Albatrosses nest in dense colonies quite unlike those of any other great albatross population. Computations from air-photos and ground counts in the mid-1990s indicate combined seasonal populations of about 5,200 pairs on the Sisters and Forty Fours islets, where formerly they were more numerous (Croxall & Gales 1997). On the adjacent New Zealand mainland, the few pairs that first bred at Taiaroa Head in the late 1930s had increased to a total of 29 pairs by 1989–90 with up to 20 breeding in any one season.

BREEDING

Biennial breeding

Successful breeding by royal and wandering albatrosses takes about a year to complete and successful pairs spend the following year at sea, thus rearing a maximum of one chick every two years. However, two pairs of Southern Royal Albatrosses that raised young at Campbell Island in 1964–65 are said to have raised chicks again the following season (Waugh *et al.* 1997). The field-notes upon which this claim was based had remained unpublished for 30 years and although the data are presented as unequivocal, they are meagre beside the considerable body of research that underpins the biennial regimes of Royal and Wandering Albatrosses (Chapter 16). In their speculation, the authors argued that annual breeding Royals would have some 33 days between successful breedings, but nothing was said about moult (Chapter 13), some of which must be accomplished before the birds attempt to breed (Weimerskirch 1991, Langston & Rohwer 1996). Few primaries could be replaced in 33 days and serious doubts must remain about this claim.

Breeding seasons

The Northern Royal Albatross returns to the breeding grounds early in October, perhaps a week before the Southern Royals. The southernmost population of the Wandering Albatross at South Georgia arrives early in November, a week or two later than the more northerly populations at the Prince Edward Islands, Iles Crozet and Iles Kerguelen. The Gough Wandering Albatross, Auckland Wandering Albatross and Antipodes Wandering Albatross return much later, perhaps towards the end of December and the Amsterdam Wandering Albatross does not arrive until late January.

Richdale (1950a) believed that mated individuals of the Northern Royal Albatross associate at sea and return to land together. Tomkins (1984) made the same claim for the Wandering Albatross at Macquarie Island, but I saw no evidence of this at South Georgia. Solitary albatrosses homing towards their islands encounter others as they get closer. At the time Richdale was working at Taiaroa Head, there were less than 16 Royals breeding there each year and during the time that Tomkins was at Macquarie, there were only ten Wanderers nesting at Caroline Cove.

Breeding grounds

Most great albatrosses breed in loose associations on the gentle gradients of open slopes, broad ridges or even on flat ground. At South Georgia, Wandering Albatrosses nest among low clumps of tussock grass, often around greens of short Antarctic hair-grass that are sometimes on quaking mires. Northerly breeding grounds are more tussocky and boggy, but the relatively flat, grassy herbfield of the Kerguelen coastal plain provides quite different nesting conditions (Derenne *et al.* 1972).

Coasts that face prevailing winds appear to have concentrations of wandering albatrosses, for example Bird Island at South Georgia, Adams Island in the Aucklands and the Péninsule Rallier du Baty of Kerguelen. Although nests are more numerous on the south-facing slopes of Adams Island, quite a lot of Wanderers nest on the northern, leeward slopes. In the Iles Crozet, Wanderers have been said to avoid the windiest places (Mougin 1970) and at Campbell Island, Southern Royal Albatrosses nest on the leeward sides of ridges in sheltered gullies and behind stunted bushes (Westerskov 1963). On the rocky islets off the Chathams Islands, Northern Royal Albatrosses appear to suffer from too much exposure; in the strongest gales incubating birds sometimes struggle to stay on their nests (Plate 33).

Wherever nests are sheltered or situated on very tussocky terrain, great albatrosses walk to some more exposed place to take off. This may be a cliff top, ridge or just flat ground where a run is possible. At South Georgia, Wandering Albatrosses habitually walk to the downwind end of flat meadows before facing into the wind and making their runs; sometimes numbers of birds congregate, apparently waiting their turn and letting the bird in front get airborne before lumbering off.

Colonies of the Northern Royal Albatross on islets of the Chatham Islands have densities of 500–800 nests per ha (Plate 32) (Robertson 1974, 1991); on the rolling hills of Campbell Island, nests of the Southern Royal Albatross are spaced at no more than 30 nests per ha.

At Macquarie Island and the Iles Kerguelen, nests of the Wandering Albatross may be far away from each other. At the Iles Crozet, average densities are less than one nest per ha. Some Wanderers nesting at South Georgia are hundreds of metres from their nearest neighbours, but within sight of larger gatherings. Densities of 20–40 nests per ha are usual and on favoured ridges densities of 106–170 nests per ha may be achieved (Tickell 1968). Similar densities evidently occur among the Auckland Wandering Albatross at Adams Island, where the tussock grass is so tall that most birds are out of sight of their neighbours (Plate 27) (Walker & Elliott 1999).

Nests

Differences between great albatross nests relate to climate, soil and vegetation rather than to species. New nests of the Southern Royal Albatross at Campbell Island and the Wandering Albatross at most sites are large mounds of loose vegetation, like truncated cones about a metre in diameter at the base and from a third to half a metre in height (Fig. 7.28). Although construction activity is most intensive at the beginning of the season (Jouventin & Lequette 1990), new material is gathered and added by both adults and the young throughout the year. By the time they have trampled the heap for a year, it is consolidated and in subsequent years becomes solid peat, sometimes columnar like mollymawk nests (Rankin 1951). The birds' excreta promote regeneration of the surrounding vegetation, often providing a sheltering ring of higher grass and the whole structure becomes a semi-permanent feature. The slopes at South Georgia are littered with such old nests which are taken over and rebuilt from time to time by both Wandering Albatrosses and giant petrels.

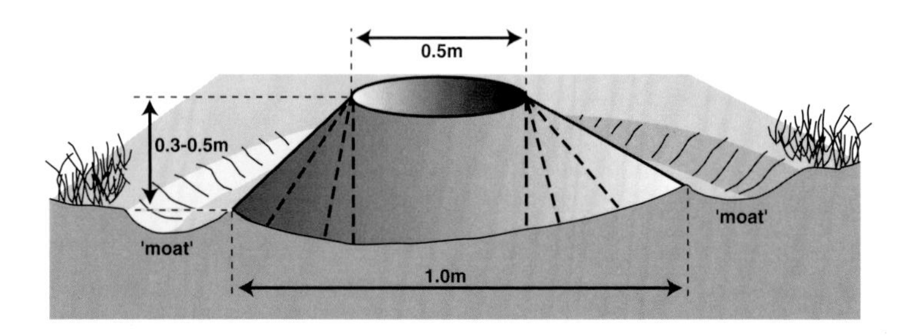

Fig. 7.28 Approximate measurements of a great albatross nest. Some may be lower than indicated and where ample vegetation is available the 'moat' is not apparent.

Where the vegetation is not so lush, it may be completely removed from a ring within reach of the sitting bird, thus forming a ditch of bare soil or peat which is sometimes imagined as a drainage 'moat'. It may function as such, but its absence elsewhere suggests that it is not a design feature. Nests like this are characteristic of the Péninsule Courbet of Kerguelen.

In drier localities, nests are much smaller. Those of the Northern Royal Albatross on the friable soil and meadow grasses of Taiaroa Head, New Zealand are only 0.6 m in diameter and 0.08 m in height (Richdale 1952); by the end of the season, they have been trampled flat by chicks (S. Webb pers. comm.). At the Chatham Islands, coarse stems are the main nest material and in their absence nests are little more than scrapes in bare soil or rims of stones.

Wandering Albatrosses at South Georgia locate their nests an average of 7 m (0–22 m, n=65) from their previous nest and 20% re-occupy nests they have used previously (Tickell 1968). At the Iles Crozet, 23% use the nests in which they successfully reared a chick two years previously while 38% of pairs use the same nest in which they lost an egg or chick the previous year. Auckland Wandering Albatrosses nest on average 21 m from their last breeding, whether successful or not; only 1% of successful breeders use the same nest two years later compared with 18% of failed pairs that do so one year later (Walker & Elliott 1999). Re-occupation of the same nest declines with time; of 236 successful pairs at Ile de la Possession, 34% re-occupied the same nest after two years, 17% after three and 12% after four years. (Fressanges du Bost & Segonzac 1976). Site tenacity in wandering albatrosses is evidently to area more than to nest site and re-occupation is a function of time since last breeding, nest density and thickness of vegetation.

Female Wandering Albatrosses normally locate their mates of former seasons and adopt the site he has selected, but occasionally synchrony fails and the female fixes on a site different from that of her mate. In one instance at South Georgia, a female laid in a nest 36 m away from her mate's site; she deserted after three days, the egg was lost and she later joined her mate. Similar instances have occurred at Macquarie Island, one of which resulted in a male moving to the site chosen by his female (Tomkins 1984).

Southern Royal Albatrosses at Campbell Island nest on average 54 m (10–81 m, n=11) from their previous nest (Sorensen 1950a).

Emigration/immigration

In spite of the strong pull of native islands, a few subadults are attracted to similar islands, especially when they find their own species active there. Of several thousand Wandering Albatrosses ringed as fledglings on the Iles Crozet, seven have been seen among the resident birds of the Prince Edward Islands, 1,000 km to the west of their natal islands and one bred there at the age of nine (Weimerskirch *et al.* 1985).

An Antipodes Wandering Albatross was ashore at Macquarie Island in 1996, and among the two immigrant Wandering Albatrosses that colonised Heard Island, at least one was from Macquarie (p.145).

Immigrant Northern and Southern Royal Albatrosses have settled at Taiaroa Head, New Zealand (Robertson & Richdale 1993) and at least one bird reared at Taiaroa Head has emigrated to the Chatham Islands.

Pre-egg periods

Wandering Albatrosses return to their nests at South Georgia and the Iles Crozet about 27 days before laying. Northern Royal Albatrosses at New Zealand average 34 days, and the longest recorded pre-egg period in both has been 45 days.

Males arrive first. At this time, most royal albatrosses find their breeding grounds vacated, but wandering albatrosses may encounter fledglings from the previous season, some of whose parents may still be bringing in food.

Wandering Albatross males occupy sites throughout the pre-egg period, leaving only for short trips to sea. Their females join them briefly at intervals. In 2,007 daily observations at 81 nests where an egg was later laid, males at South Georgia were present on 74% and females on 7% of occasions (Tickell 1968). Northern Royal males at New Zealand were present on 44% and females on 22% of days during the pre-egg period (Robertson & Richdale 1993).

Copulation

In the Wandering Albatross, copulation may occur on the day a female first returns to the breeding ground and may be repeated at intervals throughout the pre-egg period. The peak of copulation activity in both Wandering and Northern Royal Albatrosses is about ten days before laying (Tickell 1968, Tomkins 1983, Grau 1984). Apparently successful copulations were seen among Wanderers up to 24 days after laying, while failed breeders and subadults made attempts up to 65 days after the mean date of laying.

Female Wanderers walking on the breeding ground during the pre-egg period were almost always pursued and overpowered by solitary males occupying sites (Pickering in Birkhead 1988) (p.304).

Eggs

Eggs of great albatrosses are much larger than those of other albatross species. They are formed in about 40 days, the yolk being laid down in 30 days followed by 10 days for the albumin, membranes and shell (Grau 1984).

The mean calculated mass of 1,607 freshly laid Wandering Albatross eggs at South Georgia was 490 g and 95% were within the range 393–561 g. The size and mass of eggs laid by individual female Wandering Albatrosses in consecutive layings are more similar than those laid by different females. Individual difference accounts for 55% of variation in egg size at any one time within the South Georgia population (Croxall *et al.* 1992). Females which had bred on three or more occasions lay significantly larger eggs than those with less experience (Berrow *et al.* (in press)).

Egg size and mass also increase with the age of Northern Royal and Wandering Albatross females laying them, but this contributes to only a small proportion of the overall variation. In the Wandering Albatross at South Georgia, a maximum (asymptote) of about 500 g is reached between the age of 24 and 40 years. There is evidently an environmental influence that contributes to mean egg size being larger in some years than others (Croxall *et al.* 1992).

At the Iles Crozet, Weimerskirch (1992) demonstrated a similar increase in Wandering Albatross egg volume with age up to 23 years, but found that females older than 24 years laid smaller eggs.

There appears to be no marked difference in size between the eggs of the Northern and Southern Royal Albatrosses and although there may be variation between those from different populations of the Wandering Albatross, none of them have been adequately sampled or treated statistically.

Anomalous eggs, perhaps symptomatic of disease, are sometimes found. In 1926, an entirely brown Wandering Albatross egg was seen by Harrison Matthews (1929) among more than 2,000 collected by whalers from Wandering Albatross nests at South Georgia.

Robert McCormick (1884) found a two 'exceptional' shaped eggs in the nest of a Southern Royal Albatross on Auckland Island; one was long and the other rounded. In 1963, an experienced Wandering Albatross at South Georgia, laid two undersized, defective eggs that broke on handling (Tickell & Pinder 1966), and in the 1980s, among 1,137 eggs, two were undersized, weighing 109 g and 230 g (Croxall *et al.* 1992).

Laying

The date of laying varies from island to island (Table 7.4). The mean laying date of Wandering Albatrosses at South Georgia in consecutive seasons varied by up to two days.

Individual females of both the Wandering and Northern Royal Albatrosses mated to the same males may lay on the same day in separate years, but differences average about four days with ranges of up to 20 days. Some may be consistently early or late layers and this persists even when they lose partners and re-mate (Robertson & Richdale 1993). Age, experience, the outcome of previous breeding (Weimerskirch 1992) and nutrition during the pre-egg period (p.35) also influence the time of laying.

	Date of laying			
	Mean±SD (median*)	Range	n (layings)	n (years)
Northern Royal Albatross (New Zealand)	11 Nov.	27 Oct.–28 Nov.	424	55
Northern Royal Albatross (Chatham Islands)	14 Nov.±4.8	27 Oct.–30 Nov.	929	1
Southern Royal Albatross (Campbell Island)	3 Dec.*	20 Nov.–28 Dec.	107	4
Wandering Albatross (South Georgia)	24 Dec.±5.7	10 Dec.–17 Jan.	260	3
Auckland Wandering Albatross (Auckland Islands)	7 Jan.*	29 Dec.–5 Feb.		
Amsterdam Wandering Albatross (Ile Amsterdam)	28 Feb.±6.0	12 Feb.–8 Mar.	17	4

Table 7.4 Laying dates of great albatrosses: (Sorensen 1950a, Tickell 1968, Jouventin *et al.* 1989, Robertson 1993c, Waugh *et al.* 1997, Walker & Elliott (1999).

Northern Royal Albatrosses at the Chatham Islands lay on average about 19 days earlier than Southern Royals at Campbell Island, but the ranges overlap by about ten days. In New Zealand, where the colony at Taiaroa Head is mainly of Northern Royals, the progeny of an immigrant female Southern Royal and a Northern Royal male have an average laying date 11 days later than that of the Northern birds (Robertson 1993b).

At South Georgia, three times more Wandering Albatross eggs were laid during the day than at night. Robert Tomkins (1984) saw two Wandering Albatrosses laying eggs at Macquarie Island. Both were experienced females; they laid whilst standing, with their feet on the rim of the nest. The eggs were ejected quickly.

> 'Immediately before laying the first bird stood hunched, body leaning forward as if unbalanced, her head slightly forced forward and lower to the ground than was the other birds' usual upright stance. She remained in this strained, motionless position for at least three minutes, and then she laid...The second bird began to rotate her body about her pelvis, rhythmically rocking back and

forth for several minutes before laying. Her body movements were as if she was defecating in slow motion, i.e. a backward swing lifting tail and cloaca high. The frequency of her swings increased until she laid. Both birds stood motionless for a few seconds after laying, inspected and touched the egg, then settled on it and commenced to incubate'.

A Northern Royal Albatross filmed when laying at the Chatham Islands was also standing; at the moment the egg emerged, the tail was depressed under the cloaca to guide it into the nest. (Wright 1980).

Incubation

The eggs of most great albatrosses are incubated for 78–79 days (Table 7.5). Male and female Wandering Albatrosses share incubation at South Georgia in four to 14 shifts (n=166). Northern Royals at New Zealand complete incubation in 6–19 shifts.

In any one season, either male or female Wanderers may take the larger share of incubation, but on average males incubate for several days longer than females. In both the Northern Royal and Wandering Albatrosses, the first shift, taken by the female, is usually very short, sometimes lasting only an hour or so, before the male takes over. This is consistent with the male's role in occupying territory throughout the pre-egg period. However, if the male is not present at the time of laying or soon after, females can sit for much longer. One Wanderer female stayed on her egg for 20 days before deserting. Subsequent shifts are longer; Wanderer males have been known to sit for up to 38 days and Northern Royals for 17 days before being relieved, while male and female Wanderers have sat for 50 days and 48 days respectively before deserting. The mean lengths of Wanderer shifts tend to shorten as incubation progresses (Fig. 7.29).

	Incubation (days)		
	Mean±SD	range	n
NRA (New Zealand)	78.8±1.5		101
SRA (Campbell Island)	78.5±2.8	74–85	10
WA (South Georgia)	78.4±1.2	75–82	163
WA (Iles Crozet)	79	77–83	

Table 7.5 Length of incubation in great albatrosses (Tickell 1968, Robertson & Wright 1973, Fressanges du Bost & Segonzac 1976, Waugh *et al.* 1997). Northern Royal Albatross (NRA), Southern Royal Albatross (SRA), Wandering Albatross (WA).

At Bird Island, South Georgia in the late 1980s, two female Wandering Albatrosses laid eggs in separate nests built close to each other. A single male shared incubation with each female in turn, but this intriguing natural experiment failed when the male sitting on one nest excavated the side of the adjacent nest. The egg eventually rolled out between the two nests and was abandoned (M. Jones pers. comm.).

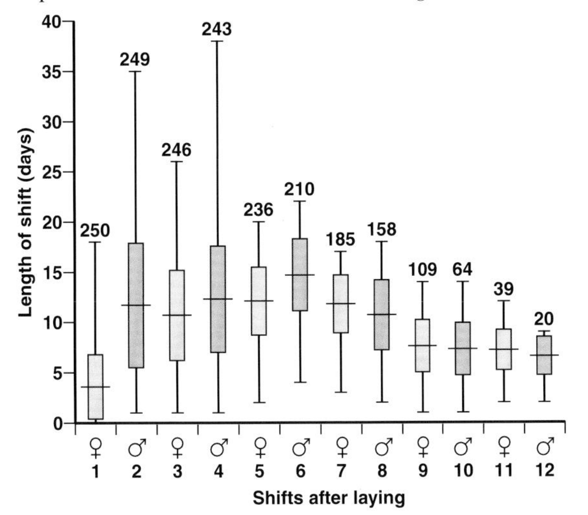

Fig. 7.29 The lengths of incubation shifts (mean±SD, range, n) by male (♂) and female (♀) Wandering Albatrosses at South Georgia in 1963 (Tickell 1968).

Hatching

Inter- and intra-specific differences in dates of hatching should correspond to those of laying and location, but hatching among Wandering Albatrosses at South Georgia is influenced by the size of eggs. Larger eggs take longer to incubate. In some years, all females apparently benefit from better nutrition in the pre-egg period and, as a consequence, lay larger eggs which take slightly longer to incubate and contribute to an improved hatching success (Croxall *et al.*1992).

Of 38 Wandering Albatross eggs which failed in one season at South Georgia, 37% were apparently infertile, 26% were broken in the first week after laying and 37% were deserted by parents sitting for longer than the average length of shift.

Chicks and Fledglings

Wandering and Northern Royal Albatross chicks take one to five days to break out of their eggs and new chicks may appear under either male or female parents. During their first days, the body temperatures of Wandering Albatross chicks are relatively low and variable. On warm days without wind, they can endure exposure while slowly losing heat, but in wind and rain body temperature falls dangerously in a few minutes. Thermoregulatory ability improves fairly slowly (Mougin 1970). Chicks of all great albatrosses are brooded for 21–43 days in shifts averaging two or three days; the males' average share tends to be longer than that of females. As chicks grow larger, brooding becomes more intermittent, and parents eventually get off the nest altogether and sit alongside. The young of all great albatrosses are in or about the nest as chicks and fledglings throughout the southern winter. These are usually not extreme, but at South Georgia, the breeding grounds are swept by drifting snow, covering nests and obscuring terrain (Plate 23).

The fledging periods of all wandering albatrosses are longer than those of royal albatrosses (Table 7.6), but there are geographic differences in the times taken for wandering albatrosses to rear young. At South Georgia, Wandering Albatross chicks fledge in 278 days. At the Iles Crozet they take about a week less (Voisin 1969) and three Amsterdam Wandering Albatross chicks took 253, 257 and 274 days respectively (Jouventin *et al.* 1989).

	Fledging (days)		
	Mean±SD	range	n
NRA (New Zealand)	240±9	216–257	64
WA (South Georgia)	278±17	263–303	35
WA (Iles Crozet)	271	258–288	

Table 7.6 Length of the fledging period among great albatrosses (Tickell 1968, Robertson & Wright 1973, Fressanges du Bost & Segonzac 1976). Northern Royal Albatross (NRA), Wandering Albatross (WA).

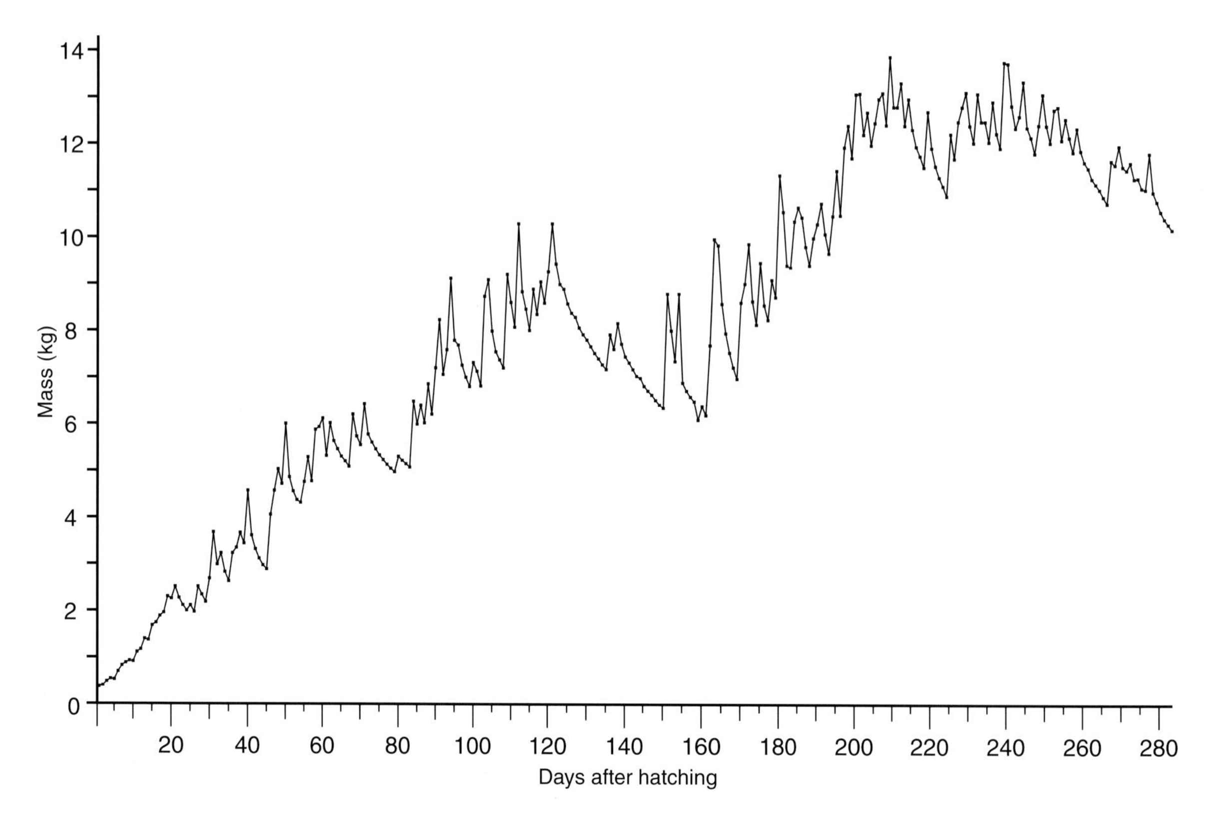

Fig. 7.30 The growth of a young Wandering Albatross at Bird Island, South Georgia. The daily mass of one chick which hatched on 29 March 1963 and was fully fledged when last seen on 6 January 1964 (Tickell 1968).

Newly hatched Wandering Albatross chicks are covered in a loose, white, primary down. Those of the Northern Royal Albatross have shorter, sparse down, particularly about the head, and areas of bare dark skin extend from the lores to the ears. As the chicks grow, secondary down soon covers these areas. Royal and Wanderer chicks have white eyelids when they hatch, but those of Royals begin to darken at about 30 days and later turn jet black (Richdale 1939).

Wandering Albatross chicks weigh about 300–400 g at hatching and do not grow very much during the first three days, thereafter they put on mass rapidly and within 15 days weigh about one kilogram (Fig. 7.30). By the time they are left alone in their nests at South Georgia they are covered in a denser, pale grey down and average about two kilograms (1.5–4.4 kg).

The larger eggs of more experienced Wandering Albatrosses produce heavier chicks; at first they grow relatively faster than those from the smaller eggs of inexperienced females, but later the difference disappears. Although first-time breeders are already experienced foragers for themselves, they have to acquire skill at saving and transporting enough extra food for the growing chick. This they appear to accomplish with their first or second chicks. First-time breeders visit their young chicks less often than experienced parents, who may be delivering smaller feeds more frequently. As the chicks get older, experienced birds visit less often and overall there is no difference in provisioning rates and meal sizes. The fledglings of inexperienced and experienced parents achieve the same size and mass by the time they fly (Lequette & Weimerskirch 1990, Berrow *et al.* in press).

Northern Royal and Wandering Albatross chicks are fed in the nest right up to the time they fly. They achieve peak mass in excess of adult mass; Wanderer chicks at South Georgia average almost 11 kg (8–15 kg, n=190) at the age of 220–230 days, towards the end of September. By the time they leave in December, they have lost much of this excess and are about as heavy as adults.

In 1963, Wanderer parents at South Georgia delivered 51–82 kg of food to their chicks between the end of brooding and their departure as fledglings (Table 7.7). Average quantities remain fairly constant from month to month, but in the final weeks ashore rather less is delivered (Fig. 7.31).

Fledglings may be fed the day before they depart, but others fast for up to three weeks. The last fast of Wanderer fledglings at South Georgia (mean 7 days, 0–21, n=19) is only half as long as the longest fast (mean 14 days, 10–24, n=19) previously endured by them. Nevertheless they lose mass at 65–150 g per day during the last 50 days in the nest (Fig.7.30), and mean daily feeds decline to less than 150 g per day during the last 20 days before departure.

	Mean±SD	Range
Number of days	246±12	231–270
Number of days when fed	81±8	67–96
Feeding frequency (1 feed per x days)	3.1±0.2	2.6–3.5
Food fed per day (g)	264±33	203–328
Total weight of food (kg)	65±7.6	51–82

Table 7.7 The mass of food and feeding frequency of 15 Wandering Albatross chicks at South Georgia between the end of brooding and their departure as fledglings (Tickell 1968).

Older fledglings become more active, not only standing and exercising their wings, but moving away from the nest. They are already familiar with their surroundings and can return to their nests from 100 m. In their last months, fledglings shuffle and walk about over a wide area, sometimes making nests. At Amsterdam Island, fledglings make up to ten nests as far as 30 m from their parents nest (Jouventin *et al.* 1989). In the last weeks at South Georgia they move around even more, extending their wings, jumping into the wind and making practice take-off runs.

Energy costs of activity might thus be expected to increase while feeds decrease and so contribute to decline in mass. In large colonies, starving young in different stages of development are always to be seen from late winter into mid-summer and emaciated individuals linger several weeks after the last adequately fed fledglings have gone.

In the Northern Royal Albatross, the bill of chicks grows faster around the base than at the tip (Richdale 1939, 1950). Bills, tarsi, middle toes and wing bones all reach full size well before fledgling Northern Royals and Wanderers depart. Flight feathers begin appearing about half way through the growth period. Those of young Northern Royal Albatrosses emerge significantly earlier than those of Wandering Albatrosses (Fig. 7.32) (Tickell 1968).

At South Georgia, 24% of 62 Wandering Albatross chicks grew complete sets of primaries in less than 150 days and only one took less than 135 days. The heavier the chicks at primary emergence, the sooner they fledge, but even underweight, chicks complete all their primaries before flying. The four outer primaries have logistic-shaped growth-curves, but they do not all grow at the same rate at the same time. Asymptote lengths are reached in the sequence, p7, p8, p9, p10 and growth rates increase from p7 to p9 to a maximum of 4.75 mm per day. There is a pattern in the growth of these primaries that leaves the completion of the longest, p10 until last. By the time fledglings fly, p7 and p8 have stopped growing, p9 is growing at only 0.1 mm per day and p10 at 0.5 mm per day. The first complete set of flight feathers may have to serve juvenile Wandering Albatrosses for five or six years before they have all been replaced. Berrow *et al.* (1999) have suggested that this pattern of growth ensures that the outermost primaries are the least abraded (by flapping among tussock grass) when fledglings start flying.

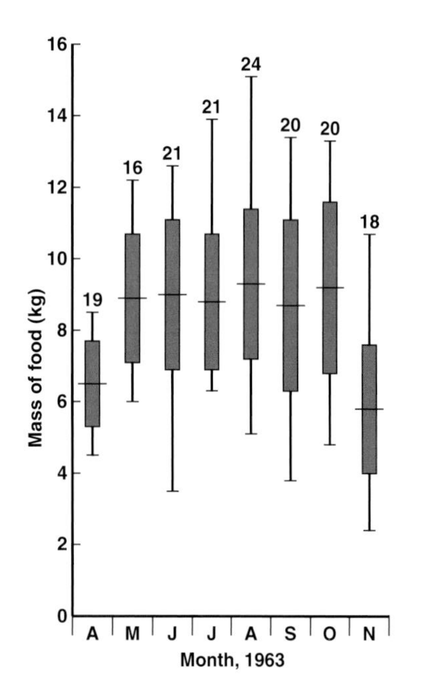

Fig. 7.31 The mass of food (mean±SD, range, n) delivered each month to Wandering Albatross chicks at South Georgia throughout the southern winter of 1963 (Tickell 1968).

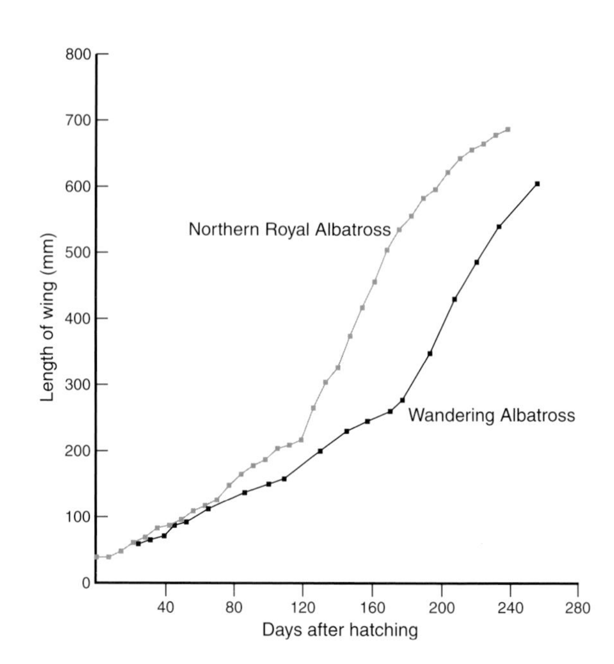

Fig. 7.32 Growth of the wings in chicks of the Northern Royal Albatross and Wandering Albatross (Richdale 1952, Tickell 1968).

Wandering Albatross fledglings leave South Georgia and Macquarie Island shortly after breeding adults of the following season begin returning. During December, large numbers of fledglings are on the breeding grounds at the same time that males are occupying sites. Fasting fledglings that approach territory-holding males to beg for food are driven off aggressively (Tomkins 1984). Although most Northern Royal Albatross fledglings have gone by the time the next season's breeding adults start returning to New Zealand, very late chicks may sometimes still be present.

Northern Royal fledglings go to sea between late September and early October while Southern Royals leave between early October and early December. Wandering Albatross fledglings are later; the mean date of departure from South Georgia is 10 December (17 November to 8 January). Amsterdam Wandering Albatross fledglings go to sea between mid-January and late February.

Juveniles

Juveniles spend all their time at sea. Almost nothing is known about these several years other than the locations of a few ringed birds.

Subadults and former breeders

Almost all young Wandering Albatrosses return to the islands upon which they were reared. A few come back at the age of three but most return as five to seven-year-olds and some not until they are ten or 11. There are significant differences between cohorts, and females tend to return slightly younger than males (Table 7.8) (Pickering 1989, Croxall *et al.* 1990b, Weimerskirch 1992, Jouventin *et al.* 1989). Subadults of the Northern Royal Albatross are also first seen in New Zealand at three years of age, but the average age of return is four years and 93% are back by the age of five (Robertson & Richdale 1993).

	Years	Sex	Mean (years)	Range	n
South Georgia	1972–77	M	6.0	4–10	466
		F	5.7	3–11	533
Iles Crozet	1982–90	M	6.2	3–10	195
		F	5.7	3–10	193

Table 7.8 The age at which subadult Wandering Albatrosses first return to the breeding islands. The differences between males and females at both locations are significant (Pickering 1989, Weimerskirch 1992).

Tomkin's marathon – one observation every 30 minutes of daylight for 103 days – on the few Wandering Albatrosses at Caroline Cove, Macquarie Island revealed a diurnal pattern of subadult activity. They came in throughout the morning and early afternoon, then left in the late afternoon, spending the night at sea; the sequence was most noticeable among the four to eight-year-olds. There was also a tendency for those birds to come in more often after an abrupt increase in wind speed and they were absent overnight more often just before full moon (Tomkins 1985).

Subadults of the Wandering Albatross apparently favour areas of higher nest density, even individuals who have themselves been reared in areas of low nest density (Weimerskirch & Jouventin 1987). Although they have already achieved adult dimensions, their body mass continues increasing up to the age of seven years (Weimerskirch 1992).

The youngest subadult Wandering Albatrosses at South Georgia and the Iles Crozet arrive from late January to early March, they spend only a few days on land and are not seen again after the end of February or early March (Fig 7.33). At this time, breeding birds have been back about ten weeks and are incubating eggs. In the following two to eight years, subadults arrive progressively earlier and leave later.

The younger they are when they first come back, the more years they spend visiting their breeding grounds before finding a mate (Fig. 7.34). This corres-

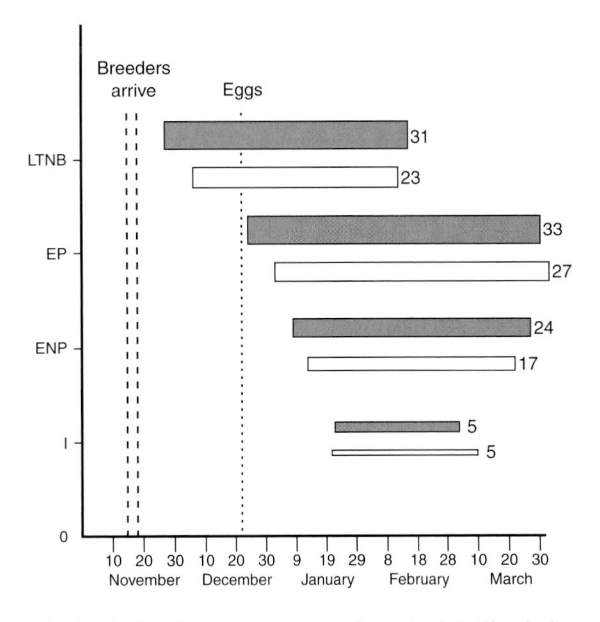

Fig. 7.33 The time spent ashore by subadult Wandering Albatrosses at Bird Island, South Georgia in relation to the stages of pair-bond formation, the median dates of arrival of breeding birds and of egg laying. I = inexperienced subadults during their first season(s) on the breeding ground, ENP = experienced subadults that have not yet started pairing, EP = experienced subadults that are pairing, LTNB = last time non-breeders in the season before they lay their first egg. The thickness of bars is proportional to the number of days ashore and the median number of days is shown at the end of each bar. Shaded bars are males and clear bars females (after Pickering 1989). Similar diagrams have been constructed for the Northern Royal Albatross (Robertson & Richdale 1993) and the Sooty Albatross (Jouventin & Weimerskirch 1984a).

ponds to the time taken to attain breeding mass. To be successful, subadult females at the Iles Crozet have to attain a threshold body mass of about eight kilograms (Weimerskirch 1992).

Each year, males spend more days ashore than females of similar experience. On average, subadult Wanderers visit Bird Island for four seasons before breeding. In the season during which they pair, subadults arrive earlier than those of the same age and experience that do not pair. In the next season they arrive at the same time as breeders, but themselves do not breed and they depart in February, long before unpaired birds. The following season they breed (Pickering 1989). First-time breeders have spent proportionally more time ashore than typical subadults of the same age that do not breed. A similar sequence of events occurs in the Northern Royal Albatross at New Zealand (Robertson & Richdale 1993).

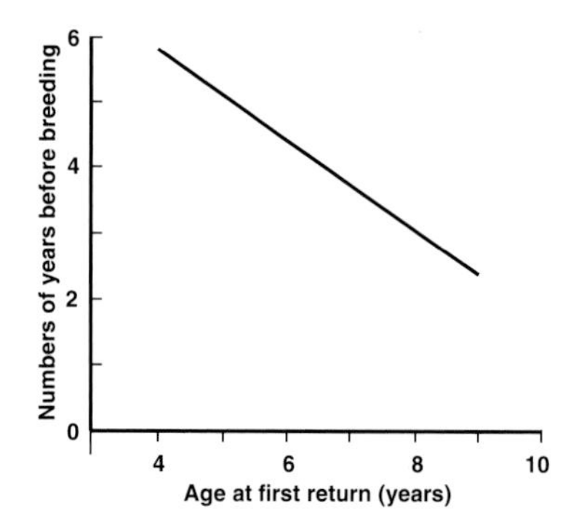

Fig. 7.34 The approximate relationship between the age at which Wandering Albatross subadults first return to Bird Island, South Georgia and the time it takes for them to acquire a mate and breed for the first time (after Pickering 1989).

Inexperienced Wanderers spend about three-quarters of their time sitting alone. By their third season, most males have a nest-site where they spend about 30% of their time, significantly more than females. In contrast, while females of similar experience spend equal proportions of their time alone or at a nest-site with a male, they also spend about 30% of their time dancing, which is significantly more than males. After dancing, a female sometimes sits with a male at a nest-site before moving on to perform with another male. Gradually, a female spends longer with just one male until eventually the two birds spend most of their time together, or alone on the male's chosen nest-site.

Subadults of the Wandering Albatross at Bird Island may have up to seven temporary partners; the number is partly a function of the time spent ashore dancing. Females dance with more partners than males; by the time the pair bond is consolidated, the number has declined to a new partner for every ten days ashore. Most subadult pairs spend about 14 days together over two seasons before breeding; some pairs can breed after only one season, providing they spend about the same number of days together (Table 7.9).

No. of seasons together as subadults before breeding	n (pairs)	Days together 3	2	1	Total
1	5	-	-	15	15
2	21	-	6	8	14
3*	1	8	8	1	17
3*	1	4	8	12	24
Former breeders	15	-	-	8	8

Table 7.9 The mean number of days on which Wandering Albatross pairs were seen together in each season before the one in which they bred (figures rounded to the nearest whole day), * values for each pair given separately (after Pickering 1989).

Experienced breeders usually re-mate with other experienced breeders. They have to spend time establishing new pair-bonds, but once that is achieved they spend only eight days together in just one season before breeding. In the 1980s, 49% of females and 15% of males had re-paired one season after losing their mate, while 98% of females and 78% of males re-paired within three seasons. There were 11 former breeding males still alive in 1986 that had not bred for five years, while all females had re-paired within four years. Only 8% of former breeders re-paired with subadults (Pickering 1989).

In the large population at South Georgia, few subadults came in on the rare calm days. Numbers increased in winds of up to 15–20 knots (28–37 kph) and decreased above 25 knots (46 kph). There were also more ashore when the winds were from the SE and SW. Careful analysis revealed no lunar influence (Pickering 1989).

First breeding

About 70% of the Wandering Albatross adults at South Georgia breed for the first time between eight and eleven years of age and about 90% have bred at least once by 13 years; all are believed to have done so by 20 years of age (Croxall 1991, Croxall *et al.* 1990b). With decline in the population there has been a progressive change in the age of first breeding discussed in Chapter 16 (Croxall *et al.* 1997b). Males breed slightly earlier than females (Table 7.10), but the difference is not significant in Northern Royals.

	Cohorts	Males			Females		
		Mean (years)	Range	n	Mean (years)	Range	n
NRA (New Zealand)	many (pre 1975)	10.6	8–22	20	9.9	8–14	19
NRA (New Zealand)	many (post 1974)	8.7	6–14	24	8.5	7–10	25
WA (South Georgia)	1 (1972)	11.4	8–15	184	10.9	8–15	184
WA (Iles Crozet)	13 (1966–78)	12.1	9–16	24	11.2	7–16	64
WA (Iles Crozet)	9 (1982–90)	10.4	7–18	183	9.6	7–18	198

Table 7.10 Age of first breeding in the Northern Royal (NRA) and Wandering (WA) Albatrosses (Weimerskirch & Jouventin 1987, Croxall *et al.* 1990b, Weimerskirch 1992, Robertson 1993c).

In very small colonies, imbalance in the sex ratio may prevent potential breeders from finding mates. This was apparent in the early decades of the Northern Royal Albatross colony at New Zealand (Robertson & Richdale 1993). At Macquarie Island, where the few Wandering Albatross nests are sometimes kilometres apart, some birds over 22 years old have not yet bred. The small numbers of Gough Wandering Albatrosses on Inaccessible Island, Amsterdam Wandering Albatrosses at Amsterdam Island and Antipodes Wandering Albatrosses at Campbell Island may all suffer from the same problem.

Adults

At the Iles Crozet, Wandering Albatrosses with breeding experience are significantly heavier than subadults. Males gradually increase in mass throughout their breeding life while females reach their maximum (asymptotic) mass at about 20 years of age. Improved physical condition is presumably associated with higher feeding efficiency among more experienced birds. Low annual mortality rates, however, means that many of these Wanderers survive into old age, when some may cease to breed altogether (Weimerskirch 1992).

Breeding success

The proportions of great albatross eggs that hatch and of chicks that develop successfully into fledglings vary from year to year and between different populations (Table 7.11).

The Wandering Albatross and Southern Royal Albatross generally have high hatching rates and very high fledging rates. Among Wanderers at the Iles Crozet, most egg-loss occurs during the first 10–20 days of incubation and most chicks are lost in the ten days following hatching (Fressanges du Bost & Segonzac 1976). At the Iles Crozet, breeding success in males and females increases progressively with age up to 20 years, but goes into decline among birds older than 20 years.

At the Chatham Islands, the Northern Royal Albatross is far more vulnerable to adverse nesting conditions; crowding, poor nests and exposure all contribute to lower breeding success. Birds incubating in high winds may be blown off nests, allowing their eggs to roll out or break in the nests. Nine percent of eggs may be lost in a single storm. In years of high density, egg-shells were at least 20% thinner than in years of low nest density (Robertson

1997). In 1993–94, 53% of eggs laid on the Forty-Fours had failed by 10 December (Robertson & Sawyer 1994). From 1990 to 1996, breeding success averaged only eight percent and in one year it was as low as three percent (Robertson 1997). Breeding success is an important parameter and its role in the computation of population size will be discussed in Chapter 16.

In very small colonies such as those at Macquarie Island, New Zealand and Ile Amsterdam, losses of just a few eggs or chicks generate erratic extremes in success rates.

	No. of eggs	No. of seasons	% eggs hatched (range)	% chicks fledged (range)	% eggs fledged (range)
NRA (New Zealand)	383	49		47	
SRA (Campbell Island)	297	4	76 (71–80)	75 (65–91)	57 (46–74)
	1 567	16		67 (43–88)	
WA (South Georgia)		13	73 (67–82)	86 (73–92)	64 (52–73)
WA (Prince Edward Islands)		10		(63–96)	
WA (Is. Crozet)	262	5	72 (54–84)	89 (84–97)	(53–74)
AKWA (Auckland Islands)	88–221	5			67 (61–78)

Table 7.11 Breeding success in the great albatrosses – Northern and Southern Royal Albatrosses (NRA, SRA), Wandering Albatross (WA) and Auckland Wandering Albatross (AKWA) (Williams & Imber 1982, Fressanges du Bost & Segonzac 1976, NZ DoC unpublished, Croxall *et al.* 1990b, Cooper & Brown 1990, Moore *et al.* 1997b, Waugh *et al.* 1997, Walker & Elliott 1999).

Survival and longevity

The mean annual adult survival rate of the Wandering Albatross at Bird Island, South Georgia over the years 1976–84 was 95% for males and 93% for females (Croxall *et al.* 1990b). It was rather higher among the Auckland Wandering Albatross in 1992–94; 98% for males and 96% for females (Walker & Elliott 1999). In 1997, there were nineteen 39-year-old Wandering Albatrosses breeding at Bird Island (Croxall *et al.* 1999).

Pooled data for the Northern Royal Albatross at New Zealand (1937–93) yielded mean survivals of 96% for males and 94% for females. One female bred in the season before she was presumed to have died at sea, 55 years after she had been first ringed as an adult (Robertson 1993c).

FOOD

The Royal and Wandering Albatrosses obtain their food mainly by seizing and scooping from the surface. Carrion is probably an important resource underestimated in most analyses. I have encountered Wandering Albatrosses off South Georgia, feeding on blubber from strips of skin torn from infant humpback whales by killer whales and at the carcasses of many penguins killed by leopard seals or furseals.

Compared with other albatrosses, they are not remarkable divers (Prince *et al.* 1994), but they will attempt to do so when an opportunity occurs:

> 'A Wanderer cruising at about 15 feet above the water suddenly baulked and dropped to the surface. It immediately plunged below, except for its extremities, to retrieve a writhing cuttle which was vigorously ejecting sepia ink into the bird's face.' (Gibson & Sefton 1959)

For a long time it was concluded that the Wandering Albatross fed at night, but in the early 1990s, sealed temperature-sensing instruments[9] became available. Free living Wandering Albatrosses were persuaded to swallow the 100 g capsules which recorded abrupt changes in internal body temperature every time the bird ingested cold prey. These direct measurements of feeding have established that the Wandering Albatross of the Iles Crozet

[9] Einkanalige Automatische Temperatur Logger (ETAL).

feed mostly during the day; only 11% of the total prey mass was caught at night and there was no indication of preferential feeding at dawn and dusk (Weimerskirch & Wilson 1992).

The diet of great albatrosses is made up of squid, fish, crustaceans and carrion (Fig. 7.35). The proportions for the Wandering Albatross at the Iles Crozet in the southern winters of 1982 and 1992 were: squid (77%, 72%), fish (15%, 24%) and carrion (8%, 3%) (Ridoux 1994, Weimerskirch *et al.* 1997b).

Undigested squid beaks regurgitated by wandering albatrosses and their chicks at six island groups included more than 60 species, many of which were common to the diets at several islands. Large samples from higher latitudes contained fewer species (South Georgia – 37) than from lower latitudes (Antipodes Island – 52) (Imber 1992) (Appendix 8).

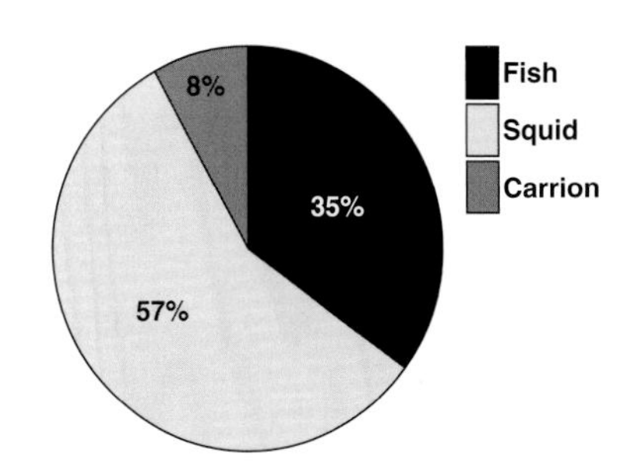

Fig. 7.35 Food delivered to 50 Wandering Albatross chicks at Marion Island. Pooled data from two cohorts collected at different stages in chick growth: September–December 1988 and June–August 1989. Segments of the diagram indicate the proportions by weight. Crustaceans in the diet amounted to only a trace (Cooper & Brown 1990).

Squid comprised about 35% by mass of the food delivered to Wandering Albatross chicks at South Georgia (Rodhouse *et al.* 1987) and eight of the 20 squid families represented contributed more than 95% of the food consumed. The most important single species of squid prey consumed was the onychoteuthid *Kondakovia longimana*, which is the biggest species with a mean mass of over three kilograms, but its relative importance varied seasonally. Although it was the most numerous (40%) of the 25 species taken in the summer of 1976–77, (when it amounted to 81% of the total fresh mass of squid eaten), during the winters of 1983 and 1984 it was less numerous (13% and 8%), but still dominant (46% and 55%) among the 46 species eaten. Smaller *Histioteuthis eltaninae, Illex* sp. and *Galiteuthis glacialis* were more numerous (Rodhouse *et al.* 1987).

Squid diets of wandering albatrosses vary between breeding areas. Some of the squid eaten by the Wandering Albatross of South Georgia, including *K. longimana*, are known to come from the cold seas south of the Antarctic Polar Frontal Zone (APFZ), while others are characteristic of warmer waters to the north. The diet of Wandering Albatrosses at several islands suggests changes in geographic feeding range or of water masses within flight ranges from year to year, and perhaps between summer and winter (Croxall & Prince 1980, Rodhouse *et al.* 1987). Cooper *et al.* (1992) concluded that the four most abundant squid taken at the Prince Edward Islands and the Iles Crozet (*K. longimana, Moroteuthis knipovitchi, H. eltaninae* and *G. glacialis*) were of similar sizes to those taken at South Georgia, but there were few similarities between the fish species taken at the Prince Edward Islands and South Georgia.

Squid taken by Wandering Albatrosses include deep-living species, some so large that they could not have been captured and killed by albatrosses. There has been much speculation as to how surface- feeding albatrosses came by them (Imber 1992). Sperm whales, long-finned pilot whales and southern bottlenose whales are all squid eaters capable of diving to great depths. They discharge fragments at the surface, attracting many seabirds including albatrosses (Clarke *et al.* 1981). Some species of squid naturally float to the surface when dead or moribund, particularly after spawning (Lipinski & Jackson 1989).

Squid fed to chicks on the Ile de la Possession in the winter of 1982 was mainly the deep-sea *K. longimana* (61%) followed by *M. knipovitchi* (3%), but in 1992, the main prey was *Moroteuthis ingens* (57%) followed by *K. longimana* (11%) (Ridoux 1994, Weimerskirch *et al.* 1997b).

Squid prey of the Northern Royal Albatross at the Chatham Islands was mainly *M. ingens* (80%), but more than half the cephalopod food samples from New Zealand birds were bottom-living octopus. Imber (1991) believed they may have been the buoyant fragments of seal prey or discards from crayfish traps. Croxall & Prince (1994) have calculated that South Georgia Wanderers could obtain 80% of their squid requirement by scavenging fish and squid that had died or been killed naturally and brought to the surface in one way or another. Cooper *et al.* (1992) came to a similar conclusion with respect to Wanderers from other islands.

In the southern winter, the Wandering Albatross of South Georgia, Prince Edward Islands, Iles Crozet, Iles Kerguelen and Macquarie Island as well as the Gough Wandering Albatross, Auckland Wandering Albatross and Antipodes Wandering Albatross, all visit the western Tasman Sea. They feed on the spawning grounds of the the giant cuttlefish *Sepia apama*, an area of about 200 km^2 off the New South Wales coast of Australia, near Bellambi. After spawning in July and August, the dead or moribund cuttlefish come to the surface where schools of dolphins sometimes play a role in breaking up the bodies. The albatrosses have been seen there since 1912, but have probably been visiting for much longer (Gibson 1963).

For over 50 years (1916–72), untreated sewage from the Sydney suburb of Malabar was discharged into the sea close inshore, under 50 m cliffs. It contained carcass trimmings of fat and meat from a nearby abattoir amounting to at least 1.8 tonnes per day. It is not known when wandering albatrosses began feeding at this outfall, but unlike cuttlefish, it was available throughout the year. By the early 1950s, up to 400 albatrosses could be seen between May and November within a kilometre of the shore; often coming in under the cliffs. By the end of the decade numbers in excess of 700 were being recorded. During the late 1960s, sewage treatment was introduced and by 1972 abattoir scraps were no longer being discharged. Within a few years, wanderers ceased to visit Malabar (Battam & Smith 1993). Bellambi and Malabar are only about 55 km apart, but wanderers that habitually fed at Malabar rarely visited Bellambi and vice-versa; many ringed and dyed birds have never been seen again at either location, even though some are known to be alive at their breeding grounds some 25 years later (Smith 1994).

Fish occurred in 93% of 80 food samples collected from Wandering Albatrosses at South Georgia in 1983 and 1984. They comprised 56% of prey items (excluding carrion) and contributed 45% of the diet by mass. Six out of ten species were identified by means of their otoliths (Appendix 8). There were differences in the relative quantities taken each year, but the approximate contributions by mass of the three main species were *Pseudochaenichthyes georgianus* (54%), *Chaenocephalus aceratus* (25%) and *Muraenolepis microps* (13%). These are characteristic species of the South Georgia continental shelf, but the latter two are demersal fish, not available to birds on the surface, so they may have been obtained as discards from trawling or perhaps as dead or moribund floaters after spawning (Croxall *et al.* 1988). At the Ile de la Possession, Patagonian toothfish from the Crozet shelf appear to be the most important fish in the Wanderer diet (Weimerskirch *et al.* 1997a).

Occasional large pelagic decapods or amphipods are conspicuous in regurgitations, but crustaceans are usually of minor importance in the diets of the Wandering Albatross at South Georgia. In 1974, however, 50 wandering albatrosses were seen in a large flock of mixed seabirds off the Falklands, feeding on shoals of lobster krill (Holloway in Bourne 1975).

Jellyfish, Portuguese man-o'-war *Physalia physalia* and pelagic tunicates are eaten by great albatrosses, but rarely detected in diets. The earliest observations are as curious today as they were to Joseph Banks in the Tasman Sea, when he rowed out in his small boat from the *Endeavour* to collect marine animals with a dip-net (Beaglehole 1962). On the 11 and 12 April 1770 he wrote in his journal:

> '...an Albatross that I had shot dischard a large quantity, incredible as it may appear that any animal should feed upon this blubber, whose stings innumerable give a much more Acute pain to a hand which touches them than Nettles.'
>
> '...I again saw undoubted proof that the Albatrosses eat ....Portuguese men of War as the sea men call them.'

Banks's albatrosses could have been either wanderers or royals. No-one else has described either of them eating these conspicuous marine animals, but Weimerskirch *et al.* (1986) have described Wandering Albatrosses eating jellyfish off the Iles Crozet and they may all be more significant as food than previously supposed.

PARASITES AND DISEASE

Large chewing lice are often conspicuous on the heads of great albatrosses. Six species are all common to wandering and royal albatrosses. Five of the species are confined to great albatrosses, but *Episbates pederiformis* occurs also on Galápagos and Black-footed Albatrosses (Appendix 5).

One species of flea, *Parapsyllus longicornis*, has been collected from the Northern Royal Albatross in New Zealand (Dunnet 1964). The tick, *Ixodes uriae* has been found in nest material of the Wandering Albatross at South Georgia (Wilson 1970), embedded near the eyes and bill of an Antipodes Wandering Albatross (Waite 1909) and on the Northern Royal Albatross at the Chatham Islands (Robertson & Sawyer 1994). Sorensen (1950) reported that chicks of the Southern Royal Albatross at Campbell Island used their feet to dislodge bloated ticks from the head and neck, leaving bleeding wounds. Infestations with these ectoparasites, although widespread, appear to remain low and are rarely debilitating. Blowflies *Lucilia sericata* have caused fatal myiasis (known to farmers as 'fly strike') in newly hatched chicks of the Northern Royal Albatross on the New Zealand mainland, but not at the Chatham Islands (Robertson 1997) (p.56).

Haematozoa have been identified in the blood of the Wandering Albatross at South Georgia, where they are more prevalent than in mollymawk hosts. Infection is evidently non-symptomatic (Peirce & Prince 1980).

Minor blisters and growths, characteristic of fowl pox lesions, have been seen on the feet of Wandering Albatrosses at South Georgia and the Iles Crozet (Voisin 1969).

INJURIES

Accidental injuries are most apparent on the feet. The webs between the toes of great albatrosses are comparatively thin; those of the Northern Royal Albatross at New Zealand are said to be irritated or pierced by coarse grasses. At South Georgia, those of the Wandering Albatross are often torn and sometimes missing altogether.

Occasionally, a Wandering Albatross at South Georgia is seen limping due to minor leg injury, perhaps the consequence of a hard landing. Sorensen (1950) described a Southern Royal Albatross crashing and breaking a humerus when landing in a Force 6 wind at Campbell Island. This bird did not survive, but another that severely fractured its breastbone (sternum) lived and apparently flew long enough for an extensive natural repair to be completed.

One adult Wandering Albatross at South Georgia had a long hole in the upper bill caused by a missing latericorn and perforation of the underlying maxillary bone; this could have been sustained during a bill-gripping struggle with another bird.

PREDATORS

Piracy on the wing (kleptoparasitism) is the trade of skuas, but it usually requires several of them to intimidate a great albatross. On 16 March 1773, on board the *Resolution*, George Forster (1777) wrote in his journal:

> ' We were much amused by a singular chace of several skuas or great grey gulls, after a large white albatross. The skuas seemed to get the better of this bird notwithstanding the length of its wings, and whenever they overtook it, they endeavoured to attack it under the belly, probably knowing that to be the most defenceless part, the albatross on these occasions had no other method of escaping, than by settling on the water, where the formidable beak seemed to keep them at bay...'

At South Georgia, particularly after the island has become snow-covered, overwintering Sheathbills scavenge for spilled food or faeces. They are no threat to well grown Wanderer chicks which respond with vigorous aggressive snapping. Antarctic Skuas that breed on the same slopes are more dangerous. When they return in the spring, they gather with intent at some Wanderer nests. Healthy chicks face-off skuas with watery sounding gulps, signalling that there is a foul oil/water mix in reserve. Anything less will encourage these predators. Feeble, undernourished chicks are harassed by several skuas, which drag them from their nests.

The sealer, George Comer claimed that 95% of chicks at Gough Island were killed by skuas and giant petrels in 1888–89 (Verrill 1895). At Tristan da Cunha, Friedrich Stoltenhoff (1873) reported that feral pigs took Gough Wandering Albatrosses and their eggs on Inaccessible Island. Pigs were last seen on that island in the 1950s, so they are hardly likely to have prevented the later recovery of the remnant albatross population (Croxall 1997) (p.68). Feral pigs are also said to restrict breeding success at a few nests of the Auckland Wandering Albatross on Auckland Island (Croxall & Gales 1997), but Challies (1975) thought that none of the albatross nests he found near pig tracks had been interfered with or abandoned.

In May 1968, a black rat slipped in under a brooding Wanderer on the Ile de la Possession and gnawed a bit of the chicks foot before it was discovered and killed by the adult. The wound healed and the chick eventually fledged (Mougin 1970).

On mainland New Zealand, several introduced alien mammals have killed chicks of the Northern Royal Albatross (Robertson 1997) (p.56).

TROPICAL ALBATROSSES

Galápagos Albatross in flight (H. Douglas).

8
THE EQUATORIAL PACIFIC OCEAN

'Practically all the oceanic area bounded by latitudes 5°S. and 3°N., and longitudes 85° and 100°W., is one in which opposing currents fight out a varying but indecisive battle for supremacy.'

R.C. Murphy (1936)

Southeasterly winds along the coast of Peru drive surface water away from the land and draw up deep water to take its place (upwelling). The temperature of this deep water averages 14°C–18°C, several degrees cooler than the surrounding surface water. It brings nutrients to the surface, and in tropical sunlight, phytoplankton can maintain primary production of 45–200 mg carbon per m^3 per day. In the adjacent, warmer ocean, primary production rates are less than 45 mg. Zooplankton supports a phenomenal biomass of medium-sized squid and fish, notably the Peruvian anchoveta that forms dense shoals and is the basis of a fishery that peaked at more than 12 million tonnes in 1970. These fish are highly sensitive to zooplankton density and their shoals break up if the total biomass of zooplankton is even moderately reduced. Among the natural predators are whales, seals, huge numbers of cormorants and other seabirds.

Every December, warm equatorial water of the Pacific Countercurrent reaches the coast of Ecuador and northern Peru. The sailors of Paita traditionally called this water El Niño[1] because it came at Christmas, but at intervals of 3–5 years much greater volumes flow farther south off Peru from January to April (Fig. 8.1) and westwards across the Pacific.

These events are linked to the Southern Oscillation. When atmospheric pressure is low over SE Asia, it is high over the southeastern Pacific Ocean (Easter Island) and vice versa. Trade wind forcing keeps more water in the western Pacific, but when trade winds weaken, surface water flows back into the eastern ocean, pushing down the thermocline off Peru and impeding the upwelling of cool deep water. Periodically this allows warm water from the Pacific Countercurrent to flow south – the El Niño/Southern Oscillation (ENSO)[2].

Ecuador and Peru

Dry air descending over Western Peru normally gives the country an arid climate. At the beginning of the year the SE trade winds weaken, but during extreme El Niño events the atmospheric circulation is reversed. The normal southeasterly airstream off the land is replaced by moisture-laden onshore northerlies and northwesterlies bringing rains between January and April. Crops grow and the whole country flourishes, but torrential rains can be devastating (Murphy 1926, Philander 1990, Suplee 1999).

The characteristic features of El Niño at sea off Ecuador and Peru are:

1. An extremely sudden rise in sea-surface temperature at the beginning, which may be as early as October or as late as March (Fig. 8.2).
2. A repetition in two successive years; overall it lasts for about 16 months including two summers and the intervening winter. From 98 years of records, scientists have identified 23 El Niño events. The most extreme have been those of 1982–83 and 1997–98 (Fig. 8.2).
3. The first temperature peak is not necessarily the highest, but it is usually the longest, and between the two peaks there is a return to almost normal surface temperature.

[1] The opposite extreme (cool, dry years off Peru) have recently become known as La Niña (baby daughter) (Philander 1990, Suplee 1999).
[2] ENSO is a major climatic anomaly that sometimes causes perturbations of catastrophic proportions on a world scale. After the four yearly seasons, it is the biggest variable in the Earth's climate (Philander 1990, Suplee 1999).

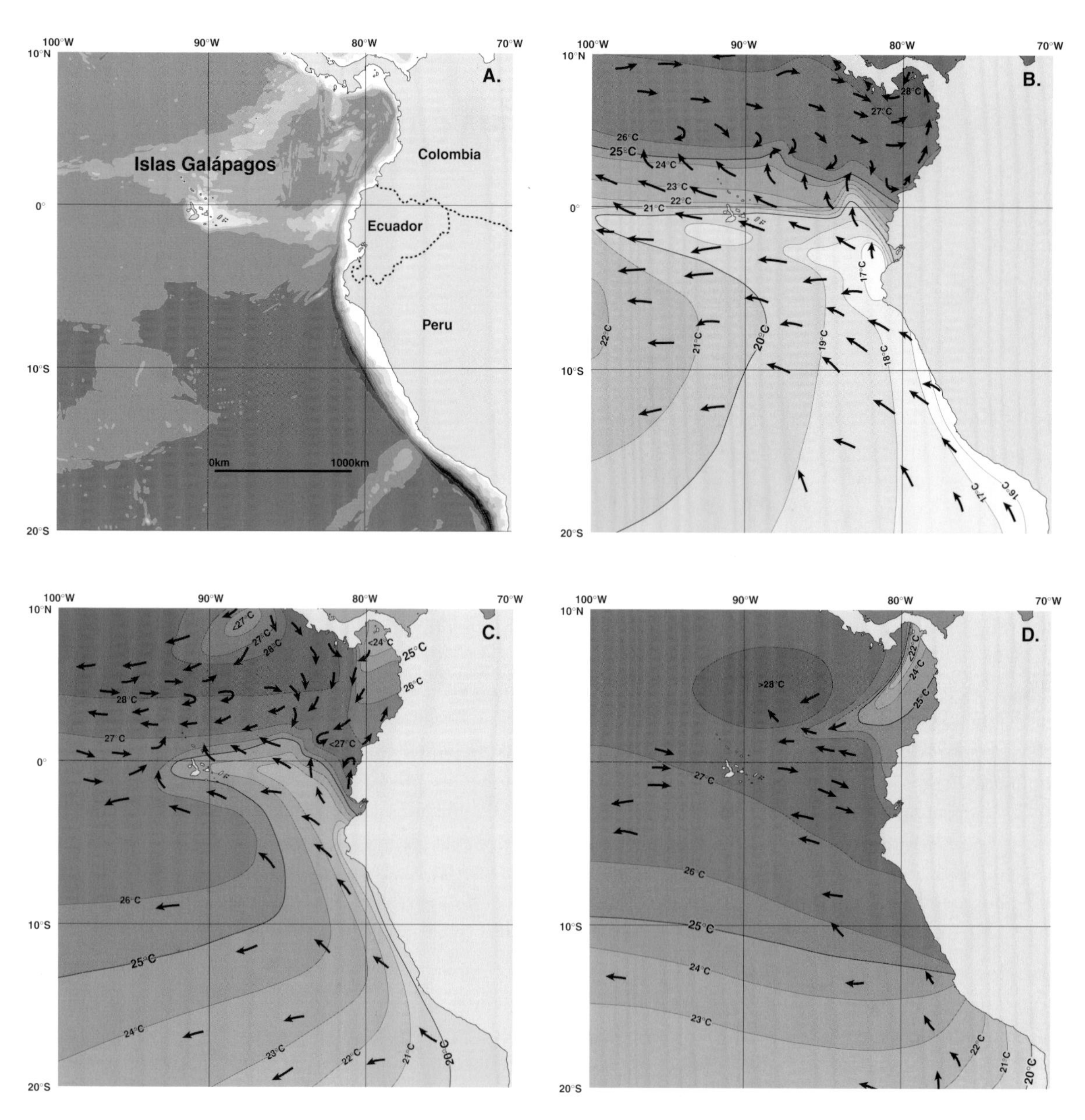

Fig. 8.1 Maps showing bathymetry (A), sea surface temperatures and currents of the equatorial Pacific Ocean off South America. During normal years, SE Trade Winds off the land force water to move northwest, causing upwelling of cool, deep water along the coast of Peru and preventing warm, easterly flowing water of the Equatorial Countercurrent from reaching the coasts of Ecuador and Peru (B). Towards the end of the year, when the SE Trade Winds weaken, warm water flows south along the coast of Ecuador. It usually reaches northern Peru during the early months of the following year (C), but if the SE Trades are replaced by NW winds, it floods much farther south (D) (after Schott 1931).

In warm surface water (30°C), plankton production along the Peruvian Coast is disrupted and the anchoveta scatter. The effect on predators can be devastating; fish, birds and marine mammals die, decomposition sets in and toxic dinoflagellates proliferate. At the Peruvian guano islands during the 1957–58 El Niño, the Guanay population crashed from 27 millions to 5.5 millions. Over the next seven years, the birds increased to 17 millions, but during the 1965 El Niño they again decreased to 4.3 millions (Idyll 1973).

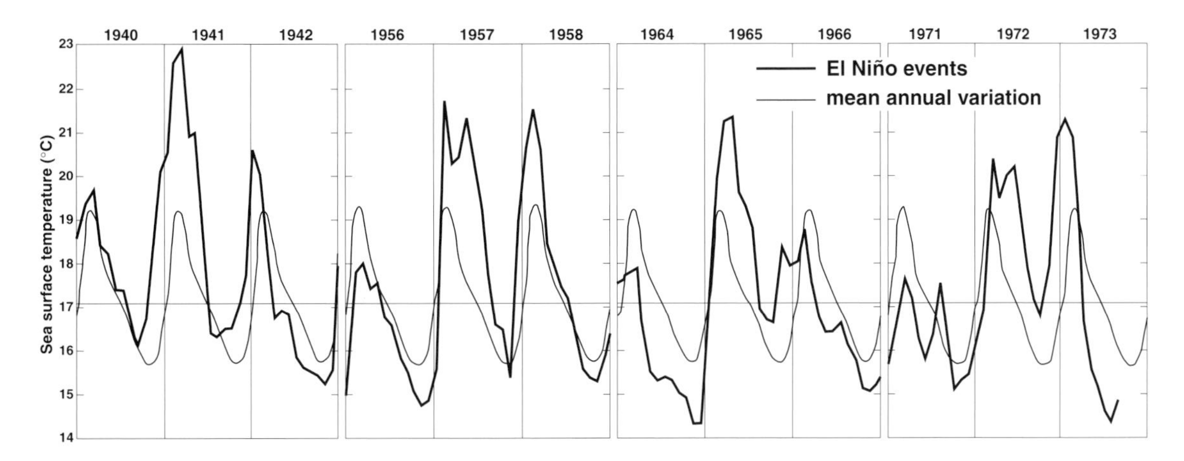

Fig. 8.2 Sea surface temperatures at Puerto Chicama, Peru (Latitude 7° 48' S) during four El Niño events, compared with the mean annual variation, 1925–1973 (after Wyrtki 1975).

Islas Galápagos

The Peru Coastal Current flowing northwest away from the mainland and south of the Islas Galápagos eventually bears west as the South Equatorial Current. The surface temperature of the sea about the islands varies seasonally from a mean of 19°C in September to 25°C in March (Philander 1990, Houvenaghel 1984), but it is composed of mixed waters that arrive by several currents from different sources. Captain Robert FitzRoy (1839) noticed this when the *Beagle* was among the islands. On 20 October 1835 he wrote:

> 'The currents about these islands are very remarkable, for in addition to their velocity, which is from two to five miles an hour, and usually towards the north-west, there is such a surprising difference in the temperature of the bodies of water moving within a few miles of each other....On one side of an island (Albermarle Island) we found the temperature of the sea, a foot below the surface, 80°Faht.; but at the other side it was less than 60°. In brief, those striking differences may be owing to the cool current which comes from the southward along the coasts of Peru and Chile, and at the Galapagos encounters a far warmer body of water moving from the Bay of Panama, a sort of "gulf stream".'

The major source of productivity at the Galápagos, not noticed by FitzRoy, is the easterly flowing Equatorial Undercurrent that sends streams of cool water among the islands, upwelling to the west of Isla Fernandina and elsewhere including Isla Española. Seabirds dependent upon local resources breed close to these upwellings (Fig. 8.3).

Although the Islas Galápagos lie on the geographic equator, the climatic equator or Inter-tropical Convergence Zone (ITCZ) is usually to the north. It is characterised by a band of warm water between latitudes 3°N and 10°N extending across the Pacific Ocean and driven by the Equatorial Countercurrent; this is one of the notorious windless Doldrums. A thermal front at 1°–3°N usually separates this water from that around the islands, but in extreme El Niño years the ITCZ moves south of the geographical equator. In 1983, the mean

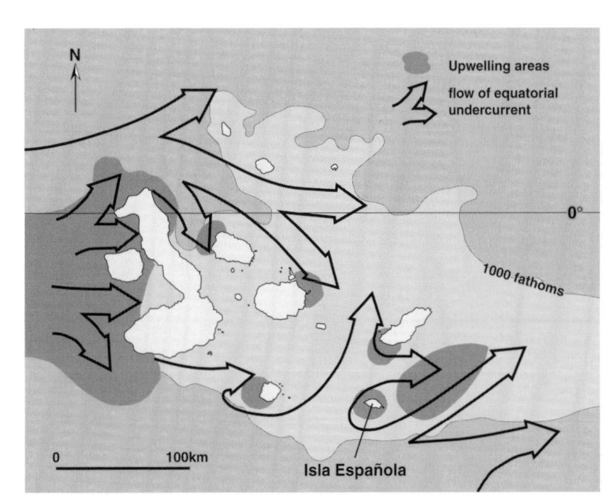

Fig. 8.3 Islas Galápagos showing the main flow of water from the Equatorial Undercurrent over the Galápagos shelf, and the location of the most prominent upwellings (shaded) (after Houvenaghel 1984).

sea temperature at the Galápagos reached 28.6°C and was accompanied by heavy rains.

Throughout the comparatively small triangle of ocean traversed by albatrosses, between the Islas Galápagos and South America (Fig. 9.2), both wind speed and wave height are below average for the south Pacific. In the three years 1985–88, satellite measurements indicated a mean wind speed about 14 km per hr (0–40 km per hr) and mean wave height of 1.8 m (0.5–3.0 m).

9
THE GALÁPAGOS ALBATROSS

'the air is full of weird noises – whoops, grunts, rattles and clunks and mad laughter.'

Bryan Nelson (1968)

The Galápagos or Waved Albatross[1] is a strange bird. Some of its features and aspects of its natural history instantly distinguish it from all the other albatrosses – sombre plumage, long bill and jutting eyebrows combined with displays that are sudden and exaggerated even by albatross standards. Set amongst dry rock and thorny scrub, the breeding ground with its scattered deserted eggs presents an antediluvian scene in keeping with the reptilian propensities of the Galápagos fauna.

Peruvian and Ecuadorian fishermen have long known it as *pájarote, pajarón* or *pajaro carnero* (Murphy 1936), but in today's Spanish, *albatros* has taken over (Harris 1974). It is the only tropical albatross; breeding almost exclusively on one island and locked into a narrow upwelling ecosystem that is immensely productive, but from time to time subject to catastrophic failures.

The Islas Galápagos were annexed by Ecuador in 1832, soon after the country had gained independence; settlement of a few islands dates from about that time. Agriculture and domestic stock did not thrive on the poor soil, so wild and feral animals contributed to the peasant economy. In 1934, a law was enacted to protect some of the indigenous animals and birds, among them the albatross, but the ravages of 200 years continued. A German biologist, Irenäus Eibl-Eibesfeldt (1959) who landed briefly in 1954, was so alarmed by what he found that he alerted international conservation agencies. In 1957, he returned in charge of the UNESCO[2] group sent to report on the state of the flora and fauna. As a result, the Charles Darwin Research Station (CDRS) began work at Academy Bay, Santa Cruz in 1964, and in 1969 the Galápagos National Park Service was founded.

While funding GNPS and fostering conservation, Ecuadorian policy has been motivated by economic objectives linked to tourism. This received international approval when the islands became one of the first four World Heritage sites declared by UNESCO in 1975. The accolade has helped to attract tourists, but has evidently done little for some of the intractable problems that afflict these islands.

The sociopolitical scenario is complex. By 1995, 60,000 to 70,000 tourists were enjoying island cruises each year, but many islanders were not sharing the prosperity created by the industry. The marine environment was threatened by several fisheries, some of them from mainland Ecuador, but the proposed exclusion zone deprived Galápagos fishermen of their livelihood and they opposed legislation (Powell & Gibbs 1995). Since 1997, however, the islanders have been brought into consultations and have been persuaded that controlling fishing is in their own interest. A legal framework for conservation and the sustainable use of resources will extend the Galápagos Marine Reserve to 40 nautical miles (74 km) and exclude large-scale fisheries. Sixty percent of the visitor fees are to be retained for management of the Park and Marine Reserve (*Oryx* 1988). A new solidarity appears to be emerging, but for long term security there has to be a reliable return for the islanders (Sitwell 1997–98).

Galápagos Albatrosses have escaped disasters on land, but there remain anxieties about their safety at sea (Merlen 1996, Sitwell 1997–98).

[1] It is difficult to imagine what inspired F.C. Du Cane Godman (1907–10) at the British Museum, to think up the name 'Waved'. L.M. Loomis (1918) of the California Academy of Sciences, more appropriately named the endemic species after its breeding islands. Alexander (1928) perpetuated Godman's curiosity and Murphy (1936) followed Loomis.

[2] United Nations Economic Cultural and Scientific Organisation.

DESCRIPTION

The white head and neck of the Galápagos Albatross is suffused with yellow, but the colour is variable and somewhat patchy (Plate 37). Prominent ridges overhang the large dark brown eyes (Fig. 9.1). The grooves (fossae) above the orbits of the skull are noticeably smaller than in other albatrosses, which implies that the salt-glands are smaller (Watson & Divoky 1971). The bill is golden yellow and proportionally much longer than in other species of similar size.

Over much of the body, the contour feathers have wavy black pencilling on their terminal fringes (vermiculations). Where these are faint and less numerous, as on the lower neck, the overall effect is pale grey, but where they become thicker, darker and closer together, as on the breast, belly and flanks, a grey-brown colour predominates. Similarly on the mantle and upper tail-coverts, they grade into more or less uniform dark brownish plumage over the back and upper surface of the wings. The underwings are whitish in the centre with fairly broad grey-brown margins and tips. The tail is brown and contrasts with a pale rump and under tail-coverts. The tarsi are pale bluish-grey as are the feet which project beyond the tail when in flight. The iris is dark brown and the orbital ring black.

Fig. 9.1 Portrait of a Galápagos Albatross showing the prominant brows (S. Halvorsen).

L.M. Loomis (1918), who examined 65 albatross skins collected at Isla Española by the California Academy of Sciences Expedition (1905–06) remarked on the variability of the plumage:

> 'The scapulars and interscapulars are plain smoky-brown in certain specimens. The light and dark markings of the upper back, sides of breast, and upper and lower tail-coverts are much coarser in some examples than in others. Their relative prominence also varies, giving the surfaces as a whole a lighter or darker appearance. The general aspect of the lower parts is lighter in some specimens and darker in others. In the extreme light manifestations, the breast, abdomen and sides are finely vermiculated with white and gray, growing darker laterally and posteriorly. In the extreme dark style the vermiculations are coarser and the flanks and abdomen are nearly uniform dark gray. The tail is smoky-brown, becoming white on the concealed portion.'

Males are recognisably larger than females with wing spans of about 2.23 m, more massive bills (Appendix 3) and conspicuous eyebrows. Chicks in their first down vary in colour from white to dark brown with whitish tips, but become more uniform brown in their second down. Fledgling plumage is much the same as that of adults except that the bill lacks the bright yellow colour. Rare mutants fledge into a uniform pale brown plumage (Harris 1973).

HISTORY

From the earliest times, the islands on the equator 800–1,000 km west of Ecuador have been known for their giant tortoises[3]. They became the haunt of buccaneers in the late 17th century and acquired a galaxy of English names. The islands were charted in 1793–94 by Captain James Colnett of the Royal Navy from the whaler *Rattler*. On 1 July 1793, he approached within about 20 km of a small island in the south which he named Hood Island after his renowned commander[4]. In 1789, the islands received their first Spanish names. About a hundred years

[3] They first appeared on a map in 1570 as Insulae de los Galápagos. In 1832 they were named officially the Archipiélago de Colón but common usage prevailed for on the 1892 Carta Geografica del Ecuador they were still the Archipiélago de Galápagos.

[4] Admiral, the Viscount Hood (1724–1816).

later, in celebration of the 400th anniversary of Columbus's discovery of the Americas, all the islands were given names associated with that historic event. Hood Island was named after Spain – Isla Española, but it remained known as Hood, even among the local Spanish speaking people. Not until the late 1970s did the Spanish names suddenly become politically more correct – and the admiral is now almost forgotten!

The *Beagle* twice approached Isla Española. On 16 September 1835 as the ship entered Gardner Bay, Fitzroy (1839) described it as:

> '...small – neither high nor low – rugged, covered with small sunburnt brushwood and bounded by a bold rocky shore. Some small beaches of white sand are visible here and there.'

Charles Darwin was on board, but he did not go ashore; the ship stayed only long enough to launch a survey boat. He was not on board when the *Beagle* returned on 14 October to make a few more soundings.

By the mid-19th century it was well known locally that albatrosses occurred only on this one island. When Simon Habel visited Isla Española in 1868, he imagined there were two species:

> 'one had a dark blackish breast and a white band crossing the head from one eye to the other; the breast of the other was grey, and the head black...' (Salvin 1876).

This was the first description of the birds on the ground and it has remained a mystery. It led Robert Ridgeway (1897), Curator of the US National Museum, to list the albatrosses of the Islas Galápagos as *Diomedea exulans* and *D. nigripes*!

Theodor Wolf, the geographer and geologist, was prevented from landing on Española in 1875. He was told that a party of more than 60 workers, who collected a lichen *Roccella babingtonii* used for extracting the dye orchil, lived there for more than a month on albatross eggs, even though each female laid just one.

In December 1881, when the *Triumph* was visiting Peru, Captain Markham shot an albatross in Callao Bay. The specimen was sent to the British Museum where Osbert Salvin (1883) described it as *Diomedea irrorata*[5]. Many years later, George Watson and George Divoky (1971) discovered a rather small and incomplete albatross skull of unknown origin in the US National Museum. In 1866, it had been provisionally named *Diomedea leptorhyncha*[6] by Elliott Coues. Watson and Divoky recognised that it was a Galápagos Albatross. Although Coues had published a description, he had not labelled the specimen nor put it in the type collection. It had been consigned to some lesser cabinet where it remained until queried many years later and catalogued 'doubtful species'. Unknown and unused for more than a hundred years, it had been overtaken by other events and was rejected (*nomen oblitum*). At the time Salvin named the species, its breeding ground was unknown. In 1896, Walter Rothschild persuaded a wealthy American, Frank B. Webster, to send a collecting expedition to the islands. Charles Miller Harris was appointed naturalist-in-chief and, accompanied by a sailing master and three collectors, set out in March 1897. At Panama three of them died of yellow fever and, after retreating to San Francisco, the fourth withdrew. Harris assembled a new team of collectors[7], chartered a schooner and sailed again in June 1897. When they returned in February 1898, they had secured a huge collection, including the albatross which was confirmed as the same *Diomedea irrorata* collected by Markham off Peru (Rothschild & Hartert 1899).

The vessel was at Gardner Bay 22–30 October and on the 26th Harris wrote in his diary:

> 'We found large colonies... from the centre of the island, south and west. These albatross have evidently used this island as a breeding ground for many years, the out-cropping rocks being worn smooth by the feet of the birds. The birds are not breeding now, as the eggs which we secured had all been deserted. There are numbers of young albatross, about the size of a "grown" goose; feathers appearing much like those of an ostrich. There must be thousands of birds on the island.'

[5] Latin: *irrorare* – to sprinkle with drops of dew.
[6] Greek: (*leptos*) – slender, (*rhynchos*) – snout.
[7] One of them was Rollo Beck who returned in 1905–06 as leader of the California Academy of Sciences Expedition (Loomis 1918).

Later visitors found more colonies. The first photographs of the birds among thorn scrub and cacti were taken during the expedition of the California Academy of Sciences (1905–06), which was also remarkable for the large number of specimens collected (Loomis 1918). In April 1925, after searching in vain for the Humboldt Current, William Beebe (1926) landed from the *Arcturus.* He saw thousands of albatrosses; two were shipped alive to the New York Zoo where they survived 'for months'.

While the genus *Diomedea* was still predominant, new albatross genera did not attract much of a following. The Galápagos Albatross was included in unsuccessful attempts to separate the Pacific albatrosses systematically from the southern hemisphere albatrosses (Mathews 1934, 1948), and von Boetticher's (1949) proposed genus *Galapagornis* was ignored. The arrival of molecular systematics has again split *Diomedea* which has been reserved exclusively for great albatrosses (Nunn *et al.* 1996), and *Phoebastria* (p.236) has returned as a proposed genus for all four Pacific albatrosses. However, although the molecular evidence confirms a separate Pacific phylogeny, the Galápagos Albatross remains so different from the three North Pacific species that it appears out of place in the same genus.

OCEANIC DISTRIBUTION

The Galápagos Albatross occurs over a comparatively small triangle of ocean between the Islas Galápagos and the adjacent mainland of South America (Fig. 9.2). Loomis (1918) predicted this and by the 1970s Harris (1973) was confident enough to claim that there had been no unequivocal record of these albatrosses to the west of the Galápagos. Subsequent records collated by Pitman (1986) support that assertion.

They are rarely encountered far out to sea (D. Day, T. Arnbom, B. Haase pers. comm.). Most sightings have been of single birds off the coasts of Colombia, Ecuador and Peru from 4°N to 15°S and quite large numbers have sometimes been seen between 5°S and 8°S.

> '...on one trip made in a small boat from the island of Lobos de Tierra to Eten on the coast the pajaros-carneros [Galápagos Albatrosses] were nearly always present in numbers; often 30 or more were in view at one time. One was observed devouring a fish about a foot in length.' (Coker 1919).

S.E. Chapman saw aggregations of up to 200 in September 1967 near the Islas Lobos de Tierra (6° 30'S) and P. & K. Meeth saw 450 in March 1980 off the Isla Lobos de Afuera (7° 00'S) where they obtained a daily count of 580. In roughly the same area, R.L. Pitman recorded a daily total of 700 in December 1985 (Tickell 1996).

They are rare north of the equator (Harris 1973, Pitman 1986, D. Day, T. Arnbom & B. Haase pers. comm.) with only two records off the coast of Colombia (Wetmore 1965). Joe Jehl (1973) saw none south of 26°S off Chile in June 1970 and Ben Haase sailing north from Iquique in July 1988, did not encounter Galápagos Albatrosses until almost 10°S off Peru. Three have been seen in February off Punta Doña Maria, Peru (14° 54'S).

In 1905–06, Gifford noticed that albatrosses were absent from the vicinity of Galápagos in December, January and February. During the rest of the year they were quite common at sea in the south of the archipelago (Loomis 1918).

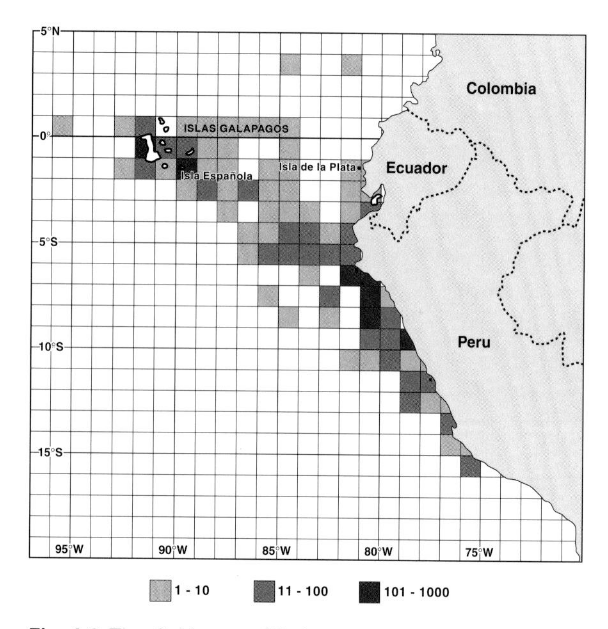

Fig. 9.2 The Galápagos Albatross at sea; 412 observations between 1881 and 1995. The large numbers seen off the coast of northern Peru between 6°S and 7°S were near the Isla Lobos de Afuera and Isla Lobos de Tierra (Tickell 1996).

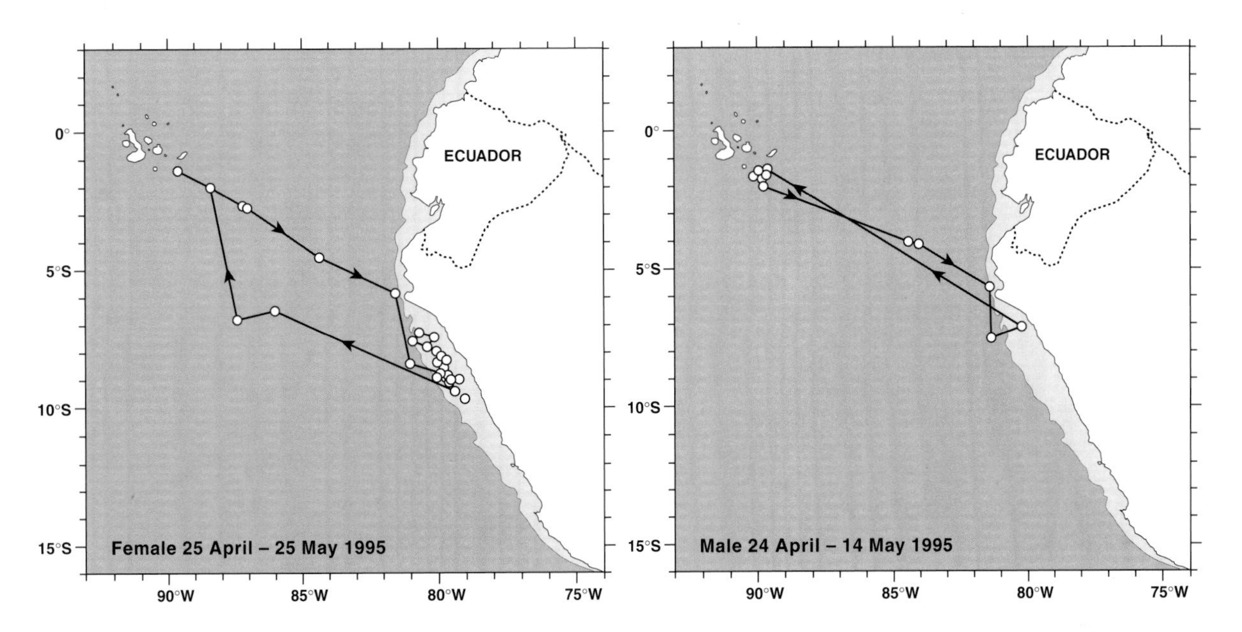

Fig. 9.3 The tracks of a female and male Galápagos Albatrosses foraging between incubation shifts on nests at Isla Española (Anderson *et al.* 1997).

> 'An odd fact about the albatrosses is the direction of their flight from the island. They fly straight out to the southward and none are seen about the north side of the island nor about any of the other islands.' (Beck 1904).

Adults tracked by satellite between early incubation shifts confirmed Beck's observation. All flew directly from the island to foraging areas in the coastal waters of Peru and Ecuador. Flights of approximately 1,200 km were made in two to seven days (Fig. 9.3) (Anderson *et al.* 1997). Comparatively few albatrosses have been seen between the islands (Loomis 1918, Nelson 1968, Harris 1973, Duffy & Merlen 1986, Hayes & Baker 1989) (Fig. 9.4). R. Pitman (pers. comm.) recorded a daily count of 61 north of Isla Isabela in November 1981. In September 1995, Godfrey Merlen (1996) saw 120 northwest of Isabela, 389 west and 239 south of Isla Fernandina, all among other seabirds feeding over dolphins and sea lions. On 27 October 1989, southeast and east of Isla Española, Don Roberson (pers. comm.) encountered aggregations of 450 and 550 albatrosses among huge numbers of other birds associated with dolphins. This was at the time most fledglings leave the island so they may have included many juveniles.

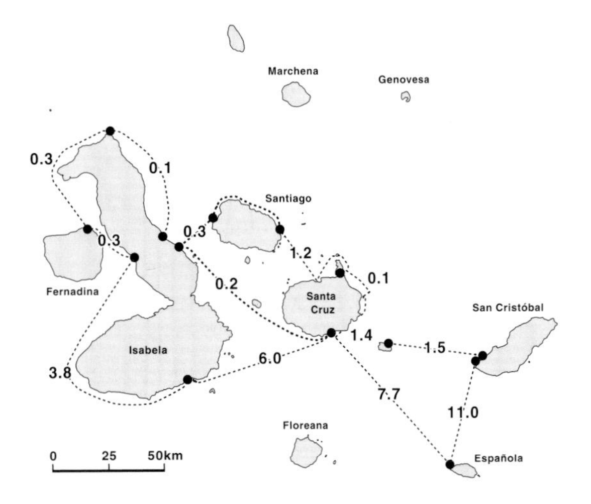

Fig. 9.4 The numbers of Galápagos Albatrosses seen at sea between the Islas Galápagos. Black spots indicate the ends of transects with the number of birds seen per 30 minutes, in an arc of 180° to the horizon on the seaward side of the vessel. The overall period of observation between 15 June and 25 July 1984 amounted to almost 60 hours (Hayes & Baker 1989).

BREEDING ISLANDS

Most Galápagos Albatrosses breed on one island, Isla Española at the southeast of the group[8], but there is another very small colony on Isla de la Plata, at almost the same latitude, 27 km off the coast of Ecuador.

Isla Española (1° 22'S, 89° 42'W)

This uninhabited island is 14.3 km x 7.6 km with hills of ancient volcanic rock, smoothed by the sea and uplifted to 200 m (Fig 9.5). There is no freshwater and at lower levels the arid *Acacia* and *Prosopis* scrub is dense and often impenetrable, but on higher ground it is more open. Drought keeps the stony soil dry for much of the year, but during rains (December–March) grasses and herbs flourish briefly.

Early visitors all appeared to have landed from Gardner Bay or on the beaches near Punta Cevallos. During the second World War, when the United States maintained an air base on Seymour Island, a radar station with a landing strip was constructed on high ground at the east end of Isla Española.

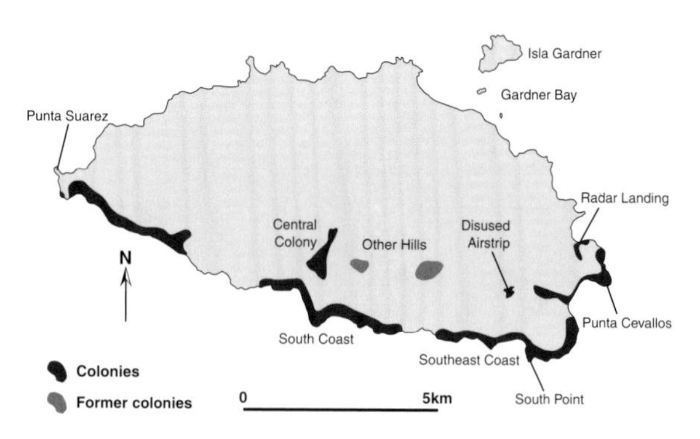

Fig. 9.5 Isla Española with Galápagos Albatross colonies and former colonies (Harris 1973, Douglas 1998).

Española tortoises *Geochelone elephantopus hoodensis* had been almost exterminated by 19th century whalers and prolonged searching in 1903 had revealed only three (Snow 1964). When Eibl-Eibesfeldt (1959) landed at Gardner Bay in 1957, he saw goats and devastated vegetation. Later searches in 1963 and 1964 yielded only two tortoises, with the result that the island acquired a high priority in conservation planning. The last tortoises were removed from the island to pens at the Charles Darwin Research Station and bred with others acquired from distant zoos. In the 1970s, all goats on Isla Española were shot and the vegetation recovered. Young tortoises were re-introduced into the island in 1975.

The Galápagos Albatross nests within walking distance of the coast or open ground free of scrub (Plate 38). After the war, they took over the abandoned runway (Brosset 1963) and are still breeding there after other inland colonies have disappeared. They are most numerous along the south and east coasts which provide convenient take-off places from the tops of steep cliffs. The Central and Punta Cevallos colonies extend inland up to two kilometres and at the former Radar Landing they nest and fly from the beach (Douglas 1998).

In 1961, Raymond Lévêque, the first Director of CDRS started counting and ringing albatrosses in colonies behind Punta Suaréz; it was soon apparent that there were many of them. Bryan and June Nelson lived there for three months in the 1964 breeding season and made the first study of Galápagos Albatross behaviour (Nelson 1968). They were followed by Michael Harris who was on the island in May 1966 and July 1967, then from February 1970 to September 1971. His field-work provided the basis for all subsequent studies and, with assistants from CDRS, he completed the first census of the whole island (Harris 1973). Catherine Rechten spent a total of 13 months at Punta Suaréz during the 1981, 1982 and 1983 breeding seasons. Her planned behavioural studies were disrupted by an El Niño event, but she was able to document its dramatic effect on the breeding biology of the albatrosses (Rechten 1985). At about the same time, Chela Vásquez investigated egg-moving behaviour and in 1988, I examined nest-building behaviour. Since 1994, David Anderson, Hector Douglas and others have completed a second census and introduced satellite tracking.

Isla Española, with its albatrosses, is one of the major attractions of the Galápagos tourist circuit, but most of the island is never visited. Punta Suaréz with its convenient landing and entertaining sea-lions, has become the show-place for visiting boats. Guided parties follow a well-marked nature trail through the edge of a small albatross colony. This same colony has been a scientific study area since 1961 and researchers sometimes camp above a

[8] In 1984, seven albatrosses were ashore on Isla Genovesa (Tower I); courtship behaviour and copulation were seen but apparently no eggs were laid (Perry 1985).

nearby beach. They are encouraged to work out of sight of tourists, but conflicts of interest are inevitable. Research not requiring known-age individuals is now carried out at the other end of the island, near Punta Cevallos.

Isla de la Plata (1° 16'S, 81° 04'W)

This uninhabited island (5.5 km x 2 km) is 27 km off the coast of Ecuador (Fig. 9.2); it has an area of 14 km^2 and rises to 167 m (Fig. 9.6). There is no freshwater and the soil is mostly dry with an arid scrub that greens only during rains (December–May). The vegetation has been browsed by feral goats and there are a few cats which may prey on some of the burrowing rice rats.

In February 1925, Murphy went ashore briefly and the resident lighthouse keeper showed him the skin of a Galápagos Albatross shot there the previous November. Almost 50 years later, Oscar Owre (1976) visited the island and was also told about albatrosses. He looked for them without success, so the following season he returned with some students and discovered several pairs. Since then, the birds have been seen from time to time (Nowak 1987). They nest among dense scrub just inland of Punta Machete on the southwest coast. Members of the University of Bristol Expedition, who were ashore from 20 July to 1 September 1990, completed the first biological survey of this important island (Agnew *et al.* 1991).

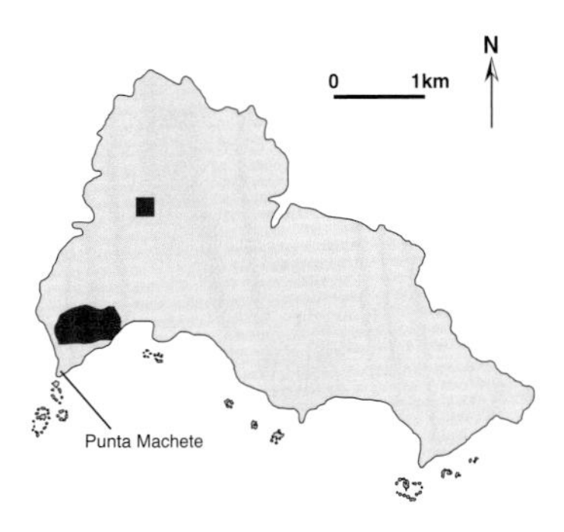

Fig. 9.6 Isla de la Plata with its small Galápagos Albatross colony behind Punta Machete and a single nest on the ridge above (Agnew *et al.* 1991).

The island has been cultivated in the past. There is an airstrip and the ruin of a fishing club that was never completed. The crews of fishing boats regularly come ashore and it is a weekend picnic beach for families from the port of Manta. It is now part of the Machalilla National Park on the adjacent mainland.

NUMBERS

Estimates of 2,000–3,000 pairs of Galápagos Albatross at Punta Suaréz and Punta Cevallos were made in 1961 (Lévêque 1963). The following year an inland colony was discovered and a total of 3,000 pairs was agreed (Brosset 1963). One of the CDRS field-workers, M. Castro who visited the island in 1964, believed that the figure might be as high as 10,000 birds (Harris 1973). However, there were several years in the 1960s when breeding success was so low that counts were not compatible.

A concerted effort was made in 1970 and 1971 to obtain a realistic figure for the whole island. Difficult terrain made it impossible to count it all at one time – many birds were nesting under trees and bushes. Study areas at Punta Suaréz were counted several times and provided the means to standardise counts from distant parts of the island at different times. It was estimated that about 10,600 pairs bred in 1970, while the following year at least 12,000 pairs were believed to have bred (Harris 1973).

At the time these counts and estimates were undertaken, feral goats maintained clearings and thinned the thorn scrub. Since 1978, when the goats were removed, the vegetation has regrown and it is now so dense that for practical purposes many areas are impenetrable and counting albatrosses has been made much more difficult. Nevertheless, in 1994 another census was undertaken. Using the same method as Harris, David Anderson estimated 15,581 breeding pairs (Anderson & Cruz 1997). From the same field data, Hector Douglas (1998) put more confidence in a single adjusted count of active nests that indicated at least 18,200 pairs. Both methods indicate that numbers have increased significantly since the early 1970s.

In May 1975, five pairs of Galápagos Albatrosses were incubating eggs in the small colony on Isla de la Plata. The University of Bristol Expedition counted 28 birds in 1990, including ten breeding pairs (Ortiz-Crespo & Agnew 1992).

BREEDING

Breeding season

Galápagos Albatrosses are absent from their breeding ground between December and April. There are recognisable colonies, but some nests may be hundreds of metres from their nearest neighbour. Successful breeding lasts seven and a half months and although they are annual breeders, pairs may periodically defer breeding (Rechten 1986). Pair-bonds survive from year to year; 12 pairs are known to have remained together for ten years (Harris 1973) and there is no known instance of divorce.

Pre-egg Period

A few albatrosses have been seen flying over the colony at the end of January (M.P. Harris pers. comm.). Towards the end of March, rafts of birds assemble on the water just off Isla Española. During the late afternoon, they fly into the island, stay overnight and go to sea again the next morning. Colonies fill up during the first week of April, mostly with males sitting at or near sites they occupied the previous season (Fig. 9.7). Females come ashore infrequently until about 15 days before laying. The median arrival dates in 1982 were 15 and 25 April for males and females respectively. There is a peak in female attendance about eight days before laying which coincides with that of male presence and is probably the time of copulation (Fig 9.8). Males are aggressively territorial throughout this period, driving off strangers from as far away as 20 m. Fugitives that cannot escape are overpowered and mounted. Harris (1973) believed males were mounted as often as females and that it was dominance behaviour rather than copulatory, but Rechten (1986) referred only to extra-pair copulations with females. Vigorous fights involving several birds occur and wounds about the head are common.

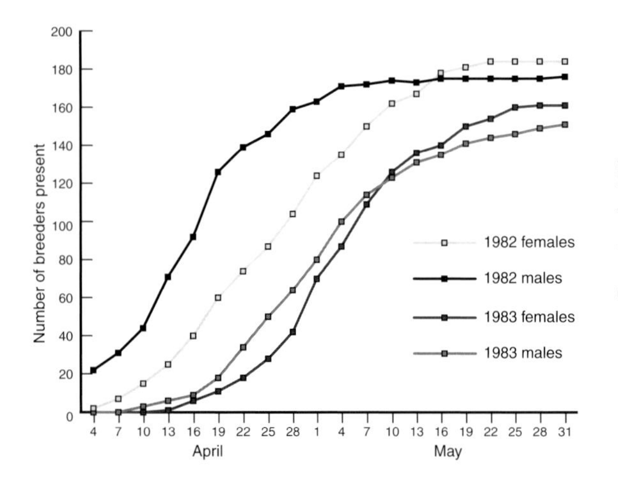

Fig. 9.7 The arrival date of ringed Galápagos Albatrosses at the Punta Suaréz breeding ground in 1982 and 1983. Only birds that bred in 1982 and were seen in both years are included. Slightly fewer males than females were ringed (Rechten 1986).

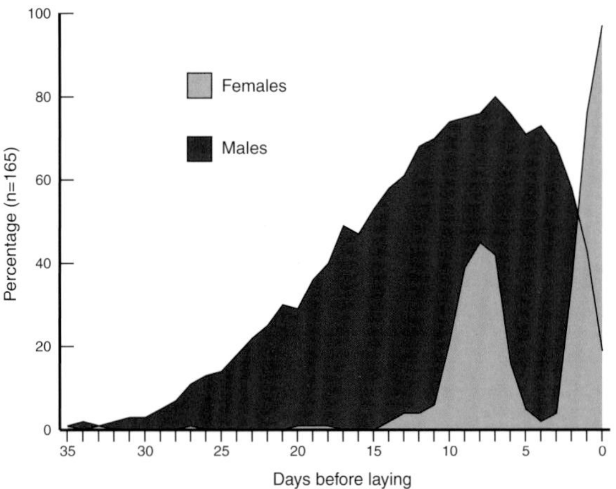

Fig. 9.8 The pre-egg period in the Galápagos Albatross. Percentage attendance at the nest by 165 males and 165 females in the 35 days before laying (Rechten 1986).

The arrival of birds is influenced by the outcome of previous breedings and environmental events, notably those associated with El Niño years. In 1983, the arrival of both sexes was delayed, males more than females, so that both arrived at about the same time. In a normal season (1982), females from only 8% of pairs were seen ashore before males, but in 1983, 40% of females arrived before their mates (Fig. 9.7). Furthermore, when males were not present on site, the normal attendance of females was disrupted. Instead of staying ashore only a short while and then going back to sea, they kept appearing on site every day for several weeks or until the male arrived (Rechten 1986). Copulation was seen at 8, 13, and 28 days before laying (Harris 1973).

Nests

Among 125 pairs that remained together in one colony at Punta Suaréz, the mean distance between their nests in consecutive seasons was 11 m (±8.7SD , 0–50). Galápagos Albatrosses pick up stems of grass and place them right or left in typical nest-building behaviour, but more often than not these are rather futile actions because there is so little grass or other material available. In 1988, 237 albatrosses sitting on eggs at Punta Suaréz were examined; 37% of them were on rock or gravel with no indication of a nest and 13% had nests of grass. The remainder were on gravel and had slight scrapes, a circle cleared of stones or a few strands of grass arranged in a ring (Table 9.1). North of Punta Cevallos, about 50 albatrosses nest in a compact colony at the back of a beach (Plate 39). Eggs are laid on coarse fragments of coral with no nest scrapes; some have been seen among boulders on a rocky shore where they would sooner or later be swamped by spring tides.

Substrate	Nest	%
rocks	none	11
gravel	none	26
gravel	slight scrape	22
gravel	ring of stones with a little grass	16
gravel	nest with grass	12
grass	nest	12
grass	moderate-large nest	1

Table 9.1 Nest characteristics of 237 Galápagos Albatross egg-sites at Isla Española, 8 May 1988.

Eggs

The average measurements and mass of Galápagos Albatross eggs are 106x70 mm and 284 g (Appendix 4). Females breeding for the first time lay longer and thinner eggs than those of older females (Harris 1973). Eggs from known-age females increase in size from 5 to 17 years of age (C. Rechten pers. comm). Deserted eggs are bleached by the tropical sun and lose their rings of peppered spots.

Laying

In 1971, eggs were laid from 9 April to mid-June, the mean date being 4 May (n=495), with half of all eggs laid during the first two weeks of May. Galápagos Albatrosses breed for the first time at four to six years of age and young birds lay later than older birds. Harris (1973) was able to show that pairs with a four-year-old partner laid after 30 May while those with a seven-year-old partner mostly laid at the beginning of May. In 1982, the median date of laying among five-year-olds was 16 May (n=16), while that of older birds was 3 May (n=189). The laying dates of seven-year-olds was no different from pooled 'older' birds (Rechten 1986).

The date of laying by individual females is not constant from year to year, but depends upon the outcome of the previous breeding. Pairs that successfully reared young in one season laid later the following year. Conversely, pairs that failed in one season return and lay earlier the next year (Rechten 1986).

Incubation

Males are sometimes present at the time of laying and take over the egg within a few hours, but on average females sit for about five days before being relieved by their mates. Thereafter, both partners share incubation in alternate shifts which average 19–22 days and are rarely less than 14 days. Just before hatching, shift lengths shorten to about eight days and by the time hatching begins partners are changing every four days. The mean incubation period is 61 days (range 59–62, n=5).

Males and females sometimes move about during incubation. The egg is held loosely by the surrounding contour feathers (brood pouch) and shuffled along the ground; it is not carried on the top of the webbed feet. In a single day, eggs were moved up to 14 m and in a few days up to 40 m (Harris 1973). In some years there is much more egg moving than in others. The behaviour is commonest in crowded colonies and is related to the proximity of neighbours. Females move eggs more than males and Vásquez noticed that movements away from neighbours are more frequent than towards them, especially if they are close (1–2 m) (Anderson & Cruz 1997).

Abandoned albatross eggs may remain viable for periods of up to six days and, if left at a recognised nest or site, a late returning partner may successfully resume incubation.

Chicks and fledglings

In 1970, the average fledging period was 167 days and no healthy young remained after 178 days.

At hatching, the chick weighs about 200 g (Nelson 1968) and is brooded almost continuously for several weeks, after which it is guarded for several more before being left on its own; many die at this time. Brooding and guarding is concerned as much with providing shade from the tropical sun as warmth. When they are left alone, chicks often suffer from the heat and cool themselves by gular ventilation and exposing the webs of their feet by sitting back on their 'heels' (tarsometatarsi/tibiotarsi).

Nelson (1968) described feeding behaviour and recorded one chick receiving feeds on seven successive days, but suggested that, on average, they were fed once every two to three days. This was later confirmed by Harris (1977), who believed that parents returning daily had been feeding nearby, but that the chicks' ability to survive on fewer, larger meals allowed foraging parents to be away for up to two weeks. He recorded that a chick could take two kilograms of food at a single feed lasting a few minutes. Adults were very aggressive to begging chicks unless the submissive 'head-away' posture was adopted.

Nelson (1968) recorded a peak mass of 5.3 kg and, in 1970, parents continued feeding chicks up to the time of their departure. Eleven fledglings averaged 3.7 kg (±0.18SD) just before they left (Harris 1973). They exercised their wings for several days and sometimes made short flights into the wind. Eventually they walked to the cliff-tops and launched themselves. Once airborne, they flew straight towards the horizon. Elsewhere on Española, some fledglings may make their first flights from 'runways' in clearings or from a beach.

Breeding success

Breeding success increases with age from 6 to 17 years, but albatrosses older than 25 years are less successful than middle-aged experienced breeders (C. Rechten pers. comm.).

Location	Year	Number of eggs	% eggs hatched	% chicks fledged	% eggs fledged
West Colony	1970	262	35	47	16
	1971	316	49		30
Hawk Rock	1970	134	51	51	26
	1971	258	56		27

Table 9.2 Breeding success of Galápagos Albatrosses at Isla Española (Harris 1973).

Cause of loss	%
Egg deformity	<1
Crushed by incubating bird	2
Chick death at hatching	3
Desertion	5
Egg rolled into cracks between rocks	10
Infertility	10
Cracking and/or breaking	18
Disappearance	52

Table 9.3 Causes of failure among 510 Galápagos Albatross eggs that did not hatch in 1970 and 1971 (Harris 1973).

There is a considerable local and annual variation in breeding success on Isla Española (Table 9.2). In 1970 at the West Colony of Punta Suaréz, 35% of eggs hatched (n=262); about half a kilometre away near Hawk Rock, 51% hatched (n=134) while five kilometres along the south coast at the Promontory only 10% hatched (n=128).

In 1971 at the West Colony, the overall breeding success was 30% (n=316), but in one colony success was as high as 82% (n= 155) .

Eggs fail to hatch for various reasons (Table 9.3). Harris (1973) believed most of those that disappeared completely were broken, probably as a result of egg-moving behaviour by incubating adults, but Vásquez (Anderson & Cruz 1997) has claimed that, in 1981, hatching success was higher among eggs that had been moved.

Most chick deaths occurred during the first month

after hatching and some died from accidents at departure, falling down cliffs or drowning in the sea (Harris 1973).

Exceptional climatic conditions associated with El Niño phenomena contribute to severe periodic constraints on breeding. The many dried out eggs that litter the breeding grounds attest to past disasters. The years 1965 and 1967 were both almost total failures, but the intervening season of 1966 was very successful. Few young were reared in 1968 or 1969 and there were large losses in 1972 and 1976. Shortage of food 900 km from the islands may have contributed to poor breeding performance, but adverse climate conditions at the breeding grounds were responsible for the dramatic disaster of 1982–83 (Rechten 1985).

At the beginning of 1982, albatrosses returned and laid normally in the study colony at Punta Suaréz, but eggs were abandoned at a conspicuously higher rate than in the previous year. Of 133 eggs laid only 14 resulted in fledged young. In the following year, there was torrential rain throughout the early months and the albatrosses were late returning (Fig 9.7). They arrived to find unusually thick vegetation, with 60% of the open spaces unsuitable for nesting; some were even swamps. Overgrown vegetation accompanied by disturbance of the normal wind regime hindered landing and take-offs. Only about 15% of females laid eggs and they were all over ten years old. Furthermore, their eggs were smaller than those which they laid the previous season. Incubation was disturbed by heavy rain and the failure of males to return. Many eggs were soon partially or completely submerged. None were incubated for longer than 21 days. All had been deserted by mid-June and searches elsewhere on Española revealed no other incubating albatrosses. In July, the rains ceased and the vegetation dried to a more characteristic state. Courtship behaviour became more frequent than usual with many failed breeders joining the normal complement of subadults during the later months of the year.

Survival

The annual survival rates of adults over nine years averaged at least 95%, and between 1970 and 1971, 97% of 288 ringed birds returned to their nests. Among 1,461 albatrosses ringed as chicks in different colonies, the minimum annual survival rate over five to nine years varied from 90.4% to 94.3% (Harris 1973).

Rechten (1985) believed that proportionately more older albatrosses may die in El Niño years. In 1983, no albatrosses older than 25 years returned compared to 19% in the previous year. Only 3.5% of albatrosses younger than 25 years were absent in 1983.

During the 1994 count, one retrapped female was at least 38 years old (Douglas & Fernández 1997).

Subadults

After spending their juvenile years at sea, some two-year-olds return and become subadults towards the end of the breeding season, but they do not remain ashore long. Many more return at the age of three and by five years of age some will already have formed pair-bonds and returned early enough to breed. Among 180 four-year-olds retrapped in 1970–71, 22% bred the following season. Most six-year-olds, if not breeding, are returning early enough to breed, although some may not do so until they are over seven years old. Harris (1973) found a tendency for individuals to pair with partners of the same age.

FOOD

Squid, fish and pelagic crustaceans make up the bulk of Galápagos Albatross food during the breeding season. On a few occasions, albatrosses have been seen in mixed species flocks associated with cetaceans and have been observed chasing prey driven to the surface by dolphins (R. Pitman pers. comm.) and plunge-diving at slow speed (B. Haase pers. comm.). They have also been seen harrying boobies (Harris 1973, Duffy 1980, Merlen 1996), but the items fed to chicks were said to be too big to have been obtained exclusively by kleptoparasitism. The Galápagos Albatross does not scavenge from ships.

In 259 stomach samples regurgitated by albatross chicks in 1970 and 1971, 53% contained squid, 41% fish and 46% crustacean remains (Harris 1973). These samples contained 299 squid beaks from seven of the eight families occurring in the eastern Pacific Ocean: Enoploteuthidae, Octopoteuthidae, Onychoteuthidae, Histioteuthidae, Ommastrephidae, Pholidoteuthidae, Chiroteuthidae. Eighty percent were from just two families, Histioteuthidae and Octopoteuthidae, with individuals ranging in mass from five grams to 450 g. Some beaks had come from squid that were much bigger than could have been captured by the albatrosses and were probably scavenged piecemeal from squid killed by larger fish or cetaceans.

The bony remains of fish in samples contained flying-fish (Exocoetidae), *Engraulis ingens*, a clupeid *Etrumeus acuminatus*, *Priacanthus cruentatus*, a carangid *Decapterus scombrinus* and a species of *Antennarius* (Cherel & Klages 1997). They ranged in length from 30 mm (20 in a single regurgitation) to 340 mm, but included the heads of much larger fish.

Pelagic crustaceans in the samples were mostly euphausians; *Benthophausia* sp. up to 120 mm in length (normally netted at a depth of 1,000 m) and *Thysanopoda monacantha*.

PARASITES

The Galápagos Albatross has a fauna of chewing lice that includes three endemic species and two shared with southern albatross hosts (Appendix 5). Palma & Pilgrim (1984) consider that host-lice characteristics are sufficient to make inclusion of the Galápagos Albatross in the genus *Phoebastria* doubtful (p.236).

Ticks are common on incubating adults and chicks; Vásquez has suggested that incubating albatrosses may move their eggs to escape from infested areas.

Mosquitoes *Aedes taeniorhynchus* occur on Isla Española. In the dry conditions that usually prevail during incubation, they are not numerous enough to disturb incubating albatrosses. When unseasonable rains leave pools of standing water, the insects proliferate and swarms seek blood meals from birds. Roger Tory Peterson (1967) suggested that such swarms caused Galápagos Albatrosses to abandon their eggs, and in 1986 Anderson & Fortner (1988) demonstrated that egg neglect increased with biting frequency by mosquitoes.

PREDATORS

Galápagos Hawks may occasionally take a young chick or scavenge a fledgling that has crashed down a cliff on take-off, but they are too few to have a significant impact.

Mockingbirds readily eat albatross eggs. Single birds rarely break through the shell unless it is already damaged, but a number of Mockingbirds hammering away together at the same spot on the shell can make a hole. They have also been seen to pick items of food from the mouths of adult albatrosses when they were regurgitating into the open beaks of their young (Brosset 1963).

NORTHERN ALBATROSSES

Steller's Albatross in flight (H. Hasegawa).

10
THE NORTH PACIFIC OCEAN

'...the SE extremity of the island[1], lying in Latitude 35°N, Long: 140°E, upon which there is a very large peak'd mountain[2] equal in height to any we have seen; The Current[3] seting round this point much stronger than before, drove us far to the eastward, that we found it vain to attempt to make the Land again; and the winter advancing, we stood to the southward for China.'

Midshipman George Gilbert on board
Discovery south of Japan, 1 November 1779[4]

The atmosphere over the North Pacific and Bering Sea is characterised by high pressure over continental Asia, low pressure over the North Pacific (Aleutian Low) and high pressure over the eastern North Pacific (Pacific High) close to the Inter-tropical Convergence Zone (ITCZ).

In the middle latitudes south of the Aleutian Islands, the day to day flight of albatrosses is influenced and perhaps determined by the tracks of cyclones and associated weather fronts travelling eastwards across the ocean. To a greater or lesser extent, these are impeded by anticyclones and ridges of high pressure. Around these systems winds blow in all directions, but when approaching and leaving the great breeding grounds of the Northwest Hawaiian Islands (NWHI), albatrosses usually experience NE trade winds.

In winter, cold northwesterly winds from continental Asia and the Arctic Ocean contribute to increased turbulence over the northwest North Pacific. A cycle of high winds and big waves in winter contrasts with breezes and moderate waves in summer over large areas of the North Pacific (Fig. 10.1).

Tropical cyclones occur on both sides of the North Pacific and are encountered by all albatrosses. Hurricanes developing over the eastern ocean off Mexico in August-October move northwest, but have usually lost most of their force by the time they approach the Northwest Hawaiian Islands. Typhoons generated over the western tropical Pacific move northwest, gathering momentum as they sweep round to the northeast over the seas south and east of Japan, where they lose energy over cooler water. They outnumber the tropical cyclones of all other oceans. Over 20 years (1940–60), an average of eight typhoons per year (range 3–20) passed Torishima in all months except March, but 73% occurred in the four months of July–October when there was an average of one per month in July, September and October with two in August. Winds sometimes exceeded 200 km per hour (Japan Meteorological Agency 1963).

Kona storms are associated with subtropical cyclones that arise in the mid-ocean northwest of the Hawaiian Islands in winter and move slowly eastwards. They have less energy than tropical cyclones but occasionally bring rain and southerly winds of over 100 km per hr to the Hawaiian Islands.

In some years, such as September 1986, cyclonic conditions in all three areas combine to produce very high winds over large expanses of ocean.

Ocean

The continents surrounding the North Pacific and the shape of the ocean floor have a profound influence upon the dynamics of the ocean and atmosphere. The Bering Sea is essentially a bay of the North Pacific Ocean (Shuntov 1993) and half of it is over the Alaskan continental shelf. The North Pacific is separated into eastern and western

[1] Honshu.
[2] Fujiyama.
[3] Kurashiwo.
[4] in Beaglehole 1967.

basins by the chain of Hawaiian and Emperor Seamounts. The ocean floor of the western basin is more rugged than that of the eastern basin (Fig 10.2).

North Pacific water circulates around the anticyclonic central North Pacific gyre and two smaller cyclonic Subarctic gyres. In mid-latitudes, the broad North Pacific Current (West Wind Drift) flows eastwards across the ocean while towards the equator the Subtropical ocean flows west under the influence of the NE trade winds. Water masses are separated by temperature and salinity fronts that meander and break off as complex mesoscale eddies, not apparent in schematic diagrams (Fig 10.3) (Schneider 1990, Roden 1991).

Bering Sea circulation is linked closely with that of the North Pacific (Dodimead *et al.* 1963). The Gulf of Alaska receives meltwater from the Alaskan mountains, which is held against the coast by onshore winds. The Aleutian Current, a westerly-flowing boundary current hugs the south coast of the Alaska Peninsula and the Aleutian Islands. Some of this relatively warm dilute water flows north through the Unimak and other passes between the islands, but most of it continues westwards until it encounters colder more saline water at about longitude 180°. East of Bowers Ridge or west of Attu Island it is deflected northwards into the Bering Sea through the Amchitka Pass.

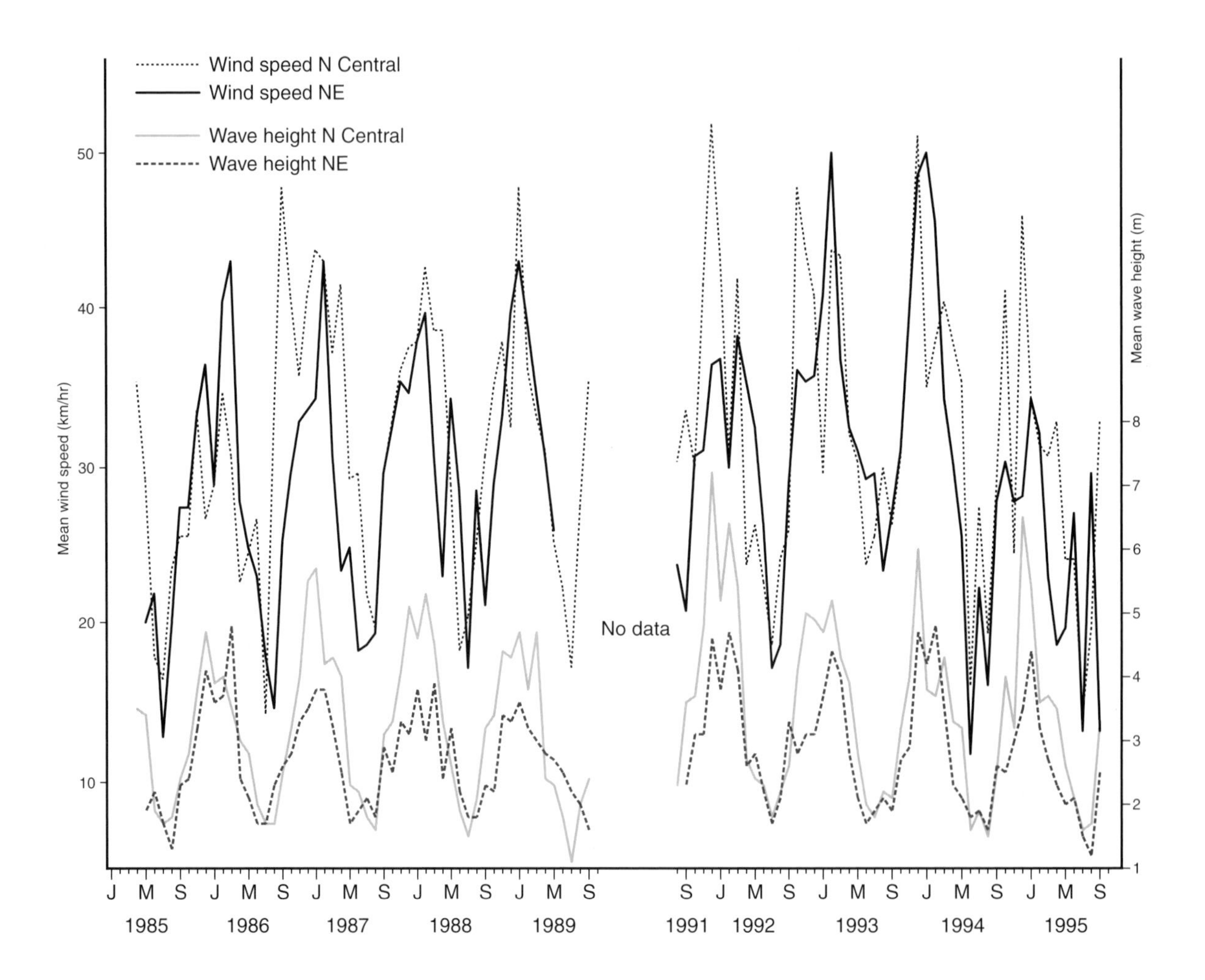

Fig. 10.1 North Pacific: mean wind speeds (upper curves) and wave height (lower curves) for the decade1985–95 from two North Pacific locations: the N Central Pacific at about 47°N 165°E and the NE Pacific at about 35°N 149°E. There is a clear annual cycle of gales in winter and breezes in summer. Note that although the NE Pacific is less rough (in terms of wave height) than the N Central Pacific, it is not so obviously less windy. There is also a significant variability between the cycles for different years (D. Cotton pers. comm.).

In winter, over the Alaskan continental shelf, Bering Sea water freezes and ice retreats only slowly from the north in spring. The climate and sea of the northwest are very cold, but become progressively milder towards the Alaska Peninsula in the southeast where large areas of Bristol Bay are covered with only thin ice or open water. The Aleutian Islands remain largely free of sea-ice throughout the winter.

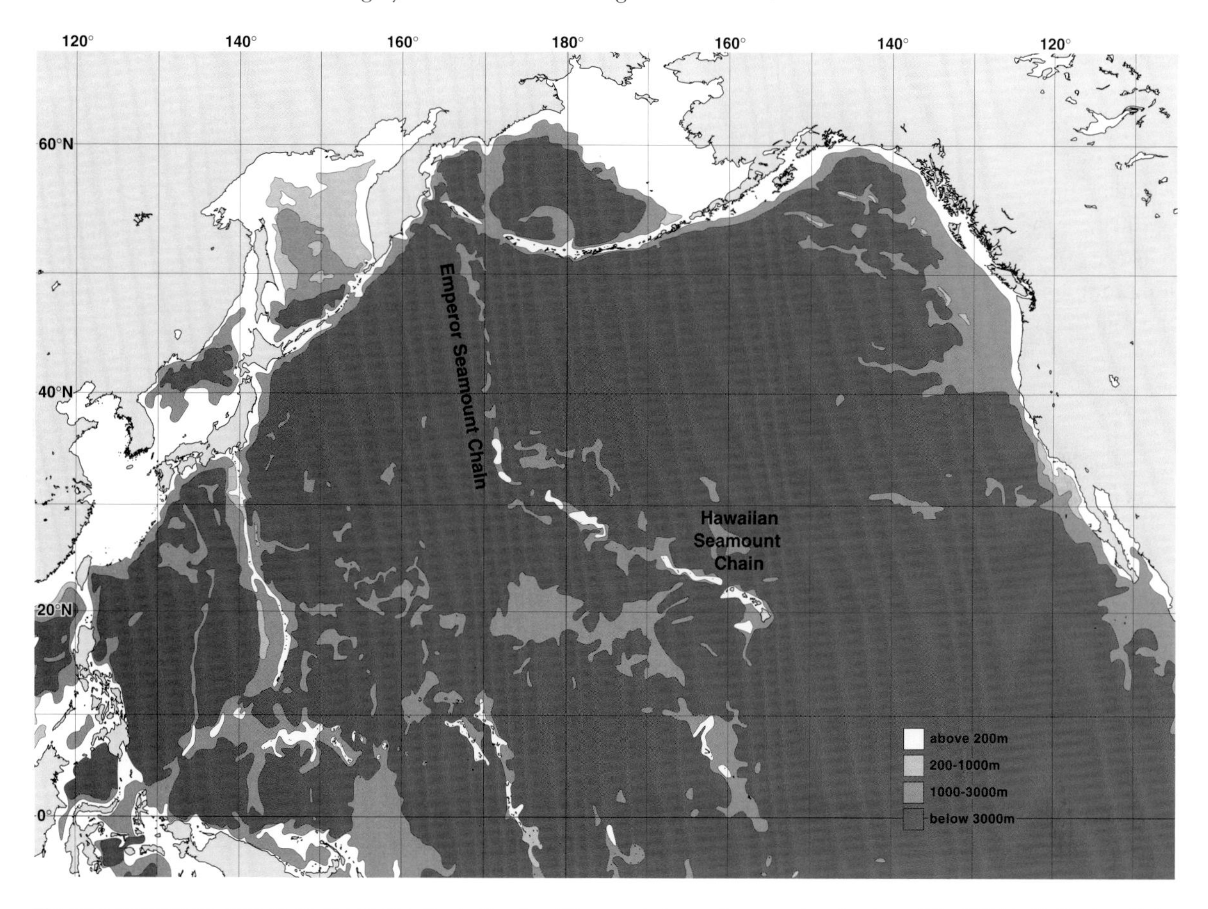

Fig. 10.2 Bathymetry of North Pacific Ocean, illustrating the difference between the complex structure of the NW Pacific Basin with its many seamounts and the less sculptured seabed of the eastern ocean.

Within the Bering Sea, cyclonic gyres circulate water in the deep-water basin, while a large anticyclonic gyre occupies most of the shelf (Fig 10.3). Upwelling of deep water and mixing along the Alaska continental slope leads to high productivity above the slope and over the margins of the shelf. Productivity is high at the perimeter of the sea, along the coast and interior part of the shelf and along the ridge between the Commander Islands[5] and the Aleutian Islands. In the deep-water basin productivity is low, but the great volume of water nevertheless holds an enormous biomass of plankton, fish and squid (Shuntov 1993). The relatively warm water of low salinity which enters between the Aleutian Islands remains within the Bering Sea for over a year and is transformed into the cold subarctic water. Some of it enters the north polar basin through the Bering Strait, but most flows back south towards the North Pacific in the western boundary current off Kamchatka (Shuntov 1993).

The Sea of Okhotsk is partially closed by the Kurile Islands[6] and lies largely over the continental shelf. It freezes in winter, being much colder in the north and in summer retains cold water at middle depths. There is a slow cyclonic circulation, with North Pacific water entering between the islands just south of Kamchatka and

[5] Russian: (*Kommandorskiye Ostrova*).
[6] Russian: (*Kurilskiye Ostrova*).

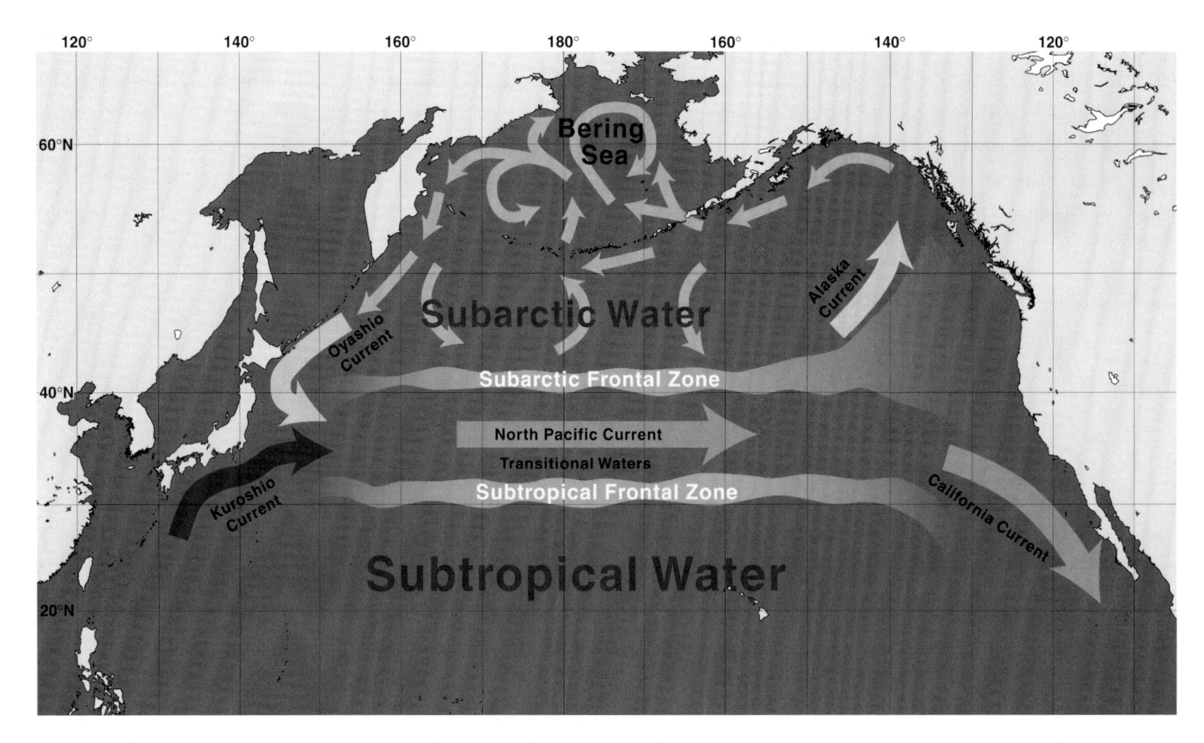

Fig. 10.3 The main hydrographic features of the North Pacific Ocean. Warm water of the Kuroshio Current and cold water of the Oyashio Current meet off Japan and flow eastwards to become the North Pacific Current. It forms a broad band of transitional water bounded by many temperature and salinity fronts that separate cold subarctic water of the north from warm subtropical water of the Central Pacific. As the North Pacific Current approaches America, it is deflected north and south. The Alaska Current flows anticlockwise then southwest down the Alaska Peninsula and along the Aleutian Islands. Some of this water enters the Bering Sea, flowing clockwise over the Alaska Continental Shelf and anticlockwise over deep water of the western basin.

flowing out between the more southerly islands off Hokkaido. It has a higher plankton productivity than the Bering Sea, but has less fish and squid (Shuntov 1993).

Arctic water flowing southwest past Kamchatka out of the Bering Sea in the western boundary current, receives additional cold water of low salinity flowing southeast out of the Sea of Okhotsk, between the Kurile Islands. Some of this water is deflected eastwards into the Subarctic Current, while the remainder continues southwest as the cold (2°C–5°C) nutrient-rich Oyashiwo[7]. It also turns east as it meets warmer water (10°C–15°C) of the Kurashiwo[8] flowing northeast past Japan. This is the northwest component of the subtropical North Pacific gyre. Small and large meanders of annual and decadal timescales off Honshu are superimposed upon fluctuations of millennial timescales, apparent in sediments over 25,000 years (Sawada & Handa 1998).

Some of this mixed water is caught up in the western subarctic gyre turning it north and northwest back towards the western Aleutian Islands, but most flows east as the North Pacific Current.

The temperature and salinity fronts that appear between the Oyashiwo and Kurashiwo to the east of Japan generate many complex meandering eddies. This is the beginning of the Subarctic Frontal Zone that extends right across the North Pacific, separating Subarctic water from Transitional water. It is a N–S gradient of several temperature and salinity fronts, whose mean position moves seasonally 100–200 km north and south of latitude 42°N. At shorter intervals, variable mesoscale meanders, eddies and jets predominate especially in the Western Basin. These are highly productive, attracting large numbers of whales, seals and seabirds. In the Eastern Basin, the Subarctic Frontal Zone is less convoluted and where it approaches North America, it is pulled apart by the Eastern Subarctic gyre (Alaska Current) and the California Current (Gould & Piatt 1993).

[7] Japanese: mother current.
[8] Japanese: black (dark blue) current.

Between the Subarctic Frontal Zone and the less clearly defined Subtropical Frontal Zone (30°N–34°N), there is a band of transitional water (42°N–32°N). It is one of the main oceanographic features of the North Pacific. In mid-ocean it is about 1,000 km broad. In the west it receives the meandering Kurashiwo extension, which may become unstable and shed large eddies. Typhoons and seamounts may also contribute similar turbulence, all capable of attracting seabirds (Haney *et al.* 1995). In the east, it feeds the California Current which encounters nutrient-rich upwelling along the continental slope. At its southern edge, the surface temperature of transitional water is about 18°C (Gould & Piatt 1993) and farther south becomes progressively warmer and less productive. Around the Northwest Hawaiian Islands, North Pacific central water remains warm throughout the year, from about 23°C in February–April to 26°C in September–October.

Climatic cycles

The El Niño/Southern Oscillation (ENSO) has long distance connections that are apparent in the northwest North Pacific Ocean (p.171). These occur at intervals of several years, but the ocean is subject to much longer cycles that are recognisable in marine deposits dating back 1,700 years (Hayward 1997). Typhoons are less frequent during El Niño events and vice versa.

From about the mid-1970s, a major climatic event affected oceanic circulation throughout the North Pacific. The Aleutian Low shifted south and deepened, causing increased westerly winds. The meeting of the Oyashiwo and Kurashiwo currents is strongly influenced by climate and wind forcing. The Oyashiwo flowed much farther south than usual and persisted until at least the late 1980s; during that time the biomass of zooplankton off Japan decreased sharply. To the east, the subarctic mixed-layer depth decreased 20–30% in 1977–88, compared with 1960–76, and zooplankton biomass in 1980–89 was double that in 1956–62.

During the mid-1970s, sea-surface temperatures along the coast of North America increased, but the central North Pacific cooled at the surface. Winter winds off Hawaii increased, the mixed-layer went deeper and productivity increased, peaking in the early 1980s and returning to pre-1975 levels by 1988. The predators of higher trophic levels, varied in their responses to these events over two to three decades; some decreased at the same time that others increased (Hayward 1997).

11

GOONEYS[1]

'as soon as the Laysan Albatross was spotted on the lawn that sun-washed noon, excited children on bicycles raced up and down the shaded streets shouting , "The goonies are here! The goonies have come back! " Small children stood in wide-eyed wonder as they watched the huge white bird...'

Mildred Fisher, Midway Atoll, 1965[2]

Laysan and Black-footed Albatrosses are both birds of the North Pacific Ocean. The Laysan is found mainly to the northwest, while the Black-footed Albatross is widespread throughout the ocean. Both breed during the northern winter, most of them on the Hawaiian Islands (Fig. 11.1).

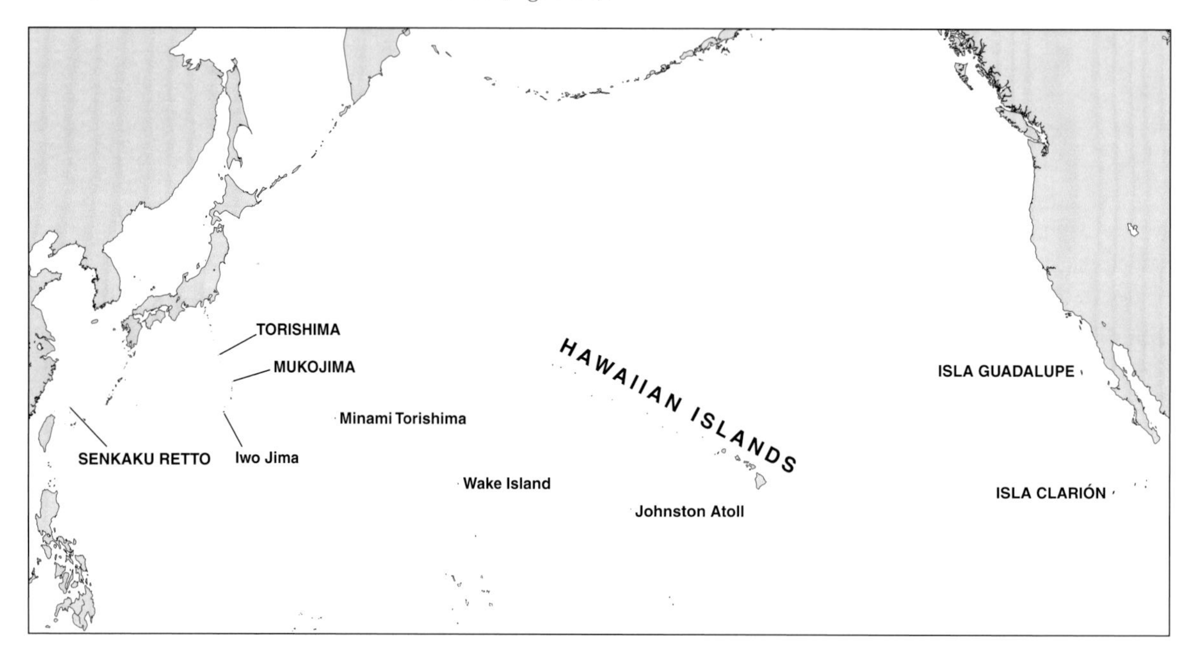

Fig. 11.1 Breeding islands of the Laysan Albatross and Black-footed Albatross in the North Pacific Ocean. There are no longer any colonies on Minami Torishima (Marcus Island), Iwo Jima or Johnston Atoll, but they may soon re-appear on Wake Island.

The Laysan Albatross in flight (Plate 40) with its wing span of little more than two metres, could easily be taken for a mollymawk, but on land its nesting habits and behaviour are very different from those of the small southern hemisphere albatrosses. The Black-footed Albatross is slightly larger and looks quite unlike the Laysan, but in spite of plumage differences, the two have more in common with each other than with any other species.

The Laysan Albatross is the most numerous albatross in the world. More of them have been killed by man than any other albatross species and several islands where they once bred no longer have colonies (Table 11.1).

Most of what we know of the biology of both species has been obtained from field-work at Midway Atoll which, because of its history and role as a naval air station, is unlike any of the other Hawaiian coral islands. Here also, albatrosses suffered horribly, but for a long time they have lived in harmony with human residents. During the

[1] p.17.
[2] Fisher M.L 1970.

1950s and 1960s, funds were available from military budgets for long-term research on the Laysan Albatrosses; the Black-footed Albatross has received less attention (Cousins 1998).

DESCRIPTION

Laysan Albatross

The Laysan is a slender white albatross with black to dark grey back, scapulars and upper wings; the mantle, rump and upper tail coverts are white, but the tail ends in a broad black to dark grey band (Plate 40).

The underparts are predominantly white, but the underwings are more patterned than in other species. The dark leading and trailing edges are irregular, with expansions at the carpal and elbow flexures, which reduce the white substantially. The darkest feathers are the ends of the primaries, which, when spread, reveal cream coloured shafts.

Most of the median primary and outer secondary under-coverts are completely or partially black-brown, as are all the lesser secondary under-coverts. At the elbows, some inner secondaries and outer humerals, together with their respective under-coverts, form dark triangles, more often than not separated from the leading edge of the underwing by a line of white or nearly white lesser under-coverts.

The forehead, crown and nape are white, but a horizontal line through the top of the eye marks a sharp transition to dark plumage around the eye below. The lores are black, fading anteriorly to pale grey, and lighter around the base of the bill. Below and behind the eye there is less black and more dark grey, fading over the cheeks below the gape and beyond the ear coverts to white (Plate 42). Immediately behind and below the eye (90°–190°)[3] there is a white post-orbital flash. The dark brown eyes are protected by long black and white eyelashes. The beak is horn-yellow/orange with black maxillary and mandibular ungues; the feet and tarsi are flesh colour (Fisher 1970).

Location	Laysan Albatross (pairs)	Black-footed Albatross (pairs)
Hawaiian Islands	558 000	59 000
Johnston Atoll	-	-
Wake Island	-	-
Minami Torishima	-	-
Iwo Jima	-	-
Mukojima	20	1 000
Torishima	-	1 000
Senkaku Retto	-	30
Islas Guadalupe & Clarión	50	-

Table 11.1 Breeding grounds of the Laysan and Black-footed Albatrosses throughout the North Pacific Ocean, with numbers (pairs) rounded to the nearest order of magnitude (Fefer *et al.* 1984, Pitman 1988, Brazil 1991, Gales 1997, Ogi *et al.* 1994b, H. Hasegawa pers. comm.).

Black-footed Albatross

A black-brown albatross, whose plumage is subject to considerable fading with age and feather wear. The inner portions of many feathers are less heavily pigmented than the exposed outer parts.

The mantle, back, scapulars, rump and tail are dark brown. In most breeding birds, the rumps and upper tail coverts are white (Bourne 1982b), but the age at which they acquire this character is not known. The upper surfaces of the wings are generally dark brown, with the exposed primaries rather darker (Plate 45).

The underpart of the body is light brown, and the belly even lighter, with white tail under-coverts and contour feathers about the vent (Plate 47). The underwings are generally darker than the body, with the primaries even darker. From above and below, the cream shafts of exposed primaries are revealed in distinct contrast to the dark brown webs.

In newly moulted birds, the crown is black-brown. It ends at a distinct post-orbital line below which the plumage is lighter, becoming slightly browner in front of the eyes and down the nape. Immediately behind the eyelid (90°–180°) there is a broad white flash that shades to light brown. The plumage surrounding the base of the bill, both upper and lower mandibles, is white, forming a narrow rim, but often spreading back over the head as brown feathers wear; in some individuals much of the head becomes pale brown or whitish (Plate 48).

Individuals are sometimes encountered with white feathers scattered among brown contour feathers of the body and/or unusual amounts of white on the belly, tail and around the base of the bill.

[3] The vertical is taken as 0°.

In juveniles, the white rim around the base of the bill is very narrow. They also lack the white upper tail coverts and white feathers around the vent; these may be acquired only very slowly and in some individuals perhaps not at all (D.W. Rice pers. comm.).

Adult Black-footed Albatrosses are almost the same mass as Laysans (Appendix 3), and with longer wings (Loomis 1918, Frings & Frings 1961, Warham 1996). The beak, tarsi and feet are black.

Sexual dimorphism

Although males sometimes appear to have flatter crowns than females, especially when pairs are seen together, it is difficult to determine the sex of individual Laysan and Black-footed Albatrosses in the field. Males of both species are slightly larger, and on average, are about 0.4 kg heavier than females, but there is overlap in all measurements. Hubert and Mable Frings (1961) found that the width of the head coupled with length of bill were the best measurements for sexing individuals. The greatest width was measured with calipers just behind the eyes. At this point the albatross skull has a blunt angle on each side that can be felt. Bill length was measured from the angle of the mouth to the tip.The Frings warned that subjective bias affects the actual measurements which should all be made by the same person with reference samples sexed by other methods. With these precautions, they were confident of sexing about 90% of individuals by the pair of measurements.

Albinos and hybrids

White plumage and partial albinos occur in both species (Bailey 1952, 1956). White Laysan chicks and fledglings have been seen regularly (Palmer in Rothschild 1893), but all-white adults have not been reported. Speckling of the dark back and upper wing surfaces of adults has been recorded. Fisher (1972) noted the following variations: 1) white or buffy bases to otherwise normal dark primary and secondary coverts, 2) all-white feathers scattered in the scapular or dorsal spinal tracts, 3) 4–14 mm fringes on flight feather coverts, 4) silvery prismatic-coloured feathers, some with brown edges on the back or mantle.

Laysan x Black-footed hybrids have been reported many times since 1891, when Henry Palmer and George Munro watched a Laysan Albatross feeding what appeared to be a Black-footed chick (Rothschild 1893). Single apparent hybrids were seen on Sand and Eastern Islands in 1939–40. The bird on Sand Island was at the same location in four consecutive seasons. In 1945, Fisher saw Laysan and Black-footed parents feeding a hybrid chick. They are treated as 'outcasts' by both species (D.W. Rice pers. comm.) and two older, ringed hybrids reached ages of at least eight to ten years of age without pairing. In all instances, the plumage of these hybrids resembled that of the Laysan Albatross (Fisher 1972):

> 'The two true hybrids we have observed are pearl gray, darker on the back, and almost white in places, particularly the ventral surfaces of the neck and belly. The smokey, pearl gray is superficial; the slightest ruffling exposes the underlying white. The shape of the bill is intermediate; its color is black in one bird, not the rosy-yellow of the adult Laysan, and it is part black and part gray in the other hybrid. Both hybrids have the Black-footed pattern of white at the base of the bill, but the white is more extensive, moving onto the forehead and supra-orbital area. One bird has one black leg and foot and one light gray leg and foot; the other bird's legs and feet are typically those of the Blackfoot. The hybrid thought to be the older is browner, less dusky gray than it was 6 years ago. The body is upright, more the posture of the Laysan Albatross, and the walking gait is more that of a Laysan than a Black-foot's head-down waddle ... Both rather obviously "thought" of themselves as Laysan Albatrosses...'

Hybrids have been reported making unsuccessful approaches to Black-footed Albatrosses (Ludwig *et al.* 1979) and John Warham (1990) saw five hybrids at Midway in 1980 which attempted unsuccessfully to dance with Laysan Albatrosses.

HISTORY

People of the North Pacific and Bering Sea would have been familiar with albatrosses long before the arrival of Europeans (Murie 1959). Off Japan the birds would have attended fishing boats and along the Aleutian Islands and Alaska Peninsula (Yesner 1976, Howard & Dodson 1933), they were known to Aleuts and Indians. Sea-going Polynesians must have found breeding grounds during and after the migrations that resulted in the settlement of the Hawaiian Islands, about a thousand years ago (Rice & Kenyon 1962). It is sometimes suggested that albatrosses were among the many birds destroyed by early settlers on the eastern high islands, but comparatively few albatrosses nest east of French Frigate Shoals.

Spanish vessels crossing the North Pacific between Mexico and the Philippines in the 16th century would also have encountered albatrosses at sea and may have found them breeding, perhaps in the Marianas Islands, which were discovered by Magellan in 1521. It was not until the late 18th century that North Pacific albatrosses began to be mentioned in the log-books and journals of European navigators.

During his third and last voyage (1776–80), Captain Cook and the crews of the *Resolution* and *Discovery* had seen many southern albatrosses by the time they sailed north from Christmas Island to discover the Hawaiian Islands, early in 1778. They were well aware that these birds were absent from the tropics and were surprised to find them at sea in the northern hemisphere. Cook himself was a meticulous observer, but there were no naturalists on board as there had been on his two earlier voyages. On 29 March 1778, when the ships were at Nootka Sound, Vancouver Island, he reviewed natural history in his journal:

> '... amongst the Oceanic tribe Albatrosses such as are seen in the Southern latitudes, but never between the Tropicks, at least I neither have seen or heard of one being seen there. And yet it should seem that they do sometimes cross these limits, if not how came they in the two Hemispheres? on the other hand if they cross the line in the Pacific Ocean why should they not do so in the Atlantic? they are numerous in the Southern Atlantic Ocean, but one was never yet seen in the Northern ...'

He had not yet suspected that the albatrosses they were seeing in the North Pacific were different from those they knew in the Southern Ocean. The following year, after the two ships had returned to the Hawaiian Islands and Cook himself had been killed in the tragic fracas at Kealakekua Bay, the answer almost presented itself. The ships were at Ni'ihau loading yams for the onward voyage, and a distant island, which the Hawaiians called Ta'oo'ra (Ka'ula), was included in the survey. They approached to within 6–8 km, and it was then, on 16 March 1779 after meeting one of the Queen's canoes, that Commander Charles Clerke guessed that the albatrosses they had been seeing might have been breeding locally. As they sailed on westwards, boobies and frigate-birds were often about the ships, but no albatrosses. The ships were far to the south of the islands when they changed course northwest towards Kamchatka and as they passed just to the west and out of sight of Kure Atoll, albatrosses were seen again and then regularly all the way to 62°N. The seabird records in the log of *Resolution* may not have been species specific, but nothing to equal them was published until modern times (Beaglehole 1967).

Many of the increasing numbers of seafarers in the North Pacific during the first half of the 19th century did not recognise albatrosses at all and often imagined them as 'large gulls'. Some adopted the whalers' name 'gooney'. On Christmas Day 1834, J.K. Townsend caught a black albatross from a ship sailing between San Francisco and Honolulu. He sent the skin to John James Audubon (1839) who named it the Black-footed Albatross *Diomedea nigripes*. It may seem perverse to name a completely black bird for its feet alone, but Audubon had already named another dark albatross for its plumage (p.115). He made no illustration from the specimen, which was never seen again (Coues 1866), but the birds themselves were easily recognised and soon became well known.

The Alaska purchase of 1867 and agreement of the Canadian border (1872), stimulated exploration of the northwest North Pacific. US Government surveys of the early 1870s provided the first opportunities for American ornithologists to visit the Alaska Peninsula and Aleutian Islands (Dall 1873, 1874). It was soon noticed that most Black-footed Albatrosses remained well to the south of the Aleutians, but among them a 'Black-browed Albatross' was reported (Baird *et al.* 1884). Although Nelson (1887) questioned whether it was possible for a southern hemisphere species to be that far north, the albatross described was clearly not a Black-footed Albatross nor a Steller's Albatross. Today we can guess that it must have been a Laysan Albatross that remained unnamed for another decade. What is remarkable is that the most abundant albatross went unnoticed for so long. As late as 1896–97, during a long voyage across the North Pacific and Bering Sea, Barrett-Hamilton (1903) wrote of many Black-footed and Steller's Albatrosses, but nothing resembling a Laysan Albatross.

The guano trade

During the early decades of the 19th century, Honolulu became the most important port in the North Pacific, catering for whalers and a burgeoning trade with China. Whaling masters soon knew all the atolls and coral islands of the Hawaiian chain (Ely & Clapp 1973) and would have returned to them whenever there were seals, turtles and birds to be had.

World demand for seabird guano increased rapidly during the 1850s into a booming industry that brought many Pacific islands under scrutiny. After the US Congress passed the Guano Islands Act of 1856 authorising American citizens to take temporary control of unclaimed seabird islands, several of the Hawaiian atolls and coral islands were claimed for the United States. The Act also drew King Kamehemeha IV's attention to the potential of islands that hitherto had been of little consequence to his people. From 1857, several were claimed for the kingdom of Hawaii and later for the provisional republic which followed[4].

Quantities of guano were exported from Hawaii between 1855 and 1861, but there is some mystery as to where it came from. On the albatross islands of the North Pacific, guano does not accumulate; it is either blown away by the wind or washed away by rain. Several ships were known to have taken guano from Johnston Atoll, but it seems to have been of poor quality, mainly sand. French Frigate Shoals may have been worked in 1859. When Captain Brooks described Laysan Island in the same year, he noticed small deposits of guano but not, he thought, in sufficient quantities to warrant any attempt to work it commercially. George Freeth, an Englishman who had visited Laysan in 1864, thought otherwise. In 1890, he secured the financial and government backing in Honolulu to set up a company[5] for the purpose of extracting guano from Laysan Island. After an inspection of the deposits with A.B. Lyons in July 1890, the latter published notes of the visit:

> 'The soil of the island consists of a peculiar kind of white sand, made up partly of fragments of sea shells, but largely of bits of egg shells and the bones of sea birds ...A rough calculation puts the bird population of the island at about 800,000; it may reach 1,000,000. They have not yet learned to fear man excessively, and are in fact no more shy than barn door fowl, so that it is very easy to study their habits.'
>
> (Lyons 1899 in Ely & Clapp 1973).

Digging began early in 1891; Laysan guano was not the dried excrement typical of arid Peruvian islands, but the product of birds, coral and rain. During seasonal rains the droppings of albatrosses and other seabirds leached into the underlying coral rock, which was transformed into a phosphate-rich lime, low in nitrogen. The resulting hard conglomerate was hacked out at the surface by gangs of immigrant Japanese labourers from Honolulu, then shovelled into trucks on rails and towed by mules to storehouses at the wharf-head. At times there were as many as 40 men working on the island (Ely & Clapp 1973).

Guano was shipped only from April to September, so the birds were not disturbed at the most sensitive time of the breeding season, during laying and incubation. The birds reoccupied former sites after excavation had ceased. Commercial extraction came to an end in 1904, but intermittent digging on a small scale continued until 1910. In one way, guano digging at Laysan Island protected the birds. Managers were appointed US officials and their presence kept Japanese fowlers away.

[4] The islands were amalgamated in 1898 when Hawaii became a Territory of the United States. In 1959, all except Midway, a Federal Territory, became the 50th state of the Union.

[5] The North Pacific Phosphate and Fertiliser Company; changed in 1894 to Pacific Guano and Fertiliser Company.

Rothschild

The Honourable Walter Rothschild was the son of the first Lord Rothschild of Tring, a renowned Victorian banker of legendary wealth and influence. In childhood, Walter conceived an ambition to create a natural history museum and by the time he went to Cambridge, his collection already included 5,000 birds. Although he did not stay at the university long enough to graduate, he was inspired by Professor Alfred Newton's talk of the Pacific islands and, while still an undergraduate he hired an Australian bird collector, Henry Palmer and sent him off to the Chatham Islands (M. Rothschild 1983).

Island faunas were then a major interest of biological science and in 1887–88, Newton's student, Scott Wilson had brought back a substantial collection of birds from Honolulu and published many papers including the beautifully illustrated *Aves Hawaiienses*. Newton had been trying to secure funds for further collecting expeditions and the frustration of his efforts was perhaps made more galling by his other former student, Rothschild, suddenly directing Palmer to the same islands. Although Rothschild belatedly offered Newton participation in the expedition, the professor favoured Wilson and declined.

Palmer reached Honolulu in December 1890 and on 27 May 1891 embarked upon a remarkable 84 day voyage with George C. Munro, along the more remote coral islands to the northwest. They stayed ashore at Laysan Island, Lisianski Island and Midway Atoll, and called at Pearl and Hermes Reef and French Frigate Shoals. Munro later described it as one of his greatest adventures, but for the collector it was as much a nightmare as a paradise:

> ' ..A small fly on the island has given me a great deal of trouble with my skins. Almost before the birds were dead they were fly-blown. It was the smallest blow-fly I had ever seen.' (Rothschild 1893).

Bird specimens reached London towards the end of 1891 and the seabirds were shown to Osbert Salvin at the British Museum, who had seen neither of the albatrosses before. Rothschild had only recently published his first short paper and was not yet a member of the British Ornithologists' Union (BOU). Having dropped out of Cambridge and alienated Professor Newton, the ornithological establishment was unlikely to see him as other than a rich dilettante, so he needed to make an impression and what better for that purpose than Laysan Island?

On this desert island, his collector had discovered an immense number of white albatrosses new to science together with a handful of endemic land birds. The subject matter was so compact that an impressive volume could be brought out very quickly. He engaged two of the best illustrators of the time (J.G. Keulemans and F.W. Frohawk) and less than two years after Palmer had landed on Laysan Island, he was ready to announce his collector's exciting discovery. On the evening of 17 May 1893, Rothschild went along to Frascati's restaurant to dine with other members of the recently formed British Ornithologists' Club (BOC), which was already beginning to take on the aura of an inner sanctum of the BOU. He had himself only just been elected to the BOU and, at the age of 25, was attending for the first time as a member, perhaps the youngest present. After dinner he presented his new albatross, *Diomedea immutabilis*[6] (Rothschild 1892–93). Three months later, the first part of *The Avifauna of Laysan* was published (Rothschild 1893), complete with photographs of the huge colony taken by J.J. Williams of Honolulu (Fig. 11.9), and extracts of Henry Palmer's journal, which gave a hint of what it was like sailing in the schooner *Kaalokai* with the irascible Captain Walker:

> '18 August 1891 At last we have reached Honolulu after a month's dreadful journey from Midway Island. I never was more pleased in my life when I found myself on shore in Honolulu once more.'

Curiously, the British Museum's seabird expert, Osbert Salvin misidentified the Black-footed Albatrosses that Palmer had sent back. He had evidently been quite unaware of Audubon's name, *Diomedea nigripes* or that Temminck's synonyms *Diomedea chinensis* and *Diomedea brachyura* which he used, had been dismissed by Elliott Coues in 1866. Laysan Island was the first documented breeding ground of this species.

Neither Audubon nor Rothschild showed any reluctance to name their specimens *Diomedea*, but long before the Laysan Albatross had been discovered, Ridgeway (1887) had grouped the Black-footed Albatross with Steller's

[6] Latin: *im-mutabilis* – unchangeable; presumably in contrast to Steller's Albatross, the other 'white gooney' which has a dark juvenile and intermediate subadult plumage.

Albatross in a subgenus based upon Reichenbach's (1850) genus *Phoebastria* (p.236). In the third Check-list of North American Birds (American Ornithologists' Union 1910) the three North Pacific albatrosses were all in this subgenus. Twelve years later, A.C. Bent (1922) raised *Phoebastria* to a genus and Gregory Mathews (1934) adopted it as the genus of all four Pacific albatrosses, including the Galápagos Albatross, on the basis of common features at the base of the bill. When resistance to this revision became overwhelming, all were returned to *Diomedea* (Mathews 1948). By the fifth edition of the Check-List (AOU 1957) *Phoebastria* had been discarded altogether. Hans von Boetticher (1949, 1955) later introduced two completely new genera, *Laysanornis* and *Penthirenia* respectively for Laysan and Black-footed Albatrosses, but both were ignored.

The arrival of molecular systematics (Nunn *et al.* 1996) split the genus *Diomedea* again and new genera had to be found for the Pacific albatrosses. Mathews was exonerated and all were returned to *Phoebastria*. Monophyly of the two gooneys was well supported by the molecular studies, so a common genus was required. Rules of priority apparently decreed that it should be *Phoebastria*, but by other criteria they are so different from Steller's Albatross that it is questionable whether a common genus is appropriate.

Fowling

Japanese fowling for seabird feathers spread across the North Pacific as islands nearer Japan were worked out. It is usually supposed that albatrosses were among the first to suffer and on some islands at least, the activity continued to be served by other species long after albatrosses had all been killed (Shelmidine 1948).

Japanese fowlers were first reported in the Northwest Hawaiian Islands (NWHI) in 1900, but they may well have been active there many years previously. In spite of US and Hawaiian opposition, they persisted intermittently for another 20 years. In response to campaigning by Audubon Societies, President Theodore Roosevelt issued Executive Order No. 1019 of 3 February 1909, that the islets and reefs of the Leeward Islands[7] were to be reserved and set apart as a preserve and breeding ground for native seabirds. The Hawaiian Bird Reservation was administered by the Bureau of Biological Survey (US Department of Agriculture). For decades there were few resources in Hawaii to enforce the order, other than the coast guard cutter *Thetis*, ever steaming up and down the islands with the watchful eyes of redoubtable commanders like W.V.E. Jacobs. Gerrit Wilder was Warden of the reservation from about 1923. He visited all of the islands from time to time, but the President's commitment to seabird conservation was to undergo many assaults before achieving the support it enjoys today.

At least some of the fowling in the NWHI was believed to have been master-minded from Honolulu. There were large numbers of immigrant Japanese labourers in Hawaii and Professor William Alanson Bryan of the O'ahu College, Honolulu, claimed that fowlers apprehended at Laysan Island had previously been digging guano on the island for Max Schlemmer, a colourful Honolulu character sometimes known as 'Admiral Max' or the 'King of Laysan'.

Surveys

In 1923, the Bureau of Biological Survey in co-operation with the Bernice P. Bishop Museum of Honolulu and the US Navy, completed the first major biological investigation of the NWHI together with Johnston Atoll and Wake Island. The expedition was led by Alexander Wetmore and became identified with the US minesweeper *Tanager* (Commander Samuel W. King) on which it sailed. Between 7 April and 6 August, almost all islands were visited (Olson 1996).

The first albatross census of the North Pacific was undertaken in 1957–1958, and was based largely upon aerial photographs of the NWHI taken from aircraft of the US Navy. Pioneer field studies of both albatrosses at Midway Atoll (Rice & Kenyon 1962) were soon followed by long-term investigations of Laysan Albatrosses directed by Harvey I. Fisher of Southern Illinois University.

A need for widespread, long-term multi-island investigations in the North Pacific had been recognised by the Smithsonian Institution and from 1963 to 1969, members of its Pacific Ocean Biological Survey Program (POBSP) visited all the Hawaiian atolls and coral islands many times, collecting a great deal of data on albatross numbers and their breeding grounds as well as making observations at sea. This was followed during the late 1970s and

[7] The Hawaiian Islands were formerly divided into the eastern, Windward Islands – big, high, volcanic, forested islands – and the western, Leeward Islands – small, low, coral atolls – now known as the Northwest Hawaiian Islands (NWHI).

1980s by the Tripartite Research Program bringing together the US Fish and Wildlife Service (USFWS), the National Marine Fisheries, the State of Hawaii Department of Lands and Natural Resources and the University of Hawaii Sea Grant Program (Harrison 1990). Since 1964, all Hawaiian albatross islands have been monitored by the Bureau of Sport, Fisheries and Wildlife (BSFW), USFWS or the National Biological Survey (NBS).

Several biologists involved with fisheries investigations had made albatross observations at sampling stations off California (Miller 1942, McHugh 1950) and Japan (Kuroda 1960), but it was not until Viacheslav Shuntov had collated years of observations from Russian fisheries vessels that distribution maps of Laysan and Black-footed Albatrosses were published (in Russian). He up-dated these maps for his impressive 1972 book on seabirds. It, too, was in Russian, but two years later it was translated into English (Shuntov 1974). In the same year, the results of POBSP were published, including other important contributions to our knowledge of albatross distributions in the North Pacific (Robbins & Rice 1974, Sanger 1974a&b).

From 1978 to 1981, biologists from the USFWS and NOAA[8] visited all the NWHI collecting marine organisms, including those in the diets of Laysan and Black-footed Albatrosses (Harrison *et al.* 1983).

OCEANIC DISTRIBUTION

Laysan and Black-footed Albatrosses have overlapping and seasonally shifting movements throughout the North Pacific Ocean. More has been written about their distributions at sea than for any other albatross species. Shuntov's (1972) maps (Figs. 11.2, 3) make some claim to precision, but do not correspond closely with Sanger's (1974) maps. Kuroda (1988) did not stray from his data and his observations are unusual in that they were made from car-transporter ships, free from direct fisheries bias.

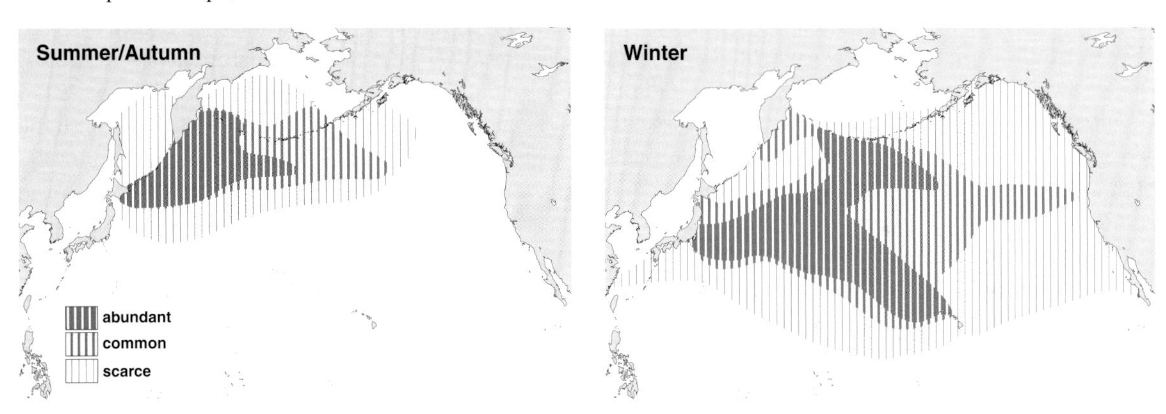

Fig. 11.2 Distribution at sea of Laysan Albatrosses in the North Pacific Ocean (Shuntov 1974).

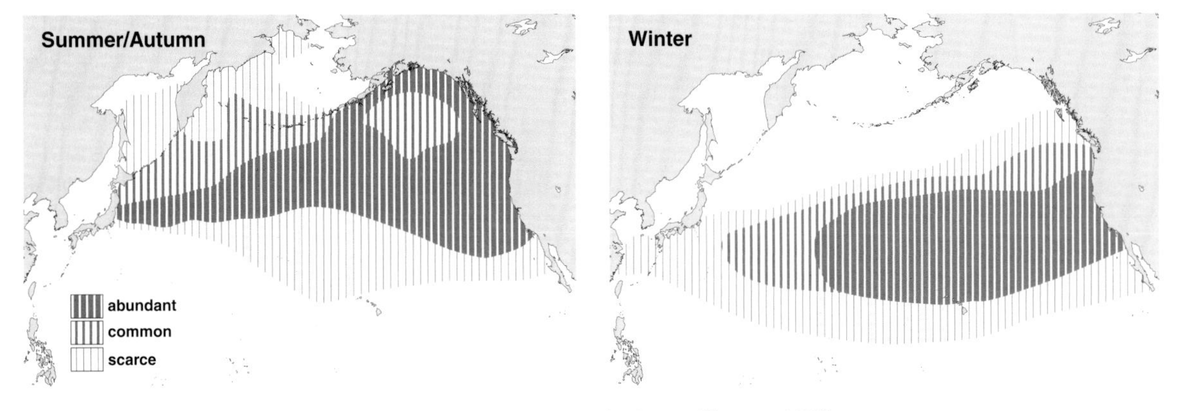

Fig. 11.3 Distribution of Black-footed Albatrosses in the North Pacific Ocean (Shuntov 1974).

[8] National Oceanic and Atmospheric Administration.

Laysan and Black-footed Albatrosses are common off Japan from Kyushu to Hokkaido. The East China Sea is open to them and occasionally albatrosses enter the Sea of Japan. Both species fly along the Kurile Islands[9], some entering the Sea of Okhotsk, while others reach the Kamchatka peninsula and as far north as the Gulf of Anadyr. To the east, they have long been known in the straits between the Aleutian Islands and they enter the deep water basin of the Bering Sea, reaching the edge of the Alaskan continental shelf. In the northeastern North Pacific, they approach the continental shelf of North America from the Aleutian Peninsula to Baja California in the south. Calm, tropical seas limit their southward movements.

Laysan Albatross

During breeding, there is constant movement between the Hawaiian Islands and the sea separating Japan from the western Aleutians. Individuals tracked by satellite, flew north to the Gulf of Alaska and the Aleutians (Anderson & Fernandez 1998). Laysan Albatrosses are capable of travelling great distances at speeds of up to 500 km per day (Kenyon & Rice 1958). This shuttle persists throughout the winter and by spring many Laysans that have lost eggs or young chicks have ceased returning to the Hawaiian Islands.

In the Gulf of Alaska, numbers begin to increase in spring and some Laysans may reach 60°N, but as summer approaches they move west along the Aleutians. Farther south off the coasts of British Columbia, Washington and California they remain rare throughout the year, except for slight increases in numbers over the continental slope in winter (Tyler *et al.* 1993, Wahl *et al.* 1993).

Adult Laysans begin leaving the Hawaiian breeding grounds in late June and July, closely followed by their young; the low latitudes are vacated and most concentrate north of 40°N, off Hokkaido, the Kurile Islands and Kamchatka as far east as 175°W. Some Laysans enter the Sea of Okhotsk and the Bering Sea, where they are regularly seen over the deep-water basins.

In late October, they begin to leave the north, and in early November, as thousands home-in on the Hawaiian Islands, the surrounding ocean is re-populated. South of the islands, they are found over the tropical ocean to 13°N and sometimes as far as 8°30'N.

Laysan Albatrosses very rarely cross the tropics into the southern hemisphere. A juvenile, ringed on a nest at Midway Atoll in March 1965 was found dead the following September on San Cristobel (10°S, 161°E) in the Solomon Islands (Robbins & Rice 1974). In April 1983, Peter Harrison (1983) saw one flying over the south Indian Ocean at 37°S, 22°E, and the following year, Jean-Paul Roux (1988) saw another at 35°S, 53°E. In October 1985, a Laysan Albatross landed on Norfolk Island (29°S, 168°E); it was probably the bird that returned to the same location in August of the following year.

Black-footed Albatross

The Black-footed Albatross is numerous in a broad band of ocean between 18°N and 45°N, from 170°E to the coast of North America where they have long been conspicuous, even over inshore waters off California (Miller 1942, McHugh 1950, Sanger 1970,1972, Gould 1983, Briggs *et al.* 1987). During the winter, while breeding in the Hawaiian Islands, Black-footed Albatrosses avoid the higher, colder latitudes of the North Pacific; when feeding chicks, individuals tracked by satellite foraged northeast, towards the coastal waters of North America (Anderson & Fernandez 1998). After they leave the breeding grounds in June, they move north but do not entirely vacate the adjacent ocean. Throughout the northern summer, they may be encountered almost anywhere north of the Hawaiian Islands. In the northeast, they are widespread, being seen often in the Gulf of Alaska, off Kodiak Island, along the Alaska Peninsula and the Aleutian Islands, although rarely near land. They enter the Bering Sea via the Aleutian passes and may be found just north of the Alaska Peninsula.

The Black-footed Albatross is numerous in the northwest North Pacific, especially in the middle latitudes. It was once abundant in the Formosa Channel (Swinhoe 1863) and has been seen off South Korea and in the strait[10] between Honshu and Hokkaido. Many occur off Hokkaido and the Kurile Islands, where they may remain south of the 11°C surface water isotherm (Shuntov 1968), even though they are found over colder water elsewhere.

[9] Russian: (*Kurilskiye Ostrova*).
[10] Japanese: (*Tsugaru-kaikyo*).

Some enter the Sea of Okhotsk and large numbers reach Kamchatka and the Commander Islands[11] in late summer. Around Japan and the East China Sea, numbers decline in autumn as birds leave these northerly regions. At this time, they withdraw to the subtropical areas of the central North Pacific some being seen south of 13°N, occasionally as far as 4° 30'N (Sanger 1974a).

In the Vienna Museum, there is a specimen of a Black-footed Albatross labelled as coming from 'New Holland' (Australia) and another said to have been collected in 1884 at Dusky Sound, New Zealand (Mathews 1930). The records have long been accepted by the Ornithological Society of New Zealand (1970) and the Royal Australian Ornithological Society (1990).

Ringing

Juvenile cohorts remain distant from their breeding islands and subadult cohorts spend less time there than breeding adults. Maps based upon at-sea observations cannot discriminate these different classes of birds. Fisher & Fisher (1972) used ringing recoveries to distinguish the oceanic movements of different age classes of Laysan Albatrosses; two years later Robbins & Rice (1974) analysed 723 recoveries of both species.

Ring recoveries at sea also reflect the distribution of fisheries. That may be acceptable if it is not assumed that the birds are absent from areas where there is no fishing. Comparative analysis of such data may also be valid if there is similar probability of the two species being captured. The numbers of the Laysan Albatross are much bigger than those of the Black-footed Albatross and more Laysan Albatrosses have been ringed so unequal numbers of recoveries should be expected. It has frequently been asserted that while the Black-footed Albatross follows ships and scavenges whatever is thrown overboard, the Laysan Albatross rarely does so. However, many Laysans have been killed in fishing gear.

Laysan Albatrosses:

All recoveries were farther north in summer than in winter. Almost all occurred west of longitude 180° and there was very little overlap between summer and winter ranges. In winter, when breeding, recoveries were between the Hawaiian Islands and Japan. In summer, almost all adults recovered at sea were north of 40°N, half of them north of 45°N and evenly scattered between Japan and the Western Aleutian Islands.

Juvenile Laysans in their first summer months after fledging were found between latitudes 40°N and 45°N, more than half of them within 500 km of Japan and the rest eastward as far as longitude 172°W. They remained in this region after adults had left for the Hawaiian Islands. Half of the recoveries were within 1,000 km of Japan, so they appeared to be spreading, but the following year all recoveries were west of 160°E and closer to Japan. Younger Laysans tended to be recovered nearer the coast than older birds.

This pattern persisted through the second winter. In their third year they appeared to move away from Japan, especially in winter, when recoveries were as far away as 134°W. There is a consistent trend for two to five-year-old Laysans to be recovered further south than the adults.

Fourth year, fifth year and adult Laysans ranged farther north, with half the summer recoveries north of 48°N, especially to the northeast, off the Aleutians.

Black-footed Albatrosses:

In winter and summer, recoveries of ringed adults were equally frequent on both sides of the North Pacific. Apart from concentrations off Japan and North America, the numbers of recoveries east and west of longitude 180° were similar. The seasonal separation in latitude was almost clear-cut, with few recoveries south of 30°N in summer or north of 35°N in winter. Adults were significantly farther north in summer than winter.

Ringed juveniles in their first months at sea scatter east and west between latitudes 35°N and 45°N, towards the coasts of Japan and North America. They move south into lower latitudes (30°N–35°N) during their first winter and recoveries off Japan and California have been numerous. First year birds were more frequent towards the western North Pacific in summer (July–August) than older birds. Second summer recoveries indicate a pronounced shift eastwards between 30°N and 47°N, and half were within 300 km of the American coast south of Alaska. All

[11] Russian: (*Kommandorskiye Ostrova*).

remained significantly farther south than adults. The few available recoveries in the second and third winters may indicate a move towards the central North Pacific, but a northward movement with increasing age in summer was indicated. In the third summer recoveries were numerous off the American coast between 40°N and 51°N, with few farther north or south. The relative number of recoveries may give an exaggerated impression of the aggregations in coastal waters. Adults (older than six years) were recovered farther west than three and four-year-olds in February in June and July.

Aggregations at sea

During 22 crossings of the North Pacific between Japan and the United States in 1983, Laysan and Black-footed Albatrosses were recorded on 1,826 and 1,237 occasions respectively. Over 80% of records of both species were of single birds. Only about 1% of Laysans were in groups larger than ten individuals, and very rarely were more than 100 seen together. Less than 1% of sightings of Black-footed Albatrosses were of groups larger than ten, and none of more than 20 birds (Kuroda 1988).

BREEDING ISLANDS

Although Laysan and Black-footed Albatrosses occur throughout the North Pacific Ocean, almost all breed at the Hawaiian Islands. The NWHI are strategically located for pelagic seabirds exploiting the North Pacific Ocean. The atolls and coral islands are within commuting distance of productive areas associated with the polar fronts off Japan, Kamchatka and the western Aleutians (Fisher & Fisher 1972). Their flat, open character is particularly well suited to accommodating large numbers; the rocky, volcanic islands east of French Frigate Shoals attract relatively few albatrosses.

Until the late 1800s, several other islands may have had substantial numbers of both species, but today relatively few breed away from the Hawaiian Islands (Table 11.1).

The Hawaiian Islands

The Hawaiian Islands form an arc of about 18 subtropical islands and atolls extending some 2,600 km across the central North Pacific, from Hawaii in the southeast to Kure Atoll in the northwest (Fig. 11.4). The mean air temperature from December to April, during the breeding season, is 21°C. After the albatrosses leave it gets much hotter; from June to October air temperatures reach 27°C to 32°C. Annual rainfall is about 100 cm. The wettest months are from December to February and the driest month is March. NE Trade Winds blow from March to October and stronger westerlies in November to February.

Howell and Bartholomew (1961) measured the microclimate in which Laysan and Black-footed Albatrosses breed at Midway Atoll and studied temperature regulation in both adults and chicks. There were no detectable differences in temperature regulating adaptations between the two species. Heat stress was reduced by radiation – exposing the feet to the air – or by evaporative cooling. Opening the mouth and 'fluttering' the expanded throat involves water-loss and chicks that are not fed often enough become dehydrated and may die.

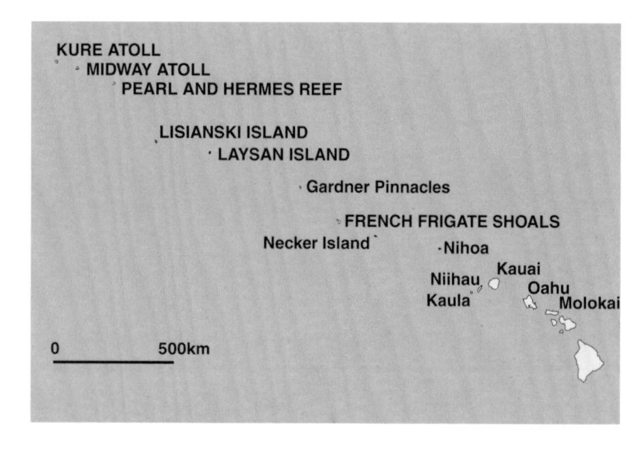

Fig. 11.4 The Hawaiian Islands: low atolls and coral islands with large colonies of albatrosses (upper case); high volcanic islands with few albatrosses (lower case).

Kure Atoll (28° 25'N, 178° 21'W)
Green Island (1 km² 8 m)

A Spanish vessel is said to have called here in 1799, but the atoll (Fig. 11.5) was eventually named after the Russian captain of the *Moller* who visited in 1827. Benjamin Morrell of the schooner *Tartar* had already (1825) noticed that Green Island[12] was covered with seafowl. In 1842, the whaler *Parker* struck the northwest reef, and

[12] The island was given several names in the early 1800s, but for more than a hundred years it was generally known as Ocean Island.

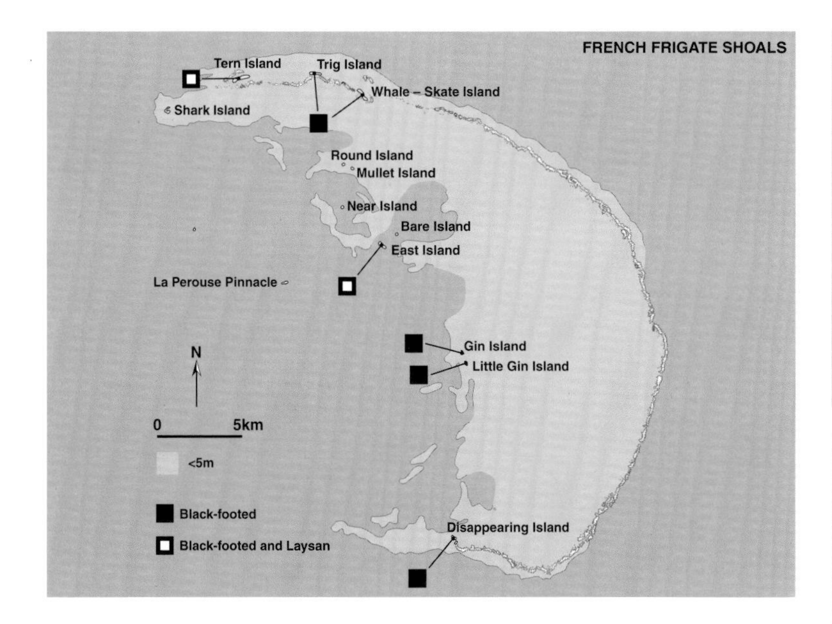

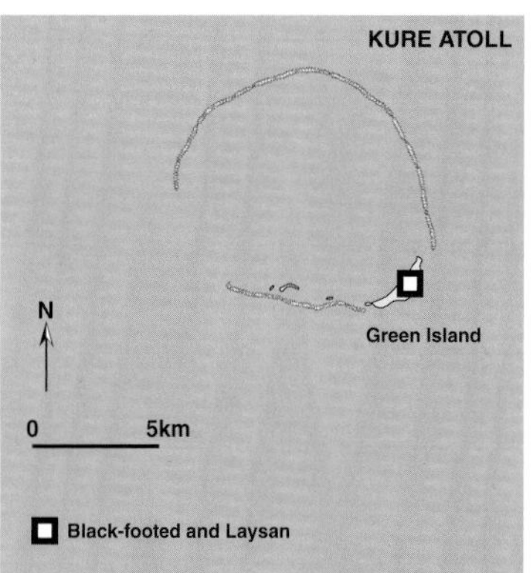

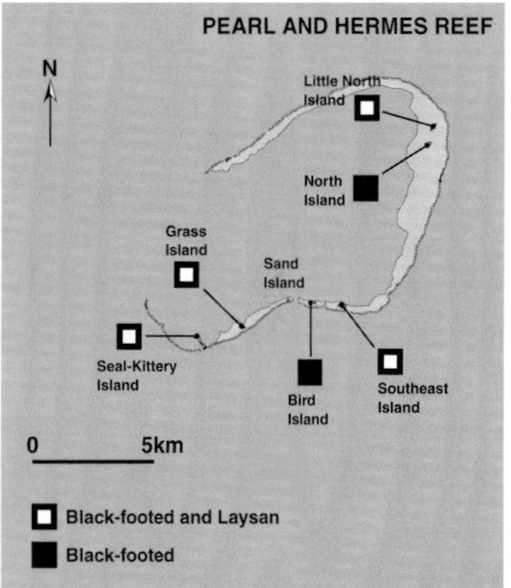

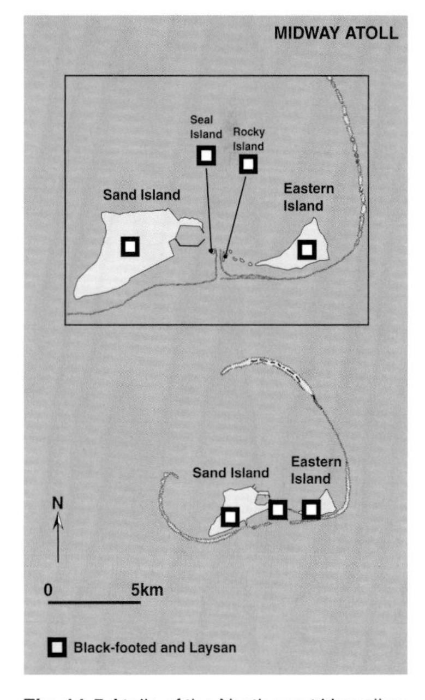

Fig. 11.5 Atolls of the North west Hawaiian Islands with the islands where both Laysan and the Black-footed Albatrosses breed and where only Black-footed Albatross colonies occur.

survivors lived on the island for seven months before being rescued. They killed about 60 seals and 7,000 seabirds, many of which would have been albatrosses. The crew of the *Saginaw*, which went aground in 1870, also depended upon seals and 'brown albatrosses'. During the first month, they killed as many as they could eat, but by the end of November, the captain set an allowance of one seal and 20 birds per day for the whole crew. About 600 albatrosses must have been taken in December, and probably more during the previous month (Woodward 1972). Pacific rats were found at this time and it has since been suggested that they reached the island long before, in Polynesian canoes.

The first seabird count was made in March 1915 by W.H. Hunter of the *Thetis*. He estimated 1,500 Black-footed Albatrosses along the north shore, and about 300 Laysan Albatrosses, with their main colony inland. The *Tanager* was at Kure in April 1923, where Alexander Wetmore (Olson 1996) found fewer of both species. Laysans were unable to penetrate the dense *Scaevola* scrub and many Black-footed Albatrosses had died.

Kure Atoll had been claimed for the kingdom of Hawaii in 1886 and biological surveys by BSFW[13] began in 1956. In 1959, a plan was devised to solve the Midway albatross problem by moving birds to Kure. Bulldozers were landed to open up a network of 15 m wide clearings in the *Scaevola* scrub of Green Island to accommodate more Laysan Albatrosses. Two years later, fledglings were translocated from Midway, but after fledging, none later returned to take up residence (Owre 1961).

In 1960, the US Coastguard installed a long range navigation (LORAN) beacon. Its 190 m mast, runway and accommodation completely changed the character of Green Island. Between 1963 and 1969, POBSP[14] and BSFW researchers spent a total of 2,327 days there in long-term studies of seabirds,

[13] US Bureau of Sport, Fisheries and Wildlife.
[14] Pacific Ocean Biological Survey Program (Smithsonian Institution).

including albatrosses (Woodward 1972). The Coastguard evacuated the island in 1992 and removed all equipment. The rats have been eradicated and Kure Atoll is now managed as a seabird sanctuary by the State of Hawaii (McDermond 1993).

Midway Atoll (28° 14'N, 177° 22'W)
Sand Island (4.9 km² 13 m)
Eastern Island (1.4 km² 10 m)

The two main islands (Fig.11.5) on the southern rim of the atoll are separated by about two kilometres of sea and three small islets. In 1859, the atoll was claimed for the United States by Captain N.C. Brooks. In 1870, an attempt was made by the US Navy to blast a harbour in its lagoon, but when this failed the atoll was abandoned. There followed a decade of shipwreck, madness and murder (Shelmidine 1948). Albatrosses were killed for food, but Japanese fowling had not started. Henry Palmer and George Munro called briefly in July 1891. They saw hardly any vegetation on Sand Island, just a little grass at one end, and some tall *Scaevola* on low mounds of sand at the other. Both men found it a desolate place and believed that in rough weather, most of the island would be swept by surf and spray. Eastern Island was quite different, being almost completely covered by ridges of green scrub almost two metres high, which were dense, tangled and difficult to traverse.

Japanese fowlers were warned off Sand Island when the USS *Iroquois* visited Midway in 1900. W.A. Bryan (1906) called briefly in August 1902. There were no albatrosses at that time of year. Sand Island was desolate and deserted, with the remains of only a few Laysan Albatrosses. On Eastern Island, he found some empty shacks and thousands of carcasses, most of them of Black-footed Albatrosses with about one third as many Laysan Albatrosses.

Following a strong note of protest to the Japanese Government, President Theodore Roosevelt placed Midway Atoll in the charge of the US Navy Department. The first trans-Pacific telegraph cable was routed via Midway and in 1903, when the Commercial Pacific Cable Company arrived to set up their station, Japanese fowlers were again at work[15]. They were evicted two months later when the *Iroquois* returned. The cable company superintendent was appointed US Navy custodian of the atoll, with specific instruction to prevent:

> '... the wanton destruction of birds that breed at Midway, and not let them be disturbed or killed except for the purpose of food supply.' (Shelmidine 1948).

The Midway albatrosses were about to experience more than three decades of protection just at the time when albatrosses on other islands were in great danger. During the initial construction there were about a hundred people on Sand Island, but by the time the Marines left in 1908, the station housed about 30 staff. A concerted effort was made to create a Pacific oasis; in 1906 several shiploads (about 9,000 tons) of soil were brought from Honolulu to create gardens and the settlement was planted with ironwood trees *Casuarina equisetifolia* and shrubs. Smaller quantities of soil continued to arrive and an island farm with ducks, chickens, turkeys and cattle made the community almost self-sufficient. Far from disturbing the albatrosses, Laysans favoured the settlement and began increasing.

The cable company welcomed visiting naturalists. Paul Bartsch was there at the beginning of November 1907. On Eastern Island, he found eight donkeys, originally introduced to Sand Island where they had become too destructive and had been been banished to Eastern. George Willett and Alfred Bailey called at Sand Island on 14 March 1913. They noted a few Laysan Albatrosses nesting near the houses, and small colonies of Black-footed Albatrosses at the other end of the island. By the time the *Tanager* arrived in April 1923, numbers of both species had increased substantially. Laysans appeared to be more numerous than Black-footed Albatrosses, especially on Eastern Island, in long openings among the *Scaevola*. There were now 16 to 18 donkeys and one was removed (Olson 1996).

In 1935, Pan American Airways began using Midway to refuel their *Clipper* flying-boats. A small hotel was built on Sand Island, and albatrosses became one of the attractions of trans-Pacific flights (Chisholm 1937). Fred Hadden (1941), who had the job of fumigating all aircraft approaching Hawaii from Asia, remained on the island

[15] There is no record that they had taken anything like the 500,000 birds claimed by Robertson and Gales (1998).

from 1936 to 1941 and took great interest in the albatrosses. His account of their behaviour was the first to benefit from such long experience of the birds. By that time, 2,000 Laysans were nesting in the PAA compound alone, but the approach of the Second World War brought the US Navy back to Midway and heralded a long period when fortune turned against the birds.

Construction of the naval airbase with runways on both Sand and Eastern islands lasted 18 months, but activity never ceased altogether throughout the war and for a decade or more thereafter. With the exception of the areas of trees around the Cable Company compound, and a few isolated areas of scrub, almost the entire surface of the two islands was either smoothed for roads, paved for runways, excavated for bunkers, or covered by buildings and dumps (Fig. 11.6).

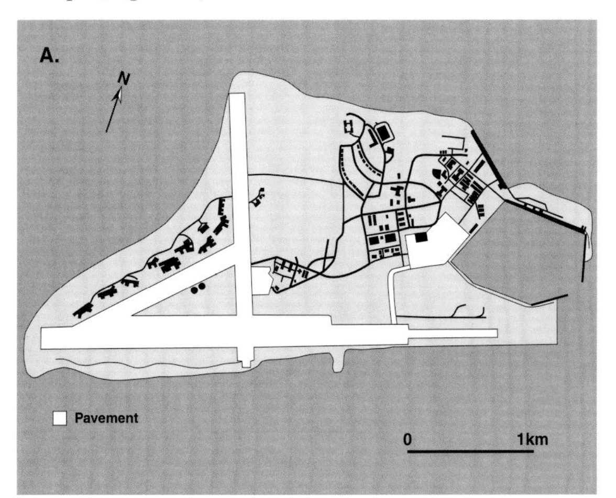

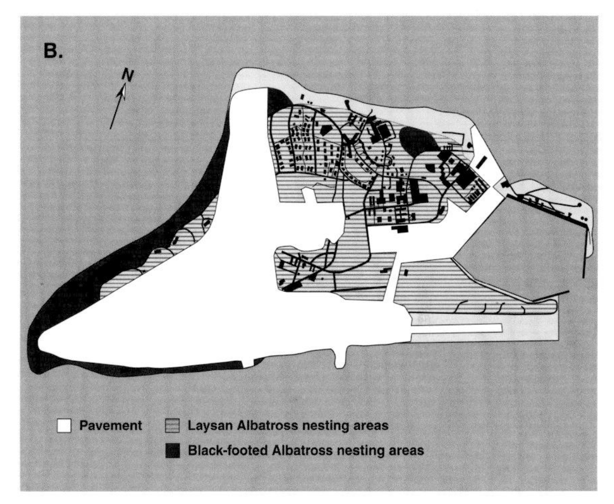

Fig. 11.6 Sand Island, Midway Atoll, showing the runways in 1956 (A) and the extent of hard-surface (pavement) in 1964 (B). The breeding areas of the Laysan Albatross and Black-footed Albatrosses are shown in B (after Fisher 1966a).

US Navy orders were emphatic about not harming birds, but everywhere a bulldozer tried to go, birds stood in the way. For a while men were detailed to walk in front of every vehicle, moving birds to one side, but this was far too slow, and eventually orders were ignored. Bulldozers ran over the birds, and in so doing created no less of a problem; the smell of decomposing dead birds and the plagues of flies made it essential to pick up the carcasses and dispose of them. Later, when aircraft started using the runways, albatrosses became such a menace that the navy ordered their extermination. Thousands were clubbed to death, with little apparent reduction in numbers (Fisher & Baldwin 1946, Fisher 1949).

The period of construction in the early years of the war was far more devastating to the birds than the brief Battle of Midway in June 1942, and afterwards albatrosses continued to be killed in great numbers. Fences, barbed-wire, poles, towers and overhead wires were hazards. The birds fell into trenches, pits and gun emplacements, and all moving vehicles, from bombers to jeeps, killed thousands. The early garrison of about 850 men rose to 15,000, and casual killing was common.

Even before the end of the war, concern was being expressed in Honolulu about the fate of seabirds at Midway. In May 1945, the National Parks Service, Board of Agriculture and Forestry of the Territory of Hawaii and BSFW, in co-operation with the US Navy, sent Harvey I. Fisher and Paul Baldwin (1946) to report on Midway. It was the first official survey of the avifauna of the atoll.

Evidence of albatross destruction was everywhere, but it was difficult to put figures to a decline because of the imprecision of prewar data. Less than half of the area formerly occupied by Laysans was still in use, and one Cable Company employee claimed there were only half as many Laysans left. The Black-footed Albatross seemed to have been much less affected because most nested around the perimeter of the island, out of the way of human activity. Fisher (1949) established good rapport with the naval commanders and was invited back the following year. He made pragmatic recommendations for monitoring and management of wildlife, but none of them was adopted.

A serious consequence of the war had been the introduction of black rats. During more than 30 years of residence by the cable company and PAA, great care had been taken to prevent the introduction of rats, but military priorities were elsewhere. After the war, an estimated 14,000 were killed (Munro 1945b), but they increased again on both Sand and Eastern Islands.

Midway remained operational for trans-Pacific military flights and by the early 1950s, about 300 aircraft were refuelling at Sand Island each month. Procedures to minimise bird strikes were remarkably successful; no aircraft crashed, but delays and replacements were expensive. In 1954, the Military Air Transport Service (MATS) requested the assistance of government biologists to reduce the number of air strikes; Philip DuMont and Johnson Neff of USFWS spent a month on the island (DuMont 1955).

The advent of long-range aircraft should have made it unnecessary to continue refuelling at Midway but, just at the time when the airbase might have become redundant, it assumed a new role in the Distant Early Warning (DEWLINE) defence of the United States. With big jets coming into service, the hazards of air-strikes became potentially more serious and costly. USFWS was contracted to carry out further investigations towards reducing this hazard.

Over two breeding seasons between 1956 and 1958, Karl Kenyon, Dale Rice, Chandler Robbins and John Aldrich carried out field-work on Laysan and Black-footed Albatrosses, mostly at Sand Island. Between November 1956 and June 1957, there were 399 albatross strikes during 1,728 aircraft landings and take-offs (Kalmbach *et al.* 1958). To discourage birds from soaring in the flight paths of approaching aircraft, the causes of updrafts – trees and revetments – were removed; but large expanses between the runways remained attractive nesting grounds. Controlled killing of Laysans continued for as long as radar surveillance aircraft operated from Midway. Gradually, as more and more of the western part of the island was flattened and hard-topped, nesting areas receded from the runways and the bird hazard gradually diminished. However, the clearing of sand binding vegetation and dunes increased drifting sand, which called for more control.

In 1965, when radar aircraft ceased flying from Midway, the killing of albatrosses came to an end (Fisher 1966a). Jets still used the island, but routine monitoring of bird movements and precautionary clearing of runways were adequate for the reduced air traffic. Most Black-footed Albatrosses had remained in the coastal dunes, but the Laysans now bred on grass and among vegetation. A rapport developed between the Navy and the birds. Albatrosses got manhandled when runways were cleared for take-off, but elsewhere they had priority. The green oasis among the buildings and under the trees was good for the Laysans and they increased (Plate 41). People became protective of the gooneys nesting about their homes and even domestic dogs learnt to ignore albatrosses and their chicks.

In 1961, the Office of Naval Research began funding investigations by Fisher (1971b) into Laysan Albatross breeding. Large numbers of fledglings were again shipped to other islands to test whether translocation might be an alternative to killing. The young that survived this experiment and the subsequent years at sea returned to Sand Island where they had been reared, thus confirming the earlier Kure Atoll experiment.

Most of Fisher's fieldwork was on Eastern Island, where there had been no aircraft since the war ended 15 years earlier and there were comparatively few people. In the early 1960s, the island became the site of Midway's main communication installation, with an array of antennas held aloft by 100 m masts. The rigging of these masts was a death-trap for flying albatrosses, especially in storms and reduced visibility. During one two-day gale, 404 Laysans were brought down, and in a single breeding season, between November 1964 and May 1965, more than 3,000 adult Laysans were killed (Fisher 1966b). In the same period, only five Black-footed Albatrosses lost their lives; they nested on exposed beaches on the seaward side, away from the masts. Years of protests and lobbying had no effect upon the Navy or the company which operated the equipment, and it was not until 1967, when satellites made the system obsolete, that the masts came down and were left to rust in a dangerous tangle on the beaches (Fisher 1970).

Most seasons between 1961 and 1973, Fisher marshalled his family and students in the field, accumulating impressive quantities of data on the breeding biology of the Laysan Albatross. Earl Meseth (1975), Eugene LeFebvre (1977) and Donald Sparling (1977) completed the first modern study of albatross behaviour and Fisher's wife, Mildred, who had been co-author in some of his classic papers wrote her book *The Albatross of Midway Island*

(1970). Physiologists later completed studies of albatrosses and their eggs at Midway that would have been difficult elsewhere (Howell & Bartholomew 1961, Pettit *et al.* 1982, Grant *et al.* 1982, Grant & Whittow 1983); but with the retirement of Harvey Fisher in 1973, long term studies of Laysan Albatrosses came to an end.

After 1977, the number of Naval personnel was reduced and by 1979, there were no longer any families living on Sand Island. USFWS contracted an independent ecological team to report on changes in vegetation and seabird numbers (Ludwig *et al.* 1979).

In 1988, Midway Atoll National Wildlife Refuge was established and managed in coordination with the Navy. USFWS personnel have been resident on the atoll since 1990. The Navy continued to reduce its operations and finally closed the station in 1993. On 31 October 1996, President Clinton suspended all Naval provisions of the 1903 and later Executive Orders pertaining to the atoll and surrounding sea. Responsibility for Midway passed from the Department of the Navy to the Department of the Interior (Clinton 1996).

Midway's two main islands had changed greatly since the Navy acquired them and the environmental clean-up lasted over three years. The breaking up of extensive areas of concrete and black-top surfacing, and the removal of many derelict buildings, fuel tanks etc. was followed by eradication of introduced rats. Control of the vegetation is planned (N. Seto pers. comm.). The atoll was finally handed-over to civilian administration in July 1997.

Pearl and Hermes Reef (27° 55'N, 175° 45'W)

The seven low islands scattered around this reef total about 34 ha (Fig. 11.5), but they change in size under the influence of wind and sea. In 1969 Grass, North, Seal and Southeast Islands had vegetation, the others were sand cays (Amerson *et al.* 1974). The reef was named after two English whaling ships, the *Pearl* and the *Hermes*, which were wrecked one night in 1822. Captain Brooks in the Hawaiian barque *Gambia* and Captain William Reynolds of the USS *Lackawanna* charted the reef in 1859 and 1867, both claiming it for their respective governments.

Among the many other vessels that visited the atoll, the Japanese schooner *Ada* worked the islands for two days in January 1882. Besides sharks, turtles and *bêche -de -mer*, they took 43 pounds of down, probably accounting for all the albatross chicks. In 1908–09, three Japanese spent seven months on Southeast Island, but no-one found any signs of fowling. In 1915, Lieutenant Munter of the *Thetis* discovered introduced rabbits, all but one of which were shot when the *Tanager* expedition arrived in 1923. Several fishing companies worked out of Pearl and Hermes in the late 1920s and a fishing station was built on Southeast Island in 1930. All such commercial ventures were opposed by the Bureau of Fisheries in favour of conservation and further research. The buildings were destroyed by the US Navy and Marine Corps in 1942.

The first estimates of albatross numbers were made at North Island in March 1915 by Willett and Bailey, and in April 1923 Wetmore counted Southeast, Grass and Seal islands. Brief, intermittent field-work in various scientific disciplines occurred throughout the 1950s and in 1963–69, POBSP and BSFW made 24 visits, amounting to 95 days in the field.

Lisianski Island (26° 04'N, 173° 58'W)

This is a flat, featureless island of about 1.8 km^2 in area, with dunes rising to about ten metres above sea level (Fig. 11.7). It was named after Urey Lisianski, captain of the Russian ship *Neva*, which approached the island in October 1805. Officers who went ashore noticed many birds, but albatrosses would have been away at that time of year. Benjamin Morrell landed from the *Tartar* at the beginning of July 1825. He mentioned several kinds of birds including rookeries of "large gulls". Gulls are rare on the Hawaiian Islands and do not breed, so he was certainly referring to the albatrosses. Another Russian ship, the *Moller* put men ashore on 3 April 1828. They collected the first specimens and made notes on the natural history (see Laysan Island below). Throughout the 19th century ships continued to be wrecked on the reef.

On 29 June 1891, Palmer and Munro landed and spent a week collecting for Lord Rothschild:

> 'The White-breasted Albatrosses are here in thousands all over the island, but the dark one is very scarce. There are only a few on the weather side of the island.'

At the time guano was worked on Laysan Island, Lisianski was also mentioned in agreements and leases, but there

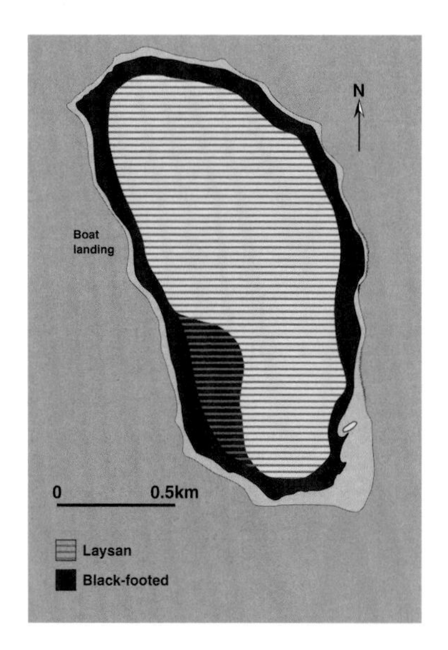

Fig. 11.7 Lisianski Island with the breeding grounds of the Laysan Albatross and Black-footed Albatross (USFWS unpublished).

is no evidence that guano was ever worked or shipped from the island. The absence of guano working left Lisianski open to fowlers seven years before they got on to Laysan. At the beginning of 1904, Japanese schooners landed 77 men on the island. They built huts and it was April before they were seen and warned. On 16 June, the *Thetis* arrived and Captain O.C. Hamlet arrested them all. By this time they had killed about 284,000 birds, but they were almost out of food, and appeared to be quite pleased to be apprehended. The men and their personal belongings were taken to Honolulu, but 326 sacks and boxes of feathers and wings were intentionally left to be collected by a Japanese vessel!

In April 1909, when Japanese fowlers set to work on Laysan Island, a party of ten was also landed on Lisianski. They were arrested by the captain of the *Thetis* shortly after he removed the men from Laysan. This time, 140,000 wings and 1.25 tons of feathers valued at $97,000 were confiscated (Clapp & Wirtz 1975). US Navy, Coastguard and Fisheries vessels continued to call at the island and estimates of albatross numbers were made from time to time.

Rabbits were put on to Lisianski Island by Max Schlemmer in 1903 at the same time that he was introducing them to Laysan Island. On 12 September 1914 Carl Elschner wrote:

> 'Surrounding the houses are small patches of tobacco, which grow wild, having been brought by Captain Schlemmer. This is in fact the only vegetation on the island, and there hardly is a blade or stalk of any other plant to be seen with the exception of perhaps two poorly looking specimens of Ipomea, which I saw ... The rabbits introduced have just exterminated the flora ...now the rest of these rabbits (we found many dead but very few living ones) will have to submit to starvation.'

The rabbit population was evidently declining in the classic manner of an alien herbivore eating out a limited habitat. By February 1916, when Stanley Parker went ashore from the *Thetis*:

> 'Every particle of vegetation, except an algae in a damp spot, has disappeared from the island ... and the rabbits have entirely disappeared.'

A field party from the *Tanager* expedition camped on the island for five days in May 1923; it was still largely devoid of vegetation.

Personnel of BSFW and POBSP visited Lisianski Island on 17 occasions between 1963 and 1969, spending a total of 27 days undertaking surveys and research on the ecology of seabirds (Clapp & Wirtz 1975).

Laysan Island (25° 47'N, 171° 45'W)

Laysan Island (Fig. 11.8) has an area of about 3.7 km^2 and its highest point is about 12 m above sea level. In the middle of the island is a large saline lagoon surrounded by ooze and covered by a hard salt crust (Table 11.2).

It appears to have been seen and named by John Briggs of New Bedford sometime in the early 1820s, but little is known about the discovery. On 28 March 1828, when the *Moller* arrived off the island, the ship's German surgeon, C. Isenbeck, landed and made the first notes on the flora and fauna which he later related to the naturalist F. von Kittlitz, who published an account of the island in 1834. In spite of being a second-hand report, von Kittlitz correctly indicated that there were two different albatrosses,

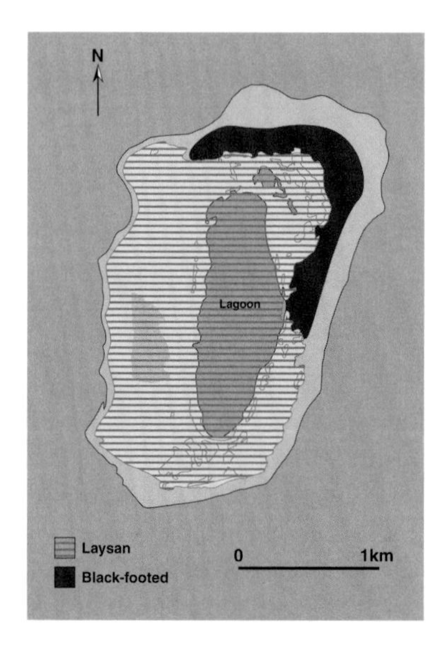

Fig. 11.8 Laysan Island with the breeding grounds of the the Laysan Albatross and Black-footed Albatross (USFWS unpublished).

one white and the other dark, nesting in considerable numbers and part of his paper was later translated for Rothschild's *Avifauna of Laysan*.

On 1 May 1857, the island was annexed by the kingdom of Hawaii and Captain John Paty of the Hawaiian schooner *Manuokawai* noted that:

> 'The island is "literally covered" with birds; there is, at a low estimate, 800,000... They were evidently unaccustomed to the sight of man, as they would scarcely move at our approach, and the birds were so tame and plentiful, that it was difficult to walk about the island without stepping upon them. The gulls[16] lay enormous large eggs, of which I have a specimen.'

	Number of islands	Area of islands (km²)
Kure Atoll	2	1.0
Midway Atoll	5	6.0
Pearl & Hermes Reef	13	0.3
Lisianski Island	1	1.8
Laysan Island	1	3.0*
Gardner Pinnacles	1	0.1
French Frigate Shoals	13	0.5
Necker Island	1	0.2
Nihoa Island	1	0.6

Table 11.2 The number and size of Hawaiian atolls and islands. * excluding 74 ha of lagoon.

Two years later Captain Brooks described the island and its vegetation.

Henry Palmer's visit to Laysan Island in 1891 coincided with the beginning of guano extraction. Rothschild (1893b) also benefited from a visit in the following year by the Honolulu photographer, J.J. Williams, who took 300 photographs while on the island. Some of these appeared in the *Avifauna of Laysan* (Fig. 11.9), and another of them gives the impression that large numbers of albatross eggs were exported. Max Schlemmer, who was then in charge of the digging, later indignantly complained to Walter Fisher that the eggs had been taken by the photographer himself, just for the picture, and maintained that no eggs had ever been shipped from the island.

Hugo Schauinsland (1899), Director of the Bremen Museum, accompanied by his wife, reached Laysan Island with guano workers at the end of June 1896 and left with them on 22 September. He completed the first comprehensive biological survey of the island. There were still a few Laysan Albatrosses and their fledglings present when they arrived, but most if not all the Black-footed Albatrosses had left. His book *Drei Monate auf einer Koralleninsel* was an important contribution that remained unavailable to most later workers until Miklos Udvardy's translation, one hundred years after Frau Schauinsland had valiantly struggled to keep the blowflies, beetles and ants from her specimens.

In 1902, when the US Fish Commission was undertaking deep sea explorations off the Hawaiian Islands, Walter Fisher and John Snyder were detached to investigate the seabirds of Laysan. They were on the island for one week (16–23 May) and although the Black-footed Albatrosses would have been about to depart, both species were present in considerable numbers. Fisher's (1903) account included the first description of Laysan and Black-footed Albatross behaviour. W.A. Bryan visited the island the following year and took away 669 specimens for the Bernice P. Bishop Museum in Honolulu.

Within a few years, the guano deposits were depleted, and in 1904, with profits declining, the company sold everything on Laysan, except the buildings, to Max Schlemmer for $1,750 and appointed him an agent of the company. He had long experience of the island, and in the manner of the time sought to diversify its economy by planting coconuts and introducing rabbits, hares and guinea pigs. With diminishing returns from guano and long delay before anything could be expected from a plantation, he was evidently anxious to participate in the profitable feather trade. In 1904, he applied to the Hawaiian Land Commission for a lease of Laysan, Lisianski and French Frigate Shoals, proposing to save the birds from extinction at the hands of the Japanese, while at the same time killing over 21,900 birds himself each year on Laysan Island alone! It was not until February 1909 that he got his lease; it stipulated that there was to be no destruction or capture of birds and Hawaii reserved the right

[16] albatrosses.

Fig. 11.9 Colonies of the Laysan Albatross (above) and Black-footed Albatross (below) at Laysan Island in 1892 (J.J. Williams).

to reclaim the islands at any time. This legislation was later declared invalid, since it post-dated President Roosevelt's Executive Order of 1909 setting up the Hawaiian Islands Reservation.

On 17 April 1909, a party of 15 Japanese landed on Laysan Island, and began collecting the wings and feathers of albatrosses and other seabirds, making use of the guano railway and buildings. By August about a ton of feathers and 70 bales of wings, representing about 64,000 birds, had been collected and shipped out by schooner. Later in the year, rumours reached Honolulu that fowlers were again in the northwest, and the *Thetis* was dispatched. It arrived at Laysan Island on 16 January 1910 to find the Japanese, who produced documents from Max Schlemmer which they claimed gave them authority to be there. Captain Jacobs noted that:

> 'One of the buildings was full of the breast feathers of birds in bulk, another was two-thirds full of loose birds' wings, and two other buildings were partly filled with bales of feathers and wings, and a number of stuffed birds of various species. On the sand adjacent to the buildings were about two hundred mats held down by rocks, under which were laid out masses of birds' wings in various stages of curing. Stretched along the beach and over the island were bodies of dead birds in large numbers from which emanated obnoxious odors.'

All the Japanese were arrested and taken on board, together with 65 bales of wings, 13 bales of feathers, 31 bags of feathers (approximately 119,000 wings and a ton of feathers) and two boxes of stuffed birds. Another 365 kg of uncured feathers and 63,500 wings were left exposed by removing mats and the sides of buildings to allow wind and weather to spoil them.

Back in Honolulu, legal suits were filed against one of the Japanese as a test case, but when it was dismissed all the Japanese were given free passages back to Japan. Max Schlemmer was apparently deeply implicated in the episode. In December 1908 he had visited Japan and persuaded a businessman, Genkichi Yamanouchi to pay him $150 monthly in gold for his PGFC agency and related documents. These all concerned guano, though it is hardly likely that the Japanese did not know that the deposits were almost exhausted. Schlemmer probably knew that PGFC was about to surrender its lease to the Hawaiian Government, making his agency worthless. He must also have known that the Japanese were really after the birds; but nothing he had written could incriminate him. Charges of poaching could not be sustained and he was also found not guilty of unlawfully bringing aliens into Hawaii (Ely & Clapp 1973).

Throughout all these events, Schlemmer continued to visit Laysan intermittently to work the remaining guano until July 1910, when the last shipment and one remaining mule left the island. He made a final visit to the island in July 1915 in a new yacht *Helene* with his son Eric and a young sailor. They were marooned after entrusting the *Helene* to the survivors of another boat, who were then wrecked a second time and did not return. After exhausting all their provisions, they were saved by the return of the albatrosses and evidently began preserving eggs in sand. The party was eventually rescued by the *Nereus*. In February 1916, when Lieutenant Munter landed from the USCG cutter, he found a barrel containing 350 eggs. The crew boiled some of them (25 minutes each) and found them still palatable (Ely & Clapp 1973).

In 1911, W.A. Bryan stayed at Laysan for a week with the Iowa State University Expedition (Dill & Bryan 1912). Evidence of the killing was everywhere, but the numbers of seabirds remaining were considered ample to enable the populations to recover. Of much greater concern was the change in the vegetation caused by the rabbits introduced by Schlemmer at the time of Bryan's earlier visit (1903). Acting on Bryan's recommendations, the Bureau of Biological Surveys promptly sent out a party to deal with the rabbits and determine the condition of the bird colonies. Between 22 December 1912 and 11 March 1913 a party of four, including George Willett and Alfred Bailey, shot 5,024 rabbits, but were not able to eradicate them; a considerable number remained. By February 1914, when Lieutenant Munter landed from the *Thetis* they were again multiplying rapidly and he strongly recommended that they be killed before the Laysan vegetation was eaten out like that of of Lisianski Island.

Unknown fowlers were again on Laysan at the beginning of 1915 and on 3 April when he went ashore, Munter reported:

> 'Dead birds were seen in piles of 10 and 15 and sometimes as many as 40 or 50 in a pile ... Only the breast feathers had been taken ... Laysan Albatross was the chief sufferer, next the Black-footed Albatross with the Frigate Bird and Blue-faced Booby following in orders of numbers killed. Between one hundred and fifty and two hundred thousand birds were found on their backs with only the breast feathers missing. In the majority of cases the feathers had been pulled out, but in some instances knives had been used and the breast had been cut away from the bodies ... As a consequence there were very few young Albatrosses and Boobies ... the western half of the island [has] only a very few young Albatrosses but there [are] hundreds of eggs with young chicks in them that never hatched ... Along the southern and southeastern part of the island quite a number of the young of the Black-footed Albatross were found ... [but] ... here as elsewhere hundreds of grown birds had been slain ... no portion of the island had been spared ...'

During the 1923 *Tanager* expedition, a field-party was ashore from 8 April to 14 May. All the rabbits were killed and the recovery of vegetation encouraged by planting large quantities of seeds including aliens. In the years following, the island was visited fairly frequently, but rabbits were never seen again.

Alfred Trempe was probably the first to ring these albatrosses using brass and blue celluloid rings made in the engine room of the *Reliance* during its 1936 patrol. The island was visited frequently during the 1950s. Rice and Kenyon (1962a) made counts of albatrosses as ground control for their 1956–57 aerial census and activity increased through the 1960s with POBSP and BSFW parties ashore on 21 occasions. In February 1967, Laysan Island was one of seven islands designated 'research natural areas' and subject to special controls (Harrison 1990). In 1991, USFWS set up a permanent camp on the island and began field-work to eradicate sand bur *Cenchrus echinatus*, an introduced grass competing with native plants.

Gardner Pinnacles (24° 42'N, 167° 54'W)

These conspicuous stacks were seen by Joseph Allen of the whaler *Maro* in 1820. Many ships have since sailed close by them, but few landings have been made. In May 1923, Alexander Wetmore (Olson 1996) climbed the main rock where he found two Laysan chicks 30 m or so above the water. A POBSP party ringed several chicks in 1963 (Clapp 1972).

French Frigate Shoals (23° 45'N, 166° 10'W)

A conspicuous stack, La Pérouse Pinnacle stands on this reef, but the 13 other named islets are low sand cays with a total area of about 45 ha (Fig. 11.5). In 1969, Tern, Trig, Whale-Skate and East Islands had appreciable vegetation while Little Gin Island had just a few stragglers (Amerson 1971). Whale-Skate Island began to erode in the early 1990s and by 1997 had almost disappeared.

In November 1796, the French warships *Broussole* and *Astrolabe,* commanded by Jean-François de la Pérouse encountered this reef, but no boat went ashore. The schooner *Fenimore Cooper* arrived in January 1859, when Lieutenant John Brooke claimed it for the United States and made a chart. He noticed guano to a depth of more than a metre on the top of La Pérouse Pinnacle; within a few years this small and difficult deposit had been removed. Ships began to be wrecked on the atoll and crews survived on one or other of the five larger islands. In the spring of 1882, the Japanese schooner *Ada* from Yokohama spent three months there and the 'bird down' said to have been obtained, would have come, at least in part, from albatross chicks.

Albatrosses were first noted on two islands of French Frigate Shoals by Henry Palmer and George Munro when the *Kaalokai* called there briefly in 1891, but it was not until the *Tanager* expedition that the 13 islands were even named. Between 22 and 28 June 1923, Alexander Wetmore (1996) explored all but one of them, and although most Black-footed Albatrosses would have departed by that time, he indicated the breeding distribution of Laysans.

Through the 1930s, the atoll was the scene of US naval-air exercises and on several occasions a tented base occupied much of East Island at times when Laysan and Black-footed Albatrosses were breeding. In 1942–43, after the Battle of Midway, coral was dredged from the lagoon and used to transform the natural curves of Tern Island (4.5 ha) into a 945m long runway supported by a steel wall, and looked like a huge (10 ha) aircraft carrier (Fig. 11.10), where fighters could refuel on flights between Honolulu and Midway. The vegetation was completely

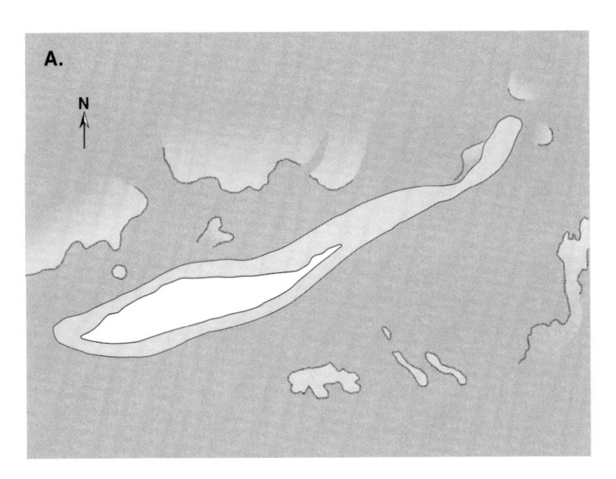

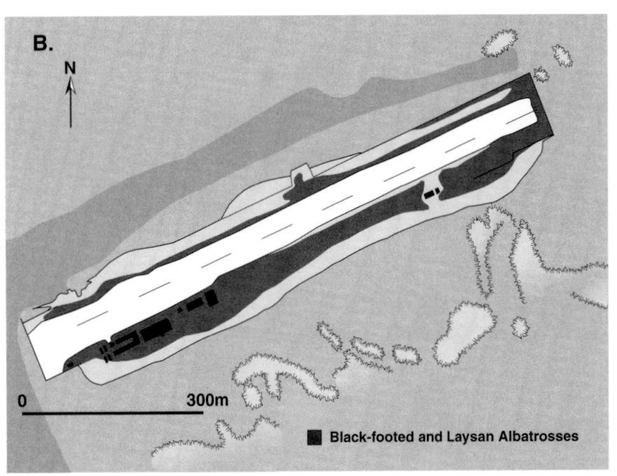

Fig. 11.10 Tern Island, French Frigate Shoals in 1932 (A) and 1966 (B); both redrawn from US Navy photographs (Amerson 1971). The mixed colony of Laysan and Black-footed Albatrosses present in 1994 is shown in B (USFWS unpublished).

obliterated, and the small numbers of Laysan and Black-footed Albatrosses that had bred on the original island were disrupted if not destroyed. The survivors would have returned, but it is hardly likely that they would have been tolerated about the new island runway while it remained operational.

In 1946, the US Coast Guard established a long-range navigation (LORAN) transmitter across the lagoon on East Island. The seven pole antenna array and 27 resident technicians were bound to have disrupted the seabirds on such a small (4.6 ha) island, but its reservation status was known and the albatrosses were still there in 1952, when the island was evacuated. The empty buildings were eventually destroyed in 1965. At the end of the war, when the Tern Island air facility was abandoned by the navy, it was used by several fishing companies until 1952, when the Coast Guard moved there from East Island. They renovated the sea-wall and buildings until 1978, when the Coast Guard withdrew altogether from French Frigate Shoals.

New beaches formed on either side of the Tern Island airstrip, increasing the area to 23 ha in 35 years. Away from the hard surface of the runway, six hectares of vegetation and sand accumulated, but in stormy weather much of the island is swept by waves. Both species of albatross had re-established themselves before 1953, when Frank Richardson (1954) checked all seven albatross islands, but nests were destroyed on Tern Island to reduce air-strikes. Albatrosses have been counted intermittently over the years. Personnel of POBSP and BSFW visited the atoll on 21 occasions between 1963 and 1969 (Amerson 1971). Since 1979, Tern Island has been a USFWS field station for conservation and research personnel within the French Frigate Shoals Wildlife Refuge.

Necker Island (23° 35'N, 164° 42'W)

In the 13th century this small rocky island (17 ha 84 m) (Fig 11.11) was briefly occupied by a small community of Polynesians, but it has remained uninhabited ever since. La Pérouse sailed past and named it after a minister in the court of King Louis XVI, on the day before he discovered French Frigate Shoals. Walter Fisher (1903) landed from the steamer *Albatross* in May 1902 and found both albatross species breeding. The first thorough investigation was by members of the *Tanager* expedition, who spent about ten days camped on the island in June 1923 (Olson 1996). More prolonged field-work was undertaken in 1965–73 by personnel of POBSP and BSFW/USFWS (Clapp & Kridler 1977, Harrison 1982).

Nihoa (23° 03'N, 161° 55'W)

In March 1789, William Douglas, captain of the *Iphegenia* sailing northwest from Ni'ihau, found this small island (63 ha 273 m Fig 11.11) which became known as Bird Island, but since the early decades of this century it has taken the Hawaiian name. Early Polynesians lived here at about the time that others were on Necker and some six hectares of slopes remain terraced from their cultivation (Harrison 1990). They are now the site of albatross colonies. Members of the *Tanager* expedition camped on the island for ten days in June 1923 (Olson 1996) and more field-work was undertaken in 1965–73 by personnel of POBSP and BSFW/USFWS (Clapp *et al.* 1977, Harrison 1982).

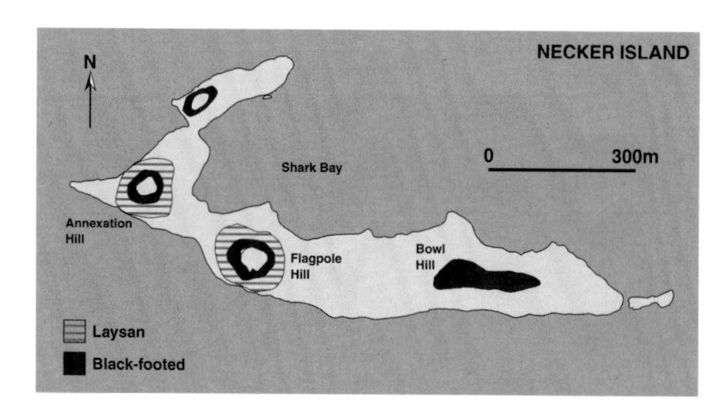

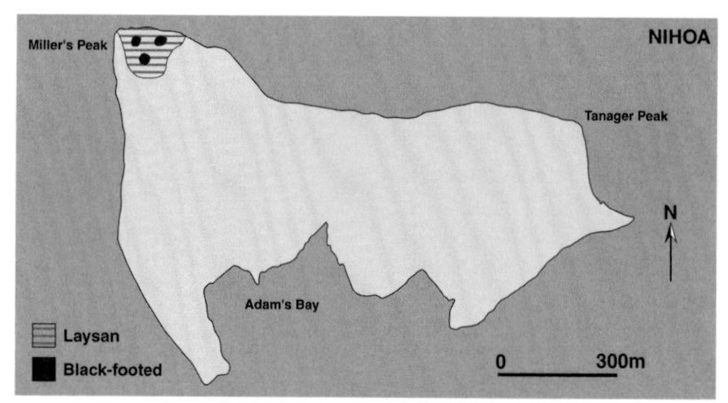

Fig. 11.11 Necker Island and Nihoa are small volcanic islands with no beaches or coral reefs. Laysan and Black-footed Albatrosses nest mostly in small, mixed colonies (C.S. Harrison pers. comm.).

Ka'ula (21° 40'N, 160° 36'W)

Edward Caum (1936) landed on this small steep island (50 ha 168 m) in August 1932. He did not see any albatrosses, which would have been away at that time of year, but he found an old egg. It was described as that of a Black-footed Albatross, but may well have been from a Laysan Albatross. Since 1952, Ka'ula has been used as a bombing target by aircraft of the US Navy and Marine Corps. Rice & Kenyon (1962a) saw only Black-footed Albatrosses when they flew over the island, but Craig Harrison (1990) recorded small numbers of both species.

Ni'ihau (21° 55'N, 160° 10'W)

A few Laysan Albatrosses still manage to breed at Ni'ihau (Rice & Kenyon 1962a) although there are no longer any on the offshore islet, Lehua (Caum 1936, Harrison 1990). During the Second World War, numbers increased in spite of wild pigs, which are still a problem (C.S. Harrison pers. comm.).

Kaua'i (21° 58'N, 159° 30'W)

Kaua'i and the other high islands to the east have been inhabited and cultivated for hundreds of years. Accessible colonies of large seabirds would have been eliminated long ago. Whether they ever included appreciable numbers of albatrosses is unknown, but with such enormous colonies on the atolls and coral island to the northwest, it is reasonable to assume that over long periods, albatrosses have from time to time prospected eastwards.

The Laysan Albatross that landed on Kaua'i in 1945 was said to have been the first within living memory (Munro, 1945a). They began increasing in the mid-1970s, but Black-footed Albatrosses have been ashore only rarely (Harrison 1990). The first Laysan chick at Kaua'i was reared on Crater Hill in 1976 and by the late 1970s, nests were becoming common along Barking Sands, although the birds there were attacked by feral dogs. In the early 1980s, chicks were being reared on the Pacific Missile Range, Mana, and by the late 1980s fledglings were flying each year from Kilauea Point, the northernmost point of Kaua'i (Harrison 1990). This was made a National Wildlife Refuge and received chicks translocated from the Pacific Missile Range. Attempts have been made to attract prospecting Laysans to a flat area maintained as a lawn (Warham 1990), but immigrant birds have been laying at sites away from the refuge, some among nearby houses (Pyle 1991).

O'ahu (21° 27'N, 158° 00'W)

A Laysan Albatross appeared on Moku Manu off O'ahu in 1946 (Fisher 1948). They began landing at Waikane and Kahuku in the late 1970s and by the late 1980s groups of 30 or more were attempting to breed along the north shore from the Kaneohe Marine Corps Air Station to Ka'ena Point. Two pairs laid in 1991 and many have been seen displaying from sea level to almost 500 m. At Dillingham Airfield, numbers have increased, but nesting has been discouraged because of the risk to aircraft. Prospecting Laysan Albatrosses have been disturbed by people and killed by dogs, so in 1993 work began on a co-operative project to attract them to Kaohikaipu Island, a state seabird reserve off the Makapu Peninsula, O'ahu. The following year, decoys with audio playback attracted subadult Laysan Albatrosses to the island (Warham 1990, Pyle 1991, McDermond 1993, 1994, Lovvorn 1994).

Moloka'i (21° 08'N, 157° 00'W)
Craig Harrison (1990) reported a nest at Kawakiu Nui (Bay), but gave no indication as to whether any albatross was actually seen.

Johnston Atoll

The only evidence of albatrosses at Johnston Atoll (16°45'N, 169° 32'W) is a second-hand report of one Black-footed Albatross with an egg on the beach of Sand Island in November 1922, one fledgling Laysan Albatross on Johnston Island in July 1923 (Olson 1996) and the recollection of both species in the late 1930s by an unknown Polynesian construction worker, 25 years after the event. Rice & Kenyon (1962a) thought that Japanese fowlers must have killed off all albatrosses many years before the *Tanager* expedition, but Wetmore (1996) searched the two islands very carefully between 10 and 20 July 1923. He found a large mound of tern and shearwater bones, but his notes on guano include a statement that deposits were due mainly to albatrosses, frigatebirds and boobies. Amerson & Shelton (1976) believed that there were never more than a few albatrosses at Johnston Atoll.

Wake Island

Titian Ramsey Peale was the first naturalist to visit an albatross colony in the North Pacific. On 20 December 1841, together with other scientists of the United States Exploring Expedition (1838–42), he spent a day on Wake Island (19° 18'N, 166° 36'E) where albatrosses were sitting on eggs. Nothing he wrote about these birds provides a clue to their identity. All North Pacific albatrosses at that time were assumed to be of one species and Peale's later description combines characteristics of all three species (Cassin 1858). A Black-footed Albatross specimen in the US National Museum (USNM 15552), labelled as coming from Wake Island, is attributed to the expedition, but there are some grounds for wondering whether it is indeed the original expedition specimen (No 745) (Olson 1996). An egg collected by Peale and also in the museum (USNM 949) was believed by Aldrich to be that of a Laysan Albatross (Rice & Kenyon 1962a), but the eggs of Laysan and Black-footed Albatrosses are difficult to distinguish (Frings 1961, Fisher 1969).

Wake Island was annexed by the United States in 1899. In 1923, when the *Tanager* and *Whippoorwill* were at the atoll, Wetmore inspected the deserted camps of Japanese fowlers occupied sometime between 1908 and 1915. They contained large quantities of bones from frigatebirds and boobies. He made no mention of albatrosses, but in 1968, he identified seven specimens collected at the time as Laysan Albatrosses (Olson 1996). The island remained uninhabited until the late 1930s, when Pan American Airways began using it for refuelling their *Clipper* flying boats on trans-Pacific flights. Several Black-footed Albatrosses were seen in 1938 and 1940 (Hadden 1941). During the Second World War, Wake Island was the scene of fierce battles, and the blockaded Japanese garrison of 4,000 starving troops would hardly have let any large seabird escape. There were no sightings of albatrosses in the late 1940s (Bailey & Niedrach 1951), but two Laysans began visiting the island in the early 1980s and by 1996, several of each species were visiting four locations (Louis E. Hitchcock pers. comm.).

Minami Torishima (Marcus Island)

During the 1890s, Japanese fishermen occasionally visited Minami Torishima (24° 18'N, 153° 58'E) for guano and birds. In 1896, the island began to be worked regularly and two years later the Japanese Government incorporated it into the Ogasawara (Bonin) Islands. W.A. Bryan arrived there on 30 July 1902, and spent a week ashore. By that time of year most albatrosses would have been at sea, but he did see one Laysan and found large quantities of bones. He learned from a Japanese resident that both Laysan and Black-footed Albatrosses had been abundant a few years earlier, but in different sentences he says 'fairly abundant' and 'exceedingly abundant'. Another indication of significant numbers is the claim by the Japanese that a few years before, one man could kill 300 Laysan Albatrosses in a day, and that almost all had been wiped out in six years. Black-footed Albatrosses had apparently been almost as abundant as Laysans, but in 1902, the fowlers had not obtained a single one (Bryan 1903). Nothing more was heard about the island until after the Second World War, when the US Air Force established a meteorological station there; no albatrosses were to be found (Kuroda 1954). There has never been any reliable account of albatrosses at Minami Torishima and the claim that the the two species totalled more than a million birds (Rice & Kenyon 1962a) is not convincing. The area of Minami Torishima (ca. 299 ha) is about the same as Laysan Island

(297 ha, excluding the lagoon) (Table 11.2; Fig 11.8), which has a high nesting density of about 300,000 Laysan and 60,000 Black-footed Albatrosses. It seems unlikely that Minami Torishima ever had larger numbers.

Ogasawara Gunto (Bonin Islands)

A few Black-footed Albatrosses were known to have nested on a small islet (Torishima) off Mukojima (27° 40'N, 142° 07'E) in the Ogasawara Gunto (Bonin Islands) before 1930 (Rice & Kenyon 1962a). They had increased to about 80 birds by 1974 and included one ringed as a chick on Pearl & Hermes Reef in the NWHI (Harrison 1990). The first Laysan Albatross chick in modern times was reared in 1976–77 (Kurata 1978) and by 1979–80, 14 eggs were laid (Hasegawa 1984).

Torishima

The Black-footed Albatross formerly bred with Steller's Albatross on Torishima (30° 29'N, 140° 19'E, 5 km^2 403 m) in the Izu Shoto. Both species were almost exterminated by fowlers in the mid-1930s. Laysan Albatrosses also began to breed in the late 1920s and increased until the mid-1930s, when they too were eliminated.

In the early 1950s, the Black-footed Albatross re-colonised Tsubame-zaki at about the same time as the Short-tailed Albatross. It has been increasing ever since, and in the early 1980s founded a new colony, two kilometres away at Hatsune-zaki (Hasegawa 1982, Ogi *et al.* 1994b).

Senkaku retto

The Black-footed Albatross was formerly well established in the Senkaku retto (25° 45'N, 123° 30'E) and in 1970–71 a few were found breeding on two adjacent rocky islets, Kita-kojima (128 m) and Minami-kojima (148 m) (Hasegawa 1984).

Iwo Jima

P.A. Holst landed on Iwo Jima (24° 47'N, 141° 19'E) in the Kazan retto (Volcano Islands) in June 1890 and described the centre of the island as very barren.

> 'A species of Black Albatross [*Diomedea nigripes*] was breeding in great numbers on the sandy ground, many rotten eggs were lying about, and there were plenty of young in down, some almost full-grown, standing like so many soldiers on guard.' (Seebohm, 1891).

Nothing more was ever heard about this colony. If it survived to the Second World War the birds would certainly have perished during the Japanese fortification of the island and the devastation wrought by the 1945 battle.

Marianas Islands

In December 1888, seven Black-footed Albatrosses and their eggs were collected by Alfred Marche on the coast of Agrihan Island (18° 45'N, 145° 45'E. 965 m) (Oustalet 1896, Jouanin 1959). This steep island is surrounded by trees and capped by the tall, dense grass *Imperator cylindrica* (Borror 1947).

A fragment of an albatross long bone was found during archaeological excavations on Pagan (18° 11'N, 145° 42'E) (Egami & Fumiko 1973).

Marshall Islands

Rice & Kenyon (1962a) cited a 19th century hearsay report that the Black-footed Albatross once bred on what was imagined to be Taongi Atoll in the Marshall Islands. Amerson (1969), who has since undertaken studies of the Marshall Islands, believes that albatrosses have never nested there and that the report can be discounted.

Mexico

Laysan Albatrosses were seen off California in the 19th century (Anthony 1898) and by the 1970s were prospecting several islands and rocks off Baja California. In May 1986, six chicks were discovered on Isla de Guadalupe (29° N, 118° 15'W), 290 km from the coast (Dunlap, 1988). The following year, they were seen in the Islas Revillagigedo, and in February 1988, two pairs were found breeding on Isla Clarión (18° 22'N, 114° 45'W) (Howell & Webb, 1990).

NUMBERS

Almost all Laysan Albatrosses (about 558,000 pairs) and 97% of Black-footed Albatrosses (about 59,000 pairs) breed in the Northwest Hawaiian Islands; most of them on Kure Atoll, Midway Atoll, Pearl and Hermes Reef, Lisianski Island, Laysan Island and French Frigate Shoals (Table 11.3). The numbers nesting is, to some extent, a function of area (Table 11.2); Sand and Eastern Islands at Midway are almost twice the area of Laysan Island and in the late-1990s, 69% of all Laysans bred at Midway compared with 22% on Laysan Island. In contrast, 35% of Black-footed Albatrosses bred on Laysan Island and and 35% on Midway.

The numbers of Laysan and Black-footed albatrosses breeding undisturbed would vary from year to year, but the past 100 years of fowling and culling have contributed to even more erratic fluctuations. The few early estimates of abundance were made at different times of the year, often when one or both species were away at sea, and at very large colonies estimates were crude. Even in recent times, remoteness has inhibited regular surveillance and quick, uncontrolled counts of individuals, rather than of eggs or chicks often defy interpretation.

Over several decades the figures indicate considerable changes. At Midway, castaways and fowlers had apparently killed albatrosses in the latter years of the 19th century, but it cannot be assumed that they were ever very numerous. From 1904, the protection afforded by the US Navy and the Pacific Cable Company on Sand Island, provided a new environment in which Laysans flourished. By the late 1930s, when preparations for the Second World War began, numbers of both species were increasing dramatically (Fig 11.12). The massive losses experienced by the birds during the war and its aftermath had no lasting effect. The continued planting of trees, shrubs and lawns appears to have favoured Laysans and compensated for their exclusion from the open areas near runways. Increase in Black-footed Albatrosses peaked in the 1930s; numbers declined in the 1940s, but later increased again to present levels (Fig. 11.12).

Location	Laysan Albatross (pairs)	Black-footed Albatross (pairs)
Kure Atoll	5 000	2 000
Midway Atoll	388 000	21 000
Pearl & Hermes Reef	11 000	7 000
Lisianski Island	27 000	4 000
Laysan Island	124 000	21 000
Gardner Pinnacles	10	-
French Frigate Shoals	2 000	4 000
low (coral) islands	**557 000**	**59 000**
Necker Island	500	100
Nihoa	50	10
Niihau	200	-
Kaula	60	10
Kauai	100	-
Oahu	10	-
high (volcanic) islands	**900**	**100**

Table 11.3 Numbers of the Laysan and Black-footed Albatrosses on the Hawaiian Islands: rounded medians (Harrison 1983, Gales 1997, USFWS unpublished) (Appendix 15).

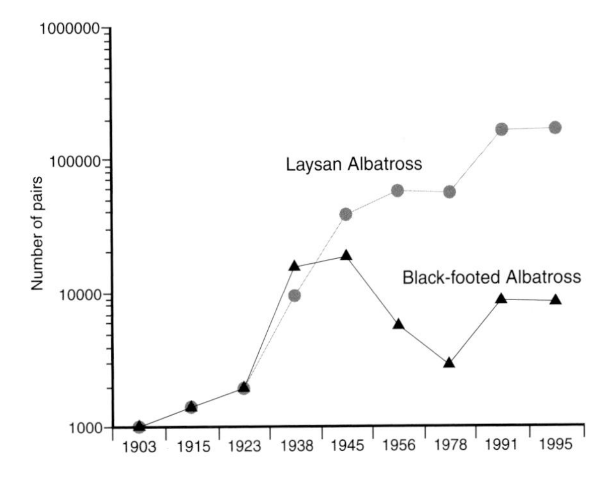

Fig. 11.12 The numbers of Laysan and Black-footed Albatrosses breeding on Sand Island, Midway Atoll, 1903–1995 (Rice & Kenyon 1962a, USFWS unpublished).

Rice & Kenyon's (1962a) census in 1957–58 was the beginning of albatross demography in the Hawaiian Islands. Kure and Midway were counted from the ground, but all other islands from aerial photographs with sample ground counts for control only at Laysan Island. Dense vegetation obscured albatrosses on parts of the Laysan and Lisianski photographs, so estimates of these areas were computed from known abundance in similar habitat on Eastern Island, Midway Atoll. A quarter of all albatrosses visible in the photographs were said to have been 'unemployed' and that proportion was deducted from the counts. There have been other counts since (Appendix 15), but none of comparable extent in a single season.

Fisher (1966c) counted the Laysan Albatross at Midway, also using aerial photographs and sample ground counts. He claimed that 25% was too high a proportion for 'unemployed' birds and that in any case the term was misleading. These and other imprecisions underline the early difficulties of obtaining consistent counts of Laysan and Black-footed Albatrosses.

With the progressive decline in human activity since 1977, disused areas have been occupied by albatrosses and the removal of naval installations will open up more ground for colonisation. In 1979, a ground count was computed from sample nest densities in measured areas of Sand Island, but was less effective on Eastern Island (Ludwig *et al.* 1979). In 1991, computation from random transects by USFWS field-workers on both islands initiated a new regime of surveillance. Eastern Island retained more of its indigenous vegetation and its albatrosses have had longer to recover from aircraft and antennas. Laysan Albatrosses on Eastern are now much more numerous than on Sand Island and Black-footed Albatrosses slightly more numerous (Appendix 15).

Through the 1930s and 1940s, Laysan Island was recovering from the devastation of the fowlers and by the 1950s to 1960s, both species reached numbers roughly equivalent to those of today. In 1991, USFWS field-workers estimated the numbers of both albatrosses from counts of birds on random transects.

A progressive increase in the number of albatrosses on islands of French Frigate Shoals followed the end of USN and USCG activity. Since the early 1980s, annual counts at Tern Island by resident staff of the FFS National Wildlife Refuge have revealed substantial increases in the numbers of both species. They are far more numerous than they ever were on the original island and at any time since they re-established (McDermond & Morgan 1993) (Appendix 15).

Increase in the numbers of Laysan Albatrosses throughout the NWHI has been reflected in the numbers of birds settling on Kaua'i, but they have not been accompanied by Black-footed Albatrosses. Off Mexico, colonising Laysan Albatrosses on Isla de Guadalupe have increased from several pairs in 1983 to 45 pairs in 1991 (Gallo-Reynoso & Figueroa-Carranza 1996).

The Black-footed Albatross has been increasing in the northwest North Pacific. In 1956–57, six fledglings were reared at Torishima, by 1988–89 there were over 400 and a new colony was started. In 1997–98 the two colonies reared a total of 914 chicks (H. Hasegawa 1982, pers. comm., Ogi *et al.* 1994b). This population may well have been founded by migrants from the Hawaiian Islands (Silva & Edwards 1999), but there is evidence that the species was well established there in the last century, before fowling began.

BREEDING

Colonies

Most Hawaiian albatross colonies are on low coral and sand islands, no more than a few metres above sea level, and often capable of accommodating huge numbers of large surface-nesting seabirds. To the east, though, volcanic islands like Necker and Nihoa, rising 100 metres or more above the sea, have no reefs or sand and the small colonies they attract indicate that they are not typical nesting grounds. On their rocky platforms and terraces, segregation is not so apparent, with both species often breeding near to each other (C. Harrison pers. comm.). The islands are exposed to the NE Trade Winds for most of the year, with frequent storms from November to March. Temperatures in winter (Nov–Feb) average 20°–25°C and in summer 25°–27°C (Harrison *et al.* 1983). All islands except the most mobile sand spits and bare rock acquire a vegetation of grasses, herbs and shrubs.

The natural vegetation of the Hawaiian coral islands is a scrub association dominated by *Chenopodium oahuense* and *Scaevola taccada* which form dense, impenetrable thickets three to four metres high. Laysan Albatrosses nest in the interior of islands, but they need spaces among this vegetation. The size and frequency of such spaces determines the number of nests. There may be no place to land and take-off. At Green Island on Kure Atoll, *Scaevola* was once thick enough to entangle albatrosses which later starved to death (Rice & Kenyon 1962a). Even the bunchgrass *Eragostris variabilis* on Lisianski Island has been thick enough to prevent Laysans from nesting.

On Midway Atoll, the habitat has been drastically modified by man (Plate 41). In 1958, territories in the most

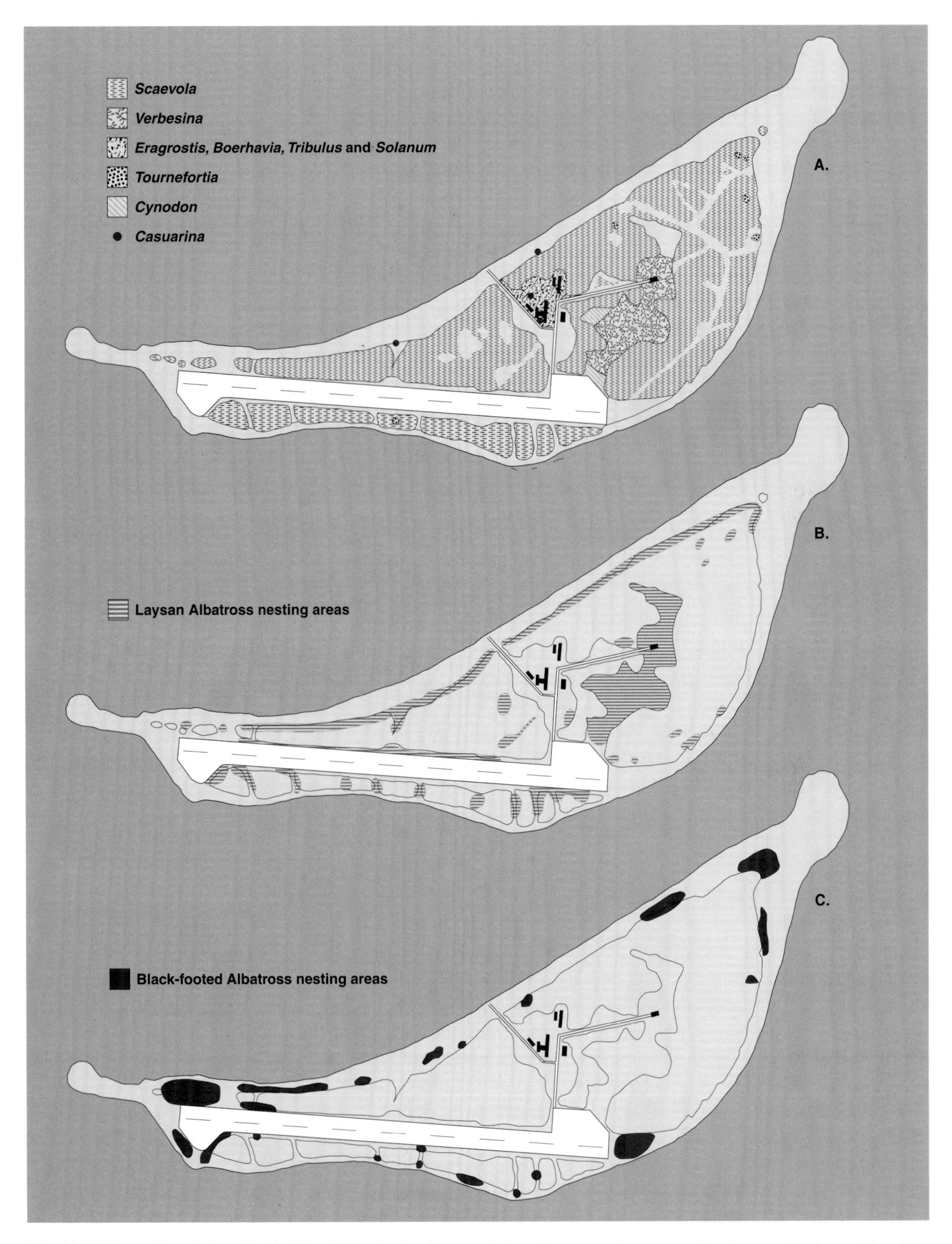

Fig. 11.13 Green Island, Kure Atoll in May 1967, showing the vegetation (A) and nesting areas of the Laysan Albatross (B) and the Black-footed Albatross (C) (Woodward 1972).

crowded area on Eastern Island occupied 2.23 m^2 per nest (Frings & Frings 1961), equivalent to 4,545 nests per hectare; this was considered to be 'saturation'. On Sand Island, maximum densities at that time were rather less and in 1979 counts of almost seven hectares (excluding scattered beach nests) gave densities of 492 to 2,398 nests per hectare (Ludwig *et al.* 1979).

In 1990, chain link fencing was removed from around a 2,800 m^2 plot in the middle of a breeding ground. Laysans that had been excluded from this area immediately occupied it. In the first year, there were 21 nests, in the second 92 and by the fourth year there were 150. They had moved in from sites immediately surrounding the vacant area with the result that nest density there decreased, no doubt to be filled up again in due course (Richardson & Sigman 1995).

On hard pan areas of Laysan Island devoid of vegetation, nest densities in 1991 reached 1,920 nests per hectare or 5.2 m^2 per nest (Rothschild 1893).

The Black-footed Albatross is an early coloniser of sand islands. As vegetation becomes established, they remain in open habitat around the periphery of islands. At Midway in the 1960s, Black-footed Albatrosses occupied areas within the Naval base, but by 1979 these had been taken over by Laysans (Ludwig *et al.* 1979).

Black-footed Albatrosses usually breed on the upper slopes of sandy beaches or dunes and flats behind, sometimes among shrubs. They are typically exposed to sun and wind, which often buries chicks in sand or blows away the sand from around or even from under them. At Sand Island, densities of 37 to 897 nests per hectare have been counted, but the distribution is patchy and in small groups, nests may be the equivalent of over 2,470 nests per hectare, or four square metres per nest (Ludwig *et al.* 1979).

The segregation of the two species in relation to vegetation was mapped at Green Island, Kure Atoll (Fig 11.13).

Nests

Both the Laysan and Black-footed Albatrosses often nest in places where there is very little nest building material and the quality of the nests depends very much on their situation. In sandy locations that the Black-footed Albatross prefers, they can often do no more than make a shallow scrape, but when there is dry grass around, as on Torishima, they are markedly attracted to it. On hard pan near the lagoon on Laysan Island, a scoop and low rim of soil is all that Laysan Albatrosses can accomplish, but on the well kept turf on the golf course and in the gardens of Midway they dig and pull grass until there is a bare 'moat' about five centimetres deep. The nest and moat is about a metre in diameter and in the most crowded areas the average distance between moats was 66 cm (Frings & Frings 1961). Later in the season all nests become more or less flattened by the trampling of chicks.

The better constructed nests provide some protection, but heavy rains cause flooding and many nests become completely submerged. The birds themselves show no avoiding behaviour; adults and chicks remain sitting in water, even when there is dry land only a metre or so away. During storms and tidal waves (*tsunamis*), huge seas break through beaches sweeping birds, eggs, chicks and debris before them (Fisher 1971a). In January 1954, almost all albatross eggs on Tern Island, FFS were lost and in 1982 hundreds of Black-footed Albatross nests at Midway were destroyed (Harrison 1990).

Pre-egg periods

Midway Atoll is free of gooneys for almost three months during the summer. The Black-footed Albatross is the first to return, in mid-October, while the Laysans begin arriving 10–14 days later, during the first days of November (Frings & Frings 1961). At first, only a few are to be seen, but within two or three days hundreds and then thousands start coming in (Fig. 11.14). The pre-egg period in the Black-footed Albatross (18–21 days) is slightly longer than that of the Laysans (14–16 days) (Rice & Kenyon 1962b). The Black-footed Albatross colonies fill up by about the third week of November and Laysans are all in by about the end of the first week of December. Laysan males predominate in the early days and Fisher & Fisher (1969) have queried the suggestion that Black-footed males and females arrive more or less together (Frings & Frings 1961).

Among Laysan Albatrosses, breeders return first, followed by successive cohorts of younger birds (Fig 11.15). Some returning subadults may be temporarily attracted to other albatross islands they pass en route to their own islands; a few ringed as fledglings at Midway have been recaptured at ages three to eight years-of-age, 84 km away

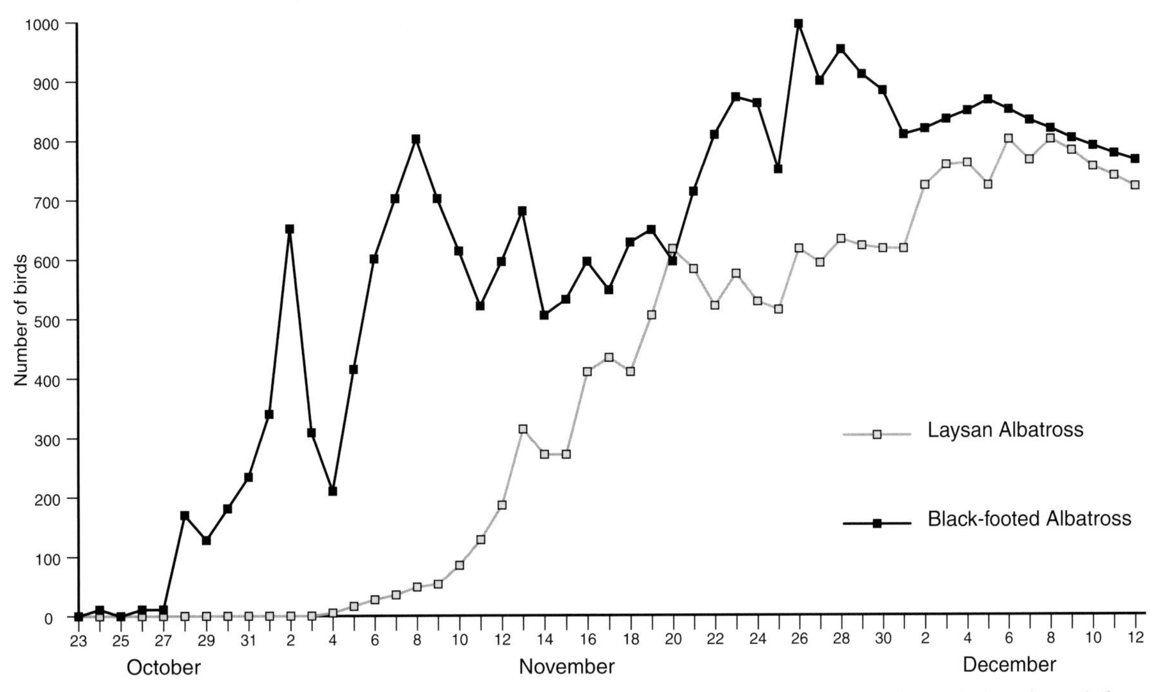

Fig. 11.14 Daily counts of Laysan and Black-footed Albatrosses on study plots at Midway Atoll from the first arrivals to the end of egg laying in 1957 (After Rice & Kenyon 1962b).

on Kure Atoll, but only two were ever confirmed as breeding and they represented a minute proportion of the number that were recruited to their natal location (Fisher & Fisher, 1969, Fisher 1976).

Adults return habitually to the same colony and nest at or near their former sites. Of 1,288 ringed Laysan Albatrosses on Fisher's study plot at Midway between 1960 and 1963, less than one percent were later found on nests outside the plot, and most moves (5–90 m) were made by widowed females. Most of these birds will have been reared from eggs laid in the same colony. Of 1,078 returned subadults formerly reared in the study plot, only three were later found nesting outside. On average, males breed 19 metres (n=56) and females 26 metres (n=49) from the nests in which they themselves were reared. Most Black-footed Albatrosses nest within five metres of the previous year's nest (BFAPBW[17] unpublished).

After an analysis of huge samples, Fisher (1976) was emphatic that Laysan Albatrosses were physiologically capable of breeding each year. In those seasons when experienced, intact pairs did not attempt to breed, there was evidently some extrinsic constraint. In such seasons, birds returned to their colonies frequently enough to maintain their nest territories. Even on islands where birds nest at high density, he thought it improbable that space was a limiting factor in Laysan colonies.

Most adult Laysan Albatrosses have long standing pair-bonds. In a sample of 3,094 nestings over 13 years

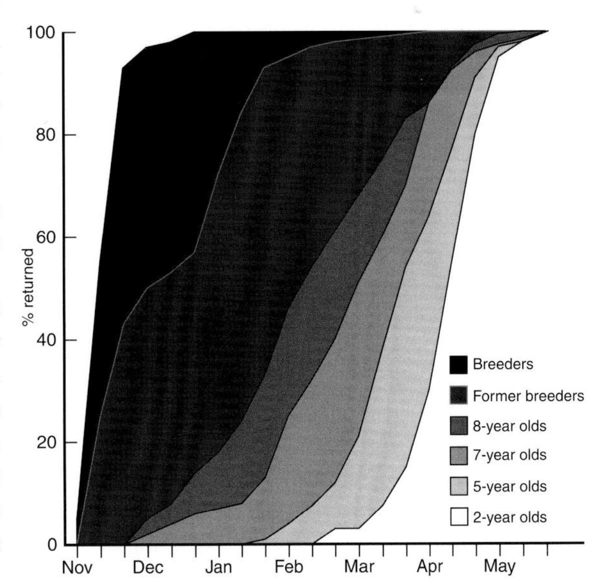

Fig. 11.15 Cumulative arrival of ringed Laysan Albatrosses at Eastern Island, Midway Atoll, 1964–65. Percentages of the total recaptures of breeders, former breeders and younger cohorts (Fisher & Fisher 1969).

[17] Black-footed Albatross Population Biology Workshop.

by 366 individuals of both sexes, most stayed with the same partner throughout this period; 10% had two partners and 2%(females) had three partners, while one female bred with five different males (Fisher 1976). If a female does not return, a few males (2%) can find another female in the same season, but most miss one breeding season. Females lose on average two seasons before re-mating and a few (3%) go five seasons without breeding (Fisher & Fisher 1969). Some Black-footed Albatrosses can also find a new mate without missing a season, but most miss one season before re-mating (BFAPBW unpublished).

Experienced males occupy their nest-sites until their mates arrive 6–16 days later. Within hours, copulation occurs then, after one or two days, both go off to sea and the nest is left empty for about ten days (2–17 days n=34). Females usually return first; 85% (n=34) of them at least 12 hours before their mates. The egg is usually laid within 24 hours of the female's return and 50% of females are alone at this time (Fisher 1969).

Eggs

Gooney eggs measure about 108 x 70 mm and weigh about 300 g. It is not possible to identify the species with confidence from single eggs, but on average Black-footed are slightly broader and heavier than Laysan eggs (Appendix 4). This difference is not equivalent to that in body mass, so Laysan eggs are relatively heavier (11.9% of female body mass; n=8) than those of Black-footed Albatrosses (9.7% of female body mass; n=10). Occasionally very small eggs, about half normal size, are found in Laysan colonies, but they do not hatch.

Laying

The Laysan Albatross at Midway lays its egg between 20 November and 24 December (n=3,540). The peak (mode) of laying is 1 December and by 10 December 97% of females have laid (Fig. 11.16). The Black-footed Albatross lays 7–10 days earlier (Fisher 1969). These statistics, however, obscure important detail; experienced Laysan females (known to have bred for at least six seasons) make their major contribution to laying between 24 and 30 November. By 4 December, when 90% of these females have laid (Fig. 11.16), only half the younger females have dropped their eggs.

Young Laysan females breeding for the first time lay from 2 to 11 December. The following year they lay two to four days earlier and in the next season a further four days earlier. Provided they keep the same mates throughout, 80% of layings thereafter will be between 27 and 28 November. The age of the female's mate also influences her date of laying. Experienced females (see above) mated to males of the same age and paired for several seasons, tend to lay before similar females mated to younger, less experienced males. Conversely, younger females usually lay earlier with older males than with mates of their own age (Fisher 1969).

Individuals, cohorts or groups with similar breeding experience all lay within time spans much shorter than that of the population. After about four years of breeding with the same mate, a female Laysan Albatross should lay each year within five days (2–7 days; n=127) of previous laying dates. This constancy is likely to be upset by loss of the mate and re-pairing (Fisher 1969).

At Midway, Rice & Kenyon (1962b) reported that both Laysan and Black-footed Albatrosses on Sand Island laid throughout the night and day. A few kilometres away on Eastern Island, Fisher (1969) found that 75% of 78 Laysan Albatrosses laid between 0500 and 1200 hours, and that laying was rare during the hours of darkness.

There are more descriptions of laying by Laysan Albatrosses than for all other species.

> 'It is easy to tell when a female is ready to lay an egg. She stands over the nest with her wings drooping at her sides and looks under herself. She dips her tail towards the nest at intervals, accompanying this with squeaks and groans. Suddenly the egg appears in the nest after one or two tail dips. The female stands a moment and then sits down on the egg.' (Frings 1961).

> 'Female at a 60° angle, tail raised slightly, feet flat and legs halfway flexed. Wings drooped slightly; quivered as she strained. Male standing alongside. She made the usual straining motions and looked down beneath her at intervals of 4 or 5 seconds, between contractions. Between her looks her bill pointed down and forward at 45° and her eyes were partly closed. The male almost motionless, but watching. When the egg dropped, after about 5 minutes, the female touched it with her bill and the male moved up to nest, looked, touched and tried to move onto egg, but female squatted on it

without getting it in incubation pouch. She raised up again and both birds looked at mucous-covered shiny egg. She then simply flattened out on it and remained resting, eyes partly closed and seemingly exhausted.' (Fisher 1969).

'17:03 – Female standing on nest; wings drooping; Cloaca alternately dilating and closing, so that the egg was visible in the cloaca when the latter was dilated. The interval between successive dilations was about seven seconds at first, later slowing to 10 and 11 seconds.
17:10 – Cloaca open about 30 mm; egg continuously visible.
17:12:00 – Female (still standing) leans back, raising breast, and lowering posterior end of body almost to surface of nest.
17:12:20 – Egg drops. Female resumes previous normal standing position, bends head down, and looks at egg. Silent: does not touch egg with beak.
17:12:50 – Female settles on egg and begins incubation.' (Rice & Kenyon 1962b).

'Abandoning the effort of further nest building, she stood up, turned and twisted. As she strained to expel the egg, the front of her body rose high and her tail flattened on the ground. Successive shudders rippled along the length of her body. Air hissed out of her open mouth as the powerful muscles of her abdomen contracted. She grunted twice. Her wings, held loosely at her side, quivered as the egg rolled into the sand and leaves which covered the floor of the nest.' (M.L. Fisher 1970).

Incubation

Adult Laysan Albatrosses lose down feathers from the abdomen at the beginning of the breeding season. The sac-like cavity just behind the sternum measures about 50 x 95 mm and is surrounded by contour feathers stiff enough to hold the egg tightly in place, sometimes even after the bird has risen to its feet. There is no room at all for a second egg.

Thickening of the skin with proliferation and engorgement of the capillaries begins before the loss of down is complete and many birds begin to re-feather the area towards the end of incubation, before the egg has hatched. There is considerable variation in this process; among old females which laid on 27 November some have completely re-feathered brood patches by 1 February while others still have bare skin. Laysans younger than four years old reveal no development of a brood pouch, but thereafter progressive development occurs during the years up to and including first breeding (Fisher 1971a).

Selecting only pairs that 'behaved normally', Fisher (1971a) observed incubation periods of 59–72 days (n=432) in the Laysan Albatross between 20 November and 18 February, but 93% of these were between 63 and 67 days. The few periods of 67 days or longer were of eggs laid early by experienced females, while those of 62 days or less were by young pairs breeding for the first time. Although the mean length of incubation in the Black-footed Albatross at Midway is little more than one day longer than that of the Laysans (Table 11.4), the difference is significant and possibly related to the slightly larger egg (Rice & Kenyon 1962b). At Kure Atoll, slightly longer incubation periods were apparent in both species (Woodward, 1972).

	Incubation (days)		
	mean±SD	range	n
Laysan Albatross	64.4±1.0	63–68	95
Black-footed Albatross	65.6±1.2	63–68	75

Table 11.4 Incubation periods of Laysan and Black-footed Albatrosses at Midway Atoll (Rice & Kenyon 1962b).

Incubation in Laysan and Black-footed Albatrosses is shared between males and females. Although female Laysans very occasionally sit for longer than males, males usually take a larger overall share of incubation, averaging 36 days compared to 29 days for females. Laysans complete incubation in three to seven shifts and most in four to six shifts (Table 11.5). Three-quarters of eggs hatched during the fifth (female) shift (Fisher 1971a). Black-footed Albatrosses divided the time into five to eight shifts (Rice & Kenyon 1962b).

Four out of five Laysan Albatross females remain in the nest for less than two days during the first shift and most of them lay their eggs and leave within 24 hours. Experience counts for much at this time; young females laying their first eggs spend twice as long at the nest, but the ability of all females to go back to sea quickly depends upon the mate returning and here again, experience helps. The second (male) shift is long and an occasional Laysan male has sat for 58 days, twice the average length, before his female returned. The third shift is the female's longest and some have gone for 38 days before relief; thereafter shift-lengths shorten. Black-footed Albatross shifts appear to be slightly shorter than in the Laysan Albatross, but sample sizes were much smaller (Rice & Kenyon 1962b).

Shift	Sex	Mean (days)	Range (days)	n
1	Female	1.7	0–14*	593
2	Male	24.1	9–58	166
3	Female	20.3	6–38	451
4	Male	14.2	1–32	269
5	Female	7.0	1–24	109
6	Male	4.8	1–11	36
7	Female	1.5	1–2	2

Table 11.5 Length of incubation shifts in male and female Laysan Albatrosses: (Fisher 1971). * experienced birds only.

Most Laysan and Black-footed chicks break out of their eggs in about three days (1–6 days) from the time of first pipping. In both species hatching eggs are encountered for periods of 23–24 days; Black-footed Albatrosses from 15 January to 7 February and Laysans between 22 January and 20 February (Fig. 11.16) (Rice & Kenyon 1962b, Fisher 1971a).

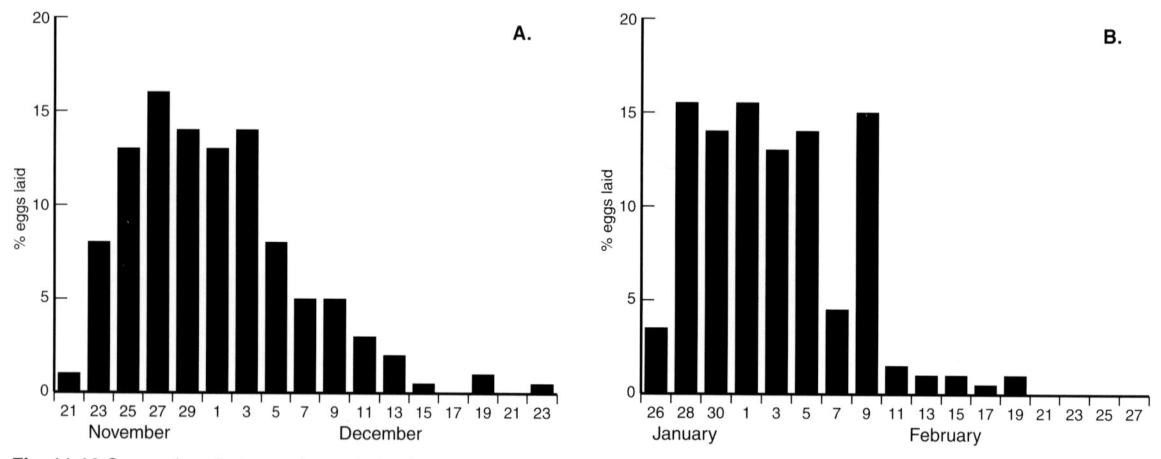

Fig. 11.16 Comparison between the periods of egg-laying (A) and hatching (B) among Laysan Albatrosses in the 1964–65 breeding season at Midway Atoll (Fisher 1971).

Chicks

Newly hatched chicks at Midway are brooded in the nest and left unattended at 12–25 days, averaging 17 days among Laysans and 19 days for Black-footed Albatrosses. As in incubation, brooding is shared equally by both sexes although shifts are much shorter, beginning at less than three days and later declining to less than two days (0.5–6 days). As chicks get too big to be covered, parents sit nearby for a further ten days or so. Eventually parents cease to guard their chicks, Laysans at about 27 days and Black-footeds at about 29 days (Rice & Kenyon 1962b). No avian predators are known to prey on these chicks, but there are environmental hazards; storms are frequent enough for small chicks to need warmth and protection from wind, drifting sand, rain and spray.

Chick feeding

At Midway, newly hatched chicks are fed on oil, but are soon weaned onto more substantial food. Laysan chicks are fed on average every one to three days (Table 11.6) and take about 165 days from hatching to departure, while Black-footed Albatrosses take only 140 days (Rice & Kenyon 1962b).

Once chicks have been left unattended, parental feeding visits are very short. Adults of both species appear to come in and feed their young by day and night, but recorded night feeds may, in fact have been delivered at dawn and sunset (Rice & Kenyon 1962b). In a single visit, which may last only 15–25 minutes, a chick may be fed three to four times. At first, feeding takes place at or near the parents' nest, but chicks soon become very mobile, shuffling up to 25 metres away and building other nests of their own, sometimes in the shade of shrubs where they may not see or even hear incoming parents. Parents do not go in search of their chicks which may thus miss feeds. If they lose touch, parents will cease to return, leaving the chick to become emaciated and die, probably more from dehydration than starvation. Abandoned chicks may be adopted by other adults, but Rice & Kenyon (1962b) believed that such young rarely received enough food to fledge. LeFebvre (1977), however, described one abandoned Laysan chick that was fed by five separate adults and survived.

Parental activity	Intervals between feeds	
	Laysan Albatross days (n)	Black-footed Albatross days (n)
Brooding	0.9 (186)	1.0 (71)
Guard stage	1.4 (79)	2.0 (29)
Foraging at sea	2.5 (67)	2.9 (57)

Table 11.6 Intervals between feeds in 12 Laysan and 5 Black-footed Albatross chicks (Rice & Kenyon 1962b).

Chick growth

Newly hatched Laysan Albatross chicks weigh about 190 g. Within a few days they are growing rapidly. By mid-May, when they are just over 100 days old, they have reached three kilograms, and are heavier than their parents. Subsequently, feeds become smaller or less frequent and body mass declines; by the time they start leaving the colonies chicks average 2.3 kg (Table 11.7). They may have been fed on their last day at the nest, but young birds on the beaches have gone 7–15 days without food and lost about half a kilogram, averaging 2.0 kg (1.6–2.6 kg n=20). One third of the body mass gained in May has been lost and by the time they leave they are considerably lighter than adults. Fledglings which do not get away by about 20 July continue to lose weight and eventually die (Fisher 1967). Black-footed chicks grow at a similar rate during the first 45 days, but after that they tend to be consistently heavier than Laysans (Rice & Kenyon 1962b).

Age (days)	Mass (g) mean±SD	n
Hatching	190±10	50
60	2 050±360	50
105	3 180±595	50
112	3 007±491	50
120	2 890±553	50
162	2 500±378	50
Leaving nest	2 310±260	16
Flying from beach	2 010±350	20

Table 11.7 Body mass of Laysan Albatross chicks (Fisher 1967).

On Eastern Island at Midway, Fisher (1967) and his students found small numbers of undersized Laysan chicks that they called 'pygmies'; in 1963 among about 8,000 fledglings handled, they recorded 62 of them. Early in July, they averaged 1.5 kg, about a kilogram below most young of the same age, but were otherwise healthy and well nourished. They often occurred in groups or 'sub-colony associations'.

Fledging

Newly hatched chicks of both species are covered by a brownish down with white-tan tips which produces a peppercorn effect (Plate 42) (Rice & Kenyon 1962b). Black-footed chicks are rather lighter due to their having more extensive white tips. By the end of the brooding period they are covered with long and very dense coats of secondary down (mesoptile), but still retain the peppercorn appearance. The bills of newly hatched Laysan chicks are blue and those of Black-footed chicks black. Primaries begin to sprout at about 35 days and feathering is proceeding under the down by March. Secondary down remains firmly attached until April–May. It begins to be lost from the back, sides and then from the belly. The upper breast, neck and head are the last areas to lose down and many fledglings still have tufts of down when they leave the island.

Fledgling plumages are almost identical to those of adults. Laysans may have a few grey flecks on the crown and upper thighs and their bills are grey rather than the orange of adults; Black-footed fledglings never have the white rumps that characterise older adults (Rice & Kenyon 1962b).

Departure

At Midway, young Black-footed Albatrosses leave the breeding grounds early in June, followed by the Laysans from about 20 June. Well fledged, but still with down about the head and neck, they start walking away from their nests towards the sea. From late June to mid-July fledgling Laysans are on the move all over the islands, walking, flapping, jumping and making short, erratic flights. They appear to head into the prevailing east-southeast winds which eventually take them to one or other of the south or east facing beaches which look out to the open sea. Only about 10% make for the beaches facing north. Once there, most remain one or two days, exercising their wings and making practice flights before leaving the island altogether. Four or five days is about the longest a fledgling might survive if it is to get away at all. Small groups of six to ten young from neighbouring nests appear to move to the beaches together (Fisher & Fisher 1969).

The departure of Laysan parents is determined largely by the behaviour of their young. They return only occasionally and briefly to the colony during the last six weeks and once they fail to find their offspring they cease coming back altogether. By the beginning of August hardly any are to be seen ashore.

Incubating Laysans that lose their eggs seldom remain on the nest longer than two more days. They go to sea and return irregularly for brief visits until late January. Likewise, if a chick dies or is lost, the parents soon cease to attend the nest regularly. Adults whose mates disappear, keep returning at infrequent intervals, but they too depart in March and April.

Breeding success

Fisher (1975a) and his students followed the fate of 2,787 Laysan Albatross eggs at Midway between 1962 and 1968; 64% (55–73%) of them hatched each year (Table 11.8). Unknown environmental factors probably contributed to the annual variation at Midway, but the experience and behaviour of females also influenced the outcome. Experienced females laying early hatched an average of 66–82% of their eggs while inexperienced females averaged slightly more than 50%. At Laysan Island in 1992–93, 81% of 205 eggs hatched (USFWS 1993).

Infertility rates of 3–16% have been recorded for Laysans in different seasons at Midway. Laysan eggs abandoned in the warm climate of Midway Atoll are known to have remained viable for as long as five days and late-returning partners can successfully renew incubation, but abandoned eggs may be taken by scavenging rats.

Year	Eggs laid	% eggs hatched	% breeding success
1961	619	-	78
1962	626	70	53
1963	708	73	70
1964	350	55	49
1966	628	61	-
1968	475	70	-

Table 11.8 Breeding success in the Laysan Albatross at Midway Atoll, 1961–64, 1966 and 1968 (Fisher 1975a).

Storms with high winds, drifting sand and flooding from heavy rain and tidal waves cause heavy losses of eggs and small chicks, and as many as 24% of all losses can occur during a single storm. Eggs may be blown out of nests and become partially or completely buried. Black-footed Albatrosses nesting on exposed beaches and flats suffer more than Laysans at inland locations. In December 1962, waves ruined 468 Laysan nests in one area of Eastern Island and on Sand Island in December 1969, large numbers of eggs were immersed for 24–48 hours; most were deserted and less than 20% produced chicks. Harrison (1990) reported a similar disaster at French Frigate Shoals in 1982. Such weather is more likely late in incubation; if catastrophic seasons are ignored, losses in the first month may be three to six percent, but nearer 25% during the second half of incubation (Fisher 1971a).

Desertion is the most frequent cause of egg loss, but there was only one instance of a pipped egg being deserted. In a study of 2,532 Laysan Albatrosses at 3,681 nests over seven years, Fisher and his students found that 21% were deserted by males and females in similar numbers; 18% deserted on only one occasion and three percent more often. In repeated desertions by 78 birds, all of which had started breeding during the seven year period of the study, one third completed incubation on only half their attempts and 60% on roughly two-thirds. Of the 78 birds, 72 abandoned their eggs during two breeding seasons, five in three seasons and one in four. Eight birds, five males and three females, abandoned all the eggs they or their mates had laid. Most desertions occurred soon after a relieving bird had taken over. In 1962–63, 26% of all desertions were associated with severe weather (Table 11.9).

Of the Laysan chicks alive immediately after hatching, 4–24% die before fledging; the mortality rate is fairly constant in any one year. In the last week of July and first week of August, hundreds of undernourished and dead Laysan fledglings are to be found in poor condition; many of them savaged and perhaps killed, when they wander away from their own nests and beg from other adults. Overall, an average 64% of Laysan Albatross eggs (49–78%, n=2,303) laid between 1961 and 1964 produced fledglings that left the nest (Fisher 1975a).

Two smaller samples of Black-footed Albatrosses at Midway in 1992 and 1993 had breeding successes of 64% and 79%, but on Laysan Island in the 1990s it was consistently lower (Table 11.10). In four seasons between 1963R;nd 1969, the Black-footed Albatross at Kure Atoll had also experienced low breeding success, 44% (38–42%) (Woodward 1972). However, on Tern Island, French Frigate Shoals Black-footed Albatrosses increased from 96 pairs in 1981 to 1,519 pairs in 1998 and throughout 17 years the mean breeding success was 69% (45–78%) (BFAPBW unpublished).

At Midway, Laysan fledglings leaving their nests walk, run and fly a kilometre or more to the beaches. They use a lot of energy, but are no longer being fed. Dehydration is common and Fisher calculated that 3.5% of those leaving in July die at this time, although the proportion may be lower for fledglings leaving earlier.

First flights over the sea end on the water, where fledglings often get into difficulties. Plumage becomes waterlogged, particularly if there is still down and many young birds drown in the surf. Fisher (1975a) believes that almost 10% of the fledglings that leave the nest never become competent juveniles flying over the ocean.

Cause	%
Nest too close to another	1–3
Egg broken by parent	2–4
Egg missing from nest	5–8
Faliure of living mate to return on time	10–20
Death of mate	1–3
Interference by second pair or subadults taking over	1–4
Mix-up with pairs from adjacent nests	0–1
Storms	0–38
Unknown	10–30
Disturbance by observers	2–4

Table 11.9 Causes of desertion in the Laysan Albatrosses (Fisher 1971).

	Year	Eggs laid	% eggs hatched	% chick fledged	% breeding success
Midway Atoll	1992	98	76	85	64
	1993	100	84	94	79
Laysan Island	1992	201	47	77	36
	1993	205	80	49	38
	1994	220	75	56	42
	1995	212	70	56	39

Table 11.10 Breeding success in the Black-footed Albatross, 1992–95 (BFAPBW unpublished).

First breeding

At Midway, Laysan Albatrosses that survived the juvenile and subadult years became breeders themselves at eight to nine years old (range 6–12 years) (Fisher 1975a). Males acquire mates about one year before females so almost all males are breeding when 12 years old and females by 13 (Table 11.11). The age at which individuals first breed is influenced by the age and experience of their first partners, as well as their own experience ashore in the years since they first returned to the colony. Subadults that pair with birds of their own experience, first breed at an age corresponding with the colony average. Those that pair with older, more experienced birds (that have lost their mates) take, on average, a year longer to breed. The modal age of first-time breeders varied, but years which appeared to favour the recruitment of younger Laysans were sometimes poor years for experienced pairs (Van Ryzin & Fisher 1976).

Fisher (1976) believed that among his Laysan Albatrosses, about 15% were 'learners' of age nine or younger, 54% prime, 10–19 years old, reproductives and 31% of birds, 20 or more years old. However, the Midway population had been recovering from former

Sex	n	Mean (years)	Range (years)
Males	471	8.4	6–16
Females	376	9.0	5–15

Table 11.11 Age of first breeding in the Laysan Albatross at Midway Atoll (Van Ryzin & Fisher 1976).

devastation throughout the 13 years of his study, and by 1972 may not have reached the stability that he assumed.

Few data are available for Black-browed Albatrosses, but the modal age of first breeding for unsexed birds is thought to be seven years (USFWS unpublished).

Survival

Among 3,305 breeding Laysan Albatrosses of all ages at Midway, over the ten years 1963–72, the mean annual survival rate was 94.5% (91.5–98.1%). There was no overall difference between the sexes, but in some years there were marked differences. About 42% of Laysans live at least of 12 years, 30% 14 years, 19% 18 years and 13% 21 years (Fisher 1975a). The mean annual survival of 224 breeding Black-footed Albatrosses (equal numbers of males and females) at Midway, over the seven years 1961–67, was 92.3%; it varied from a maximum of 99.4% in 1961–62 to 81% in the year before the 1963 El Niño (BFAPBW unpublished).

From 1,529 Laysan fledglings ringed as chicks in 1960, an average of 64% (56–77%) survive to the age of four. During most of that time they are at sea, except for their first few, brief visits to the breeding grounds in the third or fourth years. About 57% (51–64%) survive to eight years, spending progressively more of the breeding season ashore and 63% begin breeding. From 100 chicks ringed in 1944, five were still breeding at the age of 29 and one was retrapped again at the age of 42 years (Fisher 1975c).

Of 1,000 Black-footed Albatross fledglings colour-ringed at Midway in June 1957, only 27% survived to the age of five years. This seems remarkably lower than for Laysan Albatross fledglings on the same island and in almost the same years (before the later increase in fisheries-induced mortality). At least five of these birds were still returning to Eastern Island at the age of 37 years (BFAPBW unpublished).

FOOD

Laysan and Black-footed Albatrosses feed on a great variety of fish, cephalopods, crustaceans and other invertebrates, such as pelagic tunicates (Appendix 9). Stomach oil amounted to 10% of samples collected from chicks of both species at Midway in 1978–80 and is excluded from the following analyses. Although both species often feed on the same prey, differences in their diets were apparent.

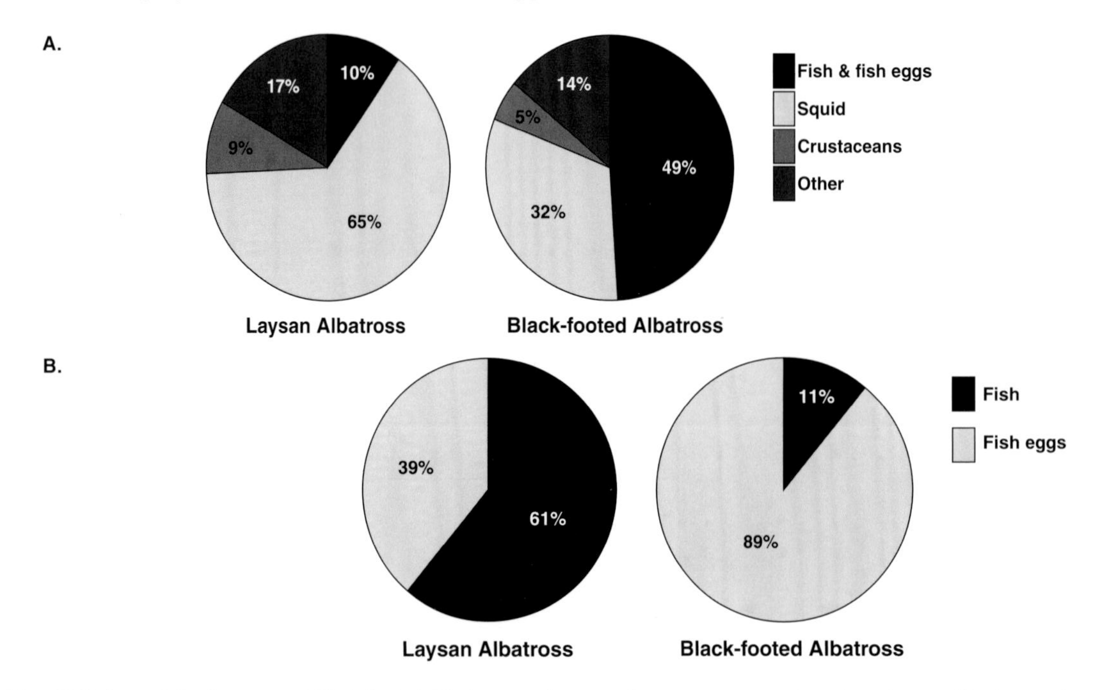

Fig. 11.17 Prey of the Laysan and Black-footed Albatrosses shown as the average percentage by volume of the four major groups (A), and the proportions of fish and fish eggs (B) (Harrison *et al.* 1983) (Appendix 9).

Laysan chicks received 10% fish, 65% squid, 9% crustaceans by volume; 93% of samples contained squid while the next most abundant item, flying fish and their eggs, were present in only 13%, which constituted 4% by volume. 'By-the-wind Sailors' *Velella* were eaten in their hundreds and amounted to 4% by volume. Remains of violet sea snails *Janthina* were found in some samples (Harrison *et al.* 1983). Black-footed Albatross chicks were fed 49% fish, 32% squid and 5% crustaceans by volume. The single most important item, the eggs of flying fish, amounted to 44%, but unexpected items like sea urchins are sometimes eaten. (Fig 11.17, Appendix 9).

Outside the breeding season, in the Transition Zone of the North Pacific during May–November 1990 and 1991, both Laysan and Black-footed Albatrosses killed in drift-nets had obtained substantial quantities of neon flying squid *Ommastrephes bartrami* and Pacific pomfret from the fisheries. Flying squid of 170–446 mm dorsal mantle length were the predominant food of both albatrosses (Fig. 11.18). By excluding these scavenged items from analyses, Gould *et al.* (1997) computed prey spectra which they believed approximated to the natural diets. Squid, particularly from the families Gonatidae and Cranchiidae, still predominated in the diet of the Black-footed Albatross, but small fish, particularly lanternfish and Pacific saury, were of equal or greater importance to the Laysan Albatross.

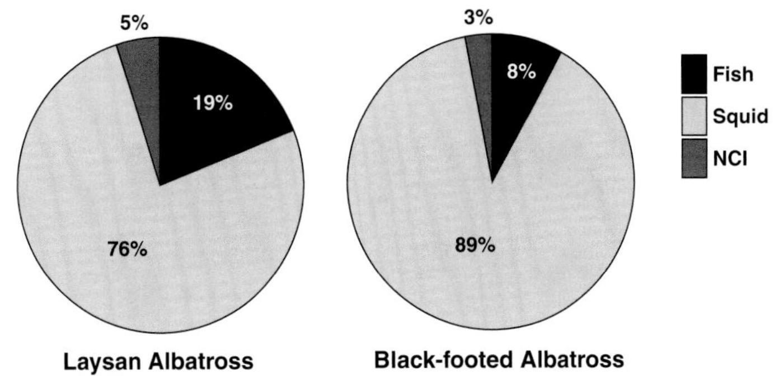

Fig. 11.18 The relative abundance (%) of identified prey taken in 1990–91 outside the breeding season by albatrosses killed when feeding at drift-nets in the Transitional Zone of the North Pacific Ocean. The high proportions of squid is a consequence of feeding at the fishery. NCI is non-cephalopod invertebrate prey (Gould *et al.* 1997).

While the diets of Black-footed Albatrosses were similar whether or not the birds were attending drift-nets, the diet of Laysans feeding at drift-nets was trophically quite different from those not scavenging. Analysis of the stable isotopes of carbon ($D^{13}C$) and nitrogen ($D^{15}N$) in muscle tissues from the two species indicated that Laysan Albatrosses fed, on average, at a lower trophic level than Black-footed Albatrosses; perhaps confirming a tendency for Laysans to consume more prey such as *Velella,* crustaceans and lantern fish that would soon disappear from the gut and may therefore be under-represented in samples. The greater tendency towards scavenging among Black-footed Albatrosses would also gain them larger, naturally dead or moribund prey from the higher trophic level.

Slight age differences were apparent; older Laysan and Black-footed Albatrosses ate more non-cephalopod invertebrates and fewer fish, while only a few newly fledged juvenile Black-footed Albatrosses scavenged marine mammal carcasses. Female Laysan and Black-footed Albatrosses both ate more squid than males. Laysan males made up the difference with increased proportions of fish and Black-footed males with non-cephalopod invertebrates (Gould *et al.* 1997).

PARASITES AND DISEASE

Nine species of chewing lice have been identified on gooneys (Appendix 5). *Docophoroides ferrisi, Harrisoniella copei* and *Paraclisis confidens* are unique to Black-footed Albatrosses and *D. niethammeri* is confined to Laysan Albatrosses. In spite of the proximity of most Laysan and Black-footed colonies, only two species of chewing lice are common to both gooneys. *Austromenopon pinguis* occurs widely on albatross hosts in both the northern and southern hemispheres, while *Perineus concinnus* is known also from Steller's Albatross, which is presently sympatric with the Black-footed Albatross on Torishima and was once sympatric with the Laysan Albatross.

A single species of feather mite, *Alloptellus pacificus*, has been found on a Black-footed Albatross. In 1993–94, soft-bodied ticks on albatross chicks at Midway Atoll were reported to be undergoing a 'population explosion' (McDermond 1994). Parasitic dermatitis was listed as one of the causes of death of 350 Laysan and Black-footed chicks autopsied at Midway (Sileo & Sievert 1988).

Introduced ants are present on Midway Atoll and Laysan Island. On Sand Island they invade nests containing newly laid eggs. They do not harm them and appear to be attracted to the water briefly available in mucous at the time of laying; as soon as the eggs dry, the ants depart.

Mosquitoes *Culex quinquifasciatus* have been implicated in the transmission of the avian pox virus that infects young albatrosses at Midway Atoll. This condition causes extensive facial lesions; there have been occasional deaths, but most infected young survived (Fefer *et al.* 1984).

Laysan Albatrosses recovered from North Pacific driftnets in 1990 and 1991 were parasitised with an oesophageal nematode *Seuratia* which may have reduced the birds' moulting efficiency (p.259).

PREDATORS

Sharks of several species including the tiger shark approach the reefs and enter the lagoons of Hawaiian atolls and coral islands every June–July to take fledgling albatrosses as they swim away from the beaches. In 1896, Schauinsland (1899) noted hundreds at Laysan Island and Fisher counted 37 large sharks from a helicopter over the lagoon at Midway. Mildred Fisher (1970) described two sharks that were killed and found to contain 7 and 13 albatrosses. At East Island, French Frigate Shoals in 1989, Rocky Strong (pers. comm.) watched 396 Laysan and Black-footed fledglings make their departure and saw 40 of them taken by tiger sharks; he noted that it often took a shark two to five attempts to capture a bird. Film of this predation reveals how the young birds have no self-preserving escape response to such large objects below them. The sharks pull the confused birds underwater and thresh off the wings before swallowing the bodies whole (DeGruy & Armstrong 1992). The number of fledglings taken in this manner appears to be considerable, but many of them die from other causes in the early years, so shark predation at this time may not significantly reduce the number that eventually breed.
At sea off California, Pribilof furseals have been seen taking Black-footed Albatrosses which were scavenging behind a fishing boat (McHugh 1952).

Pacific rats were in close contact with seabirds on Kure Atoll for over 100 years. They preyed on the eggs and young of many bird species, but albatross eggs were too big for them. Predation was heaviest in the central plain (Fig. 11.13B); Black-footed Albatrosses on or near the beaches did not suffer to the same extent as Laysans. The rats gnawed the skin and flesh of incubating adults which eventually died. In 1965–66, 57 adults were eaten and only one out of 310 young survived (Kepler 1967). Predation was most severe in years when plant food was scarce and it was when rats were most numerous that larger birds, like albatrosses, were attacked. The localised distribution of Laysan Albatross deaths suggested that individual rats or groups learnt to eat large birds when the need arose. Periodic control by poisoning began in 1966 (Woodward 1972) and there are no longer any rats on the island.

At Midway Atoll, introduced black rats devastated the colonies of smaller petrels (Atkinson 1985). They sometimes fed on deserted albatross eggs and moribund chicks, but were themselves vulnerable to albatross beaks and were not believed to cause significant mortality (Fefer *et al.* 1984, Fisher 1970). The last rat on Sand Island was killed in 1997. A few cats and dogs were once a potential hazard to albatrosses recolonising Tern Island, FFS (Richardson 1954).

Bristle-thighed Curlews use stones as tools to hammer holes in Laysan and Black-footed eggs on Laysan Island (Marks & Hall 1992) and Golden Plovers have been seen eating the contents of cracked eggs at Midway (Fisher 1971a).

12
STELLER'S ALBATROSS

'After staying in this unique southern island with the albatrosses as my friends, I have felt an intimate feeling of attachment for them, with which feeling I have written this paper.'

Toru Hattori (1889)[1]

Steller's[2] or the Short-tailed Albatross was formerly abundant in the North Pacific Ocean (Fig. 12.1). A few obscure comments from Steller's journal give us an impression of their numbers in the Bering Sea some 250 years ago; today the sight of one at sea is a rare event.

Early this century, these magnificent birds were brought close to extinction by Japanese fowlers and knowledge of the species is intimately linked with the history of the North Pacific feather trade. The only account of the fully occupied breeding ground was written in Japanese by Hattori (1889).

Concern for the species survival was voiced by the Marquis Yoshimaro Yamashina in the 1930s. Legislation, when it came, was powerless to save the last breeders and the species apparently disappeared from the breeding grounds (Yamashina 1942).

The few juveniles remaining at sea began returning as subadults in the early 1950s. Since then, numbers have increased and another breeding island has been discovered[3], so there are grounds for optimism. Steller's Albatross is still an endangered species with most of the world population breeding on one small, active volcano, but it was not eruptions that almost exterminated this species.

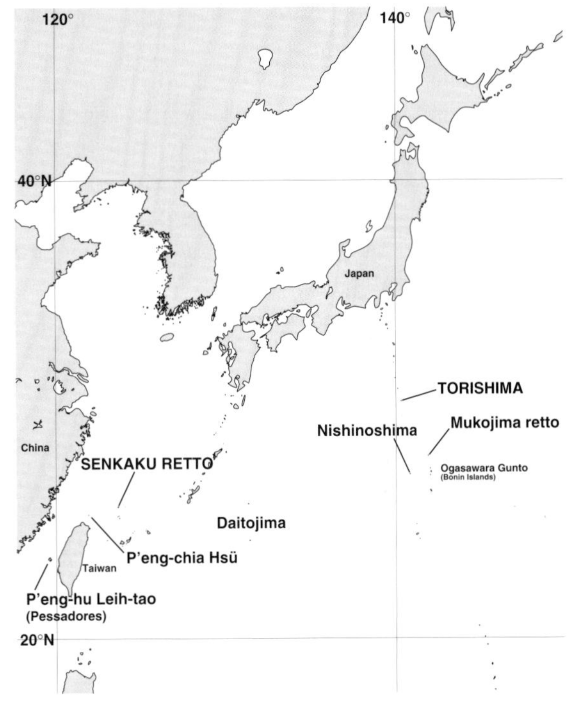

Fig. 12.1 (right) The present (upper case) and former (lower case) breeding islands of Steller's Albatross.

DESCRIPTION

Steller's Albatross (Plate 49) is the largest of the North Pacific albatrosses, but we still have to rely mostly on the measurements of a few very old museum specimens (Appendix 3). R.S. Palmer (1962) gave a wingspan of 213 cm while Swinhoe's (1873) spans of two juveniles were the equivalent of 216 and 217 cm. They seem short for albatrosses weighing between 5.1 and 7.5 kg (H. Hasegawa pers. comm., S. Rohwer pers. comm.).

The description 'short-tailed' is misleading; Laysan and Black-footed Albatrosses also have short tails, and the Galápagos Albatross has the shortest tail of all albatrosses.

[1] translated from the Japanese in Austin (1949).
[2] Georg Wilhelm Steller (1709–46) 'the illustrious traveller' (Seebohm 1890) is commemorated in five other species that bear his name. Together, they are a fitting memorial to the hard won achievements of his short life.
[3] Recent claims (Richardson 1994) that a pair of Steller's Albatrosses twice bred on Midway Atoll have been contested (Tickell 1996).

Chicks hatch into a dark brown down that appears thinner about the head; the bill and feet are black. They fledge in a dark brown plumage by which time the upper bill and feet are losing black pigment (Plate 51). Over about ten years, through successive moults, the dark feathers are gradually replaced by white ones in a manner similar to the wandering albatrosses – except that the latter start with white underparts and faces.

The underparts of juvenile Steller's start turning pale first, accompanied by a fading of the feathers around the base of the bill and across the lores, cheeks, chin, throat and neck. The breast and belly become completely white within two years. The under-surface of the wings whiten along the mid-line of the span, then towards the leading and trailing edges. Eventually, only a thin black line extends along both edges of the inner wings (secondaries and secondary under coverts). The outer underwing has rather broader, uneven black edges formed by the ends of the primaries and under primary coverts.

White feathers appear on the upper surface, first as a small patch on the mantle, then scattered on the rump and spreading forward to the back. On the upper surface of the wings, white patches first appear among the middle secondaries and then extend forwards and outwards. The back becomes mottled, then completely white, and the white feathers spread laterally and include the leading half of the inner wing. Black patches remain on the humerals and their coverts. The primaries and their coverts, outer secondaries and their coverts and a diminishing proportion of the inner secondary coverts comprise the distinctive black ends of the wings. Although there is considerable variation (Plate 50), all adults appear to achieve complete replacement plumage, but because the population is in the early stages of growth, there will probably be more individuals with intermediate plumage than in a larger, more stable population. Birds of intermediate plumage are known to incubate eggs and rear chicks.

The buff or yellow-tinged head of Steller's Albatross is a characteristic of the fully developed plumage of both males and females (Plate 52), becoming apparent at about five years of age. It resembles that on gannets (Nelson 1978). There is a progressive yellowing around the fringe of the diminishing black-brown crown, nape and hind neck (Plate 49). The yellowing appears to develop over the whole area, but to be masked by the dark pigment. Only when the brown pigment is lost is the yellow fully revealed. A similar, somewhat paler colour occurs on Galápagos Albatrosses (p.176).

Swinhoe (1873) measured the bill length of two dark-plumaged juveniles as 127 and 122 mm (Fig. 12.2). They appear to be transformed in the early years directly from black to bright pink with silver-blue-white unguis, long before the plumage has changed. The colour appears to be due to pigmentation and not not vasodilation as in great albatrosses. A conspicuous thin ridge of black skin outlines the base of the bill and is continuous with the gape stripe. This is normally hidden behind small contour feathers and, as in the mollymawks, is exposed during displays.

The tarsi and feet also become pink and in flight often project beyond the tail.

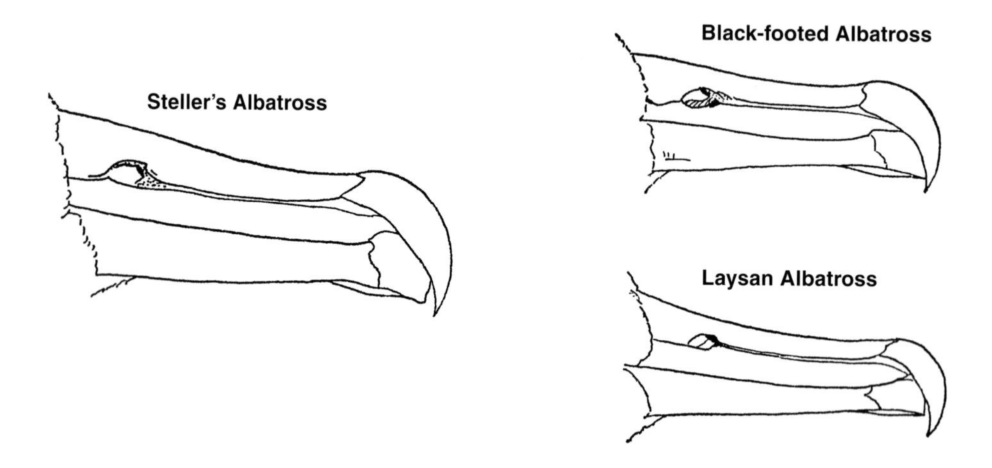

Fig. 12.2 Profiles of North Pacific albatross beaks drawn to the same scale (Palmer 1962).

HISTORY

In 1733, a Russian expedition set out from St Petersburg to explore the distant lands to the east. Three professors from the Imperial Academy of Sciences were accompanied by six student assistants, among them Stepan Petrovich Krasheninnikov. By 1736, they had reached Yakutsk and were held up. Krasheninnikov was sent ahead, but the others turned back, so he travelled on alone to Kamchatka. Two years later, Georg Wilhelm Steller, a German naturalist from the academy, was sent to Kamchatka. He joined Krasheninnikov and became the senior collaborator in a wide ranging investigation (Pearse 1968).

Steller took part in the epic voyage to Alaska of *St Peter* and *St Paul* under Captain-Commander Vitus Bering (1741–42). Returning along the Aleutian Islands, the *St. Peter* was wrecked at the Commander Islands[4], where Bering died. The survivors endured great privations. Led by Steller they built another boat from the wreckage and finally completed the voyage (Frost 1988).

Krasheninnikov set off back to St. Petersburg in 1743, but Steller spent two more years travelling as far south as the Kuril Islands[5] before leaving Kamchatka in 1744. He spent another year collecting around Yakutsk, then, tragically, after he had resumed the long, slow journey westwards, he became ill and died of a fever near Tyumen in November 1746. His specimens, journal and descriptions of birds reached St Petersburg the following year (Golder 1925).

These events took place long before Linnaeus had adopted the name *Diomedea* and neither Steller or Krasheninnikov used the name alcatraz or albatross. Steller collected albatrosses, but saw them as large gulls. Krasheninnikov (1755) may or may not have seen them, but he incorporated Steller's observations in his own work on Kamchatka and the Kurile Islands. Grieve mistranslated Krasheninnikov's *tchaiki* (gull) as cormorant, but otherwise the large reddish bill, petrel-like nostrils and long wings strongly suggest Steller's Albatross (Krasheninnikov 1755).

> ' There are two kinds of tchaiki, or [gulls] found upon this coast, which are hardly observed anywhere else. They are about the bigness of a goose, have a straight reddish bill about five inches long, and sharp on the edges, and four nostrils, such as other [gulls] have, two being near the forehead as are found in other birds which are thought to prognosticate storms, and are thence named Procellaria; their heads are of the middling size, their eyes black; their tails eight inches long; their legs are covered with hair to the knee, but below them they are bare; they have three toes of a bluish colour, and are web footed, their wings extend more than a fathom [2 m]; they are sometimes speckled; they sometimes appear near the shore, but can't stand straight upon dry ground, their feet being so near the tail that they are not able to balance their bodies; they fly slow even when hungry; but when full of meat they cannot raise themselves from the ground; and having eat too much they ease their stomachs; by throwing up...'

Peter Simon Pallas, another German naturalist, later recognised albatrosses among Steller's specimens at St Petersburg and named them *Diomedea albatrus.* His measurements established a standard for the species and clearly indicated that there were dark and white forms. Steller's journal was written in German and in his Latin translation, Pallas substituted 'albatross' and '*Diomedea*' for '*grosse Möwen*' [large gulls] (Golder 1925). Some of the gulls Steller saw may indeed have been albatrosses, but from the wording of the journal it is impossible to distinguish them from true gulls. Scholars had an early copy of the journal to work with, but none of Steller's original descriptions of birds remain. Pallas also described *Diomedea albatrus* in his great work *Zoographia Rosso-Asiatica* (1831), but no additional sources were evident. For the following translation from Pallas's (1769) Latin, I am indebted to my friend, Nan Brewer:

> 'the biggest Laros (Tschaiki) arrives about the end of the month of June in immense, and almost thousand-strong flocks, and are certainly messengers of the immediate arrival of fishes. Around

[4] Russian: (*Kommandorskiye Ostrova*).
[5] Russian: (*Kurilskiye Ostrova*).

> the end of July and before the beginning of August they leave again. But they never reach the eastern shore of Kamchatka, where they are scarcely ever seen even scattered, and like strangers; on the other hand they abound in the whole of the inner sea of Kamchatka [Sea of Okhotsk], and in the Kurile archipelago. They also visit Bering Island at the time when Steller was preparing to leave from there [early August 1742]. When they fly into these parts, the very weakest are caught, it seems, indeed, that they fly there from the south, because the first are usually seen at the end cape of the peninsula [Mys Lopatka]. Indeed, flocks have often been seen coming from the SW region and going back the same way at the time of their departure. Steller was rightly surprised that the *Diomedea* built no nest, nor looked after any offspring, when however they were present at the very time when every kind of bird was accustomed to produce young... They are the most voracious of birds, and by this reason they congregate around the river-mouths where they open their beaks wide to the shoals of salmon coming up-river. They swallow fish whole, even very large ones, of four or more pounds...'

In Daubenton's *Planches Enluminées*[6] that accompanied Buffon's *Histoire Naturelle des Oiseaux* (1786), Plate 963 was captioned *Albatros de la Chine*, but it was not mentioned at all in the text. This black bird with a greenish-grey bill was quite unlike any albatross we know today and Temminck later gave it a Linnaean binomial *Diomedea chinensis*. Another albatross specimen was then acquired from Japan. It was a largely white adult with a yellow head and sufficient traces of brown for Temminck to guess that Daubenton's black specimen had been a juvenile of the same species. In the *Planches Coloriées ...* (Temminck & Laugier 1838), this adult was illustrated (Plate 554) and the species renamed *Diomedea brachiura*, *Albatros a courte queue ou trapu*.

Robert Swinhoe was a diplomat in the British Foreign Office and from 1855–75 served at Xiamen (Amoy) and Yantai (Chefoo) on the coast of China and at Taiwan (Formosa). He was frequently at sea in the Formosa Strait and in 1858 accompanied *Inflexible* when it circumnavigated Taiwan in search of castaways. He became familiar with Steller's Albatross at sea and described it in flight, but although he was told of a breeding island, he never visited one. He acquired live birds and kept a male and female alive for some days at his home. They eventually died and were ably dissected and described (Swinhoe 1863).

The species had become widely known as *Diomedea brachyura*[7], when, in 1864, Swinhoe found Pallas's name in *Zoographia Rosso-Asiatica* (Pallas 1831) and pointed out that *Diomedea albatrus* had priority. Swinhoe knew from Temminck that juveniles were black, but in 1873 when he acquired two large black, pink-billed albatrosses he was convinced they were not juveniles. He proposed the new species *Diomedea derogata*, a name that remained in use concurrently with *D. brachyura* and *D. albatrus* for some years before being forgotten (Swinhoe 1873).

Meanwhile, Reichenbach (1850) at the Dresden Museum created a new genus, *Phoebastria*[8] for Temminck's *D. brachyura*, but with no description or justification. Apart from a brief period when it was used in America (Bent 1922), the name was not widely adopted. Recently, however, the need for a new genus for the Pacific albatrosses has led to the reinstatement of *Phoebastria* (Mathews 1934, Nunn *et al.* 1996).

The fate of albatrosses on most islands off China and Japan is forgotten history. Japanese and Chinese literature is not widely known and rarely translated. Oliver L. Austin Jr. (1949) remains an important source in this respect. As the Head of Wildlife at the Headquarters of Allied Powers occupying Japan after the second world war, he made many friends among influential Japanese ornithologists and secured useful translations, for example:

> 'I once stayed in the Pescadores [Penghu Liedao] for about one year when birds were still plentiful. During my stay I had the chance to observe albatrosses. While my observations may not be satisfactory, I regard them as valuable because wild birds have decreased so tremendously in recent Japan, and you can observe albatrosses now in Japan only on far-away Torishima.

6 Plates engraved by F.N. Martinet and coloured by hand.
7 an alternative spelling in the introduction of Temminck & Laugier (1838).
8 The meaning of this Greek name has defied translation by the scholars I have approached, but the stem it shares with the genus of sooty albatrosses *Phoebetria*, invented at the same time, implies some unknown common etymology.

> It was the last of February in 1902 when I saw the most albatrosses. The weather being calm, I had the opportunity to make a trip on the patrol ship of the local Pescadores government. During the voyage I saw albatrosses crowding on Byo-sho[9] [23° 19'N, 119° 8'E]. The hatching season was almost over, and it was the season when the parent birds do not stay on land in the daytime. Nevertheless numerous birds were seen, and I was delighted with the opportunity. But as the government officers had no business on this uninhabited island, the ship did not stay long, and I could not land on the island to collect eggs and chicks. It was hard enough to get specimens of adults.'
>
> Kaju Enomoto (1937)

Towards the end of the 19th century, Steller's Albatross was apparently breeding on Hoka-sho/Penjia Yu (Agincourt Island) (H. Hasegawa pers. comm.). Kuroda (1925) was told that the species was once abundant on Okino-daitojima (Argyle or Rasa Island) and apart from a lighthouse, it is still uninhabited (H. Hasegawa pers. comm.). There is a 1927 photograph of about 38 Steller's Albatrosses in a colony on Kitanoshima, the most northerly island of the Mukojima Retto (Parry Islands) north of the Ogasawara Gunto (Bonin Islands), but none has been seen there in recent years (Kurata 1978). The same photograph in a different publication, is said to have been taken on another island, Nishinoshima (Rosario Island), east of the Ogasawara Gunto.

Steller's Albatrosses are said to have been seen at Minami-Torishima (Marcus Island) and several have visited the Northwest Hawaiian Islands from time to time since the 1930s. In 1981, a Steller's Albatross visited Laysan Island, and Tern Island in the French Frigate Shoals (Harrison 1990). Ringed individuals from Torishima have returned repeatedly to sites among Laysan and Black-footed Albatrosses at Sand Island, Midway Atoll and there is an unsubstantiated claim that a pair bred in 1961 and 1962 (Richardson 1994, Tickell 1996). In the early 1990s, a female laid an egg in each of three seasons (E. Flint pers. comm.).

OCEANIC DISTRIBUTION

Steller's Albatrosses were once common in the the Formosa Strait (Swinhoe 1863, La Touche 1895) and the East China Sea from where they may have entered the Yellow Sea and gained access to the Sea of Japan. They have long been known about Japan (Blakiston & Pryer 1878).

In summer, large numbers moved north from Japanese waters towards Kamchatka. They may have passed between the Kuril Islands to enter the Sea of Okhotsk, but the historical evidence is ambiguous. They entered the Bering Sea, visiting the Commander Islands (Stejneger 1885) and flying as far north as Saint Lawrence and the Diomede Islands, where they used to be caught on the pack-ice (Maurie 1959). They were less frequent in Norton Sound.

Steller's Albatrosses were said to be shy, avoiding ships and difficult to approach (Nelson 1887); but in the 1860s they were described as common scavengers. Thousands attended whaling fleets at the Pribilof Islands, becoming scarce there in the 1870s after whaling had ceased (Elliott 1884).

They were common over shallow waters of Bristol Bay and along the Aleutian Islands (Dall 1874, Turner 1886). In May 1877 between the islands east of Unalaska, they were conspicuous by their white plumage and great size. During calm days, when they were most numerous, 10 to 15 were frequently seen at one time (Nelson 1887). Maurie (1959) met old Aleuts on Atka Island who insisted that small numbers had once bred during the winter on Bobrof Island and spoke of icicles on the bills of incubating birds. He thought it was an unbelievable legend. In the Gulf of Alaska, Steller's Albatrosses were considered to be most numerous at the mouth of Cook's Inlet near the Barren Islands and common about the Near Islands in March (Nelson 1887).

For most of the 20th century, Steller's Albatrosses have rarely been seen at sea. Their range overlaps that of the Black-footed Albatross; dark Steller's juveniles have been misidentified as Black-footed Albatrosses and, occasionally, vice-versa. Yesner (1976) has speculated that the Laysan Albatross may have taken over the Bering Sea niche once occupied by Steller's Albatross.

During aerial and ship surveys of over 15,000 km^2 of ocean off Alaska between 1975 and 1982, only one Steller's

[9] Chinese: Cat Island.

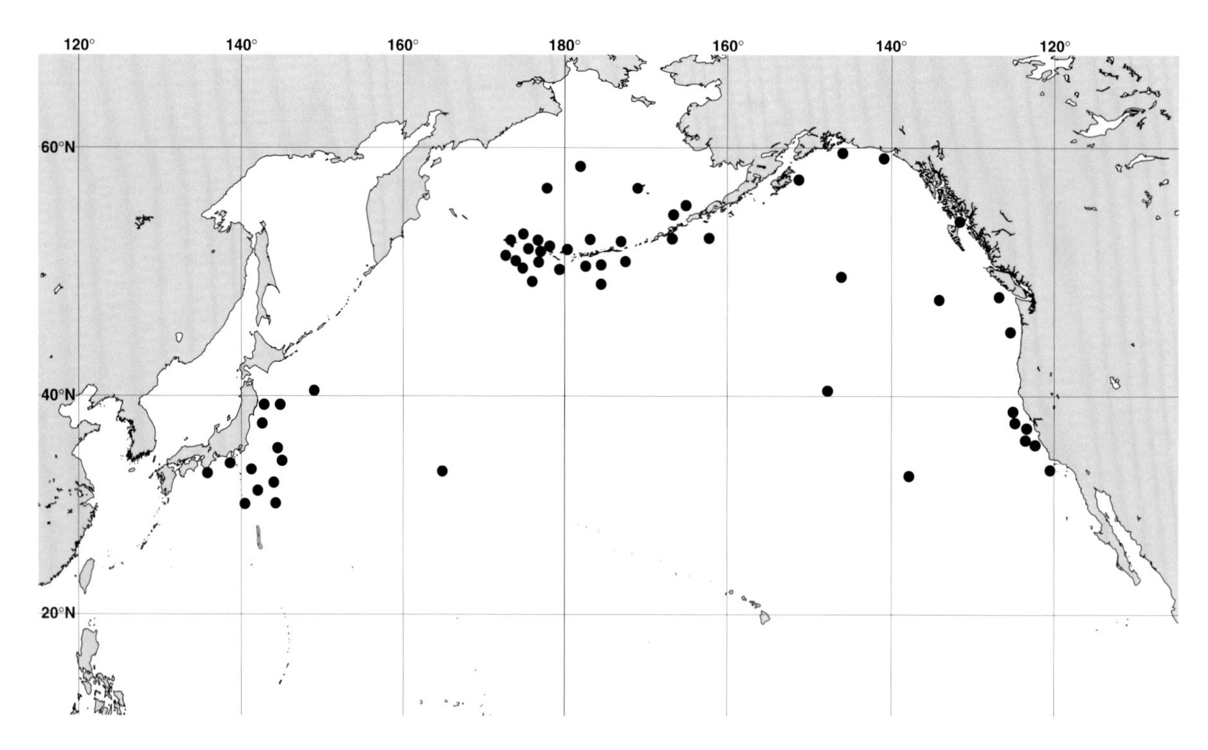

Fig. 12.3 Approximate locations of 56 Steller's Albatrosses identified at sea between 1940 and 1991 (after McDermond & Morgan 1993).

Albatross was seen. Sanger (1972) drew attention to the need for discriminating observations and evaluated the available sightings. Hasegawa and DeGange (1982) contributed many more, while Ogi *et al.* (1993) reported three over the deep basin of the Bering Sea. Another nine years of observations brought the total to 56 sightings in 52 years (Fig 12.3). McDermond & Morgan (1993) caution that extraneous influences, such as ship schedules, may influence apparent distribution patterns.

Numbers of Steller's Albatrosses increase in the northwest North Pacific during the breeding season. By December not only are the breeding birds coming to the end of incubation and probably visiting the nests more frequently, but increasing numbers of subadults are to be expected. The sharp decline in numbers in May is consistent with the end of the breeding season and departure of all birds from Torishima and the surrounding seas. The June–August peak of sightings at the Aleutian Islands and Bering Sea appears to confirm the historical significance of these waters, but the sudden decline in sightings during September may well reflect an end-of-summer decline ornithological cruises! It is difficult to make anything of the few birds seen in the eastern North Pacific and down the coast of North America. As Hasegawa & DeGange (1982) point out, much of the potential range of this species is rarely visited, let alone by ships with anyone on board capable of identifying albatrosses.

Steller's Albatross chicks were ringed at Torishima from 1962 to 1965, then every year since 1977. In addition to some local recoveries at sea off Japan, there have been several distant recoveries. In 1983, a first year juvenile was killed in fishing gear 500 km north of St. Matthew Island in the Bering Sea; it had travelled some 5,200 km in about 65 days. Ringed juveniles, seen or recovered in the Gulf of Alaska 4 to 18 months after fledging, had flown similar distances.

In spite of occasional vagrants at the Hawaiian Islands, the absence of former breeding grounds in the central North Pacific suggests that the species did not normally traverse the ocean in latitudes much lower than the Subarctic Frontal Zone. In the 19th century, they were common along the Aleutians and from the Gulf of Alaska southwards down the coast of North America as far as Southern California. Today they seen only infrequently (McLain 1898, Hasegawa & DeGange 1982, Stallcup 1990).

BREEDING ISLANDS

Torishima[10] (30° 29'N, 140° 19'E, 5 km² 403 m)

This is one of the Izu or Nampo Shoto, a widely spaced chain of islands stretching south from Japan. Torishima is 580 km south of Tokyo. It is a volcano situated on the edge of the Asiatic plate, a region of tectonic subduction. The island is 2.6 km in diameter with an outer caldera of rounded peaks and an inner cone (Fig. 12.4). The origin of the lava flows is marked by the pungent smell of sulphurous gasses from yellow fumaroles. Before the late 19th century, there appears to have been a quiescent period long enough for vegetation to have became established over parts of the island; trees and shrubs had been planted on the west and northwest. Eruptions have changed much of the island, but the forbidding landscape is still partially relieved by a segment of green vegetation on the western slopes. Elsewhere, plants are slowly recolonising the ash and lava; they include a grass *Miscanthus sinensis*, herbs *Chrysanthemum pacificum*, *Boehmeria biloba*, *Poligonum chinense* and a shrub *Elaeagnum umbellata*.

The climate is subtropical with mean temperatures of 14°C in February and 27°C in August. High winds with torrential rains are common early in the season, causing loss of nests, eggs and small chicks. In some years, typhoons hit Torishima and have resulted in the death of an occasional adult albatross (1958).

Steller's Albatross formerly nested among vegetation in several areas known as *torihara* or *torippara*[11] and the largest, on the top of the island covering about ten hectares (Fig. 12.5) (Hattori 1889). Since 1951, the growing number of albatrosses have occupied less than one hectare of fairly steep ash slope 80–110 m above the sea-cliffs of Tsubame-zaki[12] on the SE coast and backed by high cliffs above (Fig. 12.4).

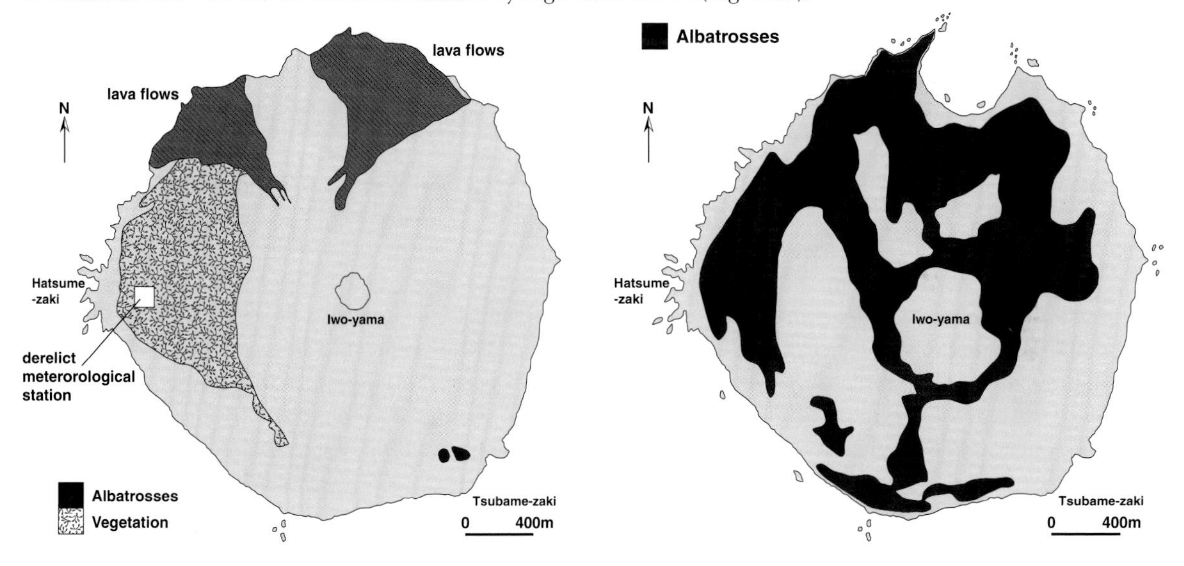

Fig. 12.4 Torishima showing the lava flows of 1939 and 1941. The main colony of Steller's Albatross is above Tsubame-zaki and the new colony is among vegetation on the west slopes above Hatsu zaki, near the derelict meteorological station (Torishima Society 1967).

Fig. 12.5 Torishima before the eruption of 1902, showing the extent of former breeding grounds of Steller's Albatross (Torishima Society 1967).

Senkaku retto

This is a group of small, uninhabited islands (Kobisho, Uotsurijima, Kita-kojima, Minami-kojima) and rocks in the East China Sea, about 150 km NE of Taiwan. They are claimed jointly by Japan (as part of the Southern Ryukyu Islands), the People's Republic of China and the Nationalist Republic of China (Taiwan).

Steller's Albatrosses formerly nested on several of the islands, but today they are found only on Minami-kojima

[10] Japanese: Bird Island. In the distant past also known as Ponafidin or St Peter's Island (Austin 1949).
[11] Japanese: bird field.
[12] Japanese: Swallow Point.

(25° 43'N, 123° 33'E). It is an islet with a conspicuous pinnacle separated by a flat shingle beach from the face of a scarp-like rock rising to 148 m (Fig. 12.6). The cliff has a broad terrace at about 100 m on which the albatrosses breed among rocks and a few sparse bushes.

Fig. 12.6 The Senkaku retto with Minami-kojima in the foreground, Kita-kojima behind and Uotsurijima in the distance. The circle marks the location of the small Steller's Albatross colony (H. Hasegawa pers. comm.).

NUMBERS

The number of Steller's Albatrosses formerly breeding on Torishima was very large. It has been said that there were more than a million birds before fowling began (Torishima Society 1967), but Hattori (1889), who appears to have been a careful observer, was on the island in the second year of fowling i.e. before serious depletion of numbers. On breeding grounds calculated as 26–32 hectares, he counted over 100,000 birds. It is impossible to say what that figure means in terms of the breeding birds, because he was present rather late in the season when most would have been away foraging and many subadults would have been ashore. The most crowded Laysan Albatross colonies have densities approaching one nest per 2.0 m^2 and most are much more widely spaced out. Steller's Albatrosses are considerably bigger birds and it's hard to imagine nest densities of more than one nest per 3.0 m^2. Hattori's area would thus have accommodated 87,000–107,000 breeding pairs, probably rather fewer.

With the killing of the last adults in the late 1930s, breeding at Torishima ceased for over a decade. During that time there remained a small number of Steller's Albatrosses at sea that had hatched and been reared during the late 1930s. These should have passed through their juvenile years and returned to Torishima as subadults in the 1940s. They probably returned to the seas about the island for some years, but without activity ashore, they evidently did not stay long. It took over ten years before an aggregation occurred at one site. Significantly, though, it occurred at the location of the last colony where, in 1929, Yamashina had seen a 1,000 birds (Fujisawa 1967).

The numbers of Steller's Albatrosses visiting Torishima in 1950–51 and the following two years were not reported, but in January 1954, 13 were photographed at Tsubame-zaki, most of them in white or nearly white plumage. Numbers increased steadily in the following years (Appendix 15). It was not recorded when the first eggs were seen, but at least seven were laid in 1954–55 and 12 the following year. Three chicks and three deserted eggs were found in January 1955. At least three fledged young were raised in 1955–56 (Ono 1955, Watanabe 1963).

The maximum number of birds counted at Tsubame-zaki in the 1950s was 30 (Watanabe 1963) and others would have been at sea. Perhaps 40 individuals were alive at the time of recolonisation. There had been no breeding for many years, so this small group would have been declining at a rate of about 5% per year, indicating that about 50 to 60 survived the last killings of the early 1940s.

The founders of the present colony were probably all back at Torishima by 1958, but numbers would have still been declining until the offspring of the first breedings began to be recruited. The maximum number of eggs laid in 1958 was 13 so the founders comprised at least 13 females and their mates, with other males. This small founder population was still declining towards the end of the 1950s, but in the early 1960s subadults from the early breedings were returning and from then on the population started increasing. By the late 1970s, the number of individual albatrosses seen each season was increasing exponentially, while the numbers of eggs laid and young fledged showed less dramatic, but significant trends (Fig. 12.7). The overall rate of increase in breeding pairs since 1954 has been 6.7% per year; since 1979 it has been 7.4% per year. In 1990–91, the number of eggs laid exceeded 100 for the first time. By the end of the 20th century, exponential increase is expected to take numbers to about 200 pairs, producing about 100 fledglings each year (Hasegawa 1995).

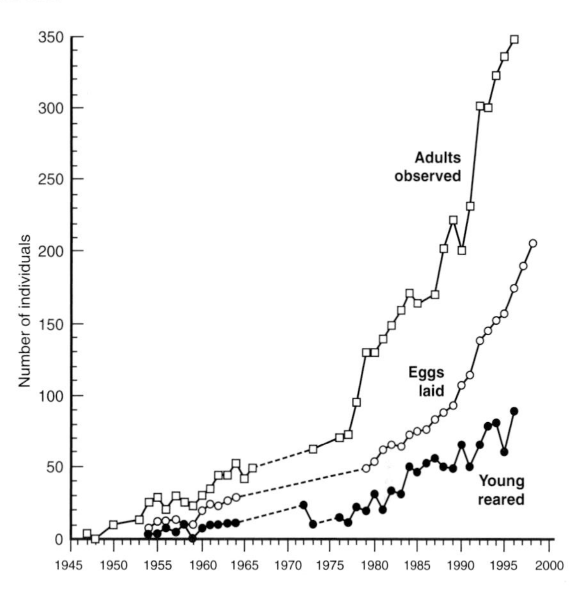

Fig. 12.7 Growth of Steller's Albatross numbers at Torishima from its re-colonisation in the early 1950s (H. Hasegawa pers. comm.).

After many years of uncertainty, about 75 Steller's Albatross on Minami-kojima in the Senkaku retto were confirmed as breeding. H. Hasegawa (pers. comm.) landed on the island in March 1991 and approached the terrace where 10 large albatross chicks were present, the following year there were 12. Altogether there may have been about 15 to 20 pairs breeding.

BREEDING

Hattori (1889) reported that Steller's Albatross returned to Torishima in September, but in recent years they have not appeared until 10 October and lay from late October to early November (Hasegawa 1980).

The favoured sites of earlier colonies at Torishima were on open ground and gentle slopes, where there was grass for nest construction. On Minami-kojima, they are on the rocky terrace of a steep cliff. These albatrosses avoid dense vegetation, but an old photograph of Kobisho shows one sitting below fig trees.

On bare soil at Torishima, nests were 0.6 m in diameter (Hattori 1889). The present day colony is on rather steep volcanic ash slopes with patchy grass and herbs. On several occasions in the last 40 years grass has been planted to improve the site. It provides nest material, shields birds from each other and permits closer spacing of nests. The colony is divided by a gully at a steeper gradient, which is probably caused by wash-outs during rains (Fujisawa 1967, Hasegawa 1978).

Five eggs obtained from the Ogasawara Gunto in November 1890 by P.A. Holst and now in the British Museum of Natural History average 119 x 74 mm. They have variable caps of reddish spots, blotches and smears over the broad end. Both parents Incubate for 64–65 days in alternate shifts that vary considerably in length (Hasegawa & DeGange 1982). Hatching occurs from late December to early January and is followed by brooding and then intermittent guarding before the chick is left alone in the nest while both parents forage at sea. By March, chicks have reached a mean body mass of 6.0±0.7 kg (range 4.4–7.5, n=53) (Hasegawa 1982).

About this time, rafts of subadults congregate on the sea off Tsubame-zaki. Fledglings are fed up to the time of their departure; all albatrosses leave by mid-June.

Breeding success

In the very early years of recolonisation, many eggs did not hatch; they were probably from younger or inexperienced pairs that produced infertile eggs. Over the ten seasons, 1955–56 to 1964–65, 49% of 180 eggs laid at Torishima hatched and of those 84% fledged, giving a breeding success of 41% (Fujisawa 1967). Since then, the colony has increased not only in numbers but also in breeding performance. In the ten seasons between 1979–80 and 1988–89 for which there are good counts, average breeding success was 57% (n=698) (Table. 12.1).

Breeding success has been depressed by egg and chick losses caused by erosion of the ash slopes, particularly during heavy rain.

Recruitment

Juveniles remain at sea for several years before returning to Torishima as subadults; the youngest birds seen ashore have been four years old. By 1981, at least five out of 23 chicks ringed in May 1973 had survived to eight-year-old and one of these had acquired a mate and bred.

Season	Eggs laid	Chicks fledged	%
1979–80	50	20	40
1980–81	54	32	59
1981–82	63	21	33
1982–83	67	34	51
1983–84	65	32	49
1984–85	73	51	70
1985–86	76	47	62
1986–87	77	53	69
1987–88	84	57	68
1988–89	89	51	57
1989–90	94	50	53
1990–91	108	66	61
1991–92	115	51	44
1992–93	139	66	47
1993–94	146	79	54
1994–95	153	82	54
1995–96	158	62	39*
1996–97	176	90	51
1997–98	194	130	67
1998–99	213	143	67

Table 12.1 Breeding success of Steller's Albatross at Torishima. * Losses due to typhoon rains (H. Hasegawa pers. comm.).

FOOD

An unidentified shrimp was once the principal item of food fed to Steller's chicks at Torishima, followed by the common Japanese squid *Todarodes pacificus*. Fish, including bonito, flying fish (Exocoetidae) and a sardine have been mentioned (Hattori 1889, Fujisawa 1967, Torishima Society 1967). Steller's Albatross may formerly have taken salmon as the fish crossed shallow estuaries to begin spawning runs up the Kamchatka rivers (Pallas 1769). They were formerly known to scavenge discarded offal from fishing and whaling vessels (Hasegawa & DeGange 1982).

PARASITES AND DISEASE

Steller's Albatross has six species of chewing lice, *Docophoroides pacificus* is endemic to the one host. Four species: *Austromenopon pinguis, Paraclisis giganticola, Perineus concinnus* and *Harrisoniella densa* are also found on the other North Pacific albatrosses; *Austromenopon navigans* is shared with the mollymawks (Appendix 5). The once large population of Steller's Albatross on Torishima was heavily infested with ticks that apparently contributed to chick mortality (Hattori 1889). They are not a problem in the small growing colony of today.

PREDATORS

Jungle Crows were apparently present when people lived on Torishima, but they have not been seen since fieldwork began in the 1970s. On one occasion, a Peregrine Falcon was believed to have taken small albatross chicks (N. Kuroda *in litt.*). Black rats are common on Torishima and a few have been trapped on the albatross nesting slopes, but they do not prey on albatross eggs or chicks. A feral cat was last seen in 1973 (Tickell 1975).

THE FEATHER TRADE

In the late 19th century, a lucrative trade in feathers developed in the Far East and seabird colonies across the North Pacific Ocean, including those of Steller's, Laysan and Black-footed Albatrosses were devastated.

Fishermen visiting remote seabird islands had traditionally gathered small quantities of feathers which could be sold in China and Japan for bedding, quilted clothing and pens (quills). In Japan, at the beginning of the Meiji era (1868), widespread encouragement of overseas trade, coupled with peasant labour and business enterprise, led to expanding markets, at first in the Far East and later worldwide. Body feathers were sold as 'swan's down', while the large feathers of the wings and tail, known in the trade as 'eagle feathers', were sold in North American and European markets for millinery.

Feather exports from all sources in Japan increased steadily from 1850 to a peak of over 383 tons per year in the period 1908–11; it then declined and no figures are available after 1915 (Austin 1949). Legal contradictions amounted to a Catch-22 situation that defied rational and practical administration. As early as 1907, the Japanese Government had prohibited the killing of albatrosses, while at the same time albatrosses were legal game birds and the Ministry of Agriculture's published statistics for 1930–42 listed 500–4,000 albatrosses taken each year (Yamashina 1942). Throughout all of this period, Japan had no laws prohibiting the traffic in or possession of protected birds or their products. The import and export of feathers flourished at Yokohama as long as birds were available to supply the demand (Austin 1949).

The killing of seabirds on numerous remote and uninhabited islands was rarely witnessed and it frequently continued, without any check or consideration of rational exploitation, until there were few birds left. We know a little of what happened to the albatrosses of the Northwest Hawaiian Islands and some of the events at Torishima have been documented in the Japanese literature; but we know nothing of what happened at other islands.

Torishima

Torishima had been known since the early 18th century. Survivors from shipwrecks drifted to the island, where they fed and clothed themselves, at least in part, from the albatrosses (Fujisawa 1967, Torishima Society 1967). Fowling at Torishima probably occurred intermittently during the 1870s, when boats from Hachijyo visited the island from time to time. It was only after the commercial potential of the international feather trade became apparent in Japan that the island was inhabited. In November 1887, Nakaemon Tamaoiki founded a settlement above Chitose Ura. In the first year there were 40 men and women, but in a few years, the work force increased to about 300 during the fowling season.

Hattori (1889), who was on a two-year survey of the Izu Shoto, arrived at Torishima two years later and stayed on the island from April to July. It was late in the albatross season, but there were still large numbers of birds and his sensitive appreciation remains a beacon of sanity at a time of destruction.

The scale of the killing created logistic problems even on so small an island. A light railway was laid from the upper slopes to the top of a cableway which ran down to the beach, where the ships were loaded. In addition to feathers, oil was boiled out of the carcasses which were then dried as fertiliser. From the beginning, fowling on Torishima had no sustained-yield rationale; adults and subadults were killed indiscriminately during the breeding season and chicks starved. Hattori (1889) said that one man could easily kill 100–200 Steller's Albatrosses in a day and an estimate of the number taken each year varied from one thousand to several hundred thousand. If we allow a two years hiatus for the 1902 eruption, killing was almost continuous for 33 years and chicks also died from ticks and scavenging crows.

Breeding success and recruitment would have started declining from 1887. We do not know the size of the original Steller's Albatross population at Torishima, but it is difficult to imagine it was big enough to sustain killing rates yielding five million birds in the 15 years (1887–1902) as claimed (Yamashina 1942).

No serious volcanic activity had occurred for more than 200 years, but on 7 August 1902, the island erupted killing all 129 islanders in a rain of hot ash. There were probably no albatrosses present at the time. After a year or so, another settlement was built on a new site called Okuyama, overlooking Hyogowan. In 1922, everyone left;

perhaps albatross numbers had become so small that there was no longer a living to be made. There followed a hiatus of five years when there was evidently no-one living on Torishima.

In 1927, people returned; a meteorological observatory was established, complete with radio communication and later Okuyama acquired a primary school with a teacher. Photographs taken during this period show cattle, cultivation and several large fishing boats moored in Hyogo wan; all of which point to a diverse economy, perhaps as the outpost of a deep-sea fishery. The inhabitants still took birds and Yamashina, who visited the island briefly on 15 February 1930, counted 1,400 albatrosses.

Yamashina arrived on the scene at a crucial time in the demise of the albatrosses, when prompt action could have saved them. He was appalled by the slaughter and left the island resolved to stop the killing, but in Tokyo even the son of a prince was not up to speeding the wheels of bureaucracy. In 1932, he sent his assistant, Nobuo Yamada to follow events on the island, but the visit was in April, far too late in the season to be effective. By then there were only a few hundred albatrosses to be seen. He went back the following year, again at the same time. The schoolteacher reported that in December 1932 and January 1933, 3,000 albatrosses had been killed and Yamada could find only 30–50.

In 1939, Torishima erupted again even more violently than in 1903; lava flowed into the sea burying Okuyama and filling Chitose ura, but all the inhabitants were rescued.

Senkaku retto

In 1896, the Senkaku retto were acquired by Tatsushiro Koga of Naha, Okinawa, who traded in seabird feathers. Over the next ten years, 157 tons of feathers were shipped from the islands. It has been said that these represented a million birds, but several species were probably included, so it is not at all clear how many albatrosses were killed. By the end of the century, a few albatrosses were still to be seen at sea in the vicinity of the islands, but only 10 to 20 could be found on Kobisho (H. Hasegawa pers. comm.).

RECOVERY AND CONSERVATION

Torishima

Throughout the Second World War there was a Japanese garrison on the Torishima. The sailors were well provisioned with 'submarine rations' and apparently did not need to take birds for food. The sighting of one 'large white bird' in April 1945 was evidently an event rare enough to be worth recording. The men were all withdrawn at the end of hostilities in August 1945.

In 1946, a new meteorological station was built on the west coast above Hatsune-zaki. Because of Torishima's location in the track of typhoons approaching Japan, it became one of the most important stations in the Japanese reporting network, with a complement of 33 scientists and supporting staff. The following winter, 1947–48, three albatrosses were reported in a Japanese newspaper (H. Hasegawa pers. comm.). Oliver Austin (1949) evidently did not know of this when he arrived off Torishima on 9 April 1949 on a Japanese whale-catcher. Heavy seas prevented him from landing, so the ship sailed around the island close inshore and the slopes were thoroughly scanned. He saw no albatrosses and after talking to the meteorologists by radio, he became convinced that the species was extinct.

On 6 January 1951, the chief meteorologist, S. Yamamoto discovered ten Steller's Albatrosses on the ash slopes, two kilometres from the weather station. The early years of recolonisation were described by Eichii Watanabe (1963). His detailed Japanese records remain the basic research on the species and deserve to be more widely known. The tremors that are a normal feature of Torishima suddenly became more frequent on 16 November 1965. It was feared that another eruption was imminent and the entire staff of the weather station was evacuated overnight. The volcano did not erupt after all, but the Japanese Meteorological Agency nevertheless abandoned the station and the island has remained uninhabited ever since.

At the end of April 1973, I was landed on Torishima by the Royal Navy and spent seven days on the island. There were 24 albatross chicks; the numbers had doubled since the meteorologists last counted them in 1965 (Tickell & Morton 1974, Tickell 1975). The following season, a Japanese television crew landed and since then several other television films have been made about the island and its albatrosses.

Hiroshi Hasegawa first visited Torishima in 1976–77. It was the beginning of a long commitment made possible by the co-operation of the Tokyo Metropolitan Fisheries Experimental Station at Izu-oshima and its research vessel *Miyako*. Since then, he has continued the research, visiting the island every year, raising public awareness through the different media and advising the Japanese government on conservation management.

When the island erupted in August 1902, the albatrosses were all away at sea. The rain of ash changed the form of the nesting grounds, notably creating the huge central crater where before there had been open slopes. It is impossible to say how much this event disturbed the breeding regime of the returning birds, but it certainly did not stop them breeding.

Today the population of Steller's Albatross continues to increase (Fig. 12.7), but it is still comparatively small compared to most other species. The ash slopes of Torishima are subject to erosion by wind and water, especially during typhoons. In the early 1950s, the founding birds favoured areas with grass whose fibrous roots stabilised the ash. This useful grass appears to be sensitive to albatross activity for it is replaced by the more robust *Chrysanthemum* shrub which flourishes in the presence of the birds; it is common on the lower gentler slopes of Tsubame-zaki, where there are many Black-footed Albatrosses.

The importance of grass was recognised in the 1950s by the meteorologists, who first collected grass from elsewhere on Torishima and transplanted it at Tsubame-zaki in summer when the birds were away at sea. This practice was repeated on a larger scale in the 1980s and average breeding success in the following years increased from 44% to 67% with more than 50 young reared. Unfortunately, between 1988 and 1992, four landslides caused more serious erosion, bringing breeding success down to 46%. In August 1993, earth-moving machinery was helicoptered in to construct protection for the slope below the colony. These expensive government works have been effective in bringing success back up to 54%, but erosion continues to be a problem at Tsubame-zaki (H. Hasegawa pers. comm.).

It has long been expected that sooner or later incoming subadults of the expanding population would attempt to nest elsewhere on the island, probably on ash. In the early 1990s, Hasegawa decided to anticipate the event by encouraging returning subadults towards the only vegetation on the island. This area is two kilometres away, almost on the opposite side of the island from Tsubame-zaki and facing west rather than southeast, but in other respects its potential as an albatross colony is superior to anywhere else on the island. Early in the 1991–92 season, about 20 lifelike replica albatrosses in two plumage phases and poses were placed on the western slopes, near the now derelict meteorological station. Recorded calls, played through loudspeakers added to the realistic imitation of a colony. Late the following season, subadults began investigating these models and in 1995–96 a single pair began breeding (H. Hasegawa pers. comm.).

The danger of another eruption at Torishima is a real one, but provided the albatross population is big enough, it need not be devastating for the birds. Steller's Albatross is an oceanic bird and its habitat is not Torishima so much as a wide expanse of ocean not influenced by one small island volcano. No albatrosses are on the island from June to September and throughout the rest of the year, a substantial proportion of the population is always at sea, even at the height of breeding. An eruption during incubation might eliminate all eggs, and the incubating birds – if they sat tight – but the odds are that many adults would fly off. During the rest of the breeding season there would be fewer adults at their nests and the many subadults would almost certainly not stay around during dangerous volcanic activity; they would assemble on the sea at a safe distance from the island.

As long as the island was not blown apart completely, an eruption would merely change the topography; much of Torishima is presently bare ash. The birds already nest on ash and as soon as it cooled they would return to the island and new colonies would form.

Senkaku retto

No albatrosses were seen at or near these islands from 1939 to 1970, then in March 1971, 12 adults were found on Minami-kojima (Ikehara & Shimojana 1971). Searches by helicopter in 1979 and 1980 failed to reveal any young. Kobisho was examined during the same flights but no albatrosses at all were seen there (Hasegawa 1982). In 1988, breeding was confirmed when seven chicks were seen on Minami-kojima from a small aircraft. In 1991 and 1992, 10 and 11 chicks were present (H. Hasegawa pers. comm.).

Legislation

Law protecting albatrosses at Torishima may have been enacted as early as 1906 (Torishima Society 1967), but it was ignored and had been forgotten by the 1920s. In 1933, Torishima was declared a *Kinryoku* (no hunting area) for a period of ten years. At a time when a sense of urgency should have prevailed, and with legislation on the statute books, interest in Torishima waned. Over the next six years, when the last remnants of the great population were being killed off, no effort was made to enforce the *Kinryoku* nor even to report on the species. Instead, throughout the North Pacific, Japanese fishing vessels were instructed by Yamashina to collect Steller's Albatrosses! None was caught and we can look back upon the episode as a perplexing finale to the incomprehensible turn of events.

The *Kinryoku* expired during the Second World War (1943) and was not renewed, so that when the albatrosses were discovered on Torishima in 1951, there was no protective legislation on the statute books. In 1953, there were 23 adults and an unknown number of eggs were laid, but no chicks were raised. Some may have been taken by feral dogs or cats, but labourers from the meteorological station may have been implicated. In 1954, the Japanese Association for the Protection of Birds (JAPB) appealed to the staff of the meteorological station, with the result that labourers were warned off the nesting slopes. The Japanese Government also forbade news photographers from approaching the birds. The albatrosses became a major concern of the Torishima meteorologists, who set about removing the cats and dogs. Guided by advice from the Yamashina Institute of Ornithology in Tokyo, they kept the growing colony under observation and collected breeding data from year to year (Watanabe 1963, Fujisawa 1967).

The Yamashina Institute also prevailed upon the Ministry of Agriculture and Forestry to designate Torishima a reservation and the staff of the meteorological station were authorised to guard the island. In 1957, the Ministry of Education named the Steller's Albatross a Natural Monument of Japan (Yamashina in Kalmbach *et al.* 1958). The conservation status was confirmed in 1958 when the Japan Government designated '*Diomedea albatrus* and its breeding habitats both on Torishima' a Natural Monument. The designation was upgraded in 1962 to a Special Natural Monument, and in 1965 the designation was shortened to read just '*Diomedea albatrus*'. Thus in Japanese law, Steller's Albatross is protected wherever it occurs in Japan (Y. Yamashina *in litt.*).

In a strange reversal of its purpose, this law was employed successfully in 1980 to delete recommendations by the Asian Continental Section of the International Council for Bird Preservation[13] that special international reserve status be negotiated for Minami-kojima in the Senkaku retto. These disputed islands lie on the edge of the continental shelf where there are undersea oil deposits and they are vulnerable to fowling and egging by fishermen.

In the 1970s, the islands were bought for 25 million yen (US$83,000) by K. Kurihara, an enterprising businessman of Saitama near Tokyo. The 9th Regional Conference of the Asian Continental Section of ICBP in 1980 was able only to urge the ICBP executive 'to endeavour, through international cooperation, to make appropriate research and monitoring, as far as possible, of the colony in question.'

Political and economic constraints remain, they still prevent more positive international protection of Minami-kojima and its albatrosses.

[13] Now known as BirdLife International.

COMPARATIVE
BIOLOGY

Light-mantled Sooty Albatrosses in flight (B. Osborne).

13
MOULT

'A bird showing wave moult appears to have a jumble of fresh, normal, worn, old and growing feathers and the greater the number of waves of moult that have passed down the primaries the greater the apparent jumble.'

Richard Brooke (1981)

Birds replace feathers by activating follicles which push out existing feathers, leaving spaces to be occupied by the new ones. Feathers grow at a fairly constant rate irrespective of size; those of large birds take longer to grow and may not all be replaced within one season. Moult studies focus upon the flight feathers of the wings (primary and secondary remiges) and tail (rectrices). Primaries are attached to the fused remnants of the bird's hand (manus), secondaries to the larger bone (ulna) of the the forearm and humerals to the upper arm (humerus). Primaries and secondaries are numbered from the wrist (carpal flexure); the primaries outwards towards the tip of the wing, while secondaries are counted inwards towards the body (Fig. 2.2, 13.1).

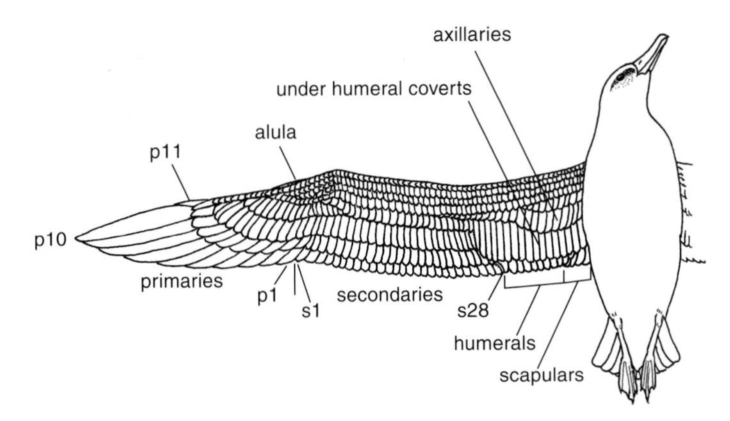

Fig. 13.1 The underwing feathering of a Laysan Albatross in flight showing the primaries and large number of secondaries, humerals and axillaries (drawn by R.J. Prytherch from photographs by S. Halvorsen).

All albatross wings have ten full sized primaries, together with a small vestige on the leading edge of each wing (10 + 1). In addition, there is a bastard wing (alula) of several small feathers. Small albatrosses have 25–29 secondaries (Kuroda 1971, Prince *et al.* 1993) and seven to ten humerals, while the much longer wings of great albatrosses have 34 secondaries (Prince *et al.* 1997) and 12–20 humerals. There are 12 paired tail feathers.

Most birds replace all their feathers seasonally in ordered sequences; the primaries outward and the secondaries inwards. Some moult them all more or less simultaneously and may be incapacitated; for example, penguins do not enter the water and geese may be flightless. In albatrosses, primaries are moulted in two opposite directions from one of the middle feathers. The outer (distal) series towards the tip of the wing (descendant) and the inner (proximal) series towards the body (ascendant). They are replaced more or less simultaneously in both wings, but not usually with exact symmetry (R. Brooke pers. comm.).

Small albatrosses do not replace any flight feathers during the breeding season, but some Wandering Albatrosses have started moulting primaries before the end of their prolonged chick-rearing period (Weimerskirch 1991). The sequence and direction of replacement can only be seen in albatrosses captured at sea some time after they have finished breeding. By the time they return to their islands, moulted feathers have grown to full size. However, the result of moult can be studied on the breeding ground, because it is possible to distinguish at least three generations of full-grown feathers present at that time. Their condition may be described by a numerical system in which individual feathers on one wing are each scored (0–5) and summed to give separate numerical assessments for all primaries and all secondaries (Ginn & Melville 1983).

Moulting is an energy demanding activity; the 20 primaries of a Laysan Albatross total 4.69 m in length (Langston & Rohwer 1996). Synthesis and elaboration of all that keratin has to be fitted into the time available between breedings. Because of their large size, albatross reproductive cycles are long and the time remaining between successive breedings is not enough to complete a moult of all flight feathers. In small albatrosses, groups of feathers are replaced in the intervals between several breedings. From time to time old, worn primaries accumulate and the birds then defer breeding in order to replace more feathers than would otherwise be possible (Langston & Rohwer 1996). In great albatrosses, the interval between the departure of fledglings and the onset of the next breeding season is too short to accommodate moulting and successful birds are obliged to defer breeding. The timing and rate of moult are related to age, breeding success, physical condition, nutrition and other seasonal imperatives such as weather. These parameters vary between species and islands so that the replacement of flight feathers in all albatrosses is prolonged, interrupted and complex. Two patterns have been recognised.

HISTORY

Stone (1900) noticed that Black-footed Albatrosses had an unusual moult:

> 'Seven of the albatrosses are molting the primaries; four of these are progressing in the usual way, the innermost quill being renewed first, but the others exhibit an exceptional order of molt. In Nos. 2 and 10 the second, third and fourth primaries are only partly grown, the old feathers having but recently cast, but the first primary (outermost) and the six inner ones are of the old plumage. In No. 3 the fourth, fifth and sixth feathers have been renewed and are only half grown, but the others have not been molted, while in No. 5 the first and second are renewed, but none of the others. Furthermore, they are full grown in one wing and only partially so in the other.'

Methods of scoring the condition of flight feathers in the field, were not worked out until the early 1960s and did not become widely known until guides to moult became available. Studies of albatross moult date from the early 1970s, when Harris (1973) scored moult in 1,487 live Galápagos Albatrosses during the breeding season, using a method adopted by Ashmole (1962) for Black Noddies. None of these albatrosses were moulting, but three generations of feathers were distinguishable and all had new outer primaries and progressively older inner primaries.

The Stresemanns did not themselves detect *staffelmauser* in the three albatrosses they examined[1], but Brooke (1981) proposed an outward wave moult of primaries in albatrosses consistent with their definition. He had only 24 assorted museum specimens, but on 26 February 1982 ten more were collected for this purpose off the Cape of Good Hope. Nine of them were Black-browed Albatrosses in which both outward and inward replacement of primaries (reversed modes) were recognised (Brooke & Furness 1982). In October 1983, Furness (1988) scored the primaries of 152 Western Yellow-nosed Albatrosses breeding at Gough Island. He again detected several moult foci which he believed further supported a wave moult hypothesis.

Seabirds driven ashore on the coasts of New Zealand during winter storms included several species of albatrosses from different islands, many of which were in moult (Kinsky 1968). The better preserved were retained as museum skins and Melville (1991) examined 57 black-browed albatross and 72 shy albatross specimens from several New Zealand collections. He scored primaries using a method devised for waders, but found the material too heterogeneous to draw any useful conclusions.

[1] 'Manche Vögel verteilen Erneuerung der zehn Handschwingen über einen langen Zeitraum. Das kann zur Folge haben, dass ein neuer Mausercyclus des Handflügels beginnt, bevor der vorhergehende Cyclus beendet ist. Wir haben diese Erscheinung als Staffelmauser, die hintereinander herlaufenden Mauserwellen als Staffeln bezeichnet und zwischen periodischer und cyclisher (besser kontinuierlicher) Staffelmauser unterschieden.' (Stresemann & Stresemann 1966).

In 1989–90, Prince and others (1993) scored moult in 168 Black-browed and 109 Grey-headed Albatrosses on the breeding grounds at South Georgia. All were live, known-aged birds between two and ten years old and none were in active moult. Three generations of feathers were identified: new feathers (fresh, unabraded and waxy in appearance), old feathers (one or two years old, abraded at the tips and no longer waxy) and third generation feathers (two or more years old, very abraded and brownish in colour) (Prince & Rodwell 1994). A two-phase model was proposed. Concurrent studies at South Georgia and the Iles Crozet revealed that the Wandering Albatross also conformed to the same model (Prince *et al.* 1997).

About this time, Henri Weimerskirch (1991) examined differences in moult and body condition among Wandering Albatrosses at the Iles Crozet. The energetic demands of moulting were believed to be substantial enough to compete with breeding, particularly in females, and strategic trade-offs were proposed between foraging success, extent of moult and decision to breed.

Opportunities for obtaining samples of moulting albatrosses at sea are rare. In the North Pacific, large numbers of seabirds have been drowned in commercial driftnets and this mortality provided an exceptional opportunity for investigating active moult. During 1990 and 1991, 308 moulting Laysan and Black-footed Albatrosses killed in this way were frozen and later made into museum specimens. Only 18 them were ringed, known-age birds, but the rest were separated into four age classes based on measurements of the bursa of Fabricius (Broughton 1994). In the laboratory, it was also possible to measure the condition of flight feathers more precisely than in previous studies. *Staffelmauser* and two other replacement hypotheses were tested and rejected in favour of an alternative, that Nancy Langston and Sievert Rohwer (1995, 1996) called the 'wrap-around model'. They concluded their research on this unique sample with an hypothesis that predicted tradeoffs between breeding and moulting.

TWO-PHASE MOULT

Juveniles and subadults

Fledgling Black-browed and Grey-headed Albatrosses leave South Georgia in April-May. They retain all their first generation juvenile feathers through their first winter and the following summer at sea. During their second winter at sea they replace all the three outer primaries (p8–10) and perhaps a few inner primaries (p1–4) which together completes PHASE 1. These feathers are retained through the third winter when many of the inner primaries are moulted (PHASE 2). Both species continue to replace primaries 8–10 in PHASE 1, but in PHASE 2, Grey-headed Albatrosses moult fewer inner primaries than Black-browed Albatrosses and several of their juvenile primaries may not be replaced until they are four to five years old (Fig 13.2). By six to seven years of age most Black-browed Albatrosses usually have an unmoulted inner (p3–4) primary at the end of the two-phase cycle, but among Grey-headed Albatrosses old worn inner primaries (p1–4/5) are more frequent. As they get older, birds of both species progress towards an annual replacement of half their primaries. The direction of PHASE 1 is outwards (descendant), probably from two or perhaps three moult centres. In PHASE 2 the moult progresses inwards (ascendant) from primary seven (Prince *et al.* 1993).

The sequence of secondary moult in Black-browed Albatrosses during the first five juvenile and subadult years is illustrated in Fig. 13.3. At two years of age, only the innermost secondaries (s25–29) have begun to be replaced, so most of the juvenile flight feathers remain. In the following year, two foci of moult are apparent and about half of the original secondaries persist. By five years of age, almost all original secondaries have been replaced and the second generation has begun to moult, again from two foci. It has been predicted that by seven years of age Black-browed Albatrosses will have achieved a two-phase cycle of secondary replacement corresponding to that of the primaries (Prince *et al.* 1993).

Juvenile Wandering Albatrosses begin moulting during their second winter at sea. The youngest return as subadults to South Georgia and the Iles Crozet at three years of age with newly moulted outer primaries (p8–10) (PHASE 1) and the following year they have new mid-inner and inner primaries (p1,2,5–7) (PHASE 2) (Fig. 13.4). By five years of age, the outer and some inner primaries are again new and the yearly alternation of PHASE 1 and 2 continues until the birds breed. This pattern is similar between the sexes at the Iles Crozet, but at South

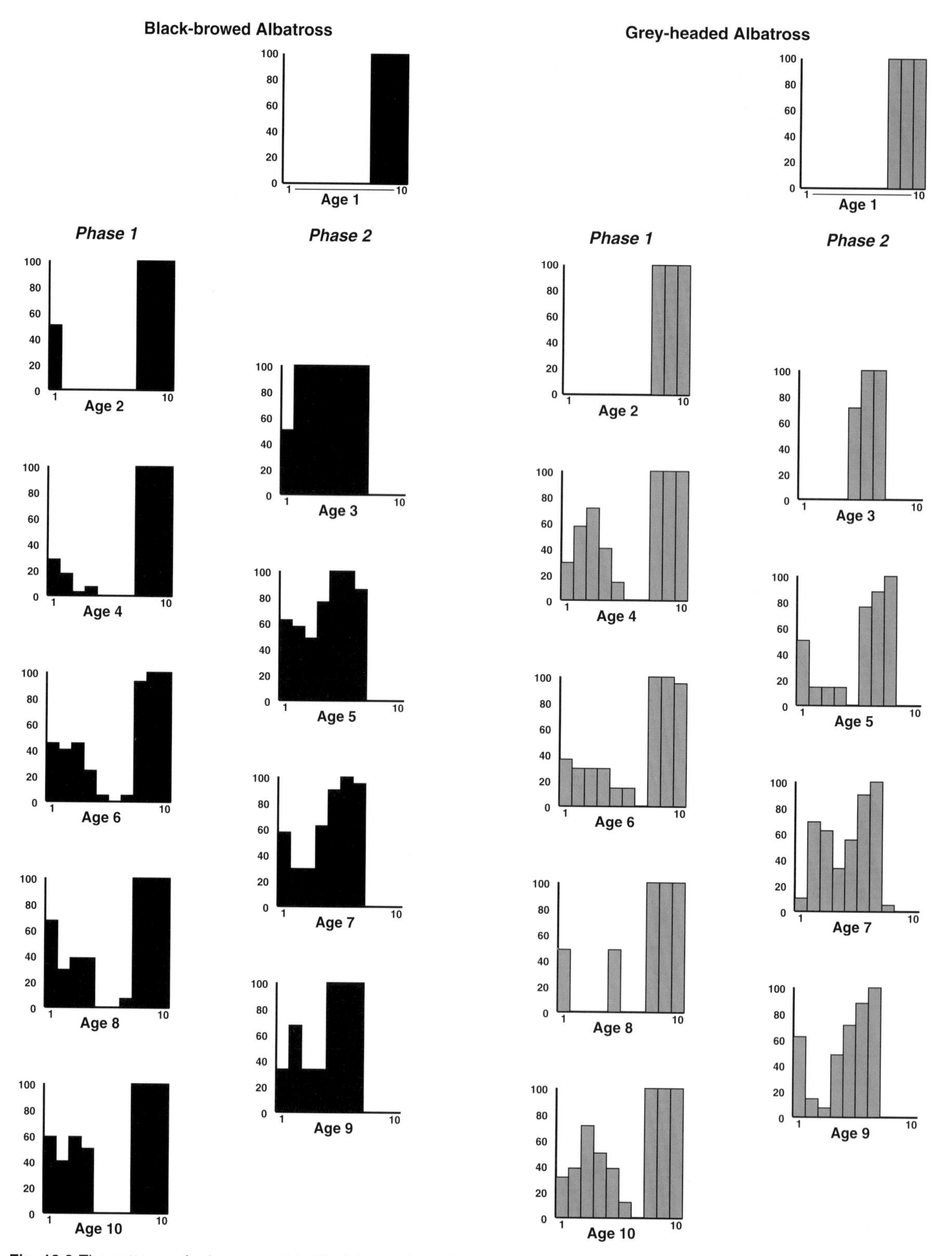

Fig. 13.2 The patterns of primary moult in Black-browed and Grey-headed Albatrosses, aged 1–10 years. In each histogram, the primaries are numbered along the x-axis with the outer (p8–10) feathers indicated at the top. The y-axis represents the percentage of birds replacing each numbered primary within the last year (Prince *et al.* 1993).

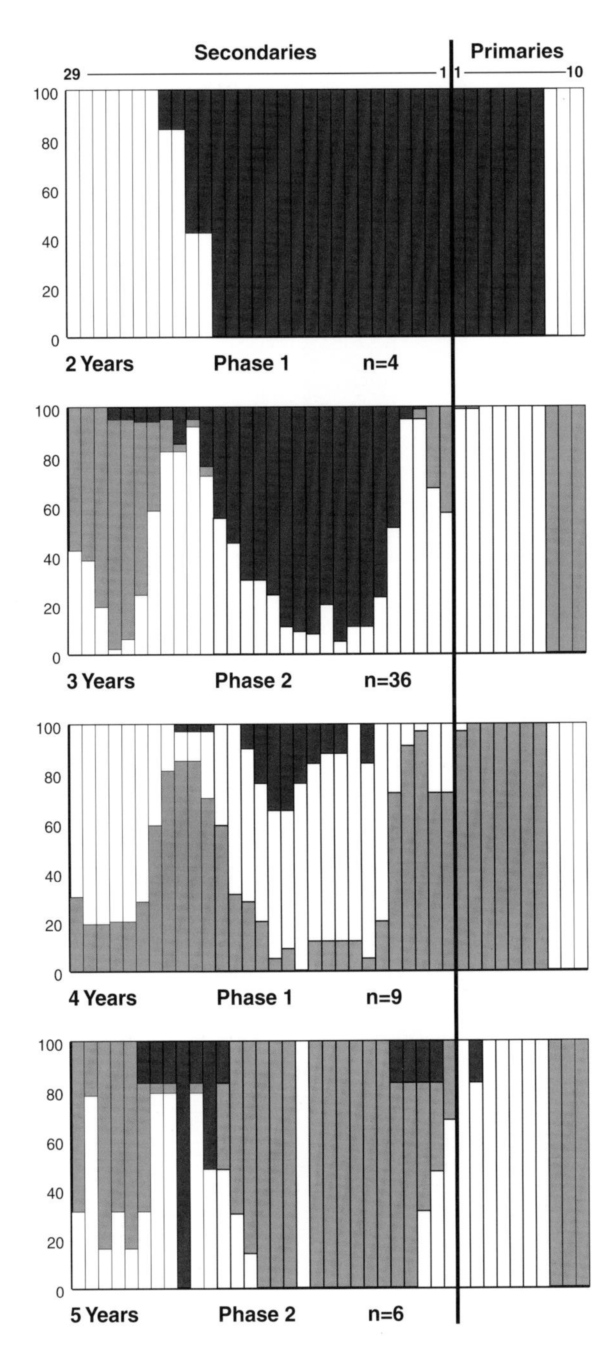

Fig. 13.3 The pattern of secondary moult in relation to primary moult in the Black-browed Albatross (n=65) between the ages of 2–5 years. Each box represents one complete wing with the proportion (%) of feathers of different ages indicated on the vertical axis. Unshaded areas indicate recently moulted feathers, grey areas one-year-old feathers and dark areas two-year-old feathers (Prince *et al.* 1993).

Georgia males replace a few more primaries than females. Four-year-old Wanderers still have 13% unmoulted fledgling primaries, but by the time they have reached five years, the primaries are almost all new or one year old.

Most three-year-old Wanderers still have all their fledgling secondaries. Moulting starts in the following winter at the inner and outer ends of the secondaries. Males always replace more feathers than females, who thus have more older secondaries. Five-year-olds of both sexes have more new secondaries than at any time in years three to nine. Together with the primaries, they make up the best set of flight feathers of their subadult years. They do not have as good a set again until they breed successfully.

Breeding adults

At South Georgia, the yearly alternation of PHASE 1 and PHASE 2 moults is retained when Black-browed Albatrosses mate for the first time and continues with slight adjustments for breeding success. Unsuccessful Grey-headed Albatrosses (Fig. 13.5) also undergo a two-phase cycle, but successful biennial breeding throws it out of synchrony.

After rearing fledglings, Grey-headed Albatrosses leave South Georgia in May and most of them do not return for 16–17 months. They spend two winters and the intervening summer at sea, without any commitment other than to feed themselves. This is much longer than the time required to regain peak fitness after breeding. In the two consecutive years most primaries are replaced at least once, although some inner primaries (notably 3) still remain unmoulted. Primaries 7, 6, 4 and 5 (in that order) are most likely to be moulted twice, some birds replacing feathers that are less than one year old. In the interval of 16–17 months, an average of 11 primaries are replaced.

Overall, a two-year moult cycle persists, but successful Grey-headed Albatrosses maintain a maximum number of functional primaries by switching the two phases and the numbers of feathers moulted at any one time.

Experienced male and female Grey-headed Albatrosses moult similar numbers of flight feathers. The physical burden of breeding is not uniform; after they have raised their chick, some individuals are in better condition than others. While they are moulting, they experience a wide range of oceanic conditions and by the beginning of the following season some are fitter

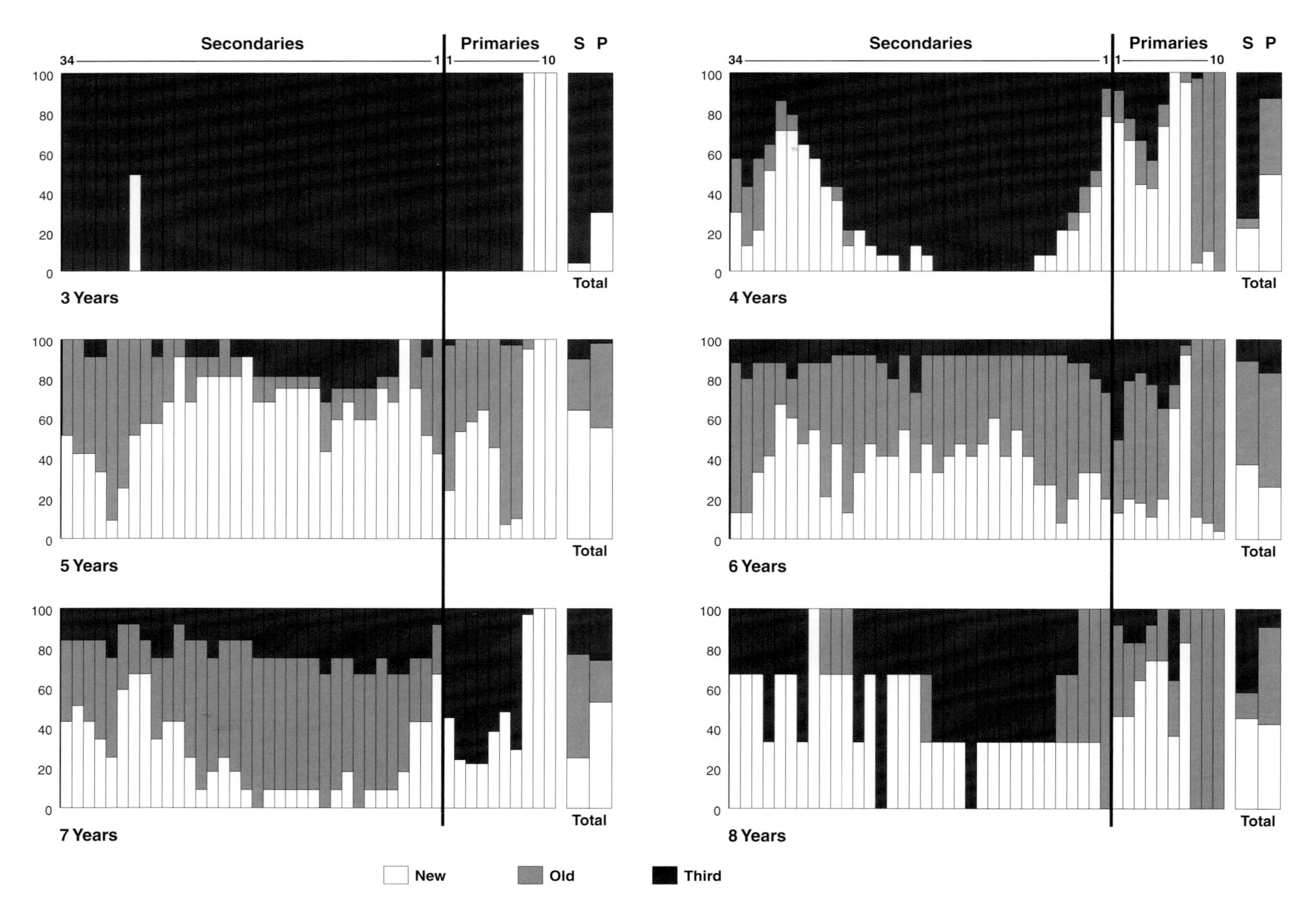

Fig. 13.4 The proportion (%) of new, old (second) and third generation feathers at each primary and secondary location in male and female Wandering Albatrosses aged 3-8, at Bird Island, South Georgia (after Prince *et al.* 1997).

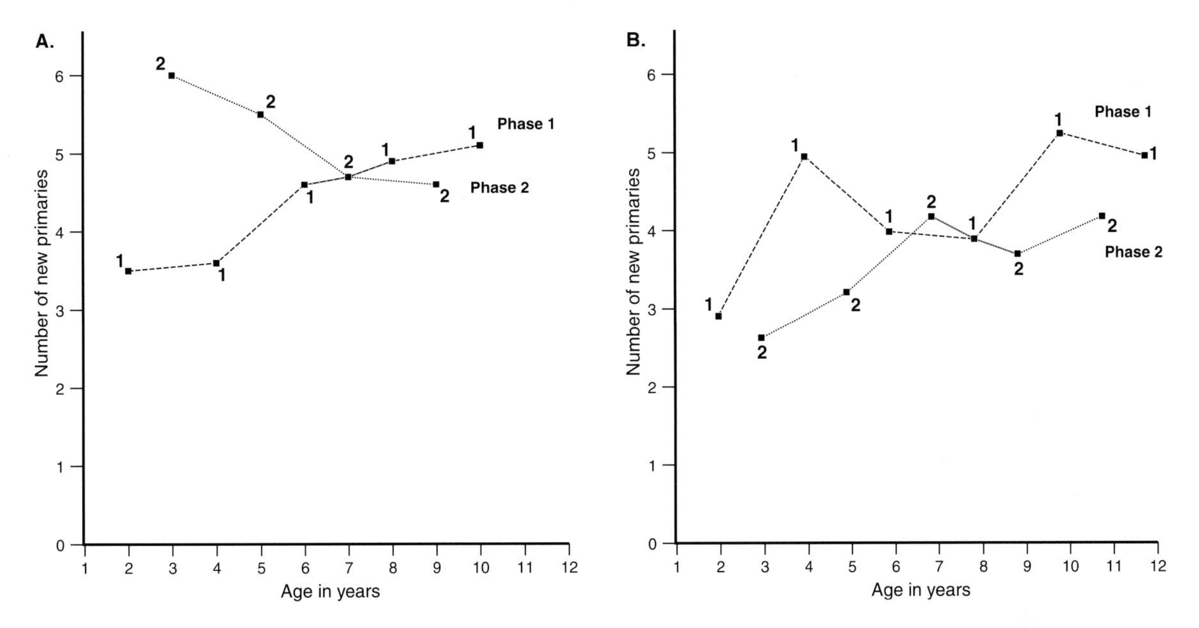

Fig. 13.5 The average number of primaries replaced each year in relation to age and phase in Black-browed Albatrosses (A) and Grey-headed Albatrosses (B). By the age at which breeding starts, about half of the primaries are replaced each year. The broken lines connect successive moults of Phase 1 (- - -) and Phase 2 (.....) (Prince *et al.* 1993).

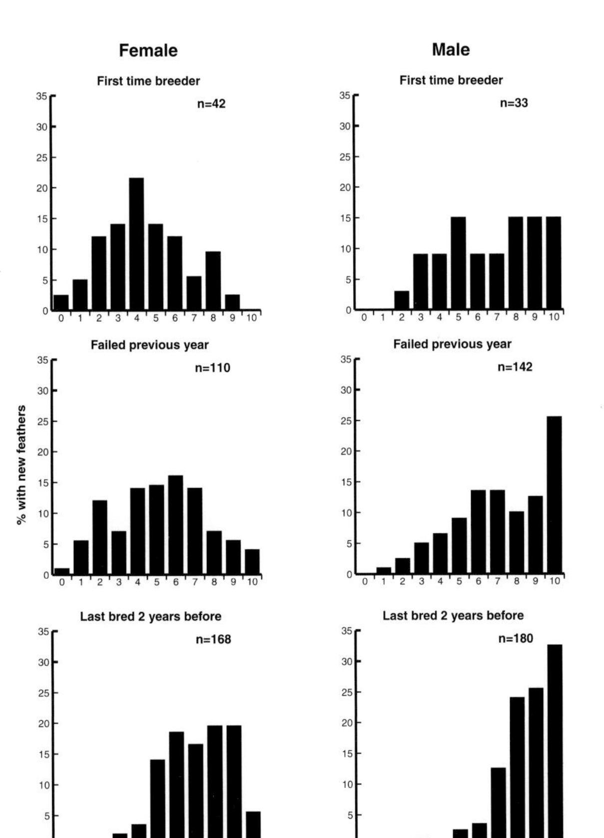

than others. Heavier males arrive with more new primaries than lighter males and in some years all return with more new primaries than in other years (Cobley & Prince 1998).

Wandering Albatrosses breeding for the first time have some new primaries from a PHASE 1 or 2 moult in the preceding winter (Fig. 13.6). Breeding interrupts the process, because successful pairs spend the next winter raising a chick, during which time moulting is suspended. They go to sea the following summer and spend 12 months away, during which time they replace their outer and some inner primaries. Males moult significantly more outer primaries than females. Pairs that lose eggs or young chicks go to sea straight away and return after five or more months with fewer new middle primaries than successful birds, especially the females (Table 13.1) (Weimerskirch 1991, Prince *et al.* 1997).

Wandering Albatross males at South Georgia moult feathers slightly faster than those at the Iles Crozet, so over the same number of years, the extra feathers replaced by South Georgia birds amount to additional moult cycles. Plumage maturation (loss of pigment) is

Fig. 13.6 (left) A frequency distribution (%) of new primaries in Wandering Albatrosses breeding at South Georgia, December 1989 – February 1990 (Prince *et al.* 1997).

Breeding success (previous season)	Moult period (months)	Number of new primaries mean±SD (range) n Males		Females	
Successful breeders	12	8.8±1.2 (5-10)	142	8.1±1.6 (3–10)	162
Failed (lost chicks)	5–8	8.6±1.5 (5-10)	33	6.4±2.4 (0–10)	38
Failed (lost eggs)	9–10	7.9±2.4 (2-10)	21	6.8±1.8 (3–10)	20

Table 13.1 The numbers of new primaries in adult Wandering Albatrosses at the Iles Crozet after moulting periods of different length (Weimerskirch 1991).

thus also slightly faster in South Georgia males, which become whiter than Crozet males of the same age. After about 20 years though, plumage maturation is almost complete; the Crozet males have lost most of their pigment and have caught up with the South Georgia birds. This mechanism is believed to be sufficient to account for observed differences in plumage at the two locations (Prince *et al.* 1997) (p.134).

WRAP-AROUND MOULT[2]

In Laysan, Black-footed and Galápagos Albatrosses, the outer primaries (usually p8–10, but sometimes p7–10 or p6–10) are moulted outwards in a single series every year, best seen in the first moult of juvenile primaries (Fig.13.7). The starting feather is not fixed, but activated so that at least the outermost are replaced (Harris 1973, Langston & Rohwer 1995).

The corresponding inner series is moulted inwards from p7, p6 or p5 as seen in older birds (Fig. 13.8). It is usually incomplete, but after a pause for breeding, replacement continues the following year at the next feather (Harris 1973). Wear of the inner primaries is uneven and the most abraded feathers are replaced first (less worn feathers are passed over). Whenever the innermost primary has been replaced, there is an immediate wrap-around to p5/6/7 and the series starts again (unless it corresponds with a pause for breeding). This is an efficient way of conserving feathers until such time as they really need replacing, then replacing them more rapidly than if all primaries remained in one series (Fig. 13.9). In any one year, 20% of Laysan Albatrosses and 29% of Black-footed Albatrosses moult all ten primaries, but most take two years to complete their primary moult.

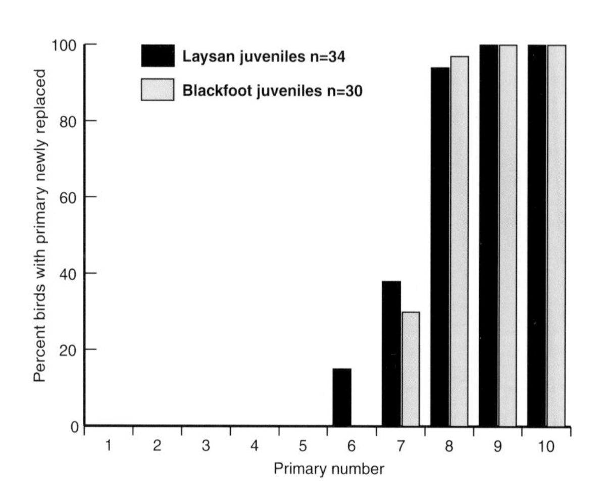

Fig. 13.7 The first primary moult in juvenile Laysan and Black-footed Albatrosses at sea, February to November 1990 and 1991 (Langston & Rohwer 1995).

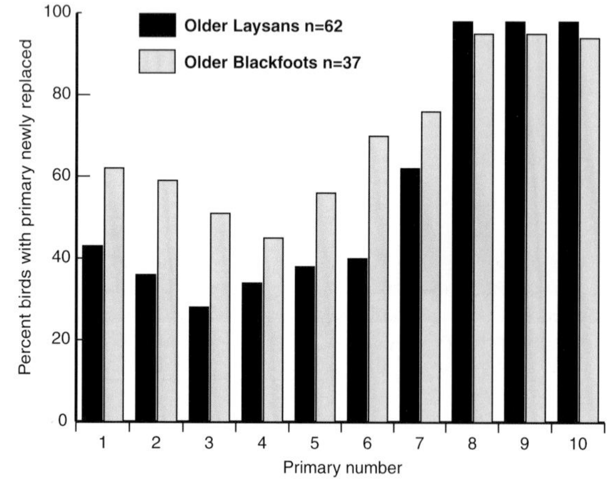

Fig. 13.8 Primary moult in subadult and adult Laysan and Black-footed Albatrosses at sea, February to November 1990 and 1991 (Langston & Rohwer 1995).

[2] A word-processing analogy — at the end of a line of text, the cursor automatically returns to the beginning of the next line.

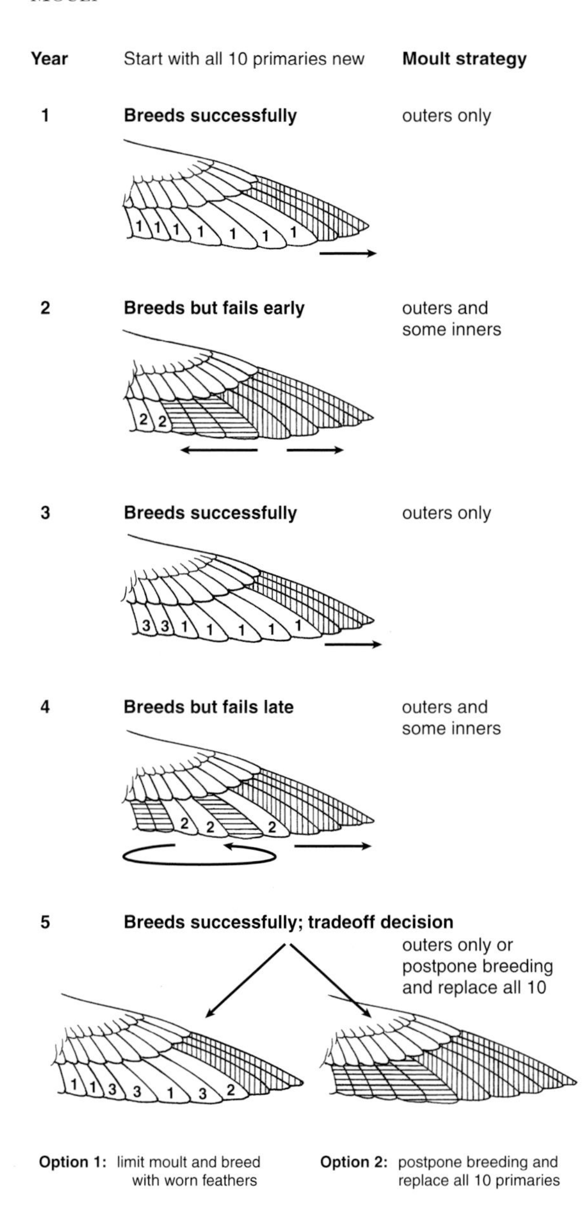

Fig. 13.9 Five years in the outer wing of an imaginary Laysan Albatross. Shaded primaries are those that have just been moulted after the breeding season. The numbers on unshaded primaries represent the years since the feather way last moulted. In year 1 the bird breeds successfully and has time to replace only the outermost three primaries. The next year it loses its egg early in the season and has time to replace eight primaries. In year 3 it is successful again and replaces the outer primaries, but old feathers begin to accumulate. In year 4 it breeds again, but loses its chick late in the season, so it replaces feathers in three locations, including the oldest, inner, and the outer primaries. It is successful again in year 5, but now has two moult options: to replace all ten primaries and not breed or to breed and allow several old primaries to accumulate (Langston & Rohwer 1996).

Curiously, the innermost primary that triggers wrap-around is not p1 but p2. Moult of p1 appears to have become linked to the adjacent secondary s1. They are moulted together more often than predicted by chance. Inner primaries are smaller than outer primaries and their wear resembles that of secondaries. In both species, moult of inner primaries has become dissociated from the outer primaries and they appear to become functionally more like secondaries.

Inward moult of secondaries among older gooneys with six or more actively growing feathers occurs in short series from several foci. On average, Laysan Albatrosses moult from 4.3±0.5 foci (n=20) and Black-footed Albatrosses from 5.6±0.5 foci (n=14) (Langston & Rohwer 1995).

Laysan Albatrosses, which are nominally annual breeders, take about 260 days to complete breeding followed by about 110 days at sea. Measurements of growth bands indicate that it would take 131 days to simultaneously replace three primaries. There is considerable variance, but few if any birds are able to keep squeezing breeding and moulting into 365 days indefinitely. Each year, about 25% of adult Laysans on Midway Atoll defer breeding (Rice & Kenyon 1962b, Fisher 1976).

Black-footed Albatrosses complete their breeding in about 245 days and have longer at sea. They replace more primaries than Laysan Albatrosses and possibly defer breeding less often.

TAIL

In both Black-browed and Grey-headed Albatrosses, tail feathers are moulted each year from the outer towards the inner pair. Juveniles probably start in their second winter at sea. Grey-headed Albatrosses moult only about half their tail feathers in the winter immediately following the departure of a fledgling in May, but in the following year all feathers are replaced (Prince *et al.* 1993).

Laysan and Black-footed Albatrosses appear to moult their tail feathers every year (Langston & Rohwer 1995). Some individual Galápagos Albatrosses still have old feathers in their tails when they return to the breeding ground (Harris 1973) and 10% of Sooty Albatrosses were still completing moult of tail feathers when they arrived at the Prince Edward Islands (Berruti 1977).

BODY

The contour feathers of sooty albatrosses arriving at the Prince Edward Islands for the beginning of the breeding season are mostly new with only a few abraded (Berruti 1977).

Galápagos Albatrosses on the breeding ground have old contour feathers among their plumage and some birds begin body moult late in the breeding season (Harris 1973). Many specimens collected in June were worn while those obtained later were very worn, with many replacing the feathers on the breast and abdomen. One taken on 28 June had apparently just finished moulting (Loomis 1918).

Shed body feathers accumulate in Black-browed Albatross colonies at South Georgia from late January. At this time, adults are foraging at sea and spending only minutes feeding their chicks, so the feathers are probably from failed breeders or subadults that remain ashore for longer periods (Tickell 1969, Tickell & Pinder 1975, Prince *et al.* 1993).

Eleven Black-footed Albatrosses captured 5–11 July 1897 in the North Pacific, about a month after leaving the breeding grounds, had belly feathers with buff edges and mottled upper surfaces due to the intermingling of feathers of different ages and different degrees of bleaching. One specimen was nearly white on the belly (Stone 1900).

BILL

Laysan Albatrosses in captivity have been seen to moult the outer edges of their bills. Ridging, perhaps due to replacement cycles, is apparent at the bases of culmens in several species.

SPECULATIONS

Although the ten large feathers inserted into the skeletal hand of albatrosses are the expected number characteristic of most seabirds, they are not a surface for delivering prolonged flying power by flapping. In gliding flight they form the end of an elongated wing that, compared to other birds, is a fairly rigid aerofoil. Albatrosses are fast gliders and the inner feathers of the hand appear to have become functional secondaries. The outer three, four or five feathers form the pointed end of the wing responsible for its trailing vortex. An aerodynamically clean, pointed wing-tip may be expected to generate less induced drag from that source than ragged feathers; which may account, at least in part, for priority in replacing them. Functionally then, small albatrosses may be imagined as having only five primaries and about 35 secondaries.

Several authors have speculated on other general principles unique to albatross feather replacement, but all have suffered from the heterogeneity of data. Moult is a common energy deficit and there are geographic variables. These have been imagined in terms of prey availability and vague notions that polar and subpolar climates are more demanding than those of the subtropical and tropical oceans. Aerodynamic implications are occasionally mentioned in passing, but wind parameters at different latitudes have never been incorporated in energy budgets.

Changes in head and neck colour occur in Grey-headed Albatrosses during the early juvenile and subadult years (Tickell 1969). This is brought about by moulting of contour feathers and Prince *et al.* (1993) have suggested that the energetic costs of these colour changes are sufficient to slow the tempo of primary replacement, compared with species such as the Black-browed Albatross that do not undergo such changes.

Langston and Hillgarth (1995) have demonstrated

		Number of feathers replaced after	
		success	**fail**
WA	South Georgia	8	5
WA	Iles Crozet	8	7
GHA	South Georgia	11	5
BBA	South Georgia	4	5
WYNA	Gough Island	5	6
GA	Islas Galápagos	8	10

Table 13.2 Comparison of primary feather replacement in albatrosses after successful and failed breeding the previous season. Figures rounded to whole feathers. Species abbreviations as in Appendix 1 (Prince *et al.* 1993).

that adult female Laysan Albatrosses parasitised with the oesophageal nematode *Seuratia* replace fewer primaries than uninfected birds. Similar correlations could probably be made with other parasites and pathogens. Albatrosses need to be healthy in order to breed successfully and complete their moult efficiently.

The two models reviewed above are exceptional in that each is based upon simultaneous samples from a pair of sibling species examined by common methodologies. Unfortunately, comparisons between these two studies cannot reflect their individual strengths, simply because the samples were taken at opposite poles of the reproductive cycles.

Table 13.2 compares the number of primaries replaced by species that have been examined during breeding (Weimerskirch 1991, Prince *et al.* 1993).

PRACTICAL APPLICATIONS

Moult scores can be used to determine the age of subadult mollymawks in the hand on the breeding ground. This has been worked out in detail for Black-browed and Grey-headed Albatrosses aged from two to five years at South Georgia (Appendix 10). By five years of age, some have acquired plumages indistinguishable from those of older birds (Prince & Rodwell 1994). The same may be achieved for subadult Wandering Albatrosses at South Georgia between the ages of three and nine years, providing the sex of individual birds is known (Prince *et al.* 1997) (Appendix 11).

14
FLIGHT

'Whilst we were heavily labouring, it was curious to see how the Albatross with its widely expanded wings, glided right up the wind.'

Charles Darwin on board HMS *Beagle* in a storm near Cape Horn, 13 January 1833[1]

Speculations about albatross flight began many years before anything was known about their natural history. Long sea voyages became more frequent in the 19th century and the spectacle of soaring albatrosses was a welcome diversion for many passengers bound for Australia and New Zealand, or rounding the Horn on the clipper route to California. The rigid wings were remembered afterwards and flight by sailing (*vol-a-voile*) was an apt description in the days of sailing ships.

Early attempts to build flying machines stimulated a widespread interest in soaring birds which persisted into the 1920s, when aircraft designers were still looking to natural fliers for ideas. Albatross aerodynamics reflect strategies for distance and endurance flying whereby energy is extracted from winds close to the ocean surface. Many aeronautical scientists and biologists have simulated albatross flight, but few have measured the birds' performance at sea and all depend upon untested models of the winds in which they fly.

For a long time, studies of albatross flight were concerned solely with how the birds use the wind and waves to stay aloft without flapping. These same sources of energy have another role; flying is about getting from one place to another and albatrosses are amazing travellers. It is only since the advent of satellite tracking that this subject has been approached and it has yet to achieve the mathematical sophistication that albatross aerodynamics has attained.

WAVES AND WIND

Albatrosses fly within a shallow layer of moving air immediately above an undulating surface. The movements of both air and water are part of the same dynamic boundary system (Roll 1965).

Sea waves are generated largely by wind action, often originating around storm centres and maintained by prevailing winds over large areas of ocean. Waves that are no longer being maintained by wind are called swell; they have great inertia and may travel for thousands of kilometres, gradually getting shallower. Waves or swell may be clearly defined as coming from one direction; or complex, combining waves and/or swell of different dimensions coming from different directions.

The dimensions of waves are determined by the properties of the wind field; its direction, speed, duration, displacement and the length over which it blows (fetch). The most characteristic feature of sea waves is their irregularity; maximum wave height (between trough and crest) is likely to be much higher than the average height[2]. The angle of up-wind slopes of sea waves increases linearly with wind speed up to 50 km per hour[3]. The down-wind slopes are generally steeper. Slopes cannot increase indefinitely and higher winds lead to a disintegration of the wave tops; in storms and hurricanes a transition zone swept by spray and foam is created between the air and sea (Roll 1965). Irregularities are less pronounced in swell.

[1] Darwin C.R. (N.Barlow Ed.), 1933.

[2] Significant Wave Height (SWH) is the average of the highest third of waves during an interval of 20 minutes.Waves of twice SWH would occur in the same interval and higher waves at much longer intervals.

[3] Ocean surface wind speed measurements are standardised as Neutral Stability Winds corresponding to a height of 19.5 m above the surface. Albatrosses fly below this height.

Vaughan Cornish (1934), who collected sea wave reports over 50 years, considered storm waves of almost 14 m to be quite common and higher ones well authenticated. One of the most impressive, measured accurately by officers of the *Rampo* in a Pacific hurricane, was 34 m high with a wavelength of 419 m and travelling at 102 km per hour (Bascom 1959). Throughout the Southern Ocean, swell of more than three metres is usual and areas with waves of 10–11 m or more occur in one place or another every few days, with swell almost as high. Mean monthly wind speeds of latitudes 40°–60°S are 20–45 km per hour, with frequent gales and storms.

In contrast, throughout the comparatively small triangle of ocean traversed by albatrosses between the Islas Galápagos and South America (Fig. 9.1), winds and waves are less pronounced. In the three years 1985–88, 1,190 satellite measurements yielded a mean wind speed of about 15 km per hr (0–40 km per hr) with a mean waves of 1.8 m (0.5–3.0 m).

HISTORY

Captain Deloitte's view that an albatross was always 'close hauled' when it flew 'in the wind's eye' was typical of a seaman's eye for detail in a natural event (Bennett 1860), but another seaman went much further. Captain Jean-Marie LeBris shot an albatross, removed a wing and held it in the breeze. The lift it generated so impressed him that he believed he had '...comprehended the whole mystery of flight.' and went home to Brittany to build a replica out of wood and Canton flannel. The aircraft was called the *Albatross*, weighed 42 kg, had a wingspan of 14 m and a wing area of 20 m^2. One Sunday in 1856, it was mounted on a horse-drawn cart and driven downhill into the wind at a gallop. It was reported that it took off not only with LeBris at the controls, but with the carter dangling below on a rope! (Stillson 1955)[4].

Theoretical interpretation of albatross flight was initiated by F.W. Hutton (1868) who proposed that the force imparted to the bird at take-off was sustained because of the low resistance offered by the albatross profile. The equation he formulated was based on ballistics. It disturbed at least one reader who dryly wrote that '...in his mathematical treatment Captain Hutton has not been happy...' and pointed to the inclined surfaces of the body as the sites of the force sustaining the bird against gravity (Webb 1869).

The 8th Duke of Argyll[5] was an influential figure in the Victorian world of science and his *Reign of Law* (1867), a response to Darwin's *Origin of Species*, ran into many editions. It contained a discussion of albatross flight using the performance of a kite as an analogue to illustrate the significance of the bird's mass. A later edition stimulated an anonymous response from a writer known only as 'R.A.' (1876) who proposed a mechanism for albatross flight that anticipated elements of modern theory. He was obviously a seafarer who had seen albatrosses and, like Hutton, believed the birds preserved take-off inertia which he called 'vis-viva'. Once airborne, though, he imagined them descending with the wind and climbing into it, in repeated conversions of 'actual' and potential energy. He foresaw that this energy would be dissipated and proposed that it was periodically restored by flapping. It was I. Lancaster (1886) who later identified gravity as the motive power for soaring flight.

In 1873, W.H. Dall (1874) noticed that Black-footed Albatrosses climbed into the wind and descended with it.

> 'It rises only against the wind, except in rare cases, when its descending momentum is sufficient to raise it slightly for a short distance, or when the reflex eddy from a high [wave] surge is strong enough to give it a slight lift.'

A significant contribution to our understanding of albatross flight was made in 1878 by William Froude. In that year, he was 69 years of age and approaching the end of a distinguished career as a naval architect, but he was still attending the sea-trials of warships and taking an active interest in the seabirds he encountered. At the 1876 meeting of the British Association for the Advancement of Science in Glasgow, there was some discussion about waves. Froude evidently said something about albatrosses and Sir William Thomson asked him to write some notes for the Mathematics and Physics Section over which he presided. Froude complied in a letter to Sir William

[4] Orville Wright doubted that this flight actually occurred!
[5] George Douglas Campbell.

written from Cape Town on 5 February 1878. He described how up-currrents could be generated on the windward and leeward slopes of waves depending upon the relative speeds and direction of waves and wind. From observations and measurements on the flight of wandering albatrosses and mollymawks at sea, he calculated the up-current needed for an albatross to fly on rigid wings and the following year discovered that soaring could occur in the absence of wind when swell slopes were moving fast enough to generate their own up-currents[6].

Froude died later during this same voyage and his 1878 letter was not published until ten years later (Froude 1888). Before he died, he was putting together a mathematical treatment of his ideas which was later edited and published by his son, R.E. Froude (1891).

Froude was meticulous in following up apparent exceptions to his hypothesis. His curiosity was aroused when he saw albatrosses flying vigorously in gale force winds where the surface of the sea was flattened and wave-slope soaring seemed impossible. He came close to, but did not make the conceptual step that provided the second theoretical mechanism of albatross flight, namely wind-shear or wind gradient soaring[7].

Lord Rayleigh (1883) and H. Airy (1883) both independently proposed wind gradient mechanisms to account for a published description of soaring birds, but it is not usually recognised that the birds referred to were pelicans, storks and vultures soaring over hills in Assam (Peal 1880) or that Rayleigh was a little sceptical:

> 'A priori, I should not have supposed the variation of velocity with height to be adequate for the purpose, but if the facts are correct, some explanation is badly wanted'

Airy did in fact briefly mention albatrosses, but in terms of slope-soaring on up-currents generated by ships and waves. It was not until after the publication of Froude's letter that A.C. Baines (1889) of New Zealand, drawing upon observations of albatrosses soaring over almost flat sea, independently concluded that albatrosses made use of

> '...the well known fact that the velocity of wind at the surface is diminished by friction so that the velocity increases with height, the rate of increase being greater near the surface.'

His contribution was later acknowledged by Lord Rayleigh (1889, 1900).

At the same time that the aerodynamics of albatross flight were attracting attention, the search for mechanisms by which albatrosses held their wings steady, had begun. In the 1870s and 1880s, James Hector (1894) dissected several albatross wings, but they were poorly preserved specimens that revealed nothing conclusive. Eventually though, a fresh specimen enabled him to demonstrate a tendon and sesamoid bone at the elbow. He believed it to be a locking device which held the wing rigid, but an alternative mechanism was later discovered (Joudine 1955, Yudin 1957) as well as sesamoid bones supporting the patagium. However, none of these was unique to albatrosses (p.28). A century later, Colin Pennycuick (1982b) discovered a shoulder lock in albatrosses and giant petrels. The humerus is normally free to be lifted above the horizontal when flapping, but pulled forward, it locks at the glenoid and cannot be raised. This, Pennycuick suggested, relieved muscles and tendons of the constant strain of suspending the body from the shoulder for long periods.

The first two decades of the 20th century, beginning with powered flight by the Wright brothers and ending with the first World War, saw rapid developments in aviation. Interest in bird flight continued and several authors turned their attention to soaring (LePage 1921).

Pierre Idrac was one of several French aeronautical scientists whose interests included birds. Government support for seabird research reflected rapid developments in aviation when it was believed that knowledge about soaring in albatrosses might benefit aircraft design. For a time, Idrac studied thermal soaring in vultures, and during voyages around Africa became interested in the different problems of soaring over the sea. From a lighthouse off Brittany he measured the characteristics of winds up to 40 m above the surface, accompanied by experiments on models in the laboratory (Idrac 1923). Together, these left him in favour of Froude's wave-slope up-currents as the means of albatross flight.

[6] This phenomenon was independently confirmed by C.J. Pennycuick (1982b).

[7] In this account I have avoided the term dynamic soaring. Pennycuick (1975) advised that the distinction between 'static' and 'dynamic' is an unsatisfactory basis for the classification of soaring, and in aerodynamics gust (pulse or turbulence) effects have also been described as dynamic.

In 1923, he went to South Georgia with an Argentine whaling company to measure the performance of albatrosses from whale-catchers at sea. The research was carefully planned; he positioned various instruments about the ship to record wind speed at different heights and floating smoke generators were released on the sea to obtain the direction of wind at the surface. Wandering and Black-browed Albatrosses were filmed with a clockwork driven ciné camera, then a technological advance over traditional hand-cranked cameras. Throughout filming, the attitude of the rolling and pitching vessel was recorded and albatross flight paths were later reconstructed, frame by frame, from the film.

The theoretical model of wind gradient soaring which he published after this expedition (Idrac 1924a&b), owes little to these empirical data. The regularity of the albatross manoeuvres he measured so carefully, appeared, a priori to exclude the use of upcurrents. He ignored all his earlier work on up-currents in favour of a single observation made during the voyage home. The ship in which he was sailing anchored in the Magellan Strait. Black-browed Albatrosses were soaring nearby over a fairly flat (0.25 m) sea and he noticed a decrease of wind speed with height on the ship. Within a year, though, French confidence was waning and alternative methods including wave-slope lift were again being discussed (Magnan 1925, Idrac 1925).

The wind gradient hypothesis led to many treatments from various aerodynamic viewpoints, all without benefit of new measurements at sea and most requiring above-average mathematical ability to comprehend (Walkden 1925, Scorer 1958, Cone 1964, Hendriks 1972, Wood 1973, Lighthill 1975, Sachs 1993). It was introduced to a wider public in William Jameson's (1958) book *The Wandering Albatross* and thus became the widely accepted and much quoted explanation of albatross flight.

In 1975, a former glider pilot, John Wilson initiated a return to the view that albatrosses used wave-slope up-currents. He created a computer simulation from which he concluded that enough energy was available in gliding descents along wave-slopes to convert into pullups without the need to invoke wind gradient effects.

Four years later, Pennycuick, another experienced flier, went to South Georgia and made a large number of measurements of albatrosses and petrels in flight over the sea. Many years before, he had built a chart-recording theodolite for measuring the flight paths of Fulmars seen from the tops of sea-cliffs. By 1979, the instrument had become an electronic 'ornithodolite' incorporating a microcomputer with data storage (Pennycuick 1982a). His measurements of albatrosses were made from a ship at sea and a headland on Bird Island. He considered the latter to be more accurate, but advised that the birds flight paths were influenced by the position and topography of adjacent land, thus introducing variables not anticipated in data collected on the open ocean (Pennycuick 1982b and pers. comm.).

Thomas Alerstam and his Swedish co-workers (1993) used tracking radar to plot the positions of albatrosses and other birds at sea, but radar signals are scattered by waves so this technique was effective only in very light winds. Since albatrosses fly most characteristically in much stronger winds, some of their measurements had to be made by optical range-finder, in the same manner as Pennycuick (1982a,b).

CHARACTERISTICS OF ALBATROSS FLIGHT

Albatrosses have narrow wings (Fig. 14.1), spanning two to three metres, with aspect ratios of 13–15 (Table 14.1). Those of great albatrosses are proportionally shorter at the ends (manus plus primaries) than those of mollymawks (Table 2.2). Narrow wings minimise drag; albatrosses have lift/drag ratios of 22–23 (compared to 10–15 in large raptors), but relatively high speeds have to be maintained. They are heavy birds, with wing loadings of 82–140 Newtons per m^2 (Table 14.1). Over the sea, lift coefficients average 0.9 –1.0 (Pennycuick 1982b). Two mollymawks at sea off South Africa were said to be not seriously inconvenienced by moult of the terminal three primaries which effectively shortened the wingspans by half a metre (Brooke 1981, Brooke & Furness 1982).

Fully extended albatross wings droop. The inner wings with their long rows of secondary feathers are usually held horizontal, but the outer wings (manus plus primaries) hang down conspicuously (see jacket cover). Robert Swinhoe (1863) noticed this in the Strait of Formosa:

> 'The Albatross on wing is never figured correctly. When flying, the wings are curved like the head of a pickax...'

Unlike the upswept (dihedral) wings of aircraft which are self-righting (counteracting sideslip by causing a roll in the opposite direction), horizontal or downswept wings (anhedral) increase the tendency to sideslip (Rüppell 1980). This may be counteracted by sweep-back and/or differential twisting of the wings (Pennycuick & Webbe 1959), but Pennycuick (1975) has also suggested that in normal gliding, extreme rolling stability would be an 'embarrassment', so this particular instability may enhance manoeuvrability. On the other hand, downward deflected (diffuser) wing tips are said to contribute to longitudinal and directional stability in tail-less aircraft at low speeds (Pennycuick 1972).

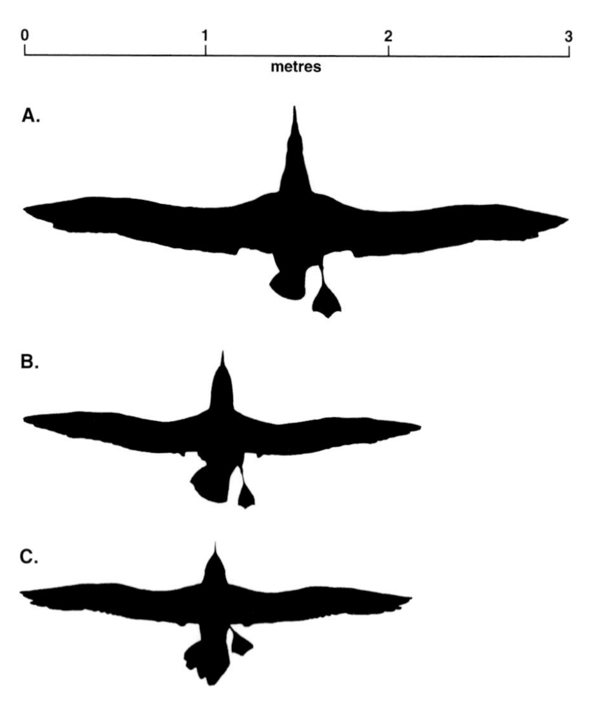

Fig. 14.1 Flight silhouettes drawn from fresh specimens of a Wandering Albatross (A), Black-browed Albatross (B) and Light-mantled Sooty Albatross (C). In each drawing, the left side of the tail is fanned with the right tarsus extended and the toes spread (after Pennycuick 1982b).

Between the very long humerus and radius of an albatross wing, the patagium extends more than half way along the leading edge and almost a third of the width at the elbow. During manoeuvres over land, when wings often assume high angles of attack, the leading edges of the patagia turn down to enclose hollow undersurfaces capable of generating high lift coefficients (Pennycuick 1982b).

In most albatrosses, the tail is short or very short and square. It can be twisted through large angles about the longitudinal axis and rotated up and down, providing additional minor control in yaw and pitch. At slow speeds, it may be fanned to provides extra lift, while the large feet are dropped as air-brakes. When manoeuvres are imminent, the feet are trailed under the tail, but in cruising flight they are tucked forward within the contour feathers (Pennycuick & Webbe 1959). The literature is

	Mass (kg)	(n)	Wing span (m)	Wing area (m²)	(n)	Wing loading N*/m²	Aspect ratio**
SRA	8.70	(19)	2.82	0.540	(8)	158	14.7
ATWA	6.64	(14)	2.71	0.471	(6)	138	15.6
CBBA	2.91	(16)	2.09	0.293	(6)	97	14.9
GHA	3.30	(21)	2.00	0.261	(6)	124	15.3
SBA	3.01	(31)	2.05	0.277	(17)	107	15.2
TSA	4.03	(36)	2.47	0.443	(1)	89	13.8
SA	2.50	(176)	2.03	0.304	(3)	81	13.5
LMSA	2.92	(28)	2.15	0.341	(3)	84	13.5
GA	2.92	(1)	2.21	0.345	(1)	82	14.2
LA	3.09	(367)	2.04	0.310	(19)	98	13.5
BFA	3.14	(306)	2.16	0.338	(11)	91	13.8

Table 14.1 Wing measurements of Southern Royal (SRA), Antipodes Wandering (ATWA), Campbell Black-browed (CBBA), Grey-headed (GHA), Southern Buller's (SBA), Tasmanian Shy (TSA), Sooty (SA), Light-mantled Sooty (LMSA), Galápagos (GA), Laysan (LA) and Black-footed Albatrosses (BFA) (Warham 1996, Spear & Ainley 1997a).
* Newton (N) = the force producing an acceleration of 1 m/s² on a mass of 1 kg.
** the aspect ratio of a wing is approximately span/chord (width).

	n	Percentage of flying time spent: Gliding (%)	Flap-gliding (%)	Flapping (%)
WA	76	93	7	0
BBA	256	82	18	0
GHA	143	94	6	0
LMSA	17	77	23	0

Table 14.2 Flight styles of albatrosses over the sea off South Georgia. Flap gliding occurred only in winds of <3 m per s (Pennycuick 1982b). Wandering Albatross (WA), Black-browed Albatross (BBA), Grey-headed Albatross (GHA), Light-mantled Sooty Albatross (LMSA).

full of anecdotes claiming that the two sooty albatrosses are more accomplished fliers than other species. Differences in some aerodynamic parameters are apparent (Pennycuick 1982b), but it is not at all clear whether the long cuneate tails contribute to performance in the air or are hydrodynamic adaptations to diving and swimming.

In spite of perennial eulogies to effortless flight, there are energetic costs and albatrosses do flap their wings (Rayner 1983), mollymawks more often than great albatrosses (Table 14.2). Pennycuick (1982b, 1987) has pointed out that although the comparatively small flight muscles (pectoralis major and supracoracoideus) attest to their minor role, they are red as expected of power muscles[8]. The lift coefficients corresponding to mean airspeeds by flapping and flap-gliding albatrosses are less than those attainable by gliding. Flap-gliding usually occurs in light winds, and in windy conditions a few flaps occasionally accompany manoeuvres. Nevertheless, albatrosses remain essentially heavy, fast gliders that extract energy from the wind.

Speed

Times for ocean passages can be obtained from ringing and homing data. Towards the end of 1956, Kenyon & Rice (1958) took 18 Laysan Albatrosses from their nests on Midway Atoll where they were incubating or brooding and transported them in US Navy aircraft to distant places in the surrounding Pacific Ocean, where they were released. Fourteen of them returned, the fastest crossed 5,149 km of ocean in ten days. It is reasonable enough to express this geographic movement as an average of 515 km per day, but it would be misleading to imagine that the bird flew at a constant speed of 21 km per hour or 6 m per second throughout the passage.

Series of locations fixed by satellite are now the most frequent source of albatross flight speeds. They are usually expressed as kilometres per hour, but the distances over which the birds are timed vary greatly. The orbits of satellites converge towards the geographic poles and are within range more frequently in higher latitudes. Furthermore, the accuracy of fixes vary and in different experiments micro-transmitters on birds have been set to signal at different intervals (Weimerskirch *et al.* 1992). The number of locations fixed each day has thus varied greatly. Walker *et al.* (1995b) attempted to compensate for these differences, and as satellite tracking becomes more readily available, further attention to such detail will be needed.

In the south Indian Ocean, over distances of 383–949 km, when locations were fixed at intervals of about two hours, five Wandering Albatrosses flew at an avearge speed of 55 km per hr. Over shorter distances, speeds of up to 81 km per hr were attained (Jouventin & Weimerskirch 1990). Five South Georgia Wanderers flying circuits of 1,810–7,470 km in the South Atlantic, with locations fixed at intervals of about one and a half hours, flew at an average speed of 29 km per hr, achieving a maximum of 88 km per hr between two locations (Prince *et al.* 1992). Six Auckland Wandering Albatrosses tracked over distances of 673–5,601 km in the Tasman Sea, with locations fixed at intervals of almost three hours, averaged 16 km per hr, with a maximum of 35 km per hr (Walker *et al.* 1995b). Two Northern Royal Albatrosses from New Zealand, foraging mostly over the shelf-edge east of South Island, and fixed every two to three hours over distances of 287–713 km, flew at an average maximum speed of 43 km per hr (Nicholls *et al.* 1994).

Reinke *et al.* (1998) tracked a single Auckland Wandering Albatross for 71 days over at least 33,000 km. They obtained 441 usable fixes at intervals averaging four hours. Fig. 14.2 illustrates the range of mean speeds over an average distance of 87 km when this bird was apparently not foraging.

Several smaller albatrosses have been tracked by satellite. Black-browed Albatrosses on migration from South

[8] myoglobin, the oxygen carrying protein of muscles, is related to the haemaglobin of red blood cells.

Georgia towards South Africa in 1992 travelled 4,213 km at an average of 22 km per hr in 1992 and 4,533 km at 48 km per hr in 1993. In that year a Grey-headed Albatross on winter migration through the Drake Passage into the South Pacific, travelled 794 km at an average of 46 km per hr, 2,977 km at 23 km per hr and 3,415 km at 39 km per hr (Prince *et al.* 1997). Three Southern Buller's Albatrosses commuting in daylight between the Snares and the central Tasman Sea averaged 21 km per hr over distances of about 900 km (Sagar & Weimerskirch 1996). A Light-mantled Sooty Albatross commuting directly between Macquarie Island and Antarctic waters to the southwest, traversed 1,781 km at 30 km per hr on the outward flight and 1,553 km at 32 km per hr on the return. Another returned over 1,459 km at and average 55 km per hr, including one section at 84 km per hr (Weimerskirch & Robertson 1994).

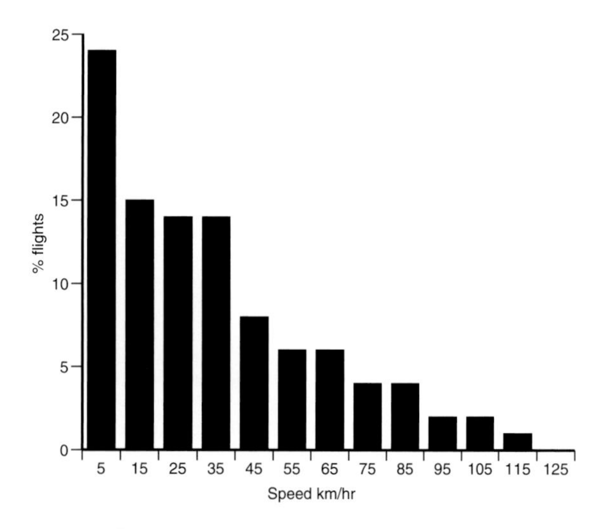

Fig. 14.2 Speeds of an Auckland Wandering Albatross on passage; computed from 303 timed flights over an average distance of 75 km between consecutive satellite fixes. Speeds are the mid-points of 10 km/hr class intervals (after Reinke *et al.* 1998).

Speed is usually imagined in relation to the ground over which a vehicle or animal is moving (groundspeed)[9], but the speed of an aircraft or bird is also measured in relation to the surrounding air (airspeed) which itself may be in motion. The maintenance of adequate airspeed is a primary condition of flight. In aircraft, airspeed is measured by on-board instruments, but no-one has yet measured airspeed on an albatross. Groundspeeds have been calculated by time/distance observations of flying birds and values for airspeeds derived by adding or subtracting the wind vectors[10]. For example, if an albatross is passing a small island at 10 m per s on a calm day its groundspeed and airspeed will be the same, but if it is flying into a wind of 5 m per s its airspeed will be 15 m per s. If the wind is coming from behind the bird, its airspeed will be only five metres per second. A wind blowing exactly from one side may cause the bird to drift sideways, but it's groundspeed and airspeed will remain the same. In practice, wind speed in albatross studies is measured at the site of observation and computed for the location of the flying bird. The mean groundspeed of Wandering Albatrosses tracked in this manner by ornithodolite just off South Georgia and mostly flying west into prevailing winds was 11 m per s, slightly faster than those of Black-browed and Grey-headed Albatrosses (10 m per s). The corresponding airspeeds were 15 m per s and 13 m per s respectively (Pennycuick 1982b) (Table 14.3).

	Mean airspeed (m/s)	**Mean groundspeed (m/s)**
	'Ornithodolite' from land in variable winds	
WA	15	11
BBA	13	10
GHA	13	10
	Optical rangefinder from ship in variable winds	
BBA	13	13
WYNA	11	13
	Radar from ship in near-calm	
BBA	13	13
GHA	13	13
WYNA	12	11

Table 14.3 Flight speeds over the sea of Wandering (WA), Black-browed (BBA), Grey-headed (GHA) and Western Yellow-nosed (YNA) Albatrosses (after Pennycuick 1982b, Alerstam *et al.* 1993).

Soaring albatrosses are constantly changing direction on a time scale of seconds, flying downwind, across wind and upwind. Their flight paths have fractal properties, revealing zigzags within zigzags the closer they are examined (Mandelbrot 1982). Wandering,

[9] The same term is used over the sea.
[10] Over long distances, speeds of albatrosses are most widely given as km per hr. Over much shorter distances and in aerodynamic calculations, metres per second (m per s) is the usual rate.

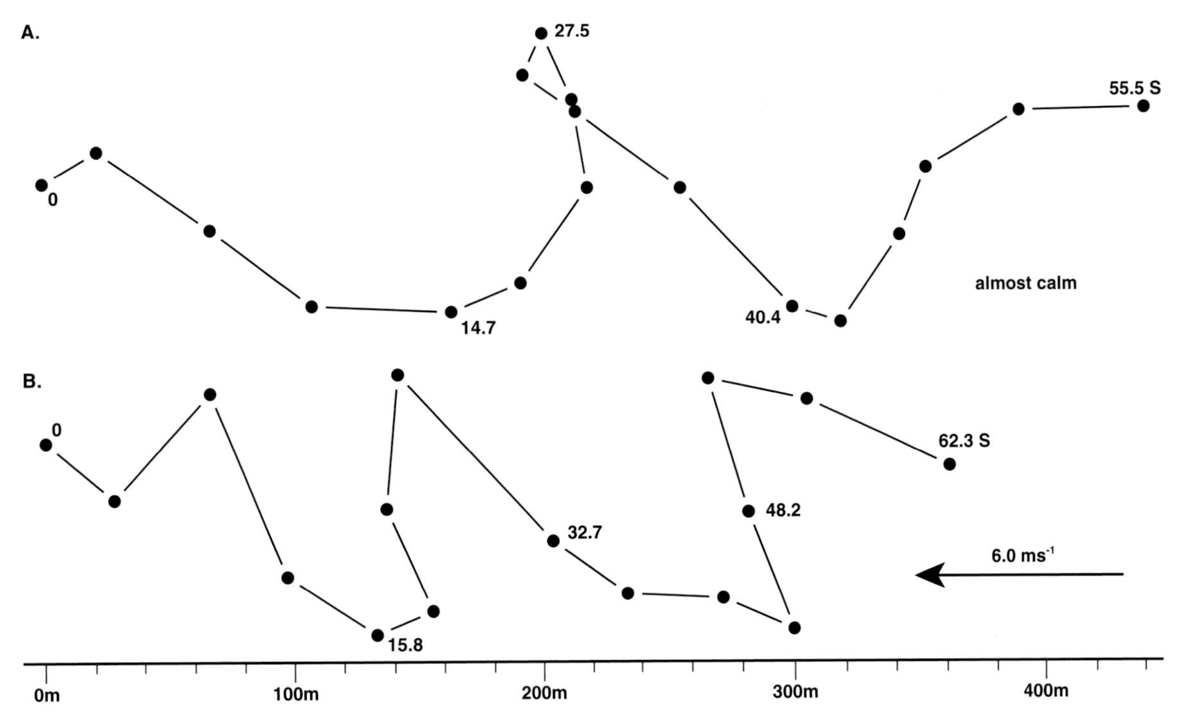

Fig. 14.3 The flight paths of two Wandering Albatrosses tracked from a ship at sea. Both birds were flying from left to right and the interval between fixes averaged 3–4 seconds. One bird (A) was slope-soaring in almost no wind along the leading edge of swell with an amplitude of about 3 m and wavelength of about 100 m. Without flapping its wings it achieved a straight-line speed of about 8 m/s. The other bird (B) was soaring into a wind of 6 m/s making cross-wind beats along the windward slopes of waves (Pennycuick 1982b).

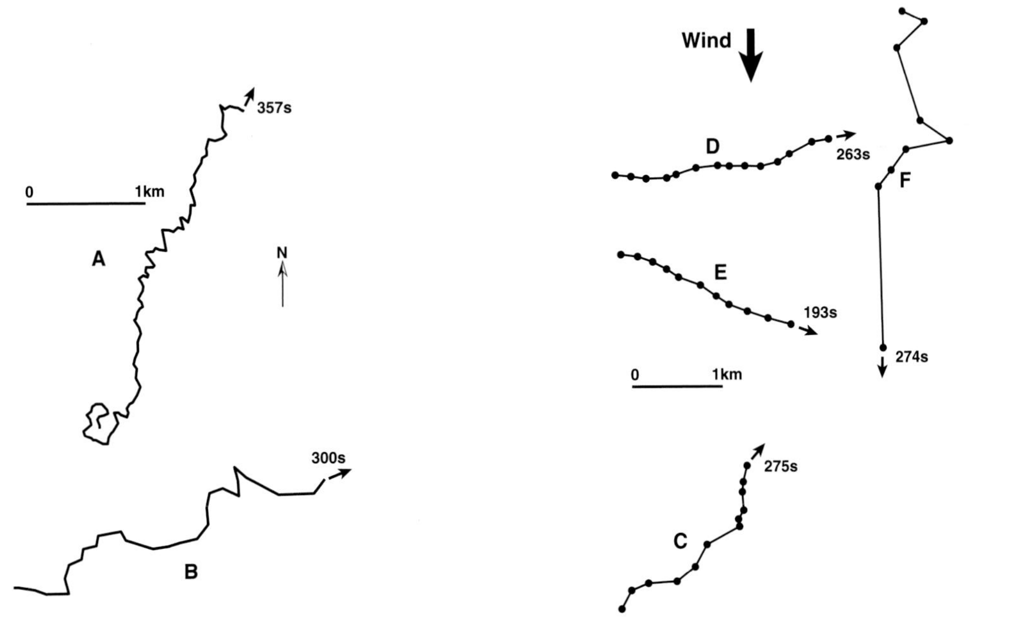

Fig. 14.4a The flight paths of two Grey-headed Albatrosses tracked by radar over open sea in near calm conditions. Positions were fixed at regular intervals of 3 seconds (A) and 10 seconds (B). The overall time is given at the end of each track (Alerstam *et al* 1993).

Fig. 14.4b The flight paths of four Black-browed Albatrosses tracked by optical range-finder in windy conditions. Positions were fixed at irregular intervals averaging 19-34 seconds. Windspeeds were 9 m/s (C), 12 m/s (D,E) and 19 m /s (F) (Alerstam *et al* 1993). The overall time is given at the end of each track.

Black-browed and Grey-headed Albatrosses tracked visually at sea by ornithodolite, over distances exceeding 700 m (Pennycuick 1982b), flew 1.47–1.54 km for every kilometre they travelled in a straight line (Fig 14.3). Several mollymawks measured by radar in near calm conditions flew 1.19–1.28 times the straight-line distance while others measured by optical range-finder, in winds of 9–19 m per s, were closer to a straight line (1.04–1.09) (Alerstam *et al.* 1993). The frequency of the radar and optical fixes from which these figures were calculated varied considerably. In Fig. 14.4a, track (A) of a Grey-headed Albatross fixed by radar at intervals of three seconds is close to the actual flight path; by increasing the interval to ten seconds (B) some of the zigzags are lost and the bird appears to fly more directly. The tracks of Black-browed Albatrosses (Fig. 14.4b) fixed optically at much longer and irregular intervals (C,D,E,F) lack detail of the actual flight path and underestimate the distances flown.

Lt D. Washer RN has described an Auckland Wandering Albatross keeping company with his helicopter at speeds of up to 110 km per hr (31 m per s) and Southern Buller's Albatrosses reached a maximum of 73 km per hr (20 m per s) (Robertson & Jenkins 1981).

Gliding

Gliding is a type of flight where no power other than that due to gravity is imparted to the bird. In still air the bird invariably follows a downward path (Fig. 14.5) and the rate at which it loses height is called the 'rate of sink' (V_s).

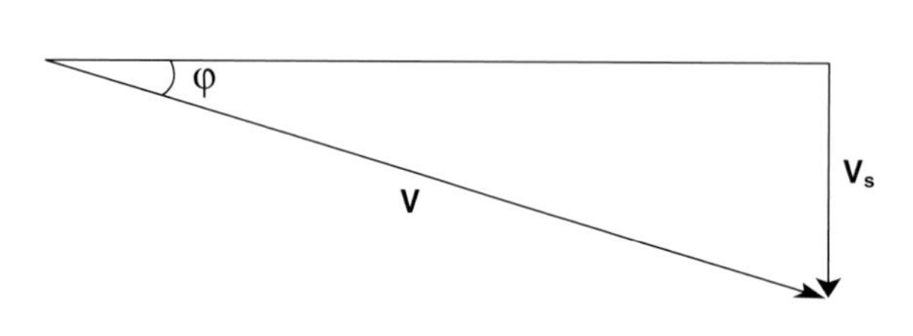

Fig. 14.5 A bird gliding at speed (V) in still air descends at an angle (φ) to the horizontal and has a sinking speed (V_S) (Pennycuick 1972).

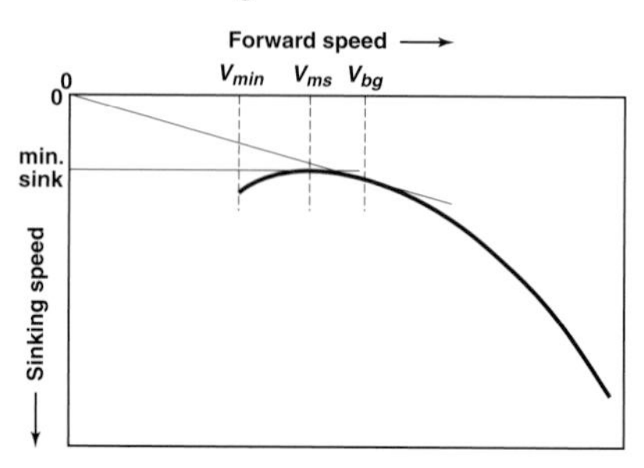

Fig. 14.6 In the glide polar, minimum gliding speed (V_{min}) is determined by the wing area and maximum lift coefficient, minimum sinking speed (V_{ms}) is achieved at a faster forward speed. The speed of best glide (V_{bg}) is the faster speed at which the bird travels the greatest distance for a given loss in height. It corresponds to the point where a tangent from the origin meets the curve (Pennycuick 1972).

The relationship between the speed of the glider (bird or aircraft) and its rate of sink is conventionally described by the glide polar (Pennycuick 1972) (Fig. 14.6). On the curve the 'minimum speed' (V_{min}) will just allow a gliding bird to keep flying, if it flies any slower it will stall. The faster speed of 'minimum sink' (V_{ms}) keeps the bird aloft longer, while the even faster speed of 'best glide' (V_{bg}) covers a greater distance at the cost of losing height more rapidly.

Nobody has measured glide polars for albatrosses, but Pennycuick (1982a) has made estimates of gliding performance (Table 14.4). These indicate that although their absolute speeds are high, albatrosses fly slowly in relation to their minimum sinking speeds, which is consistent with slope-soaring along waves.

	V_{ms} (m/s)	Min. sink (m/s)	V_{bg} (m/s)	Max. glide ratio
WA	11.9	0.59	15.7	23.2
BBA	10.7	0.57	14.0	21.7
GHA	10.6	0.56	14.0	22.0
LMSA	9.3	0.48	12.2	22.4

Table 14.4 Estimates of gliding parameters of albatrosses: the airspeed of minimum sink (V_{ms}), the rate of minimum sink and the best glide, the airspeed (V_{bg}) at which the longest distance is travelled (Pennycuick 1982b).

Wave-slope soaring

The long gliding flight of albatrosses has been a source of wonder and curiosity among seafarers for 200 years or more. Even the largest species fly so close to the surface that they disappear from sight in the troughs between the waves and it is often said that their wing-tips skim the water as they bank and turn. The observed mean heights of

Wandering and Black-browed Albatrosses gliding above the sea off South Georgia was eight metres and four metres, with pullups to 13 m and eight metres respectively (Pennycuick 1982b)[11].

Albatrosses soar along waves on winds that are rising up the slopes. Gales and huge waves are not essential; off Peru, Galápagos Albatrosses usually have only moderate winds and comparatively shallow waves (p.174), nevertheless they

> 'soar for considerable distances, keeping just above the water and rising and falling with the waves.'
>
> R.H. Coker (1919)

C.J. Wood (pers. comm.) calculated that the lowest up-current upon which wandering albatrosses can maintain level flight is about 0.6 m per s; this is close to Pennycuick's (1982b) computed minimum sinking speed. Even without surface-winds, travelling swell can generate up-currents of about one metre per second (Froude 1888).

When a gliding (descending) bird is in air rising at the same rate as its sinking speed, level flight can be maintained. If the vertical vector of the up-current is greater than a bird's sinking speed, then it can climb above the wave. Idrac (1923) measured up-currents of more than 2 m per s at eight metres above the crests of waves and they were still discernible at 15 m. Albatrosses thus have ample means to carry them over a wave crest into the next trough or to prolong a glide (Wilson 1975).

Waves and swell are also travelling – perpendicular to the bird's line of flight – so in the time it is soaring along a wave-slope, an albatross will also be drifted sideways (C.J. Wood pers. comm.). If, for example, the waves are travelling at 30 km per hr and the bird spends two thirds of its soaring circuit gliding along their slopes, it would make 20 km in an hour and 480 km in a day. Extensive wave fields that could carry birds such distances are common enough.

Pullups

In a fresh breeze, albatrosses flying close to the sea regularly climb steeply (pullup) above the waves then glide back down towards the surface. The energy (E) of a gliding albatross is partitioned such that some of it (potential) is due to its height above the water and the rest (kinetic) due to its speed of flight. Ignoring losses due to drag which are relatively small:

$$E = mgh + {}^{1}/{}_{2}mV^2$$

where m is the mass of the bird, g is the force due to gravity, h is the height above the sea and V is the speed of the bird. A gliding albatross makes tactical use of the wind and waves to convert speed into height and vice-versa by a series of climbs and descents in different directions.

John Wilson (1975) calculated that up-currents of air on wave-slopes are adequate to accelerate albatrosses in level flight to speeds sufficient to power the highest observed pullups. Acceleration towards equilibrium speed takes infinite time, so albatrosses are unlikely to achieve maximum possible speed and probably always operate at lower speeds (J. Wilson pers. comm.). The bird adjusts its glide across or down wind, increasing ground speed to V_1 close to the water (h_1); it then turns into the wind, which immediately increases its airspeed and enables it to climb steeply. In doing so the albatross slows down to some minimum speed V_2 at a height h_2 when excess kinetic energy will have been converted to potential energy:

$$h_2 - h_1 = \frac{(V_1^2 - V_2^2)}{2g}$$

The acceleration phase of the circuit is long relative to the climb, but an albatross slowing down, for example from 20 m per s to 10 m per s, could pull up 15 m (Fig. 14.7).

Energy for pullups may also be available from the wind gradient. When air flows over the sea, friction between

[11] On the night of 19 May 1982, a helicopter ferrying troops between two British warships north of the Falkland Islands crashed into the sea with the loss of 20 soldiers. Bloodstained feathers found afterwards on the sea, were said to have been those of an albatross or giant petrel that caused the accident (Geraghty 1992). Albatrosses do fly at night, but much less often than during the day (Weimerskirch & Wilson 1992), and at 300–500 feet (90–150 m), the aircraft was flying much higher than either of these birds over the sea.

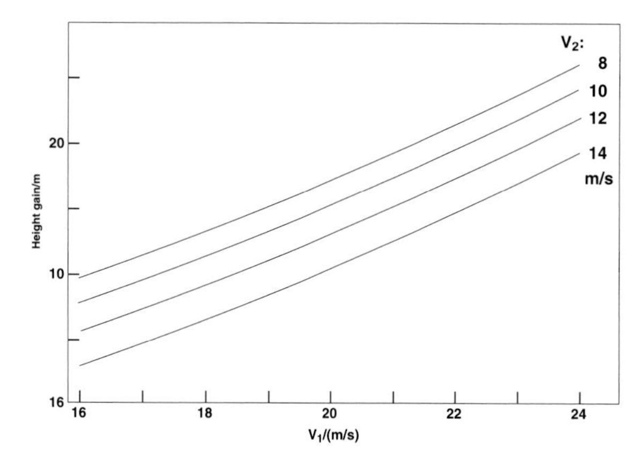

Fig. 14.7 Heights attainable by pulling up from an initial speed (V_1) just above the wave-slope to the slower speed (V_2) at the top of a climb; assuming no aerodynamic losses due to drag (Pennycuick 1982b).

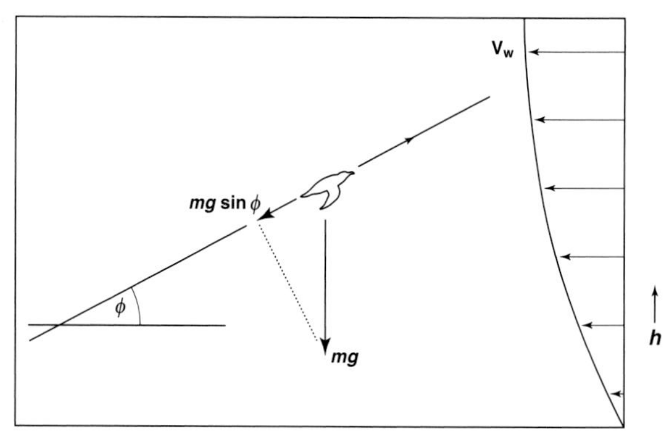

Fig. 14.8 The upward climb of an albatross into an imaginary wind-gradient (V_W), assuming no aerodynamic losses due to drag. Loss of speed proportional to *mg* sin φ may be balanced if increase in speed due to the wind-gradient approximates to wV_a sin φ, where *m* is the mass of the bird, *g* is the force due to gravity and V_a is the air speed (Pennycuick 1982b).

the surface and the moving air slows down the layer in contact with the water. That layer in turn slows down the layer immediately above it and so on. The effect diminishes with height above the surface, as the frictional force is progressively dissipated, until the velocities in adjacent layers are constant. The wind gradient (w) is defined as the rate of change (per second) of wind speed with height:

$$w = \mathrm{d}V_w / \mathrm{d}h$$

where V_w is the wind speed at height h above the sea.

An albatross climbing into the wind loses ground speed in proportion to its weight (mg) and the angle of climb (φ) — $mg \sin \varphi$ (Fig. 14.8). As it approaches its minimum glide speed, it must increase speed by flapping or descending. But when climbing into a wind gradient, it will encounter faster moving air and its airspeed (V_a) will thus increase at a rate $w\ V_a \sin \varphi$ and may balance loss of ground speed:

$$g \sin \varphi = w\ V_a \sin \varphi$$
$$w = g / V_a$$

If $w > g / V_a$, the albatross would be able to climb while conserving its airspeed and thus reach a greater height than in the absence of a wind gradient. Approaching the top of the climb, in diminishing wind gradient, energy may be further conserved by the manner of the upper turn before descending (Sachs 1993).

The highest pullups seen at sea by Pennycuick (1982b) were about 15 m (occasionally 18–20 m) among Wandering Albatrosses in reference winds of about 15 m per s measured at 25 m above sea level at the masthead of an adjacent ship. In lighter reference winds (6–8 m per s) mollymawks and giant petrels were pulling up to 13–15 m above the surface. Wind gradients required to sustain pullups to these heights were calculated to be about 0.6 per s. If the drag components of the birds had been included in the calculation, w would need to have been higher. On the logarithmic wind profiles adopted by Pennycuick (Fig. 14.9), values of this magnitude were indicated only in the first three metres above the sea.

Downwind

Flying with a wind from behind is often imagined as an easy option for birds, but they have to preserve minimum airspeed in excess of tail winds to avoid stalling. Those migrating under their own power (by flapping) can, indeed, save energy when flying downwind; they reduce flapping towards the minimum airspeed which keeps them from stalling, while the tail wind gives them faster groundspeed.

Gliding albatrosses, however, have to maintain airspeed by descending – and there is only about ten metres

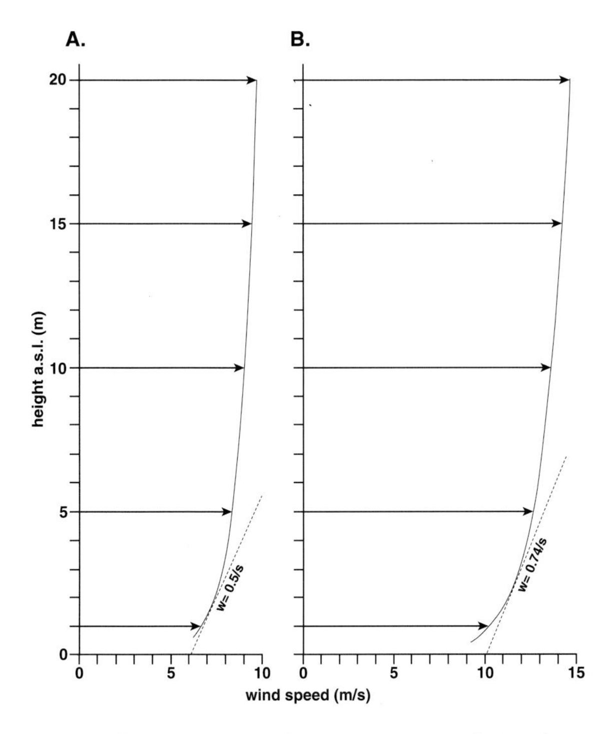

Fig. 14.9 Computed wind profiles between 1 m and 20 m above sea level (a.s.l). The reference winds at 25 m on the masthead of an adjacent ship were 10 m/s (A) and 15 m/s (B). Wind-gradients (*w*) sufficient to lift Wandering Albatrosses (0.6/s), occurred only below 3 m (after Pennycuick 1982b).

height to lose. Downwind flights are very fast and do not last long, because airspeeds decrease as tail winds increase and could soon approach stalling speed. Nevertheless, Spear & Ainley (1997b) argue that small albatrosses are able to keep flying with tail winds.

Wind gradient may help to maintain airspeed in a downwind glide. If an albatross elects to glide downwind at as flat an angle as possible (approaching minimum rate of sink), it will lose airspeed, but if it is descending in a wind gradient it will be entering progressively slower tail winds which may balance its loss of airspeed, with the net result that the glide will be prolonged. The faster the glide, the stronger the effect (Wood 1973, Pennycuick 1982b).

Weimerskirch *et al.* (1992, 1993) reported a consistent relationship between the satellite tracks of Wandering Albatrosses and the direction of the wind (Fig. 14.10). The satellite track of an Auckland Wandering Albatross analysed in greater detail by Reinke *et al.* (1998), similarly had more winds from 90° to 180° of the flight direction (Fig. 14.11). However, albatrosses do not fly in straight lines for hundreds of kilometres. The direction of the wind relative to a line between two distant satellite fixes is the resultant of many changing angles between a bird and the wind.

Several small albatrosses observed visually over straight-line distances averaging 477 m, flew mostly across and into the wind (Spear & Ainley 1997b). Over shorter distances (200–300 m), Black-browed and Grey-headed Albatrosses flew in all directions relative to light winds; moderate winds were from one side and strong winds more from behind (Alerstam *et al.* 1993).

Wind profiles

Computations of wind gradient soaring have all employed logarithmic or exponential models of wind profiles, using reference wind speeds measured at the site of observation together with assumed values for surface friction (Sutton 1953, Sachs 1993). Such models are held to be accurate only at heights above two to three times the significant wave height and the values of 10^{-2} and 10^{-3} assumed for the scale height of very rough and medium rough seas may have been overestimates for the seas observed. Below about 15 m – where albatrosses spend most of their time – there is considerable uncertainty about the shape of wind profiles, because the physical effects of individual waves upon the air flow may become dominant (Shearman, 1985).

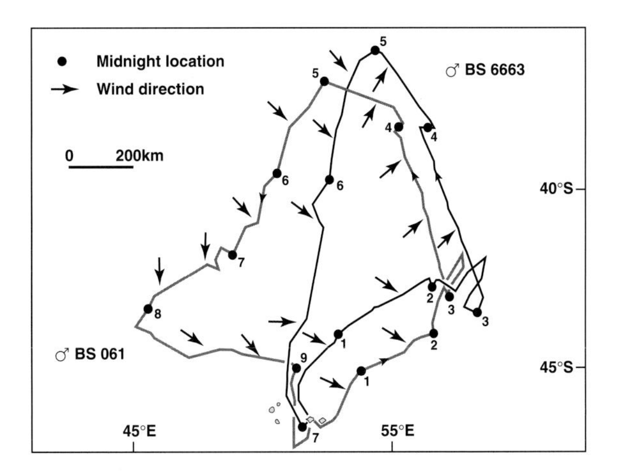

Fig. 14.10 The tracks of two male Wandering Albatrosses that both left the Iles Crozet at the same time on 28 May 1990. Successive night locations are numbered from 1 (28/29 May) to 9 (5/6 June). Arrows indicate the wind direction (Weimershirch *et al* 1993).

Ground effect

Pilots of early sailplanes landing without air-brakes were sometimes embarrassed when their aircraft remained airborne, just when touch-down was anticipated (C.J. Pennycuick pers. comm.). Trailing wing tip vortices are disrupted as wings approach a boundary, whether it is land or the surface of water; this reduces drag and increases lift (Houghton & Brock 1960). Albatross wing tips are often so close to the surface of the water that this effect could, perhaps, help them in brief turning manoeuvres or during accelerating glides close to the wave slope, as postulated by Wilson (1975).

Flight over land

Although albatrosses normally fly only a few metres above the surface of the sea, all breed on windswept islands and use slope-lift to get them to their nests. At South Georgia, albatrosses flying in over the slopes of Bird Island are mostly gliding with some flap-gliding and a few occasionally flapping (Pennycuick 1982b). Colonies are usually at altitudes within 200 m of sea level, but some are on much higher ground. Nests of Southern Royal Albatross at Campbell Island are scattered along ridges at 150–550 m, while Aucklands Wandering Albatrosses occupy slopes at about 200–600 m. Sooty Albatrosses nesting up to 800 m on Tristan da Cunha, habitually approach on the windward side of the island and avoid the lee side (Richardson 1984).

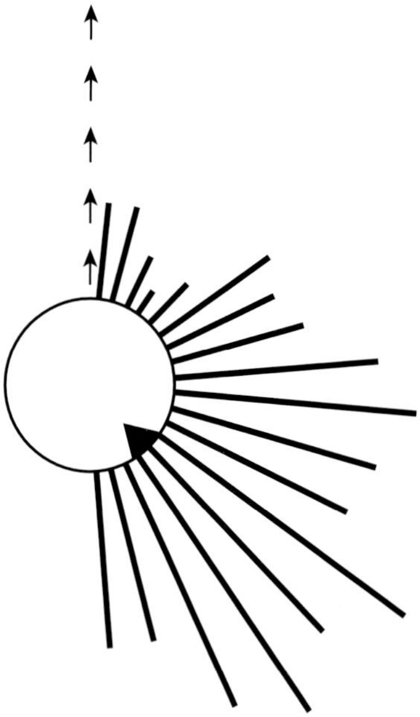

Fig. 14.11 Wind orientation in relation to the direction of migration by an Auckland Wandering Albatross. The radiating bars represent 303 incident winds from different directions (maximum of 30 on the two longest bars), at locations (circle) fixed by satellite. The broken arrows indicate the direction (0°) of the next fixed location ahead (average distance 87 km), not the second to second flight-path of the bird. The distribution of headwinds, crosswinds and tailwinds indicated on one side is cumulative for both sides. About 40% of moderate and strong winds (>14 km per hr), indicated by an arrowhead inside the circle, are at an angle of 120°–150° (redrawn from Reinke *et al.* 1998).

Cumberland Bay penetrates the north coast of South Georgia towards the high mountain range that is the backbone of the island. It is far from any albatross nests, but its whaling stations used to attract many seabirds and on the calm sunny morning of 12 November 1912, Murphy (1914) witnessed seabirds flying in a manner not described there before or since:

> '...there was a little motionless fog along the coast of the island, but it soon lifted, revealing the white-robed mountains under the brightest of skies. As my brig was being towed to its anchorage by one of the whaling steamers, innumerable birds were about, and for the first time I saw many Tubinarine species flying high. Among those which circled far up under the blue vault, until they became almost indistinguishable specks against the light cirrus clouds, were hundreds of Wandering Albatrosses. Since some of these approached, or probably surpassed, the limit of human vision, it is likely that they towered to an altitude of at least 1500 metres above the bay and the surrounding land. Two other smaller species of albatrosses [Black-browed and Light-mantled Sooty], as well as numerous individuals of the Giant Petrel, Cape Pigeon, Antarctic Whale-bird (*Pachyptila desolata*), and *Procellaria aequinoctialis*, were taking part in the same performance. Especially noteworthy is the fact that the greater part of the spiralling was taking place over water.'

It is possible that in spite of the calm at sea-level, air was moving up the steep higher slopes of the mountains surrounding Grytviken. Alternatively there may have been lee-waves down wind of the higher peaks (2900 m) (C.J. Pennycuick *pers. comm.*).

Something similar occurs in the Falkland Islands. Black-browed Albatrosses from the great colony at the west end of Steeple Jason Island sometimes soar high above the coast, just where winds coming off the sea encounter a steep ridge of sharp peaks.

Landing and Taking-off

At sea, albatrosses are in their element constantly coming and going from the surface. They have powerful leg and thigh muscles and take off by spreading their wings and running into the wind. In a strong breeze, they may be airborne within a few strides and gales guarantee almost instant lift-off from the nearest wave crest. The less wind, the harder they have to work.

> 'The Wanderer requires a tremendous effort to become air-borne in still air, and, if the unfortunate bird is further burdened with several pounds of food in its stomach, the feat becomes impossible.'
> (Gibson & Sefton 1959)

A Light-mantled Sooty Albatross has been seen to patter across the water for almost two kilometres before getting into the air (Henle 1981). However, rafts of albatrosses on calm seas in the lee of islands near Cape Horn had no difficulty taking-off (Goodman 1978).

Albatrosses may glide down and settle gently onto the surface or descend suddenly and dump themselves in the water. If there is a food available, even great albatrosses will drop a metre or two in an ungainly splash (p.164).

Approaching land is potentially dangerous and albatrosses appear to be more wary. They will often make several circuits before finally landing. There is a world of difference between the ponderous descent of a great albatross towards a clearing in the tussock grass and the panache of a mollymawk sweeping along a crowded colony and dropping neatly among the nests with a wail of warning. Off the cliffs of South Georgia, mollymawks can slow down rapidly and almost hang in the wind, tails braced up and trailing open, webbed feet. Lanchester (1910) believed the spread feet increased the tail area of albatrosses, but Pennycuick & Webbe (1959) have said that, in Fulmars, they act solely as air-brakes.

All birds approaching land use their wings to slow down. The horizontal attitude of the bird in flight changes towards the vertical as the legs are dropped and in this posture, wing-beats are directed forward and bring the bird to a stop. Albatrosses do not have strong power-strokes and their long wings are an impediment to rapid flapping. Seconds before touch-down, as posture changes, the wing bones are rotated (twisted) along their extended axis by contractions of the deep *pectoralis minor* muscles. The secondaries and primaries are rapidly fanned downwards and forwards (Scholey 1982). The humerals, adjacent to the body, remain in the horizontal plane (Fig. 14.12). Landing into brisk winds, this brings the bird to a stand-up landing, but in low winds it may not be quite enough, and as the bird touches down, it lurches into a few bounding steps or, sometimes topples forward, beak into the ground.

At Midway Atoll, Laysan Albatrosses sometimes hit the ground in sprawling bundles of wings and legs that have been said to be due to unseen pockets of warm, low-density air that the birds encounter just before they touch-down; apparently a unique hazard above the hot ground of tropical islands (Fisher 1970).

Among the dense scrub vegetation of Isla Española, most Galápagos Albatrosses nest within walking distance of the coast where they can take off down ramps on the tops of sea-cliffs or from beaches. Open spaces are scarce and birds nesting inland depend upon arid months holding back the spread of vegetation. During the unseasonal rains of the 1982–83 El Niño, some at least of these areas became so overgrown that albatrosses had great difficulty landing (Rechten 1986).

Landing can usually be accomplished in winds insufficient for take-off, but fresh winds are needed if great albatrosses are to get away from land:

> 'On the only calm day we experienced on Campbell Island the [Southern Royal] albatrosses were incapable of taking-off and landing. On such days the upper ridges present an unusual sight with all the albatrosses "grounded" although birds were seen gliding in the air above the ridges and out to sea.'
> (Westerskov 1959).

'Airfields' or 'runways' are a feature of many albatross breeding grounds. They are much used areas free of trees, shrubs, tussocks or nests where birds can take off. Departing birds habitually walk to the down-wind or up-slope end before turning, spreading their wings and making a run. Often birds from a wide area converge on one

Fig. 14.12 A Steller's Albatross landing at Torishima (H. Hasegawa).

airfield and gather in a crowd as if waiting their turn. Some of the flat meadows used by Wandering Albatrosses at South Georgia are up to 100 m long. Slopes are often preferred; on the narrow edge of the huge Black-browed Albatross colony at Steeple Jason Island short downward slopes of beaten soil are kept free of tussock by the daily traffic.

Mollymawks and sooties nesting on steep ground and cliffs have solved take-off problems. They can leap from exposed positions and dive away to gain airspeed. In large mollymawk colonies (Plate 2), most nests are some distance from cliff edges and their owners have to pass between other nests and survive the menaces of their occupants to get to the edge (Fig. 15.8). At Beauchêne and Steeple Jason Islands, Black-browed Albatrosses negotiate hundreds of nests to reach cliffs or runways. At temperate islands like Tristan da Cunha, Gough and the Snares, tall vegetation and trees often obstruct access to the cliffs.

Ships

Some albatrosses habitually follow ships and soar on air rising over the vessels. There are many accounts of birds keeping company over long distances. After sailing from Cape Town in October 1968, the *Clan Ramseu* was followed by two wandering albatrosses for 1,750 nautical miles (3,237 km) to 9° 04'S, 2° 45'W. The ship was steaming at an average speed of 17 knots (744 km per day) (L'Estrange 1972). Between Cape Town and the Prince Edward Islands, however, the average following time of 14 Wanderers was only 44 minutes and there was a tendency for the birds to stay with a ship longer in the evening than in the morning (La Cock & Schneider 1982).

In the past, sooty albatrosses appear to have been particularly attracted to sailing ships:

> '...we lay to all day under fore-staysail and trisail, in the teeth of a southwesterly gale....Dark weather alternated with bright sunshine; the temperature averaged 4.4°C.; there were several brief hail-storms; spray broke over us continually, but we shipped no heavy seas. During the morning four

> Sooty Albatrosses joined us and remained near-by for five hours, appearing to have no other purpose than to play in the howling wind for the admiration of us on board. Their ease and precision, and particularly their ability to vary their speed and to "stand still" in the air, put them in a class by themselves.' (Murphy 1936)

A decade later Harrison Matthews (1951) watched Light-mantled Sooty Albatrosses around the masts and rigging of *Discovery*:

> '...With one wing-tip almost touching a yard-arm he could hold his position while he let down one of his feet and brought it forward to scratch his head, or could tuck his head under his breast to settle a feather that he fancied was ruffled in his immaculate smoky plumage.'

When soaring alongside modern vessels, albatrosses often come very close to the superstructure. Captain R.B. Tiplady (1972) was walking around the deck of the *Pembrokeshire* one early morning watch and found a wandering albatross sitting on the forecastle head. It was later launched and flew away. However, close encounters with ships have always been potentially dangerous:

> 'In 1930 I was an AB aboard HMS *Daffodil*, a coal-burning sloop...en route from Tristan da Cunha to Simonstown and had run into a hell of a storm...I noticed an albatross battling against the winds – none of your naturalists' view of a monarch of the skies soaring along on the wind. This bird was really fighting for survival, but he didn't make it. He crashed against our main-mast and fell to the deck.' (Mullen 1982).

15
BEHAVIOUR

'...they[1] breath hard through 2 small holes in the upper part of the Bill and frequently make a sound exactly like the little toy trumpetts sold to Children at fairs...'

Lieutenant S.W. Clayton RN,
at the Falkland Islands 1773–74[2]

The conspicuous displays of albatrosses towards each other have always attracted the curiosity of those who landed on their islands. Their activities were usually seen as comic or bizarre antics and it is still difficult to avoid interpretations that endow the birds with far more than can be justified by their biology. We are still defining postures, actions and sequences while only guessing at meaning (Warham 1996).

Sex and age determine the way albatrosses behave towards one another. Most studies have managed to indicate the sex of some performing birds, but hardly any have been unequivocal about their age.

Terminology

The first names used systematically to describe albatross behaviour were those of Richdale (1949a, 1950a). The actions and vocalisations he named were mostly well defined, but some ambiguities were created between the two species he studied. Some of his names now embrace more than one action; new categories of behaviour have been defined and some common actions have acquired several names.

In this chapter I have attempted to apply the same names for common actions throughout the family. The time is long past when it might have been possible to agree priority in choosing names. Some of those used here are well established and have retained their original definition, but new ones have been created in an attempt to be unambiguous throughout the family. Synonyms have been included in parentheses and all names are tabulated in Appendix 12.

Territory and Aggression

Albatross behaviour on the breeding grounds is concerned with the acquisition and holding of territory, establishment and maintenance of a pair-bond and the satisfaction of egg and chick needs. Albatrosses defend very small territories, just the immediate surroundings of the nest or personal space of the bird itself when it is elsewhere, but competitors are recognised from a distance and forays sometimes made to drive them off. The amount of aggression is a function of the density of the birds.

In the most crowded parts of mollymawk colonies, individual nests are sometimes very close, and birds walking between them cannot avoid attracting aggressive responses from their occupants (Fig. 15.8). On steep ground the same species cannot achieve such densities and the facility to take-off from the nest or nearby, influences their behaviour.

On open slopes where great albatross pairs are widely spaced, the birds are comparatively docile and nests are occasionally alongside each other or even next to those of giant petrels.

Fights between albatrosses include charging, jabbing, thrusting, gripping or snarling with the bill. Albatrosses launch attacks from their nests, and also from a distance, towards intruders at or near their nests. On the breeding ground, attacks are seen most often during the pre-egg period and in dances, usually by nervous subadults towards inept birds attempting to join the activity. Fights occur between all ages. If the initial signal does not see off an intruder, a charge may follow and prolonged pursuits, with the participants' bills locked together. They be damaging, eyes are vulnerable and bills may be holed (Bailey 1952, Pickering & Berrow (in press)). Similar behaviour may be seen on the sea, far from breeding islands, whenever albatrosses congregate.

[1] Black-browed Albatrosses.
[2] (1774)

Chick defensive behaviour

All albatross chicks are left alone in the nest after they acquire thermal competence. Their first line of defence is their size; small chicks are vulnerable, but the earlier both parents can forage the faster chicks grow. There is evidently a selective trade-off; young chicks alone in their nests often appear quite small and it is the time of greatest mortality, but they grow rapidly and are soon heavier than most predators: hawks, caracaras, giant petrels, skuas and sheathbills. There are no natural mammalian predators on land.

A well built albatross nest is a defensive asset; it raises the chick above the level of predators. Given material, all albatrosses are able to build nests and from an early age their chicks are constantly working on them. The few chicks that fall off their nests and are unable to climb back, build new ones alongside by quarrying material from the original nests.

Albatross chicks have a common repertoire of defensive behaviour. Sitting-up they are rather like self-righting skittles. The large belly lowers the centre of gravity and makes it very difficult to dislodge the chick, especially from the hollow bowl of a well made nest. The upright posture also lifts overhanging wings which might otherwise be grabbed and used to pull the chick off its nest. Facing the danger, rhythmic snapping of the bill produces hollow *clop-clop-clop* sounds. In colonies, a threat to one chick is taken as a threat to all and the surrounding chicks all respond by facing an intruder with a chorus of *clopping* sounds. The final armament is the oily content of a large stomach, but it is usually retained until predators approach very close. As the oesophagus relaxes the *clopping* becomes a *gulping* that anticipates a nauseous deluge.

Communication

Soon after hatching a chick's soliciting calls and bill-tapping actions are reinforced by feeding and, by the end of brooding albatross chicks are also proficient at signalling their presence. This call is directed at any approaching adult and in a colony one chick calling will set off others, so that many call in chorus at a newly landed adult. Soliciting food from parents usually occurs at the nest. When old enough, the chicks of some species leave their nests and if hungry may solicit from any adult. Most albatross parents will feed young only at their own nest-site and other soliciting chicks may be savaged.

> 'One day we noticed that a white gooney had made a nest right in the middle of the road. So little by little we moved the egg, a foot at a time, over towards the side of the road. One of the parents always followed and sat on the egg, until eventually it was established in a nest at one side of the road at least 30 feet from where the egg was first laid. Both parents took turns sitting on the egg in the new locality. The change didn't seem to bother them. Then the egg hatched, and the young one when about two weeks old was left by itself. The next time we saw it,it was right back where the original nest had been made in the middle of the road. It could remember where it belonged!
>
> What really happened was the first time the old bird returned to feed it, it insisted that the young bird return to the old nest-site before it would be fed. The old bird probably went directly to the old nest and waited for the young one to come over.' (Hadden 1941).

Aggressive behaviour by a chick towards its own or closely related species is quite different from that described above towards dangerous predators. It is well developed in fledglings which can hold their own against intruding neighbours, be they chicks or adults.

An adult at the nest with an urgently begging chick may allopreen the chick's head. Chicks respond by turning their heads away and concealing their own bills. This action is both submissive and protective; it is an automatic response not only to the gentle touch of parents, but also to the fierce attack of a stranger. These actions are incorporated and become elements of more complex, later communication.

When subadult albatrosses first return to their birth-place, they bring with them the repertoire of actions and vocalisations acquired as chicks and fledglings on or near the nest several years previously. At sea, they encountered and competed with other seabirds when feeding and resting on the water with numbers of their own kind. At such times, there are random social interactions, usually assertive and not requiring the subtle postponing of gratification needed on land to acquire a mate.

The first analysis of albatross dance behaviour identified sets of related actions and sequential patterns in their performance (Meseth 1975). An albatross language was deciphered by Jouventin *et al.* (1981). The youngest birds had the characteristic postures and actions of their species, but used them in a hesitant and uncoordinated manner, with no sexual bias. It was proposed that they had a vocabulary of stereotyped postures and actions, but these signals were subject to variable expression or syntax. The behaviour of young albatrosses suggested that they could neither 'read' what was being 'said' to them nor respond appropriately. The grammar of the language rested in sequencing or synchronising the signals and could take several years to perfect.

Young albatrosses begin to learn their languages by trial and error responses during random approaches to and from other birds on the breeding grounds. These may be others of their own age, but they may learn faster from older birds. Year by year, their performances improve and their competence[3] corresponds to their age. Over several years, the number of different birds with which an individual communicates will steadily decline until it is perfecting the language of one partner. What finally emerges is a joint language of one partnership, distinguishable from all others. This synthesis of language, unique to one pair is, through repetition, synchronised with physiological cycles in the pair.

The paradox of these events is that once the mutual language is so perfected that breeding can take place, much of it ceases to be used. Breeding pairs no longer employ the full vigour of their language because they recognise each other and continually reinforce their close relationship by contact and allopreening, sharpened only occasionally by male hostility.

The striking displays of albatrosses have evolved from simple activities, but ordinary behaviour that is not stereotyped also contributes to communication, often in ways that may be difficult to recognise. For example, when confronted by con-specific activity, albatrosses may or may not appear to show interest. Standing up represents increased attention and when a standing bird becomes directly involved, the looking posture may become significant. It is debatable whether this posture is made more conspicuous as a result of selection to increase its signalling function (stereotyping) or, is merely the normal outward appearance of attention. In some species, it may be accompanied by supporting signals: a flattening of the eyebrows in Galápagos Albatrosses (Fig. 9.1); a raising of the crown feathers in Light-mantled Sooty Albatrosses (Plate 19) or an exposure of the coloured gape-stripe by mollymawks. In all these birds, an increase of emotional tension may anticipate movement or threat.

Albatrosses have well developed olfactory organs, but so far there is no reason to believe that the sense of smell is used in communication.

Autopreening

Care of the plumage is a constant concern of birds and all albatrosses spend a lot of time autopreening. Although we may guess that a solitary preening bird is just looking after its feathers, when it is in company with other birds we can never be so confident. In social situations, autopreening often occurs with regularity in encounters that lead us to suspect that it is the response to a social stimulus. It does not have to be the primary response; the bird may be nervous during an encounter and may experience associated peripheral discomfort that causes it to preen itself at an inappropriate moment. We may not read the signs, but perhaps another bird will. In all social interactions, even when the individuals are not engaged, autopreening may be considered as potential communication.

There are occasions when albatrosses open and close their mandibles suddenly to free them of contaminating food, feathers, mud or grass. The snapping sounds are audible, but may have no signalling significance and it is often difficult to distinguish such spontaneous bill sounds from stereotyped snaps. During autopreening, the mandibles work among the feathers and the sound is muffled, but when autopreening occurs during a display, bill-sounds under feathers may appear distinctly louder. A common ending to a bout of autopreening is a ruffling of the body and/or head plumage followed by vigorous shaking.

Head-shaking from side to side and sky-shaking, when the head and bill are pointed up and rotated several times, are often accompanied by a rattling of loose mandibles. These actions are not listed as stereotyped because

[3] For their analysis, Jouventin *et al.* (1981) set up four arbitrary 'niveaux' of language which may be translated as levels or standards of attainment.

they may be no more than the settling of head plumage in the same way as the body is shaken to settle contour feathers. Nevertheless, they are conspicuous in all species and sometimes occur in situations where a response to social stimuli may be indicated.

Allopreening

Allopreening is the single most important bonding behaviour among albatrosses and is a fundamental objective of much social behaviour between male and female, whether paired or together temporarily. It also occurs between individuals of the same sex. In the Sooty Albatross the rate of allopreening is higher where there are several pairs nesting close together than at solitary nests. Allopreening usually occurs between sitting individuals, but one may be standing. It may be directed towards any part of the body, but usually concentrates on the head and neck region which is not accessible to autopreening. Most albatrosses have chewing-lice in their plumage. These are particularly noticeable on the head, but there is no direct evidence that allopreening is concerned with the removal of these parasites. Recipient individuals may accept allopreening quietly, responding only by presenting new areas for attention. This is indicative of well established pairs. It is anticipated and parried by recipients that are not yet in harmony with a partner. Allopreening may be simultaneously (mutually) or alternately reciprocal and often has a dynamic rhythm of its own that stems from the shared experience of partners.

Allopreening is widespread among petrels and other seabirds (Harrison 1965); Fairy Terns, for example, allopreen like albatrosses. The concentration and mutual co-operation of the behaviour strongly suggests a bonding function reminiscent of grooming in primates.

Quiescent behaviour

Two undemonstrative birds quietly resting close together may appear to be 'non-behaviour' (Richdale 1950), but it is in fact highly significant. They are likely to be male and female, usually a pair or partners who have, for the time being at least, become comfortable together. This state may have been entered into by established breeders with few, if any, preliminaries. Less experienced partners have gone through a longer or shorter agonistic exchange beforehand. Periods of quiescent behaviour may follow other activities, such as nest-building and are usually interspersed with bouts of auto- and/or allo-preening, but for long periods the birds do no more than rest silently beside each other, often with eyelids closed and sometimes with the bill of one resting across the neck of its partner.

Individual recognition

Breeding behaviour in albatrosses rests upon the mutual recognition of individual partners who, eventually, become mated pairs. Although it may be difficult to put together a formal proof, there are ample anecdotes from ringed pairs to suggest that returning albatrosses, after being separated at sea for many months, readily recognise each other without the need to go through an elaborate courtship. Wandering Albatrosses know their own chicks before they are half grown, even though they may be diffident about feeding them away from their nests. In at least two mollymawk species, parents either do not recognise their own chicks or ignore that identity when feeding the occupants of their nests (Tickell & Pinder 1972).

Spatial activity

Change in the spatial relations between individuals is often an indicator of intention. Even among the widely scattered nests of Wandering Albatrosses, movements of one bird are carefully watched by the apparently inactive owners. The manner in which an albatross approaches, passes or withdraws may appear unremarkable and yet have meaning for another individual.

In albatross dances, there is a constant monitoring of position. The need to move around a partner without causing alarm, has led all albatross species to evolve exaggerated forms of walking or side-stepping that give advance warning. A male leading a female away, needs to be particularly sensitive to her movements behind him. All albatrosses fly over or past their nesting places and birds on the ground may be aware of intentions without any obvious signal being involved. Subtle characteristics of landings and take-offs can likewise be interpreted. Albatrosses on the ground sometimes signal to those in the air and birds in flight have been seen to communicate with each other and with those on the ground.

Nest relief

All albatrosses go through similar routines when giving-up or taking-over nest duties and these are proportionally more intense during incubation and brooding. The dynamics between the incoming bird's drive to get onto the nest immediately and the occupant's reluctance to abandon the egg or small chick, precipitates as much variation of behaviour within species as between species. It usually involves minimal stereotyped exchanges consistent with the repertoire of the species, but a common element is the ritual gathering of nest-material by the out-going bird; incoming mates collect nest material only after the prolonged refusal of a sitting bird to surrender the nest. Although the digging and placing of nest material is a practical act, the activity is highly stereotyped. When there is nothing readily available nearby, Royal Albatrosses at the Chatham Islands walk far away from their nests to pick up and place twigs that cannot possibly end up on their own nests. Although many Galápagos Albatrosses have no real nests, they still pick up a small stone or dry stem of grass before leaving.

VISUAL SIGNALS

Like most seabirds, albatrosses are devoid of prominent feathers or other structures that might disturb their clean aerodynamic lines. The best attempt at a crest, in the Light-mantled Sooty Albatross, is minimal and readily flattened. In two species, there may be some visual significance in yellow pigment on the head plumage, but in most albatrosses a sober patterning of black, grey and white is employed. Colour is conspicuous on and around some bills.

Postures and movements have been enhanced for communication. From a relaxed undemonstrative standing position with the wings and tail folded, the bill horizontal and the albatross looking forward, most signals are made by the following general movements.

Stare

Richdale's (1949a, 1950a) 'gawky look' is largely a facial expression of concentrated attention towards another bird. There is little detectable change in posture and no sound, but to a practised observer the bird is no longer relaxed. The eye expression appears to be significant, although it usually defies objective description; it seems to be a glaring, fierce look. A stare, held for two or three seconds anticipates more dynamic action, but it often relaxes without anything else happening.

Head-forward

In close social interactions, albatrosses seek to touch each other by reaching forward with their bills. The action may be accompanied by movements of the mandibles and vocalisations. The bill is directed towards any part of the body, but it is most often aimed at the bill, like a soliciting chick. When directed towards the breast, it may make contact and nibble feathers gently or quickly pull one, but touch is an intimacy permitted only to partners or mates and from others is usually rejected with thrusts, parries or clashes of the bill.

Head-down

Bowing in various styles is a repetitive component in the social interactions of all albatrosses. It may be no more than a quick nod, but deep bows can be far forward, to one side or between the legs to the belly or feet.

Head-away

All albatrosses have actions that turn the head away from their partners. These may include concealing the bill in the plumage of the neck, shoulder, breast, flanks, thighs, under the wings and along the back. This may resemble autopreening, but often the bill just touches the plumage or is even held rigidly clear of the feathers. An early observation about one such posture was that the side of the breast is the site of a dense patch of filoplumes that might be particularly sensitive to touch (Nelson 1968).

Head-up

Lifting the head and pointing the bill towards the sky is a common action among birds. In albatrosses, it may be only a quick upward flick of the head, but often it involves holding a posture and sometimes repeating the action.

The bill can tilt up fairly easily to about 60° above the horizontal, but it is often pushed vertically and some albatross species rise on tip-toe straining towards the zenith, or even leaning over backwards. Sky actions, as they are called, may be silent or accompanied by vocalisations.

Side to side movements

Exaggerated lateral swaying of the head and whole length of the neck may be slow and ponderous; quite different from the rapid shaking of the head that produces a castanet-like rattling of mandibles (Fig. 15.1A). Head-shaking is caused by a rotation on the axis vertebra; if this occurs when the head is pointing upwards in line with the vertebral axis, then the head movement also becomes a rotation (Fig. 15.1B).

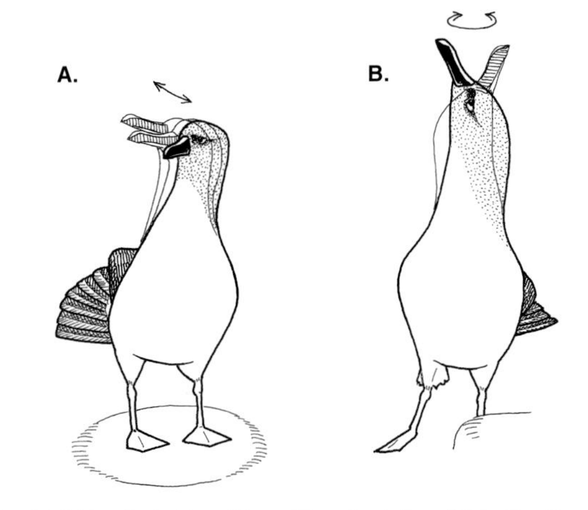

Fig. 15.1 Head shaking (A) and head rotation (B) in the Grey-headed Albatross (after Tickell 1984).

Bill-open

Opening the mandibles wide and exposing the full gape is a potent signal that may be enhanced by colour inside the mouth. In some albatross species it is effective without any sound, while in others it is accompanied by loud calls.

Wings-open

Several albatross species open their wings partially and great albatrosses assume an even more imposing presence by extending them fully.

Tail-open

Not only can the tail feathers be spread or closed like a fan but they may be cocked up and/or twisted forward on one side or the other.

SOUNDS

Albatrosses are usually silent at sea, other than when engaged in feeding frenzies and social interactions on the water. On land, sounds are essential components in the communication of all species.

Percussion sounds depend upon the force and frequency with which the mandibles close. The large bills are hollow and the skeletal spaces are connected to adjacent airways so the sound produced by a single snap is hollow, like a *clop*, especially from the rather rubbery bill of a chick or fledgling. Multiple snappings have variously been described as clopping, clacking, clappering, clattering, rattling, rapping, vibrating, drumming and buzzing. They vary in frequency from several separate snaps to a train of up to 35, like the sound of a stick being dragged along a picket fence (Meseth 1968). At the highest frequency, slight vibration of the mandibles is barely noticeable other than as a blur.

Vocalisations include a wide range of whistles, screams, shrieks, wails and trumpets. Most have many harmonics and only gooneys approach pure tones. Some characteristics of the sound may be attributed to resonating columns of air in tracheae or air sacs (Ulrich in Warham 1996).

The single calls of sooty albatrosses are isolated events on long stretches of coastal cliffs, while on spacious slopes the sounds of great albatrosses are heard here and there from time to time. In contrast, the crowded colonies of mollymawks and gooneys set up a constant noise in which the individual calls are submerged.

Groans, grunts, snorts and gurgles often precede or accompany actions. Some of these at least are reflex expulsions of air from air sacs due to muscular contractions in stretching the body; others may be quick inhalations or expirations before or after tensing actions. A grunt may precede a particular action, but if it accompanies an action that is also anticipatory, the grunt itself need have no signal function. After long screams that use most of the breathable air, wheezy or squeaky inhalations frequently follow.

Albatross vocalisations are highly variable; on sonagrams the repeated sounds of individual birds reveal distinguishing characteristics (Warham 1996) and sexual differences are apparent. On separate islands, one species may have recognisable dialects.

DANCE

Albatross displays are legendary and since the earliest descriptions, have been likened to dances. The anthropomorphic allusion may be misleading, but it is hardly likely that a pedantic quibble will dislodge such a popular image.

Walter Fisher (1904) had been watching Laysan and Black-footed Albatrosses for six days in May 1902 when he wrote:

> 'The old birds have an innate objection to idleness, and so for their diversion they spend much time in a curious dance, or perhaps more appropriately a 'cake-walk.'[4] This game or whatever one may wish to call it, very likely originated in past time during the courting season, but it has certainly long since lost any such significance. I believe the birds now practise these antics for the pure fun they derive...'

By the late 1930s, people had lived for years among the Laysan and Black-footed Albatrosses of Midway Atoll and Hadden (1941) echoed Fisher when he wrote:

> 'the gooney dance has nothing to do with mating.'

The comments had some truth and anticipated the following scenario:

1. Dances are common social interactions that occur wherever and whenever albatrosses come together on land or sea.
2. An albatross dance is a repeated or changing sequence of agonistic actions and reactions. Usually it is between a male and female, but male-male and female-female sequences occur, and dances with several participants are common.
3. Although most of the component actions are stereotyped, many are not. No two dances are the same, they are highly variable, dynamic performances.
4. Albatrosses are very slow to lose hostile responses towards strangers of the same species. Trial and error approaches during repeated dances between particular individuals synchronise the performance and may diminish nervous tension.

THE REPERTOIRES

The behaviour of albatrosses can be clearly distinguished as six repertoires consistent with other characteristics that separate the family. The repertoire of the two gooneys is more varied and the performance faster than in all other albatrosses. Throughout the mollymawks there is remarkable conformity, while differences within the great and sooty albatrosses do not disrupt their characteristic repertoires. The repertoires of Steller's Albatross and the Galápagos Albatross each have unique elements.

Actions and vocalisations common across two or more repertoires (Appendix 12) may be imagined as homologous, but the diversities of repertoires also indicates that during the course of evolution albatross behaviour has been quite labile.

MOLLYMAWKS

Richdale (1949a) defined the actions of Southern Buller's Albatross at the Snares Islands. Rowan's (1951) photographs revealed the same actions in the Western Yellow-nosed Albatross at Nightingale Island and Johnstone *et al.* (1975) noted similar behaviour among the Tasmanian Shy Albatross at Albatross Island. I compared Black-

[4] a fashionable American dance of the period.

browed and Grey-headed Albatrosses at South Georgia and proposed a characteristic mollymawk repertoire (Tickell 1984). Catherine Fitzsimons and John Warham followed-up Richdale's work at the Snares, analysed the sounds of Southern Buller's Albatross and compared them with those of other albatrosses. They also devised field experiments using cut-out models, but the results were less than expected (Warham 1996, Warham & Fitzsimons 1987). Benoît Lequette (pers. comm.) analysed the vocal signals of croaking and wailing by seven mollymawks and confirmed differences within the common repertoire.

Large colonies with crowded nests are characteristic of mollymawks, but steep slopes, rocky terrain and vegetation often keep nests well spaced. Some territories are compressed to less than one square metre and there may be interference between neighbours. From time to time, sitting birds manage to reach out and pull a neighbour's overhanging wing or tail. This is apparently a random, non-aggressive result of long periods of inactivity during incubation. The disturbed individuals usually wail in response and when released, follow it up with bouts of low intensity croaking.

At the Snares Islands, Miskelly (1984) saw a visiting Tasmanian Shy Albatross displaying to resident Salvin's Albatrosses and at Campbell Island effective communication occurs between immigrant Black-browed and endemic Campbell Black-browed Albatrosses. On the same island, Warham watched 'mutual displays' between Campbell Black-browed and Grey-headed Albatrosses (Warham & Fitzsimons 1987). My own field-notes from South Georgia, contain no records of interactions between Black-browed and Grey-headed Albatrosses (Tickell 1984).

Several species have been seen actively displaying on the sea (Cooper 1974).

Aerial activity

There is a great deal of activity in the air above or along the cliffs below crowded mollymawk colonies of South Georgia, with many prospecting subadults doing 'circuits and bumps'. Often birds hover in the wind a few metres from nests before diving away. Communication with birds on the ground may occur, but aerial postures like those described by Warham (1996) are essentially aerodynamic adjustments concerned with flight.

Western Yellow-nosed Albatrosses above sparse colonies at Tristan da Cunha and Gough Island have been reported occasionally displaying and calling in flight (Elliott 1957, Shaughnessy & Fairall 1976) rather like giant petrels (Warham 1962). Synchronised flying is rare, but has been seen in Southern Buller's and Grey-headed Albatrosses (Warham 1996).

Aggressive behaviour

Early in the breeding season, unclaimed nests in the centres of dense colonies are soon taken over by others or quarried by neighbouring birds. A male at a distance, seeing another on its nest or digging at it, will usually charge back and drive off the interloper. As long as there is an escape route, the intruder usually flees, perhaps with a loud wail, but in corners obstructed by rocks, vegetation or other nests, loud and confused melées with prolonged fights sometimes ensue. Bill-gripping is common and attacking birds may sometimes secure a grip about the neck of an opponent (Warham 1996).

In defence of occupied nests, adults tend to remain low, rising only slightly on their tarsi. Aggressive actions include pronounced swaying of the down-pointing head and neck from side to side, accompanied by irregular bill rattles and snarling sounds. Sitting birds often adopt a very low posture with the head hanging over the rim of the nest, probably because they can reach farther (Fig. 15.2). Two adjacent, antagonistic birds stretch their heads

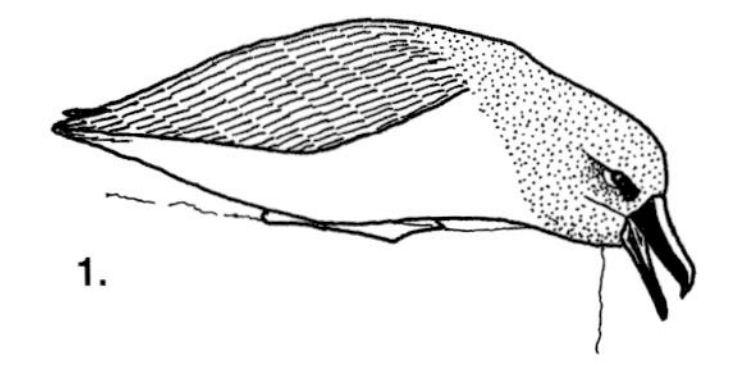

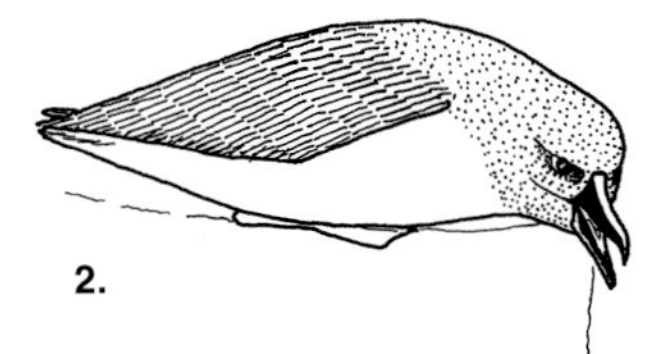

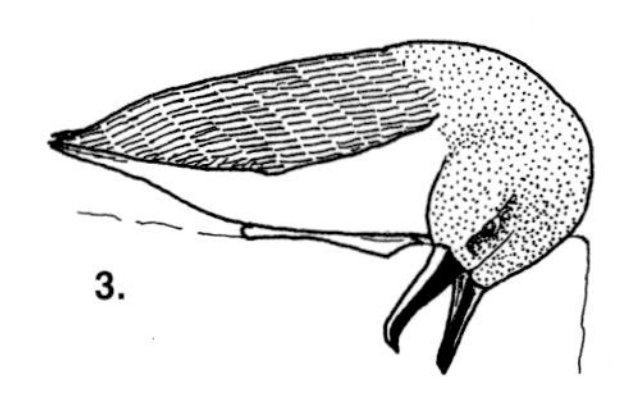

Fig. 15.2 A Grey-headed Albatross GULPING aggressively (1-3). Note the exposed gape-stripe. The sequence lasts two seconds (after Tickell 1984).

and necks horizontally towards each other, twist their heads and open bills from side to side while making snarling expressions and sounds. Usually they are just out of range of each other, but if they make contact, bills may become locked. When opponents are out of reach, redirected stabbing at the ground and/or aggressive pulling and shaking of tussock grass is fairly common.

Similar aggression is also directed towards other species, particularly scavenging skuas and sheathbills. Adult and subadult mollymawks do not normally vomit defensively like their chicks.

Stereotyped behaviour

Mollymawk social behaviour is not generally referred to by the dancing analogy. The restricted movement and absence of sky- postures do not immediately turn on the anthropomorphic imagination. There is often little space for manoeuvre between nests so displays occur on and beside the nest. Actions have to be compact and cause minimal disturbance to close neighbours. Within the common mollymawk repertoire, the vocalisations of some sympatric species may be distinguished by the practised ear.

Stare (gawky look) is often accompanied by a parting of minute contour feathers behind the mouth to expose the coloured gape-stripe. This is not an essential component of the posture, but probably signals an increased intensity.

Croaking (croaking & nodding, bowing, bowing & fanning) is typically performed by males and females on or beside the nest; sitting birds usually rise, at least onto their tarsi. The bill is opened slightly and the mandibles move with a short series of creaking or croaking sounds *eh-eh-eh-eh-eh* (Fig. 15.3). At its most vigorous, the bird appears to pivot forward in time with the sounds and the folded wings are lifted and parted very slightly, while the tail is raised and fanned widely. The depth of rocking varies; it may be deep enough for the tip of the bill to touch the feet or the ground at the base of the nest (Fig. 15.4). Throughout this action, the coloured gape-stripe is exposed.

The repetitive multi-note vocalisations have strong harmonic and temporal structures (Fig. 15. 5). Multivariate analysis of five variables in the croakings of 94 birds, indicated significant differences between the Black-browed

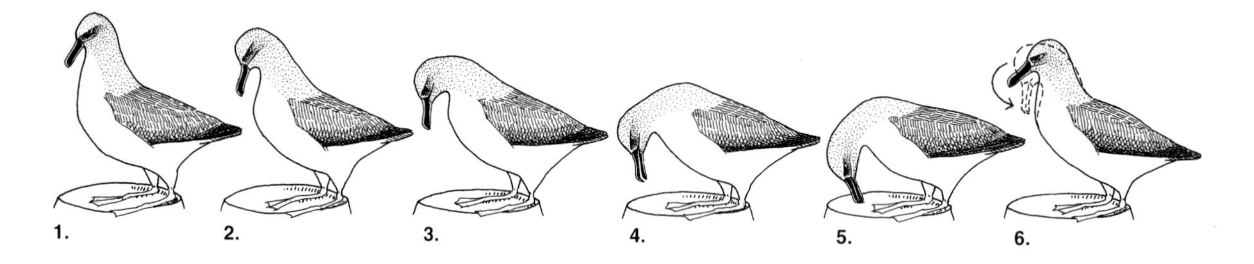

Fig. 15.3 A Grey-headed Albatross CROAKING (1–5). The head rocks into the cup of the nest and the display ends with a brief ritualised bow (6) (4 s duration) (after Tickell 1984).

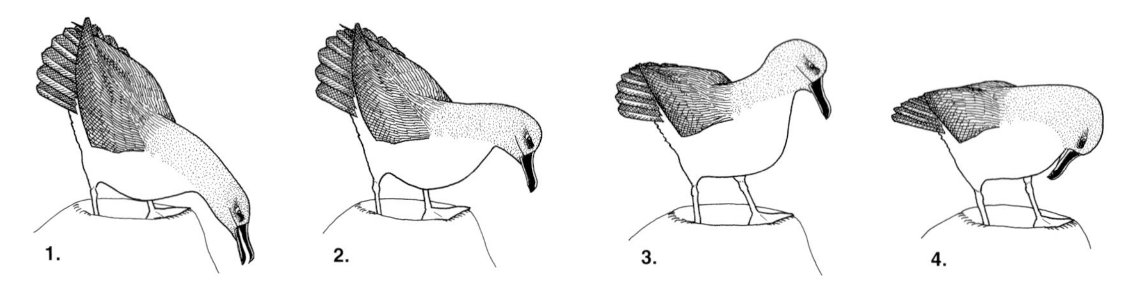

Fig. 15.4 A Grey-headed Albatross CROAKING (1–4). Note the gape-stripe and fanned tail as it recovers from deep rocking. In the concluding BOW (4), the gape-stripe has been covered (3 s duration) (after Tickell 1984).

Albatross, Grey-headed Albatross and two yellow-nosed albatrosses (B. Lequette pers. comm.). Warham & Fitzsimons (1987) concluded that croaking incorporates all the variation needed for males, females and individuals to be identified.

Croaking is a signal of identification and possession, the most common behaviour in all mollymawks. It is often repeated several times and contributes most to the background noise of crowded colonies. It is less frequent in well-spaced colonies. The Southern Buller's Albatross reunited at the beginning of the pre-egg period, croak on average five to ten times per hour, males about twice as frequently as females (Warham 1996). At their nest, a pair often croak together and set-off a chorus in surrounding nests. Next to repeating the action, it is most likely to be followed in both sexes by pointing (Warham 1996). Croaking is sometimes heard away from the nest during conflicts involving several birds and occasionally at sea among feeding frenzies.

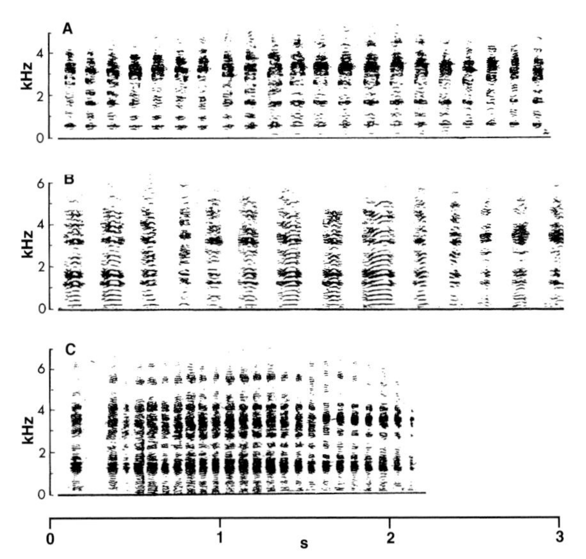

Fig. 15.5 Sonagrams of CROAKING in the Tasmanian Shy (A), Campbell Black-browed (B) and Grey-headed Albatrosses (C) (Warham & Fitzsimons 1987).

Pointing (rapier action, kiss, bill-pointing, bill-aligning, bill-touching, bill-rubbing, bill-nibbling): the lunge of a swordsman – rapier action – (Richdale 1949a) is an apt anthropomorphism; it has been abandoned here only for the sake of conformity with other species. Pointing is the common approach of one mollymawk towards another individual. The bird stretches its neck and head in a straight line towards the recipient (Fig 15.6) (Plate 3); frequently the bird lowers its centre of gravity and pivots forward to increase the upward angle (Fig. 15.7).

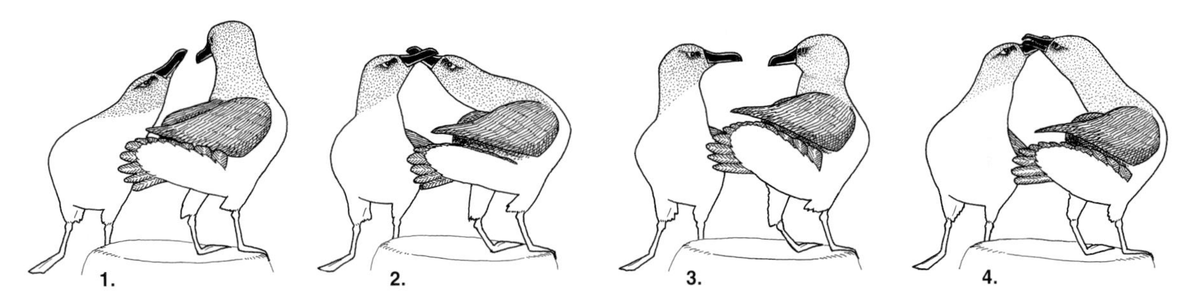

Fig. 15.6 The Grey-headed Albatross on the nest was about to step off, but was inhibited by a bird approaching and POINTING from the left (1). It responded with two POINTS; the second of these ended with a SNAP (4) (2 s duration) (after Tickell 1984).

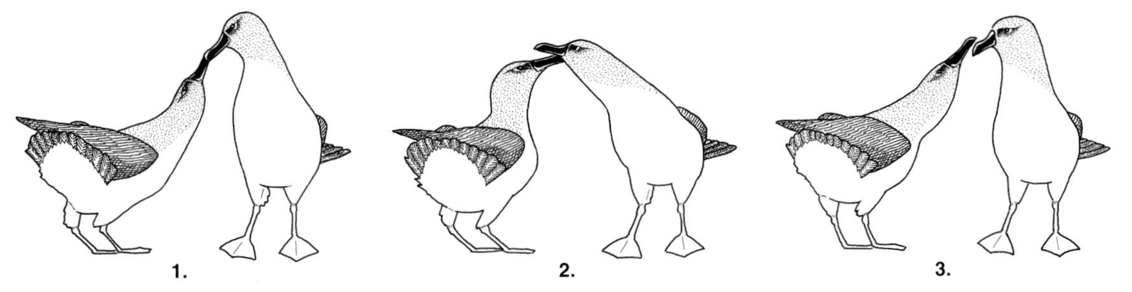

Fig. 15.7 POINTING: the Grey-headed Albatross on the left (1), making a low approach, reveals the expanded throat that accompanies throbbing vocalisation. When the right hand bird responds the bird on the left draws its head back (2) before making another, slightly higher approach (3) (2 s duration) (after Tickell 1984).

Less often, the target is some other part of the body. The tail is fanned. Withdrawal from such an approach is often accompanied by a reciprocal point from the partner and in duelling fashion, each bird points as the other withdraws, the two rocking back and forward in a rhythmic tempo. Eventually, as confidence increases, the two bills are held close together for a second before being rapidly withdrawn (bill-aligning).

While the point posture is held, the mandibles may be slightly open, with the throat visibly swollen (Fig. 15.7). This was first described in Grey-headed Albatrosses by Niall Rankin (1951) who did not hear the deep throbbing or groan audible when close to the birds. It has been recorded in both sexes of the Grey-headed Albatross and in the male Southern Buller's Albatross (Tickell 1984, Warham 1996).

Clashing (bill clashing) is a negative response to pointing. An extended bill is pushed sideways and if the pointing is repeated the clashing becomes more vigorous and noisy.

Jabbing is a more aggressive rejection of pointing. A bird standing on a nest vigorously thrusts its bill downward toward the approaching bird.

Wail (gaping, bray, trumpet): Black-browed and Grey-headed Albatrosses flying into crowded colonies often open their bills wide, seconds before touch-down, and let out a strident wail; most birds land without alarm. These single notes are also heard from birds stepping onto or off a nest, and at the moment of take-off, starting just as a bird gets airborne and sounding-off until it has cleared nearby birds on the colony edge. Wails are also made by harassed birds in the midst of others and often accompany abrupt flagging of the head from side to side (Fig. 15.8). They last several seconds and possess characteristics of species and individuals. The single prolonged note (Fig. 15.9) has many harmonics up to about six kiloHertz, often with slight frequency modulation. In the Southern Buller's Albatross wails are more frequent in males than females.

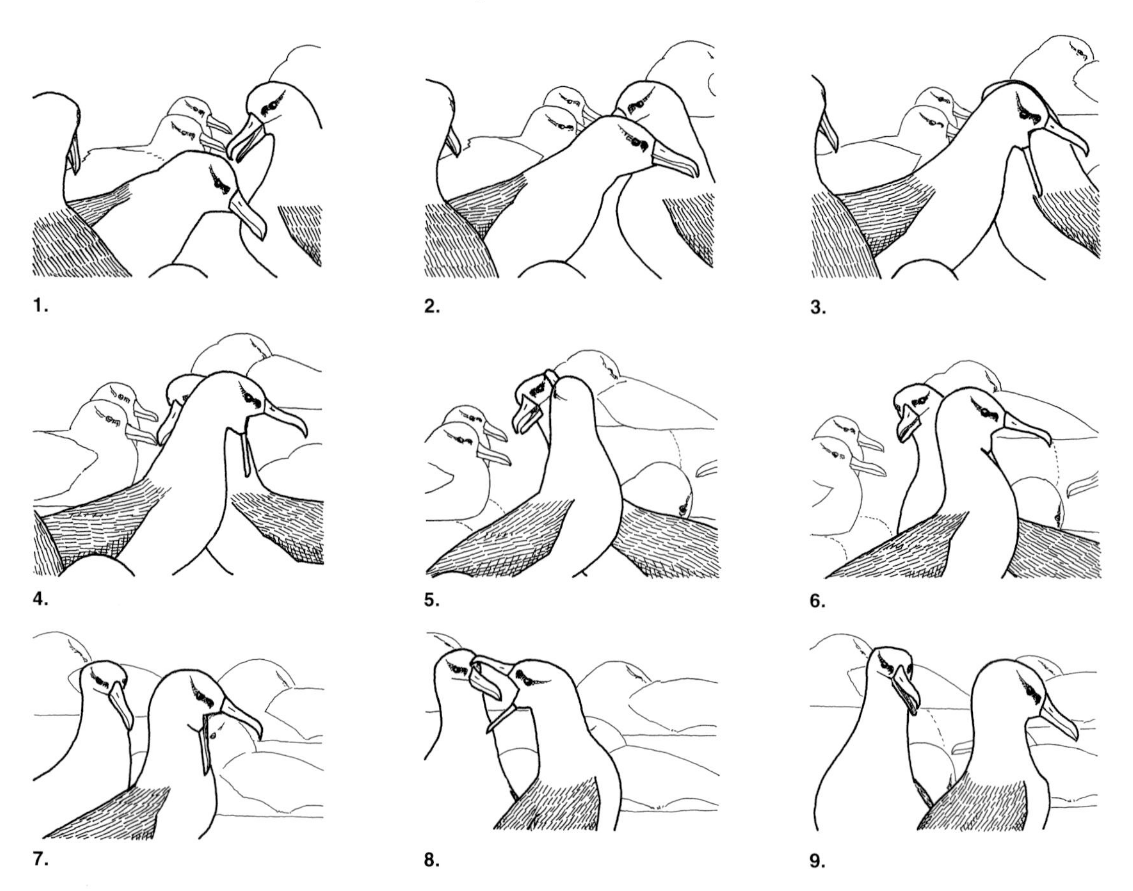

Fig. 15.8 A Black-browed Albatross WAILING and FLAGGING (3–8) as it walks through a crowded colony (4 s duration).

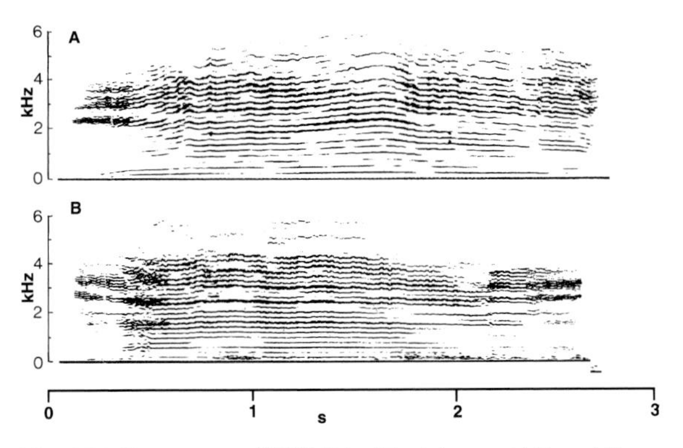

Fig. 15.9 Sonagrams of WAILS by Black-browed (A) and Grey-headed (B) Albatrosses (Warham & Fitzsimons 1987).

Nod is a quick, shallow forward bow and recovery involving just the head and upper neck.

Bow (sub-belly bow) is a deep bow with fanned tail. The bill almost touches the feet, sometimes to the left or right, even alternating; more rarely it may be directed backwards between the legs.

Neck-preen (bow, scapular action): the head alone bends down right or left and the bill enters contour feathers at the base of the neck with audible, muffled preening snaps The tail is fanned.

Scapular Action is a male action in response to a female. It is one of the most extreme examples of ritualisation in albatross behaviour and almost exclusive to mollymawks. From a standing position, the bird's neck and head swings through 180° so that the bill is pointing backwards along the mid-line of the back (Fig. 15.10A, B). As the head turns, the folded wings are lifted slightly; the tail is fanned widely and in some species twisted very slightly to one side. The bill is held rigidly just off the feathers between the lifted wings, or is pushed under the opposite wing, and is accompanied by muffled and barely audible groans. The action is quick and jerky. There is no preference as to which side it is performed, but individuals tend to repeat on the same side (Warham 1996).

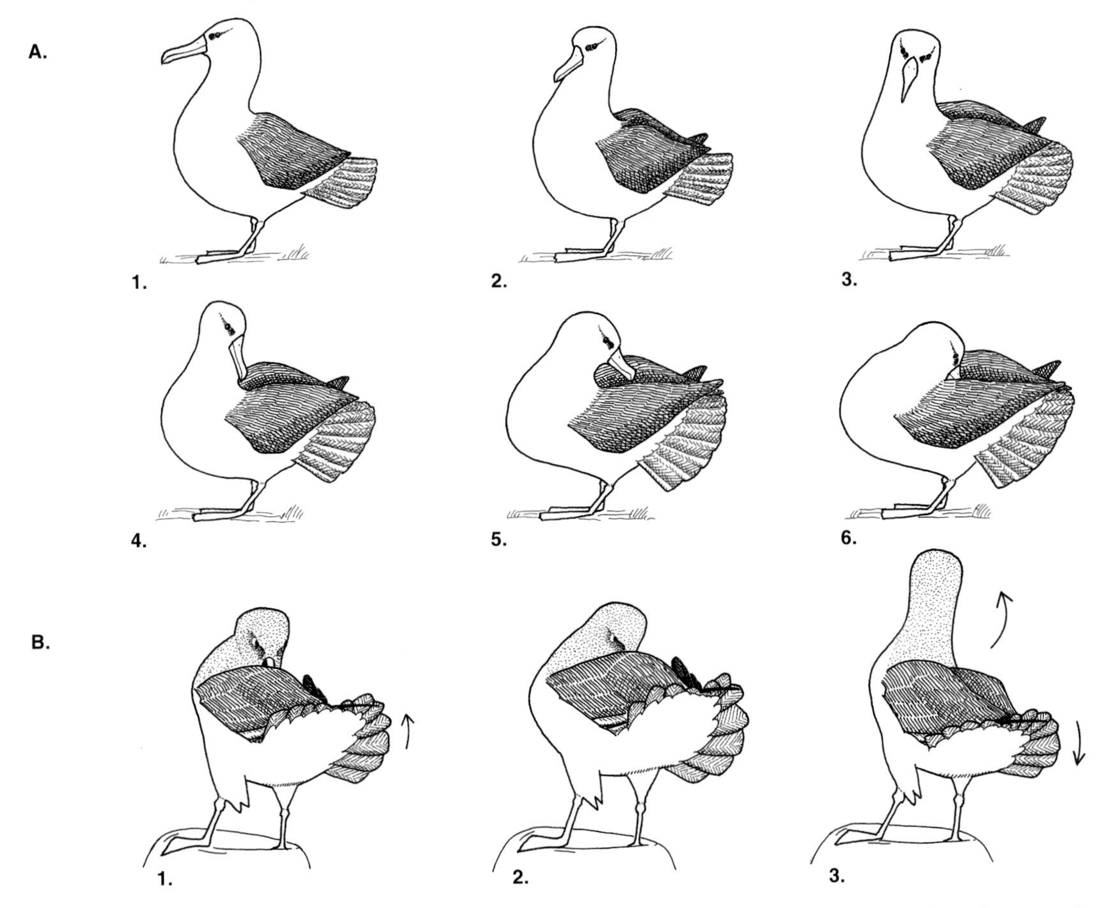

Fig. 15.10 SCAPULAR ACTION: A1–6 by a Black-browed Albatross (10 s duration) and B1–3 recovery from the posture by a Grey-headed Albatross (3 s duration) (after Tickell 1984).

Side-preen (scapular action) is also a male action. The head is lowered right or left and the bill enters contour feathers, just in front of or below the folded wing. Preening movements and sounds are made as the bill is pushed into the feathers.

Leg-preen (leg action) is a deeper version of side-preen. The standing bird reaches right down over the folded wing to preen in the feathers of the lower leg (tibiotarsus); the tail is fanned. My original description and illustration (Tickell 1984, Fig. 14c) was captioned incorrectly and refers to side-preen.

Scooping: 'The head and bill of the active bird are dipped sharply to the ground in front of the feet in the same line as the body. The bill is then pushed forward and up straight out in front to the body, thus giving a peculiar bobbing and scooping action. The tail is spread widely like a fan. The bird does not stand erect, for the legs form an acute angle at the heel so that the legs are very much bent. The movement is performed several times rapidly. As well as the bobbing and scooping of the head the body also moves up and down on bent legs.' (Richdale 1949).

Scooping is an intention movement (1) before take-off, (2) before getting onto or off a nest, (3) when approaching or passing other birds and (4) before clashing. Sometime the intention is not fulfilled. No sound is uttered. When moving, the feet make a slapping sound on the ground.

Flagging: in crowded colonies, nervous birds walking between nests abruptly turn their head from side to side. Quite often the action is accompanied by loud wails (Fig. 15.8).

Copulation

There is no obvious preliminary behaviour. A male merely steps onto a female's back and gets his claws hooked over the leading edge of her wings, which may be forced slightly open. Co-operation is essential as males may easily be shaken off, particularly from tall nests. Once mounted, the male may stretch over and around the female's head to start allopreening her breast. Allopreening always occurs in the Black-browed and Grey-headed Albatrosses at South Georgia, but was not included in descriptions of the Southern Buller's Albatross (Richdale 1949, Warham & Fitzsimons 1987). The female's bill is held horizontal and the male's, pointing downwards from above, taps it rapidly from side to side. Richdale called this action **Tattoo**; it continues while the male brushes his tail from side to side until the female lifts hers and cloacae make contact. Tattoo stops abruptly while the male pauses, then resumes. In Southern Buller's males, a low frequency, rapidly pulsed sound accompanies the final moments of copulation, ending in a much higher squeak (Warham & Fitzsimons 1987). The pause and squeak perhaps coincide with ejaculation.

The whole process lasts one or two minutes. As many as eleven mountings in two hours by a pair of Southern Buller's Albatrosses have been seen (Richdale 1949b) and in one Grey-headed pair, copulation occurred six times in half an hour (Tickell 1984). Immediately the tails disengage, the male jumps off and the two birds usually remain side by side, allopreening, although often the male forces the female off the nest.

Southern Buller's females vigorously resist attempts at forced, extra-pair copulation, sometimes made by more than one male (Warham 1996).

SOOTIES

Sorensen (1950b) made few comments on the behaviour of the Light-mantled Sooty Albatross at Campbell Island, but he quoted the keen observations from an unnamed member of the 1941 Cape Expedition.

Displays of the two sooty albatrosses were described by Berruti (1981) in a comparative study at the Prince Edward Islands (1974–75), and French scientists at the adjacent Iles Crozet (1977–78) made an innovative study of Sooty Albatross behaviour (Jouventin *et al.* 1981, Jouventin & Weimerskirch 1984b).

The Sooty Albatross and Light-mantled Sooty Albatross have similar behaviour. Both species breed at isolated nests or in small clusters on steep cliffs. At some islands, the Sooty Albatross forms loose colonies on ledges or cliff-tops, but never as dense as those of mollymawks. At most locations there are few prospecting birds and little space for the birds to walk about. Displays usually involve two individuals that come together when one lands at a nest-site occupied by the other. Potential partners often fail to stay together or even return to each other due to inept behaviour. Only rarely are three or more birds involved and the dance anthropmorphism is not as apt as in other species.

In 131 Sooty Albatross behavioural sequences, totalling 7,639 recorded events by birds of unknown age, Jouventin *et al.* (1981) defined 20 postures and actions (Table 15.1). None were exclusively male or female (Fig. 15.11).

Table 15.1 (right) Names of behaviour patterns of the Sooty Albatross with abbreviations (after Jouventin *et al.* 1981).

LANDING	LD	SKY-TWIST	ST
ADVANCE	AD	HEAD-FLICK	HF
RETIRE	RT	FLAGGING	FLG
FLY OFF	FLY	GAPE	GP
ALLO-PREEN	ALP	GAPE DOWN	GPD
AUTO-PREEN	ATP	FOOT LOOKING	FLK
WING-SHAKING	WS	BELLY-TOUCHING	BT
NEST-BUILDING	NB	CIRCLE	CIR
POINTING	PT	CLASHING	CLA
SIDE-PREEN	SP	SKY-CALL	SC
SCAPULAR ACTION	SA		

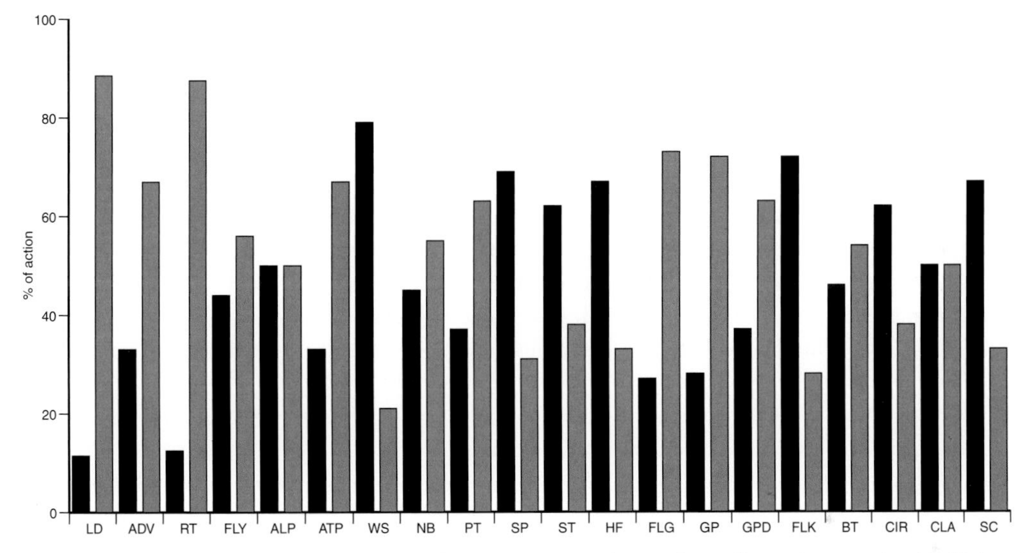

Fig. 15.11 The Sooty Albatross dance: the proportion of the component actions performed by each sex; males (black), females (grey). Abbreviations as in Table 15.1 (after Jouventin *et al.* 1981).

Aerial activity

Impressive aerial activity is characteristic of both sooty albatrosses. Close formation flying adjacent to cliffs with nest-sites is a silent male-female activity, but Richardson (1984) has reported a muted sky-call when a following bird attempted to touch the tail of a leading bird. Formation flying is not a behaviour of young, newly arrived subadults. It occurs only after early nervousness has been overcome, and before the pair-bond has become established. Breeding adults are too occupied with the nest to indulge in this activity.

Aggression

The two sooty albatrosses are territorial. Examples of overt aggression are rare, but the following episode reveals considerable aggressive potential:

> 'One adult [*P.*] *palpebrata* from an airborne group of 13 alighted near 7 settled [*D.*] *exulans*, which were investigating garbage near the stern of *Eltanin*. On approaching the larger birds, the [Light-mantled Sooty] was immediately challenged by an adult male and an adult female Wanderer, which clappered their bills and grunted at the newcomer. The [Light-mantled Sooty] abruptly seized one of its own wings at the carpal flexure and began fiercely savaging it. With its body tilted to one side and slowly beating the air with one foot, the bird battled with itself for 20–30 s, after which it swam away to preen with unusually rapid and jerky movements of its head. After drinking some water it took to the air.' (Harper 1987).

Landing and taking-off

On most albatross breeding grounds, landing and taking-off are not socially remarkable events, because walking is the mode by which birds usually encounter and leave one another. They assume significance in sooty behaviour, because landing and taking-off happens close to the activity and are so much more decisive than walking. The large proportion of landings by females (Fig. 15.11) is consistent with visiting nest-sites which are occupied by males for longer periods. Displays are usually concluded by the female taking-off and the male following; when just one bird takes-off it is twice as likely to be the female as the male.

Stereotyped actions

On their remote ledges, sooty albatrosses communicate over much greater distances than other albatrosses. Calls to birds in flight echo around the surrounding cliffs and are always heard long before the sombre birds are seen. Sky-postures have assumed a prominent role in a comparatively limited hierarchy of actions. The tails of these two species are not only fanned, like those of mollymawks, but presented to the partner by twisting forward to one side or the other. This posture may be prolonged, with the tail returning slowly to its normal position, often after the bird has progressed to another action. It appears to reinforce many actions and is most noticeable at times of excitement.

Stare: a standing bird looking intently at a partner usually raises the contour feathers of the crown, nape and mantle, creating a distinctive transverse crest which Berutti (1981) identified as threatening.

Sky-call (*chant*) is characteristic and distinct from sky-vocalisations in other albatrosses. Typically, a standing bird swings its head towards the vertical (Fig. 15.12) and, through slightly parted mandibles, gives a loud drawn out *ya-aaaaa* call. It is followed by a less shrill *eee-eeee* inhalation sound (Fig. 15.13), as the head is quickly lowered and pressed against the feathers of the neck. Duetting is characteristic of mated pairs at the nest; the male's call is followed immediately by the female's which is distinguishable by its lower, rougher tone. Individual voices are recognisable. Sitting or incubating birds perform a less intense version in which the head is not raised as high and the call is not so loud with the inhalation barely audible (Fig. 15.14). In keeping with their prolonged occupation of nest-sites, males sky-call twice as often as females.

Fig. 15.12 A Sooty Albatross making a SKY-CALL (2 s duration).

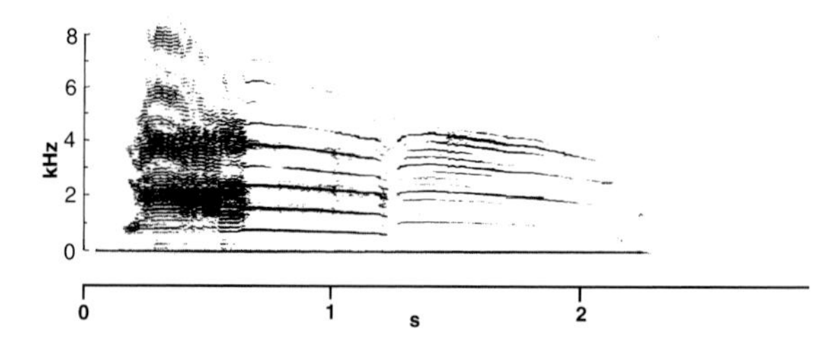

Fig. 15.13 Sonagram of a SKY-CALL from a Light-mantled Sooty Albatross at the Antipodes Islands. The call begins noisily, then several harmonics emerge until the bird is out of breath, when the fainter inhalation sound follows (Warham 1996).

Fig. 15.14 Light-mantled Sooty Albatrosses at a nest performing low intensity reciprocal SKY-CALLS (1–3) (2 s duration).

Flagging (regard): the head is turned sharply from side to side to look at the mate alternately with the right and left eye. It has been explained as a means of distance perception (Jouventin *et al.* 1981).

Pointing (bill pointing, *baiser*) in the manner common to all albatrosses, the bill is closed, but may make slight nibbling movements. Where accepted and reciprocated, contact is usually brief and gentle, leading to mutual allopreening. Rejection of the approach differs in the two species (see Clashing and Thrusting).

Gape (*bec ouvert*) begins as pointing but ends with the bird opening its bill wide before withdrawing. This may be an aggressive defence of the nest, but is also a ritualised component of the dance. A downward directed variant **Gape down** (*courbette-bec-ouvert*) occurs (Jouventin *et al.* 1981).

Clashing (bill clashing, *aiguisage des becs*) is characteristic of the Sooty Albatross. It is a joint action, but initiated and probably maintained by one of the participants. Bills that come into contact are pushed from side to side with jerky strikes and parries that can be heard some distance away. This can be aggressive combat ending in the eviction of the incomer; in displays it is ritualised and, at high intensity, the tail is fanned upwards and twisted forward to one side.

Thrusting (bill-thrusting, scooping) is characteristic of the Light-mantled Sooty Albatross. Two birds standing close together, with their crown plumage raised in distinct lateral crests, engage in unique reciprocal gaping action. No contact is made, but the pink linings of the mouth are revealed (Sorensen 1950). Both birds begin by moving their heads from side to side. The head of one is thrown up towards the partner and the bill opens wide (Fig. 15.15); the lower mandibles are held stationary and horizontal, while the upper bill is lifted. These gapes, alternating between the two birds, are synchronised to the lateral movements, such that the gape of one bird coincides with the partner looking away to one side. The tail is fanned upward and twisted forward to one side where it may be the target of points and snaps from the partner (Berruti 1981).

Side-preen (*mouvement scapulaire*): the common ritualised preening on one side of the lower neck or breast. It is usually on the opposite side to the partner and the bill may just touch the feathers (Fig. 15.16) or plunge deep into the plumage.

Scapular action: as in mollymawks the head is turned through 180° and enters plumage along the mid-line of the back between the folded wings. It is performed rarely and accompanied by a twisting of the upward fanned tail to one side (Fig. 15.17).

Foot-looking (*bec au pattes*) is a sudden deep bow towards the feet or between the legs. There may be a slight nibbling movement of the otherwise closed bill and sometimes the belly plumage is touched (bec au ventre) (Fig. 15.18).

Sky-flick (*cou de tête*): a bow is followed by throwing the head up quickly with slight backward neck-roll; there is no sound and the bill immediately returns to its normal position.

Sky-twist (*secouement de tête*): a quick, high head-shake. The head is raised to about 40° above the horizontal and rapidly rotated to right and left (Figs. 15.15(3), 15.17(1–2)).

Circle (paddle walk, *marche ressort*): when a bird moves around a nest on a ledge, the feet are lifted and placed with an exaggerated up-and-down manner and the tail is fanned. Throughout the movement, the head is turned to face the partner on the nest, who rotates to keep eye-contact.

Display

Some of the above actions appear to be associated and were grouped for analysis (Appendix 13). Jouventin *et al.* (1981) used theoretical techniques to distinguish four levels of performance – A, B, C and D – that were believed to be related to age, although the birds had not been ringed as chicks, so their real age was unknown.

Level A probably included the youngest Sooty Albatrosses who had recently returned to the island. Their displays were varied, but not yet appropriate to their sex. Responses were slow and often inappropriate, limited to long exchanges of flagging and side-preen that could be confused with common autopreening. There was little effective 'dialogue' because partners responded to each other inappropriately with the same action, for example side-preen to side-preen. Males used side-preen twice as often as females and sky-twist occurred most often at this level when young males made the greatest effort to attract females. The remarkably low incidence of gape was

Fig. 15.15 Light-mantled Sooty Albatrosses performing reciprocal THRUSTING (1–2) with SKY-TWIST at (3). SIDE-PREEN begins at (4) accompanied by twisting tail to which the partner responds by POINTING (6). The sequence ends with a brief SKY-CALL (9) (7 s duration).

Fig. 15.16 A Light-mantled Sooty Albatross seen from behind and performing SIDE-PREEN. Note the fanning and twisting of the tail to the opposite side from the preening.

caused by the quick departure of nervous females, and synchronised flying did not follow, because they took-off prematurely.

Level B comprised slightly older birds of an indeterminate age. Movements of the female played a major part at this time; she was still docile, but stayed near the male long enough for sequences to develop and when she took-off, the male was inclined to follow and fly closely in synchrony for a short while before returning to his nest-site. The female's attention at this time seemed to be captured by the male's side-preen.

Fig. 15.17 Light-mantled Sooty Albatrosses making a brief SKY-TWIST (1–2) followed by SCAPULAR ACTION (3–4) (2 s duration).

Fig. 15.18 A Light-mantled Sooty Albatross FOOT-LOOKING (2 s duration).

Exchanges were dominated by side-preening, but closer proximity led to attempted allopreening, often thwarted by clashing. Some males were only slightly more active than at level A. but others were more vigorous, with head-flicks and side-preen exchanges which accounted for a third of all male behaviour. They were twice as active as females and their propositioning was characteristic of level B.

Level C behaviour may be attributed to birds that were well on the way towards establishing a pair bond. In marked contrast to the previous level, females appeared aggressive and were far more active than males, initiating three times as many interactions and hardly ever being displaced. Flagging and pointing were alternative responses to side-preen and probably represented unsuccessful male attempts to allopreen partners. Frequent gapes indicated aggressive responses by both sexes at this time. Attempted pointing also set in motion clashing interactions which were characteristic of this level.

Level D. Breeding Sooty Albatrosses behaved in this manner. Their vocabulary declined to little more than sky-call, clashing and allopreening shared equally by both sexes. One notable difference was that females allopreened their mates only after a male sky-called, while males required no such invitation. Allopreening was also an unexpected response to clashing and probably persisted only because of the intimate relationship between established pairs.

The increase in clashing from level A to D by both sexes, and particularly the high incidence at level D, suggest that this action is not, as it appears, aggressive. Jouventin *et al.* (1981) suggest that it is an outlet for the tension generated by the proximity of the two birds. As such, it perhaps acts as a safety-valve preventing confrontation and offsetting a tendency for the female to leave every time a display degenerates into aggression. Male clashing at level B possibly reflects increased presence of females on nest-sites and male anxiety to reoccupy (Appendix 13).

GREAT ALBATROSSES

Anecdotes of great albatross behaviour were included in the early accounts of many who visited their colonies. Harrison Matthews' (1929) description of Wandering Albatrosses at South Georgia was perhaps the best, and became widely quoted (Murphy 1936).

From 1937 to 1950, Richdale (1950a) acquired an intimate knowledge of Northern Royal Albatross behaviour by watching a few ringed birds at Taiaroa Head, New Zealand. He established that once they became breeders, Northern Royals ceased to participate in the dance and that no complex behaviour was evident in the years after a pair had first bred.

In the early 1960s, I studied behaviour of the Wandering Albatross at Bird Island, South Georgia and Edouard van Zinderen Bakker (1971) followed at the Prince Edward Islands. After several seasons of field-work at the Iles Crozet and Ile Amsterdam, Pierre Jouventin and Benoît Lequette (1990) applied analytical techniques to the Wandering and Amsterdam Wandering Albatrosses that had been used previously on the Sooty Albatross. About that time (1983–86), Simon Pickering also analysed the behaviour of the Wandering Albatross at Bird Island.

Very broadly, royal and wandering albatrosses have similar repertoires of behaviour that include the full extension of wings not seen in any other albatrosses, but there are clear differences in actions and vocalisations. Vocal differences have also been recognised between three island breeding grounds of the Wandering Albatross.

Aerial activity

At Bird Island, South Georgia, Wandering Albatrosses engaged in any stage of breeding do not linger in the air over the colony. Incoming birds may make a pass before landing, especially at the very beginning of the season, but departing birds leave the island quickly. Later in the season when subadults are about, there is much more flying, particularly in the late afternoons. Young subadults may spend more time in the air, but after some experience, males land without delay. Females, however, usually make several passes over display areas before landing. In gentle breezes these passes are low and very fast, but into strong winds they can be slow. On several occasions, Pickering saw a female hang in the wind and touch the bill of a male sky-pointing from the ground. The Northern Royal Albatross flies repeated circuits over Taiaroa Head, often two or more birds together (Richdale 1950).

Sky-pointing on drooping wings, with attenuated calls, has been seen in the air over some breeding grounds of royal and wandering albatrosses (Richdale 1950, Warham 1976). It is apparently more common at some islands than others; I did not see it at all among Wandering Albatrosses at Bird Island. These displays and vocalisations in flying birds reflect those on the ground; Amsterdam Wandering Albatrosses displaying together in the air continued the activity together after they had landed (Lequette & Jouventin 1991b). Warham (1976) described similar behaviour among Antipodes Wandering Albatrosses, but in a later report of the same observation (Warham & Bell 1979), added that the head was swung from side to side in flight; this was an error (J. Warham pers. comm.).

Walking

On land the Wandering Albatross usually walks with head down and a rolling gait that gives them a peculiar hump-back appearance (Fig 15.19); but they are capable of walking upright and do so most often during a dance. The Northern Royal Albatross on the paths of Taiaroa Head usually walk upright with a slightly swaying gait, but among the tussocks of Campbell Island, Southern Royals also adopt the head-down profile (Sorensen 1950a, Warham 1996). Wings are are often extended for balance when walking.

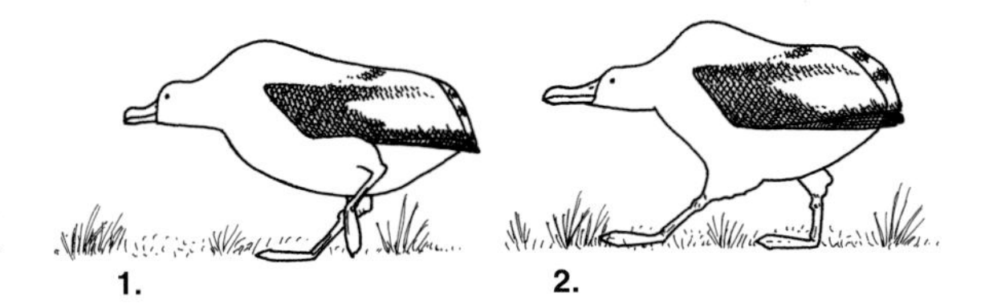

Fig. 15.19 The characteristic walk of a Wandering Albatross, at each step the head moves from side to side (1–2).

Stereotyped behaviour

Among Wandering Albatrosses at the Iles Crozet, most stereotyped actions are performed significantly more by males than females (Jouventin & Lequette 1990).

Yapping (yakker): a Wandering Albatross opens and closes its bill, uttering gruff *waa-waa-waa* vocalisations at about two to three *waas* per second, and at the same time moving its head up and down. Two birds together at a nest often duet in synchrony. They get quite excited with the head movements become faster (Fig. 15.20). The Northern Royal Albatross has more rapid yapping than the Wanderer. There are also significant difference between different populations of the Wandering Albatross. The sounds have strong harmonic structure and enough variables to signal the identity of individuals (Fig. 15.21). They are heard from single birds and more frequently among males than females at the Iles Crozet (Jouventin & Lequette 1990). Duets between pairs or subadult partners at nest-sites are frequent and female Wanderers sometimes sound noticeably higher in pitch than males. Yapping is the call that a chick uses towards a distant parent who responds with the same call.

Fig. 15.20 Wandering Albatrosses YAPPING at a nest (1–2).

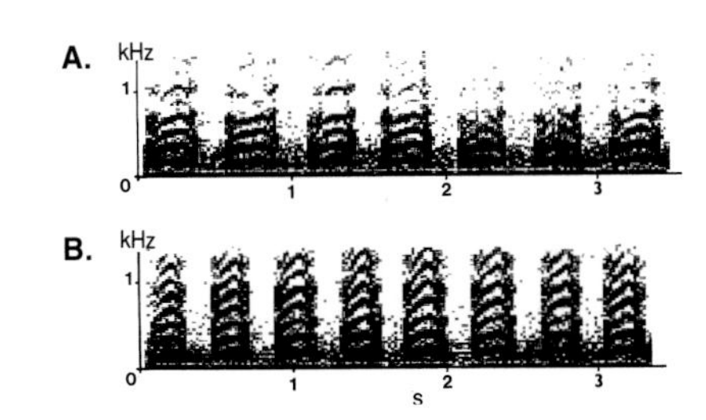

Fig. 15.21 Sonagrams of YAPPING calls by a Wandering Albatross (A) and an Amsterdam Wandering Albatross (B) (Lequette & Jouventin1991c).

Pointing (billing, head forward high, bill touching, breast billing, head forward low, greeting) is an attempt to approach and touch another individual, and is performed almost equally by males and females (Fig. 15.22). The active bird stretches its head and neck rigidly towards a partner. The angle may be up towards the bill or down towards the breast. Pointing may become synchronised with the extended bills of two birds being held close to each other tip-to-tip, crossed or parallel, and perhaps touching, while low groans may be produced (Lequette & Jouventin 1991a). Pointing frequently ends with a single pronounced snap before the bill is withdrawn. It may be reciprocal, one bird pointing

Fig. 15.22 Two Wandering Albatrosses POINTING in the high position. The left bird snaps twice before withdrawing, then both birds POINT low (3 s duration).

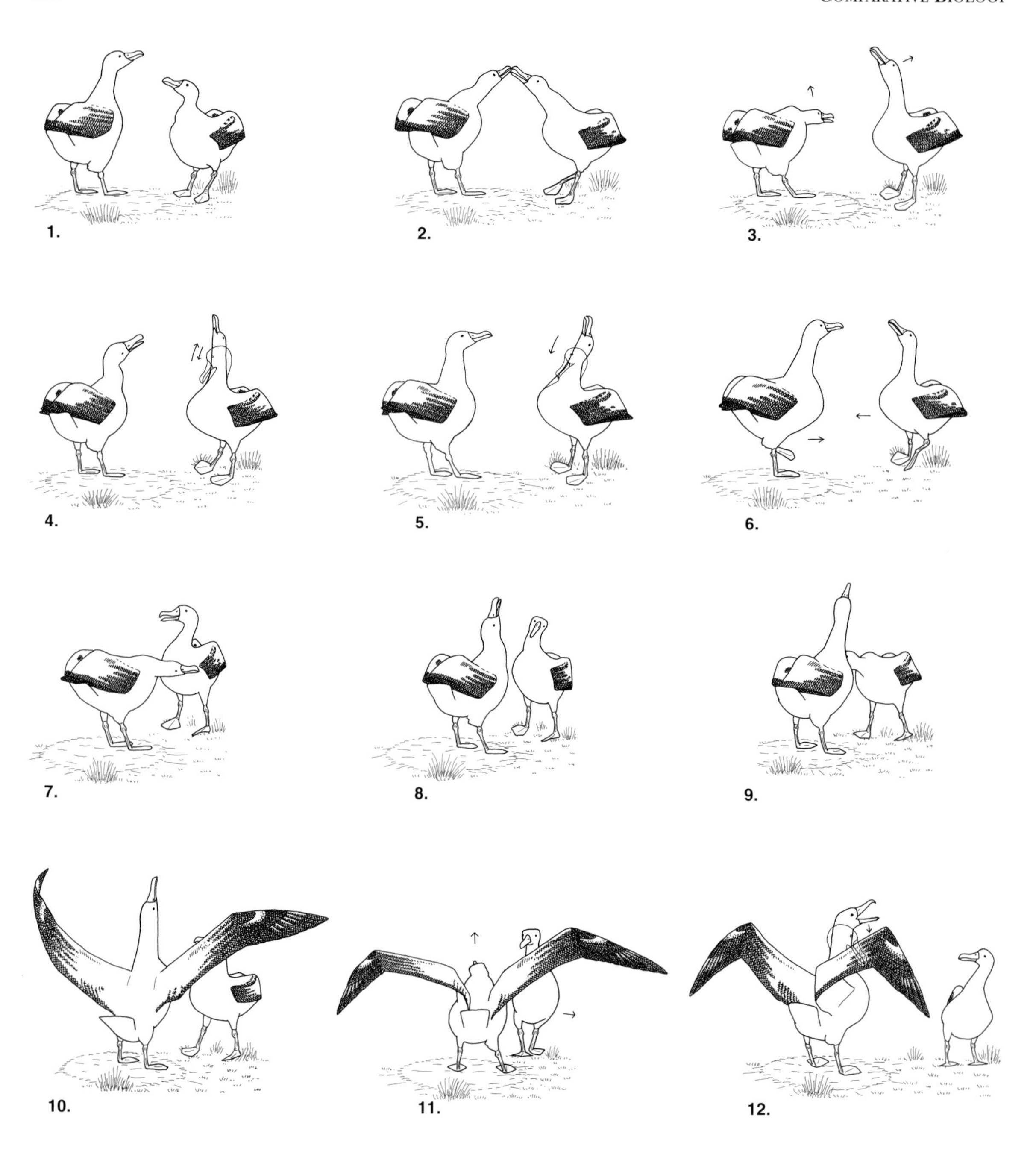

Fig. 15.23 A dance sequence between two Wandering Albatrosses (1–12) involving POINTING (2,7,9), SKY-POINTING (3-6), NECK-ROLLING/VIBRATING (8) and SKY-CALLING (10). Finally, the left bird closes its wings and snaps as the right bird circles with head FLAGGING (12) (24 s duration).

as the other is performing another action and alternating back and forth between the performers (Fig. 15.23).

In the Northern Royal Albatross, pointing low towards the breast has the objective of nipping or pulling feathers just at the moment when the recipient is looking elsewhere. Whether or not this actually happens, a return to normal stance by the recipient causes a rapid withdrawal. The pointing bird has barely a second to touch the breast, so alternation between postures is fast.

Neck roll (arched neck, head roll): a Wandering Albatross curves its neck backwards then stretches it out forwards in a point (Fig. 15.24). In the Northern Royal Albatross there is a much more pronounced backward curvature (Fig. 15.25(3–4)). In both species the action is usually accompanied by bill vibration.

Vibration (bray, rattle) of the mandibles at 15–30 times per second produces a drumming sound which gets faster towards the end and where the final percussion may be overlaid by a low vocal note rising rapidly in frequency (Warham 1996) (Fig. 15.26). It accompanies neck-roll and pointing, frequently ending with a pause and one or more single snaps before the bill is withdrawn.

Fig. 15.24 As the Wandering Albatross on the left approaches, its partner on the right performs NECK-ROLL and VIBRATE. It ends with a single snap (2 s duration).

Fig. 15.25 A dance sequence between two Northern Royal Albatrosses (1–6). The wings of the bird on the right have remained extended from a previous encounter. After POINTING (1) and BOWING (2), the bird on the left performs a characteristic NECK-ROLL and VIBRATE (3,4); note that the posture is much more exaggerated than in the Wandering Albatross (Fig. 15.24) (6 s duration).

Bow (head bob, head curl) is a rapid movement; it is highly variable sometimes involving just the head (**nod**), but usually more of the body is involved. It may be directed forwards (Fig 15.27), or a twisting of the neck presents just one side of the head to the partner (**side-bow**). Pickering has noticed Wanderers occasionally picking up and tossing vegetation when bowing, as if closeness of the head to the ground triggered the quite different action (Pickering & Berrow (in press)).

Snap (single bill snap): a bird makes one apparently spontaneous bill snap; a weaker closing of the mandibles produces a more hollow sound (**clop**). Pointing

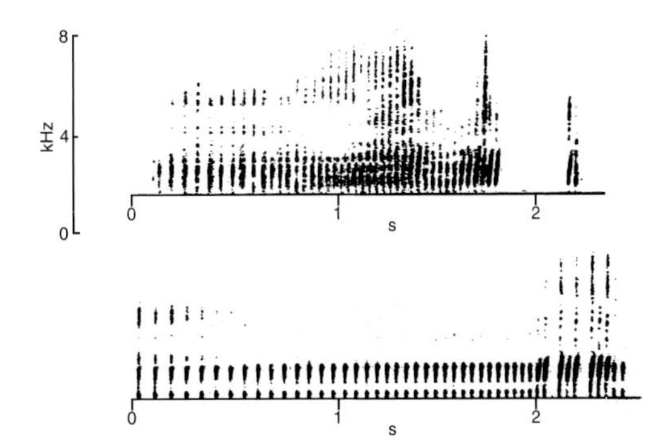

Fig. 15.26 Sonagrams of VIBRATION by Antipodes Wandering Albatrosses. Note the increasing snap rate becoming slurred towards the end and followed by double snaps (Warham 1996).

Fig. 15.27 A Wandering Albatross making a BOW (1.5 s duration).

often ends with one snap which also precedes side-preen and sky-call.

Clapper/yammer (threat) are similar aggressive bill actions. In the Northern Royal Albatross, clappering is a series of loud snaps at a frequency of about eight per second. Yammering in the Wandering Albatross combines snaps at about ten per second with a nasal tone. A bird may clapper/yammer in the direction of an intruder without any movement, but where the threat is greater, as in a dance, a Wanderer yammer becomes shrill and the bird may move towards the intruder or even charge (Fig. 15.33.7).

Side-preen (scapular-action, side-action, display preen, preen breast, snap preen): a standing bird raises its head quickly, makes a single snap then plunges its bill into the feathers on one side of the breast (Fig. 15.28A), sometimes under the carpal flexure of the wing or even lower. Several muffled preening snaps are usually audible. Sometimes the bill is held rigidly just outside the plumage.

Front-preen (preen breast) has less movement; the bill preens feathers of the upper breast or base of the neck (Fig. 15.28B). Both ritualised preening actions are reminiscent of the submissive actions of a chick (Fig. 15.29).

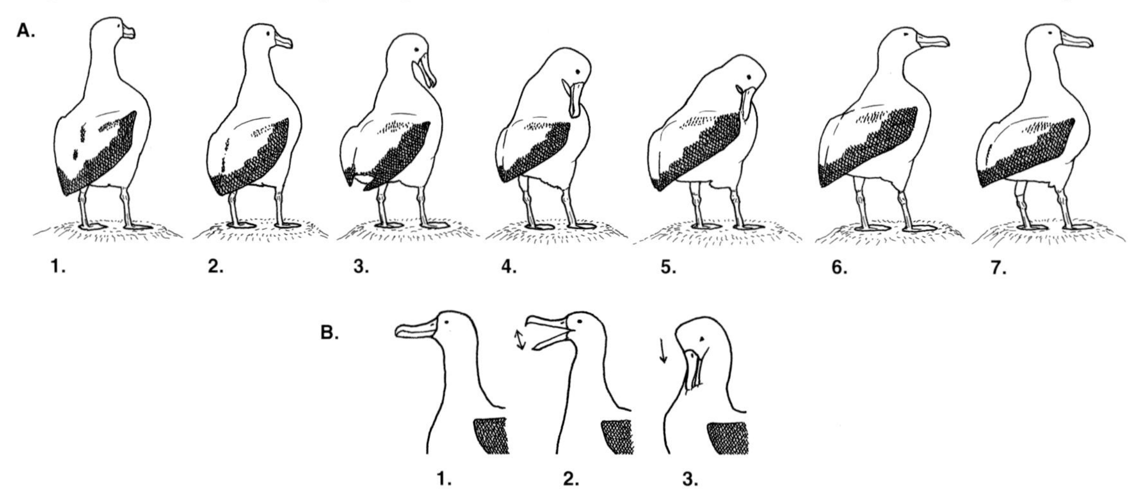

Fig. 15.28 A1–7 Wandering Albatross performing SIDE-PREEN (3.5 s duration) and B1–3 NECK-PREEN (2 s in feathers).

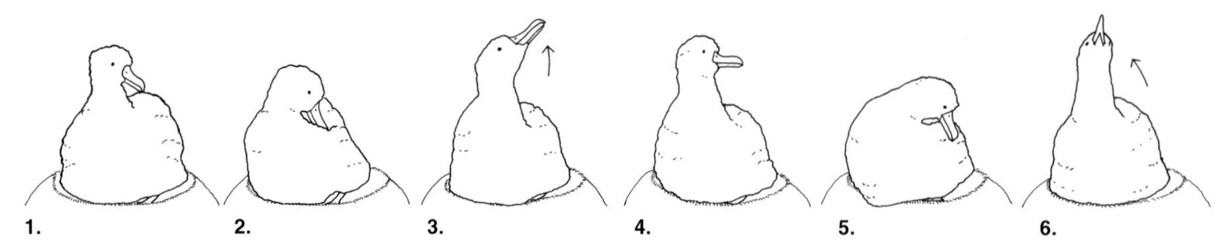

Fig. 15.29 A downy Wandering Albatross chick performing SIDE-PREEN and SKY-POINTING twice (2.5 s duration).

Sky-point (sky-call, sky-position, sky-position call): from a sitting or standing position, the head is lowered slightly then swung to the near vertical (70°–90° from the horizontal). In the Wandering Albatross, it is essentially a quiet action that develops in half grown chicks (Fig. 15.29). In full-grown birds the mandibles are usually parted and a deep groan or gurgle is sometimes heard. The sound may be caused by an involuntary squeezing of the air sacs, especially when a more motivated bird stretches beyond the vertical and leans over backwards. It may be performed more frequently by males than females (Jouventin & Lequette 1990). Several sky-points can occur in succession, but between each one, the neck usually remains stretched and just the head is lowered and raised. At South Georgia, the wings are normally folded (Fig. 15.23), but they may remain extended from the preceding behaviour. In the Amsterdam Wanderer, sky-point is normally accompanied by extended wings (Jouventin & Lequette 1990, Lequette & Jouventin 1991b).

A solitary individual may sky-point to a nearby bird or to a more distant walking individual, or to one flying overhead. It appears to be a fairly low intensity territorial and advertising signal, increasing an albatross's stature and making it more conspicuous.

Sky-trill in the Northern Royal Albatross is a sky-point accompanied by a characteristic vocalisation, altogether different from anything heard from the Wandering Albatross. As the head goes up, the hyoid apparatus is pushed down, so that the neck is visibly expanded (Plate 34). A loud bubbling sound emerges from the open bill, followed by a conspicuous nasal inhalation as the head comes down (Fig. 15.30). Each trill lasts about two seconds and several may occur in succession with the sounds coming to an end before the actions so that the last postures are silent sky-points.

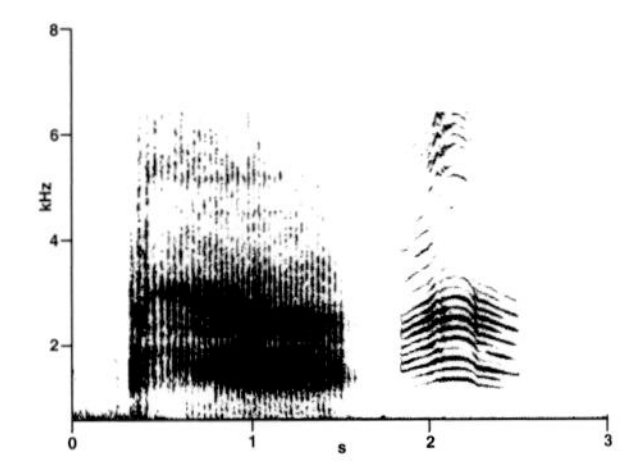

Fig. 15.30 Sonagram of SKY-TRILL by a Northern Royal Albatross. The trilling rapidly increases to about 27 pulses per s. Although two noise-like harmonics with modes at about 1.5 and 2.5 kHz dominate the trill, much higher frequencies are apparent. The pause is followed by an inhalation in which even higher frequences are noticable (from a 1991 recording by Leslie McPherson).

Sky-call is a much more excitable performance. In the Wandering Albatross it usually begins with a single snap by a standing bird before its head is lowered slightly to one side. After a quick gurgle or grunt of inhalation, the head is swung to the vertical, the breast is pushed forward, the fanned tail raised and the wings extended (Fig. 15.23) (Plate 22). Simultaneously, there is a scream lasting two to three seconds. There may be just one action with the head remaining up, but there are quite likely to be several separated by audible inhalations and snaps (Fig. 15.31). The pitch of the screams varies between island breeding grounds. Females are quite capable of screaming, but it is more frequently a male driven action, often ending in a series of yammers, with the male pressing forward and the female stepping backwards. Sometimes, under pressure from a persistent female, a screaming male will step backwards. After sky-calling, one or both birds may keep their wings extended as they continue with other actions.

Quite often a Wanderer will go though the initial head down and grunt, and then stop, as if the intention movement has served its purpose and the rest of the action and scream are no longer necessary.

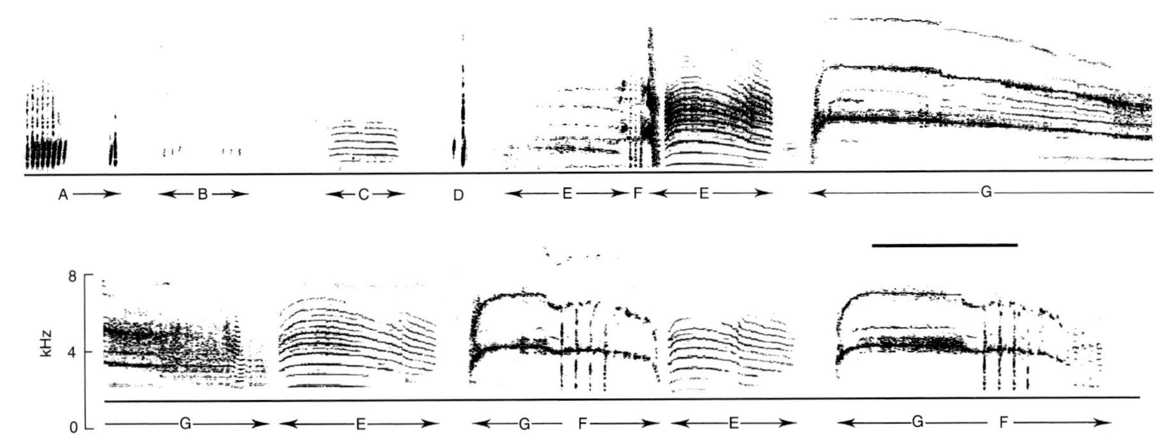

Fig. 15.31 Sonagram of a male Antipodes Wandering Albatross performing a complete vocal display from the end of a vibration (A) and comprising quiet bill clops (B), wheezing inspiration (C), double snaps (D), inspiration (E), bill snaps by female partner (F) and scream (G). The horizontal bar represents 1 s. (Warham 1996).

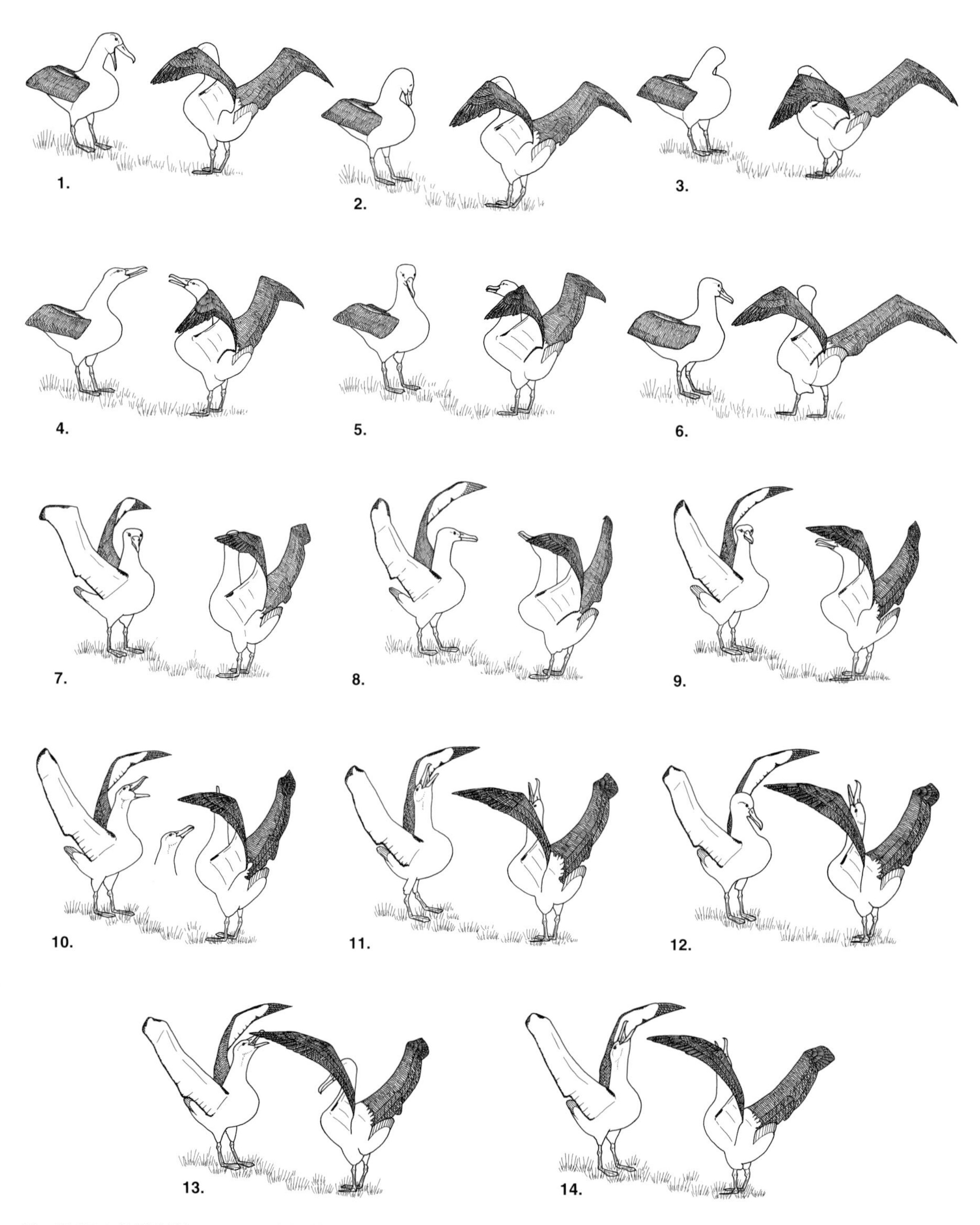

Fig. 15.32 A SKY-CALL sequence of the Northern Royal Albatross (1–14). The wings of the bird on the right have remained extended from a previous encounter. The bird on the left begins with a SNAP (1), followed by a very quick SIDE-PREEN (2,3) and SKY-POINTING (4). The head begins SWAYING at (5) and continues (6–9) with the wings opening at (6) and remaining open (7–9). SKY-CALLS begins at (10); the screams through the open bill (10–11, 13–14) cease when the head is lowered and high-pitched inspirations are heard from the closed bill (12) (16 s duration).

In the Northern Royal Albatross sky-call appears quite different (Fig. 15.32). Richdale named the sequence 'head-shake-and whine', but the main action and vocalisation is a sky-call. It begins, as it does in the Wandering Albatross, with a quick inhalation and the head lowered to one side[5], but instead of the head going straight up, it sways widely from side to side. As it lifts, the swaying diminishes and the wings are extended. At full stretch, the bird may rise on its toes, pushing its head upwards. Several screams and inhalations may follow each other as the head is raised and lowered several times.

Both partners in the display may perform together (Fig. 15.32), or one may remain less active. The screams of the Northern Royals are shorter than those of Wandering Albatrosses, lasting little more than one second (Lequette & Jouventin 1991b). Richdale (1950a) considered them ecstatic emotional states; today they may be interpreted as high intensity responses to the disturbing proximity and/or movements of partners.

In both **sky-trill** and **sky-call**, the bill is often open and it has been suggested that the black cutting-edges, visible in the two royal albatrosses and the Amsterdam Wandering Albatross, have a signal function (Lequette & Jouventin 1991b). In the absence of experimental evidence, it can be argued with equal credibility that black pigment (melanin) laid down along such precise lines on sharp edges is far more likely to have something to do with the cutting function of the bill, perhaps hardening the keratin.

Flagging (shaking head, head wag, walk): a Wandering Albatross walking upright during or near dances turns its head from side to side. These are determined movements easily distinguished from the rapid head-shaking, bill rattling action used to settle feathers.

Dance

Three-quarters of Wandering Albatrosses dancing at Bird Island are unmated subadults from different cohorts and the remainder are experienced breeders who are not currently breeding (Pickering & Berrow (in press)). Dance locations may be nest-sites, but there are spacious areas of short grass where birds gather and perform communally, rather like lekking grounds. Early in the season, breeding males sometimes leave their nests to join in, but breeding females are rarely sociable.

Wanderer dances are dynamic activities and the core performers attract other individuals, who attempt to join in and may be driven off by more aggressive participants of their own sex (Fig. 15.33). The activities of colour-ringed birds are easily observed, but are not so readily understood. In the following sequences, reproduced from field-notes, bold capitals indicate the colour rings of males (eg. **RMR** = red-mauve-red) and plain capitals females (eg. BOG = blue-orange-green).

> '16.12.1963 ...**YSM** and OBP dancing between nests 148 and 84, male energetic display with sky-calls, but female remaining calm and responding with pointing, bows, vibrating, side-preen and side-stepping. Most of the male's display is with wings extended; the female's remain closed. **YSM** very forcing, the position and performance varying as he presses forward towards the female. OBP backs away with no territorial actions and as soon as **YSM** starts sky-calling she walks around him with head down or erect and flagging... Later **YSM** leads dancing towards 148 and OBP follows. Male sat down yapping with wings still extended. Female sits beside nest, occasionally pointing. Later male still with wings out, yapping and little nest building, female also...
>
> 3.1.1964...**NOM** meets BOG and RNR near nest 218. Both females pointing and vibrating, BOG has wings extended. **NOM** sky-points and BOG sky-points and sky-calls. **NOM** presses forward pointing and BOG sky-calls more. **NOM** sky-calls. When RNR tried to join in she was yammered off so she is no longer participating. **NY** walking nearby is vigorously chased off by **NOM**....**NY** later dancing with three females RNR, YRS and BOG, but no progress because the females keep walking around the back of him. RNR and YRS both sky-calling, they blunder into and snap at each other. **NY** aggressive to all females in the trio...eventually RNR and BOG sit down and quietly watch **NY** and YRS dancing.

[5] This is the moment to which Richdale (1950a) misapplied his own (1949a) term 'scapular-action', a mistake that misled many later workers. The action in fact approaches a side-preen without the bill entering the plumage.

> SBY laid an egg 27.12.63, but her mate **RSM** was not present and after she had sat for seven days she deserted and skuas ate the egg. Five days later, **RSM** turned up and from then on kept company intermittently with two females; trouble occurred only when both females arrived together: 8.1.64...**RSM** at nest 156 with SBY and SPN...the two females attacked each other while trying to start dancing and **RSM** attacked both females in turn. At one time it seemed as if both **RSM** and SBY were together attacking SPN. Whenever **RSM** began displaying to SPN, SBY attacked and drove her off.'

The behaviour of subadults changes as they gain experience. At first they are very shy. A six-year-old male ashore at the beginning of January 1964 avoided all encounters and sat quietly, away from other birds, where he was sometimes seen bowing, yapping or sky-pointing; he would sky-point at a nearby female, but if another male approached, he got up and quickly walked away. By early February, though, a marked change had come over him. He was taking part in dances and employing the full range of behaviour. A month later he was walking after some females and keeping company with others. Females of the same age were likewise shy on arrival, but were drawn into communal dances within a day or so.

A female approaching an occupied nest-site often pauses to sky-point and is normally answered similarly by the male. At Bird Island, 80% of dances are initiated by a female walking up to a male at a nest-site (Pickering & Berrow (in press)). If she comes close enough, there will be pointing and bowing, but frequently there are only a few exchanges before she moves elsewhere.

Fig. 15.33 A dance sequence at the nest of a Wandering Albatross with a male SKY-CALLING at a circling female (1) who gets on the nest (3). The male turns to face her (4) as another female approaches from right and SKY-POINTS as the first female is POINTING (5). At the same time another bird, possibly a male, approaches from the left. The first female gets off the nest (6) and the male drives away the intruder with a YAMMER (7) and turns his attention back to the two females (8) (26 s duration).

If they remain together, the same activity will continue. Pointing and bowing by both males and females are the most frequent actions of a dance. Pointing itself was by far the commonest single stereotyped action and often led to bill touching. Females usually responded with a snap and males with a bow (Jouventin & Lequette 1990, Pickering & Berrow (in press)).

Vibrating is usually preceded by a bow, accompanies neck-roll or point and is followed by side-preen or point in both sexes (Jouventin & Lequette 1990, Pickering & Berrow (in press)). Sky-calls usually erupt in the middle of dances and afterwards the pace of activity may decline. Male Wanderers at the Iles Crozet are more active than females and almost all sky-calls are made by them (Jouventin & Lequette 1990), but at Bird Island, where sky-calls occur in a quarter of all dances (Pickering & Berrow (in press)), both sexes sky-call and there is no bias in which sex acts first or how frequently. Both partners often sky-call together and in communal dances two or three pairs may frequently be sky-calling at the same time. Dancing birds sometimes move forward, pointing towards a sky-calling partner who steps backwards maintaining distance between the two. They may also attempt to walk around or side-step while flagging the head from side to side. Sky-calling birds are disturbed by partners moving behind them and rotate on the spot, or head off the circling bird which may be frustrated and end up marking time.

The sequences of actions in Wanderer dances have been tested at South Georgia (Pickering & Berrow (in press)) and the Iles Crozet (Jouventin & Lequette 1990). Of 29 changes from one action to another (excluding repeats) in males, and 26 in females, 21 were common to both, indicating that the underlying pattern of display is similar in both sexes. In signal-response interactions between the sexes, 34 females responded to male signals and 32 males responded to female signals. Of these, 19 were common to both sexes (56% of female and 59% of male responses were the same), but significantly more females responded by repeating the same action just performed by the male (Pickering & Berrow (in press)).

A dance between one male and one female may last from a few seconds to about 15 minutes, but most last two to six minutes (n=58) (Pickering & Berrow (in press)). In communal dances, where partners are constantly changing, male-female, male-male and female-female interactions all occur, dances are longer, faster moving and chaotic. A female may walk away at any time and most dances end in this way (Table 15.2). Usually the pace of dancing gradually decreases and ends with bows before the female leaves.

How display ends		Observations n (%)
Male and female sit together at nest-site		131 (25)
Male terminates:		
male walks away	18 (4)	
male drives female(s) away	9 (2)	
total		27 (6)
Female terminates:		
female walks off	329 (64)	
another female approaches and first female walks away	11 (4)	
another male approaches and female walks away	19 (2)	
total		359 (69)
Total		517 (100)

Table 15.2 How Wandering Albatross dances end at Bird Island, South Georgia (Pickering & Berrow (in press)).

Leading away: a male who has held the attention of a female on a communal dancing ground needs to lead her away from the other birds, frequently towards a nest-site he has already claimed, but it can be anywhere free from the attentions of other individuals. It is a crucial time, as stereotyped actions subside and the male tentatively walks away. The female usually follows for a few metres, but may lose interest and go off in another direction or return to the activity. Almost two-thirds of all dances are terminated by females walking away. If she continues to follow him, they may progress towards quiescent behaviour, but more often than not something like this happens:

> '3.1.1964 subadult **NOM** and U dancing on green, both with wings extended, he leads her back to nest 219. **NOM** yapping with several sky-points. U folds her wings and sits besides nest. **NOM** suddenly gets off nest and yammers her away aggressively, but she returns and gets on the nest and **NOM** stands nearby. After half a minute he moves towards the nest and U flees.'

Dancing occurs at sea, far from the breeding grounds. On 26 November 1912 the brig *Daisy* was becalmed on a foggy ocean 65 km north of South Georgia and Murphy (1936) saw Wandering Albatrosses among thousands of seabirds on the water around the vessel.

> 'From six to about twenty of them might be together in such groups, and from time to time the huddled albatrosses would suddenly become interested in others around them and would begin to bill and bow, spread their wings, bob their heads, and caress each other by nibbling with the bill. The reaction worked rhythmically, for the initial movements of one bird would spread rapidly through a group, and within a few moments all of them would be spinning on the water, each paying attention first to one neighbor and then to another.'

Three months later (26 March 1913) hundreds of kilometres from land:

> '...the wind was again light and the ocean relatively calm. Scores of Wandering Albatrosses were about in the characteristic gams[6], and the whirlings, raising of wings, billing and squealing were in progress as lustily as during November. Yearling young, in the black plumage, were taking the same part as adult birds...'

Copulation

At Bird Island, the male Wandering Albatross chooses a nest-site and occupies it throughout the pre-egg period. The female makes occasional visits to locate the nest of the year and spends time yapping and allopreening her mate. Copulation usually occurs without any other preliminary behaviour. The male mounts and his trampling usually causes the female's wings to droop partly open at the shoulders. He reaches over the female's head, which is now under his chin, as he rattles her bill from side to side with his own. At the same time, he is brushing his tail from side to side. The movements becomes more rapid, but only when the female raises her tail can he work his tail underneath and achieve cloacal contact. He suddenly stops the bill action, raises his head and pauses for several seconds, which coincide with ejaculation. Tails then disengage and he dismounts. The pair stay quietly together, apart from more yapping or allopreening. Copulation may be repeated several times a day whenever they are together. In the Northern Royal Albatross, the behaviour is much the same, except that there is no bill rattling (Richdale 1950a).

Attempted extra-pair copulations are common. From late November and throughout December male Wandering Albatrosses on nests at Bird Island are in peak reproductive condition and waiting for their mates. Females fly in from time to time and any of them seen walking about the colony are at risk of being pursued by males from the surrounding nests; they run away and, if possible, take off. If a female cannot escape, she snaps at the base of the male's bill which causes him to turn his head submissively. If he persists and succeeds in mounting, her tactics change. Although no longer struggling, many females manage to tip their assailants off at an opportune moment. If this is not possible, a female can remain uncooperative; no matter how much a male brushes his tail she can usually keep hers down and prevent cloacal contact. If ejaculation occurs, the semen is shed onto the female's upper tail coverts or the ground.

The following extract from my field-notes at Bird Island is an example of events taking place in a Wandering Albatross colony during the pre-egg period; the bold colour-codes are again males:

> '11 December 1963: **PBY** on nest near 412 sees RNR walking near 443, he sky-points once or twice then walks and runs towards her. RNR starts running away and is also chased by **WR**; she turns and takes-off but does not have enough speed and comes down near 161, close to where SY is busy building. **PBY** and **WR** return to their nests... **WR** seen copulating with NRB and afterwards she walks towards a nearby nest but is pursued at the run by **RSM**. She reaches the bank and sits; he extends wings, bows and yapps a little, but she does not respond and he returns to his nest... **RSM** runs across the green after NRB, other males sky-pointing and yapping as she passes. When **RSM**

[6] a whalers' name for social visits between ships at sea.

gets close she makes a run to take-off, but crashes down near 430. **WSM** gets off 428 and goes after NRB who starts running. **RSM** again in pursuit, but **WS** leaves his nest and gets there first; he forcibly mounts her.

RNR is crossing the green near 443 when **PBY** sees her, sky-points several times then leaves his nest and goes in pursuit. RNR runs away then turns to make a take-off run but fails to get airborne. **PBY** goes back to his nest. RNR is walking past 81 when **WR** charges out and overpowers her in spite of resistance. **GPO** and **OBO** at different times also go after RNR but seeing **WR** in possession return to their nests. Copulation follows and after a pause, during which **WR** remains mounted, a second copulation follows. **WR** gets off and the two birds sit together occasionally yapping. RNR gets up and walks away and **WR** goes back to his nest and settles down yapping. **PBY** then sees RNR passing, gets off his nest again and goes after her, but she retreats to the bank of the stream.'

Nest relief

A relieving Wandering Albatross approaches the nest and is greeted by the incubating bird with one or two silent sky-points, it usually remains sitting down. The relieving bird may respond with a sky-point, then sits down and both start yapping together. The relief will remain by the nest and long quiet periods will follow, broken only by yapping duets and mutual allopreening. Sometimes there will be sessions of autopreening often resembling side-preen, and vibrating. The relieving bird or its incubating mate may start pulling and placing nest material, perhaps only in a rather careless manner. Sometimes, though, the relieving bird will move a few metres away and dig vigorously; perhaps because the incubating bird refuses to get off the egg. Relieving birds normally remain quietly by the nest and do not force the sitting bird off. I once displaced a sitting male to allow an incoming female to take over, but an hour later the male was back incubating. An incubating bird occasionally sits up and exposes its egg or chick. The relieving bird shows great interest, pointing its bill into the nest, but this is by no means a prelude to handing over and more often than not the incubating bird settles down again. When it does come, change-over happens quite suddenly and without ceremony. The incubating bird gets off and the waiting relief gets on. As it settles down, the new incubating or brooding bird may yap a few times and the off-duty bird immediately starts digging nest material. This may go on for a few minutes or for as long as an hour, so that the sitting bird is left with a good supply of loose grass and moss before its mate walks away and takes-off. Although this behaviour appears thoroughly functional, it is in fact highly stereotyped. A Northern Royal Albatross nesting on bare soil at the Sisters Islets, was seen to walk a long way from its nest to pick up and place a few twigs which were left far beyond the reach of its mate and closer to other sitting birds (Tickell 1991).

During brooding, shifts are much shorter and a relieving Wanderer may visit several times before a chick is handed over. The pair therefore spend much longer together at the nest.

Inter-specific behaviour

All great albatrosses interact with many species of seabirds at sea. Around pelagic food sources, feeding activity can become frenzied, but wandering and royal albatrosses often remain on the outside of the most furious scrambles. Nevertheless, when motivated, they can be as aggressive on the water as on land.

Many breeding grounds of royal and wandering albatrosses are shared with giant petrels which often nest within a metre or so of the albatrosses. At South Georgia, both species of giant petrel return earlier and sometimes take over Wanderer nests still 'occupied' by large fledglings that have walked away. When these fledglings return, they have no difficulty in displacing the giant petrels. By the time the new season's Wandering Albatrosses return, the giant petrels are sitting on eggs or brooding small chicks. The Wanderers mostly ignore them and, even when next to a Wanderer's nest, giant petrels sit tight, making only an occasional token lunge towards the huge birds tramping about them.

Chick behaviour

A very young Wandering Albatross chick does not respond to a visiting parent until it is quite near the nest, but by the age of about ten weeks they commence vigorous yapping whenever an adult lands within sight, whether or not their own parent. As an adult approaches closer, the call changes to a high-pitched whistling. Once a parent is within reach, the chick stretches up and begins begging with rapid tapping of the bill that produces a clattering sound. The target is the base of the adult's upper bill and the action is persistent, apart from pauses when the chick withdraws its head and reverts to whistling. Food is eventually delivered in the characteristic manner and between feeds the chick urgently whistles and taps. After all the food has been passed, the parent usually moves off a short distance and the chick relaxes.

Wandering Albatross parents begin to recognise their own chicks individually at an early age (Tickell & Pinder 1972). It is not so obvious when the chicks recognise their own parents because hungry chicks will continue to beg from a strange, but tolerant adult, even after being yammered off and perhaps bitten.

Long before they begin fledging, Wandering Albatross chicks reveal elements of adult stereotyped behaviour. During pauses between feeds from parents, the usual tapping and whistling of begging is sometimes interrupted by vibrating, just like the adult action. This may be incorporated into the begging sequence of some chicks. The submissive head-away posture adopted by chicks towards adults in response to the mild preening of a parent or aggressive biting from another adult, looks like the precursor of side-preen. Fledgling Wanderers and Northern Royals have both been seen to sky-call at parents (Richdale 1950a). Sky-calls occurs among fledgling Wanderers exercising their wings at Bird Island. In Wanderer chicks habituated to researchers at Bird Island, sky-point, sky call and side-preen replace the usual aggressive response to strange intruders (Fig. 15.29).

GALÁPAGOS ALBATROSS

A few Galápagos Albatrosses were displaying when F.P. Drowne landed on Isla Españole in October 1897 and his crude sketches were probably the first attempt to illustrate the displays of any albatross. Bryan Nelson's (1968) engaging account remains the best available description of the birds in action.

Galápagos Albatross displays have been filmed for scientific analysis, but the work was never completed (F. Trillmich pers. comm.). The sequences illustrated here were drawn from 16 mm film of birds whose age and sex were unknown.

Galápagos Albatross nests are widely scattered on rocky ground where dense thorny scrub covers much of the land. Most birds can land near their nest-sites, but there are various obstructions and many have to walk to cliff-tops or beaches to take-off. Social interactions occur in open clearings among the scrub. They appear to be more aggressive than other species.

> 'Nor will I soon forget the old rascal who came for me on the dead run and who, if I had not luckily cracked him on the head with the butt of my collecting pistol, would have lunged his powerful beak half way through me[7].' (Beck 1904)

Galápagos Albatrosses have distinct gape-stripes, black outside and bluish-grey inside. They are normally hidden by tiny contour feathers, but even incubating birds have a slight parting with the line visible. During agonistic encounters, black stripes are visible.

Stereotyped actions

Double call by Galápagos Albatrosses is a rather high-pitched vocalisation, sounding something like *eeelich-coo* repeated many times, but variable and incorporated into many actions. It is the most frequent call, unlike anything in other albatrosses (Fig. 15.34).

Yapping is the common albatross action by one or two birds sitting or standing at a nest. Rhythmic bowing to the ground with double calls may be accompanied by picking and placing fragments of nest material. It is a constant of all nest relief activity.

Croaking: a bird standing normally utters *uh-uh-uh...* towards the ground, accompanied by a moderate opening of the bill. This action and vocalisation bears some resemblance to actions by gooneys and mollymawks.

Pointing is the common albatross approach, stretching the neck, head and bill, high or low, towards a partner (Fig. 15.35).

[7] Whatever damage these beaks might inflict, they are not shaped to stab or pierce.

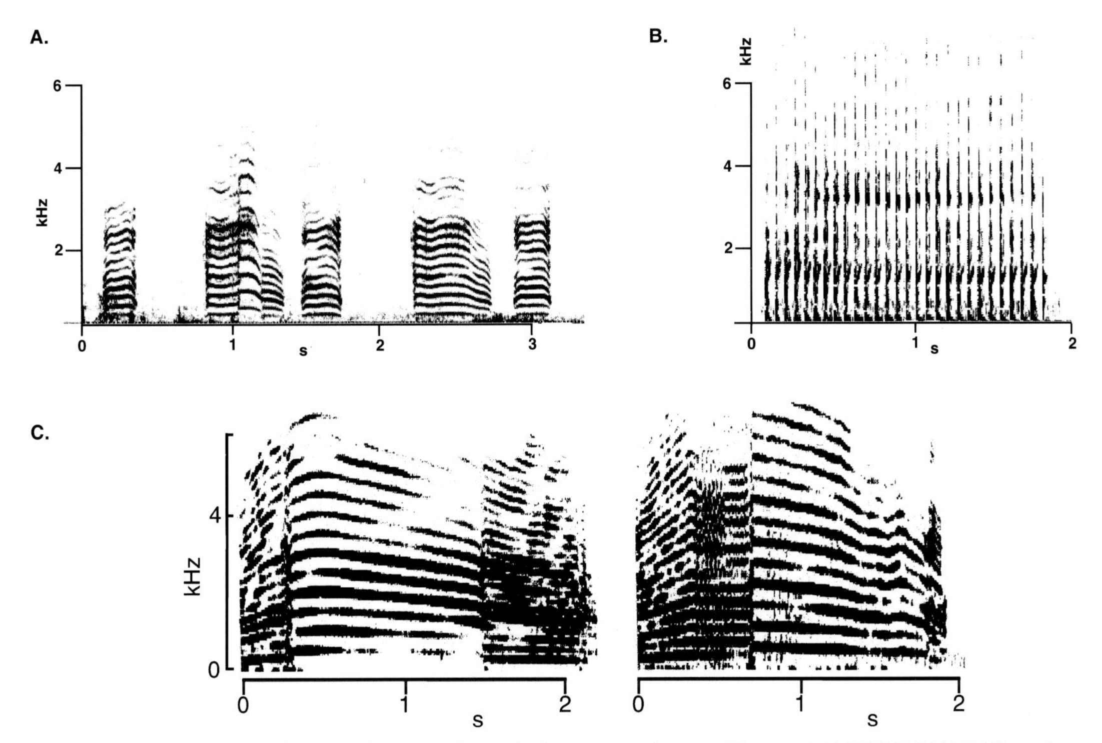

Fig. 15.34 Sonagrams of Galápagos Albatross: after a single note two phrases of harmonic DOUBLE-CALLS (A); each consists of a long and short note and in the middle of the first long note there is a halving of the number of harmonics, which gives the call a sound not heard in other albatrosses. A single burst of CLAPPERING (B). Two separate SKY-CALLS (C) each beginning and ending with instantaneous changes in the number of harmonics, another characteristic unique to Galápagos Albatrosses (from recordings by Peter Corkhill and Hector Douglas).

Clashing (bill-circling): when the bills of two birds come into contact, clashing usually follows. It may occur between birds in upright postures, but more often the pointing is nearly parallel with the ground. The bills are side by side, with both birds sliding their own sideways back and forth over the top of their partner's bill (Fig. 15.36). Each bird apparently competes to slide its own bill over the top of the other and males may be dominant, with females apparently parrying the pressure. Fluid from the nasal tubes lubricates the bills, but there is no evidence that it is secreted for that purpose. Throughout the rapid action, the two bills remain in contact, but they spring apart from time to time and clash back together with a hollow sound, like the rapping of hardwood sticks. The action is common and may erupt even between allopreening partners sitting together at a nest-site.

Clappering: in the same low posture follows clashing. When a bill suddenly loses contact with that of its partner, its mandibles clapper rapidly like castanets at a rate of about 19 claps per second (Figs. 15.34, 15.37). This may occur simultaneously, as two birds remain low but pull back from each other, or by one bird when its partner suddenly rises into an upright posture. Female clappering is usually followed by a male side-preen.

Stare is particularly striking in the Galápagos Albatross. Because of the proportionally long thin neck, the rigid standing posture seems especially tall. Small feathers erected on the projecting eyebrows give the crown its unique flat appearance (Fig. 9.1).

Bow: exaggerated silent bows to one side or the other constantly punctuate other actions.

Side-preen (scapular action, side touching): in the Galápagos Albatross this common action is highly ritualised. A standing bird suddenly bows to one side and touches a spot on the side of the breast (Fig. 15.35). The bill does not penetrate the feathers, nor are there any preening snaps.

Sky-pointing: the head is raised and the bill pointed diagonally upwards. There may be no sound or a low *oooh* or *uuuh* through slightly open mandibles; it can be so low as to be barely audible at a distance of ten metres. Sky-point may come from a solitary bird or one close to a partner (Fig. 15.35).

Fig. 15.35 Galápagos Albatrosses performing GAPE (1–2), CLASHING (3,5), SIDE-PREEN (7), SWAY-WALK (8–11) (12 s duration).

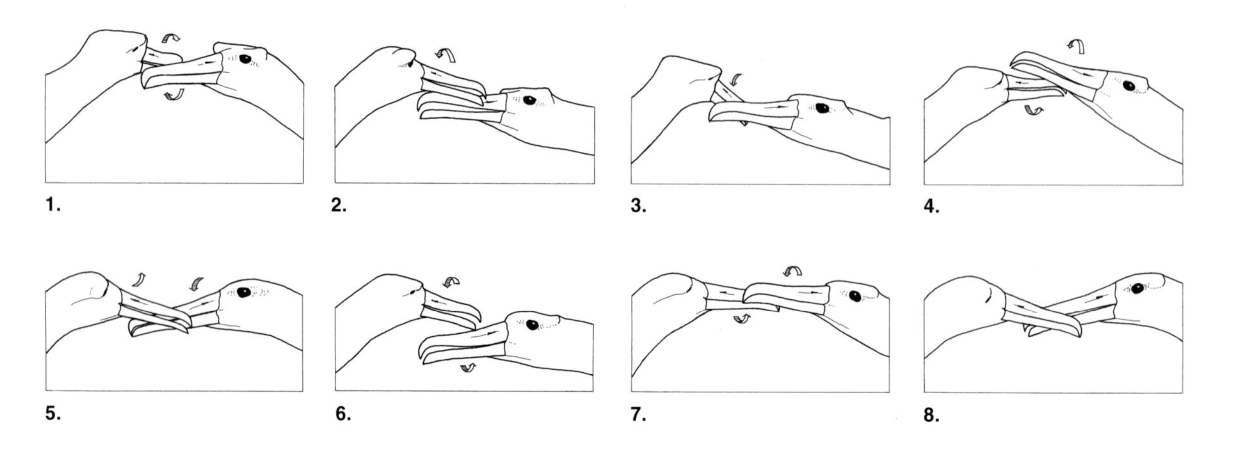

Fig. 15.36 Galápagos Albatross bill CLASHING (1.3 s duration).

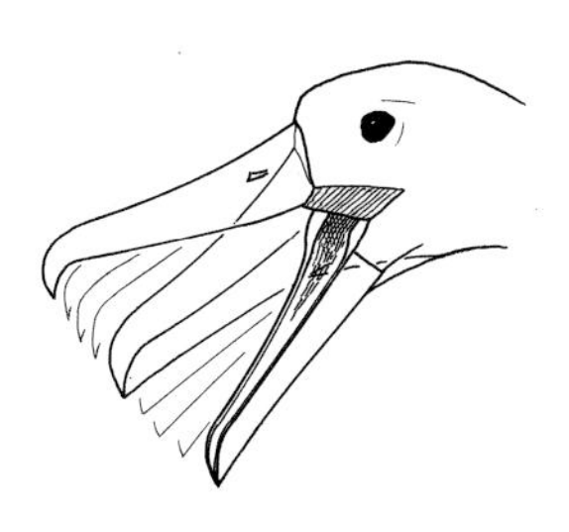

Fig. 15.37 Galápagos Albatross CLAPPERING (see Fig. 15.44).

Sky-call is preceded by side-preen after which the head is swung up to the vertical with a drawn-out *yahee-eee* sound (Fig. 15.34). It is followed by a deep side bow in both males and females.

Gape: a bird standing in the upright stare posture with the bill horizontal suddenly opens it very wide (Fig. 15.35) (Plate 38). It is usually silent, but may be accompanied by a *ya ya* sound similar to that of a low intensity yammer. Two partners in a dance may snap into synchronous gapes, but males perform this action more often than females (Nelson 1968).

Yammer: a version of the common aggressive albatross encounter with an open bill and *ya-ya-ya* or *ha-ha-ha* vocalisation. It may involve a pursuit and attack.

Sway-walk: a unique highly exaggerated lateral swaying, mostly by walking birds, but occasionally when an erect bird is standing still (Figs. 15.35, 15.38).

Fig. 15.38 Galápagos Albatrosses making a SKY-CALL (1,4), SIDE-PREEN left (3), SWAY-WALK (5–6) (7 s duration).

Dance

During all Galápagos Albatross behaviour, other than take-off and copulation, the wings remain folded and the birds do not usually stand on tip-toe. The male is responsible for the progress of the dance; in a sense he directs it, driving the tempo and switching actions. Nelson (1968) made an amusing and revealing comparison:

> ' the dance ... was like two washing machines, each with the same repertoire but out-of-phase with each other. Yet not randomly out-of-phase. The spin of washer B was more likely than any other part of the repertoire, to follow the rinse of washer A.'

Performing birds all move with sway-walk. Males are more aggressive in their performance and females must react promptly to their cues. Well practised pairs perform together very smartly. Indecisive females do not synchronise postures and males sometimes run at them, as though they were intruding males. Such tendencies then have to be appropriately placated.

Dances often include more than two birds and take place in open spaces some way from participant males' nest-sites. An important conclusion to such dances is the ability of males to lead females back to their selected nest-sites.

Copulation

A mounted male was seen to constantly slip down the sides of the female's back and push open her wings, all the time double calling through slightly open mandibles. His vigorous sideways tail movements forced him to lean over to his left and at the same time extend his right wing as a balance. The female remained quiet and leaned to her right. After ejaculation the male paused, then started again, but the female turned her head, allopreening his chin before getting up and dislodging him.

Extra-pair copulations occur among Galápagos Albatrosses: after one male was seen to pounce onto a fleeing female and copulate, she was pursued by another male and overcome. Although attempting to shake off this male, her wings were pinned to the ground and she could not move. He worked his tail sideways for a long time, but she never lifted hers and continually turned her head and disturbed him by making bill contact from below. Ejaculation occurred, but the two cloacas were not in contact. This female continued to struggle, but was still unable to escape until the male got off. She then ran away into the undergrowth.

GOONEYS

Laysan and Black-footed Albatrosses have larger repertoires of actions and vocalisations than other albatrosses and field-work at Midway Atoll during the two seasons 1956-1958, enabled Rice & Kenyon (1962) to identify and illustrate most of the separate actions and vocalisations. In the 1960s, Bartholomew & Howell (1964) undertook experimental field-work on nesting behaviour in both species and Fisher (1971,1972) described aspects of Laysan behaviour. Two of Fisher's students initiated a quantitative approach to albatross behaviour. Earl Meseth's (1975) ethological analysis of the Laysan Albatross dance was based upon 179 marked and sexed individuals, mostly subadults of unknown age and breeding experience. In 100 sexed dances, 87% were between males and females, 9% were female-female and 4% were male-male interactions. Donald Sparling (1977) recorded male and female sounds from 90 Laysan and 40 Black-footed Albatrosses for his spectrographic analysis.

Territory and aggression

Most gooneys breed on flat ground, usually with space on which to move about and display away from nest territories. In this respect they have something in common with the great albatrosses, Steller's Albatross and the Galápagos Albatross. Although the closest nests can be less than a metre apart, colonies are not as dense as those of mollymawks.

The Laysan Albatross walks with an upright posture while the Black-footed Albatross adopts a head-down gait (Fisher 1972). They differ in temperament; Black-footed Albatrosses are more aggressive, quicker to strike and are more likely to leave the nest to attack an intruder. Male Laysans defend territories three to four metres in diameter and make nest scrapes, but the female decides exactly where the egg will be laid. Subadults carefully avoid established breeders, squabbling among themselves at the boundaries of newly acquired territories until one departs or both move back.

Aggressive actions are common in crowded Laysan colonies. Sitting birds make expressive *eh* sounds, with short jabs towards the ground, and snaps or forward thrusts of their bills with biting movements. Late in the season when many young subadults[8] are wandering about, territorial males often charge up to ten metres from the nest-site, both pursuing and fleeing birds extending their wings and let out wails of distress.
Black-footed Albatrosses have been seen displaying at sea (Fisher 1904).

Stereotyped actions

Although they are so different in plumage, Laysan and Black-footed Albatrosses have similar repertoires of actions and vocalisations.

[8] Fisher's researchers called them 'walkers' (Fisher & Fisher 1969).

Double calls (Eh-eh, Eh-call) in Laysan Albatrosses are short, repeated harmonic sounds, *eh-eh, eh-eh.....* while in Black-footed Albatrosses they resemble a nasal *ha-ha, ha-ha.....* Inter-specific and individual differences are apparent in several parameters (Fig. 15.39). The notes may be compressed and low-pitched or higher-pitched harmonics, sometimes breaking up into squeaks exceeding 20 kHz (Sparling 1977). Double calls accompany a number of quite different actions, and in both species are used more than any other vocalisation.

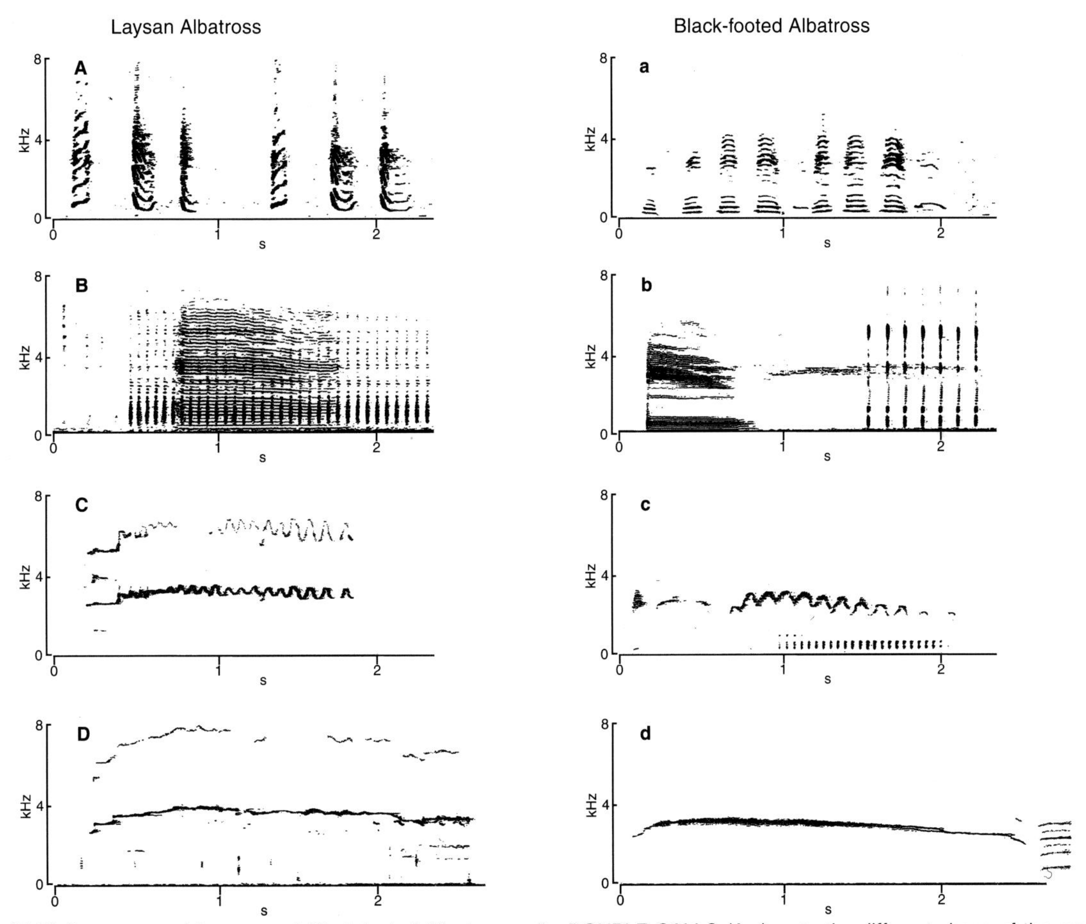

Fig. 15.39 Sonagrams of Laysan and Black-footed Albatross calls. DOUBLE CALLS (A,a); note the different shape of the many harmonics in the two species. SKY-GROANS and CLAPPERS (B,b); the clapper of a female Laysan Albatross (B) is simultaneous with the SKY-GROAN of a male partner, while the SKY-GROAN of a Black-footed Albatross (b) is followed by a BOW-CLAPPER from the same bird. WHINNY (C,c); the wavering Laysan call has two clear harmonics (C), while deeper sound of the Black-footed Albatross is accompanied by the CLAPPER of a partner (c). WHISTLES (D,d); the Laysan voice again has two harmonics (D) while the Black-footed Albatross is deeper and in this example is followed by a brief INHALATION (d) (Warham 1996).

Yapping gooneys crouch or shuffle about their nests on their tarsi uttering double calls, either alone or in duets with partners. Incubating birds have been described as 'talking to their eggs.' The behaviour is similar to the much slower action of great albatrosses.

Croaking (Eh-Eh Bow): a standing Laysan or Black-footed Albatross thrusts out its breast and arches its neck so that its bill is pointing down at about 30°. The body pivots forward in a slow bow accompanied by a succession of double calls. These are repeated several times with bows getting lower or the bird taking a few slow, deliberate steps. The pace of the movements may accelerate as the birds become more excited; the *eh-eh* sounds become higher-pitched, sometimes losing harmonics and changing into whinnies and whistles (see below). Croaking is commonly performed by males at nest-sites and less often during dances. Females perform less frequently than males (Meseth 1975). The action is similar to the croaking of mollymawks.

Pointing (billing), the universal albatross approach. Often it is not reciprocated, but when there is a response, the closed bill tips of two pointing Laysans make delicate side-to-side contact, which may shift from one side to the other. Between confident partners in dances, contact becomes more vigorous with reciprocal nibbling at the side of each others bills, very similar to the begging action of chicks. Maintaining bill contact while shifting rapidly from side to side causes the heads of the two birds to move from side to side. As the tempo increases, nibbling ceases and is replaced by bill pushing, with each partner alternately pushing sideways in the opposite direction (Meseth 1975).

Stare (gawky-look, glare, upright threat): the posture of a lone bird is erect and rigid with a glaring look. It may reveal subtle aggressive intention movements towards an intruder and is sometimes followed by a whinny (see below). During the dance, a standing bird of either sex appears taller than adjacent birds with more pronounced curves to the head, neck and breast. It may rise on its toes and slight head movements follow the tempo of a bobbing partner (Fig. 15.40).

Fig. 15.40 Laysan Albatross STARE and BOW (1–5). The right bird STARES (1) as the left bird comes down from a SKY-FLICK (1–2), then both BOW in turn (3–4) and STARE (5)

Snap(s): a single or series of repeated separate snaps (distinct from the faster clapper). They may be loud and audible at some distance. Aggressive snaps are clear warnings, sometimes accompanied by intention movements. They are frequent elements of stereotyped action sets whose meaning is often less obvious. Slight grunts may accompany these snaps.

Side-preen (scapular action) is an audible preening in the plumage on the right or left of the breast, in front of the folded wing. Both wings remain folded throughout. Before penetrating the plumage, the bill makes some slight lateral movements, perhaps anticipating the parting of feathers. Preening may progress within the plumage down the side of the breast, rearranging contour feathers over the wing. Side-preening occurs at the beginning and end of dances, and also by other birds trying to join in.

Preen-under-wing (scapular action) is performed by males and females during the dance. From a normal position the bill is suddenly swung upwards to about 60° from the horizontal and makes one sharp **air-snap** before being lowered to preen-under-wing on one side, right or left, when the bird may rise onto its toes. In the

Fig. 15.41 Laysan Albatross PREEN-UNDER-WING (2–5), SKY-POINTING (1–2) and SKY-GROAN (5–6).

Laysan Albatross one wing (on the action side) is pulled forward and flared outward with the carpal joint (wrist) held close to the body (Fig. 15.41) (Plate 44). The bird thrusts its head below the raised wing and the bill is pushed into feathers where there are preening snaps. On the opposite side, the other wing remains folded. Immediately afterwards the head and bill go up again, but higher, towards the vertical and the action ends with a hollow, resonant sky-snap.

Fig. 15.42 Black-footed Albatross performing SKY-GROAN (1,4), PREEN-UNDER-WING (2,3,9) BOW & CLAPPER (6) and POINTING (9) (4 s duration).

In the Black-footed Albatross, both wings are flared (Fig. 15.42) (Plate 46). Laysan Albatrosses always complete the action by closing the flared wing, but the Black-footed Albatross sometimes keep the wings flared throughout much of a dance (Fisher 1972).

Sky-flick (head flick): from a stare position, the head is very quickly thrown up and slightly twisted to one side while the rest of the body remains rigid. It is a silent action, but in the brief moment that the head is up, the mandibles are warped against each other from side to side. The cutting edges spring back, twice producing audible clicks. The head then comes down to stare-and-clapper (Meseth 1968).

Sky-moo is accompanied by a low vocalisation. A male after landing, emerging from a dance, or standing on a nest with folded wings raises its head and bill to about 45° (Fig. 15.43). A cow-like *moo* sound emerges from slightly parted mandibles; it is a single drawn-out note with many compressed, unmodulated harmonics.

Fig. 15.43 Laysan Albatross SKY-MOO.

Sky-groan is a slightly longer, more intense version of sky-moo that occurs during dances (Plate 43). A standing bird throws its head up and stretches skywards with its wings folded, the neck leaning slightly to one side. It may also rise on tip-toe and lean slightly backwards. Air forced through the tight airways and closed or slightly parted mandibles produces a distinctive harmonic sound, described as a strained *aww*. The Black-footed action has a more nasal timbre (Fig. 15. 39). Several bill snaps may follow before the head is lowered. Both sky-moo and sky-groan are similar to the sky-points of great albatrosses.

Sky-whistle (head-up-and whine, victory call) is more intense than sky-moo or sky-groan and characterised by loud whistles with clearly defined frequencies and few harmonics (Fig. 15. 39). After conflicts, such as a charge or bill thrust, a displacing male may suddenly fling his head vertically, stretch up on tip-toe and let out long whistles from an open bill.

Shake & whistle (head-shake-and-whine) occurs only during the dance and is similar in both sexes. The neck is curved forward in a shallow 'S' with the head down and pointing towards the partner. The head shakes from side to side accompanied by a whistle similar to that of sky-whistle. In the Black-footed Albatross (Plate 45) it is a deeper, hollow sound (Meseth 1975, Sparling 1977).

Whinny: from a rigid stare posture with almost closed bill the 'braying' of Laysans has the same frequency characteristics as the above whistles, but they are sharply modulated or broken up into a series of separate short notes (Fig. 15.39). Whistles may in fact break up into whinnies (Fig. 15. 39). During dances, whinnies are heard more often from males, and more frequently among Laysans than Black-footed Albatrosses.

Clapper (buzz, clacker, rapid bill clapper, rapping) accompanies two quite different actions in both species:

1. A rather stiff standing position similar to stare, with rapidly snapping mandibles (Fig. 15.39,44). The sound has been likened to that of a woodpecker hammering (Sparling 1977). Laysan Albatrosses usually face each other with bills tip to tip and clapper at about 20 snaps per second. Black-footed Albatrosses stand with their heads side by side (Rice & Kenyon 1962b), alternating from one side to the other (Fisher 1972), both birds lifting their wings slightly and clappering simultaneously. The action takes more than twice as long in the Laysan as in the Black-footed Albatross (Fig. 15.42) and in both species it is more frequent in males than females.

2. **Bow & clapper** (bow-clapper, bow clacker) occurs only during the dance and is often a synchronous duet of two standing birds. Here too, the

Fig. 15.44 Laysan Albatross CLAPPERING.

Laysans usually face each other while Black-footed Albatross heads are side to side. The Laysan body tilts forward and the neck is extended towards a partner. The head is dropped suddenly towards the ground, with a short burst of clappering and then raised silently to about 25° above the horizontal. Repeated actions alternate each side of the partner's bill. The Black-footed Albatross has longer bursts of clappering than the Laysan.

Bob: a watching gooney bobs up and down in tempo with the performance of others. It becomes **bob-strutting** when one moves towards or around another, with an exaggerated springing step that may become so vigorous that the bird is on tip-toe or even leaping from the ground (Fig. 15.45). The head is usually held rigid as the flexible neck bobs up and down in time with the step, but excitement may bring it down in a forward scooping movement.

Fig. 15.45 The Laysan Albatrosses on the right is circling behind the standing bird (left). Prancing on tip-toe and leaping off the ground (8,12,13) is typical BOB-STRUTTING. The standing bird on the left STARES and rotates to keep the approaching bird in view, and BOBS in time with it (3,7,8) (4 s duration).

Dance

Laysan Albatrosses approach by gentle pointing; in Black-footed Albatrosse,s pointing comes later and is much rougher, with noisy clashing of bills. Black-footed dances are faster and more vigorous. Partners often run at each other and breasts touch, while Laysan Albatrosses always stand apart. Both species stand on tip-toe, but the Black-footed Albatross stretches higher than Laysan. Furthermore, while the Laysan usually remains still when on tip-toe, the Black-footed Albatross steps forward in a characteristic mincing gait.

The wings of both species are used in postures but the tail is normally held straight in line with the sloping back and the feathers remain unflared throughout actions (Meseth 1968).

On the many islands where Laysan and Black-footed Albatrosses breed near to each other, it is not unusual to find individuals of the two species sitting together or even allopreening. Dances in which one partner is a Laysan and the other a Black-footed Albatross have been seen, but they are not common. Difference in height could be inhibiting. Laysan males may avoid Black-footed females because they are taller, and although Laysan females may be attracted to taller Black-footed males, they would probably be repelled by the vigour of the Black-footed approach (Fisher 1972).

> '...occasionally, when both species were represented in a dancing group, the odd bird would quickly get out of step and break away, seemingly much discouraged. In addition to acting more rapidly, the black fellows spread both wings from their bodies, while the Laysan usually project just one. At times the birds grew very excited, and while the postures of the performers were similar, they were rarely synchronised.' (Bailey 1952).

Dances begin slowly and end suddenly. They are usually initiated by females while males perform the core actions. At both the early and closing phases, side-preen is dominant. Females not wanting to continue just stand mute or perform side-preen. When they leave, they keep their heads lower than those of the males for about three metres and bob slightly. Males remain watching intently and may go into croaking.

Dances with three or more partners are common, but seldom start with more than two. A pair dancing attracts unoccupied birds, which often watch for some time before trying to join in. The involvement is usually brief; they are mostly ignored, lose interest and walk away, but are sometimes driven off. Violent fights may erupt without warning and take the victims by surprise. In 235 dances, 11 attacks were directed towards partners and nine at intruders.

Sets

Actions and vocalisations in the Laysan Albatross almost always occur in sets (Meseth 1975). Males signal possession of a nest-site and invite females to approach with the set: **croaking** (CR), **whinny** (W), sky-flick (SF) and **clapper** (C); which may be performed entire (CR - W - SF - C) or split into (CR - W) repeated several times before being concluded with (SF - C).

Unpaired males perform the possession/invitation action set towards passing females. Many take no notice and walk on, but if one is attracted, she will probably keep her head slightly lower than his and bob as she approaches. Actions and sets follow in an ordered pattern, usually at similar frequencies among males and females. Females may take longer to reach the tempo of the core activity, but the male will probably slow down and she will catch-up.

Three principle sets are employed during the dance. They are abbreviated as follows:

1. STARE, SKY-GROAN, = (S - SG) = [SSG]
2. AIR-SNAP, PREEN-UNDER-WING, SKY-SNAP = (AS - PUW - SS)
= [PUW]
3. SHAKE & WHISTLE, STARE, SKY-FLICK, CLAPPER
= (S&W / S - SF - C) = [S&W/SSFC]
or (S&W) (S - SF -C) = [S&W] [SSFC]

Reciprocal pointing with bow-clapper punctuate all displays and are thought to be synchronisers that can be inserted before most actions. In each individual, the performance of [PUW] alternates right and left irrespective of intervening actions or sets.

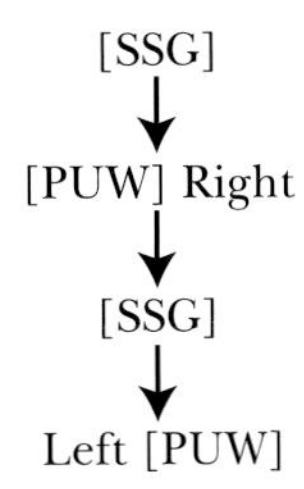

The set [PUW] is repeated more often than [SSG], which alternates with it and [PUW] follows [SSG] twice as often as it precedes it. Left/right alternation does not hold for partners. The display of [PUW] to one side does not dictate the side of the next [PUW] by its partner.

The Black-footed Albatross does not alternate left and right, and Fisher (1972) believed that their simultaneous actions reduced dependence on cues from partners and contributed to their more rapid performance.

The sets [S&W] and [SSFC] are always preceded by [PUW] and are followed by either [PUW] or [SSG] in order to maintain co-ordination of the [SSG] – [PUW] action between partners.

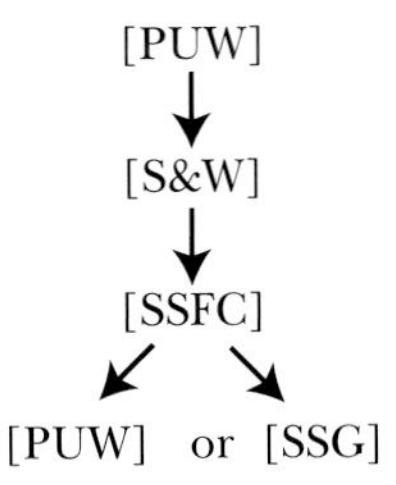

Side-preen begins and ends dances; it is also performed by other birds trying to join a dance already in progress.

Croaking is five times more frequent among males than females because it is part of the male's possession/invitation set. It may also precede various other actions and sets except [S&W] [SSFC] and perhaps causes females to quicken dancing.

Bob and bob-strut can substitute for any action.

Synchronisation

The set [PUW] by males is followed in female partners by [SSG]:

The pattern is apparent even when it is interrupted by other actions and sets (Fig. 15.46). When a good rapport has been established, the Laysan Albatross is adept at restoring the [PUW] – [SSG] alternation and make instant adjustments to regain synchrony with their partner. The anthropomorphic analogy is apt; expert dancers the world over pick-up or drop steps with ease while small breaks in continuity leave novices confused and obliged to

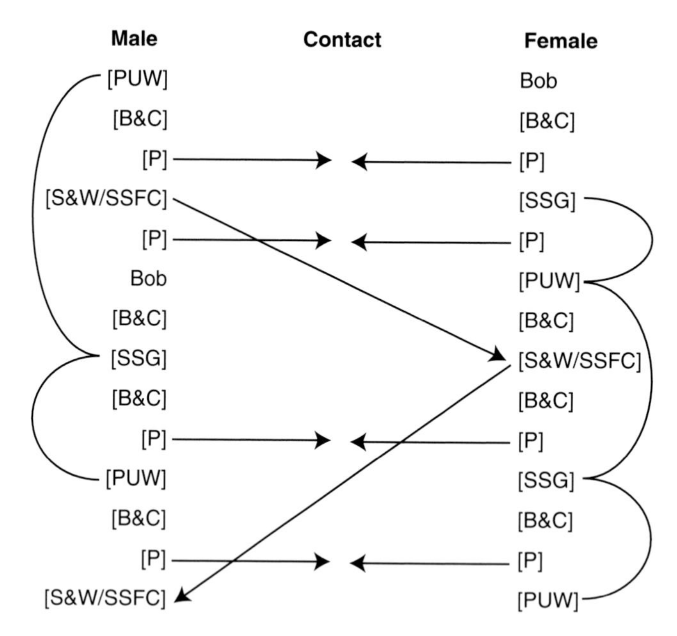

Fig. 15.46 Alternation in the Laysan Albatross dance is apparent in the [S&W / SSFC] and [SSG] – [PUW] sets. Bill-touching contact occurs during POINTING [P] (after Meseth 1975).

start again from the beginning. Likewise, premature break-up of Laysan dances often appears to correspond to 'missed steps' (E. Meseth pers. comm.).

The set [S&W/SSFC] alternates between partners, but sometimes breaks up into [S&W] [SSFC] separated by other actions, though the overall set still alternates. The set [S&W/SSFC] performed by one bird may be accompanied by any one of five simultaneous sequences by its partner. This helps the bird to maintain correspondence regardless of the displays performed before or during the [S&W/SSFC] set. Most combinations prepare for a return to the [PUW] [SSG] correspondence.

Side-preen by either partner is accompanied most often by the same action in the other, and likewise with croaking.

Copulation

Between experienced Laysan Albatrosses, copulation is a quiet act that takes place on a pair's territory undisturbed by adjacent occupants and with little if any preliminary display.

The sitting female flattens herself and allows her wings to part and droop onto the ground. The male walks up from the rear onto her back and treads until his claws hook over the leading edge of her upper wings. His wings also droop half open to make contact with the ground, somewhat steadying his precarious position. He raises a little forward and tilts his tail clockwise until the flat plane is vertical, then swishes it several times across the female's tail. She rotates her tail anti-clockwise and the cloacas come into contact as he lowers himself backwards. He makes six to eight slow thrusts then rests. The female promptly brings the activity to an end by standing and tipping him off backwards or turning her head to grab his bill aggressively and pull him off sideways. The sequence takes two to three minutes and only rarely does the male remain mounted to make a second attempt.

Afterwards, the standing female shakes and preens her plumage and both birds indulge in some yapping; the male may sky-moo or, if he has been roughly handled, sky-whistle, but well established pairs soon settle down to allopreening (Fisher 1971).

Extra-pair copulations are much more clamorous. They occur in both species, but have not been seen between species. Laysan adults and subadults may suddenly pursue and overpower a female (or male) passing through a colony. Often, many males participate and a frenzy of attempted copulations follows with several simultaneous mountings, as many as four birds deep and irrespective of sex.

> 'One female, attacked four times in 10 minutes by four different groups of males, finally escaped and, after resting for about 15 minutes, started for her mate, nearly 80 feet away. With her left eye gone and her left wing drooping she managed to run from a fifth group of males, only to be almost smothered by a group of six more, only 20 feet from her mate.' (Fisher 1971).

Males on territory were never seen going to the aid of their distant partners and, as Fisher speculated, there was nothing to inhibit them from joining in 'gang rapes' of their returning spouses when still a long way off. He concluded that the males were responding to inadvertent pre-copulatory signals:

> 'A broken leg that causes the body to droop between steps, a broken wing that depresses the primaries, or even a momentary stumble over a stone that throws the bird into a copulatory position...'

A male Laysan with a broken wing that drooped on the right side provided a convincing demonstration. As he made its way through the colony, males on its right attempted copulation while those on its left, that could not see the drooping wing showed no interest.

Fisher and his students made hundreds of cloacal examinations and smears from both males and females involved in extra-pair copulations. They found nothing to indicate that cloacal contact or insemination had occurred (Fisher 1971).

Nest relief

Incubating Laysan Albatrosses are reluctant to leave their nests and relieving mates take over only after much vocalisation and allopreening. Most males and females in 37 observed changeovers circled the nest they had just vacated, then picked up debris and dropped it or tossed it backwards. The departing bird often walked back to the nest and allopreened its incubating mate briefly. They frequently stopped for one or two minutes to pick up ironwood needles and grass for up to an hour after leaving the nest, but such behaviour did not persist after brooding newly-hatched chicks (LeFebvre 1977).

Fledgling behaviour

Fledglings have once been seen dancing; they performed croaking, preen-under-wing and sky-groan.

STELLER'S ALBATROSS

There has been no study of Steller's Albatross behaviour, but many beautiful colour photographs by Hiroshi Hasegawa (1995) have been published in Japan. Since 1976, several films and videos of these birds have also been broadcast on Japanese television. The incomplete catalogue of postures and actions illustrated here were obtained from a few sequences filmed for BBC television (Tickell 1991).

Steller's Albatrosses on Torishima nest on 30° slopes; formerly many were probably on more gentle slopes or level ground. Some nests are spaced two to three metres apart and walking subadults cause aggressive responses from the occupants of nests. Attacks occur and incubating adults sometimes rush from their nests with wide open bills to pursue intruders. In the process, eggs are sometimes pushed out of nests and are not recovered. Steller's normally walk upright although not at full height, the neck being partially withdrawn. There are no obvious spatial constraints on their displays and the dances are striking. They have conspicuous black gape-stripes that are exposed at times, but are usually concealed by tiny contour feathers, like the coloured gape-stripes of mollymawks.

Stereotyped actions

Yapping: the separate notes are longer and have more condensed harmonics than in other species (Fig. 15.47).

Pointing is usually towards the head and bill of a partner; it is often made with the neck twisted so that one side of the head looks upwards. Frequently attention is directed lower to the breast or bill of a bowing bird (Fig. 15.48).

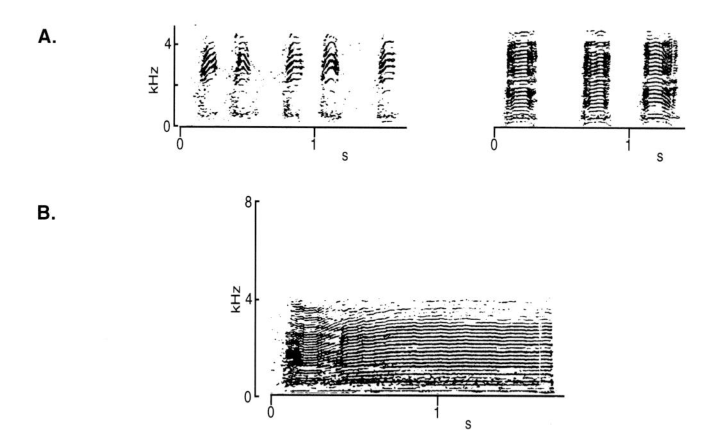

Fig. 15.47 Sonagrams of Steller's Albatross YAPPING (A) and SKY-CALL (B) (after Warham 1996).

Fig. 15.48 Steller's Albatross dance (1-12) POINTING (2,10), SKY-POINTING (3,4,11), SWAY (5-7) and PREEN-UNDER-WING (9,10,) (16 s duration).

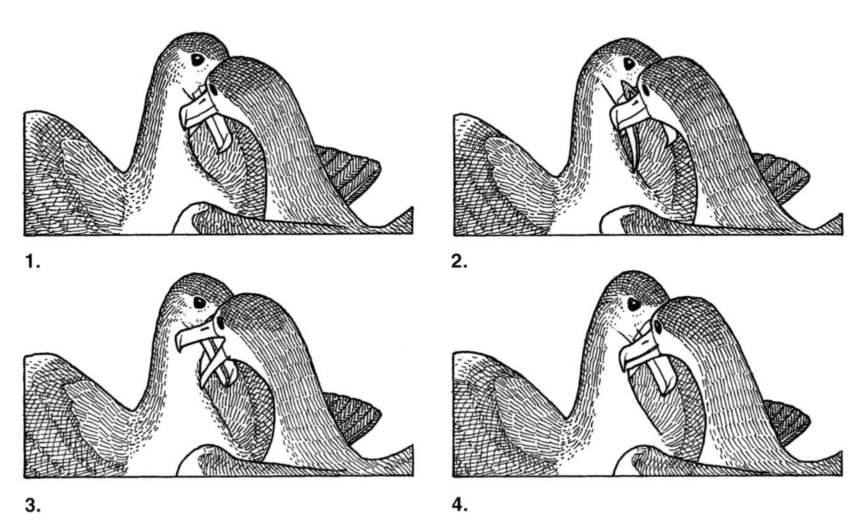

Fig. 15.49 Steller's Albatrosses bill CLASHING. Note the alternate opening of mandibles (0.2 s duration).

Clashing is the characteristic bill contact, perhaps the most common action (Fig. 15.49). The bills of two birds are side by side and in constant movement with the mandibles opening and closing regularly and, apparently, alternately. Individuals sometimes nip mildly at feathers on each other's faces. There is sometimes a reciprocal swaying forward and back with bills remaining in contact.

Scooping is a quick bow with a down-forward-and-up motion. It sometimes accompanies a few forward steps.

Head-shake is a series of tight side-to-side head movements, pivoting at the top of the neck and common to all albatrosses.

Sky-pointing: the head is raised towards the vertical with the neck twisted slightly to one side (Fig. 15.48). The mandibles are only slightly parted and there is apparently no sound. The wings may be folded, but they are usually flared open as the bird stretches upwards and fold again as it sinks onto its tarsi. The performance is often repeated in synchrony with a partner.

Swaying: the extended curved neck and head sway widely from side to side (Fig. 15.48).

Sky-call is usually preceded by swaying. The head then goes down in a deep bow as the tarsi lower the body to the ground (Fig. 15.50). The mandibles open wide and remain so, and a loud, deep groan (Fig. 15.47) is heard as the head swings up towards the vertical, twisted slightly to one side (Plate 52). At the same time, the bird rises from its tarsi and stretches upwards on tip-toe, fanning its wings and lifting the tail slightly. The head then comes down, still with an open bill, and the bird sinks onto its tarsi, closing its bill and folding its wings.

Clapper: a standing bird with a slightly lowered bill makes several separate snaps, then lowers the bill further with a short series of hollow clappers at about 20–25 per s.

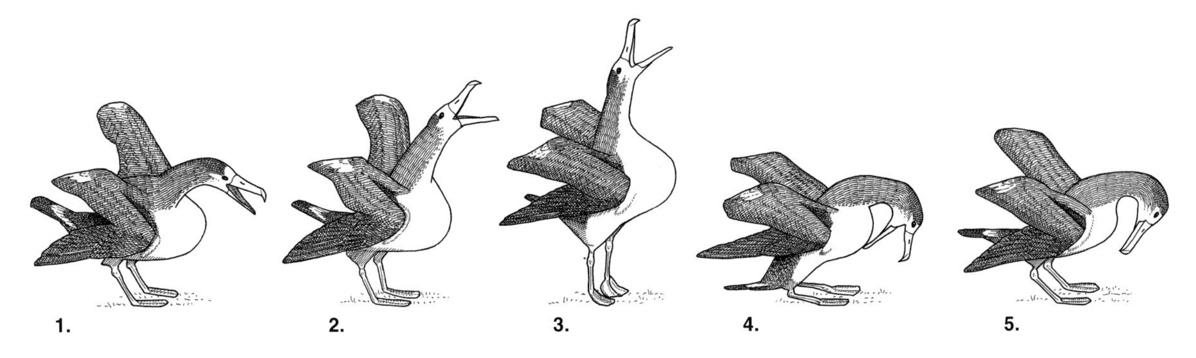

Fig. 15.50 Steller's Albatross SKY-CALL (4.5 s duration).

Preen-under-wing may begin with both wings folded. One only is flared and the bird preens deep in the feathers below on one side of the breast. If both wings are already open, the one on the preening side is lifted and flared more than the other (Fig. 15.48).

Dance

Steller's Albatross dances are vigorous activities between individuals that maintain quite close contact (Fig. 15.48). There are frequent deep grunts and bill clatters, but few loud calls.

Mutual clashing may be the most frequent action, with repeated flagging and scooping punctuating other actions.

The flared wings are similar to those of gooneys, but with more of the primaries revealed. The rhythmic opening and folding of the wings, with the lifting of the body and sky-pointing in synchrony with partners, is a unique characteristic of the species. The tail remains closed.

16
ECOLOGY

'It is difficult to assess the likely errors in the estimate of 162,360 breeding pairs[1] in an area of 230,239 m^2, at an average density of 0.70 nests m^{-2}.'

P. A. Prince (1982)

Albatross ecology began with occasional indications of numbers in seamen's journals. George Comer was said to have collected 2,000 Gough Wandering Albatross eggs in 1888 (Verrill 1895) and during the early decades of this century, whalers collected similar numbers each Christmas from Wandering Albatrosses at South Georgia (Matthews 1951). About the same time, enormous numbers of Laysan Albatross wings were confiscated from Japanese fowlers on some of the NW Hawaiian Islands (Clapp & Wirtz 1975) and islanders of Tristan da Cunha sometimes recorded the numbers of eggs and chicks taken from nests of Western Yellow-nosed Albatrosses (Rowan 1951). These old figures help us to imagine the numbers of albatrosses on islands where they have since declined.

Richdale (1939) had few Northern Royal Albatrosses when he began work at Taiaroa Head, New Zealand in the mid-1930s. Nevertheless, numbers increased slowly and records have been continued to the present (Robertson 1993c). Richdale knew whenever individual birds failed to return, but in the first 14 years of field-work he did not commit himself to population statistics. It was David Lack[2], his mentor at Oxford who later calculated that only three percent of Northern Royals died each year and that they had an average life expectation of 36 years (Lack 1954).

During the early 1940s, Jack Sorensen (1950a) counted Southern Royal Albatrosses in a sample area of about three square miles of Campbell Island ridges, they nested at an average density of two pairs per acre. From this he estimated that at least 5,000 pairs nested on the whole island each year. At the beginning of 1958, Kaj Westerskov (1963) counted 2,278 occupied nests on the island and used this number with a sample breeding success of 75%, to obtain a figure for the breeding population. Some of his assumptions were mistaken, but it was the first attempt at calculating a biennial breeding population and was the stimulus for later treatment of Wandering Albatross data (Tickell 1968).

Occupied Wandering Albatross nests were first counted on islands of the Bay of Isles, South Georgia in 1946–47 (Rankin 1951), at Macquarie Island in 1949–50 (Carrick & Ingham 1970), at Marion Island in 1951–52 (Rand 1952), at Bird Island, South Georgia in 1958–59 (Tickell & Cordall 1960) and at the Iles Crozet in 1959–60 (Voisin 1992).

Laysan and Black-footed Albatrosses had been counted on Midway Atoll in 1945 and 1946 (Fisher & Baldwin 1946, Fisher 1949). In 1956–57, biologists of the US Fish and Wildlife Service completed a survey of breeding along the whole length of the NW Hawaiian Islands. This work was based on aerial photographs taken by the US Navy together with sample counts on the ground at Kure Atoll, Midway Atoll and Laysan Island (Rice & Kenyon 1962). For some decades, the numbers of albatrosses on these islands had been increasing, but they had also been subject to considerable human disturbance. In the 13 years, 1961–73, thousands of Laysans at Midway were ringed and recaptured by Harvey Fisher and his students (Fisher 1975a,b,c, Van Ryzin & Fisher 1976). This was the first long-term study of a substantial albatross population.

Population studies at Bird Island, South Georgia began in 1958–59 and in the next six years, three cohorts of ringed known-age Wandering Albatrosses and five cohorts each of Black-browed and Grey-headed Albatrosses were established. Fertility and survival data together with field-experiments revealed that biennial breeding was not the sole prerogative of great albatrosses (Tickell & Pinder 1967) and contributed to a biennial population model (Tickell 1968). Long-term surveillance from 1975–76 to the present has revealed variable fertility while

[1] Black-browed Albatrosses at Beauchêne Island.
[2] Director of the Edward Grey Institute of Field Ornithology (1945–73).

more detailed models (Croxall *et al.* 1990b) have brought new insights to albatrosses population dynamics (Croxall *et al.* 1990b, Prince *et al.* 1994).

Intermittent counts of Wandering Albatrosses at the Iles Crozet through the 1960s became more frequent at Ile de la Possession in the 1970s, and since 1981 regular yearly counts and long-term population studies (Weimerskirch & Jouventin 1987) have continued to the present .

Distant recoveries of ringed albatrosses date from 1913 (Menegaux 1917) (p.141). Laysan Albatrosses had been ringed at Midway Atoll in 1937 and during the 1940s several hundred Laysan and Black-footed Albatrosses were ringed in most years (Robbins & Rice 1974). Southern Royal Albatrosses were ringed at Campbell Island in the 1940s (Cunningham 1951), but it was not until the 1960s that the numbers ringed were numerous enough for recoveries to accumulate (Tickell 1967, Tickell & Gibson 1968, Robertson & Kinsky 1972, Fisher & Fisher 1972, Robbins & Rice 1974). Wandering Albatrosses coloured on the breeding grounds at South Georgia in the early 1960s were reported widely by ships in the surrounding ocean (Tickell 1968), but the results were of less significance than later colouring of three species at the Iles Kerguelen (Weimerskirch *et al.* 1988).

Regular observation of ringed birds during the breeding season made it possible to be quite precise about how long off-duty birds were away at sea. The first electronic activity recorder incorporating a seawater switch (SWS) was not deployed (LeFebvre *et al.* 1967), but the design was subsequently modified and used successfully to provide measures of how long albatrosses spend on and off the surface of the sea (Prince & Francis 1984, Afanasyev & Prince 1993). External temperature loggers (XTL) have since been developed (Wilson *et al.* 1995). Short-distance micro-transmitters on birds carrying these devices, have been used to signal their arrival in colonies and facilitate the removal of loggers and other instruments.

Cheap, one-time maximum-depth gauges (MDGs) became available in the late 1980s and were fitted to many albatrosses (Prince & Weimerskirch 1994). In spite of anecdotes to the contrary, the common view that albatrosses fed by surface seizing, had long obscured alternative methods. The results obtained from MDGs were revealing and stimulated interest in diving that led to the development of more electronic instrumentation. Small time-depth recorders (TDRs) proved remarkably effective; these and MDGs contribute uniquely to our understanding of albatross feeding ecology (Hedd *et al.* 1997, Prince *et al.* 1997).

Repeated weighings of albatross chicks and adults on nests provided basic ecological data (Tickell 1968, Tickell & Pinder 1975). The early field-work was labour-intensive and time-consuming. During the 1980s, Prince & Walton (1984) invented artificial nests incorporating pneumatic and later electronic weighing machines. In Black-browed and Grey-headed Albatross colonies at South Georgia, many nests were replaced by these machines, which were connected by cables to batteries and instruments in nearby huts. The nests were activated at ten-minute intervals and the mass of occupants recorded in data loggers. Self-contained versions, with their own power supply and data storage, capable of being left unattended in the field, are now in use on several other islands.

Albatrosses were among the large birds first considered suitable for satellite tracking (Mackay 1974). Degradable harnesses were developed in the 1970s and two wandering albatrosses carried micro-transmitters over the western Tasman Sea (D. Nicholls pers. comm.), but it was another ten years before Argos CLS[3] satellites became available for biotelemetry of birds (Strikwerda *et al.* 1986). Four years after the first satellite tracking of Southern Giant Petrels[4], Wandering Albatrosses carrying 180 g micro-transmitters were tracked from the Iles Crozet (Jouventin & Weimerskirch 1990) and South Georgia (Prince *et al.* 1992). The results were so spectacular that other researchers were soon employing this technology (Nicholls *et al.* 1994, 1995, Walker *et al.* 1995b, Anderson *et al.* 1997). Smaller transmitters with extended battery-life have been developed, but attachment remains a critical problem and researchers have strong views about the relative merits of epoxy glue, adhesive tape and harnesses. Loss of transmitters will probably remain part of the cost of satellite tracking albatrosses. The addition of physiological and environmental sensors has introduced a new dimension to ecological studies and instrumental innovation is set to continue (Wilson *et al.* 1992, Bevan *et al.* 1994, 1995, Nicholls *et al.* 1997).

[3] Collecte Localisation Satellites, France.

[4] In the early months of 1985, six Southern Giant Petrels were tracked off the Antarctic Peninsula, but although several birds were followed for distances exceeding 2000 km, none of them was seen back at the breeding ground on Humble Island and six solar-powered micro-transmitters were lost (Strikwerda *et al.* 1986).

Breeding

Almost all albatrosses breed on small islands that can be approached from all directions over the sea. In groups of islands they favour remote offshore islets. Even where they are on larger islands, it is the peninsulas, capes and headlands that are usually chosen.

Breeding grounds are often said to be located in relation to their exposure to prevailing winds, but this can rarely be demonstrated other than in a very general way. Often colonies are just as conspicuous on the lee sides of islands or hills as on the windward and a few observers are convinced that some albatrosses select sheltered sites for their nests.

The circumpolar islands of the Southern Ocean, with no native trees, have peat-forming vegetation dominated by maritime tussock grasses open to all approaching albatrosses. In lower latitudes, mega-herbs, tree-ferns, shrubs and small trees become dense forests at lower altitudes. On steep ground, mollymawks and sooty albatrosses may have landing places on cliffs from where they can walk into the trees, but great albatrosses are restricted to higher, open ground.

The tropical and subtropical islands of the Pacific occupied by albatrosses may have hardly any vegetation on their sandy, volcanic or rocky surfaces. Where plants have taken hold, thickets of thorny scrub often predominate and may be dense enough to exclude albatrosses.

Most species are obviously colonial. the Laysan Albatross and Black-browed Albatross may breed in enormous colonies where nest densities can average 7,000 per hectare (Prince 1981); at other locations, the same species are much less crowded. Colonies of the same species may be on flat ground in one location and steep slopes or even cliffs in another. The Light-mantled Sooty Albatross is the nearest there is to an obligatory cliff-nesting albatross, but some nests are found on easier slopes close to the bottom of cliffs.

The nests of great albatrosses are usually so scattered that some of them hardly appear to be in colonies at all. On most breeding grounds, some nests are very close to each other, while others are so far (many kilometres) from their nearest neighbour that they might be thought solitary. Only by a broader view of a whole island may a patchy distribution become obvious. There is one remarkable exception; on the small islets off the Chatham Islands, Northern Royal Albatross nests are obviously in crowded colonies. In contrast, Southern Royals on Campbell Island have ample room and their nests are widely spaced in the manner characteristic of the five wandering albatrosses.

Unless they are ill or injured, adult albatrosses rarely encounter a natural predator on land that they cannot intimidate. By the time they are left on their own, even the chicks can see-off a lone skua, but surrounded by several determined skuas or caracaras, a small isolated chick's chance of survival is minimal. Vigilance is constant in albatross colonies, an approaching predator cannot achieve surprise. All occupants respond and thus protect their neighbour's backs, so a predator cannot go for one chick without itself being attacked from behind.The tighter the colony, the better the protection and it may be enhanced by allowing penguins to occupy spaces between the albatross nests.

Light-mantled Sooty Albatrosses, which nest alone or in twos and threes, cannot achieve this mutual help, so they protect their backs by nesting up against solid walls, where they can only be attacked from the front or side.

The social function of coloniality may be more significant among great albatrosses. Their chicks are soon big enough to look after themselves, but they too are vigilant, giving warning of potential danger to others within sight and sound. The costs of such loose coloniality may be negligible; in dense colonies there is probably competition for central sites as nests at the edge are less protected. There may sometimes be competition for nest material and birds returning early will dig at unoccupied neighbouring nests.

Sympatry

At sea different albatrosses are often found together. There have been comprehensive analyses of prey from some populations that demonstrate a partitioning of resources, but the samples from many others are small and collected by different methods, so they are less revealing.

On their breeding islands, many albatrosses are sympatric and there is potential competition for both nest-sites and resources. Colonies are mostly separate, but often those of two or more species are adjacent and may be mixed where they meet. Other ecological differences are usually less obvious.

The pelagic resources of the Patagonian Continental Shelf are large and dependable enough to sustain huge numbers of Black-browed Albatrosses. Nothing like such numbers are found on the islands off South America or at South Georgia where Black-browed are sympatric with the Grey-headed Albatross.

The two species both feed at the Antarctic Polar Frontal Zone (APFZ) to the north and west-southwest of South Georgia. Black-browed Albatrosses also feed over the shelf breaks of South Georgia, Shag Rocks and the South Orkney Islands, far to the southwest (Prince *et al.*1997) (Figs 5.33). In some seasons though, both species have to forage more extensively.

Black-browed Albatrosses at South Georgia have to obtain sufficient quantities of krill, which is an erratic and unpredictable resource. Grey-headed Albatrosses also eat krill when it is available, but by taking slightly longer over breeding, they can survive well enough on squid (Fig. 5.31). Fewer Grey-head Albatrosses breed each year, but over many years they are consistently more successful and have a total (biennial) breeding population at South Georgia roughly approaching that of the (annual) Black-browed Albatross.

On other islands, the numbers of Black-browed Albatrosses amount to only about one percent of the world population. At the Iles Kerguelen most feed in the vicinity of the island shelf and are outnumbered by Grey-headed Albatrosses which forage beyond the shelf (Weimerskirch *et al.* 1988). Campbell Black-browed Albatrosses greatly outnumber sympatric Grey-headed Albatrosses at Campbell Island (Moore & Moffat 1990).

About 20% of the two yellow-nosed albatrosses are sympatric with similar numbers of Grey-headed Albatrosses at the Prince Edward Islands, Iles Crozet and Iles Kerguelen. Sooty and Light-mantled Sooty Albatrosses are also sympatric at these islands (Weimerskirch *et al.* 1986).

Laysan and Black-footed Albatrosses are sympatric on almost all the NW Hawaiian Islands. The Black-footed Albatross colonises new sand cays but, as vegetation increases, the Laysan Albatross takes over the interior of islands, leaving the Black-footed Albatross on the dunes behind beaches. The Laysan is mainly a squid eater and has attained a world population of about 600,000 pairs. The Black-footed Albatross takes more fish, especially the egg masses of flying fish, and number about 60,000 pairs (Flint 1995).

Breeding seasons

In the southern hemisphere, most albatrosses breed during the summer (October–March); those on more northerly islands of the subtropical zone generally beginning earlier than closely related species on the more southerly subantarctic islands, two of which are just at the northern extension of winter pack-ice. Southern Buller's Albatrosses, Auckland Shy Albatrosses and Amsterdam Wandering Albatrosses all breed much later than expected from the latitude of their breeding grounds (Jouventin *et al.* 1989).

The three albatross species of the North Pacific all breed in the northern winter. Their breeding islands are in subtropical latitudes and by the time their fledglings leave and disperse northwards, it is early summer, when subarctic water becomes most productive.

The Galápagos Albatross, nesting just south of the equator, has a regime of its own. It begins breeding after the mists and rain of January to April, and towards the end of the year, their offspring disperse along the adjacent coast of South America. The Peru Coastal Current supports a huge avifauna that breeds throughout the year, but it is especially active around December. Young albatrosses evidently arrive when there is ample food.

Fecundity

The single-egg clutch (p.34) enables a pair to raise a maximum of one offspring per year, providing they are fit and have adequate resources. Sub-optimal climate or resources make increasing demands upon breeding birds that may diminish their capacity to survive at sea and/or to breed again the following season. Pairs appear to anticipate these events and reduce the costs by not attempting to breed in some seasons. There is evidently a subtle trade-off between present success and future success; natural selection may be expected to fine-tune the birds' sensitivity to their physiological condition and the environmental cues of their particular area of ocean.

An adult female albatrosses laying one egg each year has a *potential* fecundity of 1.0 (offspring) per year, but deferred breeding from time to time over many years, achieves lower *realised* fecundities. On different islands albatrosses may evolve characteristic patterns of deferred breeding, contributing to different realised fecundities.

These resemble different clutch-sizes and are probably subject to natural selection in the same manner as proposed for multi-egg clutches (Lack 1947, Ricklefs 1980). Among annual breeding albatrosses, mean realised fecundities over many years may be as high as 0.9 per year, while in biennial populations, where deferred breeding is more frequent, they may be as low as 0.4 per year.

Physiology

The gonads of adult birds increase in size at the beginning of the breeding season and regress soon after laying, accompanied by a well-known profile of hormone secretion. All albatrosses have inherent annual physiological cycles, but they may be interrupted. The annual cycle of the Black-browed Albatross may be regarded as typical. The neuro-endocrine response to seasonal changes in day-length begins long before albatrosses arrive at their breeding grounds. The photoperiodic cue for breeding in the southern hemisphere summer is *increasing* day-length. In the North Pacific, albatrosses breed in the winter so their cue is *decreasing* day-length. Two Black-browed Albatrosses from the south Atlantic that took up residence in the north Atlantic, attended Northern Gannet colonies and occupied nests in the northern summer. Within a few years of arrival in the northern hemisphere, they apparently made endocrine time shifts of six months and remained responsive to increasing day-length (B.K. Follett pers. comm.).

Male Wandering, Black-browed and Grey-headed Albatrosses arrive at their breeding islands with enlarged testes containing maturing sperm that can fertilise immediately. High testosterone levels at this time correlate with nest territoriality, agonistic behaviour towards other males, bonding behaviour towards potential mates and sometimes determined sexual assaults on other females (p.304). Testosterone levels in males (and females) decline rapidly as laying approaches, and remain low until fledglings are almost ready to depart. Interestingly, male Wanderers experience a brief but marked surge of this hormone at the time of hatching (Hector *et al.* 1986a).

Females returning shortly after males have high levels of oestrogen and several enlarged follicles in their ovaries. Copulation can take place soon after partners come together at the nest. Fertilisation may follow immediately or sperm may be stored in the female tract until after she has returned to sea to complete formation of the egg (Astheimer *et al.* 1985).

The testes of both Black-browed and Grey-headed Albatross males enlarge and atrophy each year (Fig. 16.1), but while the Black-browed males return to their islands and breed most years, Grey-headed males that reared a chick the previous season either do not return at all or make only a few late, brief visits to their nests. In Wandering Albatross males, feeding fledglings during the spring-summer inhibits maturation (recrudescence) of the testes and they leave at the same time that other males are arriving with enlarged testes, ready to begin breeding (Hector *et al.* 1986a&b, 1990).

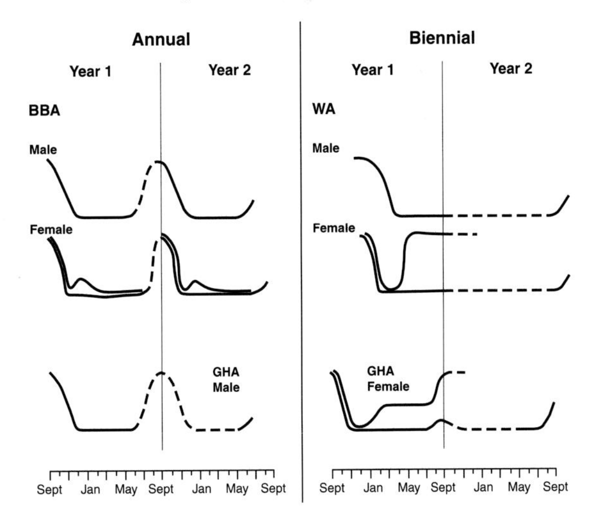

Fig. 16.1 Changes in gonad size of male and female Wandering (WA), Black-browed (BBA) and Grey-headed Albatrosses (GHA) that successfully rear chicks at South Georgia; solid curves represent actual data and dashed lines indicate periods when the birds were at sea and not accessible. In females, the additional, upper curve represent the levels of plasma progesterone in circulation (Hector *et al.* 1987).

In female albatrosses, yolk development (vitellogenesis) is promoted by a surge of oestrogen, accompanied by an increase in progesterone. Laying is followed by steep declines in the secretion of both these hormones. In the Black-browed Albatross, the levels remain low during chick rearing and throughout the following winter months at sea. In Grey-headed females, progesterone increases and fluctuates during the chick period, through the winter and into the following spring. It inhibits the release of luteinising hormone from the pituitary, which in turn blocks the surge of ovarian oestrogen essential for yolk formation. In Wandering Albatross females, a similar or more pronounced sequence prevails, but throughout the winter chick rearing, fluctuating progesterone is accompanied by significant levels of oestrogen, with an increase in the spring, even though no egg is laid that summer.

Annual environmental changes make it more or less difficult for albatrosses to find food and satisfy the energy demands of breeding and moulting, which are different for males and females (see below). Optional deferred breeding in lean years is a flexible way of responding to such constraints (Langston & Rohwer 1996). At some time in the past, disproportionately high costs of annual breeding in female Grey-headed Albatrosses evidently reduced fertility to such an extent that deferred breeding was accompanied by an improvement in lifetime fertility and natural selection eventually favoured a biennial regime. Selection pressures on males remained weak enough for them to retain the annual mechanism. In Wanderer males, the higher costs of foraging for chicks throughout the winter, ensured that they, like their females, switched to a biennial regime.

Croxall (1991) has argued that in the Grey-headed Albatross, competition for biennial females favours the retention by males of an annual endocrine cycle. After these males have bred successfully in one season (i), the 40% returning briefly, but rather late, to their colonies in the i+1 season, may acquire new partners and have a better chance of breeding in the i+2 season if their mates do not turn up (presumed dead at sea).

Less than 2% of Grey-headed males actually breed with new females in the i+1 season. If a female, absent in the i+1 season, does not die at sea, she re-assumes her role in the i+2 season. However, established females spend very little time ashore until incubating and it is quite possible for a male to consort separately with two females. Older females need not encounter a successful second female who would be absent, returning in the i+3 season, when the original female was away again. Such bigamy might prevail if both females continued to be successful, but failure of one or the other would, sooner or later result in both returning in the same season. A curious example of this occurred at Bird Island in 1987–90. A male Wandering Albatross shared incubation with two females in separate nests built next to each other. This came to an end when the male – in the normal course of nest building – excavated the side of the adjacent nest so that the egg rolled out into the cleft between the nests, where it was ignored (M. Jones pers. comm.).

Some mollymawk nests at South Georgia acquire two eggs laid by separate females (Tickell & Pinder 1966), one of them usually gets buried in the bowl of the nest, but in February 1962, a Grey-headed Albatross nest contained two healthy chicks of the same size, and at three others second chicks were found alongside occupied nests. In none of the surrounding areas were there empty nests from which chicks may have moved (Dollman 1962).

Annual breeding

Rice & Kenyon (1962b) used records of 820 ringed Laysan Albatrosses at 419 nests, to determine the proportion of birds nesting at Midway Atoll in two successive seasons, 1956–57 and 1957–58. Most pairs (87%) that lost eggs early in incubation bred the following season, but only 63% of those which successfully reared fledglings did so. Fisher (1976) later obtained figures of 81% and 68% respectively from observation of 619 ringed pairs. Among 65 pairs that remained intact and bred (successfully or unsuccessfully) 709 times in 13 years, an average of 80% bred again in the following year, 19% bred two years later and 1% three years afterwards. These figures incorporated a considerable year-to-year variation (Fig. 16.2).

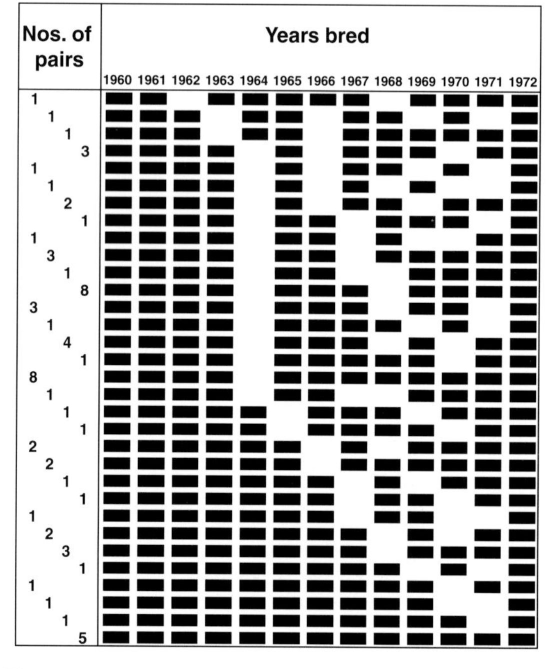

Fig. 16.2 Patterns of breeding among 65 pairs of Laysan Albatrosses that remained intact for 13 years and bred (successfully and unsuccessfully) at Midway Atoll 1960–72 (Fisher 1976).

Most black-browed and yellow-nosed albatrosses breed each season, but some pairs go for two, three or more years before their next breeding, whether or not they have been successful (Fig. 16.3). In at least one population of Black-browed Albatrosses there has been considerable yearly variation (Prince *et al.* 1994).

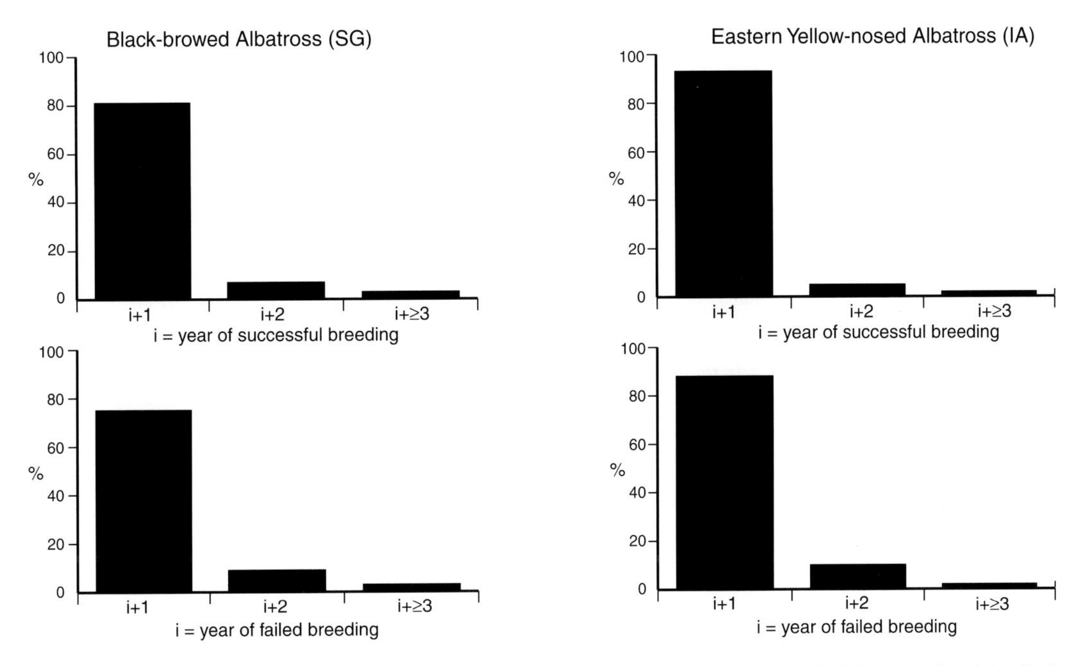

Fig. 16.3 Mean percentages of two annual albatrosses found breeding again at South Georgia (SG) (1976–87) and the Ile Amsterdam (IA) (1978–84) after breeding successfully or failing in year i (100%) (Jouventin *et al.* 1983, Prince *et al.* 1994).

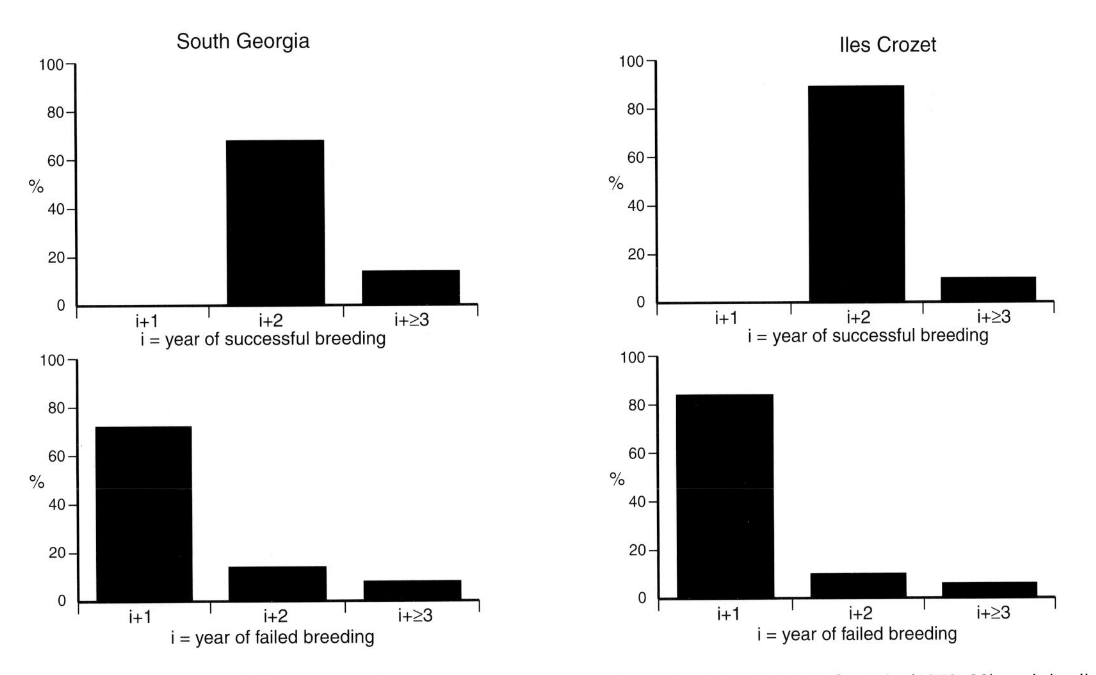

Fig. 16.4 Mean percentages of Wandering Albatrosses from two separate populations at South Georgia (1976–88) and the Iles Crozet (1966–83) found breeding again after breeding successfully or failing in year i (100%) (Jouventin & Weimerskirch 1988, Croxall *et al.* 1990b, 1997).

Biennial breeding

Fig. 16.4 illustrates the breeding frequency of Wandering Albatrosses in two separate islands. At South Georgia there is considerable year-to-year variation (Table 16.1) which may reflect differences in the dates of chick deaths and differences in the response of pairs to premature cessation of foraging. Males and females respond in much the same way. After breeding successfully, slightly fewer males return two years later, while after failed breedings rather more males returned the following year (Croxall *et al.* 1990b).

In 1963–64, Grey-headed and Light-mantled Sooty Albatrosses did not re-occupy nests at Bird Island, South Georgia in which they had successfully reared chicks the previous season (Tickell & Pinder 1967, W.L.N. Tickell unpublished), but fieldwork could not be continued into a third season and it was many years before biennial breeding was confirmed (Kerry & Garland 1984, Prince 1985). In the meantime, the Sooty Albatross was also

Year (i)	n (pairs)	After successful breeding in year i, % breeding again in year					n (pairs)	After failed breeding in year i, % breeding again in year			
		i+1	i+2	i+3	i+≥4	Total		i+1	i+2	i+≥3	Total
1976	128	0	66	13	5	84	63	71	16	6	93
1977	103	0	71	11	6	88	118	38	44	9	92
1978	146	0	59	18	7	85	64	63	9	12	84
1979	-	-	-	-	-	-	47	71	15	6	92
1980	275	0	66	7	6	80	212	67	10	6	83
1981	394	0	76	0	5	81	200	78	2	5	85
1982	326	0	74	4	5	83	174	67	16	4	87
1983	355	0	77	5	-	84	258	61	15	7	83
1984	443	0	82	-	-	82	224	85	3	-	88
1985	-	-	-	-	-	-	295	75	-	-	75

Table. 16.1 Percentage of experienced pairs of Wandering Albatrosses at South Georgia found breeding again after breeding in year i (100%) (Croxall *et al.* 1990b).

Species	No. of years	Success/failure in year i	n (pairs) in year i	% breeding in years i+1	i+2	i+3	i+≥4
			ANNUAL BREEDERS				
LA	13	unknown	644	80	19	1	
BBA	10	failure		75	9	5	
		success		81	7	4	
CBBA	4	failure	28	75	4	0	0
		success	55	80	9	0	2
EYNA	3	failure	55	98	2		
		success	95	97	3		
			BIENNIAL BREEDERS				
WA	10	failure		63	17	8	
		success		0	66	13	7
GHA	10	failure		54	24	9	
		success		1	68	10	7
GHA	4	failure	106	65	21	1	2
		success	21	0	76	5	5
SA	5	failure	176	89	10	1	
		success	72	0	83	15	1

Table 16.2 Mean percentages of pairs found breeding again after breeding in year i. Laysan (LA), Wandering (WA), Grey-headed (GHA), Black-browed (BBA), Campbell Black-browed (CBBA), Eastern Yellow-nosed (EYNA) and Sooty Albatrosses (SA) (Fisher 1976, Weimerskirch 1981, Jouventin *et al.* 1983, Croxall 1991, Waugh *et al.* 1999).

discovered to be a biennial breeder (Weimerskirch 1981) (Table 16.2). The Grey-headed Albatross is the only known biennial mollymawk and successful breeding is almost always followed by a season's absence. Some do return, and there have been rare case of Grey-headed Albatrosses breeding the next year, one pair at Bird Island bred every year for six years (J.A.L. Hector pers. comm.). No such exceptions have been documented among the two sooty albatrosses (Fig. 16.5). Whether they succeed or fail in their breeding, some Grey-headed Albatrosses put off breeding for more than one year and at Bird Island there has been considerable year-to-year variation (Prince *et al.* 1994).

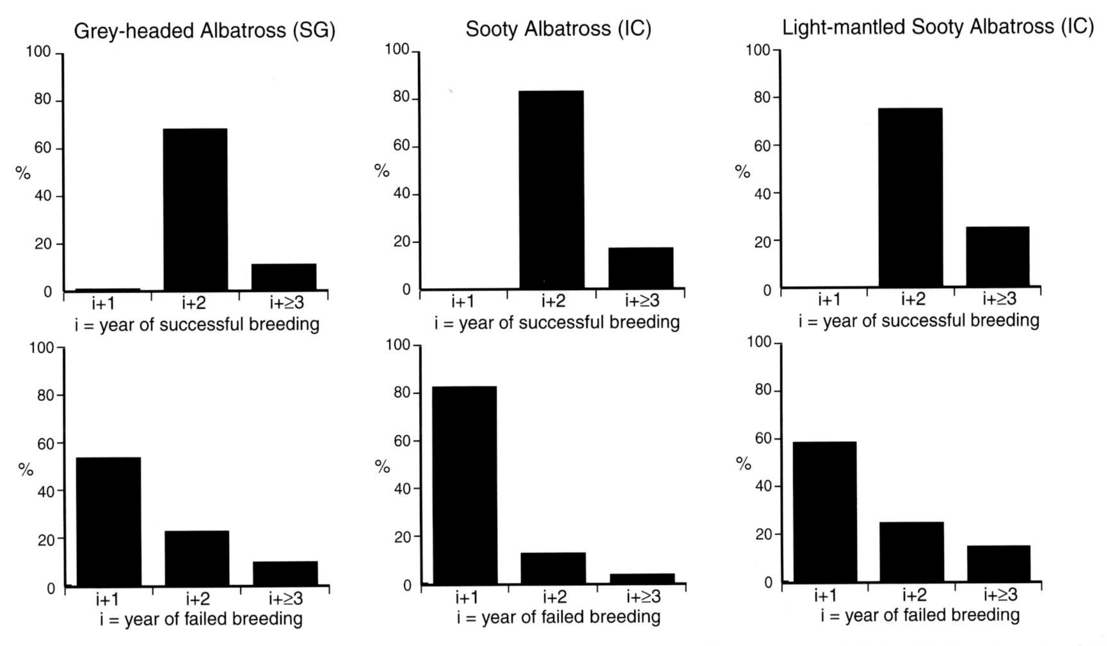

Fig. 16.5 Mean percentages of three biennial albatrosses found breeding again at South Georgia (SG) (1976-87) and the Iles Crozet (IC) (1975–84) after breeding successfully or failing in year i (100%) (Jouventin & Weimerskirch 1988, Prince *et al.* 1994).

The breeding season of Grey-headed Albatrosses at South Georgia lasts about 238 days, while that of Black-browed Albatrosses at the same island is only 204 days (Prince *et al.* 1994). Most of this difference is due to a prolonged commitment to chick rearing by Grey-headed Albatrosses (Table 5.12). It has been imagined that the biennial regime of Grey-headed Albatrosses evolved because successful breeding in one season left females less fit to breed successfully the following season. This is plausible, but it should not be assumed that there is anything broadly significant about 238 days. In one year, the breeding season of the Southern Buller's Albatross at the Snares lasted 264 days and it is an annual breeder (Sagar & Warham 1997).

Deferred sexual maturity

Juveniles remain at sea for a number of years before they return to land. In the years that follow there are discernible sexual differences; the testes of male Wandering Albatrosses enlarge between four to ten years of age, and the amount of circulating testosterone increases over that period (Fig. 16.6). By five years of age, males are producing sperm, but they do not breed before they are seven, and half of them not until they are eleven. The physiological changes associated with sexual maturation in female Wandering Albatrosses begin at sea in the juvenile years. The youngest females to return spend more years acquiring a mate than those that first arrive some years later (Fig. 7.34). In females of four to seven years old, no progressive maturation is apparent. The secretion of luteinizing hormone increases with age and maintains high levels of progesterone among younger subadults, but surges of oestradiol and enlarged ovarian follicles (yolk) are not noticed until the year of breeding (Fig. 16.6). No Wandering Albatross females have laid their first eggs at the age of six years or younger and only two percent at seven years of age (Hector *et al.* 1990).

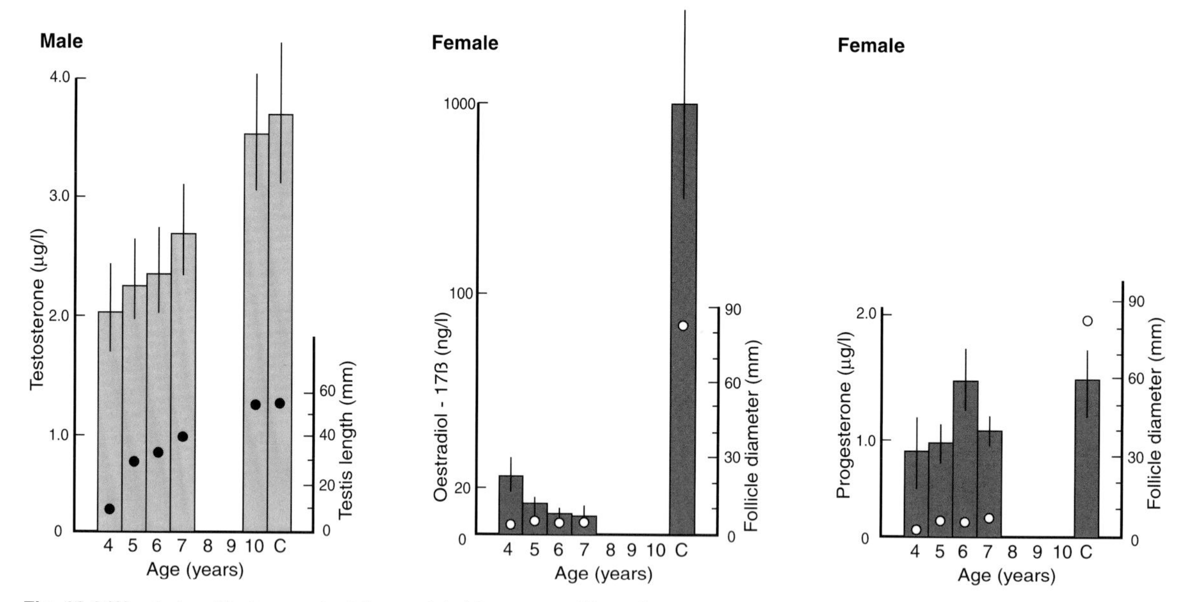

Fig. 16.6 Wandering Albatross subadults aged 4–10 years, and breeding adults at the time of copulation (C) (n=6). Males: mean testis length (n≥3) (•) and testosterone levels ± SE (n≥8). Females: maximum diameter of the largest follicle (n≥3) (o) and levels of circulating oestradiol and progesterone ± SE (n≥9) (Hector *et al.* 1986a).

POPULATION DYNAMICS

Populations have been assessed on the breeding grounds, but there has been considerable disparity in methods, often resulting in uncertainty about what has been counted. Frequently, figures have been described as 'estimates' with few if any details of how they were calculated. Gross exaggerations have been stoutly defended and counts of sympatric populations of different species have been pooled. Even where good field counts have been made, errors in other parameters have produced wildly inaccurate predictions. Rosemary Gales (1993, 1997) has boldly gone where many others might fear to tread. It is not at all clear how she arrived at her three categories of reliability. Few albatross studies predate the expansion of fisheries into the Southern Ocean and those that have been instigated in response to such hazards, suffer from lack of pre-fishing data.

Undisturbed populations of large, long-lived seabirds that invest considerable energy in rearing few offspring (K-strategy) have the potential to achieve numerical stability. Substantial populations that have been in existence for hundreds or thousands of years may be expected to have reached numbers sustainable by available resources. However, the dynamics of oceans ensure that there are substantial yearly fluctuations in the quantity and distribution of prey in surface waters and seasonal populations of albatrosses also vary. It takes many years to detect whether such fluctuations conceal trends in breeding populations or mask intrinsic stability. The physiological components of such variability are more complex than formerly imagined. Individuals have some choice about whether or not to breed in a forthcoming season and they appear to make that decision not only on the basis of the time available, but also on their current physical condition (Bevan *et al.* 1995).

At Bird Island, South Georgia, seven colonies of Black-browed Albatrosses were counted yearly over 20 years (1976–96), and within that period changes in individual colonies varied from -77% to +96% (Prince *et al.* 1994) (Fig. 16.7). The total numbers breeding on Bird Island in the seasons 1976–77 and 1989–90 were 13,328 and 14,695 pairs respectively, indicating an average increase of +0.7% per year. In subsequent years (1989–96), numbers declined at mean rates down to -9.9% per year (Croxall *et al.* 1997b). Declines among Campbell Black-browed Albatrosses over decades have also been detected, but field counts did not begin at Campbell Island until 1992, and the earlier photographic method used was not critical enough to reveal the yearly variation expected in mollymawk colonies (Waugh *et al.* 1999a).

Breeding populations of biennial species are more difficult to estimate; counts of eggs are required from at least two consecutive seasons. An early attempt to deal with Wandering Albatrosses at Bird Island (Tickell 1968) realised, however, that it was not enough just to add together two seasons' eggs. Pairs which lost eggs and young chicks did not take a year off between breeding, so the total number of pairs breeding in a population (P_i) was thought to be:

$$P_i = (1+f)\ (C_i + C_{i\text{-}1}) - Y_i \ .....^5$$

where: C_i = an egg count in year i

f = a factor for eggs broken before the count

Y_i = the number of pairs breeding in year i that also bred in year i-1.

Later, seasonal counts over 17 years (1976–91) of one Grey-headed Albatross colony at Bird Island, fluctuated between 155 and 449 pairs (Fig. 16.7) and scientists became more cautious about what they were counting. An **index** (I) of the breeding population was computed using a rather neater equation (Prince *et al* 1994):

$$I_i = N_i + qN_{i\text{-}1} .....^6$$

where: N_i is the number of eggs laid in year i, q is the proportion of birds breeding in year i-1 which did not breed in year i.

There were 9,053 pairs of Grey-headed Albatrosses on Bird Island in 1976–77 and 9,164 pairs in 1977–78. A value of q = 0.62 calculated from observations at the one study colony was applied to counts of all other colonies, with the result that the 1976–77 + 1977–78 population for the island could be considered as approximately 14,777 pairs (Prince *et al* 1994). Over the next 16 years, it declined at an average rate of -1.4% per year with individual colonies showing irregular seasonal fluctuations (Croxall *et al.* 1997b). Similar declines over decades have been claimed at Campbell Island (Waugh *et al.* 1999a).

In 1960–62, seasonal populations of Wandering Albatrosses at Bird Island were 1,554, 1,922 and 1,932 pairs (Tickell 1968). A decline became apparent in the 1970s and by the end of the decade, was confirmed as -1% per year (Croxall 1979). By 1987–89 the seasonal populations were down to 1,223, 1,366 and 1,411 pairs respectively (Croxall *et al.* 1990b) and decline was still evident at -0.8% per year in 1996 (Croxall *et al.* 1997b).

A more abrupt decline in Wandering Albatrosses was documented at the Iles Crozet and Iles Kerguelen. On the Ile de la Possession, an apparently stable population during the late 1960s went into decline at a rate of -7.0% per year (1970–76). It eased off to -1.4% per year from 1977 to 1985 then stabilised and increased at +4% per year (1986–1993) (Weimerskirch *et al.* 1997a).

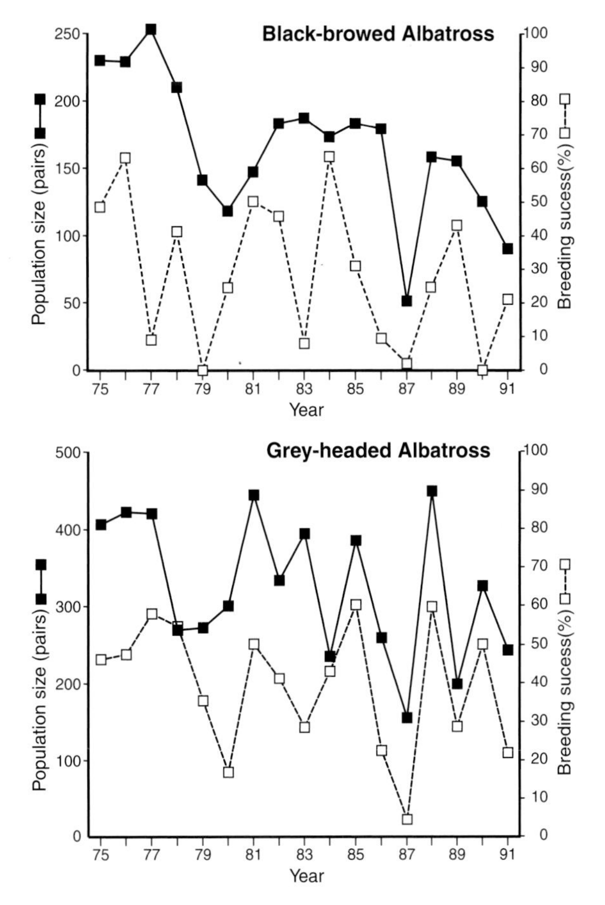

Fig. 16.7 Variation in population size and breeding success of Black-browed Albatrosses at Colony H and Grey-headed Albatrosses at Colony E on Bird Island, South Georgia 1975–91 (Prince *et al.* 1994).

[5] a more elegant version of my equation suggested by P. Rothery.
[6] a transposition of the original equation (Prince *et al.* 1994) by P. Rothery.

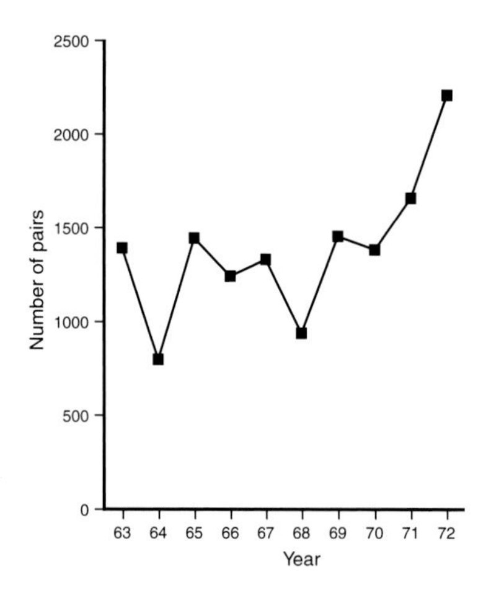

Fig. 16.8 Yearly counts of Laysan Albatrosses breeding in a study plot on Eastern Island, Midway Atoll 1963-72 (Fisher 1976).

The population of Laysan Albatrosses at Midway Atoll experienced three decades of growth followed by more than two decades of devastation and disturbance, but it was able to endure and increase (Appendix 15). In 1945, the abandoned concrete runway and surrounding stony ground of Eastern Island had only a few nests, but it had become fully occupied by the time Harvey Fisher arrived in the early 1960s. He worked with large ringed samples in plots marked out within the large colony. In ten years (1963–72) the numbers within these plots varied from 722 to 2,208 pairs (Fisher 1976) (Fig. 16.8).

Other examples lack demographic detail. In the Falkland Islands, the breeding population of Black-browed Albatrosses at West Point Island in the early 1960s was more than 4,000 pairs (R.W. Woods pers. comm.); it increased to about 12,050 pairs by 1989 (Thompson 1989) and 15,200 pairs by 1993 (Woods & Woods 1997). At the Snares Islands, Southern Buller's Albatrosses increased from 4,750 pairs in 1969 to 8,460 in 1992 (Warham & Bennington 1983, Sagar *et al.* 1994).

	Population	Years mean (range)	Males (%) mean (range)	Females (%) mean (range)	Unsexed (%)
		ANNUAL BREEDERS			
BBA*	South Georgia	1976–88	94 (84–98)	96 (86–100)	93 (82–97)
	Iles Crozet	1979–82			88 (81–96)
CBBA	Campbell Island	1989–93			95
		1984–96			95
EYNA	Amsterdam Island	1978–82			91 (75–99)
GA	Iles Galapagos	1961–70			95
		1970–71			97
LA	Midway Atoll	1961–71	95 (91–98)	95 (92–97)	95 (91–98)
		BIENNIAL BREEDERS			
WA	South Georgia	1960–63	95	97	96 (93–98)
		1976–86	95 (89–99)	93 (87–96)	93 (87–99)
	Iles Crozet	1966–69			90 (87–92)
		1970–76			89 (89–90)
		1977–85			94 (93–94)
		1986–93			96 (95–96)
AKWA	Auckland Islands	1993–95	98 (97–99)	96 (94–99)	97 (96–98)
NRA	New Zealand	1937–93	96 (94–99)	94 (89–99)	95 (93–97)
GHA	South Georgia	1977–88	96 (91–99)	96 (93–100)	95 (91–100)
	Campbell Island	1989–93			94
		1984–96			95
SA	Iles Crozet	1975–82			95 (86–100)
LMSA	Iles Crozet	1976–82			97 (83–100)

Table 16.3 Mean annual survival of adult albatrosses as a percentage of those alive in the previous year. (Tickell 1968, Harris 1973, Fisher 1975a, Croxall *et al.* 1990b, Robertson 1993c, Prince *et al.* 1994, Waugh *et al.* 1999a, Weimerskirch *et al.* 1997a, Walker & Elliott 1999). * for key to abbreviations see Appendix 1.

Survival

Albatrosses have naturally low mortality/high survival life histories with mean percentage survival of adults usually greater than 90% (Table 16.3). Inter-year variation (Fig. 16.9) is usually due to transient natural hazards and inconsistent field-work. It has taken many years for unequivocal trends to be detected (Weimerskirch & Jouventin 1987, Croxall *et al.* 1990b).

During the 1980s and 1990s, several populations of albatrosses have revealed anomalies in survival rates, with some much lower than were measured in the previous two decades. These have been attributed to fisheries-induced mortality (Weimerskirch *et al.* 1997a, Croxall *et al.* 1997b).

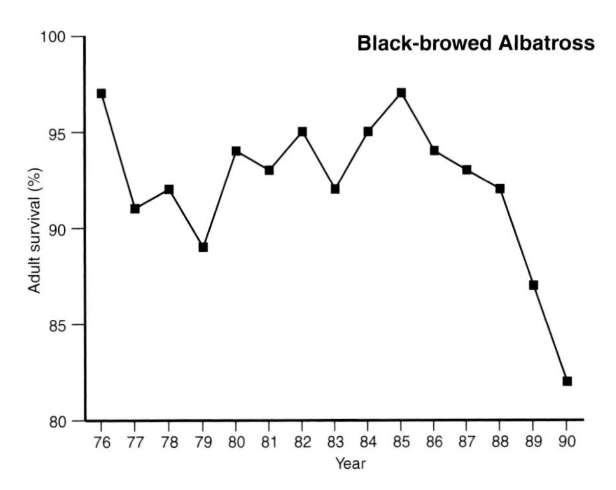

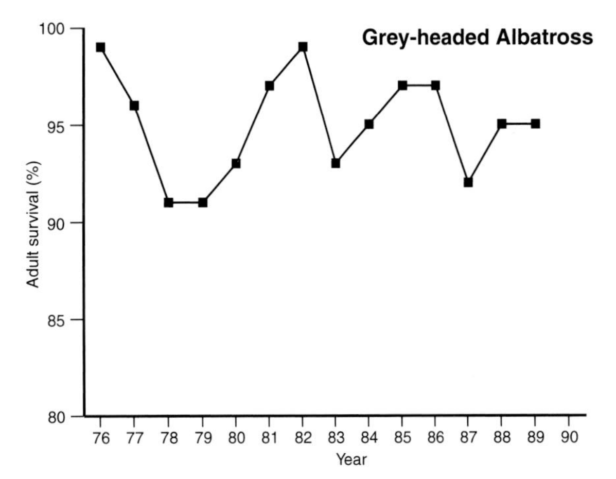

Fig. 16.9 Variation in mean annual survival rates of adult Black-browed Albatrosses at Colony H and Grey-headed Albatrosses at Colony E on Bird Island, South Georgia. The broken lines represent the overall arithmetic mean survivals 1976–88 inclusive (Prince *et al.* 1994).

Adults

In some seasons at Midway Atoll, almost twice as many male Laysan Albatrosses died as females and in other seasons vice versa; there was no consistent sexual difference in adult survival rates (Fisher 1975a). The same was true of the Grey-headed Albatross at South Georgia, but the mean annual survival rate of the Black-browed Albatross over 12 years (1976–88) was higher for females (95%) than males (93%) (Prince *et al.* 1994, Croxall *et al.* 1997b).

At the Ile de la Possession over three decades, there was no significant difference in survival rates of male and female Wandering Albatrosses (Weimerskirch *et al.* 1997a), while at Bird Island between 1958 and 1984, survival was about 95% in males and 93% in females; males exceeding females by an average of over 2% in three separate cohorts (Croxall *et al.* 1990b).

Estimates of survival rates depend upon complex interactions of ringing schedules, recapture rates and breeding success. In a capture-recapture analysis, the probability of finding biennial breeding adults during the season following a successful breeding, is less than that of recapturing birds that were unsuccessful. Bias may go undetected in cursory investigations, but it becomes apparent when survival rates in excess of 100% appear. Rothery & Prince (1990) have developed a simulation model to measure the error and make appropriate adjustments.

Juveniles

Juvenile mortality is probably high during the early years at sea, 40% of recoveries from Wandering Albatrosses, ringed as chicks at South Georgia, were obtained in the first two years, 54% in the first three years (Croxall & Prince 1990). Survivors are not available for counting until they return to land as subadults. In the 1970s and 1980s, many Wandering Albatrosses returned to South Georgia as five-year-olds, and the survival rates to this age averaged 49% for six cohorts fledged in 1972–78 and 52% for ten cohorts fledged in 1979–89 (Croxall *et al.* 1990b, 1997b). At the Iles Crozet, survival to five years of age increased steadily from 1968–69 to 1992–93 (Weimerskirch & Jouventin 1997).

Survival rates of the Grey-headed Albatross to five years of age averaged 19% for ten cohorts fledged at South Georgia in 1976–86. Survival of Black-browed juveniles from the same island declined from 23% for five cohorts fledged in 1976–81 to 15% for five cohorts fledged in 1982–86 (Croxall *et al.* 1997b).

Breeding success

The measure of breeding success is usually calculated as the proportion of eggs laid in one season that produce juveniles at sea. It is partitioned into egg and chick components, represented by hatching and fledging rates. Both relate to stochastic variation of the terrestrial and marine environments. For example, when storms sweep across Midway Atoll, Laysan Albatross nests are flooded and large numbers of eggs are lost. On the rocky islets of the Chatham Islands, the combined effect of storms and drought caused the loss of all Northern Royal Albatross eggs and chicks in 1990. Grey-headed Albatrosses sometimes arrive at South Georgia to find their nests still under ice and snow. These are intermittent events that happen at the location of the nest. Chicks are more likely to die of dehydration and starvation because of distant events at sea that prevent parents returning with enough food. An environmental perturbation may have quite different effects on closely related species. The 1986 decline in provisioning rates of 50–60% among mollymawks at South Georgia reduced the breeding success of Grey-headed Albatrosses to 45% and Black-browed to 9% (Croxall *et al.* 1999). Individual differences also contribute significantly. Some pairs of Grey-headed Albatrosses are consistently successful breeders in good and poor seasons while others, with the same experience, fail year after year (Cobley *et al.* 1998).

In the five years 1961–64, an average of 64% (49–78%) of Laysan Albatross eggs produced fledgling that left their nests on Midway Atoll, but before they could fly away over the sea, they had to walk several hundred metres to the beaches. During this journey in the summer heat, when many also made their first short flights, about 3.5% died of dehydration (Fisher 1975a).

Between 1975 and 1990, average breeding success of the Wandering Albatross at South Georgia increased from 56% to 71% then in the following five years decreased to 68%. Approximately half of the variation occurred during incubation and half during chick rearing (Croxall *et al.* 1997b). At the Iles Crozet over 28 years, Wanderer breeding success averaged 69%, but was said to have been increasing (Weimerskirch *et al.* 1997a).

Over 21 years (1976–96), Black-browed and Grey-headed Albatrosses at South Georgia experienced considerable fluctuation in breeding success, the Black-browed being the more variable. In one season, all chicks died of starvation. Disasters like this depressed the average breeding success to 27%. Although the mean hatching rates of Black-browed eggs remained fairly constant (62%), fledging rates in 1986–96 were only half of what they had been in 1976–86. Most of the variation in breeding success was due to failure during chick-rearing. Grey-headed breeding success averaged 39% with equally high hatching rates (60%) and fledging rates (65%) although, more of the variation in breeding success occurred during incubation (Prince *et al.*1994, Croxall *et al.* 1997b) (Fig. 16.10).

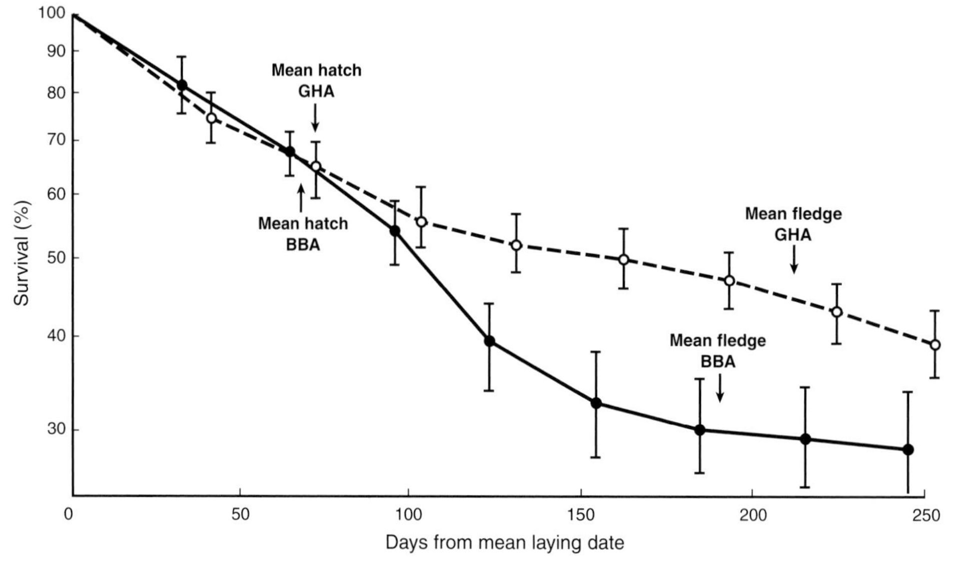

Fig. 16.10 The monthly mean survival (±SE) of eggs and chicks from eggs laid by Black-browed Albatrosses (BBA) in colony H (——) and Grey-headed Albatrosses (GHA) in colony E (- - -) at Bird Island, South Georgia, 1976–92 (Prince *et al.* 1994).

At Campbell Island over six seasons between 1984 and 1994, the mean breeding success among the Campbell Black-browed Albatross was 66% (51–84%) and in the Grey-headed Albatross, 40% (16–70%). The mean hatching rates were 86% and 80%, but the Grey-headed Albatross had a much lower fledging rate (51%), with greater variability in both hatching and fledging rates (Waugh *et al.* 1999a). During the same six years, 25 pairs of interbreeding Black-browed and Campbell Black-browed Albatrosses achieved breeding successes of only 36% to 40% (Moore *et al.* 1997).

At Kerguelen and Macquarie, annual breeding successes in Black-browed colonies, over six to seven seasons, averaged 58% (Jouventin & Weimerskirch 1988) and 67% (Copson 1988) respectively. In the Eastern Yellow-nosed Albatross at the Ile Amsterdam, it averaged 50% over seven years (Jouventin & Weimerskirch 1988).

Breeding frequency

At South Georgia over ten years (1976–86), an average of 81% (57–93%) of successful Black-browed Albatrosses bred again in the following year, compared with 75% (53–87%) of unsuccessful pairs. However, 7% of successful pairs and 9% of unsuccessful pairs next bred two years later. In the Eastern Yellow-nosed Albatross, the proportions were similar. Larger numbers of Black-browed pairs postponed breeding after seasons of low krill availability (Prince *et al.* 1994, Croxall *et al.* 1997b). In Laysan Albatrosses the reverse occurred, 68% of pairs that bred successfully in 1961–62, bred again the following year, compared with 81% unsuccessful pairs (Fisher 1976).

Less than one percent of Grey-headed pairs at South Georgia breed in two consecutive seasons. Most defer breeding for one year beyond the following season, but there is considerable variation. On average, 36% of successful and 47% of unsuccessful breeders defer breeding for two or more years (Prince *et al.* 1994, Croxall *et al.* 1997b).

Sooty and Light-mantled Sooty Albatrosses have been studied in less detail than the mollymawks and great albatrosses. Solitary nesting, high adult survival, low fertility and mass desertions have been said to indicate that these two species occupy fragile niches of low energy yield (Jouventin & Weimerskirch 1988).

Recruitment

The age of first breeding in the Wandering Albatross has been steadily declining at South Georgia and the Iles Crozet; the number of fledglings recruited into the breeding population has also fallen. In the early 1960s, at least 36% of South Georgia cohorts survived to breed, but by the 1980s it had fallen to about 27% (Croxall *et al.* 1997b, Weimerskirch *et al.* 1997a).

Recruitment rates of the Black-browed and Grey-headed Albatrosses at South Georgia in the 1970s were 27% and 36% respectively, but since 1987 they have fallen to about 7% and 5% (Prince *et al.* 1994, Croxall *et al.* 1997b). From 1975 to 1987, 25% of Campbell Black-browed and 19% of Grey-headed fledglings, returned to Campbell Island by the age of 12 years, about the average age of first breeding (Waugh *et al.* 1999a).

At Midway Atoll throughout the 1960s, at least 16% of fledgling Laysan Albatrosses returned to the colony where they were reared and 14% were known to have bred (Fisher 1975a, 1976). Over the same period, 27% of Black-footed Albatross fledglings returned to Midway by breeding age (USFWS unpublished).

Breeding equilibrium

The simplest statement of biennial breeding is that approximately half the population breeds each year. There is a constant shuffling of birds so that those breeding in alternate seasons remain identified with one homogeneous population. This is possible because biennial breeding is a behaviour-driven regime superimposed upon an annual reproductive cycle. Many pairs that are unsuccessful in a current breeding, return and try again the following year. The mechanism is tuned to breeding success. Theoretically, if all pairs in a current breeding are successful, then none of them will return next year, but if there is a total loss of eggs, all surviving birds will return and breed next year. Between these two extremes the proportion of pairs breeding in consecutive seasons will be directly proportional to the breeding success.

After a catastrophic early failure, the numbers of pairs breeding in consecutive seasons will be thrown out of balance, but if other parameters remain within the average range, balance will be restored by a series of damped oscillations (Fig. 16.11). The rate at which this is attained will also be directly proportional to breeding success (Tickell 1968, 1970).

Population regulation

Albatrosses breeding in crowded colonies may achieve maximum possible nest density in favoured central areas, but there is usually space for new nests at the periphery. Shortage of nest-sites can rarely be imagined as limiting population growth among albatrosses, but on islets of the Chatham Islands and at the Bounty Islands, competition for nest sites, and particularly for building material, is conceivable (Robertson & van Tets 1982; Robertson 1991).

The food of albatrosses may be limiting, especially during the breeding season within foraging range of nests. Afterwards, when the population is more dispersed, selection will be sensitive to different densities. Mortality from other causes will always be present and may at any time overtake food shortage as the major cause of death (Newton 1980). Density-dependent mechanisms remain, but ornithologists have been hard-pressed in their attempts to test the hypothesis and it cannot be said that albatrosses are convenient subjects.

> '...simple population models show that long-lived species, with low reproductive rates and late age at first breeding, like the fulmar, require a stronger degree of density-dependent effects to stabilise numbers than species like the shag ...if density-dependence acted only on a single parameter it would have to have an extremely large effect. In fact ...a number of breeding parameters within a single species change with increasing population size...' (Birkhead & Furness 1985).

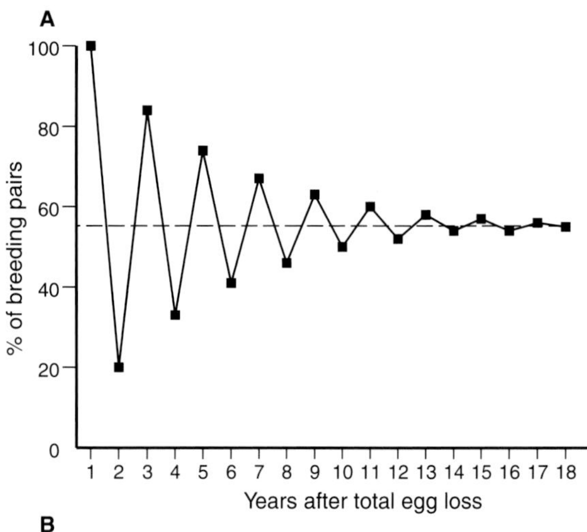

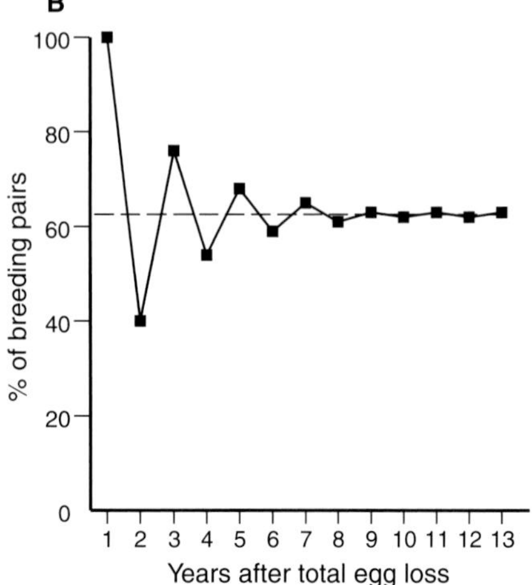

Fig. 16.11 Theoretical damped oscillations in the percentage of Wandering Albatrosses breeding each year following the loss of all eggs soon after laying in year 0. If the average breeding success is 80%, the demi-population may stabilise at approximately 55% of the total breeding population after about 15 years (A). With a lower breeding success of 60%, a demi-population of about 62% would be expected in about 9 years (Tickell 1968, 1970).

Because of their low reproductive rates, many seabird populations have not yet returned to stability after devastations caused by man (Ashmole 1963). This is obvious enough today with fisheries mortality, but albatross populations have potential for recovery.

Steller's Albatross at Torishima has been increasing exponentially since the early 1950s; from 1979 to 1995 the rate was +7.4% per year (H. Hasegawa pers. comm.) (Fig. 12.7). In the Northwest Hawaiian Islands the few irregular counts of Laysan and Black-footed Albatrosses have yielded quite varied rates of increase. In one plot among the huge population on Midway Atoll counted over 30 years, the increases were +6.0% and +9.4% per year respectively. Laysan Albatrosses on the several small islets of French Frigate Shoals increased at +4.8% per year in the years 1956–90 while on one of them, Tern Island where there were probably none in the 1940s, the increase was +13.5% per year. In 11 yearly counts on the same island (1979–90), Laysans increased at +7.9% per year and Black-footed Albatrosses at +21.3% per year (Gould & Hobbs 1993). Two counts at the Snares Islands in 1970 and 1992 indicated an increase of the Southern Buller's Albatross at a rate of about +2% per year over the 22 years.

There are no good examples of stable albatross populations. In the late 1950s and early 1960s, the numbers of

Wandering Albatrosses at South Georgia may have been stable or increasing slowly, but given the known yearly fluctuation and the high survival rates in those years, the evidence is impressionistic (Tickell 1968, Croxall *et al.* 1990b, 1997b).

Demographic models

Early attempts to deal with the arithmetic of a biennial populations have given way to more profound understanding involving cycles of several years and more accurate data from repeated resighting of ringed albatrosses.

Demographic balance in a population is achieved when:

Natality + Immigration = Mortality + Emigration

Albatrosses land on islands and in colonies other than those where they were reared, but very few settle and breed away from their natal colonies (Fisher & Fisher 1969).

Reciprocal recoveries of ringed Wandering Albatrosses between breeding grounds of the Prince Edward Islands and the Iles Crozet indicate that movements occur in both directions, over the 1,000 km between these two island groups, and that female immigrant/emigrants are twice as frequent as males (Battam & Smith 1993, Weimerskirch *et al.* 1997a). Perhaps the most remarkable example is the male Wandering Albatross, ringed as a subadult at Macquarie Island in 1967 and discovered breeding 5,800 km to the west on Heard Island, where the species had never been seen previously (p.145). Its mate must also have been an immigrant.

At least two Southern Royal Albatrosses from Campbell Island have visited the small colony of Northern Royals at Taiaroa Head, New Zealand and one of them has settled. In the other direction, a Northern Royal Albatross has been found among Southern Royals on Enderby Island in the Aucklands.

A Black-browed Albatross ringed on the Iles Kerguelen was found almost 20 years later breeding on Heard Island (Woehler 1989), and several immigrant Black-browed Albatrosses have settled in the large colonies of Campbell Black-browed Albatrosses on Campbell Island. At least one of them originated from or had visited Macquarie Island, 725 km to the southwest (Moore *et al.* 1997a). Four pairs of Salvin's Albatrosses nesting among other mollymawks on the Ile des Pingouins in the Iles Crozet, either emigrated 10,000 km from the Bounty Islands or are the progeny of immigrants; one of them was briefly ashore at South Georgia (Prince & Croxall 1983). In the ten years 1983–93, immigration is believed to have contributed significantly the 13% annual increase of Laysan Albatrosses on Isla de Guadalupe, off Mexico (Gallo-Reynoso & Figueroa-Carranza 1996). Over 50 years, Steller's Albatrosses from Torishima have visited the Northwest Hawaiian Islands (Richardson 1994, Tickell 1996).

These examples confirm gene flow between distant island populations, but they are rarely of demographic significance. Among the thousands of detailed breeding records at Bird Island, there are only two known instances of adult Black-browed Albatrosses moving from one colony to another. A few Black-browed subadults settle and breed in colonies other than those in which they were reared. In one small colony of 227 pairs, 12% of recruited breeders were immigrants, but most of them came from much larger (>2,000 pairs) colonies only 400–500 m away (Prince *et al.* 1994).

Many young albatrosses die before maturity and thus make no contribution to a breeding population, so the birth rate is replaced by recruitment, that is, the proportion of fledglings that survive to join the breeding population. Assuming there is a 1:1 sex ratio, it is convenient to represent the breeding population by the number of adult females.

A simple model for population change is:

$$N_{(i+1)} = s_A N_i + f s_J N_{(i+1-k)}$$

where: N_i = the number of females breeding in the year i

s_A = the average rate of adult survival from year i to $(i+1)$

s_J = the average rate of juvenile/subadult survival from year $(i+1-k)$ to $(i+1)$

f = the mean number of female fledglings per breeding pair

k = the age at which subadults are recruited into the breeding population.

Adult survival (%)	Breeding success (%)	Recruitment (%)
93	70	>36
94	70	34
95	60	<30

Table 16.4 Combinations of adult survival, breeding success and recruitment (% of juveniles that breed) that approach stability in a Wandering Albatross population model (Croxall *et al.* 1990b).

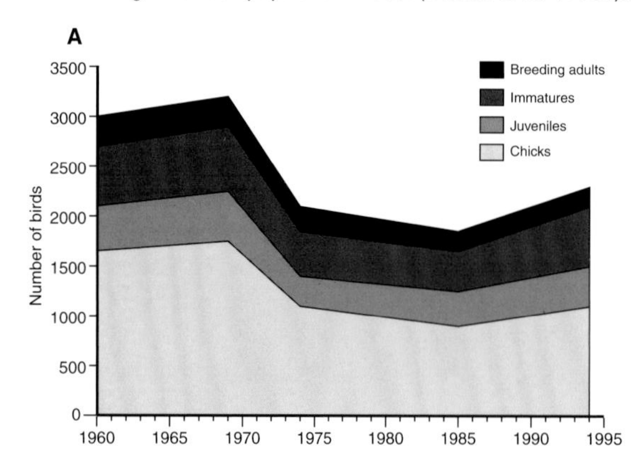

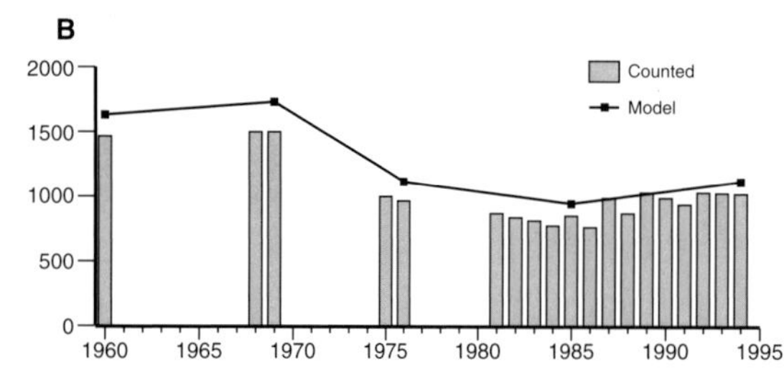

Fig. 16.12 Modelled changes in the age structure of the Wandering Albatross population at the Ile de la Possession between 1960 and 1994 (A) and the fit of the model to the numbers of birds counted (B) (Weimerskirch *et al.* 1997a).

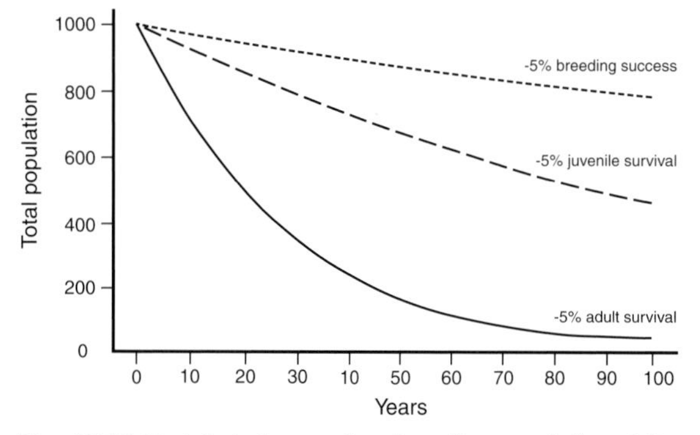

Fig. 16.13 Modelled changes in a breeding population of the Wandering Albatross at the Ile de la Possession, Iles Crozet over 100 years, following -5% decreases in the present values of breeding success, juvenile survival and adult survival (Weimerskirch *et al.* 1997a).

None of the eight species of albatrosses considered to be annual breeders adheres strictly to that regime; many intact pairs miss a season or two. Likewise biennial breeders often miss not just one, but several seasons. In the above equation, and for practical purposes, this is generally ignored, but more lengthy equations have been derived to incorporate deferred breeding (Appendix 14). One of them was constructed to distinguish the relative roles of adult survival, recruitment and breeding success in the Wandering Albatross at Bird Island, South Georgia. A population with an annual adult survival rate of 94% and a breeding success rate of 70% would need a yearly recruitment of 34% (the proportion of ringed fledglings that return to breed) to maintain numbers. If adult survival rate declined by 1% and breeding success remained the same, recruitment would have to increase to more than 36% to maintain stability. If, however, adult survival increased to 95%, the population could remain stable even though breeding success fell to 60% and recruitment to less than 30%. Wandering Albatrosses were thus five times more sensitive to changes in adult survival than juvenile survival and breeding success (Table 16.4) (Croxall *et al.* 1990b).

Two age-structured models of the Wanderer population at the Ile de la Possession, Iles Crozet have been constructed in response to fisheries-induced mortality (Moloney *et al.* 1994, Weimerskirch *et al.* 1997a). The French model agreed closely with counts on the island between 1960 and 1994 (Fig. 16.12), and was used to predict the effects of changes of breeding success, adult survival and juvenile survival (Fig. 16.13). Like the Bird Island model, it was much more sensitive to changes in adult survival (Weimerskirch *et al.* 1997a).

A similar model was created to predict the effect of driftnetting in the North Pacific over the 12 years 1978 to 1990. About 17,500 Laysan Albatrosses died each year in the nets. During that time, three breeding populations of Laysans in the NW Hawaiian Islands were growing at rates of 0.0%, +2.3% and +5.7% per year. It was concluded that the drift net mortality would cause the stable population to decline at -2.2% per year, but the other two populations would continue to increase at the lowered rates of +1.5% and +5.3% respectively (Gould & Hobbs 1993).

A model of the Northern Royal Albatross population was created to predict the demographic impact of killing albatross fledglings at Little Sister Islet (Robertson 1991). Counts of this remote and remarkable colony had been few, so population parameters from the adjacent mainland were used. Fifty years of counts at Taiaroa Head were available, but this colony had always been very small and carefully managed, quite unlike those of the Chathams. From a hypothetical stable population with an age structure of 51% adults, 22% subadults, 19% juveniles and 8% chicks, the model simulated 150 years of response to several exploitation scenarios (Robertson 1991).

FORAGING AND FEEDING

Breeding albatrosses forage over well defined areas of ocean (Figs. 5.33, 7.13, 14,), but there is considerable variation in the direction and duration of individual trips at different periods of the breeding cycle (Fig. 16.14). Birds from populations that habitually fly to more distant feeding zones tend to be smaller, have lower provisioning rates and rear smaller chicks (Waugh *et al.* 1999c).

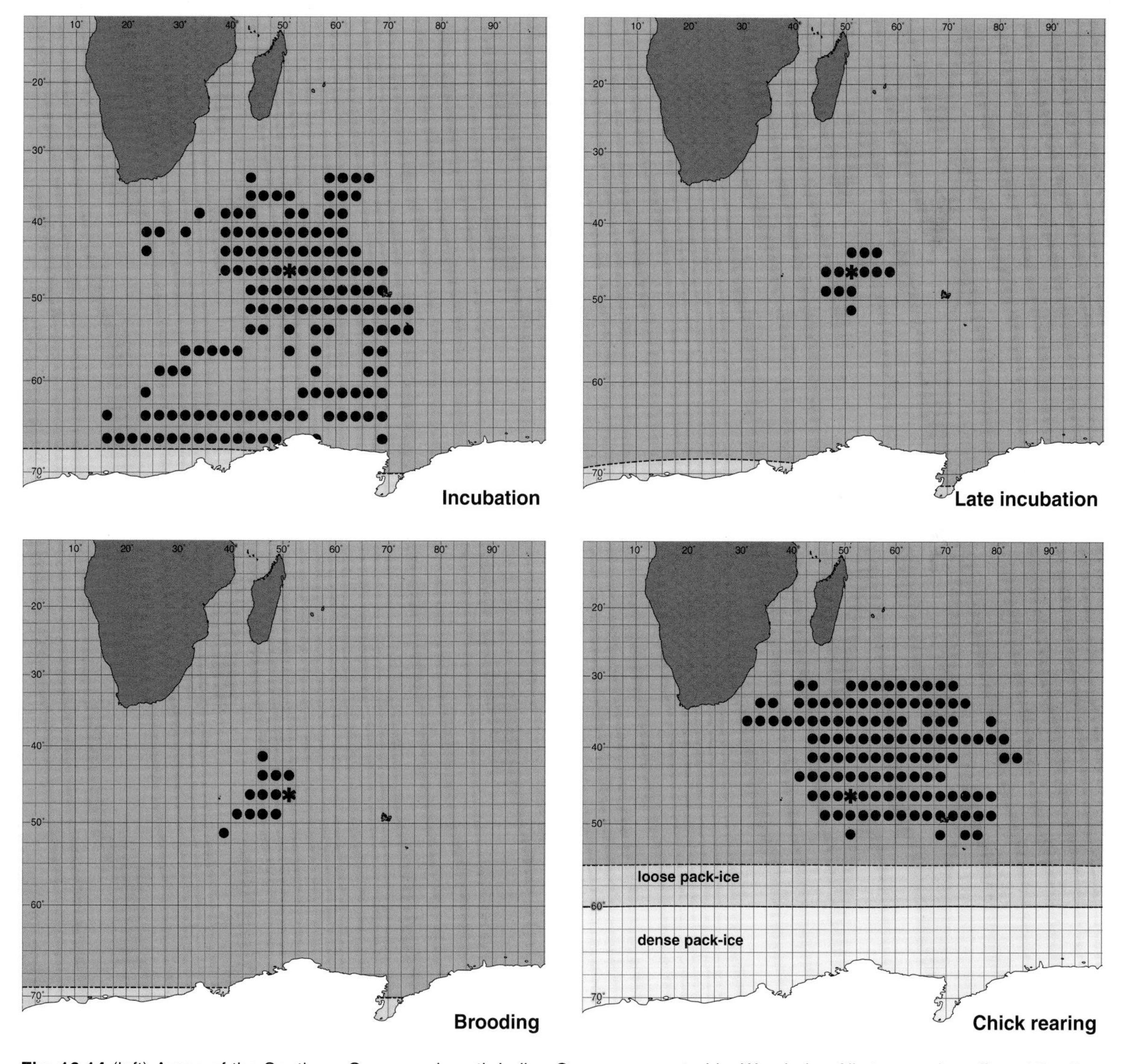

Fig. 16.14 (left) Areas of the Southern Ocean and south Indian Ocean prospected by Wandering Albatrosses breeding at the Iles Crozet (*). Pooled locations during: incubation – 11 birds on 11 flights, late incubation – five birds on five flights, brooding – 15 birds on eight flights and chick rearing – 34 birds on 16 flights (Weimerskirch *et al.* 1993). The area available during the winter is reduced by the spread of pack-ice in the south.

Observations in the south Indian Ocean during the early 1980s, indicated that darker plumaged Wandering Albatrosses were more frequently seen to the north of the Iles Crozet, while those seen far to the south were whiter.The whitest birds were likely to have been older adult males, but the darker ones could have included juveniles, subadults and adult females (Weimerskirch & Jouventin 1987). There was an apparent north-south seasonal shift in foraging zones, but in the substantial middle range both sexes were common throughout the year. Later satellite tracking revealed that males sometimes made long forays towards the Antarctic in summer, but with the approach of winter, when pack-ice occupied an increasing area of ocean, foraging circuits avoided the higher latitudes (Fig. 16.14). Females were more numerous over subtropical waters and they did not move very much farther north in winter.

It has long been known that albatross absences at sea in the breeding season vary greatly in duration and it was assumed that birds away from their nests for long periods travelled greater distances than those absent for just a few days (Tickell 1968). Satellite telemetry later confirmed that Wandering Albatrosses from the Ile de la Possession, flew long circuits over the deep ocean and short ones over the edge and slope of the Crozet shelf (Weimerskirch *et al.* 1993). However, although intervals between feeds of up to five days, measured at the nest, correlated with the duration of short foraging trips tracked by satellite, there was no such correlation between the duration of long foraging trips and the frequency of feeds (Weimerskirch *et al.* 1997d) (Fig. 16.15).

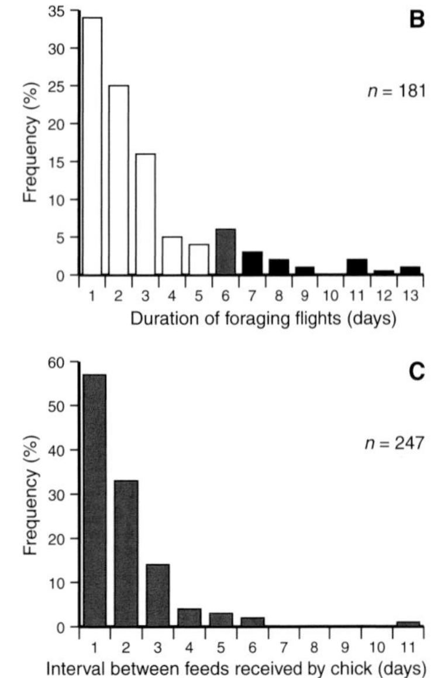

Fig. 16.15 Wandering Albatrosses foraging from the Ile de la Possession, Iles Crozet, June–July 1992, when feeding well grown chicks. The maximum foraging range has a clear bimodal distribution separating short flights (open columns) from long flights (black columns) (A). The duration of flights only just reveals a difference between short flights (0–6 days) and long flights (6–13 days) (B), but no difference is apparent in the intervals between feeding chicks (C). (Weimerskirch *et al.* 1997b).

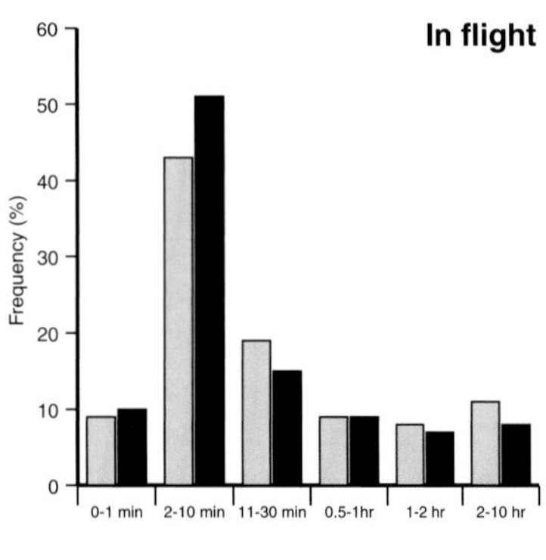

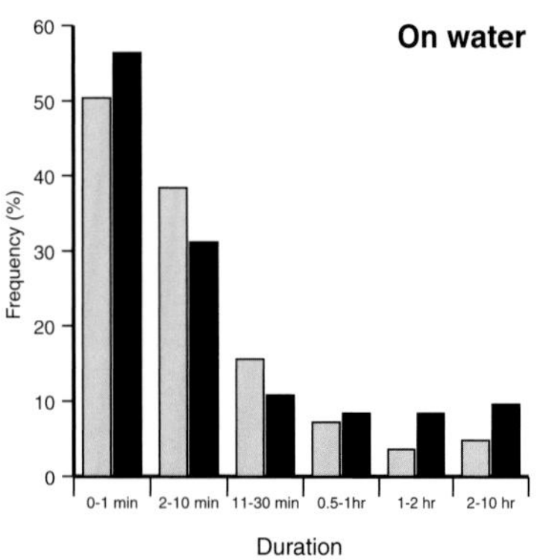

Fig. 16.16 The times spent in flight and on the water by five Wandering Albatrosses foraging from the Ile de la Possession, Iles Crozet, February-March 1994, during daylight (open columns) and at night (black columns) (Weimerskirch *et al.* 1997d).

During the first half of incubation, off-duty Wanderers from the Ile de la Possession forage up to 2,600 km from the nest with round-trips averaging 5,991 km in 14 days. By this time pack-ice has largely dispersed and some males fly south into polar waters, approaching Antarctica. Foraging birds spend 60% of their time in flight and 40% on the water. They are active in the air and on the sea throughout the day and night, landing on average 27 times, but during darkness they appear to spend more time on the water (Fig. 16.16). Some individuals fly continuously for up to 11 hours, 91% of a four-day trip. During outward flights, they spend the same amount of time in the air and on the sea as when returning to the island.

In the last weeks of incubation, trips are shorter and the foraging area is greatly reduced (Fig.16.17). Before returning to the nest, all birds spend a day or two at the edge of the Crozet shelf. Once the egg has hatched, and while the chick is being brooded, both parents forage over the Crozet shelf or off the shelf-break to an average maximum range of 256 km. Individuals tend to return repeatedly to the same area. Later, after the chick has grown enough to be left on its own, short foraging trips of about two and a half days are alternated with longer trips of about ten days.

On average one long trip is alternated with five short ones, slightly more for males than females (Weimerskirch *et al.* 1997b). On the short trips, maximum ranges average 118 km (males) and 179 km (females) and at any one time, about 60% of foraging Wanderers are within 100 km of the Iles Crozet, many on the eastern shelf-edge. Males return repeatedly to specific locations; sometimes these appear to be visited by only one individual, but others attract several or more birds. They appear to forage during the outward flight, but return directly, once they have a food load. There are indications that birds nesting on opposite sides of islands favour different foraging areas. On long trips, maximum ranges of Crozet Wanderers average 1,464 km (males) and 1,480 km (females). Individual preferences are again apparent, some move east and incorporate the Kerguelen shelf, others favour the opposite direction (Fig.16.18). They pause here and there, sometimes for several days, perhaps over upwellings or at fishing fleets, but in calm, anticyclonic conditions they may drift on the surface for several days before moving on. Feeding is not confined to particular locations and they forage throughout the trips. The greatest distances are covered during daytime and at night the birds move only short distances or remain stationary. Most individuals conform to this strategy, but alternative tactics are evident.

The physical condition of foraging Wanderers and the amount of food they deliver to their chicks varies with the apparent length of foraging trips. The food delivered after short trips is more likely to be fresh and the quantities collected per unit time are greater than on long trips, but during short trips little goes to the adults, so the energy costs of foraging are not repaid and the parents lose mass. During long trips more food is collected, but much of it becomes digested, producing concentrated loads with quantities of high energy oil. There is time for some of this to be assimilated by adults in flight, so the energy cost of foraging is recovered and parents gain mass. The relative yield of long foraging trips are thus higher than

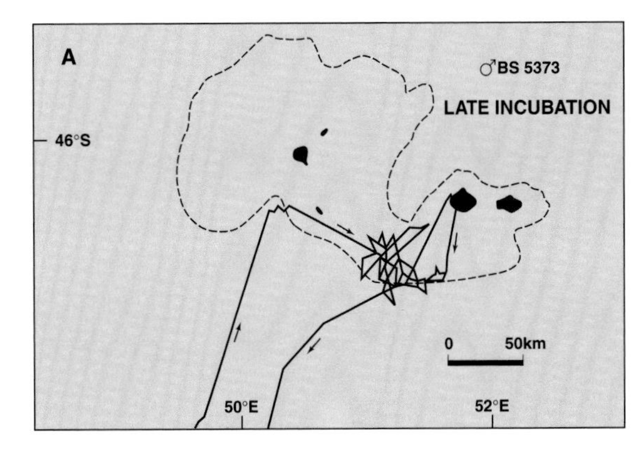

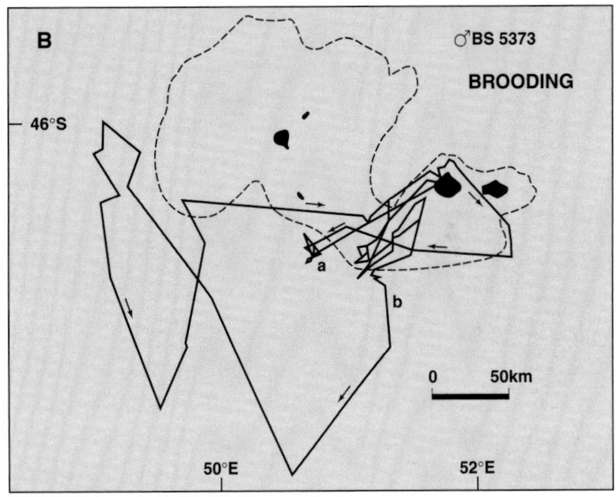

Fig. 16.17 Short foraging flights of a male Wandering Albatross tracked by satellite from the Ile de la Possession, Iles Crozet. One flight just before hatching (A) and two in succession (a & b) just after hatching (B). The broken line represents the shelf-edge surrounding the Iles Crozet (Weimerskirch *et al.* 1993).

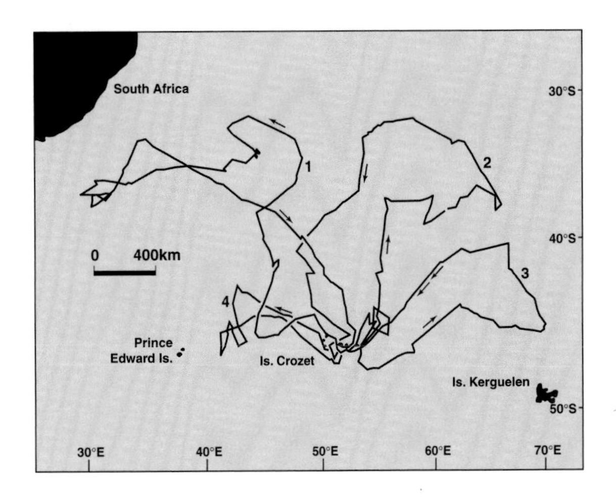

Fig. 16.18 Long foraging flights of Wandering Albatrosses tracked by satellite from the Iles Crozet during the chick-rearing period. One female (1) northwest towards South Africa and three males (2,3 & 4) (Weimerskirch *et al.* 1993).

	Trip length	WA	EYNA	BBA
Energy of diet kJ/g	short	3.5	3.8	4
	long	5.6	6.1	-
Energy to chick kJ	short	3 080	1 398	1 860
	long	5 600	2 580	-
Energy to adult* kJ	short	-7 693	-139	0
	long	4 924	920	-
Net energy kJ	short	-4 613	1 259	1 860
	long	10 524	3 500	-
Relative balance (kJ/day/g)	short	-0.14	-0.23	0.2
	long	0.11	0.21	-

Table 16.5 Estimated energy yield from short and long foraging trips by Wandering Albatrosses (WA) from the Iles Crozet, Eastern Yellow-nosed Albatrosses (EYNA) from the Ile Amsterdam and Black-browed Albatrosses (BBA) from the Iles Kerguelen (Weimerskirch *et al.* 1994). * The energy stored or lost by adults does not include the cost of flying.

those of a short ones (Table 16.5), but less goes to the chick.

There is no difference in the energy content of the meals delivered to the chick by the male and female, even though the male is ten percent larger. However, the male visits the chick more frequently and delivers 61% of the total energy it receives, compared with 39% from the female. The energy-cost of foraging may be higher for the female and she spends proportionally less of her foraging time on high-cost, short trips which are longer than those of the male. Moreover, towards the end of the chick-rearing period, the female ceases foraging for the fledgling earlier than the male. Several short trips can be made in the time taken on one long trip, so the two parents together deliver three times more energy from short trips (males 52%, females 29%) than from long trips (both 9%). A chicks benefis from frequent short trips, while the adults need long trips to maintain their own condition. Since two parents forage independently, the optimum dual strategy maximises delivery, while going some way towards randomising the detrimental effect of long absences (Weimerskirch *et al.* 1994, 1997d).

Wandering Albatrosses foraging from South Georgia have less obvious differences. The marine environment in the South Atlantic is more diverse than the ocean surrounding the Iles Crozet and appears to offer more opportunities to foraging birds (S.D. Berrow pers. comm.). Most short trips from South Georgia last one to three days and overlap with long trips of more than six days, but the maximum distances flown show no bi-modal distribution. The nature and size of meals given to chicks at the end of short and long trips are similar, but they receive fewer from long trips. The energy content of meals from long trips are similar to those from around the Iles Crozet, but Wanderers returning to South Georgia from short trips delivered three times the energy of the Crozet birds.

When their chick is recently hatched, South Georgia Wanderers fly 1,869–2,688 km to the Antarctic Polar Front and 426–732 km to the island shelf-edge (Fig. 16.19) (Arnould *et al.* 1996). Later, when foraging for a rapidly growing chick, 87% of these Wanderers make much longer trips, some more than 7,000 km. Females mostly fly north to the continental shelf of South America (Fig. 16.20) as far as 32°S off Brazil. In summer, males fly south towards the Antarctic Peninsula; in winter they go north, like the females (Prince *et al.* 1992, Prince *et al.* 1997).

Female Auckland Wandering Albatrosses incubating on Adams Island fly to the mid-Tasman Sea at about latitude 40°S, while males go west or northeast into the South Pacific as far as longitude 160°W. Most fly directly to a particular area where they cruise about for up to 15 days before returning directly. Later, when brooding small chicks, they also resort to a dual strategy.

Among the mollymawks, dual strategy is apparent in Eastern Yellow-nosed Albatrosses at the Ile Amsterdam. However, Black-browed Albatrosses at the Iles Kerguelen forage at a more uniform distance from the colony, some perhaps exclusively over and around the Kerguelen shelf (Fig. 5.35). From these comparatively short trips, adults tend to deliver more frequent, smaller meals (Weimerskirch 1997).

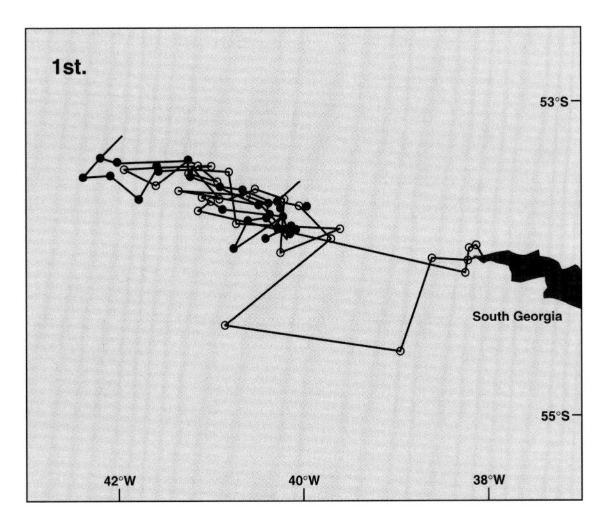

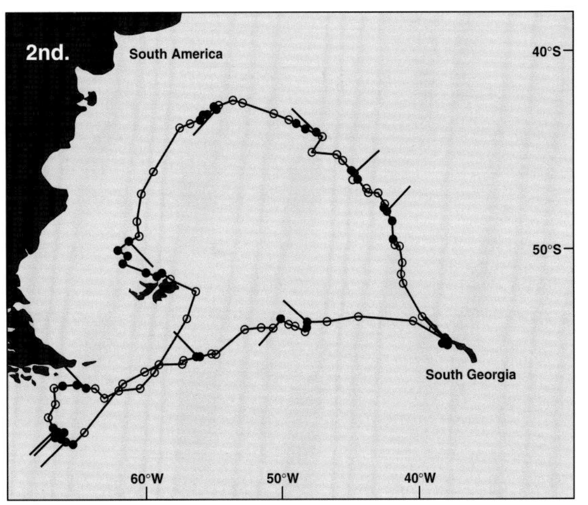

Fig. 16.19 Two consecutive foraging flights of a male Wandering Albatross tracked by satellite from Bird Island, South Georgia, where it was feeding a chick in the winter (August and September) of 1990. Daytime positions (O) and nightime positions (●) with / indicating day changes. The 1st flight, about the nearby Shag Rocks shelf, was 1,810 km in 4.5 days and the 2nd, to the Drake Passage and the edge of the continental shelf of South America, was 7,470 km in 13 days (Prince *et al.* 1992)

Searching

Wandering Albatrosses scavenge for organisms that are probably dead or moribund when they come to the surface. This natural resource is distributed over wide expanses of ocean.

The flight paths of albatrosses have fractal[7] properties and long-range correlations with Lévy[8] probability distributions (Viswanathan *et al.* 1996). Black-browed Albatrosses observed visually at intervals of seconds over several hundred metres were seen to change direction twice as frequently near krill swarms as when away from krill, while Grey-headed Albatrosses, which eat less krill, revealed no such difference in turning frequency (Veit & Prince 1997).

On a much larger scale, satellite tracks of albatrosses include intervals when fixes are separated by 100 km or more and occasions when they are much closer together in 'clusters' (Reinke *et al.* 1998), as if the birds were 'searching' in all directions. External temperature loggers (XTLs), attached to the leg of the albatross, indicate that the bird comes down onto the sea more often in clusters, but large prey are also caught between distant fixes without there being any indication of 'searching'. The XTL measurements, timed in seconds, lose precision when projected onto intervals of hours

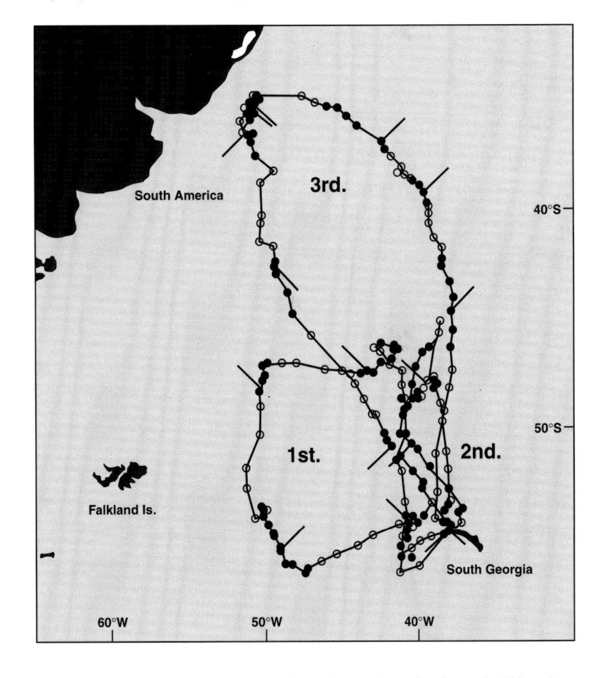

Fig. 16.20 Three consecutive foraging trips of a female Wandering Albatross tracked by satellite from South Georgia where it was feeding a chick in the winter (August and September) of 1990. Daytime positions (O) and nightime positions (●) with / indicating day changes. The 1st flight was 4,173 km in five days, the 2nd 3,298 km in four days and the 3rd 6,479 km in eight days (Prince *et al.* 1992)

[7] Fractals have the same shape whatever their scale (self-similar) and occur on different scales, but with no one scale dominating (scale-invariant) (Mandelbrot 1982, Shlesinger *et al.* 1993).

[8] Paul Lévy developed the general mathematics for determining probability distributions for the random addition of extreme variables – on long 'tails' of normal, bell-shaped curves. Lévy flights are a special class of **random walk**. The well-known model sets off with constant steps in one direction and at regular intervals changes direction right or left at random. If the length of the random steps are not constant, but chosen from a power function distribution, the track becomes very complex, even chaotic (Casti 1997).

between satellite fixes, because the albatross is probably not flying at a constant speed. These errors are compounded because the distance between fixes are shorter than the actual flight paths of the birds (p.267) (Weimerskirch *et al.* 1997b). Series of widely separated satellite fixes between the nest of the Light-mantled Sooty Albatross and the distant clusters of apparent feeding areas, have been interpreted as 'commuting' tracks (Weimerskirch & Robertson 1994), but these birds may also have paused en route to investigate and catch prey without the behaviour being registered by the technology in use at the time.

Feeding

Albatrosses are among the top predators and scavengers of the marine ecosystem. The food available at the surface of the ocean consists broadly of zooplankton, crustaceans, squid, fish and carrion. Most plankton is too small to be caught effectively by albatrosses unless it swarms, like krill in the Southern Ocean, or is deposited in masses, like flying-fish eggs in the North Pacific. Large pelagic decapods and amphipods are commonly eaten, but they never amount to an appreciable biomass. Squid is the single, most widespread and common component of all albatross diets, but in some areas shoaling fish may be as important.

Although albatrosses feed only in the top few metres of the ocean, atmospheric and hydrological perturbations influence nutrient cycles. Energy flow deep within the ocean affects marine food webs in many ways (Knox 1994) and is likely to be the distant cause of some albatross nutritional opportunities and crises. El Niño Southern Oscillation (ENSO) of the tropical Pacific (p.171) regularly affects the feeding grounds of the Galápagos Albatross off Peru, as well as the three species of North Pacific albatross off Japan. It may also influence other species in various areas of the Southern Ocean.

Range

Outside the breeding season most albatrosses travel great distances. There are enough recoveries of ringed birds to suggest that adults, subadults and juveniles follow different routes or fly in the same direction at different times. Migration towards specific distant locations is now well established in several species (Prince *et al.* 1997). One species, the Galápagos Albatross, forages in roughly the same area throughout the year (Tickell 1996).

Multi-species flocks

Although lone albatrosses are encountered often enough at sea, whenever there is any quantity of food near the surface, they are unlikely to be alone for long. Many seabirds are attracted towards distant activity. In good visibility, those rising only five metres above the water have an horizon at eight kilometres and an area of over 200 km^2 in view; higher climbs can double that area.

Abundant food at the surface may attract immense numbers of seabirds. A great deal of excitement and aggression is generated at such times and feeding frenzy is an apt description of what goes on when a trawl is being hauled[9]. During the activity, albatrosses often keep their wings open, pattering across the water to lunge and grab whatever is available. Large size is advantageous and great albatrosses can dominate the melée. Harper (1987) believes that they are more aggressive at night than during the day.

Small cetaceans, including young rorquals and seals, are prey to killer whales (Ridoux 1987, Hodges & Woehler 1994). Leopard seals kill young furseals and penguins while bull furseals also take penguins. Flocks of mixed seabirds including albatrosses are commonly encountered feeding on floating skin and blubber, or the carcasses of seals and penguins; they sometimes become feeding frenzies in which albatrosses do well, taking second place only when outnumbered by giant petrels.

Sperm and pilot whales feed in deep water and periodically expel masses of horny beaks and undigested squid fragments near the surface. Scavenging birds, including albatrosses, may thus acquire food from deep water (Clarke *et al.* 1981). Some prey species naturally float to the surface after death (Lipinski & Jackson 1989, Croxall & Prince 1994) and over squid spawning grounds, the quantities available may be considerable (Gibson & Sefton 1959, Rodhouse *et al.* 1987, Battam & Smith 1993).

Cruising whales may be followed by seabirds even when they are not feeding. Their faeces are scavenged, and

[9] feeding at commercial fisheries is discussed in Chapter 17.

organisms brought to the surface by eddies around the huge moving animals, provide additional prey; sooty albatrosses have been seen among accompanying seabirds (Routh 1949, Enticott 1986).

Whales, seals, penguins and tuna (Jehl 1974, Harrison *et al.* 1991, Swanson 1997) may all drive shoaling prey towards the surface, where they become available to seabirds. In such feeding flocks, the movements of the predators below are followed by birds above the surface in anticipation of where prey are about to appear. Some seabirds take their cue from others; multi-species flocks at krill swarms off South Georgia tend to follow Black-browed Albatrosses, which attack the surfacing krill first. In the North Pacific, Laysan Albatrosses have been seen following dolphins (Gould & Hobbs 1993). Likewise, off Peru, Galápagos Albatrosses may benefit from association with large mixed flocks of cormorants, boobies and pelicans (Duffy 1983, 1989).

The relative importance of predation and scavenging remains controversial (Croxall & Prince 1994). The distinction may have had more ecological significance before the arrival of man in the Southern Ocean, but it has now been overtaken by events. The resources released from the worldwide fishing industry are so immense that they have surely upset traditional feeding preferences. To a greater or lesser extent, all albatrosses probably scavenge from ships and fishing fleets, although some do so more than others and some species are more selective than others (Gould & Hobbs 1993, Weimerskirch 1997).

Feeding methods

Albatrosses have been characterised as surface feeders (Ashmole 1971) and during several long voyages in 1965–67 on the *Eltanin*, Harper (1987) recorded 98% of feeding Wanderers and Black-browed Albatrosses as seizing prey from the surface.

Off Australia, mollymawks glide slowly up-wind to take fish from the surface. David Barton (1979) saw Tasmanian Shy Albatrosses picking out jack mackerel as a school rippled the water. These fish may weigh up to 500 g and successful birds carried their catch away from the school to settle on the sea elsewhere and swallow them. Pilchards and anchovies are small enough to be swallowed head first by mollymawks in flight. Given enough wind, Wandering Albatrosses also stalk squid silently on the wing, just above the surface, and drop onto their prey (p.164).

As long ago as 1865, Hutton noted that Black-browed Albatrosses sometimes plunged below the surface, and in the early 1920s Harrison Matthews (1951) saw Black-browed Albatrosses submerging when scooping krill from a swarm off South Georgia. Prince (1980) watched them plunge from about nine metres and Harper (1987) from heights of two to five metres. In all these descriptions, the birds' wings opened under the water and the wing-tips remained visible, sometimes above the surface.

In the Pacific, off California, Miller (1942) watched Black-footed Albatrosses using their wings when swimming down to bait at a depth of about one metre and off South Africa, Black-browed and Shy Albatrosses were clearly seen swimming down more than two metres using their wings bent sharply at the carpal joints (Nicholls 1979). The most convincing natural experiment was related by D.A. Ogram (in Oatley 1979). With two companions, he was fishing from a small boat two miles off the coast of South Africa when they were approached by an adult Black-browed Albatross.

> 'The bird, which seemed quite fearless, rode the swell in close proximity to our craft and persistently took the small shad and sardine bait we were using. When it first joined us the bird went down to a depth of about 15 feet [4.6 m] to collect its first meal of sardine bait. Having consumed this it went down and collected one of our shad. The thieving continued even when we put our bait overboard as close as possible to the side of the boat. Eventually we became so exasperated with the albatross's interference with our sport that by throwing lines over it we caught it, drew it inboard, and shut it up in the boat's hatch. Later in the afternoon when we had done our fishing we released the bird, which seemed in no way put out and for some time still refused to leave us.'

Instrumental measurements indicate that while Wandering Albatrosses do no more than splash just below the surface, the smaller species regularly dive several metres (Prince *et al.* 1994). A study of Tasmanian Shy Albatrosses has distinguished two modes of underwater activity: short plunges where the birds dive at more than one metre per second and are underwater for five seconds or less, and slower swimming where they may be submerged for much longer and go much deeper (Fig 16.21). The average depth of dives was 1.9 ±1.7 m and 87% were less than 3.5 m (n=52). A maximum depth of 7.4 m was recorded. Tasmanian Shy Albatrosses are diurnal feeders; none dived during the hours of darkness (2200–0500) although they spent a considerable time on the water. Peaks of diving were near midday and during twilight; 64% of twilight dives were with a full moon. Diving behaviour of Tasmanian Shys from Pedra Branca differed from those breeding on Albatross Island in Bass Strait, presumably reflecting differing activity of prey (Hedd *et al.* 1997).

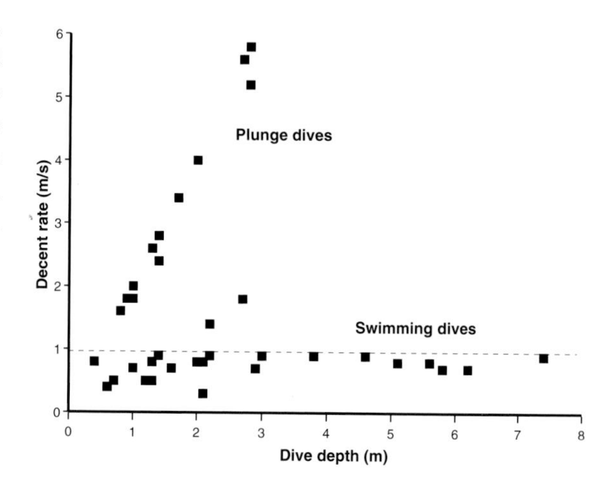

Fig. 16.21 Two types of diving by Tasmanian Shy Albatrosses. The rate of descent and depth attained by plunge dives and swimming underwater (n=52) (Hedd *et al.* 1997).

Light-mantled Sooty Albatrosses have been seen surface filtering among petrels over a swarm of krill, but on another occasion an adult was seen dropping vertically into the water like an ungainly tern, from a height of about eight metres to seize unknown prey and immediately fly off. The two sooty albatrosses have long had a reputation as superlative fliers; they are also proficient divers whose dark plumage and pointed tails may yet prove to be the attributes of accomplished swimmers. The Light-mantled Sooty Albatrosses, fitted with capillary depth gauges in 1992, recorded deeper dives than other albatrosses (mean 4.7 m, SD±3.4, range 0.7–12.4). Diving was not witnessed, but It has been suggested that plunging is unlikely to take them very deep, and that the birds achieve depth mainly by swimming (Prince *et al.* 1994).

Galápagos Albatrosses have been seen to make boobies disgorge, but gulls and jaegers may scavenge the food before the albatrosses. Whether piracy makes a significant contribution to the diet of these albatrosses remains to be established (Harris 1974, Duffy 1980, Merlen 1996).

ENERGETICS

The food and water needs of albatrosses are satisfied at sea. With no commitment ashore, they have optimum potential to maintain health within the constraints of the oceanic environment. We identify this condition with the body mass of birds first returning to land at the beginning of a breeding season (Fisher 1967, Prince *et al.* 1981). On land, albatrosses lose mass through respiration, excretion, defaecation and metabolism. Males are larger than females so such losses have allometric proportions.

Black-browed Albatrosses off duty from incubation or brooding, ingest water immediately after leaving Bird Island, but although Bevan *et al.* (1995) observed rafts of departing birds preening on the sea near the island, they did not report them drinking seawater. Several species catch rain when ashore; at Midway Atoll during squalls, hundreds of albatrosses may be seen simultaneously holding up their beaks to gently nibble or snap at falling raindrops (Howell & Bartholomew 1961).

From the time young subadults first return to the breeding grounds, progress towards optimal breeding regimes influences the time spent ashore. In all stages of acquiring mates and attaining breeding synchrony there are trade-offs between the increasing water and energy deficits of staying ashore and the imperatives of breeding. Successful breeding brings the less escapable costs of egg production, incubation, brooding and foraging for chicks. Although females alone bear the costs of egg production, they spend almost all that time at sea, where they can continue to feed, while males occupy the nest and go without food and water throughout most of the pre-egg period.

Resting metabolism

Basal metabolic rates (BMRs) in units of kiloJoules (kJ) have usually been measured on quiescent animals under strictly controlled conditions only possible in the laboratory. Over the years allometric equations have been derived by different workers for predicting BMRs from body mass (*M*) in kilograms. Those used in early albatross computations were for a wide range of non-passerines, but an equation based on seabirds (Ellis 1984) has been used in recent studies:

$$BMR = 381.8\ M^{0.721}\ kJ/day$$

The first measured metabolic rates of albatrosses were obtained in the early 1980s. Near the scientific base on Marion Island, where Wandering Albatrosses were habituated to people, Brown & Adams (1984) were able to persuade two males and two females to sit quietly inside a respiration chamber, while their oxygen uptake and carbon dioxide production were measured; one millilitre of oxygen has an energy equivalent of 20 Joules. There was no significant difference between the males and females, so the pooled resting metabolic rate of these Wandering Albatrosses was 2.5 watts[10] per kg of body mass (Table 16.6). The respirometer employed did not have constant temperature, but the result was close to the BMR predicted by the allometric method.

	Mass kg	n	Ambient temp °C	Measured MR W/kg	Predicted BMR W/kg	MR/BMR
WA	8.13	4	2–16	2.5	2.3	1.1
GHA	3.73	3	0–20	2.3	2.9	0.8
SA	2.87	4	0–20	2.9	3.1	0.9
LA	2.52	4	15–24	3.0	3.3	0.9

Table 16.6 Metabolic rates of inactive Wandering (WA), Grey-headed (GHA), Sooty (SA) and Laysan Albatrosses (LA) measured by oxygen consumption and carbon dioxide production in closed chambers away from nests (Grant & Whittow 1983, Adams & Brown 1984).

Grant & Whittow (1983) also used a simple respiration chamber with Laysan Albatrosses at Midway Atoll. These albatrosses were also well-habituated to the many people who lived among them, but they were less than half the size of the Wandering Albatross and the subtropical ambient temperature was much higher than at Marion Island. Five birds incubating on nests had a lower average metabolic rate of 2.3 W per kg. Measurements on other incubating albatrosses by different methods (see below) yielded similar values. Incubation may not be a highly energetic activity, but from time to time, albatrosses sit up, rotate their eggs, auto-preen, repair their nests and react to other birds. Energy expenditure is about 1.2–1.4 x BMR[11] (Croxall & Ricketts 1983, Brown & Adams 1984).

Energy content of diets

In 1975 and 1976, Black-browed and Grey-headed Albatross chicks aged 30–90 days at South Georgia received feeds of 573±417 g (mean±SD, n=160) and 596±206 g (n=195) respectively. Meals sometimes appeared to be mainly squid, fish, crustaceans or carrion (Fig. 5.31), but usually contained mixtures (Prince 1980).

Approximately half of these feeds were liquid, but the prey types differed in their liquid component: squid (51%), fish (43%), lamprey (70%) and krill (31%). In apparent fish and squid meals, most of this liquid was water, with seven to nine percent dissolved solids and about one percent lipids. The energy value was 1.7–2.3 kJ per g. Liquid drained from krill meals was much higher in lipids (27%) with an energy value of 12 kJ per g (Warham 1976, Prince 1980).

[10] The time dimension is now conventionally incorporated in the unit of power: watts (*W*) = joules per second (J/s).
[11] Koteja (1991) questioned BMR correlations.

The orange colour of oil from meals of Black-browed and Grey-headed Albatrosses obtained in the South Atlantic is due to the presence of astaxanthin and its esters, together with another unknown carotinoid, probably originating in pelagic crustaceans (Clarke & Prince 1976).

The constituents of oil from albatross diets are largely wax ester, triglyceride and squalene yielding 39 to 42 kJ per g. Wax ester and triglyceride occur in crustaceans, fish, squid and the blubber of marine mammals that may be consumed by albatrosses as carrion. They have been isolated from food samples taken from great albatrosses, mollymawks and sooty albatrosses, but inter- and intra-specific variation occurs. In one study, Royal and Black-browed Albatrosses had respectively 34% and 17% wax ester while Southern Buller's, Grey-headed and Light-mantled Sooty Albatrosses had none at all (Warham *et al.* 1976). In another analysis, the lipid content of food from an adult Black-browed Albatross about to feed its chick at South Georgia, comprised 49% wax ester and 32% triglyceride, while that in four chicks had only 2% wax ester and 74% triglyceride (Clarke & Prince 1976).

Squalene is functional in sharks and an intermediate in the biosynthesis of cholesterol. It has been isolated from the meals of both great albatrosses, but in significant amounts only from Royal Albatrosses (Warham 1977).

The mean energy values of the solid component of mixed meals varied from 4.5 to 5.1 kJ per g wet mass (Clarke & Prince 1980). Meals apparently containing mostly fish were also rich in lipids, although the notothenid fish about South Georgia, which mollymawks sometimes eat, contain only 2.3% lipids. Squid, as a group, have even less lipid (0.9%) and a correspondingly low energy value of around 3–4 kJ per g wet mass (Croxall & Prince 1982).

The energy used by free-living albatrosses was first measured by the method of injecting doubly-labelled water (DLW) (Adams *et al.* 1986, Costa & Prince 1987, Pettit *et al.* 1988, Arnould *et al.* 1996). This provides a single measurement of expended energy (EE) between the time a blood sample is drawn before the bird leaves the nest and a second sample taken after it returns.

Expended energy measurements on Wandering Albatrosses foraging from Marion Island and South Georgia yielded 4.6 and 3.9 W per kg respectively (1.6–1.8 x BMR), while mollymawks and Laysan Albatrosses, less than half the size of Wanderers, averaged 7.6–7.8 W per kg (2.4–2.6 x BMR). DLW methods have been useful for comparing gross energy budgets, but they cannot measure the energy costs of different activities away from the nest.

The oxygen uptake of an animal can be correlated with the amount of blood delivered to muscles; this is under autonomic control involving several circulatory parameters. The rate of heartbeat has been used to study the energetics of free-flying Black-browed Albatrosses. Heart rate data loggers (HRDLs) surgically implanted within the body cavities of Black-browed Albatrosses recorded the number of heartbeats every 30 seconds over periods of 13–33 days. The heart rate measured by these instruments had first been calibrated against oxygen uptake in the birds walking at different speeds on a treadmill in a respirometer (Bevan *et al.* 1994). On the nest, incubating birds had an average energy expenditure of 2.2 W per kg compared to 2.4 W per kg when brooding small chicks, indicating slightly more parental activity. Away from the nest, the average expended energy was 6.2 W per kg.

Salt-water switches and satellite transmitters were used in conjunction with the HRDLs to synchronise measurements of heart rate with time on the water and distance travelled. Temperature sensors recorded abrupt falls in abdominal temperature following ingestion of food and/or seawater. During sustained flight in one direction, the energy cost of flying was 2.4 W per kg, no more than for brooding a chick. Albatrosses are buoyant birds that sit high on the water and one can imagine that on a calm sea, resting birds may be using little energy, but on cold water they must lose more heat than on land. On moderate to rough seas, albatrosses almost always maintain heading on waves by footwork. Feeding involves planing, paddling, competition with other birds, diving, swimming and taking-off, all of which contribute to an overall energy expenditure averaging 5.8 W per kg (Bevan *et al.* 1995).

Change in mass during incubation

Given a known or assumed post-absorbtive loss in mass and the energy equivalents of fat and protein, metabolic rates may also be predicted from measured losses in mass.

Resting albatrosses on cold-temperate islands of the Southern Ocean may be expected to lose heat, but they have efficient insulation. Brown & Adams (1984) considered that Wandering Albatrosses at Marion Island during February were within their thermal neutral zone when the mean air temperature was 11.6°C (2.2–16.0°C), but in an environment where winds are such a dominant component of the weather, heat flux must be related to wind-chill rather than to ambient temperature.

On Midway, in ambient temperatures of 15°C to 24°C, Grant & Whittow (1983) believed that Laysan Albatrosses were within their thermal neutral zone most of the time. The birds used no more energy incubating eggs than when resting[12]. On the other hand, when exposed to the sun, they experienced heat stress and may have used energy to cool eggs overheated by the surrounding sand. Gular panting (evaporative cooling) also uses water. Galápagos Albatrosses do the same when overheated.

Observed losses in mass, however, point to much higher incubation costs and water losses probably make up the balance (Prince *et al.* 1981). Incubating albatrosses may abandon eggs more because of dehydration than starvation. Furthermore, the capacity of individual albatrosses to continue sitting under the stress of water and food deprivation determines not only whether an egg will hatch, but also influences an individual's later ability to continue feeding its chick throughout the winter (Croxall & Ricketts 1983).

Both sexes share incubation and chick rearing. At South Georgia males do more than females, but females contribute equally in proportion to their size. Incubation in Wandering Albatrosses (78 days) is shared by male and female partners in a number of shifts. Altogether, males take 54% and females 46% of the total. Sitting birds lose mass; males at a rate of 85 g per day and females at 80 g per day or 0.87% and 0.93% of their respective body mass per day. When relieved by their partners, males feeding at sea regain mass at a rate of 135 g per day, which greatly exceeds their losses, but foraging females gain 75 g per day, only just enough to balance their losses (Croxall & Ricketts 1983).

It is not known why increments in female mass between shifts should be so much lower than those of males, but the unequal division of labour between males and females offsets the difference, so that throughout incubation, partners sharing the task in average proportions can gain mass. In pairs that are not fully synchronised, individuals of either sex taking a disproportionate share of incubation may be unable to regain mass completely when off duty. Marked departure from the average division of incubation duties reduces the chance of hatching and perhaps, also, the fitness of the parent taking the greater share, especially if it is the female. Nevertheless, eggs are hatched and chicks reared by pairs with great disparity in the sharing of incubation. One female that completed 64% of the incubation, including one shift of 36 days in which she lost 23% of her body mass, managed to regain 89% of her initial body mass by the time of hatching. At the Iles Crozet the scenario is similar, but the details different. Off-duty males forage over greater areas of ocean than females (Weimerskirch 1995).

The rules that determine when a Wandering Albatross returns to its nest and resumes incubation differ between sexes. In females, body mass at the end of the previous shift is the most important factor; lightweight females from the Iles Crozet spent more time away than heavier females. Males evidently experienced no such time constraints on collecting enough food (Weimerskirch 1995).

Wandering Albatrosses appear to have a wide margin of safety that allows them to limit the risk of failure. Most incubating birds are relieved before their body reserves are seriously depleted. Desertions during incubation are mostly by inexperienced females (Weimerskirch 1995).

Following incubation, the need to bring increasing quantities of food to growing chicks places additional demands upon parents. Average feeds of 264 g per day for 246 days are little more than the daily mass increments of males in the intervals at sea between brooding shifts. These statistics obscure the enormous meals of up to two kilograms that are sometimes brought to great albatross chicks. Parents that embark upon this task while

[12] Earlier estimates by Rice & Kenyon(1962b) and Fisher (1967) did not exclude losses through defaecation (Grant 1984).

underweight may be unable to perform adequately. There is evidence that females taking 47% of incubation, later lose more chicks than those completing 43% (Croxall & Ricketts 1983). Males may collect more food than females and have a better chance of rearing a chick without assistance, providing the chick is more than half grown when the female defects. There is some evidence to support this, but there is enormous variation in parental performance, even in a single season, and one Wandering Albatross female has reared a chick unaided for 256 days at Marion Island (Brown & Adams 1984).

Adult Black-browed and Grey-headed Albatrosses are at their heaviest when they first return to South Georgia in October. Males are heavier than females and both have already lost some mass by the time they embark upon incubation. Daily loss in mass during 83 incubation shifts averaged 1.2% (0.5–3.5%) of body mass. Variation was similar in both sexes of the two species. Males remained significantly heavier than females from the beginning to the end of shifts, and in spite of considerable loss of mass during incubation shifts of 1–19 days, neither species showed any decline in mass over the whole incubation period (Prince *et al.* 1981). Both species lose mass at about the same rate while feeding their chicks, but this takes Grey-headed Albatrosses 25 days longer than Black-browed Albatrosses. By the time their fledglings depart, Grey-headed parents have each lost 800 g compared with 500 g by Black-browed parents. Furthermore, since they finish earlier, Black-browed Albatrosses have 165 days at sea to make up a deficit of 13% of body mass while Grey-headed Albatrosses would have to make good 20% in 126 days if they are to breed the following season. This is possible, but very rarely accomplished (Tickell & Pinder 1975, Croxall 1991).

In Laysan Albatrosses, males are likewise heavier than females and at peak mass on arrival. Fisher (1967) determined that it took a male three weeks foraging at sea to make up the mass lost while attending a nest during the pre-egg period. Losses of 1.08% to 1.25% of body mass per day occurred during incubation (Croxall & Ricketts 1983) and overall loss in mass during Laysan incubation continued through the 'intensive stage' of chick care and amounted to 24% of body mass. Some of this was recovered in the later weeks of the chick period, so that by the end of the breeding season, the net loss in body mass was about 9–12%. Laysan Albatrosses have about 130 days to regain breeding condition before the next season. Fisher pointed out that parallel losses in body mass are also experienced by subadults on the breeding grounds and advised that the energy cost of breeding incorporates elements that are only indirectly concerned with chick rearing.

Role of fat

Fat has long been supposed to fuel body metabolism of chicks during intervals of fasting. However, Reid *et al.* (2000) have pointed out that very little fat is deposited in Grey-headed Albatross chicks before they are 60 days old. Up to that age, growth is directed towards organ development, especially those involved in processing food (e.g. the liver) and the skeletal structure. Only after 60 days, when the flight feathers start growing, does the deposition of fat increase. Once feathers begin growing, keratin must be laid down continuously; 20 primary, 60 inner wing and 14 tail feathers amount to about 14 m and this is the only time in the life of an albatross when all feathers are grown simultaneously. During periods of fasting, if normal body fat reserves are exhausted, protein can be used to fuel body metabolism, however such a diversion of protein away from feather growth causes defects in feather structure and produces shorter, weaker flight feathers. Extra fat reduces the risk of protein being diverted away from feather growth and allows the chick to produce good quality flight feathers, some of which will not be replaced for three years. Fledglings with large, strong flight feathers can leave with maximum payloads of fat, which then buffer variable food supply during the early days at sea, before they have become proficient at foraging.

ENVIRONMENTAL IMPACT OF ALBATROSS POPULATIONS

There have been several estimates of prey biomass taken by different seabird faunas. Some have been based on weighings of food eaten, others on calculations of energy requirements together with assessments of populations and the proportions of the prey in diets. There are problems associated with such scaling-up (van Gardingen *et al.* 1997); most workers have had reservations about the figures published, but considered them useful approximations for comparative purposes.

At sea

Woehler (1997) estimated the quantities of fish, squid and crustaceans eaten by seabirds foraging in the Prydz Bay area[13] off Antarctica by scaling up species specific metabolic rates (Koteja 1991). In the 12 year period 1980–92, 752,000±176,000 tonnes of marine resources were consumed during the six months of each summer. The quantities taken by Black-browed and Grey-headed Albatrosses (each 0.04%), Wandering Albatrosses (0.87%) and Light-mantled Sooty Albatrosses (4.76%), were small compared with the maximum quantities taken by Antarctic Petrels (16.7%) and White-chinned Petrels (16.0%) (Woehler 1997).

The immense resources of the Patagonian Shelf have been responsible for the growth of huge seabird populations in the Falkland Islands and on the coast of Argentina. Although there has been an assessment of the potential for competition between seabirds and fisheries, there has been no baseline estimation of the needs of seabird populations. The Black-browed population of the Falkland Islands is probably greater than 350,000 pairs and it has been calculated that the Beauchêne Island colony alone consumed up to 12,000 tonnes of squid and fish in ten weeks from February to April 1991 (Thompson 1992, Woods & Woods 1997).

Studies over many years at South Georgia indicate that seabirds consume almost eight million tonnes of marine prey per year (excluding carrion). Almost six million tonnes of it is krill, but the four species of albatrosses take only 0.3%. Albatrosses consume 6% of the 466,000 tonnes of squid taken by all South Georgia seabirds. They are collectively the third most important consumer of squid after White-chinned Petrels and King Penguins. They also take 6% of the 390,000 tonnes of fish consumed by all South Georgia seabirds. Although southern lampreys amount to only 6.3% of mollymawk diets at South Georgia, that figure represents over a million individuals (Prince 1980); no comparable predators of these lampreys are known.

The total mass of all prey consumed in one year by each South Georgia species was calculated as: Grey-headed Albatross 42,000 tonnes, Black-browed Albatross 22,000 tonnes, Wandering Albatross 3,000 and Light-mantled Sooty Albatross 4,900 tonnes. Seven penguin and petrel species individually consumed greater biomass than any of the albatrosses. Some of the albatross prey was taken far beyond the range of penguins, diving petrels and prions. So the overall impact of albatrosses breeding at South Georgia on the stock of krill, squid and fish in the surrounding sea may have been even less than indicated by these comparisons. Furthermore, there are differences in the lengths of seasons during which different species forage from the island (Croxall *et al* 1984a).

Of the 1,811,000 tonnes of crustaceans (mostly *Euphausia vallentini*) consumed yearly by all seabirds at the Iles Crozet, albatrosses take little more than 0.02% of the total and thus have a negligible impact on pelagic crustaceans of the south Indian Ocean. Of the 114,700 tonnes of squid eaten by Crozet seabirds, the six species of albatrosses together accounted for 5%, of which almost half was taken by Grey-headed Albatrosses. Albatrosses took 6% of the 1,078,500 tonnes of fish consumed and almost three-quarters was eaten by Eastern Yellow-nosed Albatrosses (Ridoux 1989). Most of the birds were said to be feeding 100 to 200 km from their islands, but Stahl *et al.* (1985) and Weimerskirch *et al.* (1988) indicate much greater ranges, especially for the albatrosses.

Laysan Albatrosses consume about 251,000 tonnes of squid, fish, crustaceans and unidentified food, amounting to 60% of the estimated annual consumption by all seabirds from the NW Hawaiian Islands (Fefer *et al.* 1984). Black-footed Albatrosses take about 13,000 tonnes of similar prey. Both these albatrosses forage over the North Pacific Ocean within range of Japan, the Aleutian Islands and Alaska Peninsula. Huge numbers of fulmars, auks and gulls also forage over these seas and the relative impact of albatrosses on marine resources is less than their numbers at the NW Hawaiian Islands might lead us to suppose.

[13] The southwest Indian Ocean between longitudes 60°E and 90°E, and from latitude 60°S to the Antarctic continent.

On land

Albatrosses spend up to a year visiting their breeding grounds and build nests from any workable material that is readily available at the site. The impact upon the terrestrial environment depends upon the number and density of nests. Widely spaced heaps of torn-up vegetation may become semi-permanent mounds, around which albatross excreta is a major source of minerals that promote plant growth (Burger *et al.* 1978). Over long periods of time, even a thinly scattered albatross population may contribute significantly to the vegetation of island slopes, but where they expose underlying peat, erosion by wind and rain may have destructive effects (Joly *et al.* 1987).

Dense colonies of albatrosses destroy vegetation which may be completely eliminated from around nests, as in the great Black-browed Albatross colonies of the Falkland Islands. On steep ground, erosion may cause grass tussocks to tumble down slopes and where no vegetation is left to hold friable peat whole slopes may slip (Derenne *et al.* 1972, Pascal 1979). On thin soils such as the rocky islets of the Chatham Islands, once dominant plants may become stunted, perhaps because of climatic changes and nesting albatrosses then accelerate the degradation of habitat (Robertson 1997).

VICTIMS AND VERSE

A mixed flock of petrels, albatrosses and gannets feeding behind a trawler off South Africa (W.L.N. Tickell).

17
THE SEA OF MAN

> 'In the stomach [of a wandering albatross] was an undigested Roman Catholic tract with a portrait of Cardinal Vaughan'
>
> Edward Wilson 1907

Albatrosses die at sea and few dead birds are found on the breeding grounds. We know little about the natural causes of death. Like sailing ships, seabirds sometimes have the misfortune of being caught off a lee shore when a storm blows up, but wrecks of albatrosses on coasts are rare.

Today, when most governments protect albatross breeding islands, the high seas are not safe. Comparatively few albatrosses are killed intentionally, but they are a conspicuous component of the environmental cost of fisheries. There is a sense in which the bureaucratic euphemism *by-catch* obscures that reality.

Our knowledge of albatross migrations owes much to fishermen and the companies that have returned rings from birds caught on lines and in nets; inevitably, conclusions based upon such data incorporate variable fisheries bias that is difficult to measure. Satellite tracks of albatrosses also reflect the presence of fisheries. If the locations of fishing vessels are known, interesting and useful conclusions follow (Weimerskirch 1997e), but without that information the results of telemetry may be deceptive.

Early Encounters

Albatrosses were eaten by the people who settled the Aleutian Islands and Alaska Peninsula. From St. Lawrence Island to California, bones in the middens of ancient Aleut and Indian sites indicate that Steller's Albatrosses were a frequent, and sometimes the most substantial non-mammalian item of the aboriginal diet; their remains are far more frequent than those of Laysan and Black-footed Albatrosses (Howard & Dodson 1933, Friedmann 1934, Murie 1959, Yesner 1976, Yesner & Aigner 1976). In the late 19th century, Aleuts still hunted in the inter-island passages, spearing albatrosses from kayaks (Nelson 1887), while in the far north of the Bering Sea, they were captured at the edge of the ice (Murie 1959).

As the early Polynesians migrated out of the tropical central Pacific towards Hawaii and New Zealand, they too encounted albatrosses at sea and made use of them.

'Fishing' and 'Sport'

As the *Dainty* approached the Falkland Islands in 1594, Sir Richard Hawkins caught a great albatross:

> 'I caused a hooke and lyne to be brought me; and with a peece of a pilchard I bayted the hooke, and a foot from it, tyed a peece of corke, that it might not sinke deepe, and threw it into the sea which... in a little time was a good space from us, and one of the fowles being hungry, presently seized upon it, and the hooke in his upper beake. ... being brought to the sterne of the ship, two of our company went downe by the ladder of the poope, and seized on his necke and wings;... we cast a snare about his necke, and so tryced him into the ship... By the same manner of fishing, we caught so many of them, as refreshed and recreated all my people for that day.' (Markham 1878).

The practice was rediscovered many times. On 3 October 1772, during Captain Cook's second voyage (1772–75), William Bayly, the astronomer on *Adventure* wrote:

> '... they are easily catched, by the following method, take a piece of deal board about a foot square

(mor or liss) fix a line to it so that the hook when baited with any sort of meat may lay on the board & be thereby floted, put the board on the water with the hook baited, laying on it, let the board fall a-starn by veering out your line perhaps 100 Yards the whiteness of the board attracts the bird so that he pitcheth on the water by it & swims along side of it, & seeing the bait greedily swallows it hook & all & is thereby catched.'

By 12 October the crew of the *Resolution* were also pulling the birds aboard (Beaglehole 1961–67). Cook found it repugnant, hitherto albatrosses had been obtained when gentlemen naturalists and officers put off in small boats with their guns. On 5 February 1770, during the voyage of the *Endeavour* (Beaglehole 1961–67), Joseph Banks had written in his journal:

'...eat part of the Albatrosses shot on the 3rd which were so good that every body commended and Eat heartily of them tho there was fresh pork upon the table.'

His cooking directions remain classic:

'The way of dressing them is thus. Skin them over night and soak their carcasses in salt water till morn then parboil them and throw away the water, then stew them well with very little water and when sufficiently tender serve them up with a savory sauce.'

After that first albatross dinner in the Great Cabin of the *Endeavour*, the birds appeared on the table during all three of Cook's voyages. Generations of later seafarers 'fished' for albatrosses. Baited hooks were commonly used, but devices were invented to replace hooks and perhaps make it easier to remove birds from lines after capture. One took the form of a hollow triangle of metal floated on a cork. The albatross went for the bait, got the hooked end of its bill wedged in the triangle and as long as the line was pulled tight, the bird could not escape until hauled on board. Not everyone found that this method worked (Dixon 1933) and more elaborate versions were probably even less effective (Roberts 1967). 'Fishing' for albatrosses was more successful when ships were sailing slowly or becalmed. The hooked birds resisted and when a ship was making even moderate speed, they sometimes took to the air or were dragged under the waves and filled with water; either way, they were very difficult to haul in.

'Hare hooked another [albatross] towards tea-time – drowned him in dragging him in, and finally lost him by the hook breaking – it was a grievous sight to see the splendid bird float off, dead.'
Edward Roland Sill, 23 January 1862 on board an American clipper[1]

Handy sealers and whalers made pipe stems from the long bones, skinned out the toes from the large feet to make fine tobacco pouches and sewed up the skins into foot warmers. However, not all captured albatrosses were killed; many were released with marks or messages (p.29). Herman Melville (1851) describes his captain attaching a lettered leather tally.

Catching albatrosses was popular among passengers on emigrant ships. The activity continued long after fresh meat had ceased to be a desired addition to meals on board, although eggs occasionally found in female albatrosses were reserved for some captains. In 1879, whole skins were in demand and carcasses were thrown overboard. Frequently the skins were divided up to become feather boas and tippets for ladies.

Albatrosses were also shot from ships under way, when there was no chance of recovering the dead birds. Andrew Bloxam described one such incident in 1825, on the *Blonde*:

'We saw great quantities of these birds to day. Diomedea exulans or the Wandering Albatross...Guns & rifles were in requisition to fire at them as they came close to the vessel sometimes flying over it but generally over the quarters tho' we fired at them about 50 times only one was fairly brought down & that by myself with swan shot about 20 yards distant. Shot seems scarcely to have any effect upon them. As we were going at the rate of 7 knots we were unable to lower a boat to procure it.' (Olson 1995).

[1] Sill (1944).

Albatrosses flying at the mast heads and yards of sailing ships (p.275) could be brought down on deck by skilled marksmen, but ships' masters and mates did not appreciate their sails and rigging being blasted, so restricted such shooting (Green 1887). Shooting out to sea became common on passenger ships in the southern hemisphere. It was possible when wind, big seas and the speed of vessels made it impossible to catch birds on lines.

In his identification guide *Birds of the Ocean*, W.B. Alexander (1928) stated that the number of albatrosses killed after being hooked on lines was only small. David Medway (1997b) considers this quite mistaken and quotes many eyewitness accounts of albatrosses taken on lines. While it is impossible to put figures to the numbers of birds shot or killed after capture, Medway has pointed out that between 1788 and 1880 about 1.3 million free emigrants sailed from Europe to Australia. In one year alone (1874–75), 93 ships took 30,000 emigrants to New Zealand. In the middle decades of the 19th century, American clippers sailing around Cape Horn carried thousands of passengers between the east and west coasts of North America. During one month in 1850, 33 vessels arrived at San Francisco from New York and Boston (Morison 1965). Numerous eyewitness accounts, coupled with shipping statistics make a persuasive case for believing that a substantial seabird mortality occurred during the 19th century, especially among Wandering and Black-browed Albatrosses.

Moral protests evidently had little effect. The Reverend William Scoresby (1859) was one of many who were offended:

> '... the afterpart of the poop of the *Royal Charter* was yesterday occupied with "sportsmen" and lookers on, who with rifles or other guns were every now and then firing at the unconscious elegant birds gracefully hovering about our rear. I fancy 50 to 100 shots were fired, happily with rare instances of their taking effect; but in one case I saw, on being induced to look astern by a general shout, a poor striken bird struggling on the surface of the water apparently mortally wounded. This useless infliction of injury and suffering on these noble looking birds, where there was no chance of obtaining them as specimens for the museum, nor for any other use, was to my feelings, and, I believe, the feelings of many others, particularly painful.'

We can guess that this killing ended only when ships became too big and fast to trail lines and safety regulations forbad the discharge of firearms on board, but as late as December 1909, the curator of the a London museum found a fresh wandering albatross hanging among the Christmas turkeys in Leadenhall Market. It had probably been caught from a ship en route from Australia (Stubbs 1913).

Australian fishermen still shoot albatrosses as pests (Brothers 1982) and others may be killed at sea for no apparent reason (Stagi *et al.* 1997). Some South American fishermen use their feathers for fishing lures (Plenge *et al.* 1989).

SCAVENGING

Nomadic scavenging over wide expanses of ocean may be the natural strategy for at least some albatrosses. They were ranging the oceans in that distant past when all whales, seals and fish died naturally. Although William Dampier (1967) did not name them, albatrosses were evidently among the large birds he saw as the *Roebuck* approached the Cape:

> '... we past by a dead Whale, and saw Millions (as I may say) of Sea-Fowls about the Carcass (and as far around as we could see) some feeding, and the rest flying about, or sitting on the Water, waiting to take their Turns. We first discovered the Whale by the Fowls; for indeed I did never see so many Fowls at once in my Life before, their Numbers being inconceivably great: They were of divers Sorts, in Bigness, Shape and Colour. Some were almost as big as Geese, of a grey Colour, with White breasts and with such Bills, Wings and Tails. Some were Pintado-Birds, as big as Ducks, and speckled black and white. Some were Shear-waters; some Petrels; and there were several Sorts of Large Fowls. We saw these Birds, especially the *Pintado*-birds, all the Sea over from about 200 Leagues distant from the Coast of *Brazil*, to within much the same Distance of *New-Holland*.'[2].

[2] Australia

Fisheries have probably been attended by pelagic seabirds from the earliest times. Small sampans have fished off Japan for hundreds of years and as they ventured towards the edge of the continental shelf and beyond, Laysan, Black-footed and Steller's Albatrosses would have quickly taken to extracting fish from nets or lines and scavenging discards. Similarly around South America, South Africa, Australia and New Zealand in the 19th century, the small vessels that opened up fisheries attracted several species of southern albatrosses.

It may be more profitable for some albatrosses to scavenge from ships than to forage independently. As long as there is a good chance of accessible trawls being hauled frequently enough, albatrosses employ less energy sitting in rafts on the water, waiting, than in searching. Large numbers of mollymawks attend the trawling fleet off the Cape of Good Hope and in December 1994, more than 6,000 Black-browed Albatrosses assembled behind one factory ship northeast of the Falkland Islands (Bourne 1996). Nevertheless, seabirds are sensitive to ships' offerings. If the scavenging is poor – only guts and fins – albatrosses may ignore them altogether. Some albatrosses, especially inexperienced juveniles, may benefit more from scavenging than they lose by increased mortality, particularly in years when food is otherwise scarce (Gould *et al.* 1997).

Fisheries are competitors that may deplete stocks. Around the Falkland Islands, they appear to be greater predators of finfish and squid than are Black-browed Albatrosses, so the benefit gained by the birds from scavenging at fisheries may be more than offset by a potential depletion of their natural prey (Thompson & Riddy 1995). On the other hand, albatrosses are accustomed to patchy transient resources.

A fishery that offers acceptable alternative food not normally available to albatrosses, for example benthic fish, is a preferable scavenging option providing there are no excessive costs. By chance or choice, the scavenging options may vary. Females have been more numerous among the Wandering Albatrosses at longliners in both the Indian Ocean and South Atlantic (Weimerskirch & Jouventin 1987, Croxall & Prince 1990).

THE COST OF SCAVENGING

Following ships has paid off frequently enough to have become a habit among many albatrosses. The primary cost of scavenging is the time invested for food of poor nutrient value, at worst no more than garbage. There is also the potential danger of birds getting killed. Ring recoveries from the late 1950s onwards prove that albatrosses were being caught in fishing gear 50 years ago. Some were captured alive and released, many were drowned (Sladen *et al.* 1968, Robertson & Kinsky 1972, Robbins & Rice 1974). The large scale of modern fisheries has increased this cost dramatically.

Driftnetting

This ancient method of fishing became a hazard to seabirds in the early 1950s when synthetic line made it possible to set very long and virtually invisible nets. Salmon fisheries in the northwest Pacific, setting 15 km nets, were followed by squid fisheries in the central North Pacific using nets averaging about 45 km in length. Tunas and similar fish were caught in large-mesh driftnets of up to 40 km throughout much of the temperate North Pacific and the fisheries expanded into albatross latitudes of the southern Indian Ocean. Japan was prominent in all these fisheries followed by Taiwan, Korea and other nations. Altogether, fleets of vessels set thousands of kilometres of nets that became notorious for the non-target species of fish, cetaceans and seabirds killed (Northridge 1991).

Many of these fisheries were not regulated so it is impossible to guess the total number of albatrosses that died. DeGange & Day (1991) estimated that 921 Laysan Albatrosses were killed in the 1977 Japanese land-based salmon fishery, but it had already been in decline for a decade, and by 1987 the number killed was 231. The more distant Japanese salmon fishery, that employed mother-ships (factories), each with fleets of about 40 catchers, fished the western Bering Sea and the Aleutians. This fishery peaked in 1956 then declined and contracted (Northridge 1991, DeGange & Day 1991). It was a seasonal fishery (early June to mid-July) and was probably over by the time most juvenile Laysan Albatrosses reached the area. In 1981 and 1984, 228 and 114 Laysan Albatrosses were killed, but by that time the fishery was minimal.

There were five other driftnet fisheries on the high seas of the North Pacific, setting 1.5 million km of squid nets each year, in the Transitional Zone of the subarctic frontal system from Japan to longitude 145°W. In 1989, an estimated 13,500 Laysan and 5,300 Black-footed Albatrosses were killed when scavenging neon flying squid and Pacific pomphret from these nets (Northridge 1991). Craig Harrison (1990) believed that fisheries were not yet a major problem for albatrosses, but advised that the industry needed to be closely monitored. In 1990–91, Japanese large-mesh driftnets for tuna and billfish killed an estimated 1,735 Laysan and 685 Black-footed Albatrosses and in both high seas fisheries of the North Pacific, an estimated 17,500 Laysan Albatrosses and 4,400 Black-footed Albatrosses were killed (Johnson *et al.* 1993). Assuming stable populations, the 1990 rates indicated total losses of about 213,000 Laysan Albatrosses during the 1980s (Gould & Hobbs 1993). Although more Laysans were caught than Black-footed Albatrosses, the numbers were not in the proportions known on breeding islands. The driftnet mortality appears to have had a greater impact on the Black-footed Albatrosses, which lost an estimated 2.2% of their world population in the 1980s compared with 0.7% of Laysans (Gould & Hobbs 1993).

Driftnet fishing was phased out towards the end of 1992, but illegal driftnetting still continues (Alexander *et al.* 1997, Smith 1998).

Trawling

Trawlers off South Africa, South America, the Falkland Islands and New Zealand attract large numbers of albatrosses (p. 355). Where fleets are fishing the birds congregate on the surface in rafts until a vessel brings a trawl towards the surface. The resting albatrosses then take off and scramble for fish or squid escaping from contracting nets. After the trawl is aboard, lesser numbers then scavenge offal until discharge ceases (Thompson & Riddy 1995).

Few albatrosses get caught in trawling gear. Some are drowned, but others are released alive. Morant, Brooke & Abrams (1983) believed that most Black-browed Albatrosses caught in fishing gear off South Africa were killed for food.

After the establishment of a 200 mile EEZ[3] around New Zealand in 1978, fisheries observers discovered that large numbers of albatrosses were killed by Russian mid-water squid trawlers. These vessels were using net-sonde detectors to monitor the catch. The instruments were connected to the vessels by heavy cables which were a flight hazard to the birds wheeling about behind the ship, just above the water . At least 83% of dead birds were killed by these cables and the 1,144 dead shy albatrosses, collected on 27 trawlers in 1990, were believed to represent a mortality of 2,300 birds (Bartle 1991). Net-sonde detectors were widely used in many seas, so the number of albatrosses killed in more than a decade was much greater. These instruments were phased out in the 1990s.

Whaling

Scavenging pays real dividends when ships are fishing or discharging a resource not otherwise available to the birds. In the early days of whaling, dead whales were flensed while floating alongside sailing ships and the carcasses were afterwards cut loose to drift away. Great quantities of meat, blubber and offal were available at no apparent cost and albatrosses were among the seabirds that undoubtedly benefited.

The later industry was less prodigal, but shore stations and factory ships all discarded enough grease and refuse to attract many seabirds. At sea, harpooned whales vomited great quantities of krill while gallons of blood clotted in the cold water – all scavenged by seabirds. Wandering Albatrosses were big enough to swallow the plugs of skin and connective tissue cut from the whales' flukes to take towing chains (Matthews 1929).

During the boom years of Pacific whaling in the mid-19th century, Steller's Albatrosses attended the fleet at the Pribilof Islands, but soon after it ended they were notably less common over the shallow continental shelf of the Bering Sea (Elliott 1884). In the Southern Ocean, Wandering Albatrosses were common at Deception Island during the years when the whaling station was active (Gain 1914); afterwards they were rarely seen there.

The benefits albatrosses secured from whaling may have been localised, but whale-catchers in action were conspicuous cues. Apart from occasional accidents, albatrosses were safe enough with whaling. We can imagine that albatross populations were influenced by the removal of such a huge biomass of top predators from the food chain. The received dogma is that more krill became available to other marine predators, but the trade-offs in long-lived seabirds were probably far more subtle. Albatrosses join multi-species flocks following large marine mammals to secure prey driven to the surface by the feeding behaviour or turbulence of the surrounding water, and killing whales would obviously have interfered with that resource.

[3] Exclusive Economic Zone.

Longlining

Weighted longlines were traditionally laid on the sea-bottom for demersal fish; 12 km lines with over 1,000 baited hooks were in use 200 years ago. Setting these lines always attracted seabirds, but when lines were buoyed near the surface for tuna and billfish, bait sank more slowly and became more dangerous to seabirds.

Fisheries now use lines of over 100 km with 2,500–3,000 hooks set at depths of 60–150 m. Ships steam for several hours paying out lines with baited hooks over the stern. Many seabirds follow the vessels scavenging baits. It is probably a matter of chance whether a bird removes the bait from a hook or gets hooked itself and drowned.

> '...it was rarely the bird that first seized the bait that got hooked: typically...a Cape Pigeon saw a sinking bait, dived, and retrieved it to the surface; other Cape Pigeons joined in and a commotion developed; next usually came Grey-faced Petrels, perhaps also a Grey Petrel; soon the first mollymawk came in and scattered the petrels, then more mollymawks; finally a Wandering Albatross(es) splashed into the fray and finished off the bait.' (Imber 1994).

Japanese longlining in the North Pacific began in the late 1940s and proliferated rapidly. The majority of Laysan and Black-footed Albatross ring recoveries were received from birds said to have been captured on longlines set for tuna, but there was some doubt about the reporting (Fisher & Fisher 1972, Robbins & Rice 1974). Alaskan longline fisheries for Pacific cod, sablefish, and halibut caught Black-footed and Laysan Albatrosses between 1983 and 1996. By the mid-1990s Hawaiian-based longlining for swordfish killed an estimated 4,500 Black-footed Albatrosses each year. At least three individuals of the endangered Steller's Albatross were also killed (Smith 1998).

In the 1950s, Japanese longliners moved into the southern hemisphere fishing for southern bluefin tuna and allied species. About 20,000 hooks were set around New Zealand in 1955. Ten years later, the fishery was well-established around Australia and New Zealand, setting about 14 million hooks per year, and by 1969 it had expanded throughout the Southern Ocean from the South Atlantic to the South Pacific. From 92 million hooks set in 1971, the number had increased to more than 120 million per year in the 1980s (Fig. 17.1). The Japanese Southern Bluefin fishery was the largest south of latitude 35°S in the Southern Ocean, but from 1987, it declined due to the imposition of catch limits within the New Zealand and Australian EEZs. Australian and New Zealand longline fisheries, using smaller vessels, proliferated within their own EEZs, while fewer Japanese ships were joined by others from Taiwan, Korea and elsewhere, fishing on the high seas (outside the EEZs) were they were not inspected (Fig. 17.2).

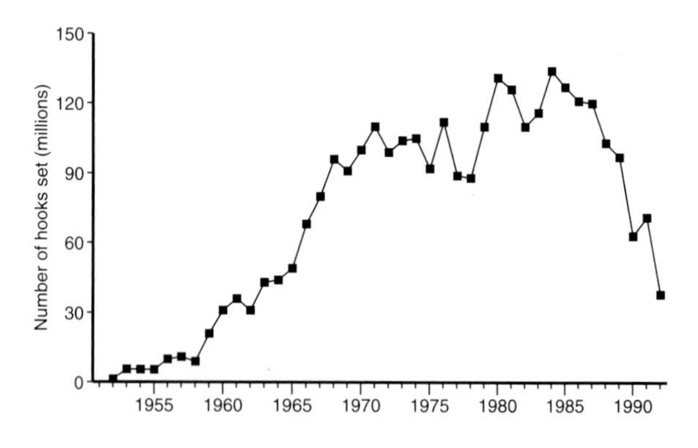

Fig. 17.1 The number of hooks set by Japanese Southern Bluefin Tuna longline fishery throughout the Southern Ocean between 1952 and 1992 (Weimerskirch *et al.* 1997a).

Albatrosses were common throughout the fishery and it is reasonable to assume that they immediately took to scavenging at these vessels. For five hours a day, each Japanese ship threw overboard a bait fish every six seconds. These bait fish measured about 30 cm in length and could be removed from the large tuna hooks by seabirds. About a third of the albatrosses that attempted to take a bait fish carried it off without being hooked.

Robert Tomkins (1985) speculated that fishing gear may have been killing increasing numbers of Wandering Albatrosses from Macquarie Island during the 1970s. BAS ringing records from 1975 to 1988 revealed that 76% of all fisheries recoveries of Wandering Albatrosses from South Georgia, over 28 years, were from longlines (Croxall & Prince 1990). In 1985, Wandering Albatrosses colour dyed on the Iles Crozet were seen in numbers by French observers on longline vessels (P. Jouventin pers. comm.) and 11 Wandering Albatrosses from the Iles Crozet were known to have been killed on Japanese longlines. Hooks were found near nests on several breeding islands and occasional live albatrosses and carcasses trailed lengths of monofilament line (Parrish 1991, Cooper 1995, H.

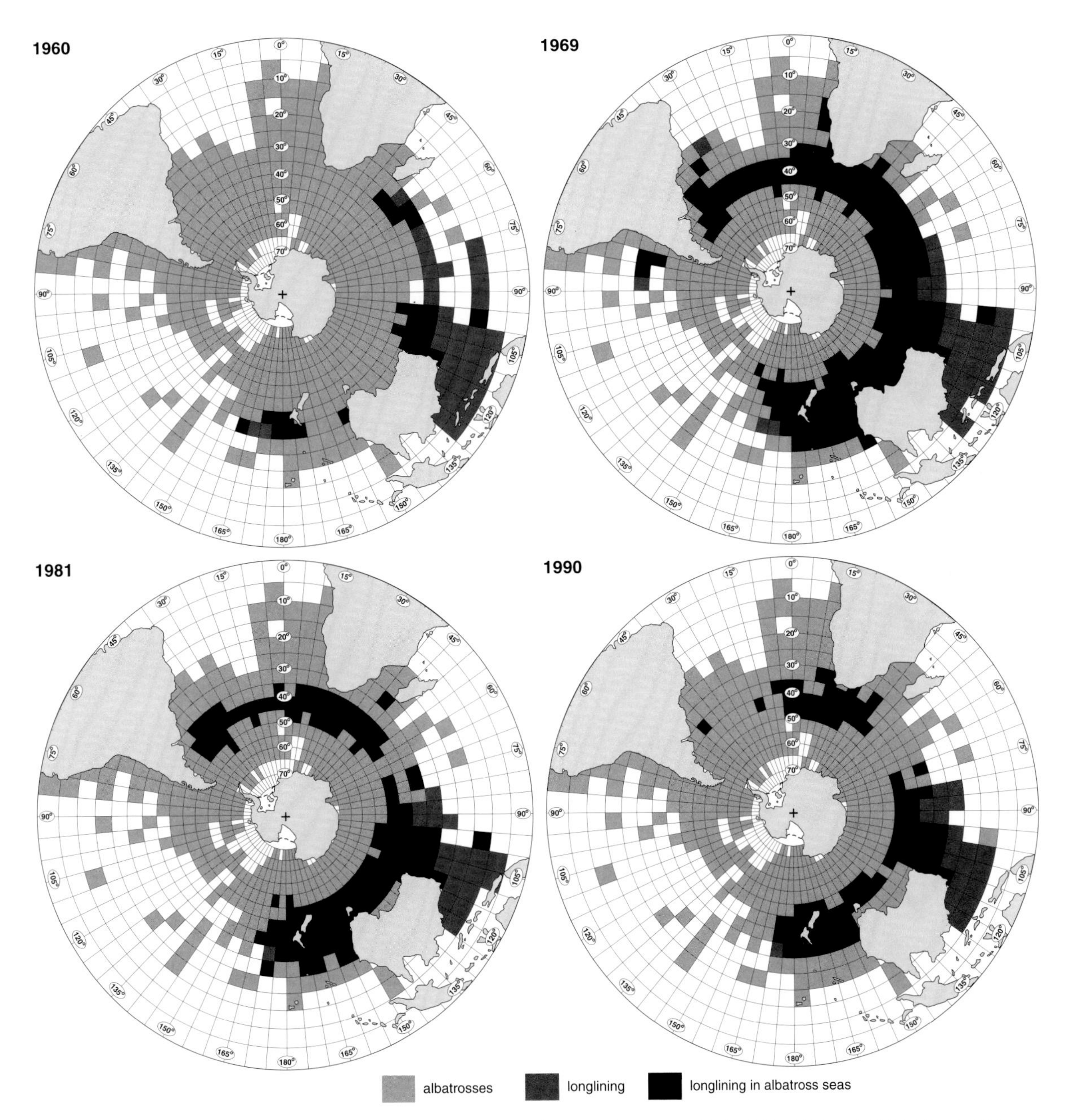

Fig. 17.2 The Japanese Southern Bluefin Tuna fishery in the southern hemisphere in 1960, 1969, 1981 and 1990. (Weimerskirch *et al.* 1997a).

Hasegawa pers. comm.). Adults with wounds to the mouth died on their nests at the Ile de la Possession (Weimerskirch & Jouventin 1987).

For decades, no-one involved in albatross research had any significant intelligence from fisheries. Officials monitoring fishing regulations on foreign vessels apparently did not notice or ignored the albatrosses that died on the gear they were observing. In the early 1980s, Nigel Brothers (1982) of the Tasmanian Parks and Wildlife Service, who was working on the breeding grounds of Tasmanian Shy Albatrosses, began trying to persuade Australian fishermen not to shoot albatrosses scavenging from trawls and droplines. By 1988, he was on board Japanese vessels off Tasmania, counting the number of seabirds hooked during the setting of longlines. He saw albatrosses killed at an average rate of 0.41 birds per 1,000 hooks set. In the period 1981–86, Japanese records showed that the Southern Bluefin Tuna fishery had set an average of almost 108 million hooks per year and simple arithmetic was enough to confirm that large numbers of albatrosses were being killed (Brothers 1991).

Over the following decade, 1988–97, Australian fisheries and wildlife biologists independently collected data from Japanese and Australian vessels longlining within the Australian Fisheries Zone (AFZ). Japanese fishing effort peaked at almost 32 million hooks in 1989 and declined steadily to less than eight million in 1997. By far the greater number were set during the southern winter (April–September), but seabirds were more likely to be caught in summer (October–March). The number of birds killed each year, estimated by wildlife observers, declined from 14,359 in 1988 to 160 in 1997. Annual catch rates varied from 0.70 to 0.02 birds per 1,000 hooks. Of 719 carcasses collected over the period 1988–96, 76% were albatrosses, identified as black-browed (26%), Grey-headed (12%), shy (12%), yellow-nosed (11%), buller's (<1%), wandering (9%), royal (3%), Sooty (1%) and Light-mantled Sooty Albatrosses (1%). Ringing returns indicated that they came from both local and distant island breeding grounds (Brothers *et al.* 1998b&c). The numbers caught varied between populations and fishing seasons. Tasmanian Shy Albatrosses from Mewstone and Pedra Branca forage throughout the year in the same areas as Japanese longliners, but Albatross Island birds encounter fewer longliners during and just after the breeding season, although they may be vulnerable in winter to Japanese as well Australian longliners and other fisheries.

Albatrosses are among many seabirds that approach the Antarctic Continent in summer as sea-ice disperses. They join huge numbers of typically polar birds foraging over the highly productive Antarctic waters. Between 1980 and 1992, Wandering Albatrosses almost ceased visiting the Prydz Bay area and no Black-browed Albatrosses have been seen there since 1984. These decreases may have been exaggerated because the species were at the southern limit of their foraging ranges, but Light-mantled Sooty Albatrosses, which normally feed in higher latitudes, also decreased by 62%. It is not possible to argue that fishery-related mortality was solely responsible for this decline. Albatrosses may have ceased visiting Prydz Bay naturally and gone elsewhere, because of changes in prey distribution and/or because foraging had become more difficult in competition with penguins and petrels from the adjacent Antarctic mainland (Woehler 1996,1997).

Seabird mortalities around New Zealand since 1988 were estimated from data collected by scientific observers of the NZ MAF[4] on Japanese longline vessels within the New Zealand EEZ. The number of ships with observers varied from less than 3% to more than 75%. Most longlines did not catch any seabirds at all and the average mortality of all seabirds declined from more than 4,000 in 1989 to less than 100 in 1994 (Murray *et al.* 1992). From 785 longlines set between 1988 and 1992, 320 dead seabirds were recovered; 34–46% of them were identified. To the north and east of North Island, 37% of those identified were albatrosses while around South Island 93% were albatrosses. Species identified included wandering, royal, black-browed, buller's, shy and Grey-headed albatrosses. In the period 1987–1994, during 1,472 days, it was calculated that longliners caught over 11,700 birds including 7,500 albatrosses. Black-browed Albatrosses hooked in 1989–1994 were predominantly juveniles (S. Waugh pers. comm.). Large differences in capture rate from year to year and between areas in any given year were attributed to differences in hydrography, phases of the moon and the effectiveness of bird lines.

In the Atlantic, off the coast of South America, trawlers have long fished for hake and other species. These vessels can be easily converted to other gear, including longlines. East of Tierra del Fuego and over the Patagonian Shelf, 19 longline ships set upwards of 25 million hooks in 1993–95. There were no independent observations of seabird mortality, but it was assumed that thousands of seabirds were killed (Schiavini *et al.* 1997). Over the shelf edge off the Rio de la Plata, in Uruguayan and Argentinean waters, large numbers of Black-browed Albatrosses are present every winter. In 1993–94, Uruguay had only five tuna boats; foreign longliners fished outside the EEZ. Flocks of ten to 400 albatrosses followed trawlers and tuna boats day and night. Many were been taken on tuna longlines and some killed by crews for food. The average mortality on longlines in 1993–94 was 4.7 albatrosses per 1,000 hooks (Stagi *et al.* 1997). This area has also been a significant source of mortality of Wandering Albatrosses (Croxall & Prince 1990) and Southern Royal Albatrosses (Robertson & Kinsky 1972). In some years, the Falkland Current flows as far as southern Brazil, where tuna boats fish between latitudes 15°S and 35°S. During 91 cruises by 14 vessels in 1994–95, 118 birds, including Wandering, Black-browed and Western Yellow-nosed Albatrosses,

[4] Ministry of Agriculture and Fisheries.

were caught on longlines (Neves & Olmos 1997).

Off South Africa, mollymawks from South Georgia have long been taken on fishing gear (Tickell 1967) and this hazard appeared to have increased through the 1970s (Croxall *et al.* 1984b).

In the 1980s, a demersal fishery for Patagonian toothfish developed on the continental slope off Chile. It employed shorter longlines which could be set more rapidly than in the tuna fishery. This fishery spread into the South Atlantic off Argentina, then around the Southern Ocean.

Fishing near the Falkland Islands had been controlled since the declaration of the FICZ[5] in 1986, but the enormous numbers of Black-browed Albatrosses remained at risk (Croxall *et al.* 1996). In November 1994, 220 Black-browed Albatrosses were killed by one toothfish longliner (Anon. 1995).

By the summer of 1988–89, Soviet longliners were fishing for toothfish off South Georgia (Dalziell & De Poorter 1993), where albatrosses are at risk. In April–May 1994, 22.5 km lines carrying 10,500 large hooks baited with squid, were set on the edge of the South Georgia shelf. It took about 90 minutes to shoot each line, at about 100 hooks per min, so scavenging birds had only a fraction of a second to grab any one bait. Few albatrosses followed the ship during night-time settings and none were caught. During daytime settings 50 km from the largest colonies, many mollymawks were in attendance. Black-browed Albatrosses persistently went for baits with fewer Grey-headed and Wandering Albatrosses approaching the lines. Seven Black-browed and nine Grey-headed Albatrosses were killed in just two settings (Ashford *et al.* 1994, Ashford & Croxall 1995). In 1993, a 200 nm EEZ was declared around South Georgia.

Over the shelf-edge to the west of the Iles Kerguelen 2.8 to 4.2 km toothfish lines carrying 2,400 to 3,600 baited hooks were set in 11 to 12 minutes and it was claimed that although considerable numbers of albatrosses gathered on the water, only two Grey-headed Albatrosses were hooked and drowned (Cherel *et al.* 1996).

Prediction

Counts of albatrosses killed on longlines have been made on few fishing vessels. Mortality rates computed from these data, have been liberally extrapolated and widely generalised in terms of the numbers of hooks set. Population models (p.340) have been used to predict the outcome of further declines or increases in survival and breeding success (Moloney *et al.* 1994, Weimerskirch *et al.* 1997a). The French model was used to examine the fate of the small, endangered Amsterdam Wandering Albatross population, given hypothetical fishing fatalities. Breeding difficulties were anticipated after 25 years and extinction in less than 50 years. The worst scenario predicted that the population could disappear in two to three years. Recovery of great albatross populations can be very slow; it is getting on for a hundred years since the number of Gough Wandering Albatrosses on Inaccessible Island was more than two or three pairs.

Different island populations are not equally at risk. The steep decline in numbers of Wandering Albatrosses on the Iles Crozet eased off in the late 1980s as fisheries effort in the area decreased (Weimerskirch *et al.* 1997a). Those breeding at South Georgia have continued to decline at about one percent per year. They are most vulnerable to toothfish longlines at the time they are brooding small chicks from March to May, when off-duty adults are foraging over the nearby island shelf where the fishery operates. At other times, they are foraging farther away and are believed to be at risk only during transit through the area of the South Georgia fishery (Croxall & Prince 1996).

The interactions between seabirds and fisheries are complex and changing. No population models have incorporated gains from fisheries. They may not have been apparent in longlining, but they were evidently present during driftnetting (Gould *et al.* 1997).

POLLUTION

Heavy metals

A number of metals, including copper and zinc, are essential trace elements for nutrition, and animals have homeostatic mechanisms to maintain them within optimum levels. Nevertheless, these mechanisms can sometimes be overloaded and produce toxic effects in wild birds.

[5] Falkland Islands Interim Conservation and Management Zone extending 200 nautical miles (370 km) from the islands.

Non-essential elements can also occur in animals; they include cadmium, lead, mercury and others. Birds acquiring these elements from their natural environments in air, water and food usually have physiological mechanisms for excreting or detoxifying them.

Pollutants released into the environment by human activity may raise environmental levels of heavy metals above normal. Animals exposed to pollutants may be unable to excrete or detoxify such high levels and suffer toxic effects. No corner of the world's oceans is now too remote to be contaminated by human pollutants, but pelagic seabirds such as albatrosses are probably less at risk than coastal species (Ohlendorf 1993).

Exposure of seabirds to trace elements can be measured by analysing their food, water and air or by assays of eggs and selected organs of the body. The liver is appropriate for assessing recent exposure of birds to most elements, but kidneys are also suitable. Differences in liver : kidney ratios of cadmium, mercury and selenium discriminate acute from chronic exposure to chemical forms of elements in the diet, and relative exposure in comparison to normal background levels. Lead is measured in blood, kidneys or bones; kidneys are good indicators of recent exposure and bones are considered the best index of lifelong exposure.

Certain elements may be deposited in growing feathers which thus reflect exposure at the time and place of feather replacement. However, the moult cycles of albatrosses are so prolonged that they would only be useful indicators of chronic or long-term exposure.

Ohlendorf's (1993) studies in the North Pacific indicated that trace elements affected seabirds indirectly through their habitat as well as causing chronic toxicity, but were unlikely to cause mortality of seabirds through their acute effects.

Cadmium: concentration in the ocean are comparatively low (Dyrssen *et al.* 1972). Very high levels of cadmium occur in *Themisto* amphipods, but although these crustaceans are eaten they are not numerous in albatross diets. Squid are more likely sources of cadmium in albatrosses. All albatross tissue samples from the South Atlantic and New Zealand seas contained cadmium. Some had total levels within the range known to have caused kidney damage in other seabirds, but with lower levels in livers than kidneys it was believed that there had been no toxic effects (Muirhead & Furness 1988, Lock *et al.* 1992).

Copper: large amounts of copper waste enter the oceans and the natural concentration is relatively high (Dyrssen *et al.* 1972). Eight albatross species from the South Atlantic and seas around New Zealand had copper in their livers and kidneys. One Grey-headed Albatross also had a high level in its feathers. There was no evidence of toxic effects (Muirhead & Furness 1988, Lock *et al.* 1992).

Zinc: large quantities of zinc waste enter the ocean and the natural concentration is higher than any of the other metals mentioned here (Dyrssen *et al.* 1972). Eight albatross species from the South Atlantic and seas around New Zealand had zinc in their livers and kidneys. One Grey-headed Albatross also had zinc in its feathers. There was no evidence of toxic effects (Muirhead & Furness 1988, Lock *et al.* 1992).

Mercury: comparatively little mercury enters the oceans, but it is widely dispersed through atmospheric transport and is an especially toxic metal. About half of the mercury in the atmosphere is released by industry and enters the sea as rain or dust. In fish, squid and other marine organisms, it is present mostly as organic methyl mercury which is many times more toxic to vertebrates than inorganic mercury (Dyrssen *et al.* 1972). Cetaceans whose diets are high in methyl mercury retain only low concentrations; most of it they detoxify into inorganic mercury that can be deposited in organs such as the liver and kidneys.

Mercury has been found in the fat of Laysan and Black-footed Albatrosses in the North Pacific (Fisher 1973) and in the livers of eight albatross species from around the Southern Ocean. The levels in Wandering and Sooty Albatrosses are very high (Thompson & Furness 1989, Lock *et al.* 1992). In birds, mercury binds to the sulphur atoms of amino acids that make up feather keratins. In this inert form it is no longer physiologically active and feathers are eventually moulted. Most birds moult annually and deposition of mercury in feathers is an effective way of excreting dietary methyl mercury, but deferred breeding in albatrosses is associated with interrupted and protracted moulting cycles. Less of their dietary mercury can be disposed of in feathers, therefore they have naturally high levels of stored inorganic mercury in their liver and kidneys. Comparisons of mercury levels in feathers of albatrosses collected up to 100 years ago with those in present day albatrosses show very little increase.

It has been suggested that albatrosses are among seabirds that are less sensitive to mercury toxicity than are terrestrial and freshwater species (Thompson & Furness 1989, Lock *et al.* 1992, Hindell *et al.*1995).

In southern hemisphere albatrosses, mercury concentrations varied between species and tended to be higher in New Zealand albatrosses and lowest in those of the Falklands and South Georgia (Thompson *et al.* 1993).

Lead: although lead is a serious pollutant of some surface waters, particularly in coastal locations of the northern hemisphere, its natural concentration in the sea is comparatively low (Dyrssen *et al.* 1972). Assays of lead have been made from the feathers of one Grey-headed Albatross and the bones of seven Salvin's Albatrosses. In all samples levels were very low (Lock *et al.* 1992). A notable exception occurred on land at Midway Atoll in 1982 and 1983. Some Laysan Albatross fledglings ingested flakes of weathered lead paint from buildings and died of lead poisoning. Their livers contained about 20 times the levels of lead found in other fledglings (Sileo & Fefer 1987).

Chlorinated hydrocarbons

Synthetic organochlorines have been widely used since the 1940s to kill terrestrial arthropods of commercial and public health importance. Residues such as DDE[6], a metabolite of DDT[7], persist in animal tissues and tend to accumulate in predators causing well known pathologies, including thinning of egg-shells which reduces breeding success in many birds. Contamination of marine food chains has been known since the 1960s. Winds carry aerosols over the sea and residues of applied compounds leach into rivers that carry them to the sea. Coastal waters are likely to be most contaminated, but ocean currents gradually distribute low concentrations of these robust compounds throughout the oceans.

Levels of PCBs[8] and the DDT group of chemicals in Laysan and Black-footed Albatrosses at Midway Atoll in 1969 were not high (Fisher 1973). By the early 1990s, more advanced analytical methods were in use, so measurements were not strictly comparable, but there is some evidence that contaminant levels were as high or higher than in 1969 (J.P. Ludwig pers. comm.).

PCBs, PCDDs[9], PCDFs[10], TCDD-EQs[11] and DDT compounds found in 1992–95 were of recent origin, not from long accumulation, and may have originated in run-off from malaria and crop-pest control in southeast Asia and China (J.P. Ludwig pers. comm.).

Average levels of DDT compounds in Black-footed Albatrosses were two to three times higher than in Laysan Albatrosses, but the variability was high. The 39-fold variation ensured that a few Black-footed Albatrosses in the population were seriously contaminated and likely to lay eggs with much higher levels of DDTs than the lowest levels known to produce deleterious effects in the eggs of 'typical' birds.

Egg-shells collected from both species in 1969 had revealed no decline in weight nor increase in breakage over museum specimens taken between 1912 and 1913 (Fisher 1973). In 1992–95, significantly more eggs of the Black-footed Albatross were crushed soon after laying than of the Laysan Albatross (Ludwig *et al.* 1997, J.P. Ludwig pers. comm.). A new technology (*beta*-backscatter) was used to measure egg-shell quality in large samples of live eggs in the field. Black-footed Albatross eggs were three to four percent thinner than the early museum specimens, but those of Laysan Albatrosses remained unchanged (Ludwig *et al.* 1997).

Some organochlorines are potent oestrogen mimics that may interfere with normal development. In large samples, low frequencies of subtle changes occur in chicks. Although there has been intensive surveillance at Midway since the late 1950s, crossed-bill abnormalities in chicks were not reported until the late 1970s (Ludwig *et al.* 1996; 1997, J.P. Ludwig pers. comm.).

Flotsam

> 'Small animal communities build up around any kind of flotsam in the open ocean, including plastic. Crabs and barnacles settle out and grow; spawning fish attach their eggs; small fish regularly associate...I have seen Laysan Albatross "surface seizing" around piles of abandoned fishing gear in the mid-Pacific, presumably taking advantage of these associations for opportunistic feeding.'
>
> (Pitman 1988).

[6] 1,1-dichloro-2,2'-bis-*p*-chlorophenyl-ethylene.
[7] 1,1,1-trichloro-2,2'-bis-*p*-chlorophenyl-ethane.
[8] polychlorinated biphenyls.
[9] polychlorinated dibenzo-*p*-dioxins.
[10] polychlorinated dibenzofurans.
[11] 2,3,7,8-tetrachlorodibenzo-*p*-dioxin equivalents.

Albatrosses pick up all sorts of floating objects from the sea. Before human artifacts appeared on the oceans, quite large indigestible objects, such as candle nuts *Aleurites molluccana*, were being swallowed by albatrosses and passed on to their chicks. Apparently healthy adults and chicks sometimes contain pieces of pumice. It has been regurgitated at the nests of the Northern Royal Albatross in New Zealand and in 1966, most of 100 Laysan Albatross chick carcasses found on Southeast Island of Pearl and Hermes Reef, had quantities of pumice, but it was not clear whether the chicks had actually died because of it (Kenyon & Kridler 1969).

Production of plastics in the United States increased from almost three million tons in 1960 to nearly 48 million tons in 1985 and reflects a much greater worldwide trend. Floating plastic on all oceans has increased accordingly and includes both raw plastic in the form of small pellets and the weathered, fragmented remains of manufactured items such as bottles, disposable cigarette lighters and childrens' toys discharged as garbage. Fisheries are responsible for discarded tangles of net and lengths of monofilament line. The North Pacific Ocean is said to have the largest quantities, mostly discharged from Japanese and US (California) petrochemical industries or vessels, and transported downstream across the ocean in the various currents of the North Pacific gyre (Robards *et al.* 1997).

All albatrosses are at risk from ingesting plastics, which contain a wide variety of toxic chemicals; grinding in the gizzard may lead to increased assimilation.

Among five species investigated off South Africa, 11% of Black-browed Albatrosses sampled had an average of 0.2 particles of plastic per bird (Ryan 1987). In the North Pacific, many more Laysan and Black-footed Albatrosses carry much larger quantities. Black-footed Albatrosses pick up smaller particles than Laysan Albatrosses. Both species ingest more plastic from Hawaiian waters than from the Gulf of Alaska and in both areas Laysans ingest four to five times the weight ingested by Black-footed Albatrosses (Gould *et al.* 1997, Ogi 1994a). When fed to chicks, the material accumulates in the gizzard (ventriculus) and stomach (proventriculus), but is eventually regurgitated about the nest before fledging. At Midway Atoll in 1987, 27% of Laysan Albatross chicks in a sample of 350 contained more than 22 cm^2 of plastic. They had significantly lower fledging weights than chicks with less plastic. In a sample of 142 Black-footed chicks, 15% had large plastic loads and their fledging weights remained normal.

There may be significant annual variation in the amount of plastic ingested. The mean volume ingested by Laysan Albatross chicks in 1986 (46 cm^2) was much greater than in 1987 (5 cm^2) and was correlated with lowered survival in 1986 (Sievert & Sileo 1993). However, there was no significant difference between 1994 and 1995 (Auman *et al.* 1997).

All 174 dead chicks examined at Midway in 1986 and 1987 contained some plastic, but only one death could be attributed directly to it (Sievert & Sileo 1993). In 1994 and 1995 at Midway, the weights of ingested plastic in 171 Laysan Albatross fledglings found dead were compared with those in 80 live fledglings (controls) injured by motor vehicles on roads of the old air base. Only six of the 251 specimens did not contain any plastic, but 90 (36%) contained the smallest measurable quantities (1–5 g), which were almost as frequent in the controls (44) as in the dead fledglings (46). In specimens containing loads of 10g and above, the dead fledglings in both years contained about twice as much ingested plastic as the controls, and the body weight of dead fledglings was significantly lower than that of the controls (Auman *et al.* 1997).

Most albatrosses naturally accumulate squid beaks which are inert, but have sharp points and edges. These are ejected around nests by adults and chicks. Hungry chicks of all species pick up objects around their nests and some get ingested. On Midway, these may include prickly ironwood cones or even coral rag. Once a colony becomes littered with plastic artifacts, they too will be picked up – in addition to whatever plastic comes with food delivered by parents – so there may be an exponential effect.

There is reason to believe that ingested plastic may reduce the amount of food and water a chick can accept. In subtropical climates, albatross chicks have to use body water for evaporative cooling and an adequate water reserve in the gut is an insurance against dehydration (Sievert & Sileo 1993).

Albatrosses do not collect nesting material at sea, so they are spared the dangers of discarded net that afflict gannets and cormorants. However, they are sometimes seen trailing lengths of monofilament line from their bills, evidently attached to embedded hooks. In 1990–91 at Midway Atoll, Black-footed Albatrosses were twice as likely to acquire lengths of line as Laysan Albatrosses (Gould *et al.* 1997).

Oil

Albatrosses, like other seabirds, have been at risk from oil pollution ever since ships started burning and transporting oil. The shores and coastal waters of South Georgia adjacent to whaling stations were once heavily oiled, but seabirds breeding on offshore islands very rarely showed any oil marks. Thirty years after the whaling industry ceased, fishing returned to the Scotia Sea. Seven oiled penguins have been found at South Georgia (Reid 1995), but albatrosses have apparently not been seriously contaminated.

CONSERVATION

In December 1989, the United Nations General Assembly adopted a consensus resolution (44/225) expressing serious concern at the indiscriminate nature of driftnet fishing (Northridge 1991). The Convention on the Conservation of Antarctic Marine Living Resources (CCAMLR), set up in 1982 to manage the marine ecosystem south of the Antarctic Convergence, endorsed the UN resolution with its own Resolution 7/IX (1990), which declared that there should be no expansion of large-mesh driftnet fishing into the Convention Area. Continued publicity and protest at the deaths of large numbers of cetaceans and seabirds sustained international diplomatic action. On 31 December 1992, a moratorium on driftnetting was signed at the United Nations. However some nations, including Taiwan, have continued to use driftnets on the high seas (Gales 1993).

In 1992, the use of net-sonde cables was banned on all trawlers fishing within the New Zealand EEZ (Commercial Fishing Regulation 1986, Amendment No.13) and from 1994–95 a similar prohibition was declared within the CCAMLR Convention Area (Conservation Measure 30/X 1991).

In 1991, after a Black-footed Albatrosses had been found with monofilament line around its legs and several ringed albatrosses had died on longlines, the US Western Pacific Regional Fisheries banned longline fishing altogether within 50 nautical miles (93 km) of the NW Hawaiian Islands (McDermond 1991). Co-operation between Australian conservation officers and Japanese longline fishermen and fisheries scientists had already produced gear and strategies for reducing seabird mortality on longlines (Brothers 1991, Murray *et al.* 1993, Cherel *et al.* 1996). Some of these measures were employed by increasing numbers of Japanese vessels during the early 1990s, but the SBT fishery had become more diverse, with less co-operative nations. Intelligence from vessels remained poor and other variables intruded (Klaer & Polacheck 1998).

Conservation Measure 29/X introduced by CCAMLR in 1991 to prevent incidental seabird mortality during longline fishing in the Convention Area was followed in 1993 by an updated 29/XI and endorsed in 1994 by 29/XIII (Sabourenkov 1994). They were notably successful in reducing albatross mortality by about 70%. Elsewhere albatrosses continue to be killed.

Soon after the confirmation of longline mortality (Brothers 1991), the Australian National Parks and Wildlife Service[12] commissioned a report on the worldwide status of albatrosses (Gales 1993). This led to the First International Workshop on Albatross-Fisheries Interactions at Hobart, Tasmania in 1995 (Alexander *et al.* 1997). It made the following urgent recommendations to reduce the numbers of albatrosses killed:

1. longlines should be set only at night
2. effective bird lines should be deployed when setting longlines
3. mechanical bait-throwing machines should be adopted
4. offal should be discharged on the opposite side from the line
5. lines should be weighted
6. only thawed bait should be used.

These measures were already in operation on some vessels and were promoted with much optimism. In the AFZ, wildlife biologists have assessed their effectiveness over several years, but the analyses suffered from inconsistent

[12] Now the Australian Nature Conservation Agency (ANCA).

methodologies (Brothers *et al.* 1998b&c). Some were unavoidable; ship-owners and masters, not researchers, decide where and when to fish. Bad weather and deck operations upset observation protocols. The design of Bird lines varied from ship to ship and during the many hours taken to set longlines, unpredictable circumstances influenced their effectiveness. They have been described as no more than glorified scarecrows by manufacturers keen to sell newly invented tube launchers (Pearce 1998).

Setting longlines at night reduces catch rates of birds, but seasonal and lunar influences are apparent (Klaer & Polacheck 1998). Well thawed baits are less buoyant than frozen fish. They catch fewer birds, but many are set when only half-thawed. Baitfish have to sink much faster to escape being taken by albatrosses. This can be achieved by attaching 4 kg weights to lines at intervals of 40 m, but raises other practical problems. Evaluation of advocated measures needs to be perfected in order to be convincing. Governments have powers to enforce the adoption of such measures within their EEZs and some have already done so. They may also temporarily close known albatross feeding grounds to fishing vessels or prohibit fishing altogether in some years.

Member nations of CCAMLR fishing in the Southern Ocean abide by agreed measures and carry independent observers on all vessels. The fisheries of non-member nations are not so constrained; there are many of them and great expanses of ocean, outside EEZs and the CCAMLR area, where seabird mortality on fishing gear is not monitored. Only about five percent of vessels in fisheries known to interact with albatrosses are monitored by independent observers.

Some of these fisheries might be influenced through the 1979 Convention on the Conservation of Migratory Species of Wild Animals, usually known as the the Bonn Convention. In 1995, there were 47 contracting parties to the Convention and nine signatories. Parties to the Convention are required to co-operate with and support research relating to migratory species. They should also work towards providing immediate protection for designated species and work towards concluding agreements for the conservation and management of listed species. By the end of 1996, Steller's Albatross and the Amsterdam Wandering Albatross were designated endangered species and given full protection in Appendix 1 of the Convention. In 1997, the rest of the southern hemisphere albatrosses were listed in Appendix 2 as having unfavourable status requiring international agreements for their conservation and management. However, the listed parties to the Convention did not include Japan, Korea, Poland, Peru, Russia, Taiwan and the Ukraine, all of whom are fishing nations active in oceans where albatrosses are common.

In 1999, the United Nations' Food and Agricultural Organisation (FAO) unanimously adopted an *International Plan of Action for Reducing Incidental Catch of Seabirds in Longline Fisheries.* It is now up to FAO member nations to put these proposals into operation (Cooper 1999).

All albatross breeding grounds are in the jurisdiction of one government or another. In international waters, the 'nationality' of some seabirds can be clearly established. Northern and Southern Royal Albatrosses, for example, all breed on New Zealand territory; they are New Zealand birds. Similarly all Steller's Albatrosses are Japanese birds. Respective governments might be encouraged to assume more direct responsibility for their seabirds when in unprotected international waters (Haward *et al.* 1997, Gales 1997).

Fisheries hazards are of grave concern. Seabirds from large populations are still being killed in dramatic numbers, but the death of one Amsterdam Wandering Albatross is more crucial than the loss of many Laysan or Black-browed Albatrosses. Sooner or later fishing technology will advance and the longline emergency will recede. However, the economic and political power of marine industries are such that hazards of one sort or another will probably reappear in different guises. Impressive efforts are being made in marine conservation, but the pristine ocean has gone; with or without help, albatrosses have to cope as best they can with seas that are increasingly managed by man and for man.

18
THE MARINER SYNDROME

'...no music could be more apt to that romantic desolation. The very spirit of the place! A chord within responds, a chord held silent for a thousand decades while the northern ice-caps slowly died, and man set foot upon the path that may lead up from barbarism. Surprising that a Sooty albatross calling to his mate in the cliffs can send such ideas running through the mind.'

Leonard Harrison Matthews[1]
at South Georgia c.1926

The *Rime of the Ancient Mariner* is one of the acknowledged classics and landmarks of English literature. At the time he wrote it, Samuel Taylor Coleridge (1798) had not yet been on a sea voyage and he never saw an albatross[2]. In his own words, the poem was a work of 'pure imagination'.

The beginning of the poem is well known. An old man waylays a young guest rushing to a wedding and makes him listen to a long fantastic tale about a tragic voyage of discovery. Although an albatross is central to the narrative, it is mentioned in only 13 out of 141 verses, so to avoid straying far from the theme of this book, I have selected just the verses that tell the albatross story:

And now there came both mist and snow,
And it grew wondrous cold:
And ice, mast-high, came floating by,
As green as emerald. (13)[3]

The ice was here, the ice was there,
The ice was all around:
It cracked and growled, and roared and howled,
Like noises in a swound ! (15)

At length did cross an Albatross,
Through the fog it came;
As if it had been a Christian soul,
We hailed it in God's name. (16)

It ate the food it ne'er had eat,
And round and round it flew,
The ice did split with thunder-fit;
The helmsman steered us through ! (17)

And a good south wind sprung up behind;
The Albatross did follow,
And every day, for food or play,
Came to the mariners' hollo ! (18)

[1] (1951)
[2] When Coleridge eventually made a sea voyage, he saw only a single wild duck on the water and dismissed the experience with some scorn!
[3] Numbers in parentheses identify verses in the original.

In mist or cloud, on mast or shroud,
It perched for vespers nine;
Whiles all the night, through fog-smoke white,
Glimmered the white Moon-shine. (19)

"God save thee, ancient Mariner !
From the fiends, that plague thee thus ! –
Why look'st thou so ?" – With my cross-bow
I shot the Albatross. (20)

And I had done a hellish thing,
And it would work 'em woe:
For all averred, I had killed the bird
That made the breeze to blow.
Ah wretch ! said they, the bird to slay,
That made the breeze to blow ! (23)

Nor dim nor red, like God's own head,
The glorious Sun uprist:
Then all averred, I had killed the bird
That brought the fog and mist.
'Twas right, said they, such birds to slay,
That bring the fog and mist. (24)

The fair breeze blew, the white foam flew,
The furrow followed free;
We were the first that ever burst
Into that silent sea[4]. (25)

Down dropt the breeze, the sails dropt down,
'Twas sad as sad could be;
And we did speak only to break
The silence of the sea ! (26)

All in a hot and copper sky,
The bloody Sun, at noon,
Right up above the mast did stand,
No bigger than the Moon. (27)

Day after day, day after day,
We stuck, nor breath nor motion;
As idle as a painted ship
Upon a painted ocean. (28)

Water, water, every where,
And all the boards did shrink;
Water, water, every where,
Nor any drop to drink. (29)

Ah ! well-a-day ! what evil looks
Had I from old and young !
Instead of the cross, the Albatross
About my neck was hung. (34)

[4] Pacific Ocean

Scholars have sought to penetrate the allegory and translate it into the emotional struggles and political events of Coleridge's life (Lowes 1927), but in P.J. Keane's (1994) words it has remained '...magnificently resistant to any attempts at reductive interpretation.'

The power of the idiom has turned the bird into a metaphor from which it has never escaped. More than any natural history it has determined the popular conception of *the albatross.* The image that has passed into the English language relates to a primitive penance imposed by fearful, superstitious seamen upon their erring shipmate. They suffer for his misdeed and he is not released from the penance until they are all dead, and in bitter loneliness he sees beauty and a common bond of life with awesome creatures of the sea:

> The self-same moment I could pray;
> And from my neck so free
> The Albatross fell off, and sank
> Like lead into the sea. (66)

At the time the poem was written, the Southern Ocean was still largely unknown, the realm of explorers, sealers and merchant venturers. Accounts of contemporary and early voyages were popular reading. The epic circumnavigations of Captain Cook were recent events; ships had sailed Antarctic seas but no-one had yet reached the icy continent. Curious seabirds figured frequently in published journals and fed the imagination of creative artists.

On the evening of 13 November 1797, Coleridge, accompanied by his friends William and Dorothy Wordsworth, set out from Alfoxden in Somerset to walk over the Quantock Hills to the fishing villages of Watchet and Dulverton. While following the moonlit moorland paths, the structure of the work came together. Coleridge left no recollections of the walk, but Wordsworth's sister, Dorothy wrote years later that her brother once said that he himself had suggested that the the poem should follow the events of some crime. Wordsworth had recently read Captain George Shelvocke's (1726) account of his voyage around the world (1719–22). In a storm near Cape Horn the following incident was related:

> '...we had not had sight of one fish of any kind, since we were come to the Southward of the streights of le Mair, nor one sea-bird, except a disconsolate black Albatross, who accompanied us for several days, hovering about us as if he had lost himself, till Hatley, (my second Captain) observing, in one of his melancholy fits, that this bird was always hovering near us, imagined, from his colour, that it might be some ill omen. That which, I suppose, induced him the more to encourage his superstition, was the continued series of contrary tempestuous winds, which had oppress'd us ever since we had got into this sea. But be that as it would, he, after some fruitless attempts, at length, shot the Albatross, not doubting (perhaps) that we should have a fair wind after it.'

Shelvocke's black albatross could have been a juvenile Wandering Albatross, a Light-mantled Sooty Albatross or even a juvenile giant petrel (Brown 1981), but the poet was unlikely to have been bound by the one event (Bourne 1982a).

From an early age, Coleridge was an avid reader. As a schoolboy at Christ's Hospital, London, he was given a subscription to a nearby library and read, he said, through the whole catalogue of books, whether or not he understood them. The Mathematical School at Christ's Hospital was run by William Wales, who had been the astronomer and meteorologist on the *Resolution,* during Captain Cook's second (1772–75) voyage. He was a dedicated, popular master. One of his tasks was to encourage boys to adopt naval careers and instruct them in navigation. Bernard Smith (1956) found that parts of the Rime corresponded to events in Cook's voyage. He believed that Coleridge remembered Wales's classroom anecdotes and found books of the voyage in Wales's library. It is a plausible, if overstated, case. Coleridge was not a King's scholar and never showed any inclination towards a career at sea. In the Upper Grammar School, he was taught by Wales one afternoon a week, but he spent most of his time reading poetry, classics and Platonic philosophy. The dominant influence in those years was not William Wales, but the sadistic headmaster, James Bowyer (Holmes 1989).

Many years later, Coleridge's publisher told him that most copies of *Lyrical Ballads* had been sold to seamen, who had heard of the *Ancient Mariner* and concluded that it was a book of seafaring songs (Patmore 1917)! The belief that seamen had an aversion to killing albatrosses dates from the poem (Jameson 1958). Sixteenth century mariners had, indeed, much to be fearful of and little to guide them, but the omens they imagined were fickle and the poem correctly reflects the swings of mood in an apprehensive crew. Albatrosses were killed with impunity often enough, before and after the poem, to demolish any taboo. Among a lusty crew on board a four-master bound for Australia in the mid-19th century a healthy disrespect was evident.

A dead loss was the albatross
The 'Ancient Mariner' slew;
The modern 'Tar'[5], acute far,
Makes him into a stew.

Day after day did that silly old man
Over his dead bird cry;
The 'Mid,'[6] more ripe, of his wings makes a pipe,
And smokes therein 'Birds-eye'[7].

(Horton 1860)

Long after the days of wind-jammers, the myth occasionally revives. In 1959, the *Calpean Star*, bound from a South Georgia whaling station to Norway, had a run of misfortunes and eventually put into Liverpool with engine trouble. A Wandering Albatross destined for a German zoo then died on board and the crew insisted on being paid off immediately (Anon 1959).

Sense of the ridiculous saves most modern readers from taking the myth too seriously. Richard Lister (1965) turned it on end in the manner of Lewis Carrol and Edward Lear.

The Albatross

I sailed below the Southern Cross
(So ran the sailor's song).
A pestilential albatross
Followed us all day long.

The creature's aspect was so grim,
And it oppressed me so,
I raised ... then, on a sudden whim,
I lowered my crossbow.

The weather grew exceeding thick,
The sullen tempest roared.
A dozen of the crew fell sick,
The rest fell overboard.

The skies were so devoid of light
We could not see to pray.
The donkeyman[8] went mad by night
The second mate by day.

[5] sailor
[6] midshipman
[7] a brand of tobacco
[8] the seaman in charge of a 'donkey-engine', a small steam engine on deck serving derricks.

We set the live men swabbing decks,
The dead men manned the pumps.
The cabin steward changed his sex,
The captain had the mumps.

The cargo shifted in the hold,
The galley boiler burst.
My hair turned white, my blood ran cold,
I knew we were accurst.

I helped the purser dig his grave
On the deserted poop.
I leaped into the foaming wave
And swam to Guadeloupe.

And there (he said) I nibbled moss
Beside the stagnant lake ...
I should have shot the albatross,
That was my big mistake.

Oliver Goldsmith included 'the Albatross, the first of the Gull Kind' in his 1774 *History of the Earth and Animated Nature.* As a contemporary of Joseph Banks and three years after Cook's first voyage, he should have been able to include something more credible than a translation of the Dutch writer, Abraham van Wicquefort, writing over a century earlier:

> 'As these birds, except when they breed, live entirely remote from land, so they are often seen, as it should seem, sleeping in the air. At night, when they are pressed by slumber, they rise into the clouds as high as they can; there, putting their head under one wing, they beat the air with the other, and seem to take their ease. After a time, however, the weight of their bodies, only thus half supported, brings them down; and they are seen descending, with pretty rapid motion, to the surface of the sea...'

This absurd notion[9] became so entrenched that it was still in circulation 87 years later and cited by a scholar of such elevated reputation as Sir John Herschel (1861) in his *Physical Geography.* Among others attracted by sleeping albatrosses was Thomas Moore (1817), who incorporated one as a single incongruous incident in his epic poem *Lalla Rookh*:

Around its base the bare rocks stood,
Like naked giants, in the flood,
As if to guard the Gulf across;
While on its peak, that braved the sky,
A ruined temple towered, so high
That oft the sleeping albatross
Struck the wild ruins with her wing,
And from her cloud-rock'd slumbering
Started – to find man's dwelling there
In her own silent fields of air!

This immensely popular romance, created out of travellers' tales, reflected a growing fascination on the part of the public for Arabian and Persian literature that eventually led to translations such as the *Rubáiyát of Omar*

[9] Still a speculation for species like swifts, martins and terns, but without strong evidence (Amlaner & Ball 1983).

Khayyám. Fantastic albatrosses were readily created from other seabirds. J.W. McCrindle (1897) identified them with the great flock of *Souspha* birds encountered by the Egyptian monk Cosmas Indiopleustes[10] during his 6th century voyage into the northwest Indian Ocean.

Annie Corder may have sailed to Australia; she knew there was tussock grass on albatross islands and had picked up the other myth – that albatrosses deserted their young during the southern winter. Wonder at females behaving in this manner pervades of her romantic poem (Layard 1897).

The Wandering Albatross

In those lone rocky islands tempests torn,
And girded by the waves and raging wind,
A melancholy constant cry forlorn
Rises like protest to a fate unkind.
Are these re-incarnate
Spirits of woeful fate,
Whose restless, roving souls must soon or late
Some peace and comfort find ? (1)

The harsh rank herbage is a sweeter rest
Than all the glorious range of growing bloom;
Those barren shelves of rigid rock their nest,
Above the sorrowful waves' portentous boom;
Seaweed and smell of brine,
Glimpse of the underline
Of seas, that lash and fret where storms combine,
This is their living room. (3)

And here they build, and here they raise their kind;
That solitary one whose early cry
For months unanswered, frets the moaning wind:
Unfed, untended, and whose lullaby
Is angry shaken thunder,
Winds that rave and sunder,
When, with a farewell solemn scream of wonder
It sees its mother fly. (5)

Who bids her start on that long pilgrimage ?
Who calls, and never will be disobeyed ?
Who holds her in this thrall of vassalage ?
That with consent and spirit undismayed
She takes to eager flight,
Through strain of day and night,
And in bewildering darkness to the light
She wanders unafraid. (6)

Whither, she cares not; where, she does not mind,
The storm her fellow and the waves her kin;
Along the furrows of the ploughing wind,

[10] Latin: Indian navigator.

On topmost heights of reeling seas, to win
Where rolling to the shore,
The headlong waters roar,
And where the sea surrounds for evermore
The whole wide world within. (8)

And in these far-off wanderings does she dream
Of that lone nestling on the sloping plain
Mid yellow tussock grass ? And does time seem
Or long, or weary ? Is it with joy or pain,
Or with constrainèd breast,
She sees once more that nest,
And leaves the tumult and the brave unrest
For home and love again ? (10)

Farewell, my friend ! I give the luck and cheer;
Seafaring hearts fare always well at sea,
And love makes all things easy. Far or near
I wish thee good, when'er I think of thee;
May winds blow strong for thee,
May waves roll long for thee,
The thunderous sea make chant and song for thee
Wherever thou may'st be. (14)

At 20 years of age, Charles Baudelaire (1821–67) was a dandy of cultivated sensibility living with other aspiring poets in the Latin Quarter of Paris. His well-to-do parents, horrified at what they perceived as a dissolute life, persuaded him to travel and in 1841 he sailed for India. On the long voyage via the Cape he was homesick, bored and isolated from his bourgeois travelling companions. In the captured albatross he saw an allegory of his own condition.

Overcome by growing melancholia, Baudelaire left the ship at the Ile Réunion and returned to France. Back in Paris, he adopted the life of style and taste for which he became legendary. His poem *L'Albatros* was written in 1843–1846 and included in the first (1857) edition of *Les Fleurs du Mal*, which was banned. It then appeared in *La Revue française* (1859) and later in the second (1861) edition of *Les Fleurs du Mal* (Baudelaire 1973, Hyslop & Hyslop 1957).

L'Albatros

Souvent, pour s'amuser, les hommes d'équipage
Prennent des albatros, vastes oiseaux des mers,
Qui suivent, indolents compagnons de voyage,
Le navire glissant sur les gouffres amers.

A peine les ont-ils déposés sur les planches,
Que ces rois de l'azur, maladroits et honteux,
Laissent piteusement leurs grandes ailes blanches
Comme les avirons traîner à côté d'eux.

Ce voyageur ailé, comme il est gauche et veule !
Lui, naguère si beau, qu'il est comique et laid !
L'un agrace son bec avec un brûle-gueule,
L'autre mime, et boitant, l'infirme qui volait !

Le Poëte est semblable au prince des nuées
Qui hante la tempête et rit de l'archer;
Exilé sur le sol au milieu des huées,
Ses ailes de géant l'empêchent de marcher[11] .

Roy Campbell was another young man in rebellion. Born to wealthy parents in Durban, South Africa, he grew up opposing the accepted values of his family while cherishing an affection for the land and African people of Natal. He went to sea with whalers and had seen albatrosses. At Oxford in the early 1920s, he affected a flamboyant 'Bohemian' way of life and his highly successful poem, *The Flaming Terrapin* (1924) included an albatross theme. Campbell returned to South Africa in 1924 as an acclaimed poet. The following year, he became editor of a new arts magazine, with William Plomer and Laurens Van der Post as contributors and assistant editors. *Voorslag*[12] was conceived with almost missionary zeal to bring modern works to an intellectually stagnant community. Nothing remotely like it had appeared before in South Africa, but after the second issue the journal's financial backers started getting worried and their interference caused Campbell to resign from the editorship (Plomer 1959, Alexander 1982).

Campbell's own poem, *The Albatross*, began the first issue of *Voorslag* (June 1926). Some of its images echo lines in *The Flaming Terrapin* and were reworked in Campbell's later poems *Adamastor* and *Tristan da Cunha.* A dying albatross recalls the noble aspirations and visions of a free ranging life. It appears to reflect the achievements, frustrations and tragedies of Campbell's own life. The following selection is from the revised 1930 edition.

Stretching white wings in strenuous repose,
Sleeving them in the silver frills of sleep,
As I was carried, far from other foes,
To shear the long horizons of the deep, (1)

A swift ship struck me down: through gusty glooms
I spun from fierce collision with her spars:
Shrill through the sleety pallor of my plumes
Whistled the golden buckshot of the stars. (2)

To the dark ocean I had dealt my laws
And when the shores rolled by, their speed was mine:
The ranges moved like long two-handed saws
Notching the scarlet west with jagged line. (5)

The cliff-ringed islands where the penguins nest,
Sheltered their drowsy legions from the foam
When evening brought the cormorants to rest,
Gondolas of the tempest, steering home. (13)

Towering far up amid the red star-sockets,
I saw deep down, in vast flotillas shoaled
The phosphorescent whales, like bursting rockets,
Bore through the gloom their long ravines of gold. (15)

[11] Freely translated: Sometimes for fun, the seamen / catch albatrosses, great birds of the ocean. / Languid companions of the voyage, who follow / the ship, gliding over the cruel waves. // When hauled on deck, / these kings of the sky become ungainly and absurd. / Sadly drooping their great white wings / like oars trailing in the water. // How awkward and feeble is this winged traveller; / Once so beautiful, now comic and grotesque. / A sailor teases its beak with a pipe, / Another, limping, mimics the pathetic walk. // The poet is like a prince of the clouds, / Riding the storm and scorning danger. / Stranded on land, among jeering crowds, / His gigantic wings are a hindrance.

[12] Afrikaans: whiplash.

My stiff quills made the hurricane their lyre
Where, pronged with azure flame, the black rain streams:
Huge brindled shadows barred with gloomy fire
Prowling the red horizon of my dreams, (21)

Thick storm-clouds threatened me with dense eclipse,
The wind made whirling rowels of the stars –
Over black waves where sky-careering ships
Gibbet the moon upon their crazy spars. (22)

Through calms that seemed the swoon of all the gales,
On snowy frills that softest winds had spun,
I floated like a seed with silken sails
Out of the sleepy thistle of the sun. (24)

Broidering earth's senseless matter with my sight,
Weaving my life around it like a robe,
Onward I draw my silken clues of flight,
Spooled by the wheeling glories of the globe. (26)

No more to rise, the last sun bombs the deep
And strews my shattered senses with its light –
My spirit knows the silence it must keep
And with the ocean hankers for the night. (32)

Curiosity about albatross death also moved Arnold Wall (1943). He was a British expatriate who, in 1898, become Professor of English and History at Canterbury College, New Zealand. Long voyages to and from Europe would have given him enough time to become captivated by albatrosses.

End of the Albatross

Balancing, swinging, swaying, poising and gliding,
Up the long currents invisible easily sliding,
Down the long currents invisible smoothly descending,
How can the beautiful creature come to his ending ?

Nowhere in air or in sea can a foe assail him;
How, in the air and that sea, can his strength ever fail him ?
How can his eyes grow dim or his wings ever falter ?
How can disease find him out in that waste of water ?

The ice-born, sleet-slinging Southerly brings him no terror;
How can his lead be at fault or his compass in error ?
Sailing before the dark storm what floe can check him ?
No loud lee-shore, nor reef, nor rock can wreck him ?

His gossips from birth are the gales and the spume flying,
And the sea roaring up at the berg and the berg replying;
Wheeling and poising, balancing, poising and swaying,
Ceaselessly, ceaselessly on, ceaselessly straying.

His end must come with a mere cessation of motion –
A long and slow volplane to the level of ocean,
A closing of eyes and a folding of wings together,
And a blending, snow to snow, of the foam and the feather.

Music hall songs of the last century thrived on the grotesque and *fin de siècle* Paris was not above embellishing a reported tragedy – ridiculing a captain and crew incapable of rescuing a fellow seaman from the water.

L'amiral dit: "Quel est l'caltat
Qu'a coupé la bouée de c'temps-là ?"

Puis il ajoutit: "Timonier !
Faites-mois vite monte l'aumônier !"

L'aumônier n'fut pas long à v'nir.
Avec tout c'qui'il faut pour bénir,

Il nous dit, face au pauv'mourant,
La prière des agonisants !

Or, pendant qu'les vieux frères pleuraient,
Les sales goinfres là-bas, s'empiffraient !

Et, quand ces voraces furent repus
Quand' du pauv'bougre y'n' restit pus,

Su' la bouée, qu'sa pauv' carcasse d'os,
Alors tout' cette bande d'albatros

Dans les gros nuages noirs s'envolit,
L'coeur gai d'avoir le ventre empli !

Mon mat'lot, les sales albatros,
Ils n'lui ont rien laissé qu'les os !

Quand su' la mer ya des gros flots,
Terriens, plaignez les pauv' mat'lots ![13]
Yan Nibor (1896)

The French yachtsman, Olivier de Kersauson (1990) disliked albatrosses. He imagined them as hyaenas of the sea, a constant reminder that disaster was never far away. The '*putain de mer*'[14], he said, had eaten the eyes of his friends who had fallen overboard. This is an ancient indictment of albatrosses (Gould 1865). A severely injured or unconscious body in the water will certainly be scavenged by seabirds, but modern yachtsmen in buoyant survival suits should be able to cope with curious albatrosses. As long ago as 1881, a sailor who fell overboard from the *Gladstone* grabbed the first albatross that approached him in the water and used it as a life-buoy to keep himself afloat until the ship stopped and lowered a boat to rescue him (Buller 1888).

[13] Freely translated: The Admiral said: "what's all the noise / who's cut the buoy this time ?" / Then he shouts "Timonier ! / help me quick, bring up the chaplain !" / The chaplain doesn't take long to come, / equipped for benediction. / He chants the last rites / in front of the poor dying man ! / While old friends were weeping, / the foul gluttons below were gorging / and by the time they were finished, / the poor fellow was no more / than a pathetic sac of bones on the buoy. / Then this flock of albatrosses / vanished into the great dark clouds, / happy with their full bellies ! / Sailor, the foul albatrosses / had left nothing but bones ! / When great waves sweep the ocean, / landsmen, remember the poor sailors !

[14] sea-whore.

In *Eight Bells,* his collection of sailors yarns and ballads, Frank Waters (1927) wrote: 'Among the numerous conceits – superstitious or otherwise – current among Mariners that sailed the seas in the days when Wind Jammers were supreme as carriers in World Trade on the seven seas, none was more evident than the pretended belief prevailing among Mariners that when old sailors died they were transmogrified into Albatross.'

Brocky Burns and old Bill Clark

In southern latitude fifty four
A thousand miles from the nearest shore,
Through foam flecked seas and dazzling spray
The bark *Triumphant* carved her way.
With every shred of canvas spread
Through the bounding billows she forged ahead
Logging twelve knots, at the very least,
And steering a course about east southeast. (1)

Early one bleak cold wintry morn,
Just as the day was beginning to dawn,
Two Albatross flew up from aft
And hovered around our gallant craft.
Sometimes low and sometimes high
Around our vessel them birds did fly.
Inspecting the rigging sails and spars
With the competent knowledge of Old Jack Tars. (2)

And as them birds went sailing by,
Something about them caught my eye,
And as I gazed I grew surprised,
For both of them I recognized.
I knew by a certain peculiar mark
That one of them was old Bill Clark.
And, by its graceful movements and rapid turns
I knew that the other was Brocky Burns. (3)

Now both of them some years ago
Were sailor men that I well did know,
Stout hearted seamen as brave and true
As ever sailed the ocean blue.
Neither of them were wont to brag
Though sailed neath many a foreign flag
On English, dago dutch, and Yanks
All wooden craft, not iron tanks. (4/6)

Said Brocky Burns, to old Bill Clark,
"Just cast your eyes on yonder bark,
And tell me Bill, does it not look good,
To see a vessel that's built of wood ?"
Says Bill to Brocky, "Times were good
When we sailed on Clippers all built of wood
With clipper bows, and handsome sterns."
"Ah them were the packets." quoth Brocky Burns. (10)

Ships nowdays, from trucks to keel,
Are built of iron, cement and steel,
From the bow-sprit, aft to the lazzarette hatch,
There is not enough wood to build a match.
Then old Bill gave his tail a swing,
And Brocky, fluttered his starboard wing,
And a doleful croak came from each mouth
As they flew away to the frigid south. (11)

John Masefield was just 16 years old when he rounded Cape Horn in the winter of 1894 as an apprentice on the four-masted barque *Gilcruix*. It was his one voyage, and part of the unforgettable experience was to see ocean birds with the eyes and imagination of 19th century seamen. *Sea-change* was one of many poems written after the bitter hardships of that voyage had mellowed and were being honed into a romantic view of the sea and seamen that became one of the hallmarks of his poetry. It was published in his first book, *Salt-water Ballads* (1902).

Sea-change

'Goneys an' gullies an' all o' the birds o' the sea
They ain't no birds, not really' said Billy the Dane.
'Not mollies, nor gullies, nor goneys at all,' said he,
'But simply the spirits of mariners livin' again. [1]

'Them birds goin' fishin' is nothin' but souls o' the drowned,
Souls o' the drowned an' the kicked as are never no more
An' that there haughty old albatross cruisin' around,
Belike he's Admiral Nelson or Admiral Noah. [2]

'When freezing aloft in a snorter, I tell you I wish –
(Though maybe it ain't like a Christian) – I wish I could be
A haughty old copper-bound albatross dipping for fish
And coming the proud over all of the birds of the sea.' [4]

Appendix 1

A checklist of albatrosses.

English names (1)	Present taxonomy	Proposed taxonomy (2)
Black-browed Albatross (BBA)	*Diomedea melanophrys melanophrys*	*Thalassarche melanophrys* (3)
Campbell Black-browed Albatross (CBBA)	*D. m. impavida*	*Thalassarche impavida*
Grey-headed Albatross (GHA)	*Diomedea chrysostoma*	*Thalassarche chrysostoma*
Western Yellow-nosed Albatross (WYNA)	*Diomedea chlororhynchos chlororhynchos*	*Thalassarche chlororhynchos*
Eastern Yellow-nosed Albatross (EYNA)	*D. c. bassi*	*Thalassarche bassi*
Southern Buller's Albatross (SBA)	*Diomedea bulleri bulleri*	*Thalassarche bulleri*
Northern Buller's Albatross (NBA)	*D. b. platei*	*Thalassarche platei*
Tasmanian Shy Albatross (TSA)	*Diomedea cauta cauta*	*Thalassarche cauta*
Auckland Shy Albatross (AKSA)	*D. c. steadi*	*Thalassarche steadi*
Salvin's Albatross (SLA)	*D. c. salvini*	*Thalassarche salvini*
Chatham Albatross (CHA)	*D. c. eremita*	*Thalassarche eremita*
Sooty Albatross (SA)	*Phoebetria fusca*	*Phoebetria fusca*
Light-mantled Sooty Albatross (LMSA)	*Phoebetria palpebrata*	*Phoebetria palpebrata*
Wandering Albatross (WA)	*Diomedea exulans exulans/chionoptera* (4)	*Diomedea exulans*
Gough Wandering Albatross (GWA)	*D. e. exulans/dabbenena* (5)	*Diomedea dabbenena*
Amsterdam Wandering Albatross (ADWA)	*D. e. amsterdamensis* (6)	*Diomedea amsterdamensis*
Auckland Wandering Albatross (AKWA)	*D. e. gibsoni*	*Diomedea gibsoni*
Antipodes Wandering Albatross (ATWA)	*D. e. antipodensis*	*Diomedea antipodensis*
Southern Royal Albatross (SRA)	*Diomedea epomophora epomophora*	*Diomedea epomophora*
Northern Royal Albatross (NRA)	*D. e. sanfordi*	*Diomedea sanfordi*
Galápagos Albatross (GA)	*Diomedea irrorata*	*Phoebastria irrorata* (7)
Laysan Albatross (LA)	*Diomedea immutabilis*	*Phoebastria immutabilis*
Black-footed Albatross (BFA)	*Diomedea nigripes*	*Phoebastria nigripes*
Steller's Albatross (STA)	*Diomedea albatrus*	*Phoebastria albatrus*

1. Some of the English names adopted in this book differ from those recently proposed elsewhere; the geographic prefixes are a pragmatic device to preserve the well understood common names of similar taxa.
2. Based upon Nunn *et al.* (1996), Robertson & Nunn (1997).
3. *Thalassarche* Reichenbach (1850) reinstated by Nunn *et al.* (1996).
4. *chionoptera* and *exulans* are both still in use as trinomials for this taxon (Prince *et al.* 1997, Alexander *et al.* 1997).
5. *dabbenena* and *exulans* are both still in use as trinomials for this taxon (Warham 1990, Bourne & Casement 1993).
6. Roux *et al.* (1983) described this albatross as a new species; but its status was later contested (Bourne 1989; Vuilleumier *et al.* 1992).
7. Genus *Phoebastria* Reichenbach (1850) reinstated by Nunn *et al.* (1996).

Appendix 2

Latin taxa of mammals, birds, fish and plants appearing in the text only as English names.

MAMMALS

rice rat	*Oryzomys albigularis*
black rat	*Rattus rattus*
brown rat	*Rattus norvegicus*
Pacific/Polynesian rat	*Rattus exulans*
house mouse	*Mus musculus*
dog	*Canis familiaris*
Arctic fox	*Alopex lagopus*
Pategonian fox	*Dusicyon griseus*
Falkland fox (Warrah)	*Dusicyon antarcticus australis*
stoat	*Mustela erminea*
ferret	*Putorius putorius*
otter	*Lutra canadensis*
cat	*Felis catus*
sealion	*Otaria* spp.
Antarctic furseal	*Arctocephalus gazella*
leopard seal	*Hydrurga leptonyx*
pig	*Sus scrofa*
guanaco	*Llama guanicoe*
reindeer	*Rangifer tarandus*
cattle	*Bos taurus*
goat	*Capra hircus*
mouflon (wild sheep)	*Ovis aries*
southern bottlenose whale	*Hyperoodon planifrons*
sperm whale	*Physeter macrocephalus*
killer whale	*Orcinus orca*
long-finned pilot whale	*Globicephala melaena*
humpback whale	*Megaptera novaeangliae*

BIRDS

King Penguin	*Aptenodytes patagonicus*
Erect-crested Penguin	*Eudyptes sclateri*
Rockhopper Penguin	*Eudyptes chrysocome*
Northern Giant Petrel	*Macronectes halli*
Southern Giant Petrel	*Macronectes giganteus*
Northern Fulmar	*Fulmarus glacialis*
Cape Pigeon (Petrel)	*Daption capense*
Snow Petrel	*Pagodroma nivea*
Grey-faced Petrel	*Pterodroma macroptera gouldi*
Bermuda Petrel	*Pterodroma cahow*
Grey Petrel	*Procellaria cinerea*
White-chinned Petrel	*Procellaria aequinoctialis*
Great Shearwater	*Puffinus gravis*
Kerguelen Petrel	*Lugensa brevirostris*
White-faced Storm Petrel	*Pelagodroma marina*
pelicans	*Pelecanus* spp.
Northern Gannet	*Sula bassana*
Australian Gannet	*Sula serrator*
boobies	*Sula* spp.
Imperial/King Cormorant	*Phalacrocorax atriceps*
Kerguelen Cormorant	*Phalacrocorax verrucosus*
Guanay	*Phalacrocorax bourgainvillii*
frigatebirds	*Fregata* spp.
Upland Goose	*Chloephaga picta*
Kelp Goose	*Chloephaga hybrida*
Mallard	*Anas platyrhynchos*
Wedge-tailed Eagle	*Aquila audax*
White-breasted Sea Eagle	*Halieaetus leucogaster*
Galápagos Hawk	*Buteo galapagoensis*
Striated Caracara	*Phalcoboenus australis*
Peregrine Falcon	*Falco peregrinus*
Weka	*Gallirallus australia*
Gough Island Moorhen	*Gallinula comeri*
Auckland Islands Rail	*Rallus pectoralis*
Pacific Golden Plover	*Pluvialis dominica*
Bristle-thighed Curlew	*Numenius tahitiensis*
Antarctic Skua	*Catharacta skua* spp.
Pacific Gull	*Larus pacificus*
Fairy Tern	*Gygis alba*
Oilbird	*Steatornis caripensis*
Galápagos Mockingbird	*Nesomimus macdonaldi*
Jungle Crow	*Corvus macrorhynchos*
Starling	*Sturnus vulgaris*
Redpoll	*Carduelis flammea*

FISH

southern lamprey	*Geotria australis*
mako shark	*Isurus* spp.
requiem shark	*Prionace glauca*
tiger shark	*Galeocerdo cuvieri*
pilchard	*Sardinops neopilchardus*
anchovy	*Engraulis australis*
Peruvian anchoveta	*Engraulis ringens*
lightfish	*Icthyococcus* spp.
lanternfish	(Myctophidae): *Electrona risso*, *Symbolophorus californiense*, *Lampanyctus jordani* & *L. ritteri*, *Diaphus gigas* & *D. theta*, *Notoscopelas japonicus*
Pacific cod	*Gadus macrocephalus*
southern blue whiting	*Micromesistius australis*
hake	*Merluccius hubbsi*
rattail (grenadier)	*Coryphaenoides* sp.
cusk-eel	*Spectrunculus grandis*
flying fish	(Exocoetidae)
Pacific saury	*Cololabis saira*
spinyfins	*Diretmus argenteus*
sablefish (blackcod)	*Anoplopoma fimbria*
horse mackerel	*Tracurus tracurus*
jack mackerel	*Trachuras declivis*
Pacific pomfret	*Brama japonica*
cod icefish	*Notothenia gibberifrons*
Patagonian toothfish	*Dissostichus eleginoides*
southern bluefin tuna	*Thunnus maccoyii*
swordfish	*Xiphias* spp.
billfish	Family: Istiophoridae
hake	*Merluccius hubbsi*
halibut	*Hippoglossus hippoglossus*

PLANTS

Throughout the Southern Ocean, tussock grasses are dominant features of the vegetation in which albatrosses breed. The distribution of several species is listed below:

Spartina arundinacea	Tristan da Cunha, Gough I., I. St Paul
Poa flabellata	South America, Falkland Is., S. Georgia, Gough I.
Poa cookii	Prince Edward Is., Is. Crozet, Is. Kerguelen, Heard I.
Poa novarae	Ile Amsterdam, I. St Paul
Poa poiformis	Albatross I. (Tasmania)
Poa tennantiana	Snares Is.
Poa astonii	Snares Is., Solander Is.
Poa foliosa	Solander Is., Auckland Is., Antipodes Is., Macquarie I.
Poa litorosa	Auckland Is., Antipodes Is., Campbell I.
Chionochloa antarctica	Auckland Is., Campbell I.

Appendix 3

Albatross measurements. Unsexed data are included only where separate male (M) and female (F) data are not available. Only samples of two or more are included. Except for Steller's Albatross all locations are breeding grounds.

Location	Sex	n	Culmen (mm) mean (range)	Tarsus (mm) mean (range)	Middle toe (mm) mean (range)	Tail (mm) mean (range)	Wing (mm) mean (range)	Mass (kg) mean (range)	Refs.
Black-browed Albatross									
South Georgia	M	5–6	119 (117–122)	83 (80–85)	138 (132–140)	221 (207–233)	541 (530–560)	3.99 (3.35–4.66)	12
	F	6–8	116 (114–121)	83 (76–88)	133 (125–142)	215 (206–222)	530 (510–545)	3.22 (2.90–3.80)	12
	M	132						3.92	17
	F	94						3.21	17
Iles Kerguelen		28–35	119 (108–124)	83 (76–89)			521 (501–550)	3.74 (2.90–4.50)	26
		123						3.68 (2.85–4.65)	29
Campbell Black-browed Albatross									
Campbell Island	M	15–26	112 (105–118)	85 (77–88)	126 (120–135)	213 (205–229)	526 (490–540)	3.10 (2.75–3.80)	16
	F	18–19	110 (105–114)	82 (75–88)	121 (116–129)	208 (200–225)	514 (495–530)	2.70 (2.20–3.15)	16
Grey-headed Albatross									
South Georgia	M	7–9	115 (113–119)	84 (79–88)	135 (128–140)	221 (219–230)	531 (512–550)	3.90 (3.10–4.35)	12
	F	10–13	115 (111–118)	83 (81–87)	132 (126–137)	210 (202–223)	517 (500–540)	3.87 (3.52–4.17)	12
	M	133						3.75	17
	F	95						3.26	17
Iles Crozet		31–41	112 (102–120)	87 (83–94)			520 (485–541)	3.48 (2.80–4.15)	24
Iles Kerguelen		33–43	111 (103–117)	81 (75–92)			515 (496–540)	3.38 (2.80–4.15)	26
Macquarie Island		28	115 (108–125)	88 (82–93)	127 (114–133)	211 (195–223)	517 (489–552)	3.24 (2.60–3.80)	27
Campbell Island	M	10	111 (108–116)	89 (83–92)	145 (139–154)	203 (197–210)	516 (499–535)		5
	M	13						3.38 (3.10–3.70)	27
	F	10						2.98 (2.80–3.20)	27
Western Yellow-nosed Albatross									
Tristan da Cunha	F	2	(109–111)	(76–81)	(107–108)	(186–199)	(456–498)	(1.87–1.93)	4
Gough Island		26–27	115 (108–122)	82 (80–86)		195 (178–214)	501 (483–520)	2.20 (1.78–2.84)	15
		11–32	114	83			473	2.24	19
Eastern Yellow-nosed Albatross									
Prince Edward Island		14–15	119 (111–124)	82 (79–87)		197 (185–210)	488 (465–499)	2.64 (2.49–2.93)	15
Amsterdam Island		33	116 (106–124)	83 (79–86)			482 (454–505)	2.06 (1.75–2.60)	10
Iles Kerguelen		13–16	116 (111–120)	77 (72–82)			476 (462–495)	2.46 (2.13–2.80)	16
Northern Buller's Albatross									
Chatham Islands	M	18	123 (117–129)	85 (82–88)	121 (117–125)	210 (201–222)	522 (507–543)	2.84 (2.50–3.30)	17
	F	18	118 (113–124)	81 (78–85)	116 (109–120)	200 (193–209)	500 (474–531)	2.43 (2.15–2.80)	17
Southern Buller's Albatross									
Snares Islands	M	6–21	121 (115–128)	85 (81–90)	124 (118–132)	202 (195–210)	529 (515–540)	3.12 (2.85–3.35)	27
	F	6–21	118 (114–123)	83 (79–86)	120 (116–125)	196 (189–211)	508 (490–525)	2.78 (2.05–3.10)	27

Location	Sex	n	Culmen (mm) mean (range)	Tarsus (mm) mean (range)	Middle toe (mm) mean (range)	Tail (mm) mean (range)	Wing (mm) mean (range)	Mass (kg) mean (range)	Refs.
Tasmanian Shy Albatross									
Albatross Island	M	18	133 (128–138)	94 (88–98)	139 (134–144)	221 (211–230)	562 (535–590)	4.35 (3.90–5.10)	27
	F	18	127 (122–132)	89 (86–92)	132 (125–138)	218 (210–225)	557 (545–570)	3.70 (3.20–4.40)	27
Auckland Shy Albatross									
Auckland Islands	M	6	139 (136–141)	97 (91–104)	144 (137–147)	233 (227–242)	607 (595–622)	4.43 (3.30–5.30)	27
	F	7	131 (126–139)	96 (92–101)	135 (134–139)	223 (213–238)	585 (569–595)	3.45 (2.60–4.20)	27
Salvin's Albatross									
Bounty Islands	M	17	129 (124–135)	92 (85–95)		222 (210–235)	577 (555–600)	4.00 (3.30–4.90)	18
	F	12	127 (123–135)	90 (87–95)		219 (210–228)	574 (555–590)	3.59 (3.30–3.70)	18
Chatham Albatross									
Chatham Islands	M	13	122 (116–130)	89 (84–96)	(120–139)	236 (221–248)	570 (550–586)	4.00 (3.60–4.70)	27
	F	10	120 (113–124)	86 (81–90)		222 (214–234)	560 (537–572)	3.77 (3.10–3.90)	27
Sooty Albatross									
Tristan da Cunha	M	5	116 (113–121)					2.60 (2.40–2.90)	20
	F	7	108 (103–109)					2.30 (1.80–2.50)	20
Gough Island		5–15	111 (102–120)	84 (81–88)			518 (503–533)	2.45 (1.98–3.01)	19
		17–29	111 (102–120)	84 (80–88)		261 (245–260)	523 (503–540)	2.41 (2.00–3.03)	21
Marion Island	M	2–5	115 (112–117)	83 (81–86)		275 (269–280)	(522–531)	2.60 (2.30–2.70)	13
	F	2–4	110 (109–111)	81 (79–84)		256 (245–267)	(508–509)	2.40 (2.30–2.60)	13
	M	11–13	116 (112–121)	84 (80–87)		273 (262–290)	524 (510–546)	2.80 (2.30–3.40)	21
	F	8–13	109 (104–113)	81 (79–83)		260 (242–267)	511 (500–522)	2.70 (2.30–3.10)	21
Iles Crozet	M	18–20	113 (109–117)				524 (490–540)	2.73 (2.20–3.25)	22
	F	12–13	107 (101–112)				501 (490–510)	2.44 (2.10–2.80)	22
Ile St Paul		5	107 (100–113)	77 (75–79)	116 (113–118)	264 (252–273)	505 (489–520)		21
Light-mantled Sooty Albatross									
Marion Island	M	3–4	109 (107–111)	88 (85–91)		297 (291–302)	559 (546–568)	(2.80–3.36)	21
	F	5	106 (104–108)	86 (84–88)		289 (279–301)	545 (527–569)	3.14 (2.62–3.70)	21
Iles Crozet		13–20	106 (98–113)	84 (79–95)			551 (520–570)	3.15 (2.85–3.60)	24
Iles Kerguelen		4	103 (97–111)	79 (75–83)	125 (123–129)	280 (265–300)	543 (520–555)		21
Macquarie Island		2–4	107 (100–112)	86 (84–88)	125 (120–130)	275 (230–320)	545 (520–560)		21
Auckland Islands		2	105 (99–111)	77 (75–80)	120 (115–125)	242 (234–250)	517 (483–550)		21
Campbell Island	M	3	106 (103–109)	86 (83–88)	139 (138–140)	299 (292–305)	549 (542–562)	3.00 (2.95–3.05)	5
South Georgia	M	11	111 (103–117)	83 (80–86)	125 (120–131)	271 (249–284)	522 (503–552)		1
	F	6	109 (98–117)	81 (78–84)	122 (116–125)	262 (236–276)	506 (490–525)		1
		8						2.87 (2.50–3.25)	31
Wandering Albatross									
South Georgia	M	52	170 (162–177)	121 (114–127)	185 (171–198)	228 (215–255)	665 (630–695)	9.11 (8.20–10.60)	31
	F	53	162 (152–172)	114 (104–119)	175 (163–186)	217 (210–230)	643 (620–675)	7.27 (6.35–8.30)	31
	M	20–21	169 (163–180)	118 (110–127)	184 (172–193)	227 (215–246)	679 (655–710)	9.77 (8.19–11.91)	8
	F	22–23	164 (155–171)	113 (106–123)	175 (165–181)	215 (206–227)	657 (630–680)	7.69 (6.72–8.71)	8

Iles Crozet		40	155 (141–165)				678 (640–788)	9.57 (6.80–11.50)	9
Macquarie Island	M	17	166 (158–181)	127 (123–132)	165 (148–172)	220 (201–228)	671 (635–701)		23
	F	10	158 (156–166)	121 (117–123)	160 (151–169)	204 (193–215)	635 (618–657)		23
	M	3						8.40 (8.00–8.60)	23
	F	3						6.90 (6.40–7.50)	23
Marion Island	M	14						9.26 (8.20–11.30)	32
	F	21						7.78 (6.40–9.00)	32
Amsterdam Wandering Albatross									
Ile Amsterdam	M	8	147 (142–156)	118 (114–121)	177 (171–183)	209 (200–215)	656 (640–674)	6.97 (6.00–8.00)	25
	F	6–8	141 (138–145)	109 (106–119)	163 (155–167)	200 (195–210)	636 (620–650)	6.12 (5.00–7.00)	25
Auckland Wandering Albatross									
Auckland Islands	M	11–13	151	117	165	200	660	6.65	28
	F	11–13	146	111	157	193	640	5.53	28
	M	35–40		119 (109–126)	161 (146–168)		659 (620–700)	6.80 (5.50–8.60)	30
	F	36–50		113 (107–117)	152 (145–162)		635 (600–670)	5.80 (4.60–7.30)	30
	M	362	151 (136–162)						30
	F	371	145 (133–157)						30
Antipodes Wandering Albatross									
Antipodes Island	M	8	148	117	177	203	655	7.35	14
	F	6	138	109	167	199	626	5.67	14
	M	10		120	167	191	664	7.46	28
	F	15		113	159	190	643	5.84	28
	M	54	151						28
	F	60	143						28
Campbell Island	M	2–8	150	118	167	207	667	6.88	28
	F	3–11	145	113	157	197	650	5.70	28
Northern Royal Albatross									
New Zealand	M	5	169 (165–172)		170 (164–176)		654 (634–669)		3
	F	5	158 (154–160)		161 (159–165)		622 (614–627)		
Southern Royal Albatross									
Campbell Island	M	5	184 (179–188)	136 (134–138)		217 (211–224)	696 (674–707)	8.84 (8.10–9.45)	5
	F	5	171 (163–177)	125 (123–126)		205 (196–210)	673 (647–686)	7.56 (6.52–9.00)	5
	M	11	186	129	178		698	10.30	33
	F	7	172	123	166		666	7.68	33
Galápagos Albatross									
Iles Galapagos	M	22	151 (142–160)	95 (91–103)	132 (125–138)	147 (137–158)	568 (550–593)		2
	F	40	141 (134–157)	92 (87–100)	125 (116–135)	139 (129–150)	547 (510–586)		2
	M	7	153					3.75	11
	F	13	142					3.04	11

Location	Sex	n	Culmen (mm) mean (range)	Tarsus (mm) mean (range)	Middle toe (mm) mean (range)	Tail (mm) mean (range)	Wing (mm) mean (range)	Mass (kg) mean (range)	Refs.
Laysan Albatross									
Hawaiian Islands	M	4	(100–112)	(80–85)	(107–119)	(133–151)	(474–500)		2
	M	152						3.31 (2.70–4.10)	6
	F	54						2.99 (2.50–3.60)	6
	M	50						3.31 (2.76–3.79)	7
	F	20						2.99 (2.39–3.44)	7
Black-footed Albatross									
Hawaiian Islands	M	3	(102–113)	(87–95)	(117–130)	(130–162)	(488–533)		2
	F	3	(94–110)	(81–93)	(109–129)	(130–150)	(485–530)		2
	M	123						3.40 (2.60–4.30)	6
	F	104						2.99 (2.60–3.60)	6
Steller's Albatross									
(British Museum)		8	135 (129–141)	89 (82–95)	133 (124–147)	167 (160–175)	550 (530–570)		31

References: 1 Nichols & Murphy (1914), 2 Loomis (1918), 3 Richdale (1942), 4 Hagen (1952), 5 Westerskov (1960), 6 Frings & Frings (1961), 7 Fisher (1967), 8 Tickell (1968), 9 Voisin (1969), 10 Segonzac (1972), 11 Harris (1973), 12 Tickell & Pinder (1975), 13 Berruti (1979a), 14 Warham & Bell (1979), 15 Brooke *et al.* (1980), 16 Robertson (1980), 17 Prince *et al.* (1981), 18 Robertson & van Tets (1982), 19 Williams & Imber (1982), 20 Richardson (1984), 21 Berruti (1984), 22 Jouventin & Weimerskirch (1984), 23 Tomkins (1984), 24 Weimerskirch *et al.* (1986), 25 Jouventin *et al.* (1989), 26 Weimerskirch *et al.* (1989a), 27 Marchant & Higgins (1990), 28 Robertson & Warham (1994), 29 Weimerskirch *et al.* (1997a), 30 Walker & Elliott (1999). Unpublished sources: 31 W.L.N. Tickell, 32 Percy FitzPatrick Institute, 33 NZ Department of Conservation.

Appendix 4

Albatross egg measurements.

Location	Length (mm) x Width (mm) mean (range)		n	Mass (g) mean (range)	n	Refs.
Black-browed Albatross						
Isles Kerguelen	103	x 66	46			32
South Georgia	104 (95–113)	x 66 (62–71)	117	257 (210–310)	117*	19
Campbell Black-browed Albatross						
Campbell Island (CBBA)	103	x 66	124	(206–256)	12	32
	102	x 66	50			31
Grey-headed Albatross						
Marion Island	108 (101–114)	x 69 (66–72)	11			8
Iles Crozet	103 (99–108)	x 69 (69–69)	3			15
Campbell Island	107 (98–116)	x 67 (61–72)	124			22
Campbell Island				(213–269)	12	10
South Georgia	106 (93–114)	x 68 (60–72)	63	276 (180–330)	63*	19
Western Yellow-nosed Albatross						
Tristan da Cunha	95 (89–101)	x 63 (57–67)	27	212 (184–238)	22	26
Gough Island	96 (83–106)	x 60 (56–65)	75			1
Eastern Yellow–nosed Albatross						
Ile Amsterdam	96 (91–105)	x 60 (56–65)	47	200 (170–225)	47	24
Southern Buller's Albatross						
Snares Islands	102 (95–112)	x 66 (61–70)	100	250 (218–290)	100	6
Tasmanian Shy Albatross						
Albatross Island	105	x 67	74			23
Auckland Shy Albatross						
Auckland Islands	106 (100–110)	x 71 (69–73)	4			17
Salvin's Albatross						
Bounty/Snares Islands	104	x 67	42			23
Chatham Albatross						
Chatham Islands	102	x 67	107			23
	100 (97–105)	x 68 (65–69)	6			4
Sooty Albatross						
Marion Island	103 (96–111)	x 65 (59–69)	67	243 (204–270)	31	27
Iles Crozet	102 (95–108)	x 65 (59–69)	74	227 (185–265)	74*	29
Tristan da Cunha	102 (94–106)	x 65 (59–70)	40	240 (216–274)	40*	26
Gough Island	102 (92–111)	x 65 (60–71)	26			27
Light-mantled Sooty Albatross						
Marion Island	102 (98–106)	x 67 (63–70)	17	258 (234–280)	17	20
Iles Crozet	104 (99–106)	x 67 (65–69)	4			32
Iles Kerguelen	102 (98–107)	x 67 (65–69)	20			32
Campbell Island	102 (98–107)	x 65 (63–67)	6	221 (210–238)	6	5
South Georgia	100	x 65	22	243	22	25
Wandering Albatross						
Marion Island	133 (123–143)	x 81 (75–85)	47	484 (437–551)	9	8
Iles Crozet	133 (123–143)	x 81 (77–85)	47			13
Iles Kerguelen	136 (122–140)	x 80 (76–83)	6			32
Macquarie Island	129 (121–139)	x 81 (78–87)	16	477 (415–520)	4	28
South Georgia	132 (114–142)	x 81 (79–86)	54			11
South Georgia				477 (425–537)	492*	33

Location	Length (mm) x Width (mm) mean (range)	n	Mass (g) mean (range)	n	Refs.
Amsterdam Wandering Albatross					
Ile Amsterdam	121 (114–126) x 76 (73–79)	7	412 (380–460)		30
Auckland Wandering Albatross					
Auckland Islands	127 (121–134) x 77 (73–82)	12			18
Antipodes Wandering Albatross					
Antipodes Islands	125 x 77	8			18
Gough Wandering Albatross					
Gough Island	127 (117–131) x 77 (72–83)	87			1
Northern Royal Albatross					
New Zealand	124 (117–132) x 78 (73–84)	53	417 (378–475)	41	7
Southern Royal Albatross					
Campbell Island	127 (118–131) x 79 (73–82)	15	425 (376–468)	15	5
Auckland Islands	129 (121–136) x 79 (73–83)	16			18
Galápagos Albatross					
Iles Galapagos	108 x 70	25			3
	105 x 69	28	284	25*	16
Laysan Albatross					
Midway Atoll	108 (101–113) x 75 (71–78)	20	278 (240–326)	20	9
	108 (100–121) x 67 (62–72)	28	278 (218–317)	26*	12
Black-footed Albatross					
Midway Atoll	107 (97–121) x 76 (68–81)	100	291 (218–335)	100	9
	108 (93–120) x 70 (64–74)	172	304 (275–333)	24*	12
Steller's Albatross					
	116 x 74	43			2
Ogasawara Gunto	119 (112–121) x 74 (73–76)	5			34
Torishima, Izu Shoto	118 (111–125) x 73 (70–79)	10	348 (310–375)	3	35

* newly laid eggs.

References: 1 Verrill (1895), 2 Bent (1922), 3 Murphy (1936), 4 Fleming (1939), 5 Sorensen (1950), 6 Richdale (1949a), 7 Richdale (1952), 8 Rand (1954), 9 Frings (1961), 10 Bailey & Sorensen (1962), 11 Tickell (1968), 12 Fisher (1969), 13 Mougin (1970), 14 Serventy *et al.* (1971), 15 Despin *et al.* (1972), 16 Harris (1973), 17 Oliver (1955), 18 Robertson (1975b), 19 Tickell & Pinder (1975), 20 Berruti (1977), 21 Sagar (1977), 22 Robertson (1980), 23 Robertson & van Tets (1982), 24 Jouventin *et al.* (1983), 25 Thomas *et al.* (1983), 26 Richardson (1984), 27 Berruti (1984), 28 Tomkins (1984), 29 Jouventin & Weimerskirch (1984a), 30 Jouventin *et al.* (1989), 31 Moore & Moffat (1990b), 32 Marchant & Higgins (1990), 33 Croxall *et al.* (1992). Unpublished sources: 34 W.L.N. Tickell.

Appendix 5

The main chewing lice [Phthiraptera: Ischnocera & Amblycera] on albatross hosts.

	Austromenopon	*Episbates*	*Docophoroides*	*Harrisoniella*	*Paraclisis*	*Perineus*
GREAT ALBATROSSES	*A. affine*	*E. pederiformis*	*D. brevis*	*H. hopkinsi*	*P. hyalina*	*P. concinnoides*
MOLLYMAWKS	*A. pinguis* *A. navigans*	–	*D. harrisoni* *D. simplex*	*H. ferox*	*P. diomedeae*	*P. circumfasciatus*
SOOTY ALBATROSSES	*A. pinguis*	–	*D. simplex*	–	*P. diomedeae*	*P. circumfasciatus*
Galápagos Albatross	–	*E. pederiformis*	*D. levequei*	*H. ferox*	*P. miriceps*	*P. oblongus*
Laysan Albatross	*A. pinguis*	–	*D. niethammeri*	*H. densa*	*P. giganticola*	*P. concinnus*
Black–footed Albatross	*A. pinguis*	*E. pederiformis*	*D. ferrisi*	*H. copei*	*P. confidens*	*P. concinnus*
Steller's Albatross	*A. pinguis* *A. navigans*	–	*D. pacificus*	*H. densa*	*P. giganticola*	*P. concinnus*

Saemundssonia sp. also occurred on two mollymawks and one of the sooty albatrosses (British Museum (Nat Hist) catalogue, Clay & Moreby 1970, Atyeo & Peterson 1970, Pilgrim & Palma 1982, Palma & Pilgrim 1984, 1987, Palma 1994, R.L. Palma pers. comm.).

Appendix 6

Prey of the two black-browed albatrosses (BBA & CBBA), the Grey-headed Albatross (GHA), the two yellow-nosed albatrosses (WYNA & EYNA), the two buller's albatross (NBA & SBA) and the two shy albatross (TSA & AKSA).

Order/Family	Species	BBA CBBA	GHA	WYNA EYNA	NBA SBA	TSA AKSA
FISH						
Petromyzontidae	*Geotria australis*	+	+++			
Engraulidae	*Engraulis australis*			*		*
Clupeidae	*Sardinops neopilchardus*			*		*
Microstomatidae	*Nansenia antarcticus*		+			
Paralepididae	*Magnisudis prionosa*	++	++			
Anotopteridae	*Anotopterus pharao*		+			
Myctophidae	*Electrona antarctica*	++	+			
	Gymnoscopelus fraseri		+			
	G. nicholsi		+			
	G. bolini	+				
	G. microlampas		+			
	Protomyctophum bolini		+			
	P. choriodon		+			
	Krefftichthys anderssoni	+	+			
Ophidiidae	*Genypterus blacodes*	+				
Macrouridae	*Coelorhynchus* sp.	+				
Moridae	*Salilota australis*	+				
	Pseudophycis bacchus				+	
Macruronidae	*Macruronus magellanicus*	+				
Muraenolepididae	*Muraenolepis microps*		++			
Merlucciidae	*Merluccius* sp.	+				
Gadidae	*Micromesistius australis*	+++				
Scomberesocidae	*Scomberesox saurus*			+		
Exocoetidae	*Cheilopogon furcatus*			+		
Melamphaidae	*Sio nordenskjöldii*		+			
Trachichthyidae	*Trachichthodes gerrardi*					+
Macroramphosidae	*Notopogon lilliei*	+				
Carangidae	*Trachurus declivis*			*		**
Pentacerotidae	*Pseudopentaceros richardsoni*				+	
Nototheniidae	*Dissostichus eleginoides*	+				
	Patagonotothen guntheri	++	++			
Channichthyidae	*Channichthys rhinoceratus*		+			
	Champsocephalus gunnari	++	++			
	Pseudochaenichthys georgianus	++	++			
Chiasmodontidae	*Dysalotus alcocki*		+			
Pinguipedidae	*Parapercis colias*					+
Gempylidae	*Thyrsites atun*			+		
	Paradiplospinus gracilis	+				
Centrolophidae	*Icichthys australis*	++				
SQUID & OCTOPUS						
Sepiidae	*Sepia* sp.	+				+
Loliginidae	*Loligo* sp.			+		
	L. gahi	+++				
	L. vulgaris	+				
	Sepioteuthis australis	+				
Lycoteuthidae	*Lycoteuthis* sp.		+			
Enoploteuthidae		+			+	
	Ancistrocheirus sp.	+			+	
Octopoteuthidae	*Octopoteuthis* sp.				+	
	Taningia danae		+			
Onychoteuthidae	*Kondakovia longimana*	+	+++	+++		
	Moroteuthis knipovitchi	+	++	++		
	M. ingens		+		+	
	M. robsoni		++	++	+	
Gonatidae	*Gonatus antarcticus*	+	++			

Order/Family	Species	BBA CBBA	GHA	WYNA EYNA	NBA SBA	TSA AKSA
Histioteuthidae	*Histioteuthis* sp.	+	++		+	
	H. atlantica				++	
	H. macrohista				+	
	H. miranda				+	
	H. dofleini				+	
	H. eltaninae		+++	+++		
Psychroteuthidae	*Psychroteuthis glacialis*	+	++			
Neoteuthidae	*Alluroteuthis antarcticus*	+	+			
Brachioteuthidae	*Brachioteuthis* sp.		+			
Ommastrephidae	*Ommastrephes bartrami*	+				+
	Martialia hyadesi	+++	+++			
	Illex argentinus	+++				
	Todarodes filippovae	+	+	+		
	Nototodarus sp.				+++	+++
	N. gouldi				+	
Batoteuthidae	*Batoteuthis* sp.		+			
Cycloteuthidae	*Discoteuthis* sp.		+			
Chiroteuthidae	*Chiroteuthis* sp.	+	++		+	
	C. veranyi	+	+			
	C. imperator		+			
Mastigoteuthidae	*Mastigoteuthis* sp.	+	+		+	
	M. psychrophila		+			
Caranchiidae				+		
	Galiteuthis glacialis	++	++			
	Mesonychoteuthis hamiltoni		+			
	Teuthowenia pellucida		+		+	
	Taonius sp.	+	+		+	
	Bathothauma sp.		+			
Octopodidae	*Octopus* sp.	+	+		+	
	O. maorum				+	
	Parelodene polymorpha	+				
Argonautidae	*Argonauta nodosa*				+	
Alloposidae	*Alloposus mollis*		+			
Ocythoidae	*Ocythoe tuberculata*				+	
CRUSTACEANS						
Calanoida				+		
	Trifur lotellae				+	
Thoracica	*Lepas australis*		+		+	+
Mysidacea	*Gnathophausia* sp.		+	+	+	+
	G. ingens			+		
	G. gigas		+	+		
	Chalaraspidum sp.			+		
Isopoda	*Livoneca raynaudi*				+	+
Euphausiacea	*Euphausia superba*	+++	+++			
	E. spinifera				++	
	E. frigida	+	+			
	Nyctiphanes australis				*	
	N. capensis				+	
Amphipoda	*Themisto guadichaudii*	+	+			
	Hyale grandicornis			+		
	Eurythenes gryllus		+	+		
	Hyperid sp.	+				
Decapoda	*Acanthephyra* sp.	+	+			
	Pasiphaea longispina		+			
	Notostomus sp.			+		
	Pandalidae					+
	Munida gregaria	++	+		+	
	Nectocarcinus integrifrons					+

Pooled stomach samples from several populations and years containing: + small to negligible quantities, ++ significant quantities, and +++ major components of diets. * Birds seen feeding at sea, some in winter far from their breeding islands (Broekhuysen & Macnea 1949, Tickell 1964, Green 1974, Prince 1980, Clarke & Prince 1981, West & Imber 1986, Brooke & Klages 1986, Weimerskirch *et al.* 1986, Ridoux 1994, Reid *et al.* 1996, Rodhouse *et al.* 1996, Croxall *et al.* 1997b, Cherel & Klages 1997 and M.J. Imber pers. comm.).

Appendix 7

Prey of the two sooty albatrosses (SA & LMSA).

Order/Family	Species	SA	LMSA
FISH			
Bathylagidae	*Bathylagus* sp.	+	
Paralepididae	*Magnisudis prionosa*	+	+
Myctophidae	*Electrona carlsbergi*	++	+
	Gymnoscopelus sp.	+	+
	G. piabilis	+	
Macrouridae		+	+
Melanonidae	*Melanonus gracilis*	+	++
Melamphaidae	*Melamphaes* sp.		+
Nototheniidae			+
	Lepidonotothen squamifrons	+	
Gempylidae	*Paradiplospinus gracilis*	+	
	Thyrsites atun	+	
SQUID & OCTOPUS			
Lycoteuthidae	*Lycoteuthis* sp.	+	+
Enoploteuthidae	*Ancistrocheirus lesueuri*	+	
Octopoteuthidae	*Octopoteuthis* sp.	+	+
	Taningia danae	+	+
Onychoteuthidae	*Onychoteuthis* sp.	+	+
	Kondakovia longimana	+++	+++
	Moroteuthis knipovitchi	+++	+++
	M. ingens	++	+
	M. robsoni	++	++
Gonatidae	*Gonatus* sp.	+	
	G. antarcticus	++	++
	G. phoebetriae	+	
Histioteuthidae	*Histioteuthis eltaninae*	+++	+++
	H. bonnellii corpuscula	++	+
	H. atlantica	+	+
	H. miranda	+	+
	H. dofleini	+	
	H. macrohista	+	
	H. meleagroteuthis	+	+
Psychroteuthidae	*Psychroteuthis* spp.	+	+++
	P. glacialis	++	+++
Architeuthidae	*Architeuthis* sp.	+	
Neoteuthidae	*Alluroteuthis antarcticus*	++	+++
Ommastrephidae	*Martialia hyadesi*	+	+++
	Todarodes sp.	+	
Lepidoteuthidae	*Lepidoteuthis grimaldii*	+	+
	Pholidoteuthis sp.		+
Batoteuthidae	*Batoteuthis* sp.	++	+
	B. skolops	+	
Chiroteuthidae	*Chiroteuthis* sp.	++	+
	C. veranyi	+	
	C. imperator	+	+
	C. macrosoma	+	+
Mastigoteuthidae	*Mastigoteuthis* sp.	+	+
Cycloteuthidae	*Cycloteuthis* sp.	+	+
	C. sirventi	+	
	Discoteuthis sp.	+	+++
Cranchiidae	*Cranchia scabra*	+	
	Galiteuthis glacialis	+++	++
	G. armata	+	
	Mesonychoteuthis hamiltoni	+	++
	Taonius pavo	++	+

Order/Family	Species	SA	LMSA
	T. belone	+	
	T. cymoctypus	+	+
	Bathothauma lyromma	+	+
	Teuthowenia sp.	+	
	T. megalops	+	
	Megalocranchia sp.		+
Octopodidae			+
Alloposidae	*Alloposus mollis*	+	
CRUSTACEANS			
Mysidacea	*Gnathophausia gigas*	+	+
Amphipoda	*Cyllopus lucasii*		+
	Themisto gaudichaudii	++	++
	Hyperia sp.	+	
	Eurythenes sp.	+	+
	E. obesus		+
	E. gryllus	+	
Isopoda			+
Euphausiacea	*Euphausia superba*	++	+++
	E. vallentini	+	+
Decapoda			+
	Pasiphaea longispina	+	+
	Acanthephyra sp.		++
	Austropandalus grayi		+

Pooled stomach samples from several populations and years containing: + small to negligible quantities, ++ significant quantities and ++ major components of diets (Imber & Berruti 1981; Thomas 1982, Imber 1991, Ridoux 1994, Cooper & Klages 1995, Croxall & Prince 1996 and Cherel & Klages 1997).

Appendix 8

Prey of the Wandering Albatross (WA) and two royal albatrosses (NRA &SRA).

Order/Family	Species	WA	NRA SRA
FISH			
Petromyzontidae	*Geotria australis*	+	+
Ophidiidae	*Spectrunculus grandis*	+	
Macrouridae		+	+
Moridae		+	+
	Halargyreus johnsonii	+	
Macruronidae	*Macruronus novaezelandiae*	+	+
Muraenolepididae	*Muraenolepis microps*	++	
Nototheniidae	*Notothenia colbecki*		+
	Gobionotothen gibberifrons	+	
	Trematomus hansoni	+	
	Dissostichus eleginoides	++	
Channichthyidae	*Chaenocephalus aceratus*	++	
	Champsocephalus gunnari	+	
	Pseudochaenichthys georgianus	+++	
Trichiuridae	*Benthodesmus elongatus*	+	
	Aphanopus sp.	+	
SQUID & OCTOPUS			
Sepiidae	*Sepia* sp.	+	
	S. apama	***	
Loliginidae	*Loligo* sp.	+	
	Sepioteuthis australis	+	
Lycoteuthidae	*Lycoteuthis* sp.	+	
Enoploteuthidae	*Ancistrocheirus lesueuri*	+	+
Octopoteuthidae	*Octopoteuthis* sp.	+	+
	O. rugosa	+	
	Taningia danae	++	+

Order/Family	Species	WA	NRA SRA
Onychoteuthidae	*Onychoteuthis* sp.	+	+
	O. banksi	+	
	Kondakovia longimana	+++	++
	Moroteuthis knipovitchi	+++	+
	M. ingens	+++	+++
	M. robsoni	++	+
Gonatidae	*Gonatus antarcticus*	++	+
	G. phoebetriae	+	
Histioteuthidae	*Histioteuthis atlantica*	++	+++
	H. dofleini	+	
	H. eltaninae	++	+
	H. macrohista	+	+
	H. bonnellii corpuscula	++	+
	H. miranda	+	+
	H. celetaria	+	
Psychroteuthidae	*Psychroteuthis glacialis*	++	
Architeuthidae	*Architeuthis* sp.	+	++
Neoteuthidae	*Alluroteuthis antarcticus*	++	+
Brachioteuthidae	*Brachioteuthis* sp.	+	
Ommastrephidae	*Illex argentinus.*	++	
	Martialia hyadesi	++	+
	Nototodarus sp.	+	++
Lepidoteuthidae	*Lepidoteuthis grimaldi*	+	
	Pholidoteuthis boschmai	+	+
Batoteuthidae	*Batoteuthis skolops*	+	+
Cycloteuthidae	*Cycloteuthis akimushkini*	++	+
	Discoteuthis sp.	+	+
Chiroteuthidae	*Chiroteuthis* sp.	++	+
	C. capensis	+	
	C. imperator	+	
Mastigoteuthidae	*Mastigoteuthis* sp.	+	+
	M. psychrophilia	+	
Grimalditeuthidae	*Grimalditeuthis bonplandi*	+	
Cranchiidae	*Cranchia scabra*	+	
	Egea sp.	+	
	Galiteuthis sp.	+	+
	G. glacialis	++	+
	Helicocranchia sp.	+	
	Megalocranchia sp.	+	+
	Mesonychoteuthis hamiltoni	+	
	Taonius sp.	++	++
	Teuthowenia pellucida	+	+
	Bathothauma sp.	+	
Vampyromorphidae	*Vampyroteuthis infernalis*	+	
Octopodidae	*Octopus maorum*		+++
	O. dofleini	+	
	Enteroctopus sp.	+	+
Argonautidae	*Argonauta nodosa*	+	+
Alloposidae	*Alloposus mollis*	+	
Ocythoidae	*Ocythoe tuberculata*	+	+
CRUSTACEANS			
Thoracica	*Lepas* sp.		+
Isopoda			+
Euphausiacea	*Euphausia superba*	+	
Decapoda	*Nematocarcinus longirostris*	+	
	Lithodes sp.	+	
	Munida sp.		+
	M. gregaria	*	

Pooled stomach samples from several populations and years containing: + small to negligible quantities, ++ significant quantities and ++ major component of diets. * Birds seen feeding at sea (Imber & Russ 1975, Clarke *et al.* 1981, Imber & Berruti 1981, Rodhouse *et al.* 1987, Croxall *et al.* 1988, Cooper & Brown 1990, Imber 1991, 1992, Cooper *et al.* 1992, Ridoux 1994, Croxall & Prince 1996, Weimerskirch *et al.* 1997, Cherel & Klages 1997 and M.J. Imber pers. comm.).

Appendix 9

Prey of the Laysan Albatross (LA) and Black-footed Albatross (BFA).

Order/Family	Species	LA	BFA
FISH			
Carcharhinidae	*Prionace glauca*	+	+
Platytrocidae	*Sagamichthyes abei*	+	+
Gonostomatidae	*Vinciguerria* sp.	+	
Sternoptychidae		+	+
	Argyropelecus sp.	+	
Photichthyidae	*Ichthyococcus* sp.	+	
Myctophidae	*Electrona risso*	+	+
	Symbolophorus californiense	+	
	Ceratoscopelus sp.	+	
	Lampanyctus sp.	+	
	L. ritteri	+	
	L. jordani	+	
	Diaphus gigas	+	
	D. theta	+	
	Notoscopelus japonicus	+	
Macrouridae	*Coryphaenoides* sp.	+	+
Moridae		+	
Scomberesocidae	*Cololabis saira*	+	+
	C. saira (ova)	+	
Exocoetidae		+	+
Exocoetidae (ova)		++	+++
	Cypselurus sp.	+	
Hemiramphidae	*Euleptorhamphus viridis*	+	
Diretmidae	*Diretmus argenteus*	+	
Scorpaenidae**			+
Carangidae	*Decapterus* sp.		+
Bramidae	*Brama japonica*	++	++
Mullidae		+	
Gempylidae		+	+
	Gempylus serpens		+
Molidae	*Ranzania laevis*	+	
MOLLUSCS			
Gastropoda	*Janthina* sp.	+	
	Janthina sp. (ova)	+	
Cephalopoda			
Enoploteuthidae	*Thelidioteuthis alessandrinii*	+	
	Abraliopsis felis		+
	Ancistrocheirus sp.	+	
Octopoteuthidae	*Octopoteuthis* sp	+	+
	O. deletron	+	+
Onychoteuthidae	*Onychoteuthis borealijaponius*	+	+
Gonatidae	*Gonatus* spp.	+	+
	Gonatopsis borealis		+
	Berryteuthis anonychus	+	+
Histioteuthidae	*Histioteuthis* sp.	+	+
	H. dofleini	+	+
Architeuthidae	*Architeuthis* sp.	+	
Ommastrephidae	*Ommastrephes bartrami**	+++	+++
	Symplectoteuthis spp.	+	+
	S. oualaniensis	+	+
	Hyaloteuthis pelagicus	+	+

Order/Family	Species	LA	BFA
Lepidoteuthidae		+	
Chiroteuthidae	*Chiroteuthis* sp.		+
	C. calyx	+	+
Mastigoteuthidae	*Mastigoteuthis* sp.	+	+
Cranchiidae	*Galiteuthis* sp.	+	+
	G. phyllura	+	+
	Leachia dislocata	+	+
	Megalocranchia sp.	+	
	Taonius spp.	+	+
	T. pavo	+	+
Octopodidae			+
Alloposidae	*Alloposus mollis*		+
Ocythoidae	*Ocythoe tuberculata*	+	+
CRUSTACEANS			
Thoracica	*Lepas* sp.	+	+
	L. fascicularis	+	
Copepoda	*Penella* spp.		+
Leptostraca	*Nebaliopsis typica*	+	+
Stomatopoda	*Pseudosquilla* sp	+	
Mysidacea	*Gnathophausia* sp.	+	+
	G. gigas	+	+
	G. ingens	+	+
Amphipoda	*Alicella* sp.		+
	Eurythenes gryllus	+	
Isopoda	*Anuropus bathypelagicus*	+	+
	A. branchiatus	+	+
Euphausiacea		+	+
Decapoda		+	+
	Notostomus sp.	+	+
	N. japonicus		+
	Acanthephyra sp.	+	
	A. exima	+	
	Planes cyaneus	+	+
	P. minutus	+	
ECHINODERMS			
Echinoidea	*Strongylocentrotus drobachiensis***		+

Pooled samples collected from 355 chicks at Midway Atoll and 216 albatrosses killed at drift-nets in the transition zone of the North Pacific Ocean outside the breeding season: + negligible quantities, ++ significant quantities and +++ major components of diets. * Also feeding at drift-nets (Harrison *et al.* 1983, Gould *et al.* 1997) and at the Aleutian Islands** (Cottam & Knappen 1939).

Appendix 10

A method of ageing subadult Black-browed and Grey-headed Albatrosses at South Georgia using moult scores, bill coloration and some general plumage characteristics (Prince & Rodwell 1994).

CLASSIFICATION OF FLIGHT FEATHERS

New feathers	0–11 months
Old feathers	second generation (12–23 months)
	third generation (24–35 months)
	fourth generation (26–47 months)
	fifth generation (48–59 months)

Distinguishing new, second generation and older feathers from each other is quite easy, but separating third, forth and fifth generations is very difficult.

New feathers are dark blackish, almost waxy in appearance. After about two months they fade to dark grey. There is no abrasion unless the feather has been damaged.

Second generation feathers are dull, matt brown; the difference being more noticable in primaries than in secondaries. There is very little abrasion, especially of secondaries. On returning birds, it is usually confined to the tip and outer webs of the outer primaries, but the tips of old outer primaries in three-year-old Black-browed Albatrosses are very abraded.

Third generation feathers are paler brown and more abraded with bleached, pale brownish-white patches in the centre. The abraded edges may be discoloured with missing notches at the tips. There are no outer primaries (8,9 & 10) of this age.

Fourth generation feathers are similar to third generation feathers, but are more bleached and abraded. Inner primaries of this age were found on three-year-old Grey-headed Albatrosses and only once on a Black-browed Albatross. Secondaries were on three-year-olds of both species; the tips and outer edges are noticably abraded with the outer webs completely bleached.

Fifth generation secondaries are retained on a few four-year-old Black-browed Albatrosses; they are indistinguishable from fourth generation feathers.

BLACK-BROWED ALBATROSS

Ages estimated using primary and secondary flight feathers.

First year (0–11 months) Bill dark bronzy-brown with blackish tip. Underwing coverts dark grey. Shoulder patch dark grey. All feathers new.

Age 1 (12–23 months) Bill dark. Underwing coverts dark grey. All feathers second generation.

Age 2 (24–35 months) Bill dark, some yellow on culmen and red on unguis. Underwing coverts off-white with brown tips. New outer primaries with large blocks of third generation feathers always present between primaries 3 and 7 and between secondaries 5 and 23.

Age 3 (36–47 months) Bill variable dark and yellow with some red on unguis. Underwing coverts white as in adult. Eyebrow dark as in adults. Outer primaries old and very abraded. Fourth generation secondaries between secondaries 6 and 24.

Age 4 (48–59 months) Bill mostly yellow with red unguis, both still with some grey. New outer primaries; no third generation primaries, only old and new.

Age 5 (60–72 months) Bill yellow with red unguis, paler than adults; most still with traces of grey. Outer primaries old, often with large blocks of second and third generation feathers in the middle of the secondaries – not to be confused with three-year-olds.

GREY-HEADED ALBATROSS

Ages estimated using only primary flight feathers.

First year (0–11 months) Bill dark with some pale, dirty brown/yellow on culmen and mandibles. Head usually grey. All feathers new.

Age 1 (12–23 months) All feathers second generation.

Age 2 (24–35 months) Bill dark with culmen beginning to turn yellow. Head white. Outer and possibly innermost primaries new; primaries 3–7 always third generation.

Age 3 (36–47 months) Bill dark laterally with culmen and lower edge of mandibles pale yellow smudged dark; unguis rosy red. Head white. Outer primaries (8–10) second generation, inner (1–4) fourth generation separated by new feathers at 5–7.

Age 4 (48–59 months) Bill culmen sometimes bright or pale yellow, but most still have some dark traces; unguis pale rosy red with some dark shading on most birds. Head in transition between white and adult grey, sometimes patchy. New outer primaries (8–10); no third or fourth generation feathers. Primaries 5–7 always second generation.

Age 5 (60–72 months) Bill culmen usually clear yellow, slightly paler than adult, merging with rosy red of unguis. Head adult grey, perhaps with some lighter patches. Outer primaries (8–10) second generation; often some third generation retained between primaries 1–5 with new ones at 4–5.

Appendix 11

A key for estimating the age of subadult Wandering Albatrosses on the breeding ground, using moult scores of primary and secondary flight feathers.

	Outer primaries (8-10)		Inner primaries (1–7)				Secondaries (34)				Age	GPI-t
			New (n)	Old (n)	Third (n)		New (n)	Old (n)	Third (n)		(years)	
Male	New	+	0	0	7	+	0	0	34	≈	3	–
	New	+	2	5	0	+	25	9	few	≈	5	12.6±1.5
	New	+	3	2	2	+	12	20	2	≈	7	13.5±1.6
	New	+	4	2	1	+	18	9	7	≈	9	14.9±1.2
	Old	+	5	1	1	+	10	2	22	≈	4	12.6±1.3
	Old	+	4	2	1	+	17	16	1	≈	6	14.1±0.8
	Old	+	6	1	few	+	16	4	14	≈	8	14.4±1.2
Female	New	+	0	0	7	+	0	0	34	≈	3	5.7±1.0
	New	+	3	4	few	+	19	9	6	≈	5	6.6±1.3
	New	+	2	2	3	+	4	16	14	≈	7	8.0±1.5
	New	+	few	few	6	+	–	–	–	≈	9	10.4±2.1
	Old	+	5	few	2	+	6	2	26	≈	4	6.5±1.9
	Old	+	2	3	2	+	11	18	5	≈	6	7.7±1.8
	Old	+	4	2	1	+	14.5	5	14.5	≈	8	9.2±1.9

The method was devised at Bird Island, South Georgia where ages could be independently confirmed from ringed, known-age birds (Prince *et al.* 1997). Proceed as follows:

1. The sex of the albatross must first be established by some other method.
2. Determine the condition of the outer three primaries (8,9 and 10): New (Phase 1) or Old (Phase 2) (see Appendix 10).
3. Examine the inner seven primaries (1–7) and count the numbers of New, Old and Third Generation feathers (see Appendix 10).
4. Examine the 34 secondaries and count the numbers of New, Old and Third Generation feathers.
5. The values in the table are means of samples, so results cannot be expected to fit exactly. Find the combination that corresponds most closely and read off the age.

The Gibson Plumage Index – used here without the tail score (GPI-t) – may help to separate non-consecutive ages and to distinguish subadults from breeding adults.

Appendix 12

Comparison of social and stereotyped actions and vocalisations in the six groups of albatrosses. Bold capitals are stereotyped postures, actions and vocalisations, with synonyms in standard lower case.

MOLLYMAWKS	SOOTY ALBATROSS	GREAT ALBATROSSES	GALÁPAGOS ALBATROSS	GOONEYS	STELLER'S ALBATROSS
Allo-preen	**Allo-preen** *Toilettage du partenaire*	**Allo-preen** Mouthing	**Allo-preen**	**Allo-preen**	**Allo-preen**
STARE Gawky look	**STARE**	**STARE** Gawky look	**STARE**	**STARE** Gawky look Glare	
POINTING Rapier action Bill aligning Throbbing (vocal)	**POINTING** Bill pointing *Baiser*	**POINTING** Billing Head forward high Head forward low Touch/nibble beaks Breast billing Billing & snap Defence billing	**POINTING**	**POINTING** Billing Bill touch	**POINTING** Pointing high Pointing low Breast billing
		VIBRATE Billing & vibrate Bill vibrate (fast) Rattle Bray		**VIBRATE**	
		NECK-ROLL Swan neck Arched neck Head roll			
			BOW-CLAPPER	**BOW-CLAPPER** Bow-clacker	
JABBING	**THRUSTING** Bill thrusting				
CROAKING Croaking & Nodding Creaking			**CROAKING** (Croaking) Gape low	**CROAKING** Double call Eh-Eh bow Eh bow	
CLASHING Fencing Beak/Bill clashing	**CLASHING** *Aiguissage de bec* Beak clashing		**CLASHING** Bill circling Fencing		**RHYTHMIC BOWING** **CLASHING** Fencing
		SNAP Single bill snap		**SNAP**	**SNAP**
	GAPE *Bec ouvert*		**GAPE** Great gape (silent)		
	GAPE DOWN *Courbette-bec-ouvert*				
WAIL(vocal) Trumpet (vocal)				**WAIL(vocal)**	
	SKY-POINT Sky (low intensity)	**SKY-POINT** **SKY-TRILL** Sky call (silent) Sky point (& gurgle) Sky position	**SKY-POINT** Sky call (silent)	**SKY-MOO** **SKY-GROAN** Sky-call	**SKY-POINT**
	SKY-FLICK *Cou de tête*			**SKY-FLICK** Head flick (& clicks)	

MOLLYMAWKS	SOOTY ALBATROSS	GREAT ALBATROSSES	GALÁPAGOS ALBATROSS	GOONEYS	STELLER'S ALBATROSS
				SKY-SNAP Head-up (& clap)	
SKY-CALL	**SKY-CALL** *Chant*	**SKY-CALL** Sky position (call) Head shake Head-shake & whine	**SKY-WHISTLE**	**SKY-CALL** Head-up & whine Victory call	
				WHINNY(vocal)	
				SHAKE & WHISTLE Head shake & whine	
		Wings extended			
	Wing flare			**Wing flare**	**Wing flare**
Tail fan	**Tail fan & twist**	**Tail fan**			
BOW Bow (nod) Bowing		**BOW** Head bob Head curl		**SIDE BOW**	**SIDE BOW** Side bow to ground
SIDE-PREEN **NECK-PREEN**	**SIDE-PREEN** [*Mouvement scapulaire*] [Scapular action]	**SIDE-PREEN** Snap preen [Scapular action] Side action Preen wing		**SIDE-PREEN** [Scapular action] Bill under wing	**SIDE-PREEN** Flank touching Side touching Preen side breast
				PREEN-UNDER-WING	**PREEN-UNDER-WING**
LEG ACTION					
		FRONT-PREEN Snap preen Preen breast			
SCAPULAR ACTION Wing action	**SCAPULAR ACTION**				
FLAGGING	**FLAGGING** *Regard*	**FLAGGING** Circling & flagging Head wag Head shake	**SWAY-WALK** sway-walk		**FLAGGING**
SCOOPING	**CIRCLE** Paddle walk *Marche ressort*		Forward bobbing	**BOB** **BOB-STRUT**	
FOOT-LOOKING Bow	**FOOT-LOOKING** *Bec au pattes*		**BOW**		
Groan		Gurgle		Grunt	
	CLAPPER	**CLAPPER**	**CLAPPER**	**CLAPPER** Clacker	**CLAPPER**
		YAMMER Threat	**YAMMER**		**YAMMER**
		YAPPING Yakker	**YAPPING**	**YAPPING**	**YAPPING**
GULPING					
GROUND-STABBING					

(Richdale 1949a, 1950a, Rowan 1951, Meseth 1975, Jouventin *et al.* 1981, Tickell 1984, Jouventin & Lequette 1990, Lequette & Jouventin 1991a, Pickering & Berrow (in press) and B. Lequette pers. comm.).

Appendix 13

Models of Sooty Albatross behaviour.

La danse de l'albatros, Phoebetria fusca (Jouventin *et al.* 1981)[1] is a landmark in the study of albatross behaviour. The numerous ritualised postures employed by the birds, in various sequences, are not random. They are imagined as letters in words of a language characterising the stages of a nuptial display lasting several years, and changing little by little as mates come to know each other better. At the beginning, females respond aggressively, but after displaying with the same male for some time, they allow closer proximity, allopreening and finally copulation.

Several axes were constructed to indicate the divergence in occurrence of pairs of actions and action sets – whether the two occurred soon after each other or at longer intervals. Seven actions contributed 84–97% of those used in the first three axes (Apx. 13.1) which were calculated to contain 60% of the available information.

Axis 1 strongly discriminated clashing and the side-preen set in both sexes, axis 2 rather less between clashing and the gape set while axis 3 separated the flagging set and gape. When sequential actions of the displaying partners were plotted as rectangular co-ordinates, a remarkable triangular array was produced, which represented progressive change in the dancing population (Apx. 13.2). It can be understood in terms of a two-dimensional model of behavioural maturation.

Side-preen versus clashing, and gape versus clashing which provide almost all contributions to the first two axes, make a triangle (Apx. 13.2) with a base formed by side-preen and gape; the two being replaced by clashings towards the apex. The base represents an extreme opposition between side-preen and gape in males and females that have apparently had little experience – subadults recently returned to land. Clashing is rare among them because they hardly ever get close enough to each other. In the middle of the triangle all three actions occur between birds that have become experienced performers. Towards the apex, pairs are forming. With progressively fewer partners, side-preens and gapes are becoming less frequent and exceeded by clashings. Maturation of behaviour may be represented as a series of bands parallel to the base and getting progressively shorter as the behaviours of male and female become compatible.

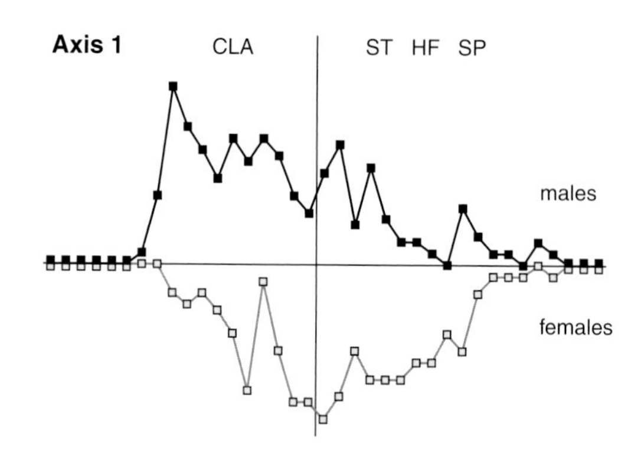

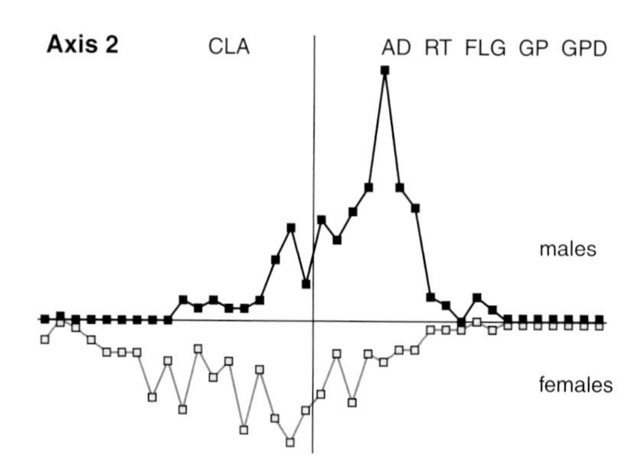

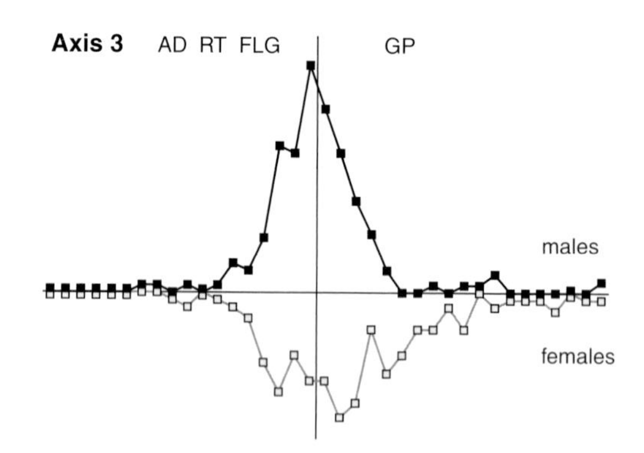

Appendix. 13.1 Pairings of actions or action sets between male and female Sooty Albatrosses. Towards the vertical zero line the two actions occur closer together. Males are above the horizontal line and females below.
Axis 1. indicates that males have a bias towards CLASHING (CLA) and females towards SKY-TWIST (ST), HEAD-FLICK (HF) and SHOULDER PREEN (SP).
Axis 2. reveals a female bias towards CLASHING (CLA) compared with male GAPE (GP), GAPE DOWN (GPD), FLAGGING (FLG), advance (AD) and retire (RT).
Axis 3. little or no difference is apparent in actions between males and females (after Jouventin *et al.* 1981).

Plotted similarly, axes 2 and 3 discriminate the gape of uneasy, aggressive females from the flagging and associated actions of more confident females motivated to approach a male more closely.

All three axes were combined to create an ingenious three-dimensional model (Apx. 13.3) in which males were represented by a trapezoid bar and females by a triangular sloping surface. The behaviour of both sexes maturates towards clashing and females have more varied displays. The base of the slope represents the five, mainly female actions; there are tensions between them in that gape is menacing while flagging and moving forward express interest in the male.

Females retain the initiative because males remain on their sites and are limited to three main actions of which only side-preen contributes strongly to all three axes. Males thus reveal fewer actions, but with highly stereotyped performance. In all sequences, side-preen is seen as an inhibitor of gape by females in need of being calmed. By averting the bill, male side-preen allows a female to approach (point) without danger. Sky-twist and head-flick are simple rapid actions used at different times by males to signal the approach of a side-preen and thus improve synchronisation between male and female. Females use sky-twist and head-flick fairly imprecisely, but males employ them more specifically. A female intent upon approaching a little-known male, perhaps watches for a head-flick which warns her of an imminent side-preen, with the result that she can point immediately it happens (Apx. 13.4).

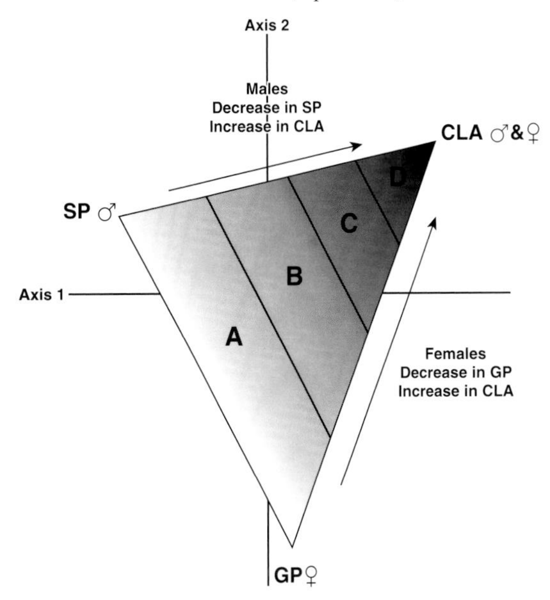

Apx. 13.2 The two-dimensional diagram combines data from axes 1 and 2. Lines connect the sequential actions of males (top) and females (below). The nuptial language progresses with time (from left to right) and four arbitrary language levels are proposed: young, newly arrived subadults (A), progressively older subadults (B, C) and breeding adults (D) (after Jouventin *et al.* 1981).

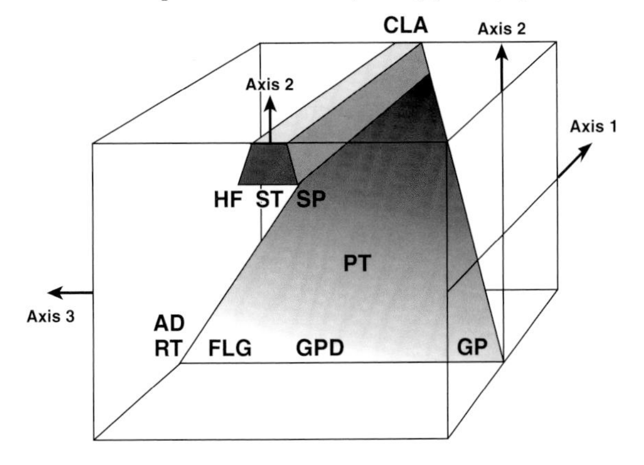

Apx. 13.3 In the three-dimensional diagram, males are represented by a trapezoid bar indicating a quick progression from HEAD-FLICK (HF), SKY-TWIST (ST) and SIDE-PREEN (SP) to CLASHING (CLA). This is repeated throughout the period of pairing. Females are represented by the sloping, triangular surface, incorporating a more varied range of behaviour that changes, over the period of pairing, from advancing (AD), retiring (RT), FLAGGING (FLG), GAPE (GP), GAPE DOWN (GPD), then POINTING (PT) and finally CLASHING (CLA) (Jouventin *et al.* 1981).

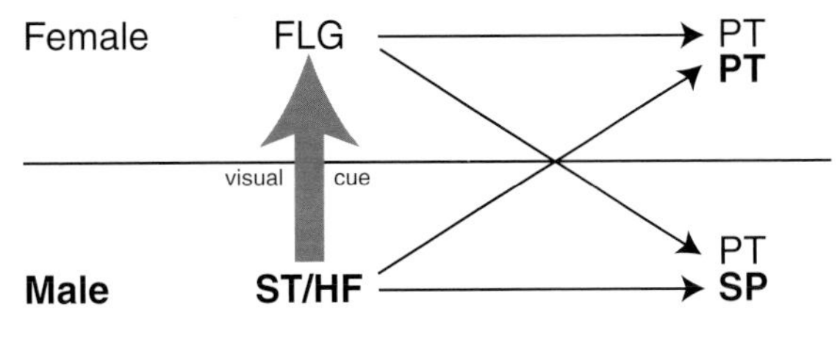

Apx. 13.4 Synchrony of POINTING (PT) and SIDE-PREEN (SP). A female looking (FLG) at a male is waiting for an opportunity to POINT. The male SKY-TWIST or HEAD-FLICK (ST/HF) is a signal that he is about to POINT or SIDE-PREEN. Either way she is ready to POINT (PT) at exactly the moment he POINTS at her, or when he SIDE-PREENS and is not looking in her direction (after Jouventin *et al.* 1981).

[1] An unpublished English translation of this work has been deposited in the library of the Edward Grey Institute, Universtiy of Oxford.

Appendix 14

A demographic model of albatross populations.

The following general formulation is a slightly expanded version of a Wandering Albatross model created in 1990 by J.P. Croxall, P. Rothery, S.P.C. Pickering and P.A. Prince. It appears rather intimidating, but it can be understood more readily if the repetitive elements are recognised.

It is assumed that the ratio of sexes is 1:1 and there is no lack of males, so only females are counted. All adult females are imagined as returning within four years of a successful breeding or within three years of an unsuccessful breeding. Variation due to breeding success is an important parameter.

The equation accommodates deferred breeding by annual and biennial breeders; in great albatrosses the obligatory pause in breeding requires a zero in one expression.

Survival is not included explicitly, but it may be implicit. Breeding success and survival are linked. If a parent dies before its chick can thrive and fledge on the food delivered by the remaining parent, then the level of breeding success will be partly due to parental mortality (survival). Deferred breeding and survival are also linked. Failed breeders are alive even if they are not visiting the breeding grounds.

Croxall *et al.* (1990b) have created a more detailed model incorporating the effects of adult survival. It is beyond the scope of this book and researchers contemplating prolonged demographic programmes will wish to study the original paper.

The number of females breeding in any one season will include adults that last bred, successfully or unsuccessfully, one or more years ago. To these are added individuals from several cohorts of subadults breeding for the first time.

$$
\begin{aligned}
N_{(i+1)} &= F_{1i}\{1-b_i\}N_i \\
&+ F_{2(i-1)}\{1-b_{(i-1)}\}N_{(i-1)} \\
&+ F_{3(i-2)}\{1-b_{(i-2)}\}N_{(i-2)} \\
&+ S_{1i}b_iN_i \ (\approx 0 \text{ in biennial species})^1 \\
&+ S_{2(i-1)}b_{(i-1)}N_{(i-1)} \\
&+ S_{3(i-2)}b_{(i-2)}N_{(i-2)} \\
&+ S_{4(i-3)}b_{(i-3)}N_{(i-3)} \\
&+ \Sigma R_{(i+1-k)}b_{(i+1-k)}N_{(i+1-k)}/2
\end{aligned}
$$

where: N_i = the number of females breeding in year i

b_i = the breeding success in year i

F_{1i}, $F_{2(i-1)}$ and $F_{3(i-2)}$ = the proportion of females that failed (lost eggs or chicks) in year i and bred again 1, 2,or ≥3 years later.

S_{1i}, $S_{2(i-1)}$, $S_{3(i-2)}$ and $S_{4(i-3)}$ = the proportion of females that successfully reared fledglings in year i and bred again 1, 2, 3, or ≥4 years later.

$R_{(i+1-k)}$ = the proportion of fledged chicks in year (i+1-k) [which are recruited at age k in year (i=1).

k = the age of subadults at the time of recruitment to the breeding population.

$R_{(i+1-k)}b_{(i+1-k)}N_{(i+1-k)}$ refers to subadults of both sexes so it is divided by 2 to make it consistent with females in the rest of the equation.

[1] Among Grey-headed Albatrosses a very small proportion of pairs are known to have bred again in the year following a successful breeding. For the purpose of this equation they may be considered negligible.

Appendix 15

Counts and estimates of breeding albatrosses.

Figures refer to the number of pairs breeding in a single season. Where the original record is a range e.g. 2,000–4,000, the mid-point (median) 3,000 has been adopted. The accuracy of figures vary, they can sometimes be assessed by referring to the original sources. The calendar year given is the one in which eggs were laid even if the count itself was of chicks made later in the same season, but in the following calendar year:

1962	the 1962–63 breeding season.
1973/4	either the1973 or 1974 breeding season where the precise season is uncertain.
1964/73	an unknown season between 1964 and 1973
1984–92	an average between 1984 and 1992 seasons

n	=	count of nests
e	=	count of eggs
c	=	count of chicks
f	=	count of fledglings (≈juveniles)
e/c	=	count at time of hatching when both eggs & chicks present
n/c	=	albatrosses present, but not counted

CAMPBELL BLACK-BROWED ALBATROSS

Location Year*	Number of pairs	Ref.
Campbell Island		
1975	74 825	85
1987	22 500	128
1988	23 400	129
1992	27 280	177

BLACK-BROWED ALBATROSS

Location Year*	Number of pairs	Ref.
South America		
Isla Diego de Almagro		
1983	10 000	100
1985	15 000	155
Islas Ildefonso		
1984	>8 500 c	115
1985	17 000	155
Islas Diego Ramírez		
1980	19 195	160
Falkland Islands		
Beauchêne Island		
	137 000	35
1980	162 360 n	88
1980	108 994 c	88
1991	134 934 n	170
1991	109 240 e/c	170
1995	149 363	155
Bird Island		
1995	17 500	155
New Island		
1960	<2 000	67
1992	>9 500	182
1992	8 910	170
1995	10 500	155
North Island		
1995	14 625	155
West Point Island		
1961	3 400 c	188
1961	>4 000	67
1989	12 050	142
1992	2 468 c	183
1994	15 400	155
Grave Cove		
1988	120	188
1992	170	155
Deaths Head		
	>10	140
South Jason Island		
1983	350	155
Elephant Jason Island		
1984	600	155
1997	250 n	188
Grand Jason Island		
1986	75 000	155
Steeple Jason Island		
1987	214 648 n	134
1987	165 804 e/c	134
1995	250 826	155
Saunders Island		
1932	4 000	162
1960s	2 000	67
1984–86	>2 000	188
1992	12 505 e	188
Keppel Island		
1987	2 085 e	120
South Georgia		
Willis Islands		
1970/75	>10 000	79
1984	33 570	146
Bird Island		
1960	5 661 c	27
1961	6 479 c	27
1962	7 747 c	183
1976	12 647	77
1990–11	4 695	146
1995	9 539	155
Jomfruene Islands		
1960	250	67
NW coast		
1970/75	3 000	79
1984	6 466	146
Elsehul		
1958	500	67
1970/75	1 500	79
Sörn & Bernt coast		
1970/75	1 000	79
Sörn & Bernt Islands		
1984	84	146
Cape North		
1984	2 050	146
Welcome Islands		
1970/75	20	79
1985	250	147
Cape Buller		
1970/75	2 500	79
1985	1 630	146
Cape Wilson		
1970/75	250	79
Cape Crewe		
1970/75	11	79
1985	322	146
Hercules Bay		
1970/75	1 000	79
Cooper Island		
1970/75	7 500	79
1985	11 785	146
Clerke Rocks		
1989	1 400	146
Rumbold Point		
1970/75	>2 000	79
1985	2 000	146

Location / Year	Number of pairs	Ref.
Green Island		
1970/75	5 000	79
1985	7 924	146
Annenkov Island		
1970/75	17 500	79
1985	17 500	146
Nuñez Peninsula		
1985	962	146
Klutschak Point		
1970/75	1 200	79
1985	850	146
Iles Crozet		
Ile de l'Est		
1981/82	350	104
Ile des Pingouins		
1981/82	300	104
1986	200	126
Iles des Apôtres		
1981/82	330	104
Iles Kerguelen		
Ile de Croy		
1982	300	104
Iles Nuageuses		
1984	1 815	121
Péninsule Loranchet		
1964	>1 000	38
Cap Français		
1985/86	250	121
Péninsule Jeanne d'Arc		
1978	1 200	97
1978	>1 750	97
1986	1 050	121
1987	905	121
Heard Island		
1947/55	234	24
1979	210	92
1986	650	136
McDonald Islands		
1979	79 c	92
Macquarie Island		
North Head		
1952	12	116
1955	11	116
1956	12	116
1959	1 f	116
1977/84	<3	116
Petrel Peak		
1959	1 f	116
1960	6 e	116
1971	10 c	116
1977/84	25	116
1992/3	29	149
1995	38	155
Bishop & Clerk Islets		
	14 c	39
	44 c	74
1992/3	91	149
1993	141	155
Snares Islands		
Toru, Western Chain		
1984	1 e	156
1985	1 c	156
Antipodes Islands		
Bollons Island		
1950	30	81
1968	6 c	82
1978	75	150
1978	111	150
	150	110
1989	86	150
1990	88	150
1992	93	150
1992	100	140
1993	75	150
1994	112	165
1995	115	165
Campbell Island		
(including BBAxCBBA pairs)		
1975	10	140
1984	3	157
1986	4	157
1992	24	157
1995	16	157

GREY-HEADED ALBATROSS

Location Year*	Number of pairs	Ref.
South America		
Islas Diego Ramírez		
1980	8 396	160
Islas Ildefonso		
1984	6	175
South Georgia		
Willis Islands		
1970/75	5 000	79
1984	15 030	146
Bird Island		
1960	4 829 c	27
1961	2 994 c	27
1962	5 766 c	183
1976	9 110	77
1977	9 146	77
1990–1	6 857	146
Jomfruene Islands		
1960	1 200	67
NW coast		
1970/75	17 000	79
1984	31 000	146
Elsehul		
1958	1 500	67
1970/75	5 600	79
Sörn & Bernt Islands		
1984	1 480	146
Cape North		
1984	415	146
Prince Edward Islands		
Marion Island		
	3 370	82
1983/90	5 037	122
1995	6 217	155
Prince Edward Island		
1972	880	87
	870	82
1979	1 500	86
Iles Crozet		
Ile de la Possession		
1981/82	10	104
Ile de l'Est		
1981/82	3 750	104
Ile des Pingouins		
1981/82	2 000	104
1986	2 000	126
Iles des Apôtres		
1981/82	180	104
Iles Kerguelen		
Ile de Croy		
1982	>5 000	104
1984/85	>7 860	121
Cap Français		
1985/86	45	121
Macquarie Island		
1984	90	116
1995	84	155
Campbell Island		
1975	11 530	62
1987	6 500	128
1988	5 800	129
1992	9 000	177
1995	6 400	155

WESTERN YELLOW-NOSED ALBATROSS

Location Year*	Number of pairs	Ref.
Tristan da Cunha		
Tristan Island		
1950/52	2 000	18
1950/52	3 000	23
1968	1 500	68
1972–3	23 000	106
	19 500	108
Inaccessible Island		
1950/52	1 400	18
1982	1 100	118
Nightingale Island		
1950/52	4 350	18
1950/52	8 000	23
1967	750	68
1972–3	4 500	106
Middle Island		
1972–3	150	106
Stoltenhoff Island		
1972–3	500	106
Gough Island		
1952–53	1 000	23
1955	<10 000	32
1972–73	7 500	106
1972/74	7 500	108

EASTERN YELLOW-NOSED ALBATROSS

Location Year*	Number of pairs	Ref.
Prince Edward Islands		
Prince Edward Island		
1964–5	2 000	47
1974/76	5 000	82
1979	7 000	86
	6 000	108
Iles Crozet		
Iles des Pingouins		
1981/82	5 800	104
1986	3 200	126
Iles des Apôtres		
1981/82	1 230	104
Iles Kerguelen		
Ile de Croy		
1984/85	50	121
Ile Amsterdam		
1970	15 000 c	51
1982	37 028	96
1981/94	25 000	155
Ile Saint Paul		
1970	7	51
1982	12	104

NORTHERN BULLER'S ALBATROSS

Location Year*	Number of pairs	Ref.
Chatham Islands		
Forty-Fours Islands		
1972	23 500	58
	16 000	140
Big Sister		
	1 500	140
Little Sister		
	500	140
1994	650	155
Three Kings Islands		
Princes Islands		
1983	5 e	109
	20	155

SOUTHERN BULLER'S ALBATROSS

Location Year*	Number of pairs	Ref.
Snares Islands		
Main & Broughton Islands		
1969/0	4 750	99
1992	8 460	148
Toru, Western Chain		
1984	5	105
Solander Islands		
Solander Island		
1985	4 500 c	112
Little Solander Island		
1985	300 c	112

TASMANIAN SHY ALBATROSS

Location Year*	Number of pairs	Ref.
Tasmania		
Albatross Island		
1959	650 c	29
1972	1 505 c	55
1993/4	5 000	163
Mewstone		
1977	1 750 e/c	75
1991	4 500	140
1993/4	7 000	163
Pedra Branca		
1978	100 e	75
1993/4	250	163

AUCKLAND SHY ALBATROSS

Location Year*	Number of pairs	Ref.
Auckland Islands		
Disappointment Island		
1972	60 000	62
1984	71 799 c	180
1992	71 683 e	180
Auckland Island		
1972	4 000	62
1989	>500	131
Adams Island		
1972	200	62
1989	50	131
1995	50	165
Antipodes Islands		
Bollons Island		
1993	6 c	150
1995	18	165
Chatham Islands		
Forty-Fours Islands		
1991	1	159
1996	1	159

SALVIN'S ALBATROSS

Location Year*	Number of pairs	Ref.
Bounty Islands		
1978	76 352	93
Snares Islands		
Toru, Western Chain		
1983	437 c	105
Rima, Western Chain		
1976	172 c	71
1983	150 c	105
Iles Crozet		
Ile des Pingouins		
1986	4	126

CHATHAM ALBATROSS

Location Year*	Number of pairs	Ref.
Chatham Islands		
The Pyramid		
1972	4 215 n	133
1973	4 000	58

SOOTY ALBATROSS

Location Year*	Number of pairs	Ref.
Tristan da Cunha		
Tristan Island		
1972/74	2 500	106
Inaccessible Island		
1950/52	2 000	23
1982	60	118
Nightingale Island		
1972/74	150	106
Stoltenhoff Island		
1972/74	37	106
Gough Island		
1972–4	7 500	106
1977	12 500	80
	5 000	139
Prince Edward Islands		
Marion Island		
	2 030	82
1983/90	2 055	122
Prince Edward Island		
1972	>1 000	87
	700	82
Iles Crozet		
Ile de l'Est		
	500	49
1981	1 300	104
Ile de la Possession		
	650	89
1980	633	187
1987	220	140
1995	273	155
Ile aux Cochons		
1973	450	65
1981	450	107
Ile des Pingouins		
1981	250	104
	200	126
Iles des Apôtres		
1981	25	104
Iles Kerguelen		
1973	3	65
1983/86	4	121
Ile Amsterdam		
	100	51
1995	350	155
Ile St Paul		
	20	51

LIGHT-MANTLED SOOTY ALBATROSS

Location Years*	Number of pairs	Ref.
South Georgia		
Bird Island		
	150	83
all islands		
	10 500	83
	5 000	98
	8 000	101
	6 250	140
Prince Edward Islands		
Marion Island		
	176	82
1983/90	201	122
Prince Edward Island		
	40	82
Iles Crozet		
Ile de l'Est		
1981	>900	104
Ile de la Possession		
1966/81	1 150	104
1995	996	155
Ile aux Couchons		
1974	75	65
1981	75	107
Ile des Pingouins		
1981	30	104
1986	300	126
Iles des Apôtres		
1981	150	104
Iles Kerguelen		
1983/86	4 000	121
Heard Island		
1947/55	350	24
Macquarie Island		
	600	50
1992	1 075	140
1994	2 000	174
Auckland Islands		
1972	5 000	59
Campbell Island		
1942/47	≥1 000	11
1995	≥1 600	152
Antipodes Islands		
1969	<1 000	81
1997	>169	175

WANDERING ALBATROSS

Location Year*	Number of pairs	Ref.
Prince Edward Islands		
Marion Island		
1951/2	700	19
1973	1 557 c	64
1976/7	1 852 f	113
1982	1 207 f	113
1983	1 137 f	113
1984	1 184 f	113
1985	1 168 f	113
1986	1 096 f	113
1987	1 066 f	113
1991	1 378	140
1995	1 794	155
Prince Edward Island		
1972	>758 c	87
1973	847 f	64
1976/7	966 f	113
1982	957 f	113
1983	1 135 f	113
1984	1 277 f	113
1985	>913 f	113
Iles Crozet		
Ile aux Couchons		
1963	2 200	36
1973	>884 c	65
1981	1 263 e	107
Ile de la Possession		
1959	475	184
1968	500	42
1981	250	104
1986	231	114
1992	312	140
1995	349	155
Ile de East		
1969	500	43
1971	437	49
1982	329	104
Iles des Apôtres		
1981	121	104
Iles Kerguelen		
Péninsule Courbet		
1971	375	57
1985	295	121
Baie Larose		
1985	9	121
Péninsule Raillier du Baty		
1987	750	121
Péninsule Joffre & Islands		
1987	15	121
Ile Howe		
1987	20	121

Iles Nuageuses		
1985	5	121
Iles Leygues		
1929	n/c	121
Heard Island		
1980	1	92
1987	0	155
Macquarie Island		
1911/13	1	4
1967	29 e	41
1974	>14 e	111
1975	16 e	111
1976	9 e	111
1977	7 e	111
1978	>11 e	111
1980	7 e	144
1986	6 e	144
1995	10	155
South Georgia		
Proud Island, Willis Islands		
1991	11 n	179
Bird Island		
1960	1 554 e	40
1961	1 922 e	40
1962	1 666 e	40
1963	2 093 e	40
1971	1 582 e	123
1972	1 477 e	123
1973	1 529 e	123
1975	1 433 e	123
1976	1 541 e	123
1977	1 382 e	123
1978	1 466 e	123
1979	1 339 e	123
1980	1 415 e	123
1981	1 404 e	123
1982	1 453 e	123
1983	1 366 e	123
1984	1 232 e	123
1985	1 491 e	123
1986	1 233 e	123
1987	1 366 e	123
1988	1 411 e	123
1993	1 329	140
1995	1 314	155
Cape Alexandra		
1958	31 f	25
1976	115	76
1982	5 f	115
1992	72 e	168
Hope Valley & coast		
1923	1 000	13
1946	298	14
1977	66	76
1982	25 f	115
1983	20 c	179
Weddell Point		
1983	30 c	179
Kade Point		
1982	26 f	115
1987	30 n	168
Saddle Island		
1982	34 f	115
1986	38 c	168
Cape Demidov		
1985	3 c	168
Samuel Islands & coast		
1983	20 c	179
Cape Rosa		
1987	7 n	168
Cape Nuñez		
1986	2 n	168
Annenkov Island		
1972	300	76
1983	168 f	115
Diaz Cove		
1986	2	168
Kupriyanov Islands		
1982	6 f	168
1986	5 n	168
Ranvik		
1986	1 n	168
Trollhul		
1983	6 c	179
Outer Lee		
1946	48	14
1993	16 e	168
Inner Lee		
1946	23	14
1976	22	76
1998	15 e	168
Petrel Island		
1946	8	14
1993	1 n	168
Prion Island		
1946	100	14
1976	61	76
1998	51 e	179
Skua Island		
1946	16	14
1993	1 e	168
Albatross Island		
1946	159	14
1976	142	76
1982	109	115
1983	100	115
1998	171 e	179
Invisible Island		
1946	15	14
1993	2 e	168
Crescent Island		
1946	31	14
1993	8 e	168
Mollyhawk Island		
1946	21	14
1993	5 e	168
Nameless Point		
1993	2 n	168

AUCKLAND WANDERING ALBATROSS

Location Year*	Number of pairs	Ref.
Auckland Islands		
Adams Island		
1972	7 000	62
1991	4 050 e	135
1995	5 762	140
1997	6 993 e	166
Auckland Island		
1972	50	62
1995	72	166
Disappointment Island		
1972	200	62
1993	254 e	185

ANTIPODES WANDERING ALBATROSS

Location Year*	Number of pairs	Ref.
Antipodes Islands		
Antipodes Island		
1969	930	81
1978	1 500	140
1994	4 522 e	172
1995	5 757 e	172
1996	5 148 e	172
Campbell Island		
1944	5 e/c	22
1958	4	26
1978	4	128
1987	5	128
1988	>4	128
1995	6	140

AMSTERDAM WANDERING ALBATROSS

Location Year*	Number of pairs	Ref.
Ile Amsterdam		
1978	5 f	119
1979	3 f	119
1980	0 f	119
1981	8 f	119

1982	1 e	119
1983	9 e	119
1984	5 e	119
1985	8 e	119
1986	8 e	119
1987	12 e	119
1988	11 e	119
1992	9	140
1995	13	155
1996	17 e	161

GOUGH WANDERING ALBATROSS

Location Year*	Number of pairs	Ref.
Tristan da Cunha		
Inaccessible Island		
1872/73	200	1
1938	2	16
1950/52	2/3	18
1981	1 c	118
1982	1 e	118
1985	3 c	118
1990	2/3	130
Gough Island		
1888	2 000	2
1955	2 000	32
1978	>792 c	94
1979	>661 c	113
1980	>431 c	113
1981	798 c	95

NORTHERN ROYAL ALBATROSS

Location Year*	Number of pairs	Ref.
Chatham Islands		
Big Sister Island		
1953	500	21
1972	1 246 e	133
1973	1 560	58
1974	1 196 e	133
1989	2 424 e	133
1990	2 315 e	133
1991	420 f	158
1992	200 f	158
1993	400 f	158
1994	720 f	158
Little Sister Island		
1953	775	21
1972	750 e	133
1973	1 150	133
1974	672 e	133
1988	201 f	133
1989	738 e	173
1990	1 065	133
1991	280 f	158
1992	240 f	158
1993	240 f	158
1994	600 f	158
Forty-Fours Islands		
1972	3 285 e	133
1973	1 749 c	133
1974	2 846 e	133
1988	719 f	147
1989	4 562	133
1990	4 520	133
1991	412	133
1992	225 f	147
1993	200 f	158
1994	740 f	158
New Zealand		
Taiaroa Head		
1935	1	17
1936	1	17
1937	2	17
1938	5	17
1939	6	17
1940	5	17
1941	6	17
1942	6	17
1943	6	17
1944	6	17
1945	5	17
1946	4	17
1947	5	17
1948	5	17
1949	6	17
1950	7	17
1951	8	17
1982	11	181
1989	20	186
1992	15	140
1994	8	181
1995	18	155

SOUTHERN ROYAL ALBATROSS

Location Year*	Number of pairs	Ref.
Campbell Island		
1957	2 278	30
1968	4 344	44
1976	4 906	128
1977	4 208	128
1979	4 575	128
1980	4 493	128
1982	4 243	128
1988	4 500	140
1994	6 308	157
1995	7 787	157
1996	7 800	155
Auckland Islands		
Enderby Island		
1972	18	62
1988	30	140
1995	46	181
1995	55	155
Auckland Island		
1972	3	62
1989	2	155
Adams Island		
1972	18	62
1990	30	140
1991	15	155

GALÁPAGOS ALBATROSS

Location year*	Number of pairs	Ref.
Islas Galápagos		
Isla Española		
1970	10 600	54
1971	12 000	54
1995	15 581	154
1995	18 200	164
Ecuador		
Isla de la Plata		
1975	5	66
1990	10	137

LAYSAN ALBATROSS

Location Year*	Number of pairs	Ref.
Izu Shoto		
Torishima		
1930	25	9
1933	1	9
Ogasawara Gunto		
Torishima, Mukojima Retto		
1976	1 f	73
1977	3 c	73
1979	14 e	103
1992	30	176
Kure Atoll		
Green Island		
1922	50 c	153
1956	270 c	28
1957	300 c	28
1959	75 c	52
1960	550	52
1961	1 080	52
1962	1 450	52
1978/81	3 500	102
1996	4 154 c	169

Midway Atoll

Sand Island

1922	≥1 000	153
1936/41	10 000	5
1944	37 500	7
1956	59 699	28
1956	60 000	28
1959	24 287 n	34
1962	46 000	33
1978	58 350	78
1981	26 060 c	171
1991	178 979 e	171
1991	>180 662	171
1995	182 574	171

Eastern Island

1922	>1 500	153
1944	17 500	7
1956	44 332	28
1957	42 625	28
1962	44 150	33
1978	175 000	78
1981	159 702 c	171
1991	247 412 e	171
1991	>247 665	171
1995	203 871	171

Seal & Rocky Islands

1957	55	28
1962	65	33

Spit Island

1991	1 021 e	171
1995	1 409	171

Pearl & Hermes Reef

North Island

1912	300 c	56
1962	384 f	56
1964	750 c	56
1966	525 c	56

Southeast Island

1922	300	153
1962	9 000 f	56
1963	7 500 c	56
1964	5 000 c	56
1966	5 300 c	56
1967	11 269 c	56
1968	6 075 c	56

Grass Island

1922	100 c	153
1962	607 f	56
1963	750 c	56
1964	337 c	56
1966	220 c	56
1967	1 120 c	56

Kittery Island

1956	0	28
1957	0	28
1962	27 c	56
1963	4 c	56
1964	7 c	56
1966	48 c	56
1967	52 c	56

Seal Island

1922	150	153
1962	186 f	56
1963	75 c	56
1964	91 c	56
1966	40 c	56
1967	399	56

Lisianski Island

1922	≤800 c	153
1953	2 000 c	60
1956	29 141	28
1963	2 600 c	60
1964	2 750 c	60
1965	4 000 f	60
1966	2 500 c	60
1979	17 626 c	171
1982	26 500 e	169

Laysan Island

1910	90 000 c	3
1912	12 312 c	15
1922	3 400 c	153
1950	103 900 f	12
1956	215 000 f	53
1957	130 554	28
1958	69 000 c	53
1963	150 000 c	53
1964	22 500 c	53
1965	150 000 f	53
1966	45 000 c	53
1967	80 000 c	53
1968	77 300 c	53
1978/81	118 500	102
1979	132 000	142
1991	145 947 e	171
1992	146 222 e	171
1994	135 252	155
1998	124 113 e	169

Gardner Pinnacles

1922	2 c	153
1957	2	28
1962	7 f	48
1966	1 f	48
1968	0	48
1978/81	12	102
	15	155

French Frigate Shoals

Tern Island

1953	28	45
1957	24	28
1959	33	45
1960	24	45
1961	34 f	45
1962	41 f	45
1965	200 c	45
1966	200 c	45
1967	156	45
1968	37 f	45
1976	186	171
1978	295	171
1979	505	171
1982	797	142
1983	854	142
1986	1 032	142
1987	990	142
1988	1 060	142
1989	1 303	142
1990	1 570	142
1991	1 570 e	171
1992	1 407 e	171
1993	1 652 e	171

Trig Island

1922	1 f	153
1953	100	20
1957	100	28
1962	8 f	45
1965	44 c	45
1966	10 c	45
1967	16 f	45
1968	17 f	45

Whale-Skate Island

1922	26 f	153
1953	300	20
1957	260	28
1962	33 f	45
1965	95 f	45
1966	75 f	45
1967	75 f	45
1968	70 f	45

East Island

1922	100 f	153
1948	>25	45
1953	200	45
1957	200	28
1962	247 f	45
1965	274 c	45
1966	339 c	45
1967	325 f	45
1988	981 e	171

Gin Island

1953	2	20

Necker Island

1902	1 500	28
1922	600 f	153
1957	2 495	28
1965	550 c	69
1969	>510	69
1978	308 c	171
1994	375 c	169

Nihoa

1922	2 c	153
1957	500	28

1968	4 c	70
1973	20 f	70
1978/81	3	102
1995	0	169
Niihau		
1957	500	28
1987	175	124
Kauai		
1944	1 c	6
1987	40	124
1990	74	132
1993	189	171
1995	100 e	169
Kaula		
1981	50	102
	63 e	169
Oahu		
1946	1	8
1990	>2	132
1994	10	155
Johnston Atoll		
1922	1 f	153
Mexico		
Isla Clarión		
1988	2	125
Isla San Benedicto		
1987	1	125
Isla de Guadalupe		
1983	4	151
1984	5	151
1985	6 c	117
1987	12	151
1988	11 e/c	151
1989	18 c	151
1990	21 c	151
1991	45	151

BLACK-FOOTED ALBATROSS

Location Year*	Number of pairs	Ref.
Senkaku Retto		
Kita-kojima		
1970	1 e	46
1991	20	155
1991	25 e	169
Minami-kojima		
1970	n/c	103
Izu Shoto		
Torishima		
1931	100	9
1956	6 c	176
1957	6 c	176
1958	11 c	176
1959	10 c	176
1960	11 c	176
1961	18 c	176
1962	13 c	176
1963	20 f	72
1964	20 c	176
1972	51 f	183
1976	126 c	72
1978	>100	176
1979	160 c	176
1981	200 c	103
1982	218 c	176
1984	347 c	176
1985	333 c	176
1988	405 c	176
1989	536 c	176
1990	571 c	176
1991	581 c	176
1992	643 c	176
1993	578 e/c	145
1993	716 c	176
1994	785 c	176
1994	1 056	178
1995	1 300	176
1995	641 c	176
1996	597 c	176
1997	914 c	176
Ogasawara Gunto		
Nakondojima, Mukojima Retto		
1974	n/c	103
Torishima		
1976/7	n/c	103
1993	1 000	176
Kure Atoll		
Green Island		
1922	300 c	153
1956	42 c	28
1957	50 c	28
1959	95 c	52
1960	160	52
1961	65	52
1962	200	52
1963	235 e	52
1964	271 e	52
1968	331 e	52
1978/81	1 000	102
1996	1 240 c	171
Midway Atoll		
Sand Island		
1922	1 000 c	153
1936/41	15 000	5
1944	17 500 c	7
1954	7 700	34
1956	3 659 e	28
1957	6 188	28
1959	1 435 n	34
1960	4 100	34
1961	4 050	34
1963	2 700	34
1978	3 000 c	78
1991	9 227 e	171
1994	8 117	171
1995	8 852	171
1996	10 107	171
Eastern Island		
1922	1 000 c	153
1944	9 000 c	7
1954	2 000	34
1956	2 286	28
1957	2 333	28
1959	2 835	34
1960	2 750	34
1961	2 900	34
1963	2 000	34
1991	>10 529	171
1994	10 614	171
1995	10 397	171
1996	11 530	171
Spit Island		
1991	1 e	171
1994	0	171
1995	6	171
1996	8	171
Pearl & Hermes Reef		
North Island		
1912	325 c	56
1956	1 900	28
1962	751 f	56
1964	550 c	56
1966	1 150 c	56
1981	1 633 c	171
Little North Island		
1962	>64 f	56
1964	16 c	56
1968	0 c	56
1981	0 c	171
Southeast Island		
1922	1 000 c	153
1956	2 300	28
1962	5 000 c	56
1963	>2 500 c	56
1964	2 000 c	56
1966	1 560 c	56
1967	2 002 c	56
1968	>1 964 f	56
1981	2 184 c	171
Bird Island		
1963	14 c	56

1966	19 f	56
1981	1	171
Grass Island		
1922	800	153
1956	1 900	28
1963	1 275 c	56
1964	480 c	26
1966	1 466 c	26
1967	1 190 c	26
1981	1 330 c	171
Kittery Island		
1956	450	28
1962	124 f	56
1963	126 c	56
1964	213 c	56
1966	353 c	56
1967	228 c	56
Seal Island		
1922	1 200	153
1956	370	28
1963	275 c	56
1964	42 c	56
1966	329 c	56
1967	219 c	56
Lisianski Island		
1922	1 000 c	153
1953	2 000 c	60
1956	3 665 e	28
1957	3 490 e	28
1957	2 618	28
1963	1 750 c	60
1964	1 400 c	60
1965	993 f	60
1966	1 161 f	60
1978	1 891	171
1981	2 109 c	171
1982	2 800	155
1996	2 683 c	171
Laysan Island		
1910	42 500	3
1912	7 722	15
1914	20 000 c	53
1922	4 700 c	153
1950	18 240 f	12
1956	32 128	28
1957	33 523	28
1963	>10 700 c	53
1964	12 500 c	53
1965	10 000 c	53
1966	15 000 c	53
1967	7 500 c	53
1968	14 694 c	53
1979	21 000	102
1991	25 109 e	171
1992	29 559 e	171
1993	32 414 e	171
1994	22 805 e	171
1995	24 813	155
1996	26 723	171
1997	24 519 e	171
1998	27 472 n	171
French Frigate Shoals		
Tern Island		
1922	8 f	153
1953	6	20
1957	2	28
1959	1	45
1965	10 c	45
1967	4 c	45
1968	2 c	45
1976	17 c	171
1978	14	171
1979	82	142
1980	96	171
1981	149	171
1982	193	142
1983	221	142
1984	292	171
1985	304	171
1986	448	142
1987	451	142
1988	516	142
1989	618	142
1990	690	142
1991	767	171
1992	895	171
1993	918	171
1994	1 034	171
1995	1 048	171
1996	1 304	171
1997	1 519	171
Trig Island		
1922	50 f	153
1953	200	20
1957	130	28
1962	39 f	45
1965	20 f	45
1966	51 c	45
1967	30 f	45
1968	80 f	45
1976	17 c	171
1978	92 c	171
1979	111 e	171
1980	66 n	171
1981	75 c	171
1982	77 n	171
1983	81 n	171
1985	0 n	171
1986	125 n	171
1987	31 c	171
1989	87 n	171
1990	38 n	171
1991	12 n	171
1992	66 n	171
1993	36 n	171
1994	10 n	171
1995	10 n	171
1996	24 n	171
Whale-Skate Island		
1922	120 f	153
1953	800	20
1957	840	28
1962	390 f	45
1965	528 c	45
1966	431 c	45
1967	>300 c	45
1968	>550 f	45
1976	300 c	171
1978	880 c	171
1979	1 130 e	171
1980	1 076 n	171
1981	1 172 n	171
1982	1 408 n	171
1983	2 046 n	171
1984	797 n	171
1985	576 n	171
1986	1 844 n	171
1987	1 190 c	171
1988	1 735 n	171
1989	1 473 n	171
1990	1 148 n	171
1991	556 n	171
1992	643 n	171
1993	818 n	171
1994	158 n	171
1995	97 n	171
Round Island		
1922	90 f	153
1957	12	28
East Island		
1922	75 f	153
1953	200	20
1957	170	28
1962	420	45
1965	547 c	45
1967	>600 f	45
1968	800 f	45
1976	450 c	171
1978	500 c	171
1979	2 500 n	171
1981	1 970 n	171
1982	2 167 n	171
1983	2 007 n	171
1984	1 536 c	171
1985	1 823 n	171
1986	2 541 n	171
1987	1 660 c	171
1988	2 009 n	171
1989	2 118 c	171
1990	1 853 n	171
1991	2 052 n	171
1992	1 974 n	171
1993	1 838 n	171
1994	1 881 n	171
1995	1 587 n	171

1996	1 750 n	171
Gin Island		
1953	14	20
1957	3	28
1982	0 c	171
1990	0 c	171
1992	1 n	171
1993	2 c	171
1994	2 n	171
1996	2 n	171
Little Gin Island		
1922	150 f	153
1953	350	20
1957	340	28
1962	17 f	45
1965	252 c	45
1966	75 f	45
1967	23 f	45
1968	>125 f	45
1979	103 c	171
1982	199 c	171
1983	206 n	171
1984	183 c	171
1986	82 c	171
1987	190 c	171
1988	146 c	171
1989	206 c	171
1990	179 c	171
1991	160 c	171
1992	238 n	171
1993	204 n	171
1994	195 n	171
1996	241 n	171
Disappearing Island		
1957	2	28
Necker Island		
1922	100 f	153
1957	368	28
1964	94 c	69
1966	27 c	69
1968	187 c	69
1978	164 c	171
1985	>288	171
1986	>259	171
Nihoa		
1922	60 f	153
1957	50 n	28
1963	60 c	70
1964	45 c	70
1967	60 c	70
1980	26 n	171
1985	0	171
1986	>28	171
Kaula		
1983	2 c	169
1993	5	171
Johnston Atoll		
1922	1	153

STELLER'S ALBATROSS

Location Year*	Number of pairs	Ref.
Senkaku Retto		
Minami-kojima		
1987	7 c	176
1990	10	176
1991	11	176
Izu Shoto		
Torishima		
1954	7 e	37
1955	12 e	176
1956	12 e	176
1957	13 e	176
1958	10 e	176
1959	10 e	176
1960	19 e	176
1961	24 e	176
1962	23 e	176
1963	26 e	176
1964	28 e	176
1972	24 f	63
1973	11 f	176
1976	15 c	72
1977	16 c	84
1978	22 c	84
1979	50 e	90
1980	54 e	90
1981	63 e	91
1982	67 e	176
1983	65 e	176
1984	73 e	176
1985	76 e	176
1986	77 e	176
1987	84 e	176
1988	89 e	176
1989	94 e	176
1990	108 e	176
1991	115 e	176
1992	139 e	176
1993	146 e	176
1994	153 e	176
1995	158 e	176
1996	176 e	176
1997	194 e	176
1998	213 e	176
1999	220 e	176

References: 1 Stoltenhoff (1873), 2 Verrill (1895), 3 Dill & Bryan (1912), 4 Mawson (1915), 5 Hadden (1941), 6 Munro (1945a), 7 Fisher & Baldwin (1946), 8 Fisher (1948), 9 Austin (1949), 10 Fisher (1949), 11 Sorensen (1950a&b), 12 Brock (1951), 13 Matthews (1951), 14 Rankin (1951), 15 Bailey (1952), 16 Hagen (1952), 17 Richdale (1952), 18 Elliott (1953), 19 Rand (1954), 20 Richardson (1954), 21 Dawson (1955), 22 Oliver (1955), 23 Elliott (1957), 24 Downes *et al.* (1959), 25 Tickell & Cordall (1960), 26 Westerskov (1960), 27 Dollman (1962), 28 Rice & Kenyon (1962a), 29 Macdonald & Green (1963), 30 Westerskov (1963), 31 Strange (1965), 32 Swales (1965), 33 Fisher (1966c), 34 Robbins (1966), 35 Woods (1966), 36 Dreux & Milon (1967), 37 Fujisawa (1967), 38 Tollu (1967), 39 MacKenzie (1968), 40 Tickell (1968), 41 Carrick & Ingham (1970), 42 Mougin (1970), 43 Prevost (1970), 44 Taylor *et al.* (1970), 45 Amerson (1971), 46 Ikehara & Shimojana (1971), 47 van Zinderen Bakker Jr. (1971), 48 Clapp (1972), 49 Despin *et al.* (1972), 50 Kerry & Colback (1972), 51 Segonzac (1972), 52 Woodward (1972), 53 Ely & Clapp (1973), 54 Harris (1973), 55 Green (1974), 56 Amerson *et al.* (1974), 57 Derenne *et al.* (1974), 58 Robertson (1974), 59 Bell (1975), 60 Clapp & Wirtz (1975), 61 Johnstone *et al.* (1975), 62 Robertson (1975b), 63 Tickell (1975), 64 Williams *et al.* (1975), 65 Derenne *et al.* (1976), 66 Owre (1976), 67 Tickell (1976), 68 Wace & Holdgate (1976), 69 Clapp & Kridler (1977), 70 Clapp *et al.* (1977), 71 Sagar (1977), 72 Hasegawa 1978, 73 Kurata (1978), 74 Lugg *et al.* (1978), 75 Brothers (1979b&c), 76 Croxall (1979), 77 Croxall & Prince (1979), 78 Ludwig *et al.* (1979), 79 Prince & Payne (1979), 80 Voisin (1979), 81 Warham & Bell (1979), 82 Williams *et al.* (1979), 83 Croxall & Prince (1980), 84 Hasegawa (1980), 85 Robertson (1980), 86 Berruti *et al.* (1981), 87 Grindley (1981), 88 Prince (1982), 89 Weimerskirch (1981), 90 Hasegawa (1982), 91 Hasegawa & DeGange (1982), 92 Johnstone (1982), 93 Robertson & van Tets (1982), 94 Williams & Imber (1982), 95 Cooper (1983), 96 Jouventin *et al.* (1983), 97 Thomas (1983), 98 Thomas *et al.* (1983), 99 Warham & Bennington (1983), 100 Clark (1984), 101 Croxall *et al.* (1984a), 102 Fefer *et al.* (1984), 103 Hasegawa (1984), 104 Jouventin *et al.* (1984), 105 Miskelly (1984), 106 Richardson (1984), 107 Voisin (1984), 108 Williams (1984), 109 Wright (1984), 110 Robertson (1985), 111 Tomkins (1985), 112 Cooper *et al.* (1986), 113 Watkins (1987), 114 Weimerskirch & Jouventin (1987), 115 Clark (1988), 116 Copson (1988), 117 Dunlap (1988), 118 Fraser *et al.* (1988), 119 Jouventin *et al.* (1989), 120 Thompson (1989), 121 Weimerskirch *et al.* (1989), 122

Cooper & Brown (1990), 123 Croxall *et al.* (1990b), 124 Harrison (1990), 125 Howell & Webb (1990), 126 Jouventin (1990), 127 Marchant & Higgins (1990), 128 Moore & Moffat (1990a), 129 Moore & Moffat (1990b), 130 Ryan *et al.* (1990), 131 Buckingham *et al.* (1991), 132 Pyle (1991), 133 Robertson (1991), 134 Thompson & Rothery (1991), 135 Walker *et al.* (1991), 136 Woehler (1991), 137 Ortiz-Crespo & Agnew (1992), 138 Voisin (1992), 139 Cooper & Ryan (1993), 140 Gales (1993), 141 McDermond (1993), 142 McDermond & Morgan (1993), 143 Robertson & Richdale (1993), 144 de la Mare & Kerry (1994), 145 Ogi *et al.* (1994b), 146 Prince *et al.* (1994), 147 Robertson & Sawyer (1994), 148 Sagar *et al.* 1994, 149 Scott 1994, 150 Clark & Robertson (1996), 151 Gallo-Reynoso & Figueroa-Carranza (1996), 152 Moore (1996), 153 Olson (1996), 154 Anderson & Cruz (1997), 155 Gales (1997), 156 Miskelly (1997), 157 Moore *et al.* (1997b), 158 Robertson (1997), 159 Robertson *et al.* (1997), 160 Schlatter & Riveros (1997), 161 Weimerskirch & Jouventin (1997), 162 Woods & Woods (1997), 163 Brothers *et al.* (1998b&c), 164 Douglas (1998), 165 Tennyson *et al.* (1998), 166 Walker & Elliott (1999). Unpublished sources: 167 Australian Antarctic Division, 168 British Antarctic Survey, 169 Black-footed Albatross Population Biology Workshop, 170 Falkland Islands Seabirds Monitoring Programme, 171 United States Fish & Wildlife Service, 172 J. Amey, 173 R. Chappell, 174 T. Disney, 175 G. Glark, 176 H. Hasegawa, 177 P.J. Moore, 178 H. Ogi, 179 J. & S. Poncet, 180 A. Rebergen, 181 C.J.R. Robertson, 182 K.R. Thompson, 183 W.L.N. Tickell, 184 R. Tufft, 185 K. Walker, 186 S. Webb, 187 H. Weimerskirch, 188 R.W. Woods, 189 A. Wright.

PHOTOGRAPHIC SECTION

Plate 1 A Black-browed Albatross on a newly renovated nest in early October, soon after it had returned to Bird Island, South Georgia (W.L.N. Tickell).

Plate 2 Part of the great Black-browed Albatross colony on Steeple Jason Island, Falkland Islands (W.L.N. Tickell).

Plate 3 Black-browed Albatross pointing display. (W.L.N. Tickell).

Plate 4 A Campbell Black-browed Albatross on its nest at Campbell Island. Note the pale iris characteristic of this population (G.A. Taylor).

Plate 5 Bull Rock South colony, above the sea cliffs of Campbell Island. Most of these birds are Campbell Black-browed Albatrosses, but there are also Grey-headed Albatrosses in this colony (G.A. Taylor).

Plate 6 A Western Yellow-nosed Albatross on a nest at Gough Island (R.W. Furness).

Plate 7 An Eastern Yellow-nosed Albatross. Note the pointed end of the yellow culminicorn (H. Weimerskirch).

Plate 8 A pair of Grey-headed Albatrosses at a nest on Bird Island, South Georgia (B. Osborne).

Plate 9 (above) A Northern Buller's Albatross. Note the very broad yellow culminicorn and conspicuous yellow stripe on the lower edge of the bill (G.B. Nunn).

Plate 10 (left) A Northern Buller's Albatross above the sea cliffs of a rocky islet in the Chatham Islands (G.B. Nunn).

Plate 11 (below) A Southern Buller's Albatross on a nest in the *Olearia* forest of the Snares Islands (J. Warham).

Plate 12 A Tasmanian Shy Albatross (G.B. Nunn).

Plate 13 An Auckland Shy Albatross in one of the great colonies on Disappointment Island (C. Fraser).

Plate 14 Salvin's Albatrosses and their chicks in one of the many mixed colonies with Erect-crested Penguins on the Bounty Islands. Note the lack of vegetation (R.H. Taylor).

Plate 15 Two Chatham Albatrosses on the Pyramid, Chatham Islands (G.B. Nunn).

Plate 16 A pair of Sooty Albatrosses at a nest on Gough Island (R.W. Furness).

Plate 17 A Sooty Albatross with its chick on the Ile de la Possession, Iles Crozet (H. Weimerskirch).

Plate 18 A Light-mantled Sooty Albatross flying along the cliffs of Bird Island, South Georgia (B. Osborne).

Plate 19 A Light-mantled Sooty Albatross. Note the blue sulcus on the bill and the transverse crest (B. Osborne).

Plate 20 A pair of Wandering Albatrosses, South Georgia (B. Osborne).

Plate 21 A Wandering Albatross at Bird Island, South Georgia, showing pink ear stain extending around the back of the head and neck (B. Osborne).

Plate 22 Wandering Albatrosses dancing in summer, at Bird Island, South Georgia (B. Osborne).

Plate 23 A Wandering Albatross chick at Bird Island, South Georgia in mid-winter (W.L.N. Tickell).

Plate 24 A Wandering Albatross male visiting its fledgling in winter, at Bird Island, South Georgia (W.L.N. Tickell).

Plate 25 A Wandering Albatross fledgling at Bird Island, South Georgia in the late winter (W.L.N. Tickell).

Plate 26 An Auckland Wandering Albatross male on a nest among megaherbs at Adams Island, Auckland Islands (K. Walker).

Plate 27 An Auckland Wandering Albatross female on a nest among tussock grass at Adams Island, Auckland Islands. Note the slightly darker plumage (K. Walker).

Plate 28 A Gough Wandering Albatross feeding its fledgling on Gough Island (R.W. Furness).

Plate 29 A pair of Antipodes Wandering Albatrosses at Campbell Island. Note the darker plumage of the female and the black crown of the male (G.A. Taylor).

Plate 30 An Amsterdam Wandering Albatross at the Ile Amsterdam. Both males and females breed in dark plumage. Note the black cutting-edge of the bill (J.-P. Roux).

Plate 31 A pair of Northern Royal Albatrosses. The female still has a few dark feathers on its crown. The black cutting-edges of the bills are characteristic of Northern and Southern Royal Albatrosses (G.B. Nunn).

Plate 32 The crowded colony of Northern Royal Albatrosses on Little Sister Islet in the Chatham Islands (K. Westerskov).

Plate 33 Northern Royal Albatrosses breeding on the rocky islets of the Chatham Islands are very exposed. In gales, many have difficulty hanging on to their inadequate nests and some may lose eggs. The faces of these birds on Little Sister Islet are dirty with wind-blown soil (K. Westerskov).

Plate 36 Southern Royal Albatross (K. Westerkov).

Plate 34 (opposite top) Northern Royal Albatrosses dancing at night on Little Sister Islet in the Chatham Islands (K. Westerkov).

Plate 35 (opposite bottom) A Southern Royal Albatross on its nest at Campbell Island (D. Allan).

Plate 37 A pair of Galápagos Albatrosses at Isla Española (S. Halvorsen).

Plate 38 Galápagos Albatross gape display on the breeding grounds – rocky clearings among thorn scrub – of Isla Española (H. Douglas).

Plate 39 An unusual colony of Galápagos Albatrosses on a beach near the Radar Landing of Isla Española (W.L.N. Tickell).

Plate 42 Laysan Albatross with its chick at Midway Atoll (S. Halvorsen).

Plate 40 (opposite top) Laysan Albatross flying over the coral reef of Midway Atoll (W.L.N. Tickell).

Plate 41 (opposite bottom) Laysan Albatrosses with chicks on Sand Island, Midway Atoll (S. Halvorsen).

Plate 43 Laysan Albatross display at Midway Atoll: sky-groan (W.L.N. Tickell).

Plate 44 Laysan Albatross display at Midway Atoll: preen-under-wing (W.L.N. Tickell).

Plate 45 Black-footed Albatross display at Midway Atoll: shake and whistle (S. Halvorsen).

Plate 46 Black-footed Albatross display at Midway Atoll: preen-under-wing (S. Halvorsen).

Plate 47 Black-footed Albatross. Note the white plumage around the vent (S. Halvorsen).

Plate 48 Black-footed Albatross (S. Halvorsen).

Plate 49 A pair of Steller's Albatrosses at Torishima (H. Hasegawa).

Plate 50 Steller's Albatrosses on the sea off Torishima. Note the various stages of plumage development (H. Hasegawa).

Plate 51 (right) A Steller's Albatross fledgling at Torishima (H. Hasegawa).

Plate 52 (below) Steller's Albatross display at Torishima: sky-call (H. Hasegawa).

Bibliography

Abbott, C.C. 1861. Notes on the birds of the Falkland Islands. *Ibis* 1: 149-167.

Adams, N.J. & Brown, C.R. 1984. Metabolic rates of sub-antarctic procellariiformes: a comparative study. *Comp. Biochem. Physiol.* 77A: 169-173.

Adams, N.J., Brown, C.R. & Nagy, K.A. 1986. Energy expenditure of free-ranging Wandering Albatrosses *Diomedea exulans*. *Physiol. Zool.* 59: 583-591.

Afanasyev, V. & Prince, P. A. 1993. A miniture storing activity recorder for seabird species. *Ornis Scand.* 24: 243-246.

Agnew, P. 1990. *Isla de la Plata Ecuador 1990.* Unpublished expedition report, University of Bristol, Bristol.

Airy, H. 1883. The soaring of birds. *Nature* 27: 590-592.

Albin, E. 1738. *A Natural History of Birds.* London.

Alerstam, T., Gudmundssen, G.A. & Larsson, B. 1993. Flight tracks and speeds of Antarctic and Atlantic seabirds: radar and optical measurements. *Phil. Trans. R. Soc. Lond.* 340B: 55-67.

Alexander, K., Robertson, G. & Gales, R. 1997. *Incidental Mortality of Albatrosses in Longline Fisheries.* Australian Antarctic Division, Kingston.

Alexander, P. 1982. *Roy Campbell A critical biography.* Oxford University Press, Oxford.

Alexander, W.B. 1928. *Birds of the Ocean.* Putnam, New York.

Alexander, W.B., Falla, R.A., Jouanin, C., Murphy, R.C., Salomonsen, F., Voous, K.H., Watson, G.E., Bourne, W.R.P., Fleming, C.A., Kuroda, N.H., Rowan, M.K., Serventy, D.L., Tickell, W.L.N., Warham, J. & Winterbottom, J.M. 1965. The families and genera of the petrels and their names. *Ibis* 107: 401-405.

Amadon, D. 1963. The Name for the Royal Albatross. *Emu* 63: 264.

Amadon, D. 1974. In memoriam: Robert Cushman Murphy, April 29, 1887 – March 20, 1973. *Auk* 91: 1-9.

American Ornithologists' Union 1910. *Check-List of North American Birds* 3rd edition. AOU, New York.

American Ornithologists' Union 1957. *Check-List of North American Birds.* Port City Press, Baltimore.

Amerson, A.B. Jr. 1969. Ornithology of the Marshall and Gilbert Islands. *Atoll Res. Bull.* 127: 1-348.

Amerson, A.B. Jr. 1971. The Natural History of French Frigate Shoals, Northwestern Hawaiian Islands. *Atoll Res. Bull.* 150: 1-383.

Amerson, A.B. Jr., Clapp, R.B. & Wirtz, W.O. II 1974. The Natural History of Pearl and Hermes Reef, Northwestern Hawaiian Islands. *Atoll Res. Bull.* 174: 1-306.

Amerson, A.B. Jr. & Shelton, P.C. 1976. The Natural History of Johnston Atoll. Central Pacific Ocean. *Atoll Res. Bull.* 192: 1-479.

Amiet, L. 1958. The distribution of *Diomedea* in eastern Australian waters: north of Sydney. *Notornis* 7: 219-230.

Amlaner, C.J. Jr. & Ball, N.J. 1983. A synthesis of sleep in wild birds. *Behaviour* 87: 85-119.

Andersen, K. 1895. *Diomedea melanophrys* in the Faeroe Islands. *Proc. R. Phys. Soc. Edinb.* 13: 91-114.

Anderson, D.J. & Cruz, F. 1997. Biology and management of the Waved Albatross at the Galápagos Islands. In *Albatross Biology and Conservation* (Robertson, G. & Gales, R., eds 1998): 105-109. Surrey Beatty, Chipping Norton.

Anderson, D.J. & Fernandez, P. (in press). Movements of Laysan and Black-footed Albatrosses at sea, January-August 1998. In Black-footed Albatross Population Biology Workshop.

Anderson, D.J. & Fortner, S. 1988. Waved Albatross Egg Neglect and Associated Mosquito Ectoparasitism. *Condor* 90: 727-729.

Anderson, D.J., Schwandt, A.J. & Douglas, H.D. 1997. Foraging ranges of Waved Albatrosses in the eastern Tropical Pacific Ocean. In *Albatross Biology and Conservation* (Robertson, G. & Gales, R., eds 1998): 180-185). Surrey Beatty, Chipping Norton.

Angot, M. 1954. Notes sur quelques oiseaux de L'Archipel de Kerguelen. *Oiseau Revue fr. Orn.* 24: 124-127.

Anon. 1959. *Daily Telegraph* 7 July.

Anon. 1995. High albatross mortality from one Confish vessel, says fisheries report. *Penguin News* 28 January. Falkland Islands. [reply 4 February 1995]

Anon. 1997. Britain's oldest bird presumed lost at sea. *Times* 12 November.

Anthony, 1898. Four seabirds new to the fauna of North America. *Auk* 15: 38-39.

Argyll, 8th Duke. 1867. *The Reign of Law.* Murray, London.

Armstrong, E.A. 1958. *The Folklore of Birds.* Collins, London.

Arnould, J.P.Y., Briggs, D.R., Croxall, J.P., Prince, P.A. & Wood, A.G. 1996. The foraging behaviour and energetics of wandering albatrosses brooding chicks. *Antarct. Sci.* 8: 229-236.

Ashford, J.R. & Croxall, J.P. 1995. Seabird interactions with longlining operations for *Dissosticus eleginoides* around South Georgia, April to May 1994. *CCAMLR Sci.* 2: 111-121.

Ashford, J.R., Croxall, J.P., Rubilar, P.S. & Moreno, C.S. 1994. Seabird interactions with longlining operations for *Dissosticus eleginoides* at the South Sandwich Islands and South Georgia. *CCAMLR Sci.* 1: 143-153.

Ashmole, N.P. 1962. The Black Noddy *Anous tenuirostris* on Ascension Island. Part 1. General Biology. *Ibis* 103b: 235-273.

Ashmole, N.P. 1963. The regulation of numbers of tropical oceanic birds. *Ibis* 103b: 458-473.

Ashmole, N.P. 1971. Seabird ecology and the marine environment. In *Avian Biology* (Farner, D.S., King, J.R. & Parkes, K.C., eds) 1: 223-286. Academic Press, New York.

Astheimer, L.B., Prince, P.A. & Grau, C.R. 1985. Egg formation and pre-egg period of Black-browed and Grey-headed Albatrosses *Diomedea melanophris* and *D. chrysostoma* at Bird Island, South Georgia. *Ibis* 127: 523-529.

Atkinson, I.A.E. 1985. The spread of commensal species of *Rattus* to oceanic islands and their effects on island avifaunas. In *Conservation of Island Birds* (Moore, P.J., ed.): 35-81. ICBP, Cambridge.

Atyeo, W.T. & Peterson, P.C. 1970. Acarina: Astigmata: Analgoidea: Feather mites of South Georgia and Heard Islands. *Pac. Insects Monogr.* 23: 121-151.

Aubert de la Rüe, E. 1959. Quelques observations faites aux iles Diego Ramirez (Chile). *Bull. Mus. 2e sér.* 31: 387-391.

Aubert de la Rüe, E. 1964. Observations sur les caractères et la répartition de la végétation des Iles Kerguelen. *Comm. Natn. Fr. Rech. Antarct.* 10: 1-60.

Audubon, J.J. 1839. *Ornithological Biography; or, an Account of the Birds of the United States of America...* 5. Black, Edinburgh.

Auman, H.J., Ludwig, J.P., Summer, C.L., Verbrugge, D.A., Froese, K.L., Colborn, T. & Giesy, J.P. 1997. PCBS, DDE, and TCDD-EQ in two species of albatross on Sand Island, Midway Atoll, North Pacific Ocean. *Environ. Toxicol. Chem.* 16: 498-504.

Austin, O.L. Jr. 1949. The status of Steller's Albatross. *Pacific Sci.* 3: 283-295.

Bailey, A.M. 1952. Laysan and Black-footed Albatrosses. *Denver Mus. Nat. Hist. Pict.* 6: 1-79.

Bailey, A.M. 1956. Birds of Midway and Laysan Islands. *Denver Mus. Nat. Hist. Pict.* 12: 1-130.

Bailey, A.M. & Niedrach, R.J. 1951. Stepping Stones across the Pacific. *Denver Mus. Nat. Hist. Pict.* 3: 1-63.

Bailey, A.M. & Sorensen, J.H. 1962. Subantarctic Campbell Island. *Proc. Denver Mus. Nat. Hist.* 10: 1-305.

Baines, A.C. 1889. The sailing flight of the albatross. *Nature* 40: 9-10.

Baird, S.F., Brewer, T.M. & Ridgeway, R. 1884. The Water Birds of North America II. *Mem. Mus. Comp. Zool. Harvard* 13: 344-361.

Bang, B.G. 1966. The olfactory apparatus of tubenosed birds (Procellariiformes). *Acta. Anat.* 65: 391-415.

Bang, B.G. & Wenzel, B.M. 1985. Nasal cavity and olfactory system. In *Form & Function in Birds* (King, A.S. & McLelland, J. eds) 3: 195-225. Academic Press, London.

Barrat, A., Despin, B., Mougin, J.-L., Prevost, J., Segonzac, M. & van Beveren, M. 1973. Note sur les baguage des oiseaux de l'archipen Crozet de 1968 à 1971. *Oiseau Revue fr. Orn.* 43: 32-50.

Barrett-Hamilton, G.E.H. 1903. Remarks on the flight and distribution of albatrosses in the north Pacific. *Ibis* (8)3: 320-324.

Barrow, K.M. 1910. *Three years on Tristan da Cunha.* Skefflington, London.

Bartholomew, G.A. & Howell, T.R. 1964. Experiments on nesting behaviour of Laysan and Black-footed Albatrosses. *Animal Behaviour* 12: 549-559.

Bartle, J.A. 1974. Seabirds of eastern Cook Strait, New Zealand, in autumn. *Notornis* 21: 135-166.

Bartle, J.A. 1991. Incidental capture of seabirds in the New Zealand subantarctic squid trawl fishery, 1990. *Bird. Conserv. Inter.* 351-359.

Barton, D. 1979. Albatrosses in the western Tasman Sea. *Emu* 79: 31-35.

Bascom, W. 1959. Ocean Waves. *Sci. Am.* 201: 74-84.

Battam, H. & Smith, L.E. 1993. *Report on Review and Analysis of: Albatross Banding Data held by the Australian Bird and Bat Banding Schemes.* Australian National Parks and Wildlife Service, Canberra.

Baudelaire, C. 1973. *Les Fleurs du Mal.* Librairie Larousse, Paris.

Beaglehole, J.C. ed. 1961-67. The Journals of Captain James Cook on his voyages of discovery. Vol. 1 *The Voyage of the Endeavour, 1768-1771*; Vol. 2 *The Voyage of the Resolution and Endeavour, 1772-1775*; Vol. 3 *The Voyage of the Resolution and Discovery, 1776-1780*; Vol. 4 *Cook's Life and Voyages: essays and lists.* Hakluyt, Cambridge.

Beck, R.H. 1904. Bird life among the Galapagos Islands. *Condor* 6: 5-11.

Beebe, W. 1926. *The Arcturus adventure.* Putnam, London.

Bell, B.D. 1975. Report on the Birds of the Auckland Islands Expedition 1972-73. In *Preliminary Results of the Auckland Islands Expedition 1972-73* (Yaldwyn, J.C. ed.): 136-142. NZ Lands & Survey, Wellington.

Bell, L.C. 1955. Notes on the birds of the Chatham Islands. *Notornis* 6: 65-68.

Bennets, J.A. 1949. Island home of "Wandering Albatross". *Afr. Wildl.* 2: 22-29,31

Bennett, A.G. 1926. A list of the birds of the Falkland Islands and dependencies. *Ibis* (12th Ser) 2: 306-333.

Bennett, G. 1860. *Gatherings of a naturalist in Australia...* Van Voorst, London.

Bent, A.C. 1922. Life histories of North American petrels, pelicans and their allies. *US Natl. Mus. Bull.* 121.

Berg, L. 1926. Russian discoveries in the Pacific. In *The Pacific*: 1-26. Academy of Sciences of the USSR. Leningrad.

Berrow, S.D., Huin, N., Humpidge, R., Murray, A.W.A. & Prince, P.A. 1999. Wing and primary growth of the Wandering Albatross. *Condor* 101: 360-368.

Berrow, S.D., Humpidge, R. and Croxall, J.P. (in press). Influence of adult breeding experience on growth and provisioning of Wandering Albatross chicks at South Georgia. *Ibis.*

Berruti, A. 1977. *Co-existence in the Phoebetria albatrosses at Marion Island.* unpublished M.Sc. Thesis. University of Cape Town, Rondebosch.

Berruti, A. 1979a. The breeding biologies of sooty albatrosses *Phoebetria fusca* and *P. palpebrata. Emu* 79: 161-175.

Berruti, A. 1979b. The plumage of fledgling *Phoebetria* albatrosses. *Notornis* 26: 308-309.

Berruti, A. 1981. Displays of the sooty albatrosses *Phoebetria fusca* and *P. palpebrata. Ostrich* 52: 98-103.

Berruti, A. 1984. *Fusca.WP* (unpublished).

Berruti, A., Griffiths, A.M., Imber, M.J., Schramm, M. & Sinclair, J.C. 1981. The status of seabirds at Prince Edward Island. *S. Afr. J. Antarct. Res.* 10/11: 31-32.

Berruti, A. & Harcus, T. 1978. Cephalopod prey of the sooty albatrosses *Phoebetria fusca* and *P. palpebrata* at Marion Island. *S. Afr. J. Antarct. Res.* 8:99-103.

Bevan, R.M., Butler, P.J., Woakes, A.J. & P.A. Prince. 1995. The energy expenditure of free-ranging black-browed albatrosses. *Phil. Trans. R. Soc. Lond. B* 350: 119-131.

Bevan, R.M., Woakes, A.J., Butler, P.J. & Boyd, I.L. 1994. The use of heart rate to estimate oxygen consumption of free-ranging Black-browed Albatrosses *Diomedea melanophrys. J. Exp. Biol.* 193:119-137.

Birkhead, J.R.. & Furness, R.W. 1985. Regulation of Seabird populations. In *Behavioural Ecology* (Sibley, R.M. & Smith, R.H. eds): 145-167. Blackwells, Oxford.

Birkhead, T.R. 1988. Behavioural Aspects of Sperm Competition in Birds. *Adv. Stud. Behav.* 18: 35-72.

Birkhead, T. & Møller, A. 1993. Female control of paternity. *Trends Ecol. Evol.* 8:100-104.

Blaber, S.J.M. 1986. The distribution and abundance of seabirds south-east of Tasmania and over the Soela Seamount during April 1985. *Emu* 86: 239-244.

Blakiston, T. & Pryer, H. 1878. A catalogue of the birds of Japan. *Ibis* (4th Ser.) 2: 209-250.

Booy, D.M. 1957. *Rock of Exile: a narrative of Tristan da Cunha.* Dent, London.

Borror, D.J. 1947. Birds of Agrihan. *Auk* 64: 415-417.

Bourne, W.R.P. 1966. Observations of seabirds. *Sea Swallow* 18: 9-36.

Bourne, W.R.P. 1967. Long distance vagrancy in petrels. *Ibis* 109: 141-167.

Bourne, W.R.P. 1975. Birds feeding on lobster-krill off the Falkland Islands. *Sea Swallow* 24: 22-23.

Bourne, W.R.P. 1977a. Half a pair of Black-browed Albatrosses. *Br. Birds* 70: 301-303.

Bourne, W.R.P. 1977b. Albatrosses occurring off South Africa. *Cormorant* 2: 7-10.

Bourne, W.R.P. 1982a. The Ancient Mariner's Albatross. *Sea Swallow* 31: 56-57.

Bourne, W.R.P. 1982b. The colour of the tail coverts of the Black-footed Albatross. *Sea Swallow* 31: 61.

Bourne, W.R.P. 1989. The evolution, classification and nomenclature of the great albatrosses. *Le Gerfaut* 79: 105-116.

Bourne, W.R.P. 1993. The early specimens of the Wandering Albatross. *Notornis* 40: 314-316.

Bourne, W.R.P. 1996. Observations of seabirds received in 1995. *Sea Swallow* 45: 33-46.

Bourne, W.R.P. & Casement, M.B. 1993. RNBWS Checklist of Seabirds. *Sea Swallow* 42: 16-27.

Bourne, W.R.P. & Casement, M.B. 1996. RNBWS Checklist of Seabirds (Revised). *Sea Swallow* 45 Supp. pp.12.

Bourne, W.R.P. & Curtis, W.F. 1985. South Atlantic Seabirds. *Sea Swallow* 34: 18-28.

Bourne, W.R.P & David, A.C.F. 1981. Nineteenth century bird records from Tristan da Cunha. *Bull. Br. Ornithol. Club* 101:247-256.

Bourne, W.R.P. & Warham, J. 1999. Albatross Taxonomy. *Birding World* 12: 123-124.

Brackenridge, W. 1971. Yellow-nosed Albatross off Tobago, West Indes. *Sea Swallow* 21: 36.

Bretagnolle, V. 1990. Behavioural affinities of the Blue Petrel *Halobaena caerulea. Ibis* 132: 102-123.

Brazil, M.A. 1991. *The Birds of Japan.* Helm, London.

Briggs, K.T., Tyler, W.B., Lewis, D.B. & Carlson, D.R. 1987. Bird communities at sea off California: 1975 to 1983. *Stud. in Avian Biol.* 11: 1-74.

Brisson, M.J. 1760. *Ornithologie...* 6. Paris.

Brock. V.E. 1951. Laysan Island bird census. *Elepaio* 12: 17-18.

Brodkorb, P. 1963. Catalogue of Fossil Birds, Pt. 1, (Archaeopterygiformes through Ardeiformes). *Bull. Florida State Mus.* 7: 179-293.

Broekhuysen, G.J. & Macnae, W. 1949. Observations on the birds of the Tristan da Cunha islands and Gough Island in February and early March 1948. *Ardea* 37: 97-113.

Brooke, M.de L. & Klages, N. 1986. Squid beaks regurgitated by Grey-headed and Yellow-nosed Albatrosses *Diomedea chrysostoma* and *D. chlororhynchos* at the Prince Edward Islands. *Ostrich* 57: 203-206.

Brooke, R.K. 1981. Modes of moult of flight feathers in albatrosses. *Cormorant* 9: 13-18.

Brooke, R.K. & Furness, B.L. 1982. Reversed modes of moult of flight feathers in the Black-browed Albatross *Diomedea melanophris. Cormorant* 10: 27-30.

Brooke, R.K., Sinclair, J.C. & Berruti, A. 1980. Geographical variation in *Diomedea chlororhynchos* (Aves: Diomedeidae). *Durban Mus. Novit.* 12: 171-180.

Brooks, A. 1937. The Pategial fan in the tubinares. *Condor* 39: 82-83.

Brooks, P. 1984. Just a Country Boy Who found his Niche. *Audubon mag.* May 1984: 48-59.

Brosset, A. 1963. La reproduction des oiseaux de mer des îles Galapagos en 1962. *Alauda* 31: 81-109.

Brothers, N.P. 1979a. Seabird Islands No. 74 Pedra Branca, Tasmania. *Corella* 3: 58-60.

Brothers, N.P. 1979b. Seabird Islands No. 78 Mewstone, Tasmania. *Corella* 3: 88-69.

Brothers, N. 1982. Shot an albatross lately? *FINTAS*: 36.

Brothers, N. 1988. Albatross search successful thanks to help of professional fishermen. *Australian Fisheries* 47: 18.

Brothers, N. 1991. Albatross mortality and Associated Bait Loss in the Japanese Longline Fishery in the Southern Ocean. *Biol. Conserv.* 55: 255-268.

Brothers, N.P. & Davis, G. 1985. Bird Observations on Albatross Island, 1981 to 1985. *Tasmanian Bird Rep.* 14: 3-9.

Brothers, N., Gales, R., Hedd, A. & Robertson, G. 1998a. Foraging movements of the Shy Albatross *Diomedea cauta* breeding in Australia; implications for interactions with longline fisheries. *Ibis* 140: 446-457.

Brothers, N., Gales, R. & Reid, T. 1998b. Seabird interactions with longline fishing in the AFZ: 1996 seabird mortality estimates and 1988-1996 trends. *Wildl. Repts.* 98/1, Parks and Wildlife Service, Tasmania.

Brothers, N., Gales, R. & Reid, T. 1998c. Seabird interactions with longline fishing in the AFZ: 1997 seabird mortality estimates and 1988-1997 trends. *Wildl. Repts.* 98/3, Parks and Wildlife Service, Tasmania.

Brothers, N.P., Reid, T.A. & Gales, R.P. 1997. At-sea Distribution of Shy Albatrosses *Diomedea cauta cauta* Derived from Records of Band Recoveries and Colour-marked Birds. *Emu* 97: 231-239.

Broughton, J.M. 1994. Size of the bursa of Fabricius in relation to gonad size and age in Laysan and Black-footed Albatrosses. *Condor* 96: 203-207.

Brown, C.R. & Adams, N.J. 1984. Basal metabolic rates and energy expenditure during incubation in the Wandering Albatross (*Diomedea exulans*). *Condor* 86: 182-186.

Brown, R.G.B. 1981. Was Coleridge's 'Albatross' a Giant Petrel? *Ibis* 123: 551.

Bruijne, J.W.A de 1970. Black-browed Albatross (*Diomedea melanophris*) in the Carribbean. *Ardea* 58: 265-266.

Bryan, W.A. 1903. A monograph of Marcus Island. *B.P. Mus. Occas. Papers* 2: 77-124.

Bryan, W.A. 1906. Report of a Visit to Midway Island. *B.P. Mus. Occas. Papers* 2: 291-299.

Buckingham, R., Elliott, G. & Walker, K. 1991. Bird and mammal observations on Adams Island and southern Auckland Island 1989. *Sci. Res. Int. Rep.* 105. Dept. of Conservation, Wellington.

Buddle, G.A. 1948. The Outlying Islands of the Three Kings Group. *Rec. Auckland Inst. Mus.* 3: 195-204.

Buffon 1786. *Histoire Naturelle des Oiseaux.* Paris.

Buller, W.L. 1873. *A history of the birds of New Zealand.* London.

Buller, W.L. 1888. *A history of the Birds of New Zealand.* 2nd edn. London.

Buller, W.L. 1905. *Supplement to the birds of New Zealand.* London.

Burger, A.E., Lindeboom, H.J. & Williams, A.J. 1978. The mineral and energy contributions of guano of selected species birds to the Marion Island terrestrial ecosystem. *S. Afr. J. Antarct. Res.* 8: 59-70.

Carmichael, D. 1818. Some account of the island of Tristan da Cunha and its natural productions. *Trans. Linn. Soc. Lond.* 12: 483-513.

Carrick, R. & Ingham, S.E. 1970. Ecology and population dynamics of Antarctic seabirds. In *Antarctic Ecology* (Holdgate, M.W. ed.): 505-525. Academic Press, London.

Carter, M.J. 1977. A Buller's Albatross in Victorian waters. *Aust. Bird Watcher* 7: 47.

Cassin, J. 1858. *US Exploring Expedition during the years 1838, 1839, 1840, 1841, 1842 under the command of Charles Wilkes USN.* 3 Mammalogy and Ornithology: 398-401.

Casti, J. 1997. Flight over Wall Street. *New Scientist* 2078: 38-41.

Catard, A. & Weimerskirch, H. 1998. Satellite tracking of petrels and albatrosses: From the tropics to Antarctica. In *22nd Int. Orn. Congr.* (Adams, N.J. & Slotow, R.H. eds) Abstracts. *Ostrich* 69: 152.

Caum, E.L. 1936. Notes on the Flora and Fauna of Lehua and Kaula Islands. *B.P. Bishop Occas. Papers* 11(21): 1-17

Cawkell, M.B.R., Maling, D.H. & Cawkell, E.M. 1960. *The Falkland Islands.* Macmillan, London.

Cerny, V. 1973. Parasite-host relationships in feather-mites. In *Proc. 3rd. Int. Congr. Acarology 1971* (Daniel, M. & Rosicky, B. eds): 761-764. Academia, Prague.

Challies, C.N. 1975. Summary Report on the problem of pigs on the main Auckland Island. In *Preliminary Results of the Auckland Islands Expedition 1972-73* (Yaldwyn, J.C. ed.): 225-232. NZ Lands & Survey, Wellington.

Chandler, A.C. 1916. A study of the structure of feathers, with reference to their taxonomic significance. *Univ. Calif. Publ. Zool.* 13: 243-246.

Chapman, F.R. 1891. The outlying islands south of New Zealand. *Trans. N.Z. Inst.* 23: 491-522.

Chastel, O., Weimerskirch, H. & Jouventin, P. 1993. High annual variability in reproductive success and survival of an Antarctic seabird, the Snow Petrel *Pagadroma nivea*: a 27-year study. *Oecologia* 94: 278-285.

Chebez, J.C. & Bertonatti, C.C. 1994. *La Avifauna de la Isla de los Estados, Isla de Año Nuevo y Mar circumdante (Tierra del Fuego, Argentina).* LOLA, Buenos Aires.

Cherel, Y. & Klages, N. 1997. A review of the food of albatrosses. In *Albatross Biology and Conservation* (Robertson, G. & Gales, R. eds): 113-136. Surrey Beatty, Chipping Norton.

Cherel, Y., Weimerskirch, H. & Duhamel, G. 1996. Interactions between longline vessels and seabirds in Kerguelen waters and a method to reduce seabird mortality. *Biol. Conserv.* 75: 63-70.

Cheshire, N.G. & Carter M.J. 1992. A Wandering Albatross with abnormal underwing plumage. *Notornis* 39: 98-99.

Chilton, C. ed. 1909. *The Subantarctic Islands of New Zealand.* Government, Wellington.

Chisholm, D.R. 1937. The majestic albatross. *Asia* 37: 84-89.

Clapp, R.B. 1972. The Natural History of Gardner Pinnacles, Northwestern Hawaiian Islands. *Atoll Res. Bull.* 163: 1-25.

Clapp, R.B. & Kridler, E. 1977. The Natural History of Necker Island, Northwestern Hawaiian Islands. *Atoll Res. Bull.* 206: 1-102.

Clapp, R.B., Kridler, E. & Fleet, R.R. 1977. The Natural History of Nihoa Island, Northwestern Hawaiian Islands. *Atoll Res. Bull.* 207: 1-147.

Clapp, R.B. & Wirtz, W.O. II. 1975. The Natural History of Lisianski Island, Northwestern Hawaiian Islands. *Atoll Res. Bull.* 186: 1-196.

Clark, G.S. 1984. Extension of the known range of some seabirds on the coast of Southern Chile. *Notornis* 31: 320-334.

Clark, G.S. 1986. Seabirds observed in the Pacific Southern Ocean during Autumn, *Australas. Seabird Grp. Newsl.* 23: 1-35.

Clark, G. 1988. *The Totorore Voyage.* Century, Melbourne.

Clark, G. & Robertson, C.J.R. 1996. New Zealand White-capped Mollymawks (*Diomedea cauta steadi*) breeding with Black-browed Mollymawks (*Diomedea melanophrys melanophrys*) at Antipodes Islands, New Zealand. *Notornis* 43: 1-6.

Clarke, A. & Prince, P.A. 1976. The origin of stomach oil in marine birds. *J. Exp. Mar. Biol. Ecol.* 23: 15-30.

Clarke, M.R., Croxall, J.P. & Prince, P.A. 1981. Cephalopod remains in regurgitations of the wandering albatross *Diomedea exulans* at South Georgia. *Brit. Antarct. Surv. Bull.* 54: 9-21.

Clarke, M.R. & Prince, P.A. 1980. Chemical composition and calorific value of food fed to mollymawk chicks *Diomedea melanophris* and *D. chrysostoma* at Bird Island, South Georgia. *Ibis* 122: 488-494.

Clarke, M.R. & Prince, P.A. 1981. Cephalopod remains in regurgitations of black-browed albatross *Diomedea melanophris* and grey-headed albatross *Diomedea chrysostoma* at South Georgia. *Brit. Antarct. Surv. Bull.* 54: 1-7.

Clarke, W.E. 1913. Ornithology of the Scottish National Antarctic Expedition. *Sci. Res. Scottish Natl. Antarct. Exped.* 4: 255-270.

Clay, T. & Moreby, C. 1970. Mallophaga and Anoplura of Subantarctic Islands. *Pac. Insects Monogr.* 23: 216-220.

Clayton, S.W. 1774. *A Short Description of Falklands Islands, their produce Climate and Natural History.* Adm 7/704. Public Records Office, London.

Clinton, W.J. 1996. Midway Atoll National Wildlife Refuge created. *Pac. Seabirds* 23(2): 39-40.

Cobb, A.F. 1933. *Birds of the Falkland Islands.* Witherby, London.

Cobley, N.D., Croxall, J.P. & Prince, P.A. 1998. Individual quality and reproductive performance in the Grey-headed Albatross *Diomedea chrysostoma. Ibis* 140: 315-322.

Cobley, N.D. & Prince, P.A. 1998. Factors affecting primary molt in the Gray-headed Albatross. *Condor* 100: 8-17.

Coker, R.E. 1919. Habits and Economic Relations of the Guano Birds of Peru. *Proc. US. Natl. Mus.* 56: 449-511.

Coleridge, S.T. 1798. The Rime of the Ancient Mariner. In *Lyrical Ballads* (Wordsworth, W. & Coleridge, S.T.). Cottle, Bristol.

Condon, H.T. 1939. The cranial osteology of certain Tubinares. *Trans. R. Soc. S. Aust.* 63: 311-328.

Cone, C.D. Jr. 1964. A mathematical analysis of the dynamic soaring of the albatross with ecological interpretations. *Virginia Inst. Mar. Sci. Spec. Scient. Rep..* 50: 1-108.

Cooper, J. 1974. Albatross displays off S-W coast of South Africa. *Notornis* 21: 234-238.

Cooper, J. 1983. Bird ringing on Gough Island 1977-82. *S. Afr. J. Antarct. Res.* 13: 47-48.

Cooper, J. 1999. News from Rome is Good for the World's Seabirds. *Australas. Seabird Grp. Newsl.* 35: 22.

Cooper, J. & Brown, C.R. 1990. Ornithological research at the sub-Antarctic Prince Edward Islands: a review of achievements. *S. Afr. J. Antarct. Res.* 20: 40-57.

Cooper, J., Henley, S.R. & Klages, N.T.W. 1992. The diet of the wandering albatross *Diomedea exulans* at sub-antarctic Marion Island. *Polar Biol.* 12: 477-484.

Cooper, J. & Klages, N.T.W. 1995. The diets and dietry segregation of sooty albatrosses (*Phoebetria* spp.) at subantarctic Marion Island. *Antarct. Sci.* 7: 15-23.

Cooper, J. & Ryan, P.G. 1993. *Management Plan for the Gough Island Wildlife Reserve.* University of Cape Town, Rondebosch.

Cooper, J.M. 1995. Fishing hooks associated with albatrosses on Bird Island, South Georgia. *Mar. Orn.* 23: 17-21.

Cooper, W.J., Miskelly, C.M., Morrison, K. & Peacock, R.J. 1986. Birds of the Solander Islands. *Notornis* 33: 77-89.

Copson, G.R. 1988. The status of Black-browed and Grey-headed Albatrosses on Macquarie Island. *Proc. R. Soc. Tas.* 122: 137-141.

Copson, G.R. 1995. An Integrated Vertebrate Pest Strategy for Subantarctic Macquarie Island. *10th. Aust. Vert. Pests. Conf..* Hobart.

Cornish, V. 1934. *Ocean Waves.* Cambridge University Press, Cambridge.

Costa, D.P. & Prince, P.A. 1987. Foraging energetics of Grey-headed Albatrosses *Diomedea chrysostoma* at Bird Island, South Georgia. *Ibis* 129: 149-158.

Cottam, C. & Knappen, P. 1939. Food of some uncommon North American birds. *Auk* 56: 138-169.

Coues, E. 1866. Critical Review of the Family Procellariidae: - Part V; embracing the Diomedeinae and Halodrominiae. *Proc. Acad. Nat. Sci. Phila.* 18: 172-197.

Cousins, K.L. 1998. Black-footed Albatross Population Biology Workshop. *Elepaio* 58: 47,52-53.

Cox, J.B. 1973. The identification of the smaller Australian Diomedea and the status of Diomedea in South Australia. *S. Aust. Orn.* 26: 67-75.

Cox, J.B. 1976. A review of the Procellariiformes occurring in South Australian waters. *S. Aust. Orn.* 27: 26-82.

Cox, J.B. 1978. Albatross killed by Giant Petrel. *Emu* 78: 94-95.

Cracraft, J. 1981. Toward a phylogenetic classification of the recent birds of the world (class Aves). *Auk* 98: 681-714.

Croxall, J.P. 1979. Distribution and population changes in the Wandering Albatross *Diomedea exulans* at South Georgia. *Ardea* 67: 15-21.Croxall, J.P. 1982. Aspects of the population demography of Antarctic and Sub-antarctic seabirds. *Com. Natn. Fr. Rech. Antarct.* 51: 479-488.

Croxall, J.P. 1991. Constraints on Reproduction in Albatrosses. *Acta XX Congr. Int. Orn.* 1: 281-302.

Croxall, J.P. 1997. Research and conservation: a future for albatrosses? In *Albatross Biology and Conservation* (Robertson, G & Gales, R. eds 1998): 269-290. Surrey Beatty, Chipping Norton.

Croxall, J.P., Black, A.D. & Wood, A.G. 1999. Age, sex and status of wandering albatrosses *Diomedea exulans* L. in Falkland Island waters. *Antarct. Sci.* 11: 150-156.

Croxall, J.P. & Gales, R. 1997. An assessment of the conservation status of albatrosses. In *Albatross Biology and Conservation* (Robertson, G. & Gales R. eds 1998): 46-65. Surrey Beatty, Chipping Norton.

Croxall, J.P., North, A.W. & Prince, P.A. 1988. Fish prey of the Wandering Albatross *Diomedea exulans* at South Georgia. *Polar Biol.* 9: 9-16.

Croxall, J.P., Pickering, S.P.C. & Rothery, P. 1990a. Influence of Increasing Fur Seal Population on Wandering Albatrosses *Diomedea exulans* Breeding on Bird Island, South Georgia. In *Antarctic Ecosystems* (Kerry, K.R. & Hempel, G. eds): 237-240. Springer-Verlag, Berlin.

Croxall, J.P. & Prince, P.A. 1979. Antarctic seabird and seal monitoring studies. *Polar Rec.* 19: 573-595.

Croxall, J.P. & Prince, P.A. 1980. Food, feeding ecology and ecological segregation of seabirds at South Georgia. *Biol. J. Linn. Soc.* 14: 103-131.

Croxall, J.P. & Prince, P.A. 1982. Calorific content of squid (Mollusca: Cephalopoda). *Br. Antarct. Surv. Bull.* 55: 27-31.

Croxall, J.P. & Prince, P.A. 1990. Recoveries of Wandering Albatrosses *Diomedea exulans* ringed at South Georgia 1958-86. *Ringing and Migration* 11: 43-51.

Croxall, J.P. & Prince, P.A. 1994. Dead or alive, night or day: how do albatrosses catch squid? *Antarct. Sci.* 6: 155-162.

Croxall, J.P. & Prince, P.A. 1996. Cephalopods as prey. I. Seabirds. *Phil. Trans. R. Soc. Lond. B* 351: 1023-1043.

Croxall, J.P., Prince, P.A., Hunter, I., McInnes, S.J. & Copestake, P.G. 1984a. The seabirds of the Antarctic Peninsula, Islands of the Scotia Sea, and Antarctic Continent between 80°W and 20°W: their status and conservation. In *Status and Conservation of the World's Seabirds* (Croxall, J.P., Evans, P.G.H. & Schreiber, R.W. eds): 637-666. ICBP, Cambridge.

Croxall, J.P., Prince, P.A. & Reid K. 1997a. Dietary segregation of krill-eating South Georgia seabirds. *J. Zool. Lond.* 242: 531-556.

Croxall, J.P., Prince, P.A., Rothery, P. & Wood, A.G. 1997b. Population changes in albatrosses at South Georgia. In *Albatross Biology and Conservation* (Robertson, G & Gales, R. eds 1998): 69-83. Surrey Beatty, Chipping Norton.

Croxall, J.P., Prince, P.A., Trathan, P.N. & Wood, A.G. 1996. Distribution and movements of South Georgia Black-browed Albatrosses. *Warrah* 9: 7-9.

Croxall, J.P., Reid, K. & Prince, P.A. 1999. Diet, provisioning and productivity responses of marine predators to differences in availability of Antarctic Krill. *Mar. Ecol. Prog. Ser.* 177: 115-131.

Croxall, J.P. & Ricketts, C. 1983. Energy costs of incubation in the Wandering Albatross *Diomedea exulans. Ibis* 125: 33-39.

Croxall, J.P., Ricketts, C. & Prince, P.A. 1984b. Impact of seabirds on marine resources, especially krill of South Georgia waters. In *Seabird Energetics* (Whittow, G.C. & Rahn, H. eds): 285-317. Plenum, New York.

Croxall, J.P., Rothery, P. & Crisp, A. 1992. The effect of maternal age and experience on egg-size and hatching success in Wandering Albatrosses *Diomedea exulans. Ibis* 134: 219-228.

Croxall, J.P., Rothery, P., Pickering, S.P.C. & Prince, P.A. 1990b. Reproductive performance, recruitement and survival of Wandering Albatrosses *Diomedea exulans* at Bird Island, South Georgia. *J. Anim. Ecol.* 59: 775-796.

Croxall, J. & Saether, B.-E. 1998. Population dynamics of fulmarine petrels and albatrosses. In *Proc. 22 Congr. Ornith. Int.* (Adams, N.J. & Slotow, R.H. eds.) Abstracts. *Ostrich* 69: 65.

Cunningham, J.M. 1951. Ringing in New Zealand. *Notornis* 4: 77-82.

Curtis, W.F. 1988a. Highlights of a South Atlantic tour - 1986/87. *Sea Swallow* 37: 3-8.

Curtis, W.F. 1988b. First occurrence of Buller's Albatross in the Atlantic Ocean. *Sea Swallow* 37: 62-63.

Curtis, W.F. 1993. Yellow-nosed Albatross *Diomedea chlororhynchos* off Cornwall. *Sea Swallow* 42: 63-65.

Curtis, W.F. 1994. Further South Atlantic records. *Sea Swallow* 43: 19-28.

Dabbene, R. 1926. Los petrels y los albatros del Atlantico austral. *Hornero* 3: 311-348.

Dall, W.H. 1873. Notes on the Avi-fauna of the Aleutian Islands from Unalaska, Eastwards. *Proc. Calif. Acad. Sci.* 5: 25-35.

Dall, W.H. 1874. Notes on the Avi-fauna of the Aleutian Islands especially those West of Unalaska. *Proc. Calif. Acad. Sci.* 5: 270-281.

Dalziell, J. & De Poorter, M. 1993. Seabird mortality in longline fisheries around South Georgia. *Polar. Rec.* 29: 143-144.

Dampier, W. 1697. *A Voyage to New Holland* 1939 edition (Williamson, J.E. ed.). Argonaut, London.

Darby, A. 1996. Researchers follow a wandering albatross for secret flight. *The Age* 13 August 1996, New Zealand.

Darby, M.M. 1970. Summer seabirds between New Zealand and McMurdo Sound. *Notornis* 17: 28-55.

Darwin, C.R. 1933. *Diary of the voyage of HMS Beagle.* (Barlow, N. ed.) Cambridge University Press, Cambridge.

Davis, P. 1957. The breeding of the Storm Petrel. *Br. Birds* 50: 85-101, 371-384.

Dawson, E.W. 1955. The Birds of the Chatham Islands 1954 expedition. *Notornis* 6: 78-82.

Dawson, E.W. 1973. Albatross populations at the Chatham Islands. *Notornis* 20: 210-230.

Deacon, G. 1984. *The Antarctic circumpolar ocean.* Cambridge University Press, Cambridge.

DeGange, A.R. & Day, R.H. 1991. Seabird mortality in Japanese gill nets. *Condor* 93: 251-258.

DeGruy, M. & Armstrong, M. 1992. *Sharks on their best behaviour.* BBC, Bristol (TV film).

de Kersauson, O. 1990. Face a face avec Dieu dans les quarantiemes rugissants. *Paris Match* 2139: 3-10.

de la Mare, W.K. & Kerry, K.R. 1994. Population dynamics of the wandering albatross (*Diomedea exulans*) on Macquarie Island and the effects of mortality from longline fishing. *Polar Biol.* 14: 231-241.

Denton, G.H. & Hughes, T.J. 1981. *The last great ice sheet.* Wiley, New York.

Derenne, P., Lufbery, J.X. & Tollu, B. 1974. L'avifaune de l'archipel de Kerguelen. *Comm. Natn. Fr. Rech. Antarct.* 33: 57-87.

Derenne, P., Mougin, J.-L., Steinberg, C. & Voisin J.F. 1976. Les Oiseaux de l'île aux Cochons, archipel Crozet (46°06'S. 50°14'E.). *Comm. Natn. Fr. Rech. Antarct.* 40: 107-148.

Derenne, P., Prevost, J. & Van Beveren, M. 1972. Note sur le baguage des oiseaux dans l'Archipel de Kerguelen depuis 1951. *Oiseau. Revue fr. Orn.* 42: 111-130.

Despin, B., Mougin, J.-L. & Segonzac, M. 1972. Oiseaux et mammifères de l'Ile de l'Est. Archipel Crozet (47°25'S. 52°12'E.). *Comm. Natn. Fr. Rech. Antarct.* 31: 1-106.

Dill, H.R. & Bryan, W.A. 1912. *Report of an Expedition to Laysan Island in 1911.* US Dept. of Agriculture, Washingron.

Diment, J.A. & Wheeler, A. 1984. Catalogue of the natural history manuscripts and letters by Daniel Solander (1733-1782), or attributed to him, in British Collections. *Arch. Nat. Hist.* 11: 457-488.

Dixon, C.C. 1933. Some observations on the albatrosses and other birds of the southern oceans. *Trans. R. Canad. Inst.* 19: 117-139.

Dodimead, A.J.F., Favorite, F. & Hirano, T. 1963. Salmon of the North Pacific Ocean - II - Review of oceanography of the Subarctic Pacific region. *Int. North Pac. Fish Comm. Bull.* 13: 1-195.

Dollman, H. 1962. *United States Antarctic Research Program. Birdbanding Expedition to Bird Is. South Georgia.* (unpublished) Br. Antarct. Surv. Archiv. 1999/102.

Dougall, W. 1888. *Far south: Stewart Island, The Snares, Auckland, Campbell, Antipodes and Bounty Islands.* Soutland Times, Invercargill.

Douglas, H.D. 1998. Changes in the distribution and abundance of Waved Albatrosses at Isla Española, Galápagos Islands, Ecuador. *Condor* 100: 737-740.

Douglas, H.D. & Fernández, P. 1997. A longevity record for the Waved Albatross. *J. Field. Orn.* 68: 224-227.

Downes, M.C., Ealey, E.H.M., Gwynn, A.M. & Young, P.S. 1959. The Birds of Heard Island. *Aust. Natn. Antarct. Res. Exp. Rep.* B1 *Zool.* 1-135.

Drake, F. 1589. (The circumnavigation of 1577-1580). In *The Principal Navigations, Voyages, Traffiques and Discoveries of the English Nation* (Hakluyt, R. ed.): 101-133. McLehose 1904 reprint of 2nd edition, Glasgow.

Dreux, P. & Milon, P. 1967. Premières observations sur L'Avifaune de l'Ile aux Cochons (Archipel Crozet). *Alauda* 35: 27-32.

Dyrssen, D., Patterson, C., Ui, J. & Weichart, G.F. 1972. Inorganic Chemicals. In *A guide to marine pollution* (Goldberg, E.D. ed.): 41-58. Gordon & Breach, New York.

Duffy, D.C. 1980. Galápagos seabirds: opportunities for research. *Bull. Pac. Seabird Grp.* 7: 76-78.

Duffy, D.C. 1983. The foraging ecology of Peruvian seabirds. *Auk* 100: 800-810.

Duffy, D.C. 1989. Seabird foraging aggregations: a comparison of two southern upwellings. *Col. Waterbirds* 12: 164-175.

Duffy, D.C. & Merlen, G. 1986. Seabird densities and aggregations in the 1983 El Niño in the Galapagos Islands. *Wilson Bull.* 98: 588-591.

DuMont, P.A. 1955. Gooneybird studies on Midway Island. *Elepaio* 15: 52-55.

Dunlap, E. 1988. Laysan Albatross nesting on Guadalupe Island, Mexico. *Am. Birds* 42: 180-181.

Dunmore, J. 1965. *French Explorers in the Pacific.* Oxford University Press, Oxford.

Dunnet, G.M. 1964. Distribution and host relationships of fleas in the Antarctic and Subantarctic. In *Biologie Antarctique* (Carrick, R., Holdgate, M.W. & Prevost, J. eds): 579-589. Hermann, Paris.

Earle, A. 1832. *A narrative of a nine month residence in New Zealand in 1827 together with a journal of a residence in Tristan d'Acunha.* Longmans, London.

Eaton, E.A. 1875. First report of the Naturalist attached to the Transit-of-Venus Expedition to Kerguelen's Island 1874. *Proc. R. Soc. Lond.* 23: 351-356, 501-504.

Edwards, G. 1747. *A Natural History of Birds.* London.

Egami, T & Fumiko, S. 1973. Archaeological excavations on Pagan in the Mariana Islands. *J. Anthropol. Soc. Nippon* 81: 203-226.

Eibl-Eibesfeldt, I. 1959. *survey on the Galapagos Islands.* UNESCO, Paris.

Elliott, G. 1997. Wandering Albatross Research at the Auckland and Antipodes Islands. *Australas. Seabird Grp. Newsl.* 32: 52.

Elliott, H.F.I. 1953. The fauna of Tristan da Cunha. *Oryx* 2: 41-53.

Elliott, H.F.I. 1957. A contribution to the ornithology of the Tristan da Cunha group. *Ibis* 99: 545-586.

Elliott, H.W. 1884. *Report on the seal islands of Alaska.* US Government, Washington.

Ellis, H.I. 1984. Energetics of free-ranging seabirds. In *Seabird Energetics* (Whittow, G.C. & Rahn, H. eds.): 203-234. Plenum, London.

Ely, C.A. & Clapp, R.B. 1973. The Natural History of Laysan Island. *Atoll Res. Bull.* 171: 1-361.

Enomoto, K. 1937. (My memories of birds). *Yacho* 4: 7-11. (In Japanese; translated in Austin, O.L. Jr. 1949).

Enticott, J.W. 1986. The pelagic distribution of the Royal Albatross *Diomedea epomophora. Cormorant* 13: 143-156.

L'Estrange, D. 1972. (northerly record of albatrosses in the South Atlantic) *Sea Swallow* 21: 31.

Falla, R.A. 1933. note proposing *Thalassarche cauta steadi. Rec. Auck. Inst. Mus.* 1: 179.

Falla, R.A. 1937. Birds. *BANZ Antarct. Res. Exped. 1929-1931. Rep. Ser. B* 2: 1-288.

Falla, R.A., Sibson, R.B. & Turbott, E.G. 1966. *A Field Guide to the Birds of New Zealand.* Collins, London.

Fanning, E. 1833. *Voyages round the world, with selected sketches...* New York.

Fefer, S.I., Harrison, C.S., Naughton, M.B. & Shallenberger, R.J. 1984. Synopsis of results of recent seabird research in the Northwestern Hawaiian Islands. In *Proceedings of the 2nd Symposium on Resource Investigations in the Northwestern Hawaiian Islands* (Grigg, R.W. & Tanoue, K.Y. eds) 1: 9-76. University of Hawaii Sea Grant, Honolulu.

Filhol, H. 1885. Oiseaux. In *Mission de L'Ile Campbell* 1: 35-64. Academie des Sciences, Paris.

Fisher, H.I. 1948 Laysan Albatross nesting on Moku Manu Islet, off Oahu, T.H. *Pac. Sci.* 2: 66.

Fisher, H.I. 1949. Populations of Birds on Midway and the Man-Made Factors Affecting Them. *Pac. Sci.* 3: 103-110.

Fisher, H.I. 1966a. Airplane-Albatross Collisions on Midway Atoll. *Condor* 68: 229-242.

Fisher, H.I. 1966b. Midway's Deadly Antennas. *Audubon Mag.* 68: 220-223.

Fisher, H.I. 1966c. Aerial Census of Laysan Albatrosses breeding on Midway Atoll in December 1962. *Auk* 83: 670-673.

Fisher, H.I. 1967. Body weights of Laysan Albatrosses *Diomedea immutabilis. Ibis* 109: 373-382.

Fisher, H.I. 1968. The "two-egg clutch" in the Laysan Albatross. *Auk* 85: 134-136.

Fisher, H.I. 1969. Eggs and Egg-laying in the Laysan Albatross, *Diomedea immutabilis. Condor* 71: 102-112.

Fisher, H.I. 1970. The Death of Midway's Antennas. *Audubon Mag.* 72: 62-63.

Fisher, H.I. 1971a. The Laysan Albatross: its incubation, hatching and associated behaviors. *Living Bird* 10: 19-78.

Fisher, H.I. 1971b. Experiments on homing in Laysan Albatrosses, *Diomedea immutabilis. Condor* 73: 389-400.

Fisher, H.I. 1972. Sympatry of Laysan and Black-footed Albatrosses. *Auk* 89: 381-402.

Fisher, H.I. 1973. Pollutants in the North Pacific Albatrosses. *Pac. Sci.* 27: 220-225.

Fisher, H.I. 1975a. Mortality and Survival in the Laysan Albatross, *Diomedea immutabilis. Pac. Sci.* 29: 279-300.

Fisher, H.I. 1975b. The relationship between deferred breeding and mortality in the Laysan Albatross. *Auk* 92: 433-441.

Fisher, H.I. 1975c. Longevity of the Laysan Albatross, *Diomedea immutabilis. Bird-banding* 46: 1-6.

Fisher, H.I. 1976. Some dynamics of a breeding colony of Laysan Albatrosses. *Wilson Bull.* 88: 121-142.

Fisher, H.I. & Baldwin, P.H. 1946. War and the Birds of Midway Atoll. *Condor* 48: 3-15.

Fisher, H.I. & Fisher, J. 1972. The oceanic distribution of the Laysan Albatross, *Diomedea immutabilis. Wilson Bull.* 84: 7-27

Fisher, H.I. & Fisher, M. 1969. The Visits of Laysan Albatrosses to the Breeding Colony. *Micronesica* 5: 173-221.

Fisher, J. & Lockley, R.M. 1954. *Sea-Birds.* Collins, London.

Fisher, M.L. 1970. *The Albatross of Midway Island.* Southern Illinois University Press, Carbondale.

Fisher, W.K. 1903. Birds of Laysan and the Leeward Islands, Hawaiian group. *Bull. US Fish Commission* 23: 767-807.

Fisher, W.K. 1904. On the habits of the Laysan Albatross. *Auk* 21: 8-20.

Fitzroy, R. 1839. *Narrative of the surveying voyages of HMSs Adventure and Beagle between the years 1826 and 1836.* London.

Fleming, C.A. 1939. Birds of the Chatham Islands. *Emu* 38: 380-413, 492-509.

Fleming, C.A. 1984. Lancelot Eric Richdale 1900-1983. *Proc. R. Soc. NZ* 112: 27-37.

Fletcher, F. 1854. *The World encompassed by Sir Francis Drake...* Hakluyt, Cambridge.

Flinders, M. 1801. *Observations on the Coasts of Van Diemen's Land, on Bass's Strait and its Islands, and on part of the Coasts of New South Wales.* Nichol, London.

Flint, J.H. 1967. Conservation problems on Tristan da Cunha. *Oryx* 9: 28-32.

Forbes, W.A. 1882. Report on the Anatomy of the Petrels (Tubinares) collected during the voyage HMS Challenger. *Zool. Challenger Exped.* 4: 1-64.

Forster, G. 1777. A Voyage Round the World. In *Georg Forsters Werke* (Kahn, R.L. ed. 1968). Akademie-Verlag, Berlin.

Forster, J.R. 1785. Mémoire sur les Albatros. *Mém. Math. Phys. Paris (Acad. Sci.)* 10: 563-572.

Fraser, C. 1986. *Beyond the Roaring Forties.* NZ Government Press, Wellington.

Fraser, M.W., Ryan, P.G. & Watkins, B.P. 1988. The Seabirds of Inaccessible Island, South Atlantic Ocean. *Cormorant* 16: 7-33.

Fressanges du Bost, D. & Segonzac, M. 1976. Note complimentaire sur le cycle reproducteur du Grande Albatross *Diomedea exulans* de L'Ile de la Possession, Archipel Crozet. *Comm. Natn. Fr. Rech. Antarct.* 40 :53-60.

Friedmann, H. 1934. Bird bones from eskimo ruins on St Lawrence Isand, Bering Sea. *J. Wash. Acad. Sci.* 24: 83-96.

Frings, C. 1961. Egg sizes of Laysan and Black-footed Albatrosses. *Condor* 63: 263.

Frings, H. & Frings, M. 1959. Observations on salt balance and behavior of Laysan and Black-footed albatrosses. *Condor* 61: 305-314.

Frings, H. & Frings, M. 1961. Some biometric studies of the albatrosses of Midway Atoll. *Condor* 63: 304-312.

Frost, O.W. ed. 1988. *Journal of a Voyage with Bering 1741-42* (transl. Engel, M.A. & Frost, O.W.). Stanford University Press, California.

Froude, R.E. 1891. On the Soaring of Birds. *Proc. R. Soc. Edinb.* 18: 65-72.

Froude, W. 1888. On the Soaring of Birds. *Proc. R. Soc. Edinb.* 19 March, 256-258.

Fryer, J. 1909. *A New Account of East India and Persia...* (Crooke, W. ed.). Hakluyt, Cambridge.

Fujisawa, K. 1967. *(Ahodori) Diomedea-albatrus.* Toko Shoin, Tokyo (In Japanese).

Fürbringer, M. 1902. Beitrag zur Genealogie und Systematik der Vögel. *Jenaische Zeitschrift für Naturwissenschaft* 34: 587-736.

Furness, R.W. 1988. Influences of status and recent breeding experience on the moult strategy of Yellow-nosed albatross *Diomedea chlororhynchos. J. Zool. Lond.* 215: 719-727.

Gadow, H. 1892. On the Classification of Birds. *Proc. Zool. Soc. Lond.* 1892: 229-256.

Gain, L. 1914. Oiseaux antarctique. *Deux. Expéd. Antarct. Fr. 1908-1910 Sci. Nat.* 5: 1-200.

Gales, R. 1993. *Co-operative Mechanisms for the Conservation of Albatross.* Australian Nature Conservation Agency, Hobart.

Gales, R. 1997. Albatross populations: status and threats. In *Albatross Biology and Conservation* (Robertson, G. & Gales, R. eds 1998): 20-45. Surrey Beatty, Chipping Norton.

Gales, R. Brothers, N. & Reid, T. 1998. Seabird mortality in the Japanese tuna longline fishery around Australia, 1988-1995. *Biol. Conserv.* 86: 37-56.

Gallo-Reynoso, J.P. & Figueroa-Carranza, A.L. 1996. The breeding colony of Laysan Albatrosses on Isla de Guadalupe, Mexico. *Western Birds* 27: 70-76.

Gamberoni, L., Geronimi, J., Jeannin, P.F. & Murail, J.F. 1982. Study of frontal zones in the Crozet-Kerguelen region. *Oceanol. Acta* 5: 289-299.

Geraghty, T. 1992. *Who Dares Wins* 3rd. edition. Little Brown, London.

Gibson, J.D. 1963. Third Report of the New South Wales Albatross Study Group (1962) Summarizing Activities to Date. *Emu* 63: 215-223.

Gibson, J.D. 1967. The Wandering Albatross (*Diomedea exulans*): results of banding and observations in New South Wales coastal waters and Tasman Sea. *Notornis* 14: 47-57.

Gibson, J.D. & Sefton, A.R. 1959. First Report of the New South Wales Albatross Study Group. *Emu* 59: 73-82.

Gibson, J.D. & Sefton, A.R. 1960. Second Report of the New South Wales Albatross Study Group. *Emu* 60: 125-130.

Gibson-Hill, C.A. 1950. Birds of the Southern Ocean. *Field* 196: 1046.

Ginn, H.B. & Melville, D.S. 1983. *Moult in Birds.* Guide 19. Br. Trust. Orn., Tring.

Gmelin, J.F. 1788-93. C. a. Linne... *Systema Naturae...* editio decima tertia.

Godley, E.J. 1975. Report on Activities of the Adams Island Party of the 1966 Auckland Islands Expedition. In *Preliminary Results of the Auckland Islands Expedition 1972-1973* (Yaldwyn, J.C. Ed.): 370-380.

Godman, F. du Cane. 1907-10. *A Monograph of the Petrels.* Witherby, London.

Golder, F.C. 1925. Bering's Voyages. *Am. Geog. Soc. Res. Ser. 2.* New York.

Goldsmith, O. 1774. *An History of the Earth and Animated Nature.* Nourse, London.

Goodman, F.R. 1978. Excursion to Cape Horn. *Geog. Mag.* 51: 104-112.

Goodridge, C.M. 1832. *Narrative of a Voyage to the South Seas.* Hamilton & Adam, London.

Gould, J. 1865. *Handbook to the Birds of Australia 2.* London.

Gould, P.J. 1983. Seabirds between Alaska and Hawaii. *Condor* 85: 286-291.

Gould, P.J. & Hobbs, R. 1993. Population Dynamics of the Laysan and other Albatrosses in the North Pacific. *Bull. N. Pac. Comm.* 53: 485-497.

Gould, P., Ostrom, P., Walker, W. & Pilichowski, K. 1997. Laysan and Black-footed Albatrosses: trophic relationships and driftnet fisheries associations of non-breeding birds. In *Albatross Biology and Conservation* (Robertson, G. & Gales, R. eds 1998): 199-207. Surrey Beatty, Chipping Norton.

Gould, P.J. & Piatt, J.F. 1993. Seabirds of the central North Pacific. In *The status, ecology and conservation of marine birds of the North Pacific.* (Vermeer, K., Briggs, K.T., Morgan, K.H. & Siegel-Causey, D. eds). Wildl. Serv. Spec. Publ., Ottawa.

Gower, E. 1803. The Loss of the Swift. *Falkland Islands J.* 1970: 24-29.

Grant, G.S. 1984. Energy cost of incubation to the parent seabird. In *Seabird Energetics* (Whittow, G.C. & Rahn, H. eds): 59-71. Plenum, New York.

Grant, G.S., Pettit, T.N., Rahn, H., Causey Whittow, G & Paganelli, C.V. 1982. Water loss from Laysan and Black-footed Albatross eggs. *Physiol. Zool.* 55: 405-414.

Grant, G.S. & Whittow, G.C. 1983. Metabolic cost of incubation in the Laysan Albatross and Bonin Petrel. *Comp. Biochem. Physiol.* 74A: 77-82.

Grau, C.R. 1984. Egg Formation. In *Seabird Energetics* (Whittow, G.C. & Rahn, H. eds): 33-57. Plenum, New York.

Graves, R. 1955. *The Greek Myths.* Penguin, London.

Green, J.F. 1887. *Ocean Birds.* London.

Green, R.H. 1974. Albatross Island. *Rec. Queen Victoria Mus.* 51: 1-17.

Greenwood, J.J.D. 1997. Introduction: the diversity of taxonomies. *Bull. Br. Ornithol. Club* 117: 85-96.

Grew, N. 1681. *Museum regalis societatis. Or a Catalogue & Description Of the Natural and Artificial Rarities Belonging to the Royal Society And preserved at Gresham College.* London.

Grindley, J.R. 1981. Observation of seabirds at Marion and Prince Edward Islands in April and May 1973. In *Proc. Symp. Birds Sea & Shore* (Cooper, J. ed.): 169-187. African Seabird Group, Cape Town.

Guthrie-Smith, H. 1936. *Sorrows and joys of a New Zealand naturalist.* Reed, Wellington.

Haase, B. 1994. A Chatham Island Mollymawk off the Peruvian coast. *Notornis* 41: 50.

Hadden, F.C. 1941. Midway Islands. *Hawaiian Planters' Record* 45: 179-221.

Hagen, Y. 1952. Birds of Tristan da Cunha. *Results Norw. Scient. Exped. Tristan da Cunha 1937-1938* 20: 1-248.

Hall, R. 1900. Field notes on the birds of Kerguelen Island. *Ibis* 7th Ser. 6: 1-34.

Hamlet, M.P. & Fisher, H.I. 1967. Air sacs of respiratory origin in some procellariiform birds. *Condor* 69: 586-595.

Haney, J.C., Haury, L.R., Mullineaux, L.S. & Fey, C.L. 1995. Seabird aggregations at a deep North Pacific seamount. *Mar. Biol.* 123: 1-9.

Hansbro, P. 1996. Seabird observations around southeast Australia, 1994-1995. *Australas. Seabird Grp. Newsl.* 30: 1-9.

Harper, P.C. 1978. The plasma proteins of some albatrosses and petrels as an index of relationships in the Procellariiformes. *N.Z. J. Zool.* 5: 509-549.

Harper, P.C. 1987. Feeding behaviour and other notes on 20 species of Procellariiformes at sea. *Notornis* 34: 169-192.

Harper, P.C. & Kinsky, F.C. 1978. *Southern Albatrosses & Petrels.* Price Milburn, Wellington.

Harris, M.P. 1973. The biology of the Waved Albatross *Diomedea irrorata* of Hood Island, Galapagos. *Ibis* 115: 483-510.

Harris, M.P. 1974. *A Field Guide to the Birds of the Galapagos.* Collins, London.

Harris, M.P. 1977. Comparative ecology of seabirds in the Galapagos archipelago. In *Evolutionary Ecology* (Stonehouse, B. & Perrins, C.M. eds): 65-76. Macmillan, London.

Harrison, C. 1982. Fragile Islands. *Geo* 4: 114-129.

Harrison, C.J.O. 1965. Allopreening as agonistic behaviour. *Behaviour* 24: 161-209.

Harrison, C.J.O. 1967. Sideways-throwing and sideways-building in birds. *Ibis* 109: 539-551.

Harrison, C.S. 1990. *Seabirds of Hawaii.* Cornell, Ithaca.

Harrison, C.S., Hida, T.S. & Seki, M.P. 1983. Hawaiian seabird feeding ecology. *Wildl. Monog.* 85: 1-71.

Harrison, J.A., Allan, D.G., Underhill, L.G., Herremans, H., Tree, A.J., Parker, V. & Brown, C.J. 1997. *The Atlas of Southern African Birds* 1. Birdlife South Africa, Johannesburg.

Harrison, L. 1916. The genera and species of Mallophaga. *Parasitology* 9: 1-156.

Harrison, N.M., Whitehouse, M.J., Heinemann, D., Prince, P.A., Hunt, G.L. Jr., & Viet, R.R. 1991. Observations of multispecies seabird flocks around South Georgia. *Auk* 108: 801-810.

Harrison, P. 1978. At sea identification of Wandering and Royal albatrosses. *Australas. Seabird Grp. Newsl.* 11: 14-21.

Harrison, P. 1983. *Seabirds.* Croom Helm, London.

Harrop, H. 1994. Albatrosses in the Western Palearctic. *Birding World* 7: 241-245.

Hartert, E. & Venturi, S. 1909. Notes sur les oiseaux de la République Argentine. *Novit. Zool.* 16: 159-267.

Hasegawa, H. 1978. Recent Observations of Short-tailed Albatross *Diomedea albatrus* at Torishima. *Misc. Rep. Yamashina Inst. Orn.* 10: 58-69.

Hasegawa, H. 1980. Observations on the status of the Short-tailed Albatross *Diomedea albatrus* on Torishima in 1977/78 and 1978/79. *J. Yamashina Inst. Orn.* 12: 59-67.

Hasegawa, H. 1982. The Breeding status of the Short-tailed Albatross *Diomedea albatrus*, on Torishima, 1979/80-1980/81. *J. Yamashina Inst. Orn.* 14: 16-24.

Hasegawa, H. 1984. Status and conservation of seabirds in Japan, with special attention to the Short-tailed Albatross. In *Status and Conservation of the World's Seabirds* (Croxall, J.P., Evans, P.G.H. & Schreiber, R.W. eds): 487-500. ICBP, Cambridge.

Hasegawa, H. 1995. (*The Short-tailed Albatross: A love symphony.* Kodansha, Tokyo). (In Japanese).

Hasegawa, H. & DeGange, A.R. 1982. The Short-tailed Albatross, *Diomedea albatrus*, its status. distribution and natural history. *Am. Birds.* 36: 806-814.

Hatch, S.A. 1987. Copulation and mate guarding in the Northern Fulmar. *Auk* 104: 450-461.

Hattori, T. 1889. (The story of the Albatross of Torishima). *Dobutsugaku Zasshi* 1: 405-411. (In Japanese; translated in Austin, O.L. Jr. 1949).

Hayes, F.E. & Baker, W.S. 1989. Seabird Distribution at Sea in the Galapagos Islands: Environmental Correlations and Associations with Upwelled Water. *Col. Waterbirds* 12: 60-66.

Hayward, T.L. 1997. Pacific Ocean Climate change: atmospheric forcing, ocean circulation and ecosystem response. *Trends Ecol. Evol.* 12: 150-154.

Haward, M., Bergin, A. & Hall, H.R. 1997. International legal and political bases to the management of the incidental catch of seabirds. In *Albatross Biology and Conservation* (Robertson, G. & Gales, R., eds 1998): 255-266. Surrey Beatty, Chipping Norton.

Headland, R. 1984. The Island of South Georgia. Cambridge University Press, Cambridge.

Headland, R.K. 1989. Chronological List of Antarctic Expeditions and related historical events. Cambridge University Press, Cambridge.

Hector, J. 1894. On the Anatomy of Flight of certain Birds. *Trans. NZ Inst.* 27: 284-287.

Hector, J.A.L. 1984. Techniques for the serial collection of blood samples and inspection of gonads in free-living albatrosses. *Br. Antarct. Surv. Bull.* 63: 127-133.

Hector, J.A.L., Croxall, J.P. & Follett, B.K. 1986a. Reproductive endocrinology of the Wandering Albatross *Diomedea exulans* in relation to biennial breeding and deferred sexual maturity. *Ibis* 128: 9-22.

Hector, J.A.L., Follett, B.K. & Prince, P.A. 1986b. Reproductive endocrinology of the Black-browed albatross *Diomedea melanophris* and the Grey-headed albatross *D. chrysostoma*. *J. Zool. Lond.* 208: 237-253.

Hector, J.A.L., Pickering, S.P.C., Croxall, J.P. & Follett, B.K. 1990. The endocrine basis of deferred sexual maturity in the wandering albatross, *Diomedea exulans* L. *Funct. Ecol.* 4: 59-66.

Hedd, A, Gales, R., Brothers, N., & Robertson, G. 1997. Diving behaviour of Shy Albatrosses (*Diomedea cauta*) in Tasmania: Initial findings and Dive Recorder Assessment. *Ibis* 139: 452-460.

Hendriks, F. 1972. *Dynamic soaring.* Unpublished Ph.D. Thesis, UCLA, Los Angeles.

Henle, K. 1981. Remarkable observations on the flight of albatrosses. *Ecology of Birds* 3: 311-313. (In German with English summary).

Herschel, J.F.W. 1861. Physical Geography: 347. *Encylop. Britannica.* Edinburgh.

Hicks, G.R.F. 1973. Latitudinal distribution of seabirds between New Zealand and the Ross Sea, December 1970. *Notornis* 20: 231-250.

Hill, L & Wood, E. 1976. *Penguin Millionaire.* David & Charles, Newton Abbot.

Hindwood, K.A. 1955. Seabirds in sewage. *Emu* 55: 212-216.

Hoare, M.E. ed. 1982. *The Resolution Journal of Johann Reinhold Forster 1772-1775.* Hakluyt, Cambridge.

Hodges, C.L. & Woehler, E.J. 1994. Associations between seabirds and cetaceans in the Australian sector of the southern Indian Ocean. *Mar. Orn.* 22: 205-212.

Holdgate, M.W. 1958. *Mountains in the Sea.* Macmillan, London.

Holgersen, H. 1957. Ornithology of the "Brategg" Expedition. *Scient. Results. "Brategg" Exped. 1947-48.* 4. Griegs, Bergen.

Holmes, R. 1989. *Coleridge.* Hodder & Stoughton, London.

Hopkins, G.H.E. & Clay, T. 1952. *A checklist of the genera and species of Mallophaga.* British Museum, London.

Horning, D.S. Jr. & Horning, C.J. 1974. Bird records of the 1971-'73 Snares Islands, New Zealand, Expedition. *Notornis* 21: 13-24.

Horton, E. 1860. Use of the Albatross. *Zoologist* 18: 6981.

Hough, R. 1971. *The Blind Horn's Hate.* Hutchinson, London.

Houghton, E.L. & Brock, A.E. 1960. *Aerodynamics for Engineering Students.* Arnold, London.

Houston, D.C., 1978. The effect of food quality on breeding strategy in Griffon vultures (*Gyps* spp.). *J. Zool. Lond.* 186: 175-184.

Houvenaghel, G.T. 1984. Oceanographic Setting. In *Galapagos* (Perry, R. ed.): 43-54. Pergoman, Oxford.

Howard, H. & Dodson, L. 1933. Bird remains from an Indian shell mound near Point Mugu, California. *Condor* 35: 235.

Howard, P.F. 1954. Banding of Black-browed Albatross at Heard Island and Macquarie Island. *Emu* 54: 256.

Howell, S.N.G. & Webb, S. 1990. The seabirds of Las Revillagigedo, Mexico. *Wilson Bull.* 102: 140-146.

Howell, T.R. & Bartholomew, G.A. 1961. Temperature regulation in Laysan and Black-footed Albatrosses. *Condor* 63: 185-197.

Huin, N. 1997. Prolonged incubation in the Black-browed Albatross *Diomedea melanophris* at South Georgia. *Ibis* 139: 178-180.

Huin, N., Prince, P.A. & Briggs, D.R. (in press). Chick provisioning rates and growth in Black-browed Albatross *Diomedea melanophris* and Grey-headed Albatross *D. chrysostoma* at Bird Island, South Georgia. *Ibis.*

Hunter, F.M., Burke, T. & Watts, S.E. 1992. Frequent copulation as a method of paternity assurance in the Northern Fulmar. *Animal Behaviour* 44: 149-156.

Hunter, R. 1997. Shark Attack. *S. Oceans Seabird Study Ass. Nltr.* 15: 9.

Hunter, S. 1984. Breeding biology and population dynamics of giant petrels *Macronectes* at South Georgia (Aves: Procellariiformes). *J. Zool. Lond.* 203: 441-460.

Hunter, S. & Klages, N.T.W. 1989. The diet of Grey-headed albatrosses at the Prince Edward Islands. *S. Afr. J. Antarct. Res.* 19: 31-33.

Huntley, J. 1971. Vegetation. In *Marion and Prince Edward Islands* (Van Zinderen Bakker, E.M. Sr., Winterbottom, J.M. & Dyer, R.A. eds). Balkema, Cape Town.

Hutton, F.W. 1865. Notes on some birds inhabiting the Southern Ocean. *Ibis* New Ser. 1: 276-298.

Hutton, F.W. 1867. Notes on the birds seen on a voyage from London to New Zealand in 1866. *Ibis* New Ser. 3: 185-193.

Hutton, F.W. 1868. On the mechanical principles involved in the flight of the albatross. *Trans. NZ Inst Sci.* 1: 58-60.

Hutton, F.W. 1871. On the Sailing Flight of the Albatross; a Reply to Mr J.S. Webb. *Trans. NZ Inst Sci.* 4: 347-350.

Hyslop, L.B. & Hyslop, F.E. Jr. 1957. *Baudelaire: A self-portrait.* Greenwood, Connecticut.

Idrac, P. 1923. Remarques sur le vol de l'albatros et sur un procédé simple d'étudier les variations d'inclinations du vent. *Techq. aéronaut.* 19: 602-606.

Idrac, P. 1924a. Contributions à l'étude du vol des albatros. *C. r. hebd. Séanc. Acad. Sci., Paris.* 179: 28-30.

Idrac, P. 1924b. Étude théorique des manoeuvres des albatros par vent croissant avec l'altitude. *C. r. hebd. Séanc. Acad. Sci., Paris.* 179: 1136-1139.

Idrac, P. 1925. Étude expérimentale et analytique du vol sans battements des oiseaux voiliers de mers australes d'albatros en particulaire. *Techq. aéronaut.* 16: 9-22.

Idyll, C.P. 1973. The Anchovy Crisis. *Sci. Am.* 228: 22-29.

Ikehara, S. & Shimojana, M 1971. The terrestrial animals of the Senkaku Islands. In *Report of scientific research on the Senkaku Islands* (Ikehara, S. ed.): 85-114. University of Ryukyus, Okinawa. (In Japanese with English summary).

Illiger, C. 1811. *Prodromus Systematis Mammalium et Avium.* Berlin.

Imber, M.J. 1991. Feeding ecology of Antarctic and subantarctic Procellariiformes. *Acta XX Congr. Int. Orn.* 3: 1402-1412.

Imber, M.J. 1992. Cephalopods eaten by wandering albatrosses (*Diomedea exulans* L.) breeding at six circumpolar localities. *J. R. Soc. NZ* 22: 243-263.

Imber, M.J. 1994. Report on a Tuna Long-lining fishing voyage aboard *Southern Venture* to observe by-catch problems. *NZ DoC Sci. Res. Ser.* 65:1-12.

Imber, M.J. & Berruti, A. 1981. Procellariiform seabirds as squid predators. In *Proc. Symp. Birds Sea & Shore* (Cooper, J. ed.): 43-61. African Seabird Group, Cape Town.

Imber, M.J. & Russ, R. 1975. Some foods of the Wandering Albatross (*Diomedea exulans*). *Notornis* 22: 27-36.

Jameson, W. 1958. *Wandering Albatross.* Hart-Davis, London

Japan Meteorological Agency 1963. *Meteorological data and report of Marcus and Torishima Islands.* Tokyo. (In Japanese).

Jehl, J.R. Jr. 1973. The distribution of marine birds in Chilean waters in Winter. *Auk* 90: 114-135.

Jehl, J.R. Jr. 1974. The distribution and ecology of marine birds over the continental shelf of Argentina in winter. *Trans. San Diego Soc. Nat. Hist.* 170: 217-234.

Jenkins, J.A.F. 1986. The Seabirds of Fiji. *Australas. Seabird Grp. Newsl.* 25: 1-70.

Jobling, J.A. 1991. *A Dictionary of Scientific Bird Names.* Oxford University Press, Oxford.

John, D.D. 1934. The second Antarctic commission of RRS *Discovery* II. *Geog. J.* 83: 381-398.

Johnson, D.H., Shaffer, T.L. & Gould, P.J. 1993. Incidental catch of marine birds in the North Pacific high seas driftnet fisheries in 1990. *N. Pac. Fish. Comm. Bull.* 53: 473-483.

Johnstone, G.W. 1980. Australian Islands in the Southern Ocean. *Bird Observer* 586: 85-88.

Johnstone, G.W. 1982. Zoology. In *Expedition to the Australian Territory of Heard and McDonald Islands 1980* (Veenstra, C. & Manning, J. eds): 33-39. Techn. Rep. 31, Dept. Natn. Development, Canberra.

Johnstone, G.W. & Kerry, K.R. 1976. Ornithological observations in the Australian sector of the Southern Ocean. *Acta XVI Congr. Int. Orn.* 725-738.

Johnstone, G.W., Milledge, D. & Dorwood, D.F. 1975. The White-capped Albatross of Albatross Island: numbers and breeding behaviour. *Emu* 75: 1-11.

Joly, Y., Frenot, Y. & Vernon, P. 1987. Environmental modifications of a subantarctic peat bog by the Wandering Albatross (*Diomedea exulans*): a preliminary study. *Polar Biol.* 8: 61-72.

Jouanin, C. 1959. Une colonie méconnu d'Albatros a pieds noirs, *Diomedea nigripes,* dans les Iles Mariannes. *Bull. Mus. Natl. Hist. Nat.* Paris (2) 31: 477-480.

Jouanin, C. & Mougin, J.-L. 1979. Order Procellariiformes. In *Checklist of Birds of the World* (Mayr, E. & Cottrell, G.W. eds) 1: 48-121. Harvard University Press, Cambridge.

Joudine, K. 1955. A propros du mécanisme fixant l'articulation du coude chez certaines oiseaux (Tubinares). *Acta XI Congr. Int. Orn.* 279-283.

Jouventin, P. 1990. Shy Albatross *Diomedea cauta salvini* breeding on Penguin Island, Crozet Archipelago, Indian Ocean. *Ibis* 132: 126-127.

Jouventin, P. & Lequette, B. 1990. The dance of the Wandering Albatross *Diomedea exulans. Emu* 90: 123-131.

Jouventin, P., Martinez, J. & Roux, J.-P. 1989. Breeding biology and current status of the Amsterdam Island Albatros *Diomedea amsterdamensis. Ibis* 131: 171-182.

Jouventin, P., de Monicault, G. & Blosseville, J.M. 1981. La dance de l'albatros, *Phoebetria fusca. Behaviour* 78: 43-80.

Jouventin, P., Roux, J.-P., Stahl, J.-C. & Weimerskirch, H. 1983. Biologie et frequence de reproduction chez L'Albatros a bec jaune (*Diomedea chlororhynchos*). *Gerfaut* 73: 161-171.

Jouventin, P., Stahl, J.-C., Weimerskirch, H. & Mougin, J.-L. 1984. The seabirds of the French subantarctic islands & Adelie Land, their status and conservation. In *Status and Conservation of the World's Seabirds* (Croxall, J.P., Evans, P.G.H. & Schreiber, R.W. eds): 609-625. ICBP, Cambridge.

Jouventin, P. & Stonehouse, B. 1985. Biological Survey of Ile de Croy, Iles Kerguelen. *Polar Record* 22: 688-691.

Jouventin, P. & Weimerskirch, H. 1984a. L'Albatros Fuligineux a Dos Sombre *Phoebetria fusca,* example de strategie d'adaptation extreme a la vie pelagique. *Rev. Ecol. (Terre & Vie)* 39: 401-429.

Jouventin, P. & Weimerskirch, H. 1984b. Les Albatros. *La Recherche* 15: 1228-1240.

Jouventin, P. & Weimerskirch, H. 1988. Demographic strategies of Southern Albatrosses. *Acta XIX Congr. Int. Orn.* 1: 857-865.

Jouventin, P. & Weimerskirch, H. 1990. Satellite tracking of Wandering Albatrosses. *Nature* 343: 746-748.

Jury, M.R. 1991. Anomalous winter weather in 1984 and a seabird irruption along the coast of South Africa. *Mar. Orn.* 19: 85-89.

Kalmbach, E.R., Delacour, J., Gabrielson, I.N., McCabe, R.A. & Munro, D.A. 1958. Report to the American Ornithologists' Union of the Committee on Bird Protection, 1957. *Auk* 75: 81-85.

Keage, P.L. 1982. *The Conservation Status of Heard Island and the McDonald Islands.* University of Tasmania, Hobart.

Keane, P.J. 1994. *Coleridges's Submerged Politics.* University of Missouri Press, Columbia.

Kenyon, K.W. & Kridler, E. 1969. Laysan Albatrosses swallow indigestible matter. *Auk* 86: 339-343.

Kenyon, K.W. & Rice, D.W. 1958. Homing of Laysan Albatrosses. *Condor* 60: 3-6.

Kepler, C.B. 1967. Polynesian rat predation on nestling Laysan Albatrosses and other Pacific seabirds *Auk* 84: 426-430.

Kerry, K.R. & Colback, G.C. 1972. Follow the Band! *Austr. Bird Bander* 10: 61-62.

Kerry, K.R. & Garland, B.R. 1984. The breeding biology of the Light-mantled Sooty Albatross *Phoebetria palpebrata* on Macquarie Island. *Tasmanian Naturalist* 79: 21-23.

Kidder, J.H. 1875. Contributions to the natural history of Kerguelen Island, made in connection with the United States Transit-of-Venus Expedition, 1874-75. 1 Ornithology. *Bull. US Natn. Mus.* 1: 1-47.

Kinsky, F.C. 1968. An unusual seabird mortality in the southern North Island (NZ), April 1968. *Notornis* 15: 143-155.

Kirkwood, R.J. & Mitchell, P.J. 1992. The Status of the Black-browed Albatross *Diomedea melanophys* at Heard Island. *Emu* 92: 111-114.

Klaer, N. & Polacheck, T. 1998. The Influence of Environmental Factors and Mitigation Measures on By-Catch Rates of Seabirds by Japanese Longline Fishing Vessels in the Australian Region. *Emu* 98: 305-315.

Klein, J.T. 1750. *Historiae Avium Prodromus...* Schmidt, Lubeck.

Koteja, P. 1991. On the relation between basal and field metabolic rates in birds and mammals. *Funct. Ecol.* 5: 56-64.

Krasheninnikov, S.P. 1755. *The history of Kamschatka and the Kurilski Islands...* (Grieve, J. transl. 1764). Jefferys, London.

Kurata, Y. 1978. Breeding records of the Laysan Albatross *Diomedea immutabilis* on the Ogasawara Islands. *Misc. Rep. Yamashina Inst. Orn.* 10: 185-189.

Kuroda, N. 1925. *Avifauna of the Riukiu Islands.* Tokyo.

Kuroda, N. 1954. *On the Classification and Phylogeny of the Order Tubinares, particularly the Shearwaters (Puffinus)...* Tokyo.

Kuroda, N. 1960. Notes on the breeding seasons in the Tubinares (Aves). *Jap. J. Zool.* 12: 449-464.

Kuroda, N. 1961. A note on the pectoral muscles of birds. *Auk* 78: 261-263.

Kuroda, N. 1971. Some external and anatomical measurements of a Laysan Albatross with new naming to parts of the pectoral muscles and under-wing coverts. *Misc. Rep. Yamashina Inst. Orn.* 6: 316-320.

Kuroda, N. 1988. A distributional analysis of *Diomedea immutabilis* and *D. nigripes* in the North Pacific. *J. Yamashina Inst. Orn.* 20: 1-20.

Lachmund, D.F. 1674. *De Ave Diomedea dissertatio...* Amsterdam.

La Cock, G. & Schneider, D.C. 1982. Duration of ship following by Wandering Albatrosses *Diomedea exulans. Cormorant* 10: 105-107.

La Grange, J.J. 1962. Notes on the birds and mammals of Marion Island and Antarctica (SANAE). *J. S. Afr. Biol. Soc.* 3: 27-84.

Lack, D. 1947. The significance of clutch-size. *Ibis* 89: 302-352.

Lack, D. 1954. *The Natural Regulation of Animal Numbers.* Oxford University Press, Oxford.

Lancaster, I. 1886. Gravitation and the soaring bird. *Am. Nat.* 20: 514-521.

Lanchester, F.W. 1910. *Aerodonetics.* Constable, London.

Langston, N.E. & Hillgarth, N. 1995. Moult varies with parasites in Laysan Albatross. *Proc. R. Soc. Lond.* B. 261: 239-243.

Langston, N.E. & Rohwer, S. 1995. Unusual patterns of incomplete primary molt in Laysan and Black-footed Albatrosses. *Condor* 97: 1-19.

Langston, N.E. & Rohwer, S. 1996. Molt – breeding tradeoffs in albatrosses: life history implications in big birds. *Oikos* 76: 498-510.

Lashmar, A.F.C. 1990. Albatross studies – Kangaroo Island, South Australia: A complete summary 1971-1988. *Corella* 14: 44-50.

Layard, E.F. 1867. Note on skins and eggs brought back from the Crozette Islands by Captain Armson... *Ibis* 3 (New Ser.): 457-461.

Layard, N.F. 1897. *Songs in Many Moods.* Longman, London.

Latham, J. 1785. *A General Synopsis of Birds* 3. London.

La Touche, J.D.D. 1895. Notes on South Formosa and its birds *Ibis* (1895): 305-338.

LeFebvre, E.A. 1977. Laysan Albatross breeding behaviour. *Auk* 94: 270-274.

LeFebvre, E.A., Birkebak, R.C. & Dormann, F.D. 1967. A flight-time integrator for birds. *Auk*: 84:124-128.

LePage, L. 1921. Note on the soaring flight of birds. *Aeronautical Res. Council Rep. Memo.* 742: 917-923.

Lequette, B. & Jouventin, P. 1991a. The dance of the Wandering Albatross II: Acoustic Signals. *Emu* 91: 172-178.

Lequette, B. & Jouventin, P. 1991b. Comparison of visual and vocal signals of Great Albatrosses. *Ardea* 79: 383-393.

Lequette, B., Verheyden, C. & Jouventin, P. 1989. Olfaction in subAntarctic seabirds: its phylogenetic and ecological significance. *Condor* 91: 732-735.

Lequette, B. & Weimerskirch, H. 1990. Influence of parental experience on the growth of Wandering Albatross chicks. *Condor* 93: 726-731.

Lesson, R.P. 1825. Distribution géographique de quelques Oiseaux marins observés dans le Voyage autour du Monde de la Corvette la Coquille. *Ann. Sci. Nat. Paris* 6: 88-103.

Lévêque, R. 1963. Le statut actuel des vertébrés rare et menacés de l'archipel des Galalagos. *Terre & Vie* 110: 397-430.

Lichtenstein, M.H.C. ed. 1844. *Descriptiones animalium quae in itinere ad maris australis terras par annos 1772 1773 et 1774 ... Ionnes Reinoldus Forster...* Berlin.

Lighthill, M.J. 1975. Aerodynamic aspects of animal flight. In *Swimming and Flying in Nature* (Wu, T.Y.T., Brokaw, C.J. & Brennen, C. eds): 423-491. Plenum, New York.

Linnaeus, C. 1758. *Systema Naturae...* 1. 10th edition, Stokholm.

Lipinski, M.R. & Jackson, S. 1989. Surface-feeding on cephalopods by procellariiform seabirds in the southern Benguela region, South Africa. *J. Zool. Lond.* 218: 549-563.

Lister, R.P. 1965. The Albatross. *Punch* 248: 502.

Liversidge, R. & Le Gras, G.M. 1981. Observations of Seabirds off the eastern Cape, South Africa. 1958-63. In *Proc. Symp. Birds Sea & Shore* (Cooper, J. ed.): 149-167. African Seabird Group, Cape Town.

Lock, J.W., Thompson, D.R., Furness, R.W. & Bartle, J.A. 1992. Metal concentration in seabirds of the New Zealand region. *Environ. Pollut.* 75: 289-300.

Loomis, L.M. 1918. A review of the albatrosses, perels and diving petrels. *Proc. Calif. Acad. Sci.* (4th Ser.) 2: 1-187.

Loranchet, J. 1915. Observations biologiques sur les oiseaux des Iles Kerguelen. *Revue fr. Orn.* 4: 240-242.

Lovvorn, J. 1994. (untitled regional report). *Pac. Seabirds* 21: 28.

Lowe, P.R. & Kinnear, N.B. 1930. Birds *Br. Antarct. ("Terra Nova") Exped. 1910 Nat. Hist. Rep. Zool.* 4(5): 103-193.

Lowes, J.L. 1927. *The Road to Xanadu.* Constable, London.

Lubbock, B. 1937. *The Arctic Whalers.* Brown & Ferguson, Glasgow.

Ludwig, J.P., Auman, H.J., Giesy, J.P., Sanderson, J.T., DeDoes, J.M. & Jones, P. 1996. Reproductive Hazards to North Pacific Albatrosses from PCBs and TCDD-EQS. *Pac. Seabirds* 23: 41.

Ludwig, J.P., Ludwig, C.E. & Apfelbaum, S.I. 1979. *Midway Island Survey 1-24 February 1979.* unpublished report, USFWS, Honolulu.

Ludwig, J.P., Summer, C.L., Auman, H.J., Gauger, V., Bromley, D., Giesy, J.P., Rolland, R. & Colborn, T. 1997. The roles of organochlorine contaminants and fisheries bycatch in recent population changes of Black-footed and Laysan Albatrosses in the North Pacific Ocean. In *Albatross Biology and Conservation* (Robertson, G. & Gales, R. eds 1998): 225-238. Surrey Beatty, Chipping Norton.

Lugg, D.J., Johnstone, G.W. & Griffin, B.J. 1978. The outlying islands of Macquarie Island. *Geog. J.* 144: 277-287.

Lysaght, A. 1959. Some eighteenth century bird paintings in the library of Sir Joseph Banks (1743-1820). *Bull. Br. Mus. Nat. Hist.* (hist. Ser.) 1: 251-371.

Macdonald, D. & Green, R.H. 1963. Albatross Island. *Emu* 63: 23-31.

Mackay, R.S. 1974. Field Studies on Animals. *Biotelemetry* 1: 286-312.

MacKenzie, D. 1968. Birds and seals of Bishop and Clerk Islets, Macquarie Island. *Emu* 67: 241-245.

McCallum, J. 1985. Buller's Mollymawks on Rosemary Rock, Three Kings Islands in 1985. *Notornis* 32: 257-259.

McCormick, R. 1884. *Voyages of discovery in the Arctic and Antarctic Seas...* Sampson Low, London.

McCrindle, J.W. (transl.) 1897. *The Christian Topography of Cosmas, an Egyptian Monk.* Hakluyt, London.

McDermond, K. 1991. *Bull. Pac. Seabird Grp.* 18: 8.

McDermond, K. 1993. *Bull. Pac. Seabird Grp.* 20: 49-51.

McDermond, K. 1994. *Bull. Pac. Seabird Grp.* 21: 31-33.

McDermond, K. & Morgan, K.H. 1993. Status and conservation of North Pacific albatrosses. In *The status, ecology, and conservation of marine birds of the North Pacific* (Vermeer, K., Briggs, K.T., Morgan, K.H. & Siegel-Causey, D. eds): 70-81. Environment, Canada.

McFadgen, B.G. & Yaldwyn, J.C. 1984. Holocene Sand dunes on Enderby Island, Auckland Islands. *NZ J. Geol. Geophy.* 27: 27-33.

McHugh, J.L. 1950. Increasing abundance of albatrosses off the coast of California. *Condor* 52: 153-156.

McHugh, J.L. 1952. Fur-seals preying on Black-footed Albatrosses. *J. Wildl. Manag.* 16: 226.

McLain, R. 1898. Capture of the Short-tailed Albatross on the coast of Southern California. *Auk* 15: 267.

McLelland, J. 1989. Larynx and trachea. In *Form and Function in Birds* (King, A.S. & McLelland, J. eds) Academic, London 4: 69-103.

McNab, R. 1909. *Murihiku.* Whitcombe & Tombs, Christchurch.

Magnan, A. 1925. *Le Vol a Voile.* Roche d'Estrez, Paris.

Mallet, J. 1995. A species definition for the Modern Synthesis. *Trends Ecol. Evol.* 10: 294-299.

Mandlebrot, B.B. 1982. *The fractal geometry of nature.* W.H. Freeman, San Francisco.

Marchant, S. & Higgins, P.J. (Co-ordinators) 1990. *Handbook of Australian, New Zealand and Antarctic Birds.* Oxford University Press, Oxford.

Markham, C.R. ed. 1878. *The Hawkins' Voyages.* Hakluyt, London.

Marks, J.S. & Hall, C.S. 1992. Tool use by Bristle-thighed Curlews feeding on albatross eggs. *Condor* 94: 1032-1034.

Marples, B.J. 1946. List of the Birds of New Zealand. *Bull. Orn. Soc, NZ* 1: (supp.).

Martin, G.R. 1998. Eye structure and amphibious foraging in albatrosses. *Proc. R. Soc. Lond.* B. 265: 665-671.

Masefield, J. 1902. *Salt-water Ballads.* Grant Richards, London.

Mathews, G.M. 1912-13. *The Birds of Australia* 2. Witherby, London.

Mathews, G.M. 1927. *Systema Avium Australasianorum.* British Ornithologists Union, London.

Mathews, G.M. 1930. *Bull. Br. Ornith. Club* 51: 29.

Mathews, G.M. 1934. Remarks on Albatrosses and Mollymawks. *Ibis* (13th Ser.) 4: 807-816.

Mathews, G.M. 1936. The ossification of certain tendons in the patagial fan of Tubinares. *Bull. Br. Orn. Club* 56: 45-50.

Mathews, G.M. 1948. Systematic Notes on Petrels. *Bull. Br. Orn. Club* 68: 155-170.

Mathews, G.M. & Hallstrom, E.J.L. 1943. *Notes on the order Procellariiformes.* Canberra.

Matthews, L.H. 1929. The Birds of South Georgia. *Discovery Rep.* 1: 561-592.

Matthews, L.H. 1951. *Wandering Albatross.* MacGibbon & Key, London.

Mawson, D. 1915. *The Home of the Blizzard.* Heinemann, London.

Mawson, D. 1934. The Kerguelen Archipelago. *Geog. J.* 83: 18-29.

Mayr, E. & Amadon, D. 1951. A Classification of Recent Birds. *Am. Mus. Novit.* 1496: 1-42.

Medway, D.G. 1993. The identity of the Chocolate Albatross *Diomedea spadicea* of Gmelin, 1789 and of the Wandering Albatross *Diomedea exulans* of Linnaeus, 1758. *Notornis* 40: 145-162.

Medway, D.G. 1997a. Type specimens of albatrosses collected on Cook's second voyage. In *Albatross Biology and Conservation* (Robertson, G & Gales, R. eds 1998): 3-12. Surrey Beatty, Chipping Norton.

Medway, D.G. 1997b. Human-induced mortality of Southern Ocean Albatrosses at sea in the 19th century: a brief historical review. In *Albatross Biology and Conservation* (Robertson, G & Gales, R. eds 1998): 189-198. Surrey Beatty, Chipping Norton.

Meeth, P. & Meeth, K. 1983. Seabird observations from six Pacific Ocean crossings. *Sea Swallow* 32: 145-162.

Meeth, P. & Meeth, K. 1988. A Shy Albatross off Somalia. *Sea Swallow* 37: 66.

Melville, D.S. 1991. Primary moult in Black-browed and Shy Mollymawks. *Notornis* 38: 51-59.

Melville, H. 1851. *Moby-Dick.* London & New York.

Menegaux, A. 1917. Reprise d'un albatros bague. *Revue fr. Orn.* 5: 64.

Merlen, G. 1996. Scavenging Behavior of The Waved Albatross in Galapagos: A Potential Problem With Increasing Longlining? *Pac. Seabirds* 23: 10-12.

Meseth, E.H. 1968. *The behaviour of the Laysan Albatross, Diomedea immutabilis, on its breeding ground.* Unpublished thesis Ph.D. S. Illinois Univ. Carbondale.

Meseth, E.H. 1975. The dance of the Laysan Albatross *Diomedea immutabilis. Behaviour* 54: 217-257.

Micol, T. & Jouventin, P. 1995. Restoration of Amsterdam Island, South Indian Ocean, following control of feral cattle. *Biol. Conserv.* 73: 199-206.

Miller, L. 1942. Some tagging experiments with Black-footed Albatrosses. *Condor* 44: 3-9.

Miskelly, C.M. 1984. Birds of the Western Chain Snares Islands, 1983-84. *Notornis* 31: 209-223.

Miskelly, C.M. 1997. Biological Notes on the Western Chain, Snares Islands, 1984-85 and 1985-86. *Conservation Advisory Notes* 144: 1-10.

Moloney, C.L., Cooper, J., Ryan, P.G. & Siegfried, R. 1994. Use of a population model to assess the impact of longline fishing on Wandering Albatross *Diomedea exulans* Populations. *Biol. Conserv.* 70: 195-203.

Moore, P.J. 1996. *Light-mantled sooty albatross on Campbell Island, 1995-96: a pilot investigation.* NZ Dept. of Conservation, Wellington.

Moore, P.J. & Moffat, R.D. 1990a. *Research and Management Projects on Campbell Island 1987-88.* NZ Dept. of Conservation, Wellington.

Moore, P.J. & Moffat, R.D. 1990b. *Mollymawks on Campbell Island.* NZ Dept. of Conservation, Wellington.

Moore, P.J., Scott, J.J., Joyce, L.J. & Peart, M. 1997b. *Southern Royal Albatross Diomedea epomophora epomophora census on Campbell Island 4 January-6 February 1996, and a review of population figures.* NZ Dept. of Conservation, Wellington.

Moore, P.J., Taylor, G.A. & Amey, J.M. 1997a. Interbreeding of Black-browed Albatross *Diomedea m. melanophris* and New Zealand Black-browed Albatross *D. m. impavida* on Campbell Island. *Emu* 97: 322-324.

Moore, T. 1817. *Lalla Rookh an Oriental Romance.* Longmans, London.

Morant, P.D., Brooke, R.K. & Abrams, R.W. 1983. Recoveries in southern Africa of seabirds breeding elsewhere. *Ringing & Migration* 4: 257-268.

Morison, S.E. 1965. *The Oxford History of the American People.* Oxford University Press, Oxford.

Morlan, J. 1994. Light-mantled Sooth Albatross (*Phoebetria palpebrata*) over Cordell Banks, Marin County, California. *Australas. Seabird Grp. Newsl.* 27: 5-8.

Moseley, H.N. 1879. *Notes of a Naturalist on the "Challenger".* Macmillan, London.

Mougin, J.-L. 1970. Observations ecologiques sur les grands albatros (*Diomedea exulans*) de l'Ile de la Possession (Archipel Crozet) en 1968. *Oiseau Revue fr. Orn.* 40:16-36.

Mougin, J.-L. Jouanin, C. & Roux, F. 1997. Intermittent breeding in Cory's Shearwater *Calonectris diomedea* of Selvagem Grande, North Atlantic. *Ibis* 139: 40-44.

Muirhead, S.J. & Furness, R.W. 1988. Heavy metal concentrations in the tissues of Seabirds from Gough Island, South Atlantic. *Mar. Pollut. Bull.* 19: 278-283.

Mullen, J. 1982. Tail Piece. *Sea Swallow* 31: 71.

Munch, P.A. 1946. Sociology of Tristan da Cunha. *Res. Norweg. Sci. Exped. Tristan da Cunha 1937-38* 13: 1-330.

Munro, G.C. 1945a. The Laysan Albatross on Kauai. *Elepaio* 5: 70.

Munro, G.C. 1945b Notes on the bird life of Midway Islands. *Elepaio* 6: 22-26.

Murie, O.J. 1959. *Fauna of the Aleutian Islands and Alaska Peninsula.* Govt. Print, Washington DC.

Murphy, E.J., Watkins, J.L., Reid, K., Trathan, P.N., Everson, I., Croxall, J.P., Priddle, J., Brandon, M.A., Brierley, A.S. & Hoffmann, E. 1998. Interannual Variation of the South Georgia marine ecosystem: biological and physical sources of variation in the abundance of krill. *Fish. Oceanogr.* 7: 381-390.

Murphy, R.C. 1914. Cruise of the "Daisy": observations of birds in the south Atlantic. *Auk* 31: 439-457.

Murphy, R.C. 1917. A new albatross from the west coast of South America. *Bull. Amer. Mus. Nat. Hist.* 37: 861-864.

Murphy, R.C. 1930. Birds collected during the Whitney South Seas expedition. XI. *Am. Mus. Novit.* 419: 1-15.

Murphy, R.C. 1936. *Oceanic Birds of South America.* Macmillan, New York.

Murphy, R.C. 1938. Wandering Albatross in the Bay of Panama. *Auk* 11: 126.

Murphy, R.C. 1947. *Logbook for Grace.* Macmillan, New York.

Murray, D. 1989. The Gibson Code. *Corella* 13: 104.

Murray, T.E., Bartle, J.A., Kalish, S.R., & Taylor, P.R. 1993. Incidental capture of seabirds by Japanese southern bluefin tuna longline vessels in New Zealand waters 1988-1992. *Biol. Conserv. Int.* 3: 181-210.

Murray, T.E., Taylor, P.R., Greaves, J., Bartle, J.A. & Molloy, J. 1992. *Seabird bycatch by Southern Fishery longline vessels in New Zealand waters.* NZ Fisheries Assessment Doc. 92/93. Wellington.

Nelson, B. 1968. *Galapagos, Islands of Birds.* Longmans, London.

Nelson, E.W. 1887. Birds of Alaska I. In *Report upon the natural history collections made in Alaska between the years 1877 and 1881.* Government Printer, Washington DC.

Nelson, J.B. 1978. *The Sulidae.* Oxford University Press, Oxford.

Neves, T. & Olmos, F. 1997. Albatross mortality in fisheries off the coast of Brazil. In *Albatross Biology and Conservation* (Robertson, G & Gales, R. eds 1998): 214-219. Surrey Beatty, Chipping Norton.

Nevitt, G.A., Veit, R.R. & Kareiva, P. 1995. Dimethyl sulphide as a foraging cue for Antarctic Procellariiform seabirds. *Nature* 376: 680-682.

Newton, A. 1896. *A Dictionary of Birds.* Adam & Black, London

Newton, I. 1980. The role of food in limiting bird numbers. *Ardea* 68: 11-30.

Nibor, Y. 1896. In L'Albatros: Mythes et réalites (Vincent, T. 1990). *Chasse-marée* 52: 22-37.

Nicholls, D., Murray, D., Battam, H., Robertson, G., Moors, P., Butcher, E. & Hildebrandt, M. 1995. Satellite tracking of the Wandering Albatross *Diomedea exulans* around Australia and in the Indian Ocean. *Emu* 95: 223-230.

Nicholls, D.G., Murray, M.D., Butcher, E. & Moors, P. 1997. Weather Systems Determine the Non-breeding Distribution of Wandering Albatrosses over Southern Oceans. *Emu* 97: 240-244.

Nicholls, D.G., Murray, M.D., Elliott, G.P. & Walker, K.J. 1996. Satellite tracking of a Wandering Albatross from the Antipodes Islands, New Zealand, to South America. *Corella* 20: 28.

Nicholls, D.G., Murray, M.D. & Robertson, C.J.R. 1994. Oceanic flights of the Northern Royal Albatross *Diomedea epomophora sanfordi* using satellite telemetry. *Corella* 18: 50-52.

Nicholls, G.H. 1979. Underwater swimming by albatrosses. *Cormorant* 6: 38.

Nichols, J.T. & Murphy, R.C. 1914. A review of the genus *Phoebetria. Auk* 31: 526-534.

Nitzsch, C.L. 1867. *Nitzsch's Pterylography* (transl. Sclater, P.L.). Ray Society, London.

Northridge, S.P. 1991. Driftnet fisheries and their impact on non-target species: a worldwide review. *FAO Fisheries Tech. Paper* 320: 1-115.

Nowak, J.B. Isla de la Plata and the Galapagos. 1987. *Noticias de Galápagos* 44: 17.

Nunn, G.B., Archander, P., Bretagnolle, V., Cooper, J., Robertson, C. & Robertson, G. 1994. A Mitochondrial DNA Phylogeny Petrels *Procellariiformes. J. Orn.* 135 (supp.): 34.

Nunn, G.B., Cooper, J., Jouventin, J., Robertson, C.J.R. & Robertson, G.G. 1996. Evolutionary relationships among extant albatrosses (Procellariiformes: Diomedeidae) established from complete Cytochrome-*b* gene sequences. *Auk* 113: 784-801.

Nunn, G.B. & Stanley, S.E. 1998. Body Size Effects and Rates of Cytochrome *b* Evolution in Tube-nosed Seabirds. *Mol. Biol. Evol.* 15: 1360-1371.

Nuttall, Z. (transl. & ed.) 1914. New light on Drake. A collection of documents relating to his voyage of circumnavigation 1577-1580. Hakluyt, London.

Oatley, T.B. 1979. Underwater swimming by albatrosses. *Cormorant* 7: 31.

Ogi, H., Hattori, Y., Yoshino, T., Yamahara, T., Sawada, N. & Furukawa, T. 1993. Sightings of the Short-tailed Albatross Over the Bering Sea. *J. Yamashina Inst. Orn.* 25: 102-104.

Ogi, H., Momose, K., Sato, F. & Baba, N. 1994a. Plastic particles found in the gizzard of a starved Black-footed Albatross (*Diomedea nigripes*). *J. Yamashina Inst. Orn.* 26: 77-80.

Ogi, H., Sato, F., Mitamura, A., Baba, T. & Oyama, H. 1994b. A Survey of Black-footed Albatross Breeding Colonies and Chicks on Torishima, January 1994. *J. Yamashina Inst. Orn.* 26: 126-131.

Ogilvie, M.A. 1978. *Wild Geese.* Poyser, Berkhamstead.

Ohlendorf, H.M. 1993. Marine birds and trace elements in the temperate North Pacific. In *The Status, Ecology and Conservation of Marine Birds of the North Pacific* (Vermeer, D., Briggs, K.T., Morgan, K.H. & Siegel-Causey, D. eds): 232-240. Canad. Wildl. Serv., Ottawa.

Oliver, W.R.B. 1930. *New Zealand Birds.* Reed, Wellington.

Oliver, W.R.B. 1955. *New Zealand Birds* 2nd edition. Reed, Wellington.

Ollason, J.C. & Dunnet, G.M. 1978. Age, experience and other factors affecting the breeding success of the fulmar *Fulmarus glacialis* in Orkney. *J. Anim. Ecol.* 47: 961-976.

Olmos, F., Martuscelli, P., Silva e Silva, R. & Neves, T.S. 1995. The Sea-birds of São Paulo, southeastern Brazil. *Bull. Br. Ornithol. Club* 115: 117-128.

Olson, S.L. 1983. Fossil seabirds and changing marine environments in the late Tertiary of South Africa. *S. Afr. J. Sci.* 79: 399-402.

Olson, S.L. 1984. Evidence of a large albatross in the Miocene of Argentina (Aves: Diomedeidae). *Proc. Biol. Soc. Wash.* 97: 741-743.

Olson, S.L. 1985. The fossil record of birds. In *Avian Biology* 8 (Farner, D.S. & Parkes, K.C. eds): 79-252. Academic Press, London.

Olson, S.L. 1995. Birds observed at sea during the voyage of HMS *Blonde* to Hawaii (1824-1826). *Sea Swallow* 44: 38-43.

Olson, S.L. 1996. History and Ornithological Journals of the *Tanager* expedition of 1923 to the Northwest Hawaiian Islands, Johnston and Wake Islands. *Atoll Res. Bull.* 433: 1-210.

Ono, Y. 1955. Status of birds on Torishima, particularly of Steller's Albatross. *Tori* 14: 24-32. (In Japanese with English summary).

Open University, 1989. *Ocean Circulation.* Pergamon, Oxford.

Orlando, C. 1958. Cattura di un albatro ulatore (*Diomedea exulans exulans*, Linnaeus) in Sicilia. *Riv. Italiano Orn.* 28: 101-113.

Ornithological Society of Japan 1974. *Check-list of Japanese Birds* 5th edition. Gakken, Tokyo.

Ortiz-Crespo, F.I. & Agnew, P. 1992. The birds of La Plata Island, Ecuador. *Bull. Br. Ornithol. Club.* 112: 66-73.

Oryx 1988. Galápagos Funding. *Oryx* 32: 105.

Oustalet, M.E. 1896. Les Mammifères et les oiseaux des Iles Mariannes. *Nouv. Arch. Mus. Hist. Nat.* (3 er Ser.) 7: 141-228.

Owre, O. 1976. A second breeding colony of the Waved Albatross *Diomedea irrorata. Ibis* 118: 419-420.

Pallas, P.S. 1769. *Spicilegia Zoologica* 5: 28-32.

Pallas, P.S. 1831. *Zoographia Rosso-Asiatica...* (Academy of Sciences, St Petersburg).

Palliser, T. 1999. The Ship's Log. *Australas. Seabird Grp. Newsl.* 35: 17.

Palma, R.L. 1994. New synonymies in the lice (Insecta: Phthiraptera) infesting albatrosses and petrels (Procellariiformes). *NZ Entomologist* 17: 64-69.

Palma, R.L. & Pilgrim, R.C.L. 1984. A revision of the genus *Harrisoniella* (Mallophaga: Philopteridae) *NZ J. Zool.* 11: 145-166.

Palma, R.L. & Pilgrim, R.C.L. 1987. A revision of the genus *Perineus* (Phthiraptera: Philopteridae) *NZ J. Zool.* 14: 563-586 (1988).

Palmer, R.S. 1956. *Egg Profiles.* Am. Ornithol. Union, Washington.

Palmer, R.S. 1962. *Handbook of North American Birds* 1. Yale University Press, New Haven.

Parrish, R. 1991. Buller's Mollymawk hooked. *Notornis* 38: 344.

Pascal, M. 1978. Note sur *Phoebetria fusca, Diomedea chlororhynchos* et *Diomedea chrysostoma* aux Iles Kerguelen. *Oiseau Revue fr. Orn.* 48: 69-71.

Pascal, M. 1979. Données ecologiques sur l'albatros a sourcils noirs *Diomedea melanophris* (Temminck) dans l'archipel des Iles Kerguelen. *Alauda* 47: 165-172.

Paterson, A.M., Gray, R.D., & Wallis, G.P. 1993. Parasites, petrels and penguins: does louse presence reflect seabird phylogeny? *Int. J. Parasitology* 23: 515-526.

Patmore, C. (ed) 1917. *Table Talk and Omnia of Samuel Taylor Coleridge*. Oxford University Press, Oxford.

Paulian, P. 1953. Pinnepèdes, Cétacés, Oiseau des Iles Kerguelen et Amsterdam. *Mem. Instit. Sci. Madagascar* A, 8: 111-234.

Paxton, R.O. 1968. Wandering Albatross in California. *Auk* 85: 502-504.

Peal, S.E. 1880. Soaring of Birds. *Nature* 23: 10-11.

Pearce, F. 1998. Thrown a lifeline. *New Scientist* 160: 13.

Pearse, T. 1968. *Birds of the Earlier Explorers in the Northern Pacific*. Centennial Canadian Confederation Project.

Peirce, M.A. & Prince, P.A. 1980. *Heptazoon albatrossi*, new species (Eucoccida: Heptozoidae) from *Diomedea* spp. in Antarctica. *J. Nat. Hist.* 14: 447-452.

Pennycuick, C.J. 1972. *Animal Flight*. Edward Arnold, London.

Pennycuick, C.J. 1975. Mechanics of Flight. In *Avian Biology* 5 (Farner, D.S., King, J.R. & Parkes, K.C. eds): 1-75. Academic Press, New York.

Pennycuick, C.J. 1982a. The ornithodolite: an instrument for collecting large samples of bird speed measurements. *Phil. Trans. R. Soc. Lond.* B 300: 61-73.

Pennycuick, C.J. 1982b. The flight of Petrels and Albatrosses (Procellariiformes), observed in South Georgia and its vicinity. *Phil. Trans. R. Soc. Lond.* B 300: 75-106.

Pennycuick, C.J. 1987. Flight in Seabirds. In *Seabirds* (Croxall, J.P. ed.): 43-62. Cambridge University Press, Cambridge.

Pennycuick, C.J. 1989. *Bird Flight Performance*. Oxford University Press, Oxford.

Pennycuick, C.J. 1999. *Measuring birds' wings*. Boundary Layer publications, Bristol.

Pennycuick, C.J. & Webbe, D. 1959. Observations on the Fulmar in Spitzbergen. *Br. Birds* 52: 321-332.

Perrins, C.M., Harris, M.P. & Britton, C.K. 1973. Survival of Manx Shearwaters *Puffinus puffinus*. *Ibis* 115: 535-548.

Perry, R. ed. 1985. *Galapagos*. Pergamon, Oxford.

Peters, J.L. 1931. *Check-list of Birds of the World* 1. Harvard University Press, Cambridge.

Peterson, P.C. & Atyeo, W.T. 1972. the Feather mite family Alloptidae, Gaud. II – The Oxalginae, New subfamily (Analgoidea). *Acarology* 8: 651-668.

Peterson, R.T. 1967. The Galapagos. *Natl. Geogr. Mag.* 131: 541-585.

Pettit, T.N., Grant, G.S., Whittow, G.C, Rahn, H. & Paganelli, C.V. 1982. Embryonic oxygen consumption and growth of Laysan and Black-footed Albatrosses. Am. J. Physiol. 242: 121-128.

Pettit, T.N., Nagy, K.A., Ellis, H.I. & Whittow, G.C. 1988. Incubation energetics of the Laysan Albatross. *Oecologia* (Berlin) 74: 546-550.

Philander, S.G. 1990. *El Niño, La Niña and the Southern Oscillation*. Academic Press, New York.

Phillips, R.A. & Hamer, K.C. 1999. Lipid reserves, fasting capability and the evolution of nestling obesity in procellariiform seabirds. *Proc. R. Soc. Lond.* B 266: 1329-1334.

Pickering, S.P.C. 1989. Attendance patterns and behaviour in relation to experience and pair-bond formation in Wandering Albatross *Diomedea exulans* at South Georgia. *Ibis* 131: 183-195.

Pickering, S.P.C. & Berrow, S.D. (In press). Courtship Behaviour of the Wandering Albatross *Diomedea exulans* at Bird Island, South Georgia. *Mar. Orn.*

Pilgrim, R.L.G. & Palma, R.L. 1982. A List of chewing lice (Insecta: Mallophaga) from Birds in New Zealand. *Notornis* 29: supp.

Pitman, R.L. 1986. *Atlas of seabird distribution and relative abundance in the eastern tropical Pacific*. US National Marine Fisheries Service. Admin. Rep. LJ-86-02C.

Pitman, R.L. 1988. Laysan Albatross breeding in Eastern Pacific – and a comment. *Bull. Pac. Seabird Grp.* 38: 58.

Plenge, M.A., Parker, T.A. III, Hughes, R.A. & O'Neill, J.P. 1989. Additional Notes on the distribution of birds in West-Central Peru. *Gerfaut* 79: 55-68.

Plomer, W. 1959. Voorslag Days. *London Mag.* 6: 46-52.

Pomare, M.W.P.N. & Cowan, J. 1930. *Legends of the Maori*. Fine Arts, Wellington.

Powell, J.R. & Gibbs, J.P. 1995. A report from Galápagos. *Trends Ecol. Evol.* 10: 351-354.

Powlesland, R.G. 1985. Seabirds found dead on New Zealand beaches in 1983 and a review of albatross recoveries since 1960. *Notornis* 32: 23-41.

Powlesland, R.G. & Powlesland, M.H. 1994. Seabirds found dead on New Zealand Beaches in 1993, with a review of *Sterna albostriata*, *S. crispa* and *S. striata* recoveries, 1943-1992. *Notornis* 41: 275-286.

Prévost, J. 1970. Relation d'une visite a l'Ile de l'Est, Archipel Crozet, en 1969. *Revue fr. Orn.* 40 (supp.): 1-15.

Prince, P.A. 1980. The food and feeding ecology of Grey-headed Albatross *Diomedea chrysostoma* and Black-browed Albatross *D. melanophris*. *Ibis* 122: 476-488.

Prince, P.A. 1982. The Black-browed Albatross *Diomedea melanophris* population at Beauchêne Island, Falkland Islands. *Comm. Natl. Fr. Rech. Antarct.* 51: 111-117.

Prince, P.A. 1985. Population and energetic aspects of the relationship between Blackbrowed and Greyheaded Albatrosses and the Southern Ocean marine environment. In *Antarctic Nutrient Cycles and Food Webs* (Siegfried, W.R., Condy, P.R. & Laws, R.M. eds):473-477. Springer-Verlag, Berlin.

Prince, P.A. & Croxall, J.P. 1983. Birds of South Georgia: New records and re-evaluation of status. *Br. Antarct. Surv. Bull.* 59: 15-27.

Prince, P.A. & Croxall, J.P. 1996. The Birds of South Georgia. *Bull. Br. Ornith. Club* 116: 81-104.

Prince, P.A., Croxall, J.P., Trathan, P.N. & Wood, A.G. 1997. The pelagic distribution of South Georgia albatrosses and their relationships with fisheries. In *Albatross Biology and Conservation* (Robertson, G & Gales, R. eds 1998): 137-167. Surrey Beatty, Chipping Norton.

Prince, P.A. & Francis, M.D. 1984. Activity budgets of foraging Gray-headed Albatrosses. *Condor* 86: 297-300.

Prince, P.A. & Payne, M.R. 1979. Current status of birds at South Georgia, Antarctica. *Br. Antarct, Surv. Bull.* 48: 103-118.

Prince, P.A. & Ricketts, C. 1981. Relationships between food supply and growth in albatrosses: an interspecies chick fostering experiment. *Ornis. Scand.* 12: 207-210.

Prince, P.A., Ricketts, C. & Thomas, C. 1981. Weight loss in incubating albatrosses and its implications for their energy and food requirements. *Condor* 83: 238-242.

Prince, P.A. & Rodwell, S.P. 1994. Ageing immature Black-browed and Grey-headed Albatrosses using moult, bill and plumage characteristics. *Emu*: 94: 246-254.

Prince, P.A., Rodwell, S., Jones, M. & Rothery, P. 1993. Moult in Black-browed and Grey-headed Albatrosses *Diomedea melanophris* and *D. chrysostoma*. *Ibis* 135: 121-131.

Prince, P.A., Rothery, P., Croxall, J.P., & Wood, A.G. 1994. Population dynamics of Black-browed and Grey-headed albatrosses *Diomedea melanophris* and *D. chrysostoma* at Bird Island, South Georgia. *Ibis* 136: 50-71.

Prince, P.A. & Walton, D.W.H. 1984. Automated measurement of meal sizes and feeding frequency in albatrosses. *J. Applied Ecol.* 21: 789-794.

Prince, P.A., Weimerskirch, H., Huin, N. & Rodwell, S. 1997. Molt, maturation of plumage and aging in the Wandering Albatross. *Condor* 99: 58-72.

Prince, P.A., Wood, A.G., Barton, T. & Croxall, J.P. 1992. Satellite tracking of wandering albatrosses (*Diomedea exulans*) in the South Atlantic. *Antarct. Sci.* 4: 31-36.

Pye, T. & Bonner, W.N. 1980. Feral brown rats *Rattus norvegicus*, in South Georgia (South Atlantic Ocean). *J. Zool. Lond.* 192: 237-255.

Pyle, R.L. 1991. Albatrosses in the Hawaiian Islands region. *Am. Birds* Summer: 324.

R. A. 1876. The Sailing Flight of Birds. *Nature* 13: 324-325.

Rahn, H., Ackermann, R.A. & Paganelli, C.V. 1984. Eggs, yolk and embryonic growth rate. In *Seabird Energetics* (Whittow, G.C. & Rahn, H. eds):89-112. Plenum, New York.

Rand, R.W. 1952. Bird banding on Marion Island. *Ostrich* 23: 120-122.

Rand, R.W. 1954. Notes on the birds of Marion Island. *Ibis* 96: 173-206.

Rand, R.W. 1957. Grey-headed Albatross on Marion Island. *Bokmakierie* 9: 40-42.

Rand, R.W. 1962. Seabirds south of Madagascar. *Ostrich* 33: 48-51.

Rankin, N. 1951. *Antarctic Isle.* Collins, London.

Ray, J. 1713. *Synopsis methodica Avium.* London.

Rayleigh, Lord. 1883. The Soaring of Birds. *Nature* 27: 534-535.

Rayleigh, Lord. 1889. The sailing flight of the albatross. *Nature* 40: 34.

Rayleigh, Lord. 1900. The Mechanical Principles of Flight. *Manchester Memoirs* 44: 1-26.

Rayner, J.M.V. 1983. Form and function in avian flight. *Current Ornithology* 5: 1-66.

Rechten, C. 1985. The Waved Albatross in 1983 – El Niño leads to complete breeding failure. In *El Niño en las Galápagos: El Evento de l982-83* (Robinson, G. & del Pino, E.M. eds): 227-237. Fundacion Charles Darwin par las Islas Galápagos, Quito.

Rechten, C. 1986. Factors determining laying date of the Waved Albatross *Diomedea irrorata. Ibis* 128: 492-501.

Reichenbach, L. 1850. *Avium Systems Naturale.* Dresden.

Reid, K. 1995. Oiled penguins observed at Bird Island, South Georgia. *Mar. Orn.* 23: 53-57.

Reid, K., Prince, P.A. & Croxall, J.P. 1996. The fish diet of black-browed albatross *Diomedea melanophris* and grey-headed albatross *D. chrysostoma* at South Georgia. *Polar Biol.* 16: 469-477.

Reid, K., Prince, P.A. & Croxall, J.P. 2000. Fly or die: the role of fat stores in the growth and development of Grey-headed Albatross chicks. *Ibis* (in press).

Reid, T. & Carter, M.J. 1988. The nominate race of the Yellow-nosed Albatross *Diomedea chlororhynchos chlororhynchos* in Australia *Aust. Bird Watcher* 12: 160-164.

Reid, T. & James, D. 1997. The Chatham Island Mollymawk (*Diomedea eremita*) in Australia. *Notornis* 44: 125-128.

Reinke, K., Butcher, E.C., Russell, C.J., Nicholls, D.G. & Murray, M.D. 1998. Understanding the flight movements of a non-breeding Wandering Albatross *Diomedea exulans gibsoni* using a geographic information system. *Aust. J. Zool.* 46: 171-181.

Reischek, A. 1888a. The habits and home of the Wandering Albatross (*Diomedea exulans*). *Trans. NZ Inst.* 21: 126-128.

Reischek, A. 1888b. Notes on the islands to the south of New Zealand. *Trans. NZ Inst.* 21: 384-385.

Reynolds, P.W. 1935. Notes on the Birds of Cape Horn. *Ibis* 5: 65-101.

Rice, D.W. & Kenyon, K.W. 1962a. Breeding distribution, history, and populations of North Pacific albatrosses. *Auk* 79: 365-386.

Rice, D.W. & Kenyon, K.W. 1962b. Breeding cycles and behaviour of Laysan and Black-footed Albatrosses. *Auk* 79: 517-567.

Richards, R. 1984. The Maritime Fur Trade: Sealers and Other Residents on St Paul and Amsterdam Islands. *Great Circle* 6: 24-42, 93-109.

Richardson, F. 1954. Notes on the birds of French Frigate Shoals. *Elepaio* 14: 61-63, 73-75.

Richardson, M.E. 1984. Aspects of the ornithology of the Tristan da Cunha group and Gough Island, 1972-1974. *Cormorant* 12: 122-201.

Richardson, S.A. 1994. Status of Short-tailed Albatross on Midway Atoll. *Elepaio* 54: 35-37.

Richardson, S.A. & Sigman, M. 1995. At Midway Atoll. *Bull. Pac. Seabird Grp.* 22: 42.

Richdale, L.E. 1939. A Royal Albatross nesting on the Otago Peninsula, New Zealand. *Emu* 38: 467-488.

Richdale, L.E. 1942. Supplementary Notes on the Royal Albatross. *Emu* 41: 169-184, 253-264.

Richdale, L.E. 1949a. The Pre-egg Stage in Buller's Mollymawk. *Biol. Monogr.* 2, Otago Daily Times and Witness, Dunedin.

Richdale, L.E. 1949b. Buller's Mollymawk: incubation data. *Bird-banding* 20: 127-141.

Richdale, L.E. 1950a. The Pre-egg Stage in the Albatross Family. *Biol. Monogr.* 3, Otago Daily Times and Witness, Dunedin.

Richdale, L.E. 1950b. (review of Sorensen 1950). *Emu* 50: 142-143.

Richdale, L.E. 1952. Post-egg Period in Albatrosss. *Biol. Monogr.* 4, Otago Daily Times and Witness, Dunedin.

Richdale, L.E. & Warham, J. 1973. Survival, pair bond retention and nest-site tenacity in Buller's Mollymawk. *Ibis* 115: 257-263.

Ricketts, C. & Prince, P.A. 1981. Comparison of growth in albatrosses. *Ornis Scand.* 12:120-124.

Ricklefs, R.E. 1980. Geographical Variation in clutch size among passerine birds: Ashmole's Hypothesis. *Auk* 97: 38-49.

Ridgeway, R. 1887. *A Manual of North American Birds.* Lippincott, Philadelphia.

Ridgeway, R. 1897. Birds of the Galapagos Archipelago. *Proc. US Natl. Mus.* 19: 459-670.

Ridley, M. 1986. *Evolution and Classification.* Longman, London.

Ridoux, V. 1987. Feeding association between seabirds and killer whales, *Orcinus orca*, around subantarctic Crozet Islands. *Canad. J. Zool.* 65: 2113-2115.

Ridoux, V. 1989. Impact des oiseaux de mer sur les ressources marines autour des îles Crozet: estimation préliminaire. *Actes Colloque Rech. Fr. Terres Australes, Strasbourg, 1987*: 85-94.

Ridoux, V. 1994. The diets and dietary segregation of seabirds at the subantarctic Crozet Islands. *Mar. Orn.* 22: 1-192.

Ritchie, I.M. 1970. A preliminary report on a recent botanical survey of the Chatham Islands. *Proc. NZ Ecol. Soc.* 17: 52-56.

Robards, M.D., Gould, P.J. & Piatt, J.F. 1997. The Highest Global Concentrations and Increased Abundance of Oceanic Plastic Debris in the North Pacific: Evidence from Seabirds. In *Marine Debris* (Coe, J.M. & Rogers, D.B. eds): 71-80. Springer-Verlag, New York.

Robbins, C.S. & Rice, D.W. 1974. Recoveries of Banded Laysan Albatrosses (*Diomedea immutabilis*) and Black-footed Albatrosses (*D. nigripes*). In *Pelagic Studies of Seabirds in the Central and Eastern Pacific Ocean* (King, W.B. ed.): 232-277. Smithsonian Contr. Zool. 158, Washington.

Roberts, B.B. ed. 1967. *Edward Wilson's Birds of the Antarctic.* Blandford, London.

Robertson, C.J.R. 1974. Albatrosses of the Chatham Islands. *Wildlife – A Review* 5: 20-22.

Robertson, C.J.R. 1975a. Yellow-nosed Mollymawk (*Diomedea chlororhynchos*) recorded in the Chatham Islands. *Notornis* 22: 342-344.

Robertson, C.J.R. 1975b. Report on the Distribution, Status and Breeding Biology of Royal Albatross, Wandering Albatross and White-capped Mollymawk on the Auckland Islands. In *Preliminary Results of the Auckland Islands Expedition 1972-73* (Yaldwyn, J.C. ed.).: 143-151. Dept. Lands & Survey, Wellington.

Robertson, C.J.R. 1980. Birds on Campbell Island. *Res. Ser.* 7: 105-116. Dept. Lands & Survey, Wellington.

Robertson, C.J.R. 1985. In *Complete Book of New Zealand Birds*. Reader's Digest, Sydney.

Robertson, C.J.R. 1991. Questions on the Harvesting of Toroa in the Chatham Islands. *Sci. Res. Ser.* 35: 1-105. Dept. Conservation, Wellington.

Robertson, C. 1993a. Effects of Nature Tourism on Marine Wildlife. *Proc. Mar. Conserv. & Wildl. Protection Conf.*: 53-60. NZ Conservation Authority, Wellington.

Robertson, C.J.R. 1993b. Timing of egg laying in the Royal Albatross (*Diomedea epomophora*) at Taiaroa Head 1937-1992. *Conserv. Advisory Sci. Notes* 50: 1-8. Dept. Conservation, Wellington.

Robertson, C.J.R. 1993c. Survival and longevity of the Northern Royal Albatross *Diomedea epomophora sanfordi* at Taiaroa Head 1937-93. *Emu* 93: 269-276.

Robertson, C.J.R. 1997. Factors influencing the breeding performance of the Northern Royal Albatross. In *Albatross Biology and Conservation* (Robertson, G & Gales, R. eds 1998): 99-104. Surrey Beatty, Chipping Norton.

Robertson, C. & Jenkins, J. 1981. Birds at sea in southern New Zealand waters, February-June 1981. *Australas. Seabird Grp. Newsl.* 16: 17-29.

Robertson, C.J.R. & Kinsky, F.C. 1972. The dispersal movements of the Royal Albatross (*Diomedea epomophora*). *Notornis* 19: 289-301.

Robertson, C.J.R. & Nunn, G.B. 1997. Towards a new taxonomy for albatrosses. In *Albatross Biology and Conservation* (Robertson, G & Gales, R. eds 1998): 13-19. Surrey Beatty, Chipping Norton.

Robertson, C.J.R. & Richdale, L.E. 1993. The breeding phenology of the Royal Albatross (*Diomedea epomophora sanfordi*) 1937-74. *Conserv. Advisory. Sci. Notes* 48: 1-13. Dept. Conservation, Wellington.

Robertson, C.J.R., Robertson, G.G. & Bell, D. 1997. White-capped Albatross (*Thalassarche steadi*) breeding at the Chatham Islands. *Notornis* 44: 156-158.

Robertson, C.J.R. & Sawyer, S. 1994. Albatross research on (Motuhara) Forty-Fours Islands: 6-15 December 1993. *Conserv. Advisory. Sci. Notes* 70: 1-10. Dept. Conservation, Wellington.

Robertson, C.J.R. & van Tets, G.F. 1982. The status of birds of the Bounty Islands. *Notornis* 29: 311-336.

Robertson, C.J.R. & Warham, J. 1992. Nomenclature of the New Zealand Wandering Albatrosses *Diomedea exulans*. *Bull. Br. Ornithol. Club* 112: 74-81.

Robertson, C.J.R. & Warham, J. 1994. Measurements of *Diomedea exulans antipodensis* and *D.e. gibsoni*. *Bull. Br. Ornithol. Club* 114: 132-134.

Robertson, C.J.R. & Wright, A. 1973. Successful Hand-rearing of an Abandoned Royal Albatross Chick. *Notornis* 20: 49-58.

Robertson, G. & Gales, R. eds 1998. *Albatross Biology & Conservation*. Surrey Beatty, Chipping Norton.

Roden, G.I. 1991. Subarctic-Subtropical Transition Zone of the North Pacific: Large-Scale Aspects and Mesoscale Structure. In *Biology, Oceanography, and Fisheries of the North Pacific Transition Zone and Subarctic Frontal Zone* (Wetherall, J.A. ed.): 1-38. NOAA, Honolulu.

Rodhouse, P.G., Clarke, M.R. & Murray, A.W.A. 1987. Cephalopod prey of the Wandering Albatross *Diomedea exulans*. *Mar. Biol.* 96: 1-10.

Rodhouse, P.G. & Prince, P.A. 1993. Cephalopod prey of the black-browed albatross *Diomedea melanophrys* at South Georgia. *Polar Biol.* 13: 373-376.

Rodhouse, P.G., Prince, P.A., Clarke, M.R. & Murray, A.W.A., 1990. Cephalopod prey of the Grey-headed Albatross *Diomedea chrysostoma*. *Mar. Biol.* 104: 353-362.

Rodhouse, P.G., Prince, P.A., Trathan, P.N., Hatfield, E.M.C., Watkins, J.L., Bone, D.G., Murphy, E.J. & White, M.G. 1996. Cephalopods and mesoscale oceanography at the Antarctic Polar Front: satellite tracked predators locate pelagic trophic interactions. *Mar. Ecol. Prog. Ser.* 136: 37-50.

Roll, H.U. 1965. *Physics of the Marine Atmosphere*. Academic, London.

Rothery, P. & Prince, P.A. 1990. Survival and breeding frequency in albatrosses. *Ring* 13: 61-74.

Rothschild, M. 1983. *Dear Lord Rothschild*. Hutchinson, London.

Rothschild, W. 1892-93. (exhibition and description of *Diomedea immutabilis*). *Bull. Br. Ornith. Club* 1: 48.

Rothschild, W. 1893. *The Avifauna of Laysan*. Porter, London.

Rothschild, W. & Hartert, E. 1899. A review of the ornithology of the Galapagos Islands. *Novit. Zool.* 6: 85-142.

Routh, M. 1949. Ornithological observations in the Antarctic seas, 1946-47. *Ibis* 91: 577-606.

Roux, J.-P. 1987. Sooty Albatross *Phoebetria fusca* breeding in the Kerguelen Archipelago: a confirmation. *Cormorant* 14: 50-51.

Roux, J.-P. 1988. Second record of a Laysan Albatross *Diomedea immutabilis* in the Indian Ocean. *Cormorant* 16: 56.

Roux, J.-P., Jouventin, P., Mougin, J.-L., Stahl, J.-C. & Weimerskirch, H. 1983. Un nouvelle albatros *Diomedea amsterdamensis* n. sp. decouvert sur l'Ile Amsterdam (37° 50'S. 77° 35'E). *Oiseau Revue fr. Orn.* 53: 1-11.

Rowan, M.K. 1951. The Yellow-nosed Albatross *Diomedea chlororhynchos* Gmelin at its breeding grounds in the Tristan da Cunha group. *Ostrich* 22: 139-155.

Rowlands, B.W., Trueman, T., Olson, S.L., McCulloch, M.N. & Brooke, R,K, 1998. *The Birds of St. Helena 15° 58'S 5° 42'W*. Checklist 16, British Ornithologists' Union, London.

Rüppell, G. 1980. *Vogelflug*. Rowohlt, Hamburg.

Ryan, P.G. 1987. The incidence and characteristics of plastic particles ingested by seabirds. *Mar. Environ. Res.* 23: 175-306.

Ryan, P.G. & Avery, G. 1987. Wreck of juvenile Black-browed Albatrosses on the west coast of South Africa during stormy weather. *Ostrich* 58: 139-140.

Ryan, P.G., Avery, G., Rose, B., Ross, G.J.B., Sinclair, J.C. & Vernon, C.J. 1989. The Southern Ocean seabird irruption to South African waters during winter 1984. *Cormorant* 17: 41-55.

Ryan, P.G., Dean, W.R.J., Moloney, C.L., Watkins, B.P. & Milton, S.J. 1990. New information on seabirds at Inaccessible Island and other islands of the Tristan da Cunha group. *Mar. Orn.* 18: 43-54.

Sabourenkov, E. 1994. Information on CCAMLR measures to prevent incidental mortality of seabirds. *Mar. Orn.* 22:250-252.

Sachs, G. 1993. Minimaler Windbedarf für den dynamischen Segelflug der Albatrosse. *J. Orn.* 134: 435-445.

Saether, B.E., Laurentsen, S.H., Tveraa, T., Andersen, R. & Pedersen, H.C. 1997. Size dependent variation in reproductive success of a long-lived seabird, the Antarctic Petrel (*Thalassoica antarctica*). *Auk* 114: 333-340.

Sagar, P.M. 1977. Birds of the Western Chain, Snares Islands, New Zealand. *Notornis* 24: 178-183.

Sagar, P. 1999. Southern Buller's Mollymawks. *Australas. Seabird Grp. Newsl.* 34: 18.

Sagar, P.M., Molloy, J., Tennyson, A.J.D. & Butler, D. 1994. Numbers of Buller's Mollymawk breeding at the Snares Islands. *Notornis* 41: 85-92.

Sagar, P.M., Stahl, J.-C. & Molloy, J. 1998. Sex determination and natal philopatry of Southern Buller's Mollymawks (*Diomedea bulleri bulleri*). *Notornis* 45:271-278.

Sagar, P.M. & Warham, J. 1993. A long-lived Southern Buller's Mollymawk (*Diomedea bulleri bulleri*) with a small egg. *Notornis* 40: 303-304.

Sagar, P.M. & Warham, J. 1997. Breeding biology of Southern Buller's Albatrosses at The Snares, New Zealand. In *Albatross Biology and Conservation* (Robertson, G & Gales, R. eds 1998): 92-98. Surrey Beatty, Chipping Norton.

Sagar, P.M. & Weimerskirch, H. 1996. Satellite tracking of Southern Buller's Albatrosses from the Snares, New Zealand. *Condor* 98: 649-652.

Salvin, O. 1876. ...on the avifauna of the Galapagos Archipelago–Dr. Habel's Account. *Trans. Zoo. Soc. Lond.* 9: 456-461.

Salvin, O. 1883. A List of the Birds collected by Captain A.H. Markham on the West Coast of America. *Proc. Zool. Soc. Lond.* 1883: 419-432.

Salvin, O. 1896. Tubinares (petrels and albatrosses). *Catalogue of the Birds of the Br. Mus.* 25: 340-355.

Sanger, G.A. 1970. The seasonal distribution of some seabirds off Washington and Oregon with notes on their ecology and behavior. *Condor* 72: 339-357.

Sanger, G.A. 1972. The recent pelagic status of the Short-tailed Albatross (*Diomedea albatrus*). *Biol. Conserv.* 4: 189-193.

Sanger, G.A. 1974a. Black-footed Albatross (*Diomedea nigripes*). In *Pelagic Studies of Seabirds in the Central and Eastern Pacific Ocean* (King, W.B. ed.): 96-128. Smithsonian Inst. Press, Washington.

Sanger, G.A. 1974b. Laysan Albatross (*Diomedea immutabilis*). In Pelagic Studies of Seabirds in the Central and Eastern Pacific Ocean (King, W.B. ed.): 129-153. Smithsonian Inst. Press, Washington.

Savours, A. 1961. The wreck of *Betsy and Sophia* on Iles Kerguelen. *Geog. J.* 127: 317-321.

Sawada, K. & Handa, N. 1998. Variability of the path of the Kuroshio ocean current over the past 25,000 years. *Nature* 392: 592-595.

Schiavini, A., Frere, E., Gandini, P., Garcia, N. Crespo, E. 1997. Albatross–fisheries interactions in Pategonian shelf waters. In *Albatross Biology and Conservation* (Robertson, G & Gales, R. eds 1998): 208-213. Surrey Beatty, Chipping Norton.

Schlatter, R.P. & Riveros, G.M. 1997. Historia Natural del Archipiélago Diego Ramírez, Chile. *Ser. Cient. INARCH* 47: 87-112.

Schmidt-Nielsen, K. 1960. The salt-secreting gland of marine birds. *Circulation* 21: 955-967.

Schneider, D.C. 1990. Seabirds and Fronts: a brief overview. *Polar Res.* 8: 17-21.

Scholey, K.D. 1982. *Developments in vertebrate flight...* Unpublished Ph.D thesis. University of Bristol, Bristol.

Schott, G. 1931. Der Perú-Strom und seine nördlichen Nachbargebiere in normalei und abnormaler Ausbildung. *Annal. Hydr. Marit. Meteor.* 5: 161-169, 6: 200-213, 7: 240-252.

Schuinsland, H.H. 1899. (transl. Udvardy, M.D.F.). Three months on a Coral Island (Laysan). *Atoll Res. Bull.* 432: 1-53 (1996).

Scofield, P. 1994. Report on ornithogical observations from G.R.V *Tangaroa* during leg one of Tan93/10 on the Southern Plateau. *Australas. Seabird Grp. Newsl.* 27: 9-10.

Scopoli, G.A. 1769. *Annus I. historico-naturalis.* Lipsiae.

Scorer, R.S. 1958. *Natural Aerodynamics.* Pergamon, London.

Scoresby, W. 1859. *Journal of a Voyage to Australia and Round the World for Magnetical Research.* Longman, London.

Scott, J. 1994. *Marine conservation at Macquarie Island.* Parks and Wildlife Service, Hobart.

Seebohm, H. 1890. On the Birds of the Bonin Islands. *Ibis* (6th Ser) 2: 95-108.

Seebohm, H. 1891. On the Birds of the Volcano Islands. *Ibis* (6th Ser.) 3: 189-192.

Segonzac, M. 1972. Données récentes sur la faune des îles Saint-Paul et Nouvelle Amsterdam. *Oiseau Revue fr. Orn.* 42: (spécial) 1-68.

Selkirk, P.M., Seppelt, R.D. & Selkirk, D.R. 1990. *Subantarctic Macquarie Island: environment and biology.* Cambridge University Press, Cambridge.

Serventy, D.L. 1950. Taxonomic trends in Australian Ornithology–with special reference to the work of Gregory Mathews. *Emu* 49: 257-267.

Serventy, D.L., Serventy, V.N. & Warham, J. 1971. *The Handbook of Australian Sea-birds.* Reed, Sydney.

Shackleton, E. 1919. *South.* Heinemann, London.

Sharpe, R.B. 1879. *Zoology, Birds.* In Petrological, botanical and zoological collections made in Kerguelen's Land and Rodrigues during the Transit of Venus expeditions, 1874-75. *Phil. Trans. R. Soc. Lond.* 168: 101-162.

Shaughnessy, P.D. & Fairall, N. 1976. Notes on seabirds at Gough Island. *S. Afr. J. Antarct. Res.* 6: 23-25.

Shearman, R.J. 1985. The Mathematical Representation of Wind Profiles Over the Sea. *J. Underwater Tech.* 11: 13-19.

Shelmidine, L.S. 1948. The Early History of Midway Islands. *Am. Neptune* 8: 179-195.

Shelvocke, G. 1726. *A Voyage round the World By the Way of the Great South Sea performed in the years 1719-1722...* Senex, London.

Shlesinger, M.F., Zaslavsky, G.M. & Klafter, J. 1993. Strange kinetics. *Nature* 363: 31-37.

Shoemaker, V.H. 1972. Osmoregulation and excretion in birds. In *Avian Biology* (Farner, D.S., King, J.R. & Parkes, K.C. eds) 2: 527-574.

Shuntov, V.P. 1968. (Some regularities in the distribution of albatrosses (Tubinares, Diomedeidae) in the North Pacific.) *Zool. Zh.* 47: 1054-1064. (In Russian).

Shuntov, V.P. 1972. (*Seabirds and the Biological Structure of the Ocean.* Far East Books, Vladivostok). (In Russian).

Shuntov, V.P. 1974. *Seabirds and the Biological Structure of the Ocean.* (Irène Allardt transl. of Shuntov 1972). US Dept. Sport Fisheries and National Science Foundation, Washington.

Shuntov, V.P. 1993. Biological and physical determinants of marine bird distribution in the Bering Sea. In *The status, ecology, and conservation of marine birds of the North Pacific* (Vermeer, K., Briggs, K.T., Morgan, K.H. & Siegel-Causey, D. eds): 10-17. Environment, Canada.

Sibley, C.G. & Ahlquist, J.E. 1990. *Phylogeny and Classifications of Birds. A Study in Molecular Evolution.* Yale University Press, New Haven.

Sievert, P.R. & Sileo, L. 1993. The effect of ingested plastic on growth and survival of albatross chicks. In *The status, ecology, and conservation of marine birds of the North Pacific* (Vermeer, K., Briggs, K.T., Morgan, K.H. & Siegel-Causey, D. eds): 212-217. Environment, Canada.

Sileo, L. & Fefer, S.I. 1987. Paint chip poisoning of Laysan Albatross at Midway Atoll. *J. Wildl. Diseases* 23: 432-437.

Sileo, L. & Sievert, P.R. 1988. Causes of mortality of albatross chicks from Hawaii. *Bull. Pac. Seabird Grp.* 15: 38.

Sill, E.R. 1944. *Around the Horn.* Yale University Press, New Haven.

Silva, M.C. & Edwards, S.V. 1999. Conservation genetics of Black-footed Albatrosses: the origin of bycatch birds. *Pac. Seabirds* 26: 53.

Siple, P.A. & Lindsey, A.A. 1937. Ornithology of the Second Byrd Antarctic Expedition. *Auk* 54: 147-159.

Sitwell, N. 1997-98. Galapagos scientists in Southampton. *Galapagos Newsl.* 5: 9-11.

Sladen, W.J.L., Wood, R.C. & Monaghan, E.P. 1968. The USARP Bird banding Program, 1958-1965. In *Antarctic Bird Studies* (Austin, O.L. Jr. ed.): 213-262. American Geophysical Union, Washington.

Slipp, J.W. 1952. A Record of the Tasmanian White-capped Albatross, *Diomedea cauta cauta*, in North Pacific Waters. *Auk* 69: 458-459.

Smit, F.G.A.M. 1970. Siphonaptera of South Georgia and Heard Island. *Pac. Insects Monog.* 23: 291-292.

Smith, B. 1956. Coleridge's *Ancient Mariner* and Cook's Second Voyage. *J. Warburg Courtauld Inst.* 19: 117-154.

Smith, L.E. 1997a. An Unusual Wandering Albatross On Macquarie Island. *S. Oceans Seabird Study Ass. Nltr.* 13: 7.

Smith, L.E. 1997b. Gough Island Albatross A First for Australia. *S. Oceans Seabird Study Ass. Nltr.* 16: 5

Smith, L.E. 1998. Arthur (Arfie) Mothersdill 1915-1998. *S. Oceans Seabird Study Ass. Nltr.* 18: 5.

Smith, L.E., Battam, H. & Milburn, P.J. 1999. Amsterdam Albatross vs Antipodean Albatross. *S. Oceans Seabird Study Ass. Ntlr.* 22: 4-6.

Smith R.I.L. 1981. Terrestrial plant biology of the sub-Antarctic and Antarctic. In *Antarctic Ecology* (Laws, R.M. ed.): 61-162. Academic, London.

Smith, R.I.L & Prince, P.A. 1985. The natural history of Beauchêne Island. *Biol. J. Linn. Soc.* 24: 233-283.

Smith, T. 1998. Avoiding seabird bycatch in Alaska longline fisheries. *Pac. Seabirds* 25: 44.

Snow, D.W. 1964. The Giant Tortoises of the Galapagos Islands. *Oryx* 7: 277-290.

Snow, D.W. 1997. Should the biological be superseded by the phylogenetic species concept? *Bull. Br. Ornithol. Club* 117:110-121.

Snyder, D.E. 1958. Correcting an old albatross error. *Auk* 75: 478-479.

Sorensen, J.H. 1950a. The Royal Albatross. *Cape Expedition Series Bull.* 2: 1-39. DSIR, Wellington.

Sorensen, J.H. 1950b. The Light-mantled Sooty Albatross. *Cape Expedition Series Bull.* 8: 1-30. DSIR, Wellington.

Sparling, D.W. Jr. 1977. Sounds of Laysan and Black-footed Albatrosses. *Auk* 94: 256-269.

Spear, L.B. & Ainley, D.G. 1997a. Flight behaviour of seabirds in relation to wind direction and wing morphology. *Ibis* 139: 221-233.

Spear, L.B. & Ainley, D.G. 1997b. Flight speed of seabirds in relation to wind speed and direction. *Ibis* 139: 234-251.

Spear, L.B., Ainley, D.G., Nur, N. & Howell, S.N.G. 1995. Population size and factors affecting at-sea distributions of four endangered procellariids in the tropical Pacific. *Condor* 97: 613-638.

Squires, W.A. 1952. The Birds of New Brunswick. *Monogr. Ser. 4*, New Brunswick Mus.

Stagi, A., Vaz-Ferreira, R., Marin, Y. & Joseph, L. 1997. The conservation of albatrosses in Uruguayan waters. In *Albatross Biology and Conservation* (Robertson, G & Gales, R. eds 1998): 220-224. Surrey Beatty, Chipping Norton.

Stahl, J.-C. 1987. Distribution des oiseaux marins dans le sud-ouest de l'Océan Indien: données préliminieres de la campagne APSARA II - ANTIPROD III. In *Les rapports des campagnes à la mer MD 38/ASPARA II - ANTIPROD III* (Fontugne, M. & Fiala, M. eds.): 175-190. TAAF, Paris.

Stahl, J.-C., Bartle, J.A., Cheshire, N.G., Petyt, C. & Sagar, P.M. 1998. Distribution and movements of Buller's albatross (*Diomedea bulleri*) in Australasian seas. *NZ J. Zool.* 25: 109-137.

Stahl, J.-C., Jouventin, P., Mougin, J.-L., Roux, J.-P. & Weimerskirch, H. 1985. The foraging zones of seabirds in the Crozet Islands Sector of the Southern Ocean. In *Antarctic Nutrient Cycles and Food Webs* (Siegfried, W.R., Condy, P.R. & Laws, R.M. eds):478-486. Springer-Verlag, Berlin.

Stallcup, R. 1990. *Ocean Birds of the Nearshore Pacific.* Point Reyes Bird Observatory, California.

Stejneger, L. 1885. Results of ornithological explorations in the Commander Islands and in Kamtschatka. *US Natl. Mus. Bull.* 29: 1-382.

Stiles, F.W. 1974. Black-browed Albatross on freshwater. *Auk* 91: 844-845.

Stillson, B. 1955. *Wings.* Gollancz, London.

Stoltenhoff, F. 1873. Two years on Inaccessible. *Cape Monthly Mag.* 7: 321-337.

Stone, W. 1900. Report on the birds and mammals collected by the McIlhenny Expedition to Pt. Barrow, Alaska. *Proc. Acad. Nat. Sci. Philadelphia* 1900: 4-49.

Stone, W. 1930. Townsend's Oregon Tubinares. *Auk* 47: 414-415.

Strange, I.J. 1965. Beauchêne Island. *Polar Record* 12: 725-730.

Strange, I.J. 1969. The wise men of West Point. *Animals* 11: 458-462.

Strange, I.J. 1976. *The Bird Man.* Gordon & Cremones, London.

Strange, I.J., Parry, C.J., Parry, M.C. & Woods, R.W. 1988. *Tussac grass in the Falklands.* Falkland Islands Govt., Stanley.

Stresemann, E. 1975. *Ornithology.* Harvard University Press, Cambridge.

Stresemann, E. & Stresemann, S.V. 1966. Die mauser der vögel. *J. Orn.* 107: 291-303.

Strikwerda, T.E., Fuller, M.R., Seegar, W.S., Howey, P.W. & Black, H.D. 1986. Bird-borne satellite transmitter and their location program. *Johns Hopkins APL Tech. Digest* 7: 203-208.

Stubbs, F.J. 1913. Asiatic Birds in Leadenhall Market. *Zoologist* (4th Ser.) 17: 156-157.

Sugden, D.E. & Clapperton, C.M. 1977. The Maximum Ice Extent on Islands groups in the Scotia Sea, Antarctica. *Quaternary Res.* 7: 268-282.

Summerhayes, C.P., Hofmeyr, P.K. & Rioux, R.H. 1974. Seabirds off the southwestern coast of Africa. *Ostrich* 45: 83-109.

Suplee, C. 1999. El Niño La Niña. *Natl. Geogr. Mag.* 193: 72-95.

Sutherland, W.J. & Brooks, D.J. 1979. Nest of Black-browed Albatross in Shetland. *Br. Birds* 72: 286-288.

Sutton, O.G. 1953. *Micrometeorology.* McGraw Hill, London.

Swales, M.K. 1965. The Sea-birds of Gough Island. *Ibis* 107: 17-42, 215-229.

Swanson, N. 1973. Status, latitudinal and seasonal occurrences of albatross species in Kangaroo Island waters (South Aust.). *S Aust. Ornithologist* 26: 75-77.

Swanson, R. 1997. Kangaroo Island South Australia. Brief Report. *S. Oceans Seabird Study Ass. Nltr.* 14: 3.

Swinhoe, R. 1863. The Ornithology of Formosa, or Taiwan. *Ibis* 5: 377-435.

Swinhoe, R. 1873. On a Black Albatross of the China Sea. *Proc. Zool. Soc. Lond.* 1873: 781-786.

Szijj, L.J. 1967. Notes on the winter distribution of birds in the western Antarctic and adjacent Pacific waters. *Auk* 84: 366-378.

Taylor, G.A. 1996. Seabirds found dead on New Zealand beaches in 1994. *Notornis* 43: 187-196.

Taylor, G.A. 1997. Seabirds found dead on New Zealand beaches in 1995. *Notornis* 44: 201-212.

Taylor, R.H. 1975. The Distribution and Status of Indroduced Mammals on the Auckland Islands. In *Preliminary Results of the*

Auckland Islands Expedition 1972-73 (Yaldwyn, J.C. ed.): 233-243. Dept. Lands & Survey, Wellington.

Taylor, R.H., Bell, B.D. & Wilson, P.R. 1970. Royal Albatrosses, feral sheep and cattle on Campbell Island. *NZ J. Sci.* 13: 78-88.

Temminck, C.J. & Laugier de Chartrouse, M. 1838. *Nouveau Recueil de Planches Coloriées d'Oiseau...* Paris.

Temple, R.C. 1919. *The Travels of Peter Mundy, in Europe and Asia, 1608-1667.* Hakluyt, London.

Tennyson, A., Imber, M. & Taylor, R. 1998. Numbers of Black-browed Mollymawks (*Diomedea m. melanophrys*) and White-capped Mollymawks (*D. cauta steadi*) at the Antipodes Islands in 1994-95 and their population trends in the New Zealand region. *Notornis* 45: 157-166.

Thomas, G. 1982.The food and feeding ecology of the Light-mantled Sooty Albatross at South Georgia. *Emu* 82: 92-100.

Thomas, G., Croxall, J.P. & Prince, P.A. 1983. Breeding biology of the Light-mantled Sooty Albatross (*Phoebetria palpebrata*) at South Georgia. *J. Zool. Lond.* 199: 123-135.

Thomas, T. 1983. Données récentes sur l'avifaune des îles Kerguelen (Terres Australes et Antarctiques Françaises). *Oiseau Revue fr. Orn.* 53: 133-141.

Thomson, C.W. 1885. Report on the Scientific Results of the *The Voyage of HMS Challenger during the years 1873-76...* Part 1 Narrative. Macmillan, London.

Thompson, D.R. & Furness, R.W. 1989. The chemical form of mercury stored in South Atlantic seabirds. *Environmental Pollution* 60: 305-317.

Thompson, D.R., Furness, R.W. & Lewis, S.A. 1993. Temporal and spatial variation in mercury concentrations in some albatrosses and petrels from the sub-Antarctic. *Polar Biol.* 13: 239-244.

Thompson, K.R. 1992. Quantitative analysis of the use of discards from squid trawlers by Black-browed Albatrosses *Diomedea melanophris* in the vicinity of the Falkland Islands. *Ibis* 134: 11-21.

Thompson, K.R. & Riddy, M.D. 1995. Utilization of offal and discards from "finfish" trawlers around the Falkland Islands by the Black-browed Albatross *Diomedea melanophris. Ibis* 137: 198-206.

Thompson, K.R. & Rothery, P. 1991. A census of the Black-browed Albatross *Diomedea melanophrys* population on Steeple Jason Island, Falkland Islands. *Biol. Conserv.* 56: 39-48.

Thrower, N.J.W. ed. 1981. *The three Voyages of Edmond Halley in the Paramore 1698-1701.* Hakluyt, London.

Thurston, M.H. 1982. Ornithological observations in the South Atlantic Ocean and Weddell Sea, 1959-64. *Br. Antarct. Surv. Bull.* 55: 77-104.

Tickell, W.L.N. 1964. Feeding preferences of the albatrosses *Diomedea melanophris* and *D. chrysostoma* at South Georgia. In *Biologie Antarctique* (Carrick, R., Holdgate, M.W. & Prevost, J. eds): 383-387. Hermann, Paris.

Tickell, W.L.N. 1967. Movements of Black-browed and Grey-headed Albatrosses in the South Atlantic. *Emu* 66: 357-367.

Tickell, W.L.N. 1968. The biology of the great albatrosses, *Diomedea exulans* and *Diomedea epomophora.* In *Antarctic Bird Studies* (Austin, O.L. Jr. ed.): 1-55. American Geophysical Union, Washington.

Tickell, W.L.N. 1969. Plumage changes in young albatrosses. *Ibis* 111: 102-105.

Tickell, W.L.N. 1970. Biennial Breeding in Albatrosses. In *Antarctic Ecology* (Holdgate, M.W. ed.): 551- 557. Academic, London.

Tickell, W.L.N. 1975. Observations on the Status of Steller's Albatross (*Diomedea albatrus*) 1973. *Bull. Int. Council. Bird. Protection* 12: 125-131.

Tickell, W.L.N. 1976. The distribution of Black-browed and Grey-headed Albatrosses. *Emu* 76: 64-68.

Tickell, W.L.N. 1980. The pink ear stain of Wandering Albatrosses *Diomedea exulans. Australas. Seabird Grp. Newsl.* 13: 12-17.

Tickell, W.L.N. 1984. Behaviour of Blackbrowed and Greyheaded Albatrosses at Bird Island, South Georgia. *Ostrich* 55: 64-85.

Tickell, L. 1991. Marathon Birds. *The Natural World.* BBC, Bristol (TV film).

Tickell, W.L.N. 1993. An Atlas of Southern Hemisphere Albatrosses. *Sea Swallow* 42: 28-40.

Tickell, W.L.N. 1996. Galápagos Albatrosses at sea. *Sea Swallow* 45: 83-85.

Tickell, W.L.N. & Cordall, P.A. 1960. South Georgia Biological Expedition 1958/9. *Polar Record* 10: 145-146.

Tickell, W.L.N. & Gibson, J.D. 1968. Movements of Wandering Albatrosses *Diomedea exulans. Emu* 68: 6-20.

Tickell, L. & Morton, P. 1974. The albatross of Torishima. *Geog. Mag.* 49: 359-363.

Tickell, W.L.N. & Pinder, R. 1966. Two-egg clutches in albatrosses. *Ibis* 108: 126-129.

Tickell, W.L.N. & Pinder, R. 1967. Breeding Frequency in the Albatrosses *Diomedea melanophris* and *D. chrysostoma. Nature* 213: 315-316.

Tickell, W.L.N. & Pinder, R. 1972. Chick recognition in albatrosses. *Ibis* 114: 543-548.

Tickell, W.L.N. & Pinder, R. 1975. Breeding biology of the Black-browed Albatross *Diomedea melanophris* and Grey-headed Albatross *D. chrysostoma* at Bird Island, South Georgia. *Ibis* 117: 433-451.

Tickell, W.L.N. & Woods, R.W. 1972. Ornithological observations at sea in the South Atlantic Ocean, 1954-64. *Br. Antarct. Surv. Bull.* 31: 63-84.

Tilman, H.W. 1961. *Mischief among the Penguins.* Hart Davis, London.

Tiplady, R.B. 1972. *Sea Swallow* 21: 31.

Tollu, B. 1967. Reconnaissances systematiques des cotes sitiees a l'est du meridien 69° 05' E. *Terres Australes Antarct. Fr.* 40 (Supp.): 1-45.

Tomkins, R.J. 1983. Fertilization of Wandering Albatross eggs on Macquarie Island. *Notornis* 30: 244-246.

Tomkins, R.J. 1984. Some aspects of the morphology of Wandering Albatrosses on Macquarie Island. *Emu* 85: 29-32.

Tomkins, R.J. 1985. Reproduction and mortality of Wandering Albatrosses on Macquarie Island. *Emu*: 40-42.

Torishima Society, 1967. *Torishima.* Toko Shoin, Tokyo. (In Japanese).

Tuck, G.N., Polacheck, T., Croxall, J.P., Weimerskirch, H., Prince, P.A. & Wotherspoon, S. 1999. The potential of Archival Tags to Provide Long-term Movement and Behaviour Data for Seabirds: First Results from Wandering Albatross (*Diomedea exulans*) of South Georgia and the Crozet Islands. *Emu* 99: 60-68.

Tuck, G.S. 1956. Sea Reports. *Sea Swallow* 9: 5-8.

Tuck, G.S. & Heinzel, H. 1978. *A field guide to the Seabirds of Britain and the World.* Collins, London.

Turbott, E.G. & committee. 1990. *Checklist of Birds of New Zealand and the Ross Depencency, Antarctica.* Orn. Soc. NZ, 3rd edition.

Turner, L.M. 1886. *Contributions to the natural history of Alaska.* US Govt., Washington.

Tyler, W.B., Briggs, K.T., Lewis, D.B. & Ford, R.G. 1993. Seabird distribution and abundance in relation to oceanographic processes in the California Current System. In *The status, ecology, and conservation of marine birds of the North Pacific* (Vermeer, K., Briggs, K.T., Morgan, K.H. & Siegel-Causey, D. eds): 48-60. Environment, Canada.

Tynan, C. 1998. Ecological Importance of the Southern Boundary of the Antarctic Circumpolar Current. *Nature* 392: 708-710.

Vallentin, R. 1904. Notes on the Falkland Islands. *Manchester Memoirs* 48: 1-51.

van Gardingen, P.R., Foody, G.M. & Curran, P.J. (eds,) 1997. Scaling-up. Cambridge University, Cambridge.

Van Ryzin, M.T. & Fisher, H.I. 1976. The age of Laysan Albatrosses, *Diomedea immutabilis*, at first breeding. *Condor* 78: 1-9.

van Zinderen Bakker, E.M. Jr. 1971. The Genus Diomedea. In *Marion and Prince Edward Islands* (van Zinderen Bakker, E.M. Sr., Winterbottom, J.M. & Dyer, R.A. eds): 273-281. Balkema, Cape Town.

Veit, R.R. & Prince, P.A. 1997. Individual and population dispersal of Black-browed Albatrosses *Diomedea melanophris* and Grey-headed Albatrosses *D. chrysostoma* in response to Antarctic krill. *Ardea* 85: 129-134.

Verheyden, C & Jouventin, P. 1994. Olfactory behavior of foraging Procellariiformes. *Auk* 111: 285-291.

Verrill, G.E. 1895. On some birds and eggs collected by Mr. Geo. Comer at Gough Island, Kerguelen Island and the island of South Georgia... *Trans. Conn. Acad. Arts Sci.* 9: 430-478.

Viot, C.R., Jouventin, P. & Bried, J. 1993. Population Genetics of Southern Seabirds. *Mar. Orn.* 21: 1-25.

Viswanathan, G.M., Afanasyev, V., Buldyrev, S.V., Murphy, E.J., Prince, P.A., & Stanley, H.E. 1996. Lévy flight search patterns of wandering albatrosses. *Nature* 381: 413-415.

Voisin, J.F. 1969. L'Albatross hurleur *Diomedea exulans* a lîle de la Possession. *Oiseau Revue fr. Orn.* 39: 82-106.

Voisin, J.F. 1979. Observations ornithologiques aux îles Tristan da Cunha et Gough. *Alauda* 47: 73-82.

Voisin, J.F. 1983. Observations of birds at sea between Réunion, Kerguelen and the Crozet Islands, January-March 1982. *Cormorant* 11:49-58.

Voisin. J.F. 1984. Observations on the birds and mammals of île aux Cochons, Crozet Islands, in February 1982. *S. Afr. Antarct. Res.* 14: 11-17.

Voisin, J.F. 1992. Observations sur les oiseaux de l'île de la Possession, archipel Crozet en décembre 1959 et janvier 1960. *Oiseau Revue fr. Orn. 62: 72-77.*

von Boetticher, H. 1949. Gatlungen und Untergatlungen der Albatrosse. In Beiträge zur Gatlungssystematik der Vögel (Wolters, H.E. & Boetticher, H. von eds). Goerke & Evers, Krefeld.

von Boetticher, H. 1955. *Albetrosse und andere sturmvögel.* A ziemsen verlag, Wittenberg.

von Kittlitz, F.H. 1834. Nachricht von den Brüteplätzen einiger tropischen Seevögel im Stillen Ocean. *Senckenb. Nat. Gesell., Mus. Senckenb.* 1: 117-126.

von Tschudi, J.J. 1856. Beitrage von geographischen Verbreitung der Meersvogel. *J. Orn.* 4: 134-162, 177-191.

Voous, K.H. 1970. Moulting Great Shearwater (*Puffinus gravis*) in Bay of Biscay. *Ardea* 58: 265-266.

Vuilleumier, F., LeCroy, M. & Mayr, E. 1992. (Comment that *D. amsterdamensis* should be *D. exulans amsterdamensis*) *Bull. Br. Ornithol. Club* 112A: 269.

Wace, N.M. & Holdgate, M.W. 1976. Man and Nature in the Tristan da Cunha Islands. *IUCN Monogr.* 6: 1-107. Morges, Switzerland.

Wahl, T.R., Morgan, K.H. & Vermeer, K. 1993. Seabird distribution off British Columbia and Washington. In *The status, ecology, and conservation of marine birds of the North Pacific* (Vermeer, K., Briggs, K.T., Morgan, K.H. & Siegel-Causey, D. eds): 39-47. Environment, Canada.

Waite, E.R. 1909. Vertebrates of the subantarctic islands of New Zealand. In *The Subantarctic Islands of New Zealand* (Chilton, C. ed.): 542-600. Government, Wellington.

Walkden, S.L. 1925. Experimental study of the "soaring" of albatrosses. *Nature* 116: 132-134.

Walker, K., Dilks, P., Elliott, G. & Stahl, J.C. 1991. Wandering Albatross on Adams Island February 1991. *Sci. Res. Int. Rep.* 109: 1-14. Dept. Conservation, Wellington.

Walker, K. & Elliott, G. 1999. Population changes and biology of the Wandering Albatross *Diomedea exulans gibsoni* at the Auckland Islands. *Emu* 99 (in press).

Walker, K., Elliott, G., Davis, A. & McClelland, P. 1995a. Wandering Albatross on Adams Island: Census, nesting data and body measurements. *Sci. Res. Ser.* 78: 1-29. Dept. Conservation, Wellington.

Walker, K., Elliott, G., Nicholls, D., Murray, D. & Dilks, P. 1995b. Satellite tracking of Wandering Albatross (*Diomedea exulans*) from the Auckland Islands: Preliminary Results. *Notornis* 42: 127-137.

Wall, A. 1943. *About our Birds.* Whitcombe & Tombs, Wellington.

Waluda, C.M., Trathan, P.N. & Rodhouse, P.G. 1999. Influence of oceanographic variability on recruitment in the *Ilex argentinus* (Cephalopoda: Ommastrephidae) fishery of the South Atlantic. *Mar. Ecol. Prog. Ser.* 183: 159-167.

Wanless, S. & Harris, M.P. 1988. Seabird records from the Bellingshausen, Amundsen and Ross Seas. *Br. Antarct. Surv. Bull.* 81: 87-92.

Warham, J. 1962. The biology of the Giant Petrel *Macronectes giganteus. Auk* 79: 139-160.

Warham, J. 1967. Snares Island birds. *Notornis* 14: 122-139.

Warham, J. 1976. Aerial displays by large petrels. *Notornis* 23: 255-257.

Warham, J. 1977. The incidence, functions and ecological significance of petrel stomach oils. *Proc. NZ Ecol. Soc.* 24: 84-93.

Warham, J. 1982. A distant recovery of a Buller's Mollymawk. *Notornis* 29: 213-214.

Warham, J. 1984. Obituary–Lancelot Eric Richdale OBE. *Ibis* 126:591-594.

Warham, J. 1985. The composition of petrel eggs. *Condor* 85: 194-199.

Warham, J. 1990. *The Petrels.* Academic, London.

Warham. J. 1996. *The Behaviour, Population Biology and Physiology of the Petrels.* Academic, London.

Warham, J. & Bell, B.D. 1979. The birds of Antipodes Island, New Zealand. *Notornis* 26: 121-169.

Warham, J. & Bennington, S.L. 1983. A census of Buller's Albatross *Diomedea bulleri* at the Snares Islands, New Zealand. *Emu*: 83: 112-114.

Warham, J. & Fitzsimons, C.H. 1987. The vocalisations of Buller's Mollymawk *Diomedea bulleri* (Aves: Diomedeidae), with some comparative data on other albatrosses. *NZ J. Zool.* 14: 65-79.

Warham, J. & Johns, P.M. 1975. The University of Canterbury Antipodes Islands Expedition 1969. *J. R. Soc. NZ* 5: 103-131.

Warham, J., Watts, R. & Dainty, R.J. 1976. The composition, energy content and function of the stomach oils of petrels (Order, Procellariiformes). *J. exp. mar. Biol. Ecol.* 23: 1-13.

Warheit, K.I. 1992. A review of the fossil seabirds of the Tertiary of the North Pacific: plate tectonics, paleoceanography and faunal change. *Paleobiology* 18: 401-424.

Watanabe, E. 1963. (breeding biology of Steller's Albatross at Torishima). In *Meteorological data and report of Marcus and Torishima Islands*: 156-168. Japan Meteorological Agency, Tokyo (In Japanese).

Waters, F. 1927. *Eight Bells.* Appleton, New York.

Waterston, G. 1968. Black-browed Albatross on Bass Rock. *Br. Birds* 61: 22-27.

Watkins, B.P. 1987. Population sizes of king, rockhopper and macaroni penguins and wandering albatrosses at the Prince Edward Islands and Gough Island 1951-86. *S. Afr. J. Antarct. Res.* 17: 155-162.

Watson, G.E. 1975. *Birds of the Antarctic and Sub-Antarctic.* American Geophysical Union, Washington.

Watson, G.E., Angle, J.P., Harper, P.C., Bridge, M.A., Schlatter, R.P., Tickell, W.L.N., Boyd, J.C. & Boyd, M.M. 1971. Birds of the Antarctic and Subantarctic. *Antarctic Map Folio Series 14.* American Geophysical Union, Washington.

Watson, G.E. & Divoky, G.J. 1971. Identification of *Diomedea leptorhyncha* Coues 1866, an albatross with remarkably small salt glands. *Condor* 73: 487-489.

Waugh, S.M. 1998. Dye-marking of New Zealand black-browed and grey-headed Albatrosses from Campbell Island. *NZ J. Mar. Freshwater Res.* 32: 545-549.

Waugh, S.M., Prince, P.A. & Weimerskirch, H. 1999d. Geographical variation in morphometry of black-browed and grey-headed albatrosses from four sites. *Polar Biol.* 22: 189-194.

Waugh, S.M., Sagar, P.M. & Cossee, R.O. 1999b. New Zealand Black-browed Albatross *Diomedea melanophrys impavida* and Grey-headed Albatross *D. chrysostoma* Banded at Campbell Island: Recoveries from the South Pacific Region. *Emu* 99:29-35.

Waugh, S.M., Sagar, P.M. & Paull, D. 1997. Laying dates, breeding success, and annual breeding of Southern Royal Albatross *Diomedea epomophora epomophora* at Campbell Island during 1964 to 1969. *Emu* 97: 194-199.

Waugh, S., Sagar, P. & Stahl, J.-C. 1996. Campbell, Snares and Solander Islands albatross research – reports of dyed birds. *Australas. Seabird Grp. Newsl.* 31: 28-29.

Waugh, S.M., Weimerskirch, H., Cherel, Y., Shankar, U., Prince, P.A. & Sagar, P.M. 1999c. Exploitation of the marine environment by two sympatric albatrosses in the Pacific Southern Ocean. *Mar. Ecol. Prog. Ser.* 177: 243-254.

Waugh, S.M., Weimerskirch, H., Moore, P.J. & Sagar, P.M. 1999a. Population dynamics of Black-browed and Grey-headed Albatrosses *Diomedea melanophrys* and *D. chrysostoma* at Campbell Island, New Zealand, 1942-96. *Ibis* 141: 216-225.

Webb, D.J., Killworth, P.D., Coward, A.C. & Thompson, S.R. 1991. *The FRAM Atlas of the Southern Ocean.* NERC, Swindon.

Webb, J.S. 1869. On the Mechanical Principles involved in the Sailing flight of the Albatross. *Trans. N.Z. Inst. Sci.* 2: 233-236.

Weddell, J. 1825. *A Voyage towards the South Pole.* Longman, London.

Weimerskirch, H. 1981. La stratégie de reproduction del'albatros fuligineux a dos sombre. *Comm. Natl. Fr. Rech. Antarct.* 51: 437-447.

Weimerskirch, H. 1991. Sex-specific differences in moult strategy in relation to breeding in the wandering albatross. *Condor* 93: 731-737.

Weimerskirch, H. 1992. Reproductive effort in long-lived birds: age-specific patterns of condition, reproduction and survival in the wandering albatross. *Oikos* 64: 464-473.

Weimerskirch, H. 1995. Regulation of foraging trips and incubation routine in male and female wandering albatrosses. *Oecologia* 102: 37-43.

Weimerskirch, H. 1997. Foraging strategies of Indian Ocean albatrosses and their relationships with fisheries. In *Albatross Biology and Conservation* (Robertson, G & Gales, R. eds 1998): 168-179. Surrey Beatty, Chipping Norton.

Weimerskirch, H. 1998. Breeding and foraging decision in petrels and albatrosses. In *Proc. 22 Congr. Ornith. Int.* (Adams, N.J. & Slotow, R.H. eds.) Abstracts. *Ostrich* 69: 64.

Weimerskirch, H., Bartle, J.A., Jouventin, P. & Stahl, J.C. 1988. Foraging ranges and partioning of feeding zones in the three species of southern albatrosses. *Condor* 90: 214-219.

Weimerskirch, H., Brothers, N. & Jouventin, P. 1997a. Population dynamics of wandering albatross *Diomedea exulans* and Amsterdam albatross *D. amsterdamensis* in the Indian Ocean and their relationships with long-line fisheries: conservation implications. *Biol. Conserv.* 79: 257-270.

Weimerskirch, H., Cherel, Y., Cuenot-Chaillet, F. & Ridoux, V. 1997b. Alternative foraging strategies and resource allocation by male and female wandering albatrosses. *Ecology* 78: 2051-2063.

Weimerskirch, H., Doncaster, C.P. & Cuenot-Chaillet, F. 1994. Pelagic seabirds and the marine environment: foraging patterns of wandering albatrosses in relation to prey availability and distribution. *Proc. R. Soc. Lond.* B 255: 91-97.

Weimerskirch, H. & Jouventin, P. 1987. Population dynamics of the Wandering Albatross, *Diomedea exulans*, of the Crozet Islands: causes and consequences of the population decline. *Oikos* 49: 315-322.

Weimerskirch, H. & Jouventin, P. 1997. Changes in population sizes and demographic parameters of six albatross species breeding in the French sub-Antarctic islands. In *Albatross Biology and Conservation* (Robertson, G & Gales, R. eds 1998): 84-91. Surrey Beatty, Chipping Norton.

Weimerskirch, H., Jouventin, P., Mougin, J.L., Stahl, J.C. & van Beveren, M. 1985. Ringing recoveries and the dispersion of seabirds breeding in French Austral and Antarctic Territories. *Emu*: 85: 22-33.

Weimerskirch, H., Jouventin, P. & Stahl, 1986. Comparative ecology of the six albatross breeding on the Crozet Islands *Ibis* 128: 195-213.

Weimerskirch, H., Lequette, B. & Jouventin, P. 1989b. Development and maturation of plumage in the wandering albatross *Diomedea exulans. J. Zool. Lond.* 219: 411-421.

Weimerskirch, H., Mougey, T. & Hindermeyer, X. 1997c. Foraging and provisioning strategies of black-browed albatrosses in relation to the requirements of the chick: natural variation and experimental study. *Behav. Ecol.* 8: 635-643.

Weimerskirch, H. & Robertson, G. 1994. Satellite tracking of Light-mantled Sooty Albatross. *Polar Biol.* 14: 123-126.

Weimerskirch, H., Salamolard, M. & Jouventin, P. 1992. Satellite telemetry of foraging movements in the wandering albatross. In *Wildlife Telemetry* (Priede, I.G & Swift, S.M. eds): 185-198. Ellis Horwood, Chichester.

Weimerskirch, H., Salamolard, M., Sarrazin, F. & Jouventin, P. 1993. Foraging strategy of Wandering Albatrosses through the breeding season: a study using satellite telemetry. *Auk* 110: 325-342.

Weimerskirch, H. & Wilson, R.P. 1992. When do wandering albatrosses *Diomedea exulans* forage? *Mar. Ecol. Prog. Ser.* 86: 297-300.

Weimerskirch, H., Wilson, R. & Lys, P. 1997d. Activity pattern of foraging wandering albatross: a marine predator with two modes of prey searching. *Mar. Ecol. Prog. Ser.* 151: 245-254.

Weimerskirch, H., Zotier, R. & Jouventin, P. 1989a. The Avifauna of the Kerguelen Islands. *Emu* 89: 15-29.

Wenzel, B.M. & Meisami, E. 1990. Quantitative characteristics of the olfactory system of the Northern Fulmar (*Fulmarus glacialis*): a pattern for sensitive odor detection? In *10th Int. Symp. Olfaction & Taste, Oslo* (Doving, K.B. ed.). GCS A/S, Oslo.

West, J.A. & Imber, M.J. 1986. Some foods of Buller's Mollymawk *Diomedea bulleri. NZ J. Zool.* 13: 169-174.

Westerskov, K. 1959. The Nesting Habitat of the Royal Albatross on Campbell Island. *Proc. NZ Ecol. Soc.* 6: 16-20.

Westerskov, K. 1960. *Birds of Campbell Island.* NZ Government, Wellington.

Westerskov, K. 1961. History of Discovery and Taxonomic Status of the Royal Albatross. *Emu* 61: 153-170.

Westerskov, K. 1963. Ecological Factors Affecting Distribution of a Nesting Royal Albatross Population. *Proc. 13th Int. Orn. Congr.* 795-811.

Wetmore, A. 1965. *Birds of Panama.* Smithsonian, Washington.

Wheeler, A. 1986. Catalogue of the natural history drawings commissioned by Joseph Banks on the *Endeavour* voyage 1768-1771. *Bull. Br. Mus. Nat. Hist.* (hist. Ser.) 13: 1-172.

White, A. 1855. *A Collection of Documents on Spitzbergen and Greenland comprising a translation from F. Martens voyage to Spitzbergen...* Hakluyt, London.

White, C.M.N. 1973. *Diomedea cauta* in South African waters. *Bull. Br. Ornithol. Club* 93: 56.

Wilkins, G.H. 1923. Report on the birds collected during the voyage of the "Quest" (Shackleton-Rowett Expedition) to the southern Atlantic. *Ibis* (11th Ser) 5: 474-511.

Wilkinson, H.E. 1969. Description of an Upper Miocene albatross from Beaumaris, Victoria, Australia and a review of fossil Diomedeidae. *Mem. Natl. Mus. Victoria* 29: 41-51.

Williams, A.J. 1984. The status and conservation of seabirds on some islands in the African sector of the Southern Ocean. In *Status and Conservation of the World's Seabirds* (Croxall, J.P., Evans, P.G.H. & Schreiber, R.W. eds): 627-635. ICBP, Cambridge.

Williams, A.J., Burger, A.E., Berruti, A. & Siegfried, W.R. 1975. Ornithological Research on Marion Island. *S. Afr. J. Antarct. Res.* 5: 48-50.

Williams, A.J. & Imber, M.J. 1982. Ornithological observations at Gough Island in 1979, 1980 and 1981. *S. Afr. J. Antarct. Res.* 12: 40-45.

Williams, A.J., Siegfried, W.R., Burger, A.E. & Berruti, A. 1979. The Prince Edwards Islands: a sanctuary for seabirds in the Southern Ocean. *Biol. Conserv.* 15: 59-71.

Williamson, I.A.W. 1975. Northern limits of albatrosses at sea off west coast of Africa. *Sea Swallow* 24: 62.

Willis, E.O. & Oniki, Y. 1993. On a *Phoebetria* specimen from southern Brazil. *Bull. Br. Ornithol. Club* 113: 60-61.

Wilmore, S.B. 1974. *Swans of the World.* David & Charles, Newton Abbot.

Wilson, E.A. 1907. Aves. In *National Antarctic Expedition 1901-04, Natural History* 2: 108.

Wilson, G.J. 1973. Birds of the Solander Islands. *Notornis* 20: 318-323.

Wilson, J.A. 1975. Sweeping flight and soaring by albatrosses. *Nature* 257: 307-308.

Wilson, N. 1970. Acarina: Metastigmata: Ixodidae of South Georgia, Heard and Kerguelen. *Pac. Insects Monog.* 23: 78-88.

Wilson, P.R. & Orwin, D.F.G. 1964. The sheep population of Campbell Island. *NZ J. Sci.* 7: 460-490.

Wilson, R.A. 1959. *Bird Islands of New Zealand.* Whitcombe & Tombs, Wellington.

Wilson, R.P., Cooper, J. & Plötz, J. 1992. Can we determine when marine endotherms feed? a case study with seabirds. *J. Exp. Biol.* 167: 267-275.

Wilson, R.P., Weimerskirch, H. & Lys. P. 1995. A device for measuring seabird activity at sea. *J. Avian Biol.* 26: 172-175.

Winthrope, I. (no date). The Shearwaters of *Diomedes. Seabird Rep.* 3: 37-40.

Woehler, E.J. 1989. Resightings and recoveries of banded seabirds at Heard Island, 1985-88. *Corella* 13: 38-40.

Woehler, E.J. 1991. Status and conservation of the seabirds of Heard Island and the McDonald Islands. In *Seabird Status and Conservation* (Croxall, J.P. ed.): 263-277. ICBP, Cambridge.

Woehler, E.J. 1996. Concurrent decreases in five species of Southern Ocean seabirds in Prydz Bay. *Polar Biol.* 16: 379-382.

Woehler, E.J. 1997. Seabird abundance, biomass and prey consumption within Prydz Bay, Antarctica 1980/81-1992/93. *Polar Biol.* 17: 371-383.

Woehler, E.J., Graham, R. & Hodges, C.L. 1991. Multiple sightings of a Royal Albatross *Diomedea epomophora* in the southern Indian Ocean. *Mar. Orn.* 19: 71-72.

Woehler, E.J., Hodges, C.L. & Watts, D. 1990. *An atlas of the pelagic distribution and abundance of seabirds in the southern Indian Ocean.* Antarctic Division, Kingston.

Wood, C.J. 1973. The flight of albatrosses (a computer simulation). *Ibis* 115: 244-256.

Wood, K.A. 1992. Seasonal Abundance and Spatial Distribution of Albatrosses off Central New South Wales. *Aust. Bird Watcher* 14: 207-225.

Woods, R.W. 1966. Letter to the editor. *Animals* 8: 224-225.

Woods. R.W. 1988. *Guide to Birds of the Falkland Islands.* Anthony Nelson, Owestry.

Woods, R.W. & Woods, A. 1997. *Atlas of Breeding Birds of the Falkland Islands.* Anthony Nelson, Owestry.

Woodward, P.W. 1972. The natural history of Kure Atoll, northwestern Hawaiian Islands. *Atoll Res. Bull.* 164: 1-318.

Wordsworth, F. 1876. The Strathmore: Letter from Mrs Wordsworth, the lady who survived the wreck. *Blackwood's Edinburgh Mag.* 120: 317-342.

Wright, A.E. 1984. Buller's Mollymawks breeding on the Three Kings Islands. *Notornis* 31: 203-207.

Wright, D. 1980. *Royal Albatross.* NZ National Film Unit, Lower Hutt (16 mm colour film).

Wynne-Edwards, V.C. 1939. Intermittent Breeding of the Fulmar (*Fulmarus glacialis* (L.)), with some General Observations on Non-breeding in Sea-Birds. *Proc. Zool. Soc. Lond.* Ser. A. 109: 127-132.

Wyrtki, K. 1975. El Niño – The Dynamic Response of the Equatorial Pacific Ocean to Atmospheric Forcing. *J. Phys. Oceanogr.* 5: 572-584.

Yamashina, Y. 1942. (Birds of the seven islands of Izu). *Tori* 11: 191-270. (In Japanese).

Yesner, D.R. 1976. Aleutian Islands albatrosses: a population history. *Auk* 93: 263-280.

Yesner, D.R. & Aigner, J.S. 1976. Comparative biomass estimates and prehistoric cultural ecology of the southwest Umnak region, Aleutian Islands. *Arctic Anthropol.* 13: 91-112.

Yudin, K.A. 1957. (Certain adaptive peculiatities of the wing in the birds of the order Tubinares). *Zool. Zh.* 36: 1859-1873. (In Russian).

Zimmerman, D.A., Turner, D.A. & Pearson, D.J. 1996. *Birds of Kenya and northern Tanzania.* A. & C. Black, London.

Zink, R.M. 1981. Observations of seabirds during a cruise from Ross Island to Anvers Island, Antarctica. *Wilson Bull.* 93: 1-20.

Zink, R.M. 1997. Species Concepts. *Bull. Br. Ornithol. Club* 117: 97-109.

Zotier, R. 1990. Breeding ecology of the White-headed Petrel *Pterodroma lessoni* on the Kerguelen Islands. *Ibis* 132: 525-534.

Index

A

Abbott, C.C. 17
ACC 41, 42, 45
Adams Island 57, 90, 91
adults 36
adusta 135
aerial photographs 219, 220, 323
aggregations 204
Agincourt Island 237
Agrihan Island 218
Agulhas Current 45
air sacs 32
aircraft 208
al - câdous 14
Alaska Current 192
Alaskan continental shelf 189
Albatros a courte queue 236
Albatros de la Chine 236
Albatros mouton 15, 136
albatross destruction 207, 219
Albatross Island 54, 89
Albatrus 16
albatrus 135, 235, 383
Albin 15
albinos 196
Alcatraz 14
Aldrich, John 208
Aleutian Islands 23, 189, 202, 203
Aleutian Peninsula 202
Aleuts 357
Alexander, W.B. 18, 175, 359
alula 28
Amadon, Dean 137
Amchitka Pass 190
Amsterdam Wandering Albatross **130**, 144, 383
amsterdamensis 383
ANARE 52, 54
anatomy 26, 32
anchoveta 171
Anderson, David 180
Annenkov Island 95
annual breeding 328
Año Nuevo group 62
Antarctic Circumpolar Current 23, 41
Antarctic circumpolar ocean 41
Antarctic Continent 42
Antarctic Convergence 41
Antarctic Front 43
Antarctic Ocean 41
Antarctic Polar Frontal Zone (APFZ) 41, 44, 45
antarctica 114
antibiotics 56
anticyclones 41, 189
antipodensis 383
Antipodes Island 59
Antipodes Islands 47, 59, 91, 120, 148
Antipodes Wandering Albatross 60, **130**, 148, 383
ants 232
archival tags 21
artificial eggs 56
atmospheric pressure 41
Aubert de la Rüe, Edgar 50, 51, 63
Auckland Island 57, 90, 91
Auckland Islands 47, 57, 90, 120, 147
Auckland Islands Expedition 58
Auckland Shy Albatross **71**, 75, 90, 91, 92, 383
Auckland Wandering Albatross **130**, 147, 383
Audubon, John James 197
Austin, Oliver L. Jr. 236, 244
Australia 52
Australian Antarctic Division 21

B

Bailey, Alfred 206
bait-throwing machines 369
Baja California 202
Baldwin, Paul 207
Banks, Joseph 14, 73, 74, 114, 135, 358
BAS 20
Bass, George 54
Bass Strait 45
bassi 383
bathymetry 41, 42
Baudelaire, Charles 377
Bay of Isles 66
Bayly, William 357
Beauchêne Island 63, 64, 65, 94
Beck, Roland ('Rollo') Howard 18, 61, 62
behaviour 33, 276, 400, 402
  aggression 276
  allopreening 279
  autopreening 278
  chick defensive 277
  communication 277
  copulation 288, 304, 310, 318
  dance 282, 301, 309, 322
  Galápagos Albatross 306
  gooneys 310
  great albatrosses 294
  individual recognition 279
  inter-specific behaviour 305
  mollymawks 282
  nest relief 280, 305, 319
  quiescent behaviour 279
  sooties 288
  sounds 281
  spatial activity 279
  terminology 276

territory 276
visual signals 280
Bellingshausen Sea 44
Benguela Current 23, 45
Bering Sea 23, 189, 191, 202
biennial breeding 33, 115, 127, 153, 330
bills 25, 30, 31, 72, 73, 132, 134, 135
biological species concept 71, 130
Bird Island (Falklands) 94
Bird Island (South Georgia) 20, 64, 66, 95, 205, 215
bird lines 369
Birds of the Ocean 18
Black-browed Albatross 28, 64, **71**, 73, 86, 87, 88, 89, 90, 91, 92, 93, 94, 95, 282, 383
Black-footed Albatross 17, 31, **194**, 310, 383
blowflies 56
Bloxam, Andrew 136, 358
Bollons, Captain 56
Bollons Island 59, 91
Bonin Islands 218, 237
Bonn Convention 370
Bossière, Henri Emile 51
Bounty Islands 47, 60, 91, 92
brachiura 236
brachyura 199, 236
Bransfield Strait 44
Brazil Current 44, 76
breeding equilibrium 337
breeding frequency 337
breeding grounds 153
breeding population 37
breeding season 33
Galápagos Albatross 182
gooneys 222
great albatrosses 153
mollymawks 100, 101
sooty albatrosses 123
Steller's Albatross 241
breeding success 36, 127, 163, 184, 228, 242, 336
Brisson, Mathurin Jaques 16
Bristle-thighed Curlew 232
British Ornithologists' Club 199
British Ornithologists' Union 199
bronchi 32
brooding 103, 226
Brooke, Richard 249
Broughton Island 56, 90
Bryan, William Alanson 200, 211, 213, 217
BSFW 201, 205, 209, 210, 214, 215
Buffon 22, 74, 236
Bull Rock 91
Buller, Walter 17, 74, 75, 136
bulleri 74, 383
by-catch 357
Byo-sho 237

C

California 203
California Current 192, 193
Campbell Black-browed Albatross **71**, 90, 383
Campbell Island 47, 58, 90, 120, 148
Campbell, Roy 378
Cape Alexandra 95
Cape Buller 95
Cape Crewe 95
Cape Expedition 58, 59
Cape North 95
Cape Nuñez 95
Cape Paryadin 95
Cape Wilson 95
cats 38, 48, 49, 50, 51, 52, 53, 57, 59, 64, 67, 232, 242
cattle 50, 53, 57, 63, 67
cauta 74, 75, 383
cauta mollymawks 71
CCAMLR 369
Challenger 18
changing oceans 23
Chapman, F.R. 58
Chappell, R. 61
Charles Darwin Research Station 175
Chatham Albatross **71**, 73, 90, 383
Chatham Islands 47, 60, 92, 149
checklist 383
chicks 103, 105, 126, 133, 158, 184, 226, 234
fasting 160
feeding 109, 159, 226
fledging 227
fledging period 126
growth 36, 105, 159, 184, 227
Chile 45
chinensis 199, 236
chionoptera 383
chlorinated hydrocarbons 367
chlororhynchos 74, 383
Chocolate Albatross 135
chrysostoma 74, 383
Ciconiiformes 24
circumpolar migration 142
Clark, Gerry 20, 46
classification 24
Clayton, S.W. 17
Clerke, Commander Charles 197
cliffs 123
clusters 124
Coleridge, Samuel Taylor 371
colonies 98, 99, 124, 220, 241
Commander Islands 203, 235
communication 33
comparative anatomy 18
conservation 48, 52, 53, 54, 55, 56, 57, 58, 59, 60, 61, 64, 69, 70, 175, 180, 200, 206, 209, 215, 216, 219, 220, 244, 245, 246, 369

continental shelf 44, 45
Cook, Captain James 46, 51, 56, 65, 112, 197, 358
cooking 358
Cooper Island 95
copulation 34, 155
coral islands 220
Corder, Annie 376
Cory's Shearwater 16
Cosmas 376
Coues, Elliott 18, 24, 75, 114, 199
counts 405
Croxall, John P. 20
crustaceans 230, 231
culminata 74
culminicorn 73
cyclones 41, 189
cytochrome-*b* 14, 25

D

Dabbene, Roberto 143
dabbenena 383
Daisy 18
Daitojima 233
Damien II 20
Dampier, William 359
Darwin, Charles 63, 260
de Kerguelen-Trémarec, Yves-Joseph 51
deferred sexual maturity 331
demipopulation 37
demographic models 339, 365, 404
demography 219
d'Entrecasteaux, de Bruni 50
departure 105, 227
depth gauges 324
derogata 236
description
  Amsterdam Wandering Albatross 132, Plate 30
  Antipodes Wandering Albatross 132, Plate 29
  Auckland Shy Albatross 73, Plate 13
  Auckland Wandering Albatross Plates 26, 27
  Black-browed Albatross 72, Plate 1
  Black-footed Albatross 195, Plates 47, 48
  Campbell Black-browed Albatross 72, Plate 4
  cauta mollymawks 73
  Chatham Albatross 73, Plate 15
  Eastern Yellow-nosed Albatross 72, Plate 7
  Galápagos Albatross 169, 176, Plates 37, 39
  Gough Wandering Albatross 133, Plate 28
  Grey-headed Albatross 72, Plate 8
  Laysan Albatross 195, Plates 40, 42
  Light-mantled Sooty Albatross 113, 247, Plate 19
  Northern Buller's Albatross 72, Plates 9, 10
  Northern Royal Albatross 133, Plate 31
  Salvin's Albatross 73, Plate 14
  Sooty Albatross 113, Plate 16
  Southern Buller's Albatross 72, Plate 11
  Southern Royal Albatross 133, Plates 35, 36
  Steller's Albatross 187, 233, Plates 49, 50
  Tasmanian Shy Albatross 73, Plate 12
  Wandering Albatross 39, 132, Plates 20, 22
  Western Yellow-nosed Albatross 72, Plate 6
Dillingham 216
Diomede Islands 16
Diomedea 14, 15, 16, 17, 114, 135, 177, 197, 200, 235, 383
Diomedeidae 24
Diomedeinae 24
Diomedella 75
Diomedes 16
Disappointment Island 57, 90, 91
disease 37, 110
divers 25
diving 265, 348
diving petrels 22, 25, 29
Dixon, C.C. 19
DNA sequencing 25
DNA-DNA hybridisation 24, 25
dogs 52, 53, 216, 232
domestic fowl 50
donkeys 53, 67, 206
Dougall, William 59, 60
Drake Passage 44, 45
Drake, Sir Francis 62
driftnetting 360
  moratorium on 369
ducks 53
Dunedin 56
dynamic soaring 262

E

earthquakes 53
East Australian Current 45
East China Sea 202, 203
Eastern Island 205, 206
Eastern Yellow-nosed Albatross **71**, 73, 86, 87, 383
ecology 323
Ecuador 171
eddies 190
Edwards, George 15
egging 64, 66, 68, 69
eggs 34, 102, 125, 155, 183, 224, 241, 389
El Niño 171
El Niño/Southern Oscillation 171, 193
Elephant Jason Island 94
Elsehul 95
emigration 155
Enderby Island 57
endocrinology 20
energetics 348
environmental impact 353
epomophora 14, 136, 383
equator 23
Equatorial Countercurrent 173
Equatorial Undercurrent 173

eremita 75, 383
Euphausia superba 44
extra-pair copulation 34
exulans 135, 383
eyelids 132
eyes 72

F

Fairchild, Captain 58, 136
Falkland Current 44, 76
Falkland Islands 41, 47, 63, 93, 94
Falla, Robert 17, 49, 56, 75
Fanning, Edmund 17
Fanning, Henry 49
FAO 370
Faroes 77
fat 352
feather trade 233, 243
feathers 26
fecundity 326
feeding 346, 347
feeding chicks 35
feet 132
FICZ 65
Filhol, Henri 18, 33, 58, 136
fire 63
first breeding 105, 127, 163, 229, 242
first flight 36
First International Conference 21
First International Workshop on Albatross-Fisheries 369
fish 230, 231, 242
Fish & Wildlife Service 20
fish eggs 230, 231
Fisher, Harvey I. 20, 200, 207, 208, 323
Fisher, James 203
Fisher, Mildred L. 194, 208
Fisher, Walter 211
fisheries 21, 175, 360
fishing 50
'fishing' (for albatrosses) 358
FitzRoy, Captain Robert 173
flapping 29
fledging period 103
fledglings 126
Fleming, Charles 61
flight 29, 260
flight muscles 29
Flinders, Matthew 54
flotsam 367
flying 29
food 107
 Black-browed Albatross 106
 Black-footed Albatross 230
 Galápagos Albatross 185
 great albatrosses 394
 Grey-headed Albatross 106
 gooneys 396
 Laysan Albatross 230
 Light-mantled Sooty Albatross 128
 mollymawks 391
 Northern Royal Albatross 165
 Sooty Albatross 128
 sooty albatrosses 393
 Steller's Albatross 242
 Wandering Albatross 164, 165
foraging 35, 341
Forbes, William Alexander 18
Formosa Channel 202
Forster, George 73, 74
Forster, Johann Reinhold 16, 73, 74, 112, 114, 135
Forty Fours 61, 92
fossils 22, 23
fowl 53
fowling 68, 69, 200, 206, 209, 210, 213, 214, 217, 219, 233, 240, 243
foxes 64
FRAM 42, 43
French Frigate Shoals 198, 204, 211, 214, 219, 220, 237
frigatebirds 25
Froude, William 18, 261
fuliginosa 114, 124
Fulmar 16, 22, 28, 33, 34, 36
Fürbringer, Max 24
furseals 38
fusca 112, 115, 383

G

gadfly-petrels 22
Gadow, Hans Friedrich 24
Galapagornis 178
Galápagos Albatross 25, 29, **175**, 383
Galápagos Hawks 186
gape-stripe 72, 73
Gardner Pinnacles 204, 211, 214, 219
geese 53
giant petrels 25, 28, 29, 30, 33, 34
Gibson Code 131
Gibson Plumage Index 131, 133, 134
gibsoni 383
Giglioli, Enrico 75
Gilbert, George 189
glacials 23
glaciation 24
gliding 265, 268
Gmelin, J.F. 74, 114
goats 53, 57, 59, 63
Godman, F. Du Cane 75, 175
Golden Plover 232
Goldsmith, Oliver 375
gony 17
gooneys 17, 25, **194**, 197
Gough Island 47, 70, 95, 96, 121, 150
Gough Wandering Albatross 68, **130**, 150, 383

Gould, John 74
Gower, Erazmus 71
Grand Jason Island 64, 94
great albatrosses 16, 18, 25, 28, **130**
Green Island 95, 204, 205, 220, 221, 222
Greenwood, Jeremy 13
Grew, Nehemiah 15
Grey Petrel 28
Grey-headed Albatross **71**, 73, 74, 86, 87, 89, 90, 93, 94, 95, 283, 383
ground effect 271
guanacos 64
guano 172, 198, 209, 211, 214, 217
Gulf of Alaska 202
Gulf of Anadyr 202
gulls 16, 197, 209, 211, 235
'gulls' 17, 197, 235

H

Habel, Simon 177
Hadden, Fred 206
hallux 30
hares 63
Harris, Charles Miller 177
Harris, Michael 180
Harris, Richard 33
Harrison, Peter 18, 131
hatching 103, 126, 157
Hattori, Toru 233
Hawaiian Islands 194, 204, 232
Hawkins, Sir Richard 357
heads 72, 73
Heard Island 47, 52, 88, 119, 145
heat stress 204
heavy metals 365
Hector, Julian A.L. 327
Hilsenberg 115
Hasegawa, Hiroshi 245
Hoka-sho 237
Holst, P.A. 218
home range
  Black-browed Albatross 108, 109
  Grey-headed Albatross 108
  Tasmanian Shy Albatross 83
  Wandering Albatross 140
Honolulu 198
Hood Island 176
Hooker, Joseph 63, 87
horses 53
humeral feathers 28
humerus 28
Hutton, F.W. 33, 115
hybrids 91, 134, 196
Hydrobatidae 24
Hydrobatinae 24

I

ice-sheets 23
Idrac, Pierre 262
Ile Amsterdam 50, 86, 119, 144
Ile aux Cochons 48
Ile de Croy 52
Ile de la Possession 48
Ile de l'Est 48
Ile des Pingouins 48
Ile Saint Paul 47, 50, 87, 119
Iles Amsterdam 47
Iles Crozet 47, 48, 86, 118, 143
Iles Kerguelen 47, 51, 87, 119, 145
Illiger, Carl 24
Ilots des Apôtres 48
immigration 155
immutabilis 199, 383
impavida 74, 383
Inaccessible Island 66, 68, 95, 150
incubation 102, 125, 157, 183, 225, 351
Indian Ocean 41, 45, 48
Indians 357
injuries 167
inter-ramicorn 30
Inter-tropical Convergence Zone (ITCZ) 173
interglacials 23
irrorata 177, 383
Isla Bartolomé 63, 93
Isla Clarión 194, 218
Isla de Guadalupe 218, 220
Isla de la Plata 181
Isla de los Estados 62
Isla Diego de Almagro 47, 62, 92
Isla Española 176, 180
Isla Gonzalo 63, 93
Isla Guadalupe 194
Isla Norte 63
Isla Observatorio 62
Islands 23
Islas Año Nuevo 62
Islas Diego Ramírez 47, 63, 93
Islas Evout 62
Islas Galápagos 173
Islas Ildefonso 47, 62, 93
Islas Revillagigedo 218
Iwo Jima 194, 218
Izu Shoto 218

J

Jacobs, W.V.E. 200
Jameson, William 130
Janthina 231
Japan 202, 203
Japanese Meteorological Agency 244
John, Dilwyn 43
Johnston Atoll 194, 217

Jomfruene 95
Jouventin, Pierre 21
juveniles 36, 126, 160

K

Ka'ena Point 216
Kamchatka 191, 202, 203, 235
Kaneohe 216
Kaohikaipu Island 216
Kaua'i 204, 216, 219, 220
  Crater Hill 216
Ka'ula 204, 216, 219
Kenyon, Karl W. 20, 208
Keppel Island 64, 94
keratin 31
Kilauea Point 216
Kita-kojima 218
Kitanoshima 237
Klein, Jacob Theodore 16
Klutschak Point 95
Krasheninnikov, Stepan Petrovich 235
krill 44
Kurashiwo Current 192, 193
Kure Atoll 204, 211, 219, 220, 221, 222, 232
Kurile Islands 191, 202, 235
Kuroda N. 201

L

La Niña 171
La Pérouse Pinnacle 214
Lachmund, Friderici 16
Lack, David 323
landing 29, 273
Langston, Nancy 251
large petrels 28
Latham, John 74, 114
Laurens Peninsula 52
laying 102, 125, 156, 183, 224
Laysan Albatross 17, 30, **194**, 198, 310, 383
Laysan Island 198, 199, 204, 211, 219, 222, 237
Laysanornis 200
Leadenhall Market 359
Leeward Islands 200
LeFebvre, Eugene 208
leopard seals 38
leptorhyncha 177
Lesson, R.P. 136
Lévêque, Raymond 180
Lichtenstein 74
life expectation 323
Light-mantled Sooty Albatross 59, **112**, 288, 383
Linnaeus 16
Lisianski Island 204, 209, 211, 219
Lister, Richard 374
Little Solander Island 56
Logbook for Grace 18
longevity 37, 128, 164
longlining 362
Loomis, L.M. 175
loons 25
Lyrical Ballads 373

M

MacDonald Islands 119
Macquarie Island 45, 47, 53, 89, 119, 145
  Bishop and Clerk Islets 53, 89
  Judge and Clerk Islets 53
  Macquarie Ridge 45, 53
Mallard 53
Mallemuk 16
Man-of-War 15
man-overboard 380
Mana 216
Manx Shearwater 34
Maori 55, 57, 60
Marcus Island 217, 237
marginal ice zone 42
Marianas Islands 218
marine mammal carcasses 231
Marion du Fresne, Marc Macé 46, 48
Marion Island 46
Markham, Captain 177
Marshall Islands 218
Masefield, John 382
mass 25, 28
mast 205, 208, 215
Mathews, Gregory Macalister 18, 74, 115
Matthews, Leonard Harrison 131, 371
Mawson, Douglas 53
maxillary unguis 132
McCormick, Robert 57, 63
McDonald Islands 47, 53, 88
measurements, 385
Medway, David 359
melanophris 74
melanophrys 74
Meseth, Earl 208
mesoscale 190
mesoscale eddies 41
messages 29
Mewstone 54, 89
mice 48, 49, 50, 51, 52, 53, 57, 64, 65, 67, 70
Midway Atoll 17, 20, 35, 194, 200, 204, 205, 206, 211, 219, 220, 222, 237
migrations 29
Minami Torishima 194, 217, 237
Minami-kojima 218, 239
Mockingbird 186
molly 16
mollymawks 16, 25, 71
Moloka'i 204, 217
Moore, Thomas 375
Moriori 57, 60
mosquitoes 186, 232

Mouflon 52
moult 26, 249, 398, 399
mouth 32
Mouton du Cap 15
Mozambique Channel 45
Mukojima Retto 194, 218, 233
multi-species flocks 35, 346
Munro, George C. 199, 206, 209, 214
Murphy, Robert Cushman 13, 18, 41, 75, 171, 175
myth 381, 382
Myxoma 52

N

National Biological Survey 201
naturalists 17
NE trade winds 189
Nealbatrus 75
Necker Island 204, 211, 215, 219, 220
Nelson, Bryan 180
nest building 34
nests 98, 99, 100, 122, 125, 154, 163, 183, 222
New Island 63, 64, 65, 94
New Zealand 45, 47, 55, 89, 146
New Zealand offshore islands 47, 55
Nightingale Island 66, 68, 95
nigripes 197, 199, 383
Nihoa 204, 211, 215, 219, 220
Ni'ihau 204, 216, 219
Nishinoshima 233, 237
Norfolk Island 202
North America 202
North Atlantic 77, 79
North East Island 56
North Island 205
North Pacific 23, 189
North Pacific Current 190
Northern Buller's Albatross 55, **71**, 89, 92, 383
northern hemisphere 143
Northern Royal Albatross 13, 35, 55, 61, **130**, 146, 294, 383
numbers
  Galápagos Albatross 181
  gooneys 219
  great albatrosses 151
  mollymawks 96, 97
  sooty albatrosses 112, 121, 122, 123
  Steller's Albatross 240, 241

O

O'ahu 216, 219
  Moku Manu 216
Oceanic Birds of South America 18
oceanic distribution
  black-browed albatrosses 76
  Black-footed Albatross 201, 202
  buller's albatrosses 80, 81
  Chatham Albatross 85
  Galápagos Albatross 178
  Grey-headed Albatross 78
  Laysan Albatross 201, 202
  Light-mantled Sooty Albatross 115, 116, 117, 118
  royal albatrosses 138
  Salvin's Albatross 84
  shy albatrosses 82
  Sooty Albatross 115, 116, 117, 118
  Steller's Albatross 237, 238
  wandering albatrosses 137
  yellow-nosed albatrosses 79, 80
Ogasawara Gunto 218, 237
oil 369
olfactory lobes 31
Otago Peninsula Trust (OPT) 55, 56
Otago Harbour 55
Otago Peninsula 55
otoliths 33
Ovid 16
Oyashiwo Current 192, 193

P

Pacific Cable Company 206
Pacific Ocean Biological Survey Program (POBSP) 200, 205
pack-ice 42, 44
Pagan 218
Pallas, Peter Simon 235
Palmer, Henry 199, 206, 209, 214
palpebrata 112, 114, 115, 383
Pan American Airways 206
Panama isthmus 22
parasites 390
  blowflies 167
  chewing lice 37, 110, 129, 167, 186, 231, 242
  feather mites 37, 110, 232
  fleas 37, 110, 129, 167
  Haematozoa 110
  leeches 110
  nematode 110, 232
  ticks 37, 110, 167, 186, 232, 242
Parkinson, Sydney 114, 135
Patagonian Shelf 44
Pearl and Hermes Reef 204, 209, 211, 219
Pedra Branca 54, 89
Pelecanoididae 24
P'eng-chia Hsü 233
P'eng-hu Leih-tao 233
penguins 24
Péninsule Courbet 52
Péninsule Rallier du Baty 51
Pennycuick, Colin J. 263
Penthirenia 200
Péron, P.F. 50
Peru 171
  Islas Lobos de Afuera 178
  Islas Lobos de Tierra 178
Peru (Ocean) Current 45

Peru Coastal Current 173
Pescadores Islands 236
petrels 22, 24, 26, 27, 29, 30, 36, 37
philopatry 33, 101
Philosophical Institute of Canterbury 58
Phoebastria 178, 200, 236, 383
Phoebetria 17, 112, 114, 115, 383
physiologists 209
physiology 327, 331, 348, 351, 352
phytoplankton 42
pigs 49, 50, 53, 57, 59, 63, 67
pink-stained plumage 132, Plate 21
piracy 38, 167
plasma protein 25
platei 383
Plautus 16
Pliny the Elder 16
plumage 26
POBSP 209, 210, 214, 215
pollution 366
Polynesians 197, 215, 357
Poncet, Jérôme and Sally 20, 62, 66
population dynamics 332
population regulation 338
Port-aux-Français 52
Post-egg Period in Albatrosses 19
pre-egg period 34, 101, 125, 155, 182, 222
predators 35, 38, 111, 129, 167, 186, 232, 242
primaries 28
Prince Edward Islands 46, 47, 86, 118, 143
Prince, Peter A. 20, 251, 323
Princes Islands 55
prions 22, 28, 29
Procellaria 24
Procellariidae 24
Procellariiformes 24
Procellariinae 24
propatagium 28
pullups 269

Q

'quakerbird' 112, 114

R

rabbits 49, 52, 53, 57, 63, 210, 213, 214
radius 28
range 346
rank 71, 130
Rasa Island 237
rats 38, 49, 50, 51, 52, 53, 59, 63, 65, 67, 205, 208, 232, 242
Ray, John 16
Rayleigh, Lord 262
re-pairing 162
Rechten, Catherine 180
recruitment 337
rectrices 29
Redpoll 53
regia 136
regurgitation 32
Reichenbach, Heinrich Gottlieb Ludwig 17, 75, 114
reindeer 52, 65
rhamphotheca 30
Rhothonia 136
Rice, Dale W. 20, 203, 208
Richdale, Lancelot Eric 19, 33, 57
Ridgeway, Robert 75, 177
Rima 56, 90
Rime of the Ancient Mariner 371
ringing 203
Robbins, Chandler 203, 208
Roberts, Brian B. 87
Robertson, Christopher J.R. 61, 62
Robertson, Graham 61, 62
Rocky Island 205
Rohwer, Sievert 251
Roosevelt, Theodore 200
Rosario Island 237
Rothschild, Walter 18, 74, 75, 177, 199
Royal Navy 244
Rua 56, 57
Rumbolds Point 95
runways 205, 207, 215

S

salt glands 30
Salvin, Osbert 18, 74, 199
salvini 75, 383
Salvin's Albatross 60, **71**, 73, 86, 91, 92, 383
Sand Island 205, 206, 207
sand islands 220
sanfordi 14, 136, 383
satellite telemetry 21
satellite tracking 117, 141, 142, 179, 265, 271, 324, 357
Saunders Island 64, 94
scapulars 28
scavenging 231, 359, 360
Schauinsland, Hugo 211
Schlemmer, Max 200, 210, 211, 213
Scoresby, Reverend William 359
Scotia Arc 44
Scotia Sea 44
Sea of Japan 202
Sea of Okhotsk 191, 202, 203
sea-level 24
sea-surface temperature 193
Seabirds 18
Seal Island 205
Seal-Kittery Island 205
sealers 46, 50, 51, 52, 53, 54, 56, 58, 60, 63, 65, 67, 70, 112
sealing 20
sealions 38
seamounts 41, 190, 193

searching 345
seasonal population 37
Second World War 207, 214, 217, 219
secondary feathers 28
Senkaku Retto 194, 218, 233, 239, 244, 245
sense of smell 31
sesamoid ossicles 28
sex ratio 163
sexual dimorphism 26, 196
sharks 38
shearwaters 22, 28, 29, 30
sheep 50, 52, 53, 57, 59, 63, 67
Shelvocke, Captain George 373
Shetland 77
ships 274
shooting 358, 359
Short-tailed Albatross 233
Shuntov, Viacheslav 201
shy albatrosses
  in Gulf of Aqabah 84
  in North Pacific 84
  off East Africa 84
size 25
skins 358
'small' albatrosses 16, 28
Snares Islands 47, 56, 90
Snares Western Chain 56
  Rima 56
  Rua 56
  Tahi 56
  Toru 56, 90
  Wha 56
Snowy Albatross 130
Solander, Daniel 56, 73, 74, 114, 135
Solander Islands 55, 56, 89, 90
Solomon Islands 202
sonagrams 285, 287, 290, 295, 297, 299, 307, 311, 320
sooties 17, 112
Sooty Albatross 25, 69, **112**, 288, 383
Sorensen, Jack 323
Sørn & Bernt 95
sounds 281
South Africa 45, 46
South Equatorial Current 173
South Georgia 20, 24, 47, 65, 94, 95, 120, 149
South Jason Island 64, 94
South Korea 202
South Pacific 45
Southeast Island 205
Southern Buller's Albatross 56, **71**, 73, 89, 90, 282, 383
Southern Ocean 23, 41
Southern Royal Albatross 13, 35, 59, **130**, 147, 148, 383
spadicea 135
Sparling, Donald 208
species 13
speed 265
squid 106, 128, 165, 230, 231, 242
squid beaks 33
staffelmauser 250
standing 30
starling 53
'starvation theory' 33
Stead, Edgar Fraser 56
steadi 383
Steeple Jason Island 64, 94
Steller, Georg Wilhelm 233, 235
Steller's Albatross 25, 35, 37, **233**, 383
stomach oils 32
storm petrels 22, 24, 28, 29, 34, 36
subadults 36, 161, 162, 185
Subarctic Frontal Zone 192
Subtropical Convergence 44, 45
Subtropical Frontal Zone 192
superstition 374
surface currents 43
surface temperature 43, 45
survival 164, 185, 230, 334
swimming 30, 265, 347
sympatry 325

T

TAAF 21, 49, 50, 51, 52
Tahi 56
Taiaroa Head 19, 55, 56, 134
tail 25, 29
taking-off 273
Tanager 200, 210, 214, 215, 217
Tarsi 30, 132
Tasman, Abel Janszoon 54, 55
Tasman Sea 41, 45
Tasmania 53, 54
Tasmanian Islands 89
Tasmanian Offshore Islands 47, 54
Tasmanian Parks and Wildlife Service 21
Tasmanian Shy Albatross **71**, 89, 383
taxonomy 13, 383
Temminck, Coenraad Jacob 74, 136, 236
Tern Island 215, 220, 232
Thalassarche 16, 75, 383
Thalasseus 75
Thalassogeron 75
The Pre-egg Stage in the Albatross Family 19
The Pyramid 61, 92
The Sisters 61, 92
Three Kings Islands 55, 89, 90
tiger shark 232
Tilman, H.W. 49
Tollu, Benôit 87
tomia 132
Torishima 194, 218, 220, 233, 239, 241, 243, 244
  Hatsune-zaki 218
  Tsubame-zaki 218
toroa 55
Totorore 20, 46

Townsend, J.K. 197
trachea 32
trans-equatorial migration 23
Transitional Waters (North Pacific) 192, 193
translocation 205, 208
trawling 361
Tremiti Islands 16
Tristan da Cunha 47, 66, 68, 94, 95, 120, 150
tropical cyclones 41, 189
Tubinares 24
tubular nostrils 22, 31
Tufft, Roger 49, 144
Tussac 63
tussock grass 46, 384
two-phase moult 251
typhoons 244

U

ulna 28
United Nations 369
upwelling 171
US Coastguard 205
US Fish and Wildlife Service (USFWS) 201, 208, 214, 215, 220, 323
US Navy 17, 206, 207, 214, 215, 216
USARP 20

V

Van Diemen's Land 53
van Wicquefort, Abraham 375
vasoconstriction 135
vasodilation 135
Vásquez, Chela 180
Velella 231
Vélin, Charles 51
Vienna Museum 203
virus infections 37
volcanic eruptions 67, 239, 243, 245
volcanic islands 220
Volcano Islands 218
von Bellingshausen, Admiral 53, 145
von Kittlitz, F. 210

W

Waite, Edgar 59, 74
Wake Island 194, 217
Walker, Kath 58
walking 30
Wall, Arnold 379
Wandering Albatross 28, **130**, 131, 143, 145, 149, 294, 383
Watanabe, Eichii 244
wave height 174
wave-slope soaring 268
Waved Albatross 175
waves 260
Weddell Sea 44
weight, see mass
Weimerskirch, Henri 21
Wekas 53
Welcome Islands 95
West Point Island 63, 64, 94
West Wind Drift 41
Western Chain 56, 57, 90
Western Yellow-nosed Albatross 34, 69, **71**, 74, 94, 95, 96, 282, 383
Westerskov, Kaj 59, 136
Wetmore, Alexander 200
whaling 20, 65, 361
White-capped Albatross 74
White-chinned Petrel 35
Wilder, Gerrit 200
Willett, George 206, 213
Williams, J.J. 199, 211, 212
Willis Islands 95
Wilson, Edward A. 18, 19, 20, 357
Wilson, John 263
wind 260
wind profiles 271
wind speed 174
Windward Islands 200
wing area 28
wing measurements 264
wing span 25, 130
wings 27
Wordsworth, C.F. 49
Wordsworth, Florence 17
Wotherspoon 61
wrap-around moult 256
Wright, Alan 62
Wynne-Edwards, V.C. 33

Y

Yamashina, Marquis Yoshimaro 233, 244
yellow plumage 176, 234

Z

zooplankton 42